ALLE ZEIT WACH 1842

Volumes already published

Volume 1: Trees I (1986)
Volume 2: Crops I (1986)
Volume 3: Potato (1987)
Volume 4: Medicinal and Aromatic Plants I (1988)
Volume 5: Trees II (1989)
Volume 6: Crops II (1988)
Volume 7: Medicinal and Aromatic Plants II (1989)
Volume 8: Plant Protoplasts and Genetic Engineering I (1989)
Volume 9: Plant Protoplasts and Genetic Engineering II (1989)
Volume 10: Legumes and Oilseed Crops I (1990)

Volumes in preparation

Volume 11: Somaclonal Variation in Crop Improvement I
Volume 12: Haploids in Crop Improvement I
Volume 13: Wheat
Volume 14: Rice
Volume 15: Medicinal and Aromatic Plants III
Volume 16: Trees III

Biotechnology in Agriculture and Forestry 10

Legumes and Oilseed Crops I

Edited by Y. P. S. Bajaj

With 175 Figures

Springer-Verlag Berlin Heidelberg New York
London Paris Tokyo Hong Kong

Professor Dr. Y. P. S. BAJAJ
A-137
New Friends Colony
New Delhi 110065, India

ISBN-13:978-3-642-74450-1 e-ISBN-13:978-3-642-74448-8
DOI: 10.1007/978-3-642-74448-8

Library of Congress Cataloging-in-Publication Data. Legumes and oilseed crops I / edited by Y. P. S. Bajaj. p. cm. – (Biotechnology in agriculture and forestry; 10) Includes bibliographical references. ISBN-13:978-3-642-74450-1(U. S.:alk.paper)1. Legumes– Propagation – In vitro. 2.Oilseed plants – Propagation – In vitro. 3. Legumes – Biotechnology. 4. Oilseed plants – Biotechnology. I. Bajaj, Y. P. S., 1936– . II. Title: Legumes and oilseed crops One. III. Series. SB177.L45L45 1990 633.3'042 – dc20

Softcover reprint of the hardcover 1st edition 1990

Typesetting: K+V Fotosatz GmbH, Beerfelden.

2131/3145-543210 – Printed on acid-free paper

Dedicated to
Shamiran and Paul Bajaj

Preface

Legumes and oilseed crops are a rich source of protein and energy. The major objectives in grain-legume breeding are to increase grain yield, protein quantity and quality and digestibility, and to decrease toxic factors, seed hardness and cooking time. Similarly, oilseed crops such as brassicas and peanut share somewhat similar genetic improvement goals, but suffer from susceptibility to a number of pathogens and diseases. In spite of the best efforts through conventional breeding, their yield has been virtually static, with no significant breakthrough. The lack of genetic variability in the base population has been considered to be a major limiting factor for the slow progress made in the improvement of these crops. The desired goals can be achieved by incorporating additional genetic variability in the existing germplasm. The protein and oil contents which are genotypically oriented can be enhanced further by suitable crosses. In this regard, progress made during the last decade, in the area of in vitro manipulation and recombinant DNA technology, holds promise for the improvement of these crops.

Among the oilseed crops, *Brassica* spp., soybean and sunflower have been well studied, wherein haploids, somaclones, somatic hybrids, cybrids and transformed plants have been produced. Oilpalm is one of the best examples where micropropagation is being commercially practiced. Regarding legumes, though there is extensive work on forages, such as *Medicago* and *Trifolium*, much needs to be done on food legumes. The studies on soybean, peanut, pea, winged bean, chickpea and small grain legumes have been encouraging, but to date have not been fruitfully incorporated into breeding programs.

Taking these developments into consideration, the present volume on *Legumes and Oilseed Crops I* has been compiled. This volume presents 31 chapters on various aspects of cell and tissue culture, in vitro genetic manipulation and other biotechnological approaches for the improvement of legumes and oilseed crops contributed by international experts from 20 countries. It deals specifically with the case history, importance, distribution, conventional propagation, review of the tissue culture work, micropropagation, somaclonal variation, production of haploids, wide hybridization and other recent advances in the in vitro manipulation of crops such as soybean, pea, chickpea, winged bean, alfalfa, clovers, lupines, birdsfoot trefoil, crownvetch, goat's rue, indigofera, brassica, sunflower, linseed, coconut, oilpalm, olive oil, etc. In addition, there are three general chapters dealing with the wide hybridization, genetic transformation and the cryopreservation of germplasm. These biotechnological approaches have far-reaching implications for the much needed genetic variability in these crops.

This book may be of interest to research scientists, teachers and advanced students of plant biotechnology, botany, agriculture, plant pathology, and especially plant breeding.

New Delhi, January 1990

Y. P. S. Bajaj
Series Editor

Contents

II.6 Winged Bean [*Psophocarpus tetragonolobus* (L.) D.C.]
S. VENKETESWARAN (With 10 Figures)

Section III Forage Legumes

III.1 Alfalfa, Lucerne (*Medicago* spp.)
S. ARCIONI, F. DAMIANI, M. PEZZOTTI and E. LUPOTTO
(With 7 Figures)

III.2 Clovers (*Trifolium* spp.)
E.G. WILLIAMS, G.B. COLLINS, and J.R. MYERS
(With 9 Figures)

III.3 Lupins (*Lupinus* spp.)
C. SATOR (With 3 Figures)

III.4 Cell and Tissue Culture of *Stylosanthes* spp.
E. G. M. MEIJER and L. SZABADOS (With 3 Figures)

III.5 Birdsfoot Trefoil (*Lotus corniculatus* L.)
E. B. SWANSON, D. A. SOMERS, and D. T. TOMES
(With 7 Figures)

III.6 Crownvetch (*Coronilla varia* L.)
D. L. GUSTINE and B. G. MOYER (With 2 Figures)

III.7 Goat's Rue (*Galega officinalis* L.)
V. NAŠINEC and B. NĚMCOVÀ (With 5 Figures)

Section IV Oilseed Crops

IV.1 Interspecific Hybridization in *Brassica* Through Ovary Culture
N. INOMATA (With 7 Figures)

IV.2 Regeneration of Plants from Protoplasts of Oilseed *Brassica* Crops
K. VAMLING and K. GLIMELIUS (With 4 Figures)

IV.3 Cybrids in Oilseed *Brassica* Crops Through Protoplast Fusion
G. PELLETIER (With 3 Figures)

IV.4 Sunflower (*Helianthus annuus* L.): Establishment of Cultures, Transformation, and the Regeneration of Plants
K.E.P. ROBINSON and N.P. EVERETT (With 1 Figure)

IV.5 Sunflower (*Helianthus annuus* L.): In Vitro Interspecific Hybridization
N.E. BOHOROVA and A.I. ATANASSOV (With 7 Figures)

IV.6 In Vitro Production of Haploids in *Helianthus*
H.Y. YANG, H. YAN and C. ZHOU (With 6 Figures)

IV.7 Sunflower (*Helianthus* spp.): Anther Culture and Field Studies on Haploids
R. JONARD and A. MEZZAROBBA (With 5 Figures)

IV.8 Flax (*Linum usitatissimum* L.): In Vitro Studies
A. MCHUGHEN (With 2 Figures)

IV.9 Flax (*Linum usitatissimum* L.): Embryo Culture Studies
A. PREŤOVÁ (With 16 Figures)

IV.10 Coconut (*Cocos nucifera* L.): Micropropagation
J. BLAKE (With 3 Figures)

List of Contributors

AHMAD, M. S., Nuclear Institute for Agriculture and Biology, PAEC, P.O. Box 128, Jhang Road, Faisalabad, Pakistan

ALTAF, N., Nuclear Institute for Agriculture and Biology, PAEC, P.O. Box 128, Jhang Road, Faisalabad, Pakistan

ARCIONI, SERGIO, Centro di Studi per il Miglioramento, Genetico delle Piante Foraggere, Facultà di Agraria, Borgo XX. Giugno, 06100 Perugia, Italy

ATANASSOV, A. I., General Laboratory of Genetic Engineering, Agric. Acad., 2232, Kostinbrod-2, Bulgaria

BAJAJ, Y. P. S., Former Professor of Tissue Culture, Punjab Agricultural University Ludhiana, India. Present Address: A-137, New Friends Colony, New Delhi 110065, India

BARWALE, U. B., Department of Agronomy, University of Illinois at Urbana-Champaign, Urbana, IL 61801, USA

BLAKE, JENNET, Unit for Advanced Propagation Systems, Dept. of Horticulture, London University Wye College, Wye, Ashford, Kent TN 25 5AH, United Kingdom

BOHOROVA, N. E., Institute of Genetics, Bulgaria Academy of Sciences, Sofia 1113, Bulgaria

CHOO, WOOI KHENG, Sime Darby Plantations, P.O. Box 147, 41910 Klang, Selangor, Malaysia

COLLINS, GLENN, B., Department of Agronomy, University of Kentucky, Lexington, KY 40546, USA

DAMIANI F., Centro di Studi per il Miglioramento, Genetico delle Piante Foraggere, Facultà di Agraria, Borgo XX. Giugno, 06100 Perugia, Italy

DAVEY, M. R., Plant Genetic Manipulation Group, Department of Botany, University of Nottingham, University Park, Nottingham, NG7 2RD, United Kingdom

EVANS, P. K., Department of Biology, University of Southampton, Southampton, SO9 5NH, United Kingdom

EVERETT, N. P., Eni Chem Americas, Inc., 2000 Princeton Park Corporate Center, Monmouth Junction, NJ 08852, USA

FEDELI, E., SSOG Stazione, Sperimentale Oli e Grassi, Via G. Colombo 79, 20133 Milano, Italy

GLIMELIUS, KRISTINA, Department of Crop Genetics and Breeding, University of Agricultural Sciences, Box 7003, 75007 Uppsala, Sweden

GRANT, J. E., APPM, Forestry Research Unit, P.O. Box 63, Ridgley, TAS 7321, Australia

GRIGA, M., OSEVA – Plant Breeding and Seed Corporation, Research and Breeding Institute of Technical Crops and Legumes, Sumperk-Temenice, Czechoslovakia

GUSTINE, D. L., USDA-ARS, Regional Pasture Research Laboratory, University Park, PA 16802, USA

HAMMATT, N., Plant Genetic Manipulation Group, Department of Botany, University of Nottingham, University Park, Nottingham NG7 2RD, United Kingdom

INOMATA, N., Department of Biology, College of Liberal Arts and Sciences, Okayama University, Tsushima, Okayama 700, Japan

JONARD, ROBERT, Laboratoire de Cultures de Tissus, Groupe de Physiologie Végétale Appliquée, Université des Sciences et Techniques du Languedoc, 34060 Montpellier, France

LUPOTTO, E., Istituto Sperimentale per la Cerealicoltura, Sezione Maiscoltura, Via Stezzano 24, 24100 Bergamo, Italy

MARTINS-LOUÇÃO, M. A., Departamento de Biologia Vegetal, Faculdade de Ciencias, Campo Grande Bloco C2, 1700 Lisboa, Portugal

MCHUGHEN, ALAN, Crop Development Centre, University of Saskatchewan, Saskatoon, Saskatchewan S7N 0W0, Canada

MEIJER, E. G. M., Department of Plant Molecular Biology, Leiden University, 2333 AL Leiden, The Netherlands

MEZZAROBBA, ANTOINE, UCST Verneuil, BP 83, 11400 Castelnaudary, France

MOYER, B. G., USDA-ARS, US Regional Pasture Research Laboratory, University Park, PA 16802, USA

MYERS, JAMES R., Department of Agronomy, University of Kentucky, Lexington, KY 40546, USA

NAŠINEC, V., Institute of Experimental Botany, Czechoslovak Academy of Sciences, 37005 Ceské Budejovice, Czechoslovakia

NĚMCOVÀ, B., Institute of Experimental Botany, Czechoslovak Academy of Sciences, 37005 Ceské Budejovice, Czechoslovakia

NISBET, G. S., ICI Seeds, Jealott's Hill Research Station, Bracknell, Berkshire RG12 6EY, United Kingdom

NOVÁK, F. J., Plant Breeding Unit, Joint FAO/IAEA Programme, International Atomic Energy Agency's Laboratories, Seibersdorf, Austria

PELLETIER, G., Laboratoire de Biologie Cellulaire, INRA, 78000 Versailles, France

PEZZOTTI, M., Centro di Studi per il Miglioramento, Genetico delle Piante Foraggere, Facultà di Agraria, Borgo XX. Giugno, 06100 Perugia, Italy

PREŤOVA, ANNA, Institute of Experimental Genetics of Slovak Academiy of Sciences, Hlohovska 2, 94992 Nitra, Czechoslovakia

ROBINSON, K. E. P., DNA Plant Technology Corporation, 6701 San Pablo Avenue, Oakland, CA 94608-1239, USA

RUGINI, E., Istituto die Ortofloroarboricoltura, Facoltá die Agraria, Universitá Degli Studi della Tuscia, Via Riello, 01100 Viterbo, Italy

SATOR, C., Institut für Produktions- und Ökotoxikologie, Bundesforschungsanstalt für Landwirtschaft, Bundesallee 50, 3300 Braunschweig, FRG

SOMERS, D. A., Department of Agronomy and Plant Genetics, University of Minnesota, St. Paul, MN 55108, USA

SWANSON, E. B., Department of Plant Science, Allelix Crop Technologies, 6850 Goreway Drive, Mississauga, Ontario L4V1P1, Canada

SZABADOS, L., C.I.A.T., Cali, Colombia

THANH-TUYEN, NGUYEN, T., Tissue Culture Laboratory, Department of Horticulture, Visayas State College of Agriculture, Leyte, Philippines 6521-A

TOMES, D. T., Department of Biotechnology Research, Pioneer Hibred International, Box 38, Johnston, IA 50131, USA

VAMLING, KRISTOFER, Department of Crop Genetics and Breeding, University of Agricultural Sciences, 26800 Svalöv, Sweden

VENKETESWARAN, S., Department of Biology, University of Houston, 4800 Calhoun, Houston, TX 77004, USA

WEBB, K. J., AFRC Institute for Grassland and Animal Production, Welsh Plant Breeding Station, Aberystwyth, Wales, SY23 3EB, United Kingdom

WIDHOLM, J.M., Department of Agronomy, University of Illinois at Urbana-Champaign, Urbana, IL 61801, USA

WILLIAMS, ELIZABETH, G., Department of Botany, The University of Georgia, 2502 Plant Sciences, Athens, GA 30602, USA

YAN, H., Department of Biology, Wuhan University, Wuhan 430072, China

YANG, H.Y., Department of Biology, Wuhan University, Wuhan 430072, China

ZHOU, C., Department of Biology, Wuhan University, Wuhan 430072, China

Section I
Wide Hybridization, Transformation, Cryopreservation

I.1 Wide Hybridization in Legumes and Oilseed Crops Through Embryo, Ovule, and Ovary Culture

Y. P. S. Bajaj[1]

1 Introduction

Legumes and oilseed crops are the most important source of plant proteins and energy, and are cultivated throughout the world in wide areas (Table 1). The grain legumes are grown in the tropics, sub-tropics and temperate zones. They are amongst the earliest food crops to be cultivated by man, and constitute one of the most importent sources of dietary protein, especially in Asia, Latin America and Africa. Some species which are commonly cultivated for human consumption are peanut (*Arachis hypogaea*), pigeon pea (*Cajanus cajan*), chickpea (*Cicer arietinum*), soybean (*Glycine max*), lathyrus pea (*Lathyrus sativus*), lentil (*Lens esculenta*), lima bean (*Phaseolus lunatus*), kidney or navy bean (*Phaseolus vulgaris*), garden pea (*Pisum sativum*), winged bean (*Psophocarpus tetragonolobus*), broad bean (*Vicia faba*), blackgram (*Vigna mungo*), greengram (*Vigna radiata*), cowpea (*Vigna unguiculata*), etc. They are of paramount importance not only for their value as human food, but also because of the high protein content for livestock and fish feed. Moreover, legumes are endowed with the ability to fix atmospheric nitrogen and thus improve soil fertility. In different countries they are consumed in various forms, such as immature seed, dry grain, as condiments or roasted grain, for flour production, and as fermented products (Parpia 1973). The protein content of legumes is affected by genetic as well as environmental factors, while the nutritional quality depends on several factors such as their digestibility, balance of essential amino acids, and the presence of certain toxic substances. Legume proteins are very rich in lysine and threonine, but poor in sulfur-containing amino acids, methionine, and cystine (Bressani 1973). The high level of lysine makes grain legumes an ideal supplement for cereals which are deficient in this particular amino acid. Legumes also form a good source of vitamins, thiamine, and niacin as well as minerals, calcium, and iron. In addition to these constituents, they possess about 60% carbohydrates, mainly in the form of starch. The calorie yield of pulses per unit weight is almost the same as in cereals (Aykroyd and Doughty 1966).

The major objectives in grain-legume breeding are to increase grain yield, methionine, cystine, and protein digestibility, and to decrease anti-enzymes and seed hardness or cooking time. Similarly, oilseed crops, such as brassica and

[1] Formerly Professor of Tissue Culture, Punjab Agricultural University, Ludhiana, India. Address for correspondence: A-137 New Friends Colony, New Delhi 110065, India

Biotechnology in Agriculture and Forestry, Vol. 10
Legumes and Oilseed Crops I (ed. by Y. P. S. Bajaj)

Table 1. World production, area and average yield of the major grain legumes and oilseed crops in 1981. (FAO 1982)

Grain legume	Production 10^6 mt	Area 10^6 ha	Average yield kg/ha
Soybean (*Glycine max*)	87941	50219	1751
Groundnut (*Arachis hypogaea*)	19368	19329	1002
Dry bean (*Phaseolus vulgaris*)	14053	24805	567
Dry pea (*Pisum sativum*)	8215	7890	1041
Chick pea (*Cicer arietinum*)	6292	10104	623
Broad bean (*Vicia faba*)	4178	3619	1154
Lentil (*Lens culinaris*)	1128	1953	577
Oil seed			
Soybean	87941	50219	1751
Seed cotton (*Gossypium* spp.)	45689	33389	1369
Ground nut (*Arachis hypogaea*)	19368	19329	1002
Sunflower seed (*Helianthus annuus*)	13765	11648	1182
Rape seed (*Brassica napus*)	12147	11712	1037
Coconut (*Cocos nucifera*)	3665	–	–
Linseed (*Linum usitatissimum*)	2274	4913	463
Sesame seed (*Sesamum indicum*)	1959	6308	311
Castor bean (*Ricinus communis*)	810	1490	544
Palm kernel (*Elaeis guinensis*)	1.89		

peanut, share somewhat similar goals, but suffer from susceptibility to a number of pathogens and diseases. Diseases such as leaf spot (*Cerospora arachidicola*) and root rot (*Rhizoctonia destrens*) cause considerable damage to peanut. There are more than 100 pathogens that affect soybean, and about 35 diseases (Sinclair 1982), that result in tremendous losses. There has been no significant increase in the production of either pulses or oilseed crops during the last decade because of their low static yields and susceptibility to various viral, fungal, and bacterial diseases. For qualitative and quantitative improvement, the breeder has to rely upon the extent of genetic variability present in the base population. The lack of genetic variability has been considered to be a major limiting factor to the progress made in the improvement of these crops. The desired goals can be achieved by incorporating additional genetic variability in the existing germplasm. Protein quality and quantity which are genotypically oriented (S. Bajaj 1972; S. Bajaj et al. 1971) can be enhanced further by suitable crosses. However, in addition to the conventional methods of plant breeding, like introduction, selection, and hybridization, recent advances in biotechnology (Bajaj 1986, 1988) have opened several new possibilities for the creation of genetic variability and selection of desired traits.

Recently, there has been much interest in the transfer of genes in crops through recombinant DNA technology and genetic manipulation of cell and tissue cultures, and also in combining the characters of two crop species into one through somatic hybridization by the fusion of protoplasts. These technologies and their implementation in the induction of genetic variability and crop improvement have been recently reviewed (Schell 1987; Cocking and Davey 1987; Goodman et al. 1987; Bajaj 1989a, c). Somatic hybrids have recently been obtained in some legumes and brassica crops (see Bajaj 1989b). Considerable progress has been made in the use of various in vitro techniques in grain legumes (see Bajaj and Gosal 1982). *Medicago* and *Trifolium* are the two most important forage legumes on which extensive work has been successfully done to generate somaclonal variation (see Chap. III.1, 2, this Vol.). Among the oilseed crops, in vitro studies on peanut (Bajaj 1988b), cotton (Bajaj and Gill 1985, 1987), soybean (Hildebrand et al. 1986), brassica (Pelletier et al. 1983), sunflower (Power 1987), flax (McHughen and Rowland 1988) have also helped to induce genetic variability.

Most of these techniques are in the process of being developed, and it will be quite some time before they are available to plant breeders for routine use in crop improvement. However, some of the old and simple in vitro techniques, such as culture of excised ovaries (Bajaj 1966), ovules (Bajaj 1965) and embryos (Johri and Bajaj 1962) have already contributed a great deal in obtaining interspecific and intergeneric hybrids in otherwise incompatible crosses, and numerous examples can be cited (see Table 2).

In this chapter, literature on the use of these techniques in wide hybridization in the legumes and oilseed crops is reviewed, and our work on interspecific hybridization in *Arachis*, *Brassica*, *Gossypium*, *Phaseolus*, and *Vigna* is summarized.

2 Incompatibility Problems and in Vitro Culture

In hybridization through conventional plant breeding methods, the pollen from one plant is simply put on the stigma of the other plant, and the rest is left to nature. If the cross fails, the useful alien gene cannot be transferred, and there is not much that a traditional plant breeder can do to achieve the desired goal. The failure of hybridization may be due to various types of incompatibility; some of the common forms of pre- and post-fertilization incompatibilities are shown in Fig. 1, and listed below:

1. Non-germination of pollen on the stigma;
2. Abnormal germination of pollen;
3. Pollen tube is short, with the result that it does not reach the egg;
4. Pollen tube bursts before reaching the ovule/egg;
5. Fertilization does not take place;
6. Fertilization takes place but the zygote does not divide;
7. Zygote divides to form a few celled embryo and does not develop further, or it aborts;
8. Endosperm is not formed to support the development of embryo;
9. Embryo remains small and does not mature.

In such cases, various in vitro techniques such as somatic hybridization, uptake, and incorporation of DNA, in vitro fertilization, and the culture of excised embryos, ovules, and ovaries can be employed.

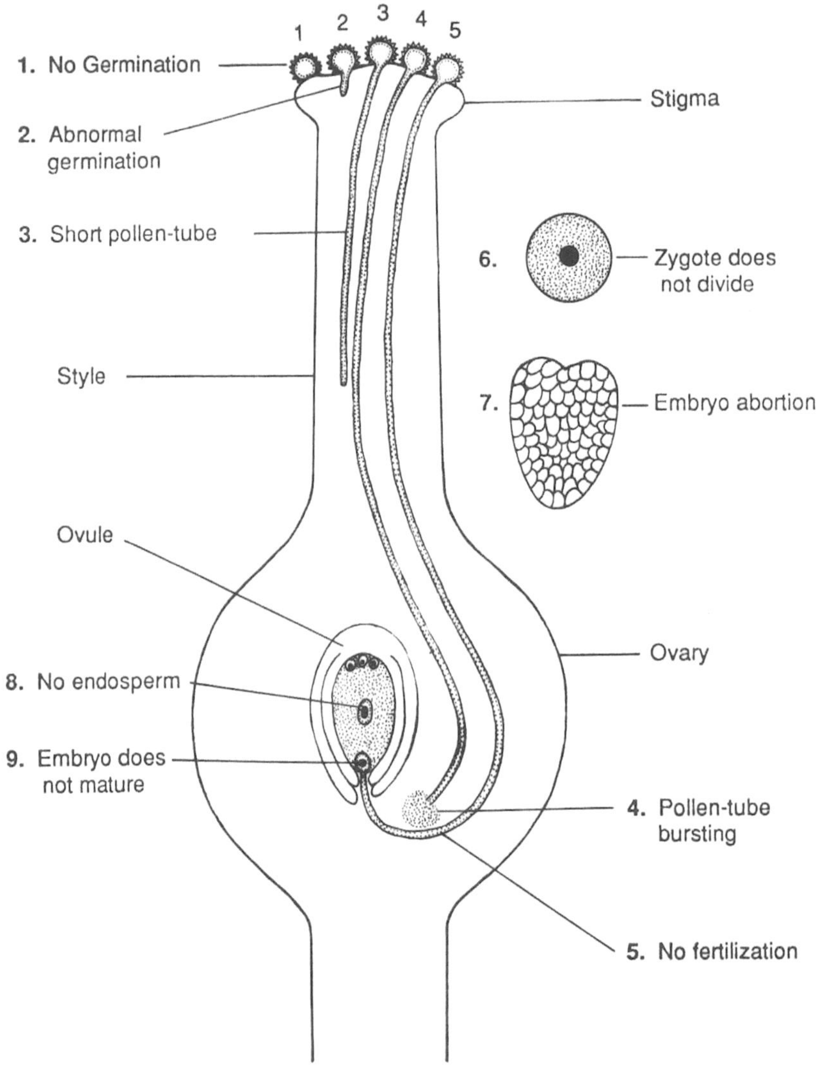

Fig. 1. Schematic representation showing various causes of incompatibility where in vitro technology can be applied for wide hybridization

3 Wide Hybridization

The literature on the in vitro production of hybrids in various legumes and oilseed crops through the culture of excised embryos, ovules and ovaries is summarized in Table 2.

Table 2. Summary of the work on wide hybridization in legumes and oilseed crops through embryo, ovule, and ovary culture

Crop	Culture/explant	Objective/response	Reference
Arachis hypogaea × *A. villosa*	Embryo	Hybrid plants	Bajaj et al. (1982)
Brassica juncea × *B. napus*	Embryo, ovule, ovary	Hybrid plants	Bajaj et al. (1986)
Brassica juncea × *B. hirta*	Ovule	Hybrid plantlets	Mohapatra and Bajaj (1984)
Brassica juncea × *B. hirta*	Ovary	Hybrid plants	Mohapatra and Bajaj (1987)
Brassica juncea × *B. campestris*	Ovary	Hybrid plants	Mohapatra and Bajaj (1988)
Brassica campestris × *B. oleracea*	Ovary	Hybrids	Matsuzawa (1978)
Brassica campestris × *B. oleracea*	Ovaries	Plants	Inomata (1979)
Brassica × *Raphanus*	Ovules	Plants	Takeshita et al. (1980)
Gossypium arboreum × *G. stocksii*	Embryos	Hybrid plants, seed formation after backcrossing	Gill and Bajaj (1984a, b, 1986)
G. herbaceum × *G. stocksii*			
G. arboreum × *G. stocksii*			
G. arboreum × *G. anomalum*			
Gossypium arboreum × *G. hirsutum*	Ovule	Hybrid plantlets	Gill and Bajaj (1987)
Glycine max × *G. tomentella*	Embryo	Hybrids	Newell and Hymowitz (1982); Broué et al. (1982)
Helianthus annuus × *H. hirsutus*	Embryo	Hybrids	Bohorova et al. (1985)
H. scaberimus × *H. annuus*			
Lathyrus clymenum × *L. articulatus*	Embryo	Hybrids	Pecket and Selim (1965)
Lens	Embryo	Hybrid	Cohen et al. (1984)
Linum perenne × *L. austriacum*	Embryo	Hybrids	Laibach (1925)
Lotus, several species	Embryos	Hybrids	Grant et al. (1962)
Medicago sativa × *M. rupestris*	Ovule-embryo	Plants	McCoy (1985)
Melilotus alba × *M. officinalis*	Embryo	Hybrids	Webster (1955)
Ornithopus pinnatus × *O. sativus*	Embryo	Tetraploid hybrids	Williams and de Lantour (1981)
Phaseolus vulgaris × *P. acutifolius*	Embryo	Hybrids	Honma (1956); Bajaj, Saettler and Adams (see Bajaj and Gosal 1982); Mok et al. (1978)
P. vulgaris × *P. ritensis*	Embryo	Hybrids	Braak and Kooistra (1975)
P. mungo × *P. calcaratus*	Embryo	Hybrid	Dana and Biswas (1975)
P. aureus × *P. angularis*	Embryo	Hybrid	Ahn and Hartman (1978)
P. vulgaris × *P. lunatus*	Embryo	Hybrid	Mok et al. (1978)
P. vulgaris × *P. angustissimus*	Embryo	Hybrid	Belivanis and Doré (1986)
Trifolium ambiguum × *T. hybridum*	Embryo	Hybrids	Keim (1953) Williams (1980)
T. pratense × *T. sarosiense*	Embryo	Hybrids	Phillips et al. (1982)
Vigna mungo × *V. radiata*	Embryo	Hybrids showing variation	Gosal and Bajaj (1983a, b)

3.1 *Arachis hypogaea* × *A. villosa* (Bajaj et al. 1982)

Arachis hypogaea, commonly known as peanut or groundnut, is one of the most important protein-rich oilseed crops, and occupies the third position in the world as a source of edible oil. For its improvement, naturally occurring wild germplasm has not been much utilized (Singh and Moss 1982), owing mainly to cross-incompatibility. In improvement programs, increased yield, high oil and protein, resistance to various pathogens, and early maturity have been the main objectives, and international efforts are being made to achieve them. Germplasm with these characters is available in the wild species (see Smartt and Gregory 1967; Spielman et al. 1979). However, in most of the cases, the crosses are either incompatible, or the frequency of hybridization is low. Interspecific hybridization in the genus *Arachis* (*A. hypogaea*×*A. correntina*), first reported in 1952 by Krapovick and Rigoni, was continued by other workers (Kumar et al. 1957; Smartt and Gregory 1967), and this subject has been reviewed (Smartt 1979). The existing variation being limited, there is a need for the induction of genetic variability through various in vitro techniques (Bajaj 1984, 1988c), including interspecific hybridization through embryo culture. Our work (Bajaj et al. 1982) on successful hybridization between the cultivated tetraploid ($2n = 40$) *A. hypogaea* and a wild diploid ($2n = 20$) species *A. villosa* (resistant to drought, pests, and the tikka disease, and containing high oil content – Cherry 1977) is summarized.

The plants of *Arachis hypogaea* L. cv. M13, and *Arachis villosa*, grown at the Experimental Fields of the Punjab Agricultural University formed the source of the material. The seeds were sown during July 1978 and 1979, and crosses were made during August–October. The flower buds of *A. hypogaea* were emasculated in the evening and pollinated on the following morning with suspensions of pollen (in sucrose 20% and boric acid 0.1%) of *A. villosa*. Thirty to 35 days after pollination, the developing pegs were surface-sterilized with 0.5% solution of mercuric chloride for 7–10 min and rinsed twice with sterile distilled water. The hybrid embryos were then excised under aseptic conditions, using a stereoscopic dissecting microscope in a Laminar Flow Chamber (Klenzaids, Bombay) and were cultured on modified Murashige and Skoog's medium (1962, indicated as MS) supplemented with various concentrations and combinations of indoleacetic acid (IAA), 2,4-dichlorophenoxyacetic acid (2,4-D), casein hydrolysate (CH) and kinetin (kin). The cultures were incubated at 23°–25 °C in diffused light.

The plantlets (4–6 weeks old) obtained from the embryos were first transferred to a basal medium containing 1 mg/l of naphthalene acetic acid (NAA). This procedure helped to induce more roots. When sufficient roots were developed, the plantlets were taken out of the tubes, the agar adhering to them was gently washed, and the plants were transferred to autoclaved soil in pots. To prevent desiccation, glass beakers were inverted over the pots (Fig. 2E). After 1 week, when the plants were established, the beakers were removed.

For chromosome counts, the root tips were treated with a saturated aqueous solution of 1-bromonaphthalene for 3–4 h, and then fixed in acetic alcohol (1:3). Squashes were made by staining with Feulgen and acetocarmine.

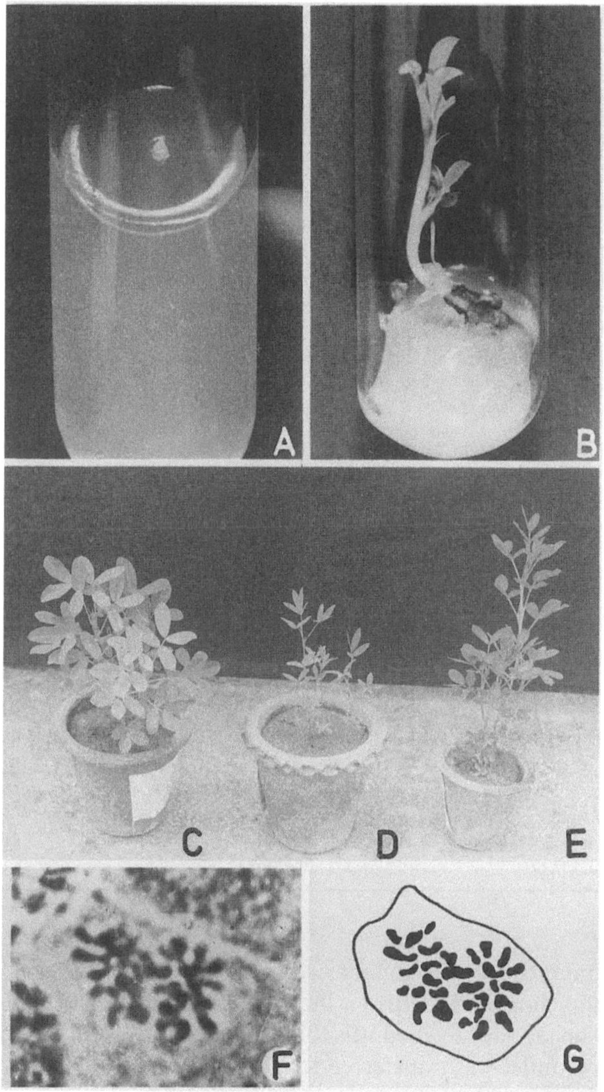

Fig. 2A–G. Interspecific hybridization between *Arachis hypogaea* × *A. villosa* through the culture of excised embryos. **A** An immature hybrid embryo (30 days after pollination) at the time of culture **B**. A hybrid plant obtained from an embryo cultured on MS+IAA (4 mg/l)+kin (2 mg/l) after 6 weeks. **C–E** The embryo derived plants of *A. hypogaea* (2 n = 40), *A. hypogaea* × *A. villosa* (2 n = 30), and *A. villosa* (2 n = 20), respectively. **F, G** Root tip squashes of the hybrid plant *A. hypogaea* × *A. villosa* with triploid chromosome number (2 n = 30) photographed under high power (**F**), and a camera lucida drawing (**G**). (Bajaj et al. 1982)

In Vitro Development of the Hybrid Embryos. Of the 9546 pollinations made (Table 3) with the pollen suspensions of *A. villosa*, 4770 (50%) gynophores penetrated the soil within 3 weeks. After the next 1 week, 1302 pegs were used for the culture of hybrid embryos, whereas the remaining 1076 pegs were left in the field. There was no further enlargement and seed-setting in these pegs.

Culture of Hybrid Embryos. The in vitro response of the hybrid embryos on various media is summarized in Table 4. The embryos cultured on MS+IAA 4 mg/l+kin 2 mg/l (MS 3 medium) produced plantlets (Fig. 2), whereas the addition of 2,4-D and casein hydrolysate caused excessive proliferation. On MS 3, the

Table 3. In vivo and in vitro studies for obtaining hybrid embryos (*A. hypogaea* × *A. villosa*). (Bajaj et al. 1982)

	No.
Flowers pollinated[a]	9546
Gynophores penetrating the soil	4770
Gynophores showing pod enlargement	2378
Gynophores left under natural conditions	1076
Pegs excised for culturing	1302
Embryos cultured on various media	669

[a] A drop of suspension of pollen of *Arachis villosa* prepared in solution of sucrose (20%) + boric acid was applied on the stigma of *A. hypogaea*.

Table 4. Growth responses of the hybrid embryos (*A. hypogaea* × *A. villosa*, 30 – 35 days after pollination) cultured on various media. (Bajaj et al. 1982)

Media	No. of embryos cultured	No. of embryos showing growth	Percent embryos showing growth	Remarks
MS + 2,4-D (1 mg/l) + kin (0.5 mg/l) + CH (500 mg/l)	156	12	7.6	Few embryos started to proliferate in the third week, and a mass of callus was formed in 6 weeks
MS + IAA (4 mg/l) + kin (2 mg/l)	306	19	6.2	The embryos developed into plants in 5 – 6 weeks. 10-week-old plants were 8 cm long with 3 – 5 pairs of leaves
MS + IAA (2 mg/l) + kin (0.5 mg/l)	207	5	2.4	10-week-old plants were 5 cm long with 2 – 3 pairs of leaves

embryos swelled and started to grow within 2 weeks. The epicotyl turned green, and elongated to form a shoot (Fig. 2B), the growth of the radicle was rather slow. The tap root formed was thick, short and very slow-growing. However, when such a plantlet was transferred to a basal medium + NAA (1 mg/l), the roots started to elongate. The hybrid plants were triploids and showed $2n = 30$ (Fig. 2F, G), i.e., an intermediate chromosome number between their parents. The plants were erect, and resembled *A. hypogaea*.

3.2 *Phaseolus vulgaris* × *Phaseolus acutifolius* (Fig. 3)

The common bean, *Phaseolus vulgaris*, is susceptible to common bacterial halo blight (*Pseudomonas phaseolicola*, *Xanthomonas phaseoli*) and root rot (*Fusarium solani*), whereas resistance to these are found in tepary bean (*Phaseolus acutifolius*). Interspecific crossing is incompatible, but hybrids can be obtained through embryo culture (Honma 1956). During 1967 – 1970, while working at the Michigan State University in collaboration with the U.S. Department of Agricul-

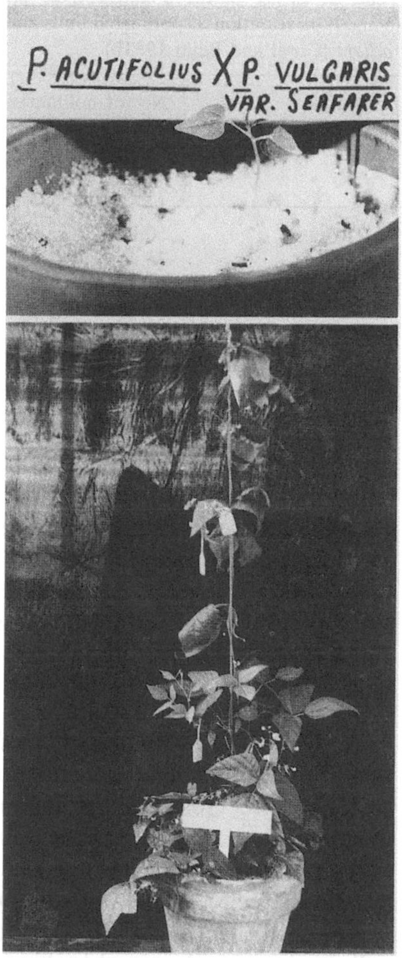

A

B

Fig. 3 A, B. Interspecific hybridization on *Phaseolus vulgaris*×*P. acutifolius* through embryo culture. **A** A hybrid plantlet transferred to a mixture of soil and perlite. **B** A mature hybrid plant undergoing flowering and occasional pod formation. (Bajaj and Gosal 1982)

ture, we (Bajaj, Saettler, and Adams) crossed these species by first applying lanolin paste containing growth regulators to the freshly pollinated flowers to prevent early shedding. This encouraged the embryos to develop to a stage when they could be excised and cultured on agar medium as well as on filter paper bridges in liquid medium (see Bajaj and Gosal 1982, Plate 1). The embryo-derived hybrids, on transfer to pots/greenhouse, continued to develop further (Fig. 3). Once again, lanolin paste was applied to the developing pods. Most of the pods abscissed and the ones which remained were sterile. Occasionally one seed per pod was formed. These seeds were generally small (for more information on interspecific hybridization in *Phaseolus* see Table 2 and Hucl and Scoles 1985).

3.3 *Vigna mungo* × *V. radiata* (Gosal and Bajaj 1983 a, b)

Hybrids were obtained in an otherwise incompatible cross involving blackgram (*Vigna mungo* ♀) and greengram (*Vigna radiata* ♂). Blackgram is resistant to

Table 5. Effect of the application of growth regulators GA (100 mg/l) + NAA (25 mg/l) + kin (5 mg/l) on retention of pollinated buds and developing pods in a cross involving *Vigna mungo* × *V. radiata* (Gosal and Bajaj 1983b)

Days after pollination	No. of pollinated buds and developing pods[a]		
	Control	Water application	Growth regulator application
0	50.00	50.00	50.00
3	36.66	40.33	42.66
6	26.33	32.33	39.33
9	11.66	21.00	24.66
12	4.66	7.33	19.00
15	4.33	5.66	10.33

[a] Values represent the mean of three replications.

yellow mosaic virus as well as to seed shattering, and greengram is very much prone to both these defects, but possesses a large number of seeds per pod and contains more easily digestible protein than that of blackgram. Whereas greengram can be easily crossed as a female to blackgram (Sen and Ghosh 1960; Boling and Matlock 1961), the reciprocal cross is not successful (Chowdhury et al. 1977; Singh 1981); however, through embryo rescue this has been attained (Gosal and Bajaj 1983a, b).

Effect of Growth Regulators on Pod Retention. The daily application of the solution of a mixture of growth regulators containing GA + NAA + kinetin to buds immediately after cross-pollination, and to the developing pods, considerably helped to retain a larger number of them longer. Whereas in the case of control only 8% of the pods were retained after 15 days, in the hormone-treated ones is increased to 20% (Table 5). In the untreated pods the number of ovules was 1–3 per pod and most of them shrivelled. On the other hand, the treated pods contained 2–5 well-developed and normal-looking ovules per pod. However, this increased number of ovules per pod was not comparable to the ones which were selfed (6–7 ovules/pod).

In Vitro Culture of Embryos. The parental as well as the hybrid embryos were cultured on MS + IAA + kinetin, fortified with either coconut water or casein hydrolysate. The parental embryos were more amenable to form complete plants than the hybrid embryos cultured at the same age and under similar conditions (Table 6). The young hybrid embryos (11 DAP) did not develop into normal-looking plants but most of them underwent callusing. The plants produced from such embryos were spindly, rootless, and albino. The maturing embryos (12–17 DAP) developed into plantlets (Fig. 4A–C). The radicle began to elongate within 4 days, followed by the unfolding of the small cotyledons. The appearance of primary leaves was followed by the elongation of the epicotyl. Plant recovery was higher in a medium containing casein hydrolysate than in a medium containing coconut water. However, the plants formed on the CW medium were healthier, quicker growing, and sturdier.

Table 6. In vitro growth response of parental as well as hybrid embryos cultured on MS + IAA (1 mg/l) + kin (0.2 mg/l) + CW (70 mg/l). (Gosal and Bajaj 1983a)

Embryo (parentage)	No. of embryos cultured	No. of embryos forming plants	Percent embryos forming plants
Vigna mungo	30	24	80.00
V. radiata	30	21	70.00
V. mungo × *V. radiata*	30	19	63.33

Raising the Parental and Their Resulting F_1 Plants in Pots. The in vitro obtained 4- to 5-week-old (Fig. 4D) parental as well as hybrid plantlets transferred to the pots were reared to maturity (Fig. 4E). In all, 287 plantlets with well-developed roots and shoots were obtained from the 498 cultured embryos. However, upon transferring to the soil, 143 plants established well, but only 89 reached maturity.

Data on various aspects of the morphological characters (Table 7) showed that the F_1 plants were intermediate between both parents with respect to leaf color, shape, number of branches, and height. Flower color is intermediate between the two parents. Whereas in the case of blackgram and greengram this is bright yellow and greenish yellow respectively, the F_1 flowers were dull yellow. The flowering was delayed by 9–15 days. Pod setting, total number of pods, and number of seeds per pod were highly reduced. However, at the same time, the 50-seed weight from the F_1 plants (F_2 seed) increased slightly. Most of the pods were incompletely filled, constricted and possessed single or, at the most, three seeds per pod. The pods generally collapsed at the distal end. The F_1 plants were healthy and green like those of the female parent, blackgram, and unlike those of greengram, which showed severe yellowing because of the yellow mosaic infection.

Cytological Studies of F_1 Plants. The chromosome number of both the parents and the F_1 plants, as determined from the pollen mother cells during diakinesis, metaphase-I, and anaphase-I was observed to be the same, i.e., $2n = 22$. The pairing behavior and the anaphasic movements of the chromosomes in the F_1 plants, however, pointed to their hybrid nature. In the parents, there was a regular formation of II at diakinesis and metaphase I. However, in the F_1 plants, out of the 119 cells studied, 30.2% showed irregular pairing, and hence most had I even at late diakinesis, 4,2% were with anaphasic bridges, and 11.7% showed unequal distribution at anaphase-I; the remaining cells showed regular meiosis.

Pollen fertility, as determined with acetocarmine staining, was 88.9%, 76.2%, and 46.2% in blackgram, greengram, and their hybrids, respectively. Moreover, the pollen grains produced by the F_1 plants were larger.

Field Studies of F_2 Plants. The seeds of the two parents as well as of the F_1 plants (Fig. 4F–H) were spaced planted in parallel rows in the field and the F_2 generation was raised (Fig. 4I). The comparative observations on morphological characters are shown in Table 7. Leaflet shape, color, size, and plant height were highly variable. However, pod-setting percentage, number of branches per plant, and number of seeds per pod were better than for the F_1 plants. There were some

Fig. 4 A–J. Interspecific hybridization between *Vigna mungo* (blackgram ♀) and *V. radiata* (greengram ♂) through, the culture of hybrid embryos. **A, B** Plantlets obtained from the hybrid embryos 3 and 4 weeks after culture on MS+IAA+kinetin+CW. **C** A hybrid plantlet from an embryo cultured on MS+IAA+kin+CH. **D** 5-week-old embryo-derived plants 4 days after transfer to soil. **E** An F_1 plant (10-week-old) at flowering stage. **F–H** Freshly harvested seeds of *Vigna mungo* (**F**), hybrid F_1 (**G**), and *V. radiata* (**H**). **I** 5-week-old plants (at vegetative stage) of *Vigna mungo*, *V. mungo* × *V. radiata* (F_2) and *V. radiata* in the experimental field. **J** 10-week-old F_2 plant in the field showing formation of pods. (Gosal and Bajaj 1983 b)

Table 7. Morphological characters of the parents, F_1 and F_2 plants of cross involving *Vigna mungo* and *V. radiata*. (Gosal and Bajaj, 1983b)

Character	*V. mungo*	*V. radiata*	F_1	F_2
Days to flower	43.80	37.80	52.00	42.20
Leaflet length (cm)	8.94	6.02	6.78	6.96
Leaflet width (cm)	2.02	5.08	2.92	3.22
Height at flowering (cm)	10.94	15.66	11.74	12.76
Number of branches	4.60	5.00	5.20	
Setting percentage	86.88	56.64	20.28	79.72
Height at maturity (cm)	16.98	29.96	19.12	18.48
Number of pods per plant	13.00	24.80	5.00	13.80
Number of seeds per pod	4.80	9.20	2.80	5.00
50-seed weight (mg)	2278.00	2103.40	2293.00 F_2 seed	2262.20 F_3 seed
Leaf color and shape	Narrow, long and dark green	Heart-shaped and light green	Intermediate to both parents in length and width and dark green	Different shapes resembling those of the F_1 plants and parents, dark green
Flower color	Yellow	Greenish yellow	Yellow	Greenish yellow to completely yellow
Pod length, color and extent of filling at maturity	3.5 – 4.5 cm long brownish and completely filled	6.5 – 7.5 cm long black and completely filled	Mostly single seed, constricted (2.5 – 3.5 cm long and incompletely filled) and black	3.5 – 5.0 cm long completely filled and brown to blackish
Seed color	Black	Green	Black and brownish black (F_2 seed)	Brownish black, brownish green and black (F_3 seed)
General appearance of the plants	Healthy and dark green throughout the growth period	Healthy and light green in the initial stages and yellowing caused by yellow mosaic virus towards rainy season	Healthy and green throughout the period, no infections were observed	Healthy and green through the period, no infections were observed

In *V. mungo, V. radiata* and F_1 the values are the mean of five different plants, but the values for F_2, are the means of 15 plants.

plants with a combination of different characters, in that few resembled greengram but showed no infection of yellow mosaic. Likewise, others were compact like blackgram, but with a branching pattern resembling that of greengram. Segregation also occurred for flower color and height of the plant, but the segregants did not excel any of the parents. The seeds produced from these field-grown F_2 plants showed large variation in color and size.

To conclude, the resulting material from an otherwise incompatible cross *Vigna mungo* (♀)×*V. radiata* (♂) possesses cytoplasm of *V. mungo*, unlike the conventionally obtained reciprocal cross. Desirable segregants can be obtained only if the genes put into the alien cytoplasm express their full potential. In the present case, the interaction between the nuclear genes and the cytoplasm appears to be favorable, as there was a good level of fertility in the F_1 plants. The F_2 segregation showed the quantitative type of inheritance pattern of characters, such as number of pods per plant, seeds per pod, and seed size. Thus the selection of better-yielding genotypes would be possible only by raising a large population for many generations.

3.4 Interspecific Hybridization in Oilseed *Brassica* Crops

The improvement of *Brassica* has mostly been confined to the exploitation of the naturally occurring genetic variation in the cultivated species. The existing variation being limited, breeders have to resort to wide hybridization, which provides an important method of incorporation of desirable characters such as resistance to disease, drought, and frost, which imbue the plant with wider adaptability. The genus *Brassica* has 159 species, which include a number of wild species of economic importance. Some of these species are difficult to hybridize due to certain crossability barriers, such as abortion of hybrid embryos. In order to enhance wide hybridization, various in vitro techniques of embryo (Matsuzawa 1978), ovule (Takeshita et al. 1980), and ovary culture (Inomata 1979), and recently protoplast fusion (see Chap. IV.2, 3, this Vol.) have been employed. Our work on interspecific hybridization in *Brassica* is summarized here.

3.4.1 Brassica napus×Brassica juncea Hybrids (Bajaj et al. 1986)

The plants of *Brassica napus* cv. Tower (2 n = 38) and *B. juncea* cv. RLM 198, RLM 514 (2 n = 36), were raised in the fields. To prevent selfing, only a few buds were retained in each inflorescence, which were emasculated and bagged. Pollinations were carried out the next morning, and the flowers were tagged.

The ovaries were excised 2–9 days after pollination (DAP), surface-sterilized with 0.25% mercuric chloride or chlorine water for 7–10 min, rinsed twice in distilled water and cultured (Bajaj, 1966). Likewise, ovules and embryos were excised from the immature pods 7–12 and 10–14 DAP respectively, and cultured. All manipulations were done under aseptic conditions in a laminar flow chamber (Bombay, Klenzaids). The basal media of Murashige and Skoog (MS 1962) and White (1943), supplemented with various combinations and concentrations of IAA, kin, CH, and CW were employed.

Table 8. Formation of hybrid seeds in excised ovaries of *B. napus* × *B. juncea* and *B. juncea* × *B. napus* cultured on various media at different days after pollination (DAP). (Bajaj et al. 1986)

Media	DAP	Size at culture		Size at maturity		No. of ovaries cultured	No. of ovaries showing seed formation	Percent ovaries showing seed formation
		Length	Breadth	Length	Breadth			
B. napus × *B. juncea*								
White's	2	0.59	0.031	1.20	0.21	74	28	37.8
MS	7	1.81	0.163	3.10	0.29	60	24	40.0
MS	9	2.69	0.210	3.90	0.30	60	48	80.0
B. juncea × *B. napus*								
CH	6	1.94	0.110	2.50	0.39	74	48	62.1
White's	8	2.56	0.120	2.98	0.32	68	46	67.6

The cultures were maintained at 25 °±2 °C. The embryo-derived plants with well-developed roots were taken out of the tubes, their roots were washed to remove agar, and the plants were then transplanted into the pots. Some of the plants were transferred to Kulu valley during May-June, and grown in the field. The F_1 hybrid seeds obtained from the in vitro cultured ovaries were sown in the field, and plants were raised to maturity.

Ovary Culture. The ovaries of the two crosses cultured from 2–9 DAP on various media (Table 8) showed various degrees of success, ranging from 37.8% to 80.0%. Whereas the simple basal media of White and MS supported the growth of older ovaries (8–9 DAP), the younger ovaries showed a low level of success (37.8%). Thus, the best stage for the culture of the hybrid ovaries is 8 or 9 DAP. The ovaries showed elongation within 4–6 days of culture. On the media supplemented with auxins and kin, the basal cut end of the ovary proliferated to form callus. In the cultured ovaries there was more increase in breadth than in length, whereas in the field there was a proportionate increase in length and breadth.

The ovaries showed considerable swelling at places where the seeds developed (Fig. 5A). The number of seeds per ovary ranged from one to four, and varied in color and size (Fig. 5B). On germination, the seeds produced fertile hybrid plants.

Ovule and Embryo Culture. The hybrid ovules and embryos cultured at various stages of development, i.e., 7–12, and 10–14 DAP respectively, showed a somewhat similar response. The ovules cultured on basal medium germinated to form plantlets, whereas on MS+IAA (1 mg/l)+kin (0.2 mg/l)+CH (500 mg/l), they proliferated and formed a mass of callus and plantlets. The in vitro response of the hybrid ovules cultured at various days after pollination is shown in Table 9. Likewise, hybrid embryos cultured on MS+IAA (1 mg/l)+kin (0.2 mg/l)+CH (500 mg/l) or CW (7%) showed swelling and elongation within 4 days, and formed multiple shoots and plantlets. In some cases, profuse proliferation of the hybrid embryo was observed and a mass of callus was occasionally formed. The comparison of the effect of media containing CH and CW on in vitro plant formation is shown in Table 10.

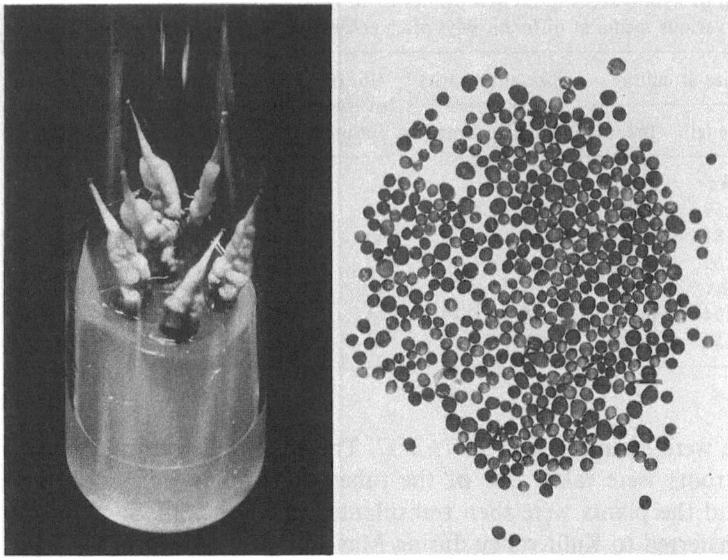

Fig. 5 A, B. In vitro formation of seeds from cultured hybrid ovaries of *Brassica napus* × *B. juncea*. **A** Ovaries 5 weeks after culture; note the bulging out of the seeds. **B** Seeds obtained from the cultured ovaries. (Bajaj et al. 1986)

Table 9. In vitro plantlet formation from hybrid ovule cultures of *B. juncea* × *B. napus* and *B. napus* × *B. juncea* on MS medium at different days after pollination. (Bajaj et al. 1986)

Days after pollination	Stage of embryo	*B. juncea* × *B. napus*			*B. napus* × *B. juncea*		
		No. of ovules cultured	No. of plants formed	Percent plant formation	No. of ovules cultured	No. of plants formed	Percent plant formation
7	Globular	210	104	49.5	–	–	–
8	Late globular	190	90	47.3	–	–	–
10	Heart-shaped	164	50	30.4	200	84	42.0
12	Late heart-shaped	–	–	–	160	60	37.5

The root tip squashes of the in vitro plants showed chromosome number 37, which is an intermediate between the two parents. The ovule- and embryo-derived hybrid plants, on transfer to the field, matured and produced seeds. The plants generally resembled their maternal parent, but some showed a wide range of morphological characters, such as height, size, and shape, and color of the leaves, which varied from green to dark green. The leaves were slightly hairy with larger lobes. The pods were mostly fertile, and the seeds also showed variation in size, color, and shape.

Table 10. In vitro response of the culture of embryos of *B. juncea, B. napus* and their hybrids on various media (embryos obtained from 10–14-day-old ovaries). (Bajaj et al. 1986)

Parent/cross	Media					
	CH 500 mg/l			CW 7%		
	No. of embryos cultured	Plantlet formation	Percent plant formation	No. of embryos cultured	Plantlet formation	Percent plant formation
B. juncea cv. RLM 198	126	98	77.7	118	104	88.1
B. napus cv. Tower	–	–	–	70	38	54.2
B. juncea × *B. napus*	126	92	73.0	–	–	–
B. napus × *B. juncea*	70	40	57.1	82	50	60.9

3.4.2 Brassica juncea × *Brassica hirta* (Mohapatra and Bajaj 1984, 1987)

Commercially grown species of oilseed *Brassica* (*B. juncea, B. napus, B. campestris*) are mostly susceptible to leaf blight (*Alternaria brassicae*). White mustard, European mustard (*Sinapis alba*, although resistant to blight, is incompatible, and so past attempts to hybridize *B. juncea* (2 n = 36) × *B. hirta* (2 n = 24) through conventional methods were unsuccessful (Downey et al. 1980). However, by employing embryo, ovary and ovule culture techniques, hybrids have been obtained (Mohapatra and Bajaj 1984, 1987).

The flower buds were emasculated, and pollinated the next day. The ovaries were excised 4–9 days after pollination, sterilized and cultured, whereas for ovule culture, 10–15 DAP ovaries were used.

Culture of Excised Ovules. The younger ovules (10 DAP), had a tendency to proliferate and formed callus, the older ovules (15 DAP), however, germinated. The parental ovules started to grow within 2 days of culture, and produced plants in 10 days. The hybrid ovules, on the contrary, required 25–30 days to germinate (Fig. 6A). The germination percentage of parental ovules in *B. juncea* and *B. hirta* was 60% and 44% respectively (Table 11), however only about 2% of the hybrid ovules germinated to form plants (Fig. 6B). Of all the media, the best response was observed on White's medium supplemented with CH (300 mg/l), and on MS + IAA (2 mg/l) + kin (0.5 mg/l) + CH (500 mg/l). Since the response of cultured ovules was not very good, further work was concentrated on the culture of excised ovaries.

Culture of Excised Ovaries. The excised ovaries cultured 4–9 DAP on various media (Table 12) yielded a varying degree of success. The best response in terms of hybrid seed set was obtained from 7-day-old ovaries cultured on White's medium containing 300 mg/l of CH. Relatively old ovaries (7 DAP) responded better than 4 DAP ovaries, the younger ovaries had a tendency to form callus at the basal cut end. In some cases shoots regenerated from the callus. The seed formation was inhibited in those cultures where callus was formed.

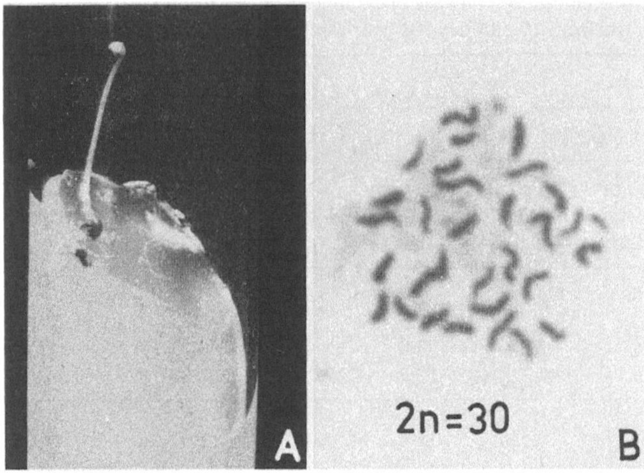

Fig. 6 A, B. In vitro culture of hybrid ovules (15 DAP) of a cross *Brassica juncea* × *B. hirta*. **A** Plantlet from a hybrid ovule 25 days after culture on MS+IAA (2 mg/l)+kin (0.5 mg/l)+CH (500 mg/l). **B** Root tip squash of a hybrid showing an intermediate chromosome number (2 n = 30). (Mohapatra and Bajaj 1984)

Table 11. In vitro growth response of parental as well as hybrid ovules (15 DAP) of *Brassica* cultured on MS + IAA (2 mg/l) + kin (0.5 mg/l) + CH (500 mg/l). (Mohapatra and Bajaj 1984)

Ovules (Percentage)	No. of ovules cultured	No. of ovules forming plants	Percent plantlet formation
Brassica juncea	60	36	60.00
Brassica hirta	45	20	44.44
B. juncea × *B. hirta*	260	6	2.31
B. hirta × *B. juncea*	210	4	1.90

In general, the ovaries started to elongate and to grow within a week of their culture. The places where the ovules grew swelled and eventually formed seeds. The number of seeds in each ovary varied (Table 12). In many cases, the growth of the ovary was normal but without seeds, perhaps due to lack of fertilization. Although a good number of seeds was obtained, most of them were poorly developed and wrinkled. Hybrid ovaries of *B. juncea* × *B. hirta* and *B. hirta* × *B. juncea* were cultured, but hybrid seeds were obtained only from *B. juncea* × *B. hirta* (Table 13). This indicates that perhaps *B. juncea*, being a digenomic (AABB) species, has some buffering effect in fertilizations with *B. hirta* pollen, whereas *B. hirta*, a monogenomic (DD) species, when taken as a female, did not produce any seed.

Germination of F_1 (Ovary-Derived) Seeds in the Field. Both parental and hybrid ovary-derived seeds were sown in the field (at the Regional Research Station, Ranital Balasore) in parallel rows with 30 cm between seeds, and 50 cm between rows, and were raised to mature plants. Thirty to 40% of the parental seeds ger-

Table 12. Effect of various media and age of the ovary on seed setting in an incompatible cross (*B. juncea* × *B. hirta*). (Mohapatra and Bajaj 1987)

Media		4 DAP		5 DAP		7 DAP		9 DAP	
		No. of ovaries cultured	Average no. of seeds per silique	No. of ovaries cultured	Average no. of seeds per silique	No. of ovaries cultured	Average no. of seeds per silique	No. of ovaries cultured	Average no. of seeds per silique
MS	Basal MS	45	0.35	45	0.85	45	0.6	45	–
MS_1	MS + IAA (2.0 mg/l) + kin (0.5 mg/l)	27	0.45	27	0.60	28	1.25	26	0.50
MS_2	MS + NAA (0.5 mg/l) + kin (2.5 mg/l)	34	0.85	35	2.60	34	1.50	33	0.25
MS_3	MS + NAA (0.5 mg/l) + kin (2.5 mg/l) + coconut milk (15%)	39	0.35	40	0.60	38	0.45	39	0.20
MS_4	MS + CH (500 mg/l)	21	0.75	21	1.20	20	1.25	22	0.40
W_1	White's basal + CH (300 mg/l)	40	1.00	40	1.90	40	1.50	40	1.60

Table 13. In vitro response of the ovaries of *B. juncea, B. hirta* and their hybrids cultured on basal MS medium supplemented with CH. (Mohapatra and Bajaj 1987)

Parents/Cross	No. of ovaries cultured	No. of ovaries growing	No. of ovaries forming seeds	Percent success
B. juncea	35	24	10	41
B. hirta	40	28	7	18
B. juncea × *B. hirta*	370	265	9	4
B. hirta × *B. juncea*	285	193	0	0

Table 14. Comparative morphology of two ovary-derived parents and a F_1 hybrid (*B. juncea* × *B. hirta*). (Mohapatra and Bajaj 1987)

Characters	*B. juncea*	*B. hirta*	F_1 hybrid
1. Plant height	122 cm	143 cm	148 cm
2. Stem	Medium thickness and stout	Thick and hollow	Medium thickness and stout
3. Number of primary branches	4	10	7
4. Leaves	Petioled, basal leaves are large with terminal round lobe and upper leaves are lyrately lobed	Petioled, hairy and pinnatipartite	Petioled, hairy and intermediate shape to both the parents
5. Days to first flowering	9 weeks	11 weeks	Flowers appeared after 10th week but did not open
6. Siliqua	Semi-erect, thin, narrow with short beak. Pod size 3 cm	Nearly horizontal, thick with long beak. Pod size 2.3 cm	Rarely present, semi-erect with intermediate beak length, pod size 2.5 cm, no seed
7. Maturity period	125 days	140 days	135 days

minated, whereas only 5 plants emerged from 30 hybrid seeds (16.6%). The parental seeds took 2–4 days for germination, but the hybrid seeds required almost 7 days. In the beginning, the growth of the hybrid plants was slow and after 1 month the plants looked sturdy.

Characters of the Hybrid Plants. The characters of the two parent plants and a hybrid plant obtained through ovary culture are shown in Table 14. Out of five hybrid plants, three bore abnormal flowers, and two remained in the vegetative phase. The hybrid plant derived through ovule culture showed certain male characters. The comparative observations on morphological characters of the ovary-derived plants indicated their hybrid nature, which was confirmed by the chromosome numbers of the root tips ($2n = 30$). This chromosome number is intermediate between the number of their parents *B. juncea* ($2n = 26$) and *B. hirta* ($2n = 24$) (Fig. 6).

The in vitro production of interspecific hybrids in an otherwise incompatible cross, *B. juncea*×*B. hirta*, and their successful transfer to the field will eventually enable the incorporation of this germplasm for further crossing and utilization in further research on oilseed *Brassica* species.

3.4.3 Brassica juncea×*B. campestris* (Mohapatra and Bajaj 1988)

Ovary culture was employed for the production of interspecific hybrids of the partially incompatible cross, *Brassia juncea* ($2n = 36$)×*Brassica campestris* ($2n = 20$). Five- to seven-day old ovaries cultured on White's medium supplemented with casein hydrolysate (300 mg/l) and sucrose (5%) produced more seeds than any other media tried, but seed development was better on media fortified with plant hormones. The seed yield was better in *B. juncea*×*B. campestris* than in their reciprocal cross. The plants obtained from ovary-derived seeds were transferred to the field; they were intermediate in some morphological characters and chromosome number ($2n = 28$) compared to their parents. The flower buds generally did not open and had poorly developed anthers with mostly sterile pollen. The pod size/setting was very much reduced, but some healthy seeds were obtained (Mohapatra and Bajaj 1988). Initially the excised ovaries were cultured on MS medium supplemented with CH (500 mg/l). Four- to nine-day-old ovaries of both the parents, when cultured on MS+CH (500 mg/l) started to grow within 1 day, whereas in those of the cross, signs of growth were visible after 2 days. The subsequent rate of growth in the parents was also faster, and 41%–46% of the ovaries set well-developed seeds as compared to poorly developed 13.8% in the cross (Table 15). Thus, in order to improve seed set and to obtain well-developed seeds, various factors, such as age of the ovary at the time of culture, and the effect of different growth regulators were studied.

Effect of Age of the Ovary. The age of the ovary at the time of culture had a profound effect on its further development and seed set. The seed set response of ovaries of various ages (4, 5, 7, 9 DAP) cultured on different media is shown in Table 16. Five and seven DAP ovaries gave the best response, and a maximum of 66% of the ovaries set seeds. Four DAP was too early for culture, as in most of the cases the ovaries did not set seed, whereas in 9 DAP the seed set was also less, it seems that perhaps because the process of abortion of the embryo had already set in, resulting in poor seed set. Thus in subsequent experiments 5 and 7 DAP ovaries were cultured.

Table 15. In vitro response of the excised ovaries of *B. juncea, B. campestris* and their hybrids cultured on basal MS medium supplemented with CH. (Mohapatra and Bajaj 1988)

Parents/Cross	No. of ovaries cultured	No. of ovaries growing	No. of ovaries forming seeds	Percent success
B. juncea	35	24	10	41.3
B. campestris	36	26	12	46.1
B. juncea×*B. campestris*	112	80	11	13.8

Table 16. Effect of age of hybrid ovaries of *B. juncea* × *B. campestris* on seed development on different media. (Mohapatra and Bajaj 1988)

Days after pollination	Media					
	MS	MS 1	MS 2	MS 3	MS 4	W 1
4	− (0)	− (0)	+ (20)	− (0)	+ (11)	+ (20)
5	+ + (66)	+ + (50)	+ (40)	+ + (37)	+ + (66)	+ + (16)
7	+ (49)	+(25)	+ (16)	+ + + (16)	+ + (44)	+ + (48)
9	+ + (16)	+ (25)	− (0)	− (0)	− (0)	+ + (20)

− No seed development; + Poorly developed seeds; + + Average developed seeds; + + + Well-developed seeds. Figures in parentheses indicate percentage of seed-forming ovaries.

Table 17. Effect of various media on seed setting in vitro of excised ovaries of an interspecific cross *B. juncea* × *B. campestris*. (Mohapatra and Bajaj 1988)

Medium	No. of capsules examined	No. of seeds obtained	Average seed set per capsule
MS	25	32	1.3
MS 1	16	14	0.9
MS 2	21	18	0.9
MS 3	24	12	0.5
MS 4	44	20	0.5
W 1	22	27	1.4

Effect of Media. A comparison of different media with growth regulators was undertaken (Table 17). The hybrid ovaries of *B. juncea* × *B. campestris*, when cultured on basal MS medium, produced small roots at the basal cut end after 7 days of culture, but without any callus formation. However, on W1 and MS4, well-developed roots were produced without callusing; although the rate of growth of the ovaries was slow, the patterns of growth were similar in MS and W1. MS2 medium formed profuse callus without any root development. The highest average seed set per capsule in medium W1 and MS was quite comparable (1.4 and 1.3 seeds per ovary respectively) followed by MS1 and MS2 (0.9 per ovary), MS3 and MS4 (0.5 per ovary). However, the seed development was better in MS3, followed by MS4, W1, MS and MS2 media.

Hybrid Seed Germination of F_1 Plants. The ovary-derived seeds of the interspecific cross showed about 12% germination in pots. Three-week-old plants, when transferred to the field, continued further growth and underwent flowering. The hybrid plants could readily be differentiated from their parents (Table 18, Fig. 7). The most conspicuous differences were observed in leaf shape and flower-bud opening. Similarly, clearcut differences were observed in floral morphology, pod setting, and seed development. The flower buds of the hybrid plants rarely opened and the anthers were small. Pod set, pod growth, pod size, and seed set were very much reduced in the hybrids. The plants were intermediate in height.

Table 18. Comparative morphology of the parents and the ovary culture derived hybrids (*B. juncea* × *B. campestris*) (results based on average of ten plants). (Mohapatra and Bajaj 1988)

Characters	*B. juncea* cv. RLM 198	*B. campestris* cv. TL 15	F_1 hybrid
1. Plant height	137.0 cm	78.0 cm	98.0 cm
2. Stem	Thick and stout	Slender and hollow	Thick and stout (like *B. juncea*)
3. Number of primary branches	8	6	6
4. Leaves	Petioled; basal leaves are large with terminal rounded lobe and upper leaves are lyrately lobed; glabrous or hairy	Auricled-stem clasping; terminal acute to obtuse; glabrous or hairy; waxy or bloomless	Basal leaves like *B. juncea* (petioled) upper leaves like *B. campestris* (auricled)
5. Days to 1st flowering	11 weeks	8 weeks	8–9 weeks
6. Inflorescence	Corymbose raceme. Petals – yellow, Androecium – tetradynamous	Corymbose raceme. Petals – bright yellow, Androecium – tetradynamous	Corymbose raceme petals – dull yellow, Anthers below the stigma (heterostyle)
7. Siliqua	Semi-erect, thin, narrow with short beak. Pod size – 3.8 cm	Large, flat with short beak. Pod size – 4.5 cm	Intermediate with long beak. Pod size – 2.0 cm
8. Maturity period	160 days	92 days	98 days
9. 1000-seed weight (in g)	3.36	3.37	3.2
10. Seed color	Brown	Dark brown	Dark brown
11. Seed setting	11.8	12.9	0.01

Table 19. Pollen fertility/sterility study of the parents and the in vitro produced hybrids. (Mohapatra and Bajaj 1988)

Parents/hybrids	No. of pollen studied	No. of pollen fully fertile	No. of pollen partially fertile	No. of pollen completely sterile	Percent fertility
B. campestris	280	265	0	15	94.6
B. juncea	220	210	0	10	95.4
B. juncea × *B. campestris* (F_1)	676	43	273	360	26.6

Study of Pollen and Chromosome Numbers. The comparison of the pollen of the hybrids and their parents revealed high sterility in the hybrid (Table 19). The anthers were poorly developed, and the pollen mostly empty and abortive. As compared to more than 94% pollen fertility in the parents, the hybrid pollen showed only 26.6% fertility, as evidenced by acetocarmine staining.

Fig. 7. Interspecific hybridization in *Brassica juncea*×*B. campestris* through ovary culture. Eight-week-old plants of *B. juncea*, *B. campestris* and their hybrid. (Mohapatra and Bajaj 1988)

3.5 Interspecific Hybridization in Cotton

Cotton is the most important source of fiber. There are four cultivated cotton species, two diploids (*Gossypium arboreum*, *G. herbaceum*, $2n = 26$) and two tetraploids (*G. hirsutum*, *G. barbadense*). In addition, there are 35 wild species, which, though shortfiberd or lintless, have a number of desirable traits, such as fiber quality and resistance to insects, drought, and diseases (Hutchinson et al. 1947; Prentice 1972). Although there exists a lot of genetic variability, yet the routine methods to transfer the desirable characters into the cultivated species have been unsuccessful due to various incompatibility barriers, such as early abortion of the hybrid embryo (Weaver 1958; Pundir 1972). Even in cases where hybridization is effective, the frequency of hybrid formation is very low, and eventually there is a hybrid breakdown (Gerstel and Philipps 1958). Through the use of embryo and ovule culture, hybrids have been obtained in a number of species of *Gossypium*.

3.5.1 *Gossypium arboreum* × *G. hirsutum* (Gill and Bajaj 1987)

Gossypium hirsutum (2 n = 52) and *G. arboreum* (2 n = 26) are the two species which occupy a major portion of the area under cotton in India. *G. hirsutum* types generally yield more, possess longer fiber of superior quality, and are resistant to diseases such as *Fusarium* wilt, whereas *G. arboreum* can better tolerate environmental stresses and is comparatively resistant to various insect pests. Efforts to combine the desirable characters of these two species have not been encouraging, due to abortion of the hybrid embryo/endosperm in the early stages of seed development (Weaver 1957; Pundir 1972; Steward and Hsu 1978). Thus in our ongoing cotton improvement program, tissue culture methods have been incorporated with a view to increasing genetic variation (Bajaj and Gill 1985, 1987).

Cultivars of *Gossypium hirsutum* (F414 and LH372) and *G. arboreum* (G27 and LD133) were grown in the experimental fields of the Punjap Agricultural University at Ludhiana and reciprocal crosses between these species were made in the field. A mixture of growth regulator solution (gibberellic acid 50 mg/l + naphthalene acetic acid 100 mg/l) was applied in cotton wool at the base of the pedicle of the pollinated flowers on the day of pollination and the day after pollination. The developing bolls (cross-pollinated as well as selfed) were excised 3 days after pollination (DAP), sterilized by dipping them in 70% ethyl alcohol, and subjected to flaming. The ovules were scooped out of the bolls aseptically in a laminar flow chamber and cultured on agar solidified MS medium or in liquid SH medium (Stewart and Hsu 1978) with or without growth regulators. The cultures were transferred to fresh medium every 4 weeks. The germinating ovules were transferred to a medium containing half the concentration of MS salts and 1% sucrose. The culture conditions and other details were as provided earlier (Gill and Bajaj 1984a).

Effect of Growth Regulators. In both parents there was normal development of the bolls; however, after interspecific pollination, the flowers generally abscised

Table 20. Effect of growth regulators (GA3 50 mg/l + NAA 100 mg/l) on boll retention in interspecific crosses in *G. hirsutum* × *G. arboreum* and their reciprocals (Gill and Bajaj 1987)

Cross	Treated			Control		
	No. of flowers crossed	No. of bolls shed	Percent retention	No. of flowers crossed	No. of bolls shed	Percent retention
G. hirsutum × *G. arboreum*						
F 414 × G 27	71	20	71.8	36	30	16.7%
F 414 × LD 133	62	13	79.0	15	14	6.6%
LH 372 × G 27	56	24	57.1	31	29	6.4%
LH 372 × LD 133	34	16	52.9	17	14	17.6%
G. arboreum × *G. hirsutum*						
G 27 × F 414	28	11	60.7	11	11	0
LD 133 × F 414	16	9	43.7	17	17	0

Table 21. Percentage of ovules giving seedlings and the callus growth response of excised ovules (3 days after pollination) of *G. arboreum* and *G. hirsutum* cultured on various media (figures in parenthesis indicated the number of ovules cultured). (Gill and Bajaj 1987)

Medium (mg/l)	*G. arboreum* germination (%)	Callus[a] development	*G. hirsutum* germination (%)	Callus development
MS basal	17.6 (125)	+	8.3 (96)	–
MS+CH 250	29.2 (96)	+	22.2 (99)	–
MS+CW 7%	21.2 (80)	+	25.4 (110)	–
MS+CH 250+IAA 0.5+kin 0.2	22.2 (108)	+	23.8 (122)	–
MS+CH 250+IAA 0.5+kin 0.5	23.7 (93)	+	25.0 (120)	–
MS+CH 250+IAA 1+kin 0.2	28.1 (121)	+	44.4 (133)	+ +
MS+CH 250+IAA 1+kin 0.5	20.3 (103)	+ + +	24.1 (108)	+ +
MS+CH 250+IAA 1.5+kin 0.5	37.5 (112)	+ + +	18.8 (117)	+ +

– no callus; + little callus; + + good callus; + + + profuse callus.

Table 22. Percentage germination and callus growth response of hybrid ovules cultured on various media (data based on 100 ovules cultured in each experiment). (Gill and Bajaj 1987)

Media (mg/l)	*G. hirsutum* cv. F414× *G. arboreum* cv. G27		*G. arboreum* cv. G27× *G. hirsutum* cv. F414	
	Germination (%)	Callus[a]	Germination (%)	Callus
MS+CH 250+IAA 1+kin 0.2	27.5	+	10.7	+ +
MS+CH 250+IAA 1.5+kin 0.5	14.3	+	34.7	+ + +
SH (basal)	0	–	0	–
SH+IAA 0.5	0	+ + +[b]	0	–
SH+IAA 1	0	+ + +[b]	0	+ +

[a] – no callus; + little callus; + + good callus; + + + profuse callus.
[b] Fibers were also formed in these cultures.

within 3 days. The application of a solution of GA (50 mg/l) and NAA (100 mg/l) to the flowers immediately after cross-pollination and the following day considerably enhanced their retention (Table 20).

In the controls (untreated) the flower retention with 3 days of pollination was only 6.4% to 17.6% in *G. hirsutum*×*G. arboreum*. When *G. arboreum* was used as the female parent, all the bolls were shed after 3 days. The process of shedding started on the second day after pollination, and by the third day the bolls were either completely shed or they fell on slight disturbance. However, with the application of growth regulators, depending on the cultivar, the retention varied between 52.9% and 79% in *G. hirsutum*×*G. arboreum*, and 43.7% and 60.7% in the reciprocal crosses.

Culture of Hybrid Ovules. The hybrid ovules were cultured on MS agar as well as on SH liquid medium (Tables 21, 22). On basal SH liquid medium there was no visible growth or germination, the ovules gradually sank to the bottom and

Fig. 8A–H. Regeneration of plants from in vitro cultured ovules and embryos of various interspecific crosses in *Gossypium*. **A–C** Growth of ovules of *G. hirsutum*×*G. arboreum* (cultured 3 days after pollination) on Stewart and Hsu's (1978) medium containing IAA (0.5 mg/l) after 14, 21 and 28 days of culture, respectively. **D** *G. arboreum*×*G. hirsutum* ovules (3 DAP) after 50 days of culture on MS+IAA (1.5 mg/l)+kin (0.5 mg/l)+CH (250 mg/l); note the profuse proliferation, and the radicular root. **E** A plantlet obtained from a *G. arboreum*×*G. hirsutum* ovule (3 DAP) after 85 days of culture on MS+IAA (1 mg/l)+kin (0.2 mg/l)+CH (250 mg/l). **F** A plantlet formed from *G. arboreum*×*G. stocksii* embryos (15 DAP) after 35 days of culture on MS+IAA (1.5 mg/l)+kin (0.5 mg/l)+CH (250 mg/l) showing root and shoot formation. **G** Diversity in plants derived from the embryos of a three-way cross (*G. arboreum* ×*G. herbaceum*)×*G. somalense* (8 months old). **H** *G. arboreum*×*G. anomalum* hybrid at flowering stage after 8 months of culture (at 3 months it was transferred to the greenhouse). (Bajaj and Gill 1985)

turned brown. When IAA (0.5 or 1 mg/l) was added to this medium, there was profuse callusing (Fig. 8A–C), but no germination even after 4 months of culture.

On MS agar medium, the parental as well as the hybrid ovules underwent callusing and their responses were similar. However, the germination of hybrid ovules was slower in comparison to their respective parents, and no germination was observed in the absence of IAA and kin, whereas the parents ovules germinated even on basal medium. *G. hirsutum*×*G. arboreum* ovules produced more vigorous seedlings than the reciprocal hybrid.

The seedlings from the in vitro germinated ovules were transferred to a fresh medium (half-strength MS with 1% sucrose) and placed in an illuminated room,

Table 23. Variability in chromosome number in hybrid callus cells obtained from embryos of a cross *Gossypium hirsutum* (2n = 52) × *G. arboreum* (2n = 26). Data based on 477 dividing cells. (Bajaj and Gill 1985)

Chromosome no.	No. of cells studied	Percent
25	7	1.47
26–29	70	14.67
30–33	105	22.01
34–37	161	33.75
38–41	123	25.79
42–45	7	1.47
46	4	0.84

where they continued further growth. The seedlings from *G. arboreum*×*G. hirsutum* crosses had a lower survival rate (40%) than the seedlings from the reciprocal crosses. The cytological examination showed a triploid chromosome number (2 n = 3x = 39) in the root tips of hybrid plantlets.

Hybrid Embryo-Derived Callus. The embryos of a cross *G. hirsutum*×G. arboreum, cultured on Smith et al. medium (see Stewart and Hsu 1978) proliferated to form callus. This was then subcultured periodically on MS+NAA (2 mg/l)+BAP (0.5 mg/l). The callus was creamy-white, soft, and friable. Cytological examination of 1-year-old callus showed a wide range of chromosome number, varying from less than 26 to more than 45 (Table 23). However, the majority of the cells belonged to the category 26–41. The cells with less than 26 and more than 45 chromosomes were rare, i.e., 1.4% and 0.84%, respectively. No polyploids were observed.

The formation of hybrid callus which shows greater variations can be utilized to generate environments for enhancing chromosome breakage and reunion events (Larkin and Scowcroft 1981). Such a hybrid callus may thus provide means for obtaining genetic exchange in the interspecific hybrids. The plants from hybrid callus are likely to show an enhanced frequency of genetic variation. This is an area which needs to be explored, and which may open wide vistas for future plant breeding programs.

3.5.2 Hybridization Among Diploid Gossypium Species (G. arboreum, G. anomalum, G. herbaceum, G. stocksii (Gill and Bajaj 1984a, b, 1986)

Effect of Growth Regulators on Boll Retention. The daily application of a mixture of growth regulators (GA_3 50 mg/l+NAA 100 mg/l) to the flowers immediately after crossing considerably enhanced their retention and ensured further development in all three crosses (Table 24).

In the controls (untreated), the boll retention after 15 days of pollination was only 11.5% in *G. herbaceum*×*G. stocksii*, 14.8% in *G. arboreum*×*G. anomalum* and 19.1% in *G. arboreum*×*G. stocksii*. The shedding started after 4 days of pollination, and the abscission occurred at a point where the pedicle is jointed to

Table 24. Effect of growth regulators (GA3 50 mg/l + NAA 100 mg/l) on boll retention in various interspecific crosses in *Gossypium*. (Gill and Bajaj 1984b)

Crosses	Control			Treated		
	No. of flowers crossed	No. of bolls shed	Percent retention	No. of flowers crossed	No. of bolls shed	Percent retention
1. *G. arboreum* × *G. stocksii*	21	17	19.1	86	32	62.8
2. *G. arboreum* × *G. anomalum*	27	23	14.8	102	35	68.3
3. *G. herbaceum* × *G. stocksii*	26	23	11.5	68	30	55.8

Table 25. In vitro growth response of the hybrid embryos cultured (15 days after pollination) on MS + IAA (1.5 mg/l) + CH (250 mg/l) + kin (0.5 mg/l) + sucrose 3 percent. (Gill and Bajaj 1984a)

Cross	No. of embryos cultured	No. of plants formed	Percent plant formation
1. *G. arboreum* × *G. stocksii*	31	22	71
2. *G. arboreum* × *G. anomalum*	74	51	69
3. *G. herbaceum* × *G. stocksii*	39	19	48

the stalk. However, when the growth regulators were applied, boll retention was increased to 55.8% in *G. herbaceum* × *G. stocksii*, 62.8% in *G. arboreum* × *G. stocksii*, and 68.3% in *G. arboreum* × *G. anomalum*. The best response was observed in the cross *G. arboreum* × *G. anomalum*, where the retention increased from 14.8% to 68.3%. The base of the pedicle became rather thick and strong, and abscission was prevented.

Culture of Hybrid Embryos. The hybrid embryos of three crosses, i.e., *G. arboreum* × *G. stocksii*, *G. arboreum* × *G. anomalum*, and *G. herbaceum* × *G. stocksii* cultured on MS + indoleacetic acid (1.5 mg/l) + kin (0.5 mg/l) + casein hydrolysate (250 mg/l) and 3% sucrose and incubated in the dark, showed an overall increase in size during the first 2 weeks. The cotyledons enlarged, the radicle underwent elongation and penetrated the medium in most of the cases. Upon exposure to light, the cotyledons turned green, the first leaf appeared in 30–35 days, and the embryos developed into small seedlings (Fig. 8F). In cases where the embryos did not produce seedlings, either the radicle failed to penetrate the medium or the root stopped growing after a few days.

The response of the cultured embryos varied in different crosses (Table 25). The best growth response was observed in *G. arboreum* × *G. anomalum*, although the number of embryos producing seedlings was slightly more in *G. arboreum* × *G. stocksii*. The *G. arboreum* × *G. anomalum* embryos exhibited fast growth and most of them produced seedlings in 30–35 days, whereas in *G. arboreum* × *G. stocksii*, it took at least 35 days. In *G. herbaceum* × *G. stocksii*, only 48% embryos produced healthy seedlings. Seedlings with first pair of leaves were obtained in 35 days (Fig. 8F).

Fig. 9. Embryo-derived hybrid plants (*Gossypium arboreum* × *G. anomalum*) 6 months after transfer to the field (total time taken from embryo culture to flowering stage is 11 months) being examined by Dr. Manjeet S. Gill (*left*) and Professor Y.P.S. Bajaj (*right*). (Bajaj and Gill 1985)

The plants were transferred into the pots at the stage with two- to three-leaf stage with well-developed roots. There was 80% – 90% survival of potted plants, which were then grown in the greenhouse and field (Figs. 8H, 9). The hybrids grew normally, *G. arboreum* × *G. anomalum* hybrids showed vigorous growth, and attained almost double the height of either parent. This hybrid also produced a large number of flowers as compared to their parents as well as to other hybrids. The hybrids in all three crosses were perennial in nature and remained green throughout the winter season (Nov–Feb). The hybrids of *G. arboreum* × *G. anomalum* flowered throughout the year. The *G. herbaceum* × *G. stocksii* hybrids, however, produced only a few flowers in winter, *G. arboreum* × *G. stocksii* did not flower in winter at all, but resumed flowering in summer.

Effect of Light and Dark. The embryos incubated in the dark for the first 15 days, followed by exposure to light, yielded normal seedlings. The effect of light vs. dark on percent plantlet formation from the embryos of *G. arboreum* and *G. herbaceum* is shown in Table 26. When the embryos were kept in the dark, most of them exhibited normal growth, and produced vigorous seedlings. These embryos increased in size during the first 2 weeks when kept in the dark, then turned green upon exposure to light, and produced plantlets. However, when the embryos were exposed to light from the very beginning, they turned green at a very early stage.

Table 26. Effect of light and dark treatments on the in vitro growth response of cotton embryos cultured (15 days after pollination) on MS + IAA (1.5 mg/l) + CH (250 mg/l) + kin (0.5 mg/l) + sucrose 3 percent. (Gill and Bajaj 1984b)

Species	Treatment	No. of embryos cultured	No. of plants formed	Percent plant formation
G. arboreum	Dark (15 days)	36	27	75.0
	Light	32	14	43.7
G. herbaceum	Dark (15 days)	33	20	60.6
	Light	34	13	38.2

The radicle failed to elongate, and did not penetrate the medium in most of the cases. The hypocotyl was thin, and consequently the seedlings produced were also comparatively weak.

Utilization of Hybrids in Backcrossing. The F_1 plants predominantly exhibited characters of the wild parent, as is expected, since wild plant characteristics are usually dominant (Mehboob Ali et al. 1964). The *G. arboreum* × *G. anomalum* hybrid showed particularly vigorous growth. The hybrids flowered after 5–6 months of transfer into the field, and all were perennial and self-sterile. The *G. arboreum* × *G. anomalum* hybrid produced a large number of empty bolls. Hybrids involving *G. stocksii* had smooth stems, showed creeping habit and were not weak, as observed by Mehboob Ali et al. (1964). In hybrids involving *G. arboreum*, red petal color and red stem were dominant. In all the hybrids, flower size and shape, bract size and shape, and pedicel length resembled the wild species.

Efforts to backcross the F_1 plants with their respective cultivated parent met with limited success. The *G. arboreum* × *G. anomalum* hybrid produced some seeds when pollinated with *G. arboreum* pollen or vice versa. Seeds set on the hybrid plant possessed strong and fine fiber as has also been reported by earlier workers (Silow 1941; Afzal et al. 1945; Kalyanaraman and Santhanam 1955; Iyengar et al. 1957). In the backcross progeny, segregation was observed for plant color, leaf shape and size, bract shape and size, growth habit, and petal color. Only one plant was green and it had most of the characters of *G. anomalum* with the exception of small flower size. Only one plant in the backcross progeny produced seeds upon self-pollination. However, second backcross generation seeds were obtained from all the plants after pollinating with *G. arboreum* pollen or vice versa (Gill and Bajaj 1986).

To summarize, interspecific hybrids in cotton can be successfully achieved through the culture of excised embryo/ovules. Normally in many crosses, especially between diploids and tetraploids, the hybrid embryos abort at a very early stage. However, this can be delayed by the application of growth hormones, and when grown to a stage where the embryo is few-celled, the ovules could be excised and cultured on agar medium to form hybrids. The hybrids (triploids) between *G. hirsutum* and *G. arboreum* were sterile. The pollen from triploid plants was used to pollinate *G. hirsutum* and *G. arboreum* and a few shrivelled seeds were obtained.

The hybrids between diploid and diploid could be easily obtained. In all the hybrids, flower size and shape, bract size and shape, and pedicel length resembled the wild species. The hybrid *G. arboreum* × *G. anomalum* showed particularly vigorous growth and produced some seeds when pollinated with *G. arboreum* pollen or vice versa. However, second backcross generation seeds were obtained from all the plants.

4 Conclusions and Prospects

Legumes are rich in proteins, whereas oilseed crops are a concentrated source of energy; crops such as soybean and peanut are rich in both. They are also the basic material for livestock feed. The main objectives in the improvement of legumes are the increase in grain yield, methionine, cystine, protein-digestibility, and anti-enzyme, and reduction of seed hardness and cooking time. Likewise, in oilseed crops, disease resistance is of primary interest. There have been very few technological breakthroughs because of the numerous diseases to which they are susceptible. During the last decade there has been no significant increase in the yield per acre of these crops, which has been mainly attributed to the lack of genetic variability in the available germplasm. Moreover, there is a continuous depletion of the existing germplasm pool. There is an urgent need to resort to in vitro methods for the induction and improvement of genetic variability. Although somatic hybridization and recombinant DNA technologies have great potentials for the future, they still have to be developed and refined, and it will be quite some time before they are routinely available for crop improvement. However, interspecific hybrids have already been obtained in crops such as peanut, cotton, brassica, soybean, *Vigna*, *Phaseolus*, etc. by the use of culture of excised embryos, ovules, and ovaries. It is recommended that plant breeders should immediately incorporate these simple and easy techniques in plant breeding programs for the induction of genetic variability.

Acknowledgments. Appreciation is expressed to my former graduate students Dr. S.S. Gosal, Dr. Manjeet S. Gill, Dr. D. Mohapatra, Mr. S.K. Mahajan, and Mr. Pradeep Kumar, whose work has been reviewed here.

References

Afzal M, Sikka SM, Rehman A (1945) Cytogenetic investigations in some *arboreum-anomalum* crosses. Indian J Genet 5:82–91

Ahn CS, Hartman RW (1978) Interspecific hybridization between mung bean (*Vigna radiata* L.) and adzuki bean (*V. angularis*). J Am Soc Hort Sci 103:3–6

Aykroyd WR, J Doughty (1966) Legumes in human nutrition. FAO, Rome

Bajaj S (1972) Biological value of legume proteins as influenced by genetic variation. In: Miller M (ed) Nutritional improvement of food legumes by breeding. Wiley, New York, pp 223–232

Bajaj S, Mickelson O, Baker LR, Markarian D (1971) The quality of protein in various lines of pea. Br J Nutr 25:207–212

Bajaj YPS (1965) Development of ovules of *Abelmoschus esculentus* L. var. Pusa Sawani in vitro. Proc Natl Inst Sci (India) 30:175–185
Bajaj YPS (1966) Growth of *Hyoscyamus niger* ovaries in culture. Phyton 23:57–62
Bajaj YPS (1984) Peanut. In: Ammirato PV, Evans DA, Sharp WR, Yamada Y (eds) Handbook of plant cell culture, vol 3. MacMillan, New York, pp 193–225
Bajaj YPS (ed) (1986) Biotechnology in agriculture and forestry 2. Crops I. Springer, Berlin Heidelberg New York Tokyo
Bajaj YPS (ed) (1988a) Biotechnology in agriculture and forestry 6. Crops II. Springer, Berlin Heidelberg New York Tokyo
Bajaj YPS (1988b) Induction and cryopreservation of somaclonal variation in peanut (*Arachis hypogaea* L.). Proc Int Congr Plant Tissue Cult – Tropical Species. Bogota, Colombia
Bajaj YPS (ed) (1989a) Biotechnology in agriculture and forestry 9. Plant protoplasts and genetic engineering II. Springer, Berlin Heidelberg New York Tokyo
Bajaj YPS (1989b) Genetic engineering and in vitro manipulation of plant cells – technical advances. In: Bajaj YPS (ed) Biotechnology in agriculture and forestry 9. Plant protoplasts and genetic engineering II. Springer, Berlin Heidelberg New York Tokyo, pp 1–25
Bajaj YPS (1989c) Recent advances in the isolation and culture of protoplasts and their implications in crop improvement. In: Bajaj YPS (ed) Biotechnology in agriculture and forestry 8. Plant protoplasts and genetic engineering I. Springer, Berlin Heidelberg New York Tokyo, pp 3–22
Bajaj YPS, Gill MS (1985) In vitro induction of genetic variability in cotton (*Gossypium* spp). Theor Appl Genet 70:363–368
Bajaj YPS, MS Gill (1987) Biotechnology of cotton improvement. In: Crocomo OJ et al. (eds) Biotechnology of plants and microorganisms. Ohio State Univ, Columbus, pp 118–151
Bajaj YPS, Gosal SS (1982) Induction of genetic variability in grain-legumes through tissue culture. In: Rao AN (ed) Tissue culture of economically important plants. COSTED, Singapore, pp 25–41
Bajaj YPS, Kumar P, Labana KS (1982) Interspecific hybridization in the genus *Arachis* through embryo culture. Euphytica 31:365–370
Bajaj YPS, Mahajan SK, Labana KS (1986) Interspecific hybridization of *Brassica napus* and *B. juncea* through ovary, ovule and embryo culture. Euphytica 35:103–109
Belivanis T, Doré C (1986) Interspecific hybridization of *Phaseolus vulgaris* L. and *Phaseolus angustissimus* A. Gray using in vitro embryo culture. Plant Cell Rep 5:329–331
Bohorova N, Atanassov A, Georgieva-Todorova J (1985) In vitro organogenesis, androgenesis and embryo culture, in the genus *Helianthus* L. Z. Pflanzenzuecht 95:35–44
Boling M, Matlock RS (1961) Mangbean hybridization technique. Agron J 53:54
Braak JP, Kooistra E (1975) A successful cross between *Phaseolus vulgaris* L. and *P. ritensis* Jones. With the aid of embryo culture. Euphytica 24:669–679
Bressani R (1973) Legumes in human diets and how they might be improved. In: Nutritional improvement of food legumes by breeding. PAG, UN, New York, pp 15–42
Broué P, Douglass J, Grace JP, Marshal DR (1982) Interspecific hybridization of soybeans and perennial *Glycine* species indigenous to Australia via embryo culture. Euphytica 31:715–724
Cherry JP (1977) Potential sources of peanut seed proteins and oil in the genus *Arachis*. J Agric Food Chem 25:186–193
Chowdhury RK, Chowdhury JB, Singh VP (1977) An amphidiploid between *Vigna radiata* var. *radiata* and *Vigna mungo*. Crop Improv 4:113–114
Cocking EC, Davey MR (1987) Gene transfer in cereals. Science 236:1259–1262
Cohen D, Ladizinsky, Ziv M, Muehlbauer FJ (1984) Rescue of interspecific *Lens* hybrids by means of embryo culture. Plant Cell Tissuc Organ Cult 3:343–347
Dana S, Biswas MR (1975) Black gram × rice bean cross. Cytologia 40:787–795
FAO (1982) Fats and oils situation. FAO, Rome
Gerstel DV, Phillips LL (1958) Segregation of synthetic amphiploids in *Gossypium* and *Nicotiana*. Cold Spring Harb Symp Quant Biol 13:225–238
Gill MS, Bajaj YPS (1984a) In vitro production of interspecific hybrids in cotton. Curr Sci 53:102–103
Gill MS, Bajaj YPS (1984b) Interspecific hybridization in the genus *Gossypium* through embryo culture. Euphytica 33:305–311
Gill MS, Bajaj YPS (1986) In vitro production of interspecific hybrids in *Gossypium* and their utilization in backcrossing. Z Pflanzenzuecht 96:174–176

Gill MS, Bajaj YPS (1987) Hybridization between diploid (*Gossypium arboreum*) and tetraploid (*Gossypium hirsutum*) cotton through ovule culture. Euphytica 36:625–630
Goodman RM, Hauptli H, Crossway A, Knauf CV (1987) Gene transfer in crop improvement. Science 236:48–54
Gosal SS, Bajaj YPS (1983a) In vitro hybridization of an incompatible cross – black gram × green gram. Curr Sci 52:556–557
Gosal SS, Bajaj YPS (1983b) Interspecific hybridization between *Vigna mungo* and *Vigna radiata* through embryo culture. Euphytica 32:129–137
Grant WF, Bullen MR, de Nettancourt D (1962) The cytogenetics of *Lotus*. 1. Embryo-cultured interspecific diploid hybrids closely related to *L. corniculatus* L. Can J Genet Cytol 4:105–128
Hildebrand DF, Phillips GC, Collins GB (1986) Soybean [*Glycine max* (L.) Merr.]. In: Bajaj YPS (ed) Biotechnology in agriculture and forestry 2. Crops I. Springer, Berlin Heidelberg New York Tokyo, pp 283–308
Honma S (1956) A bean interspecific hybrid. J Hered 47:217–220
Hucl P, Scoles GJ (1985) Interspecific hybridization in the common bean: a review, Hortscience 20:352–357
Hutchinson JB, Silow RA, Stephens SG (1947) The evolution of *Gossypium* and the differentiation of the cultivated cottons. Oxford Univ Press, London, pp 160
Inomata N (1979) Production of interspecific hybrids in *Brassica campestris* × *B. oleracea* by culture in vitro of excised ovaries. II. Development of excised ovaries on various media. Jpn J Breed 29:115–120
Iyengar NK, Santhanam V, Leela M (1957) Further studies in *arboreum-anomalum* crosses. Indian Cott Gr Rev 12:369–375
Johri BM, Bajaj YPS (1962) Behaviour of mature embryos of *Dendrophthoe falcata* (L.f.) Ettingh. in vitro. Nature (Lond) 193:194–195
Kalyanaraman SM, Santhanam V (1955) A note on the performance of some interspecific hybrids involving wild species of *Gossypium* L. *arboreum-anomalum* crosses. Indian Cott Gr Rev 11: 136–140
Keim WF (1953) Interspecific hybridization in *Trifolium* using embryo culture techniques. Agron J 45:601–606
Krapovickas A, Rigoni VA (1952) Estudios citologicos en el genero *Arachis*. Rev Invest Agric (Bueños Aires) 5:289–294
Kumar LS, Cruz RD, Oke JC (1957) A synthetic allohexaploid in *Arachis*. Curr Sci 26:121–122
Laibach F (1925) Das Taubwerden von Bastardsamen und die künstliche Aufzucht früh absterbender Bastardembryonen. Z Bot 17:417–459
Larkin PJ, Scowcroft WR (1981) Somaclonal variation – a novel source of variability from cell cultures for plant improvement. Theor Appl Genet 60:197–214
Matsuzawa Y (1978) Studies on interspecific hybridization in *Brassica*. 1. Effect of temperature on the development of hybrid embryos and improvement of the cross success rate by ovary culture in *B. campestris* × *B. oleracea* crosses. Jpn J Breed 28:186–196
McCoy TJ (1985) Interspecific hybridization of *Medicago sativa* L. and *M. rupestris* M.B. using ovule-embryo culture. Can J Genet Cytol 27:238–245
McHughen A, Rowland G (1988) Canada's first somaclonal variant crop cultivar. 2nd Canadian Plant Tissue Culture Genet Eng Workshop, May–June 1988, Ottawa, 43 pp
Mehboob Ali S, Krishnayya VV, Satyanarayana Murthy K (1964) A note on an interspecific hybrid in cotton (*Gossypium arboreum* race *indicum* × *G. stocksii*). Indian Cott Gr Rev 18:107–108
Mohapatra D, Bajaj YPS (1984) In vitro hybridization in an incompatible cross – *Brassica juncea* × *Brassica hirta*. Curr Sci 53:489–490
Mohapatra D, Bajaj YPS (1987) Interspecific hybridization in *Brassica juncea* × *Brassica hirta* using embryo rescue. Euphytica 36:321–326
Mohapatra D, Bajaj YPS (1988) Hybridization in *Brassica juncea* × *Brassica campestris* through ovary culture. Euphytica 37:83–88
Mok DWS, Mok MC, Rabakoarihanta A (1978) Interspecific hybridization of *Phaseolus vulgaris* with *P. lunatus* and *P. acutifolius*. Theor Appl Genet 52:209–215
Murashige T, Skoog F (1962) A revised medium for rapid growth and bioassays with tobacco tissue cultures. Physiol Plant 15:473–497

Newell CA, Hymowitz T (1982) Successful wide hybridization between the soybean and a wild perennial relative, *Glycine tomentella* Hayata. Crop Sci 22:1062–1065

Parpia HAB (1973) Utilization problems in food legumes. In: Nutritional improvement of food legumes by breeding. PAG, UN NY, pp 281–295

Pecket RC, Selim ARAA (1965) Embryo culture in *Lathyrus*. J Exp Bot 16:325–328

Pelletier C, Primand C, Vedel F, Chetrie P, Remy R, Rousselle P, Renard M (1983) Intergeneric cytoplasmic hybridization in Cruciferae by protoplast fusion. Mol Gen Genet 191:244–250

Philipps GC, Collins GB, Taylor NL (1982) Interspecific hybridization of red clover (*Trifolium pratense* L.) with *T. sarosiense* Hazsl. using in vitro embryo rescue. Theor Appl Genet 62:17–24

Power CJ (1987) Organogenesis from *Helianthus annuus* inbreds and hybrids from the cotyledons of zygotic embryos. Am J Bot 74:497–503

Prentice AN (1972) Cotton with special reference to Africa. Longmann, London, pp 68–76

Pundir NS (1972) Experimental embryology of *Gossypium arboreum* and *G. hirsutum* and their reciprocal crosses. Bot Gaz 133:7–26

Schell J St (1987) Transgenic plants as tools to study the molecular organization of plant genes. Science 128:451–460

Sen NK, Ghosh AK (1960) Interspecific hybridization between *Phaseolus aureus* and *P. mungo*. Bull Bot Soc Bengal 14:1–4

Silow RA (1941) The comparative genetics of *Gossypium anomalum* and the cultivated Asiatic cottons. J Genet 42:259–358

Sinclair JB (ed) (1982) Compendium of soybean diseases. Am Phytopathol Soc, St Paul

Singh AK, Moss JP (1982) Utilization of wild relatives in genetic improvement of *Arachis hypogaea* L. Part 2. Chromosome complements of species in section *Arachis*. Theor Appl Genet 61:305–314

Singh DP (1981) Breeding for resistance to disease in greengram and blackgram. Theor Appl Genet 59:1–10

Smartt J (1979) Interspecific hybridization in the grain legumes: A review. Econ Bot 33:329–337

Smartt J, Gregory WG (1967) Interspecific cross-compatibility between the cultivated peanut *Arachis hypogaea* L. and other members of the genus *Arachis*. Oleagineux 22:455–459

Spielman IV, Burge AP, Moss JP (1979) Chromosome loss and meiotic behaviour in interspecific hybrids in the genus *Arachis* and their implications in breeding for disease resistance. Z Pflanzenzuecht 83:236–250

Stewart JM, Hsu CL (1978) Hybridization of diploid and tetraploid cottons through in-ovulo embryo culture. J Hered 69:404–408

Takeshita M, Kato M, Tokumasu S (1980) Application of ovule culture to the production of intergeneric hybrids in *Brassica* and *Raphanus*. Jpn J Genet 35:373–387

Weaver JB Jr (1957) Embryological studies following interspecific crosses in *Gossypium*. I. *G. hirsutum*×*G. arboreum*. Am J Bot 44:209–214

Weaver JB Jr (1958) Embryological studies following interspecific crosses in *Gossypium*. II. *G. arboreum*×*G. hirsutum*. Am J Bot 45:10–16

Webster GT (1955) Interspecific hybridization of *Melilotus alba*×*M. officinalis* using embryo culture. Agron J 47:138–142

White PR (1943) The cultivation of animal and plant cells. Ronald, New York

Williams EG (1980) Hybrids between *Trifolium ambiguum* and *T. hybridum* obtained with the aid of embryo culture. N Z J Bot 18:215–220

Williams EG, de Lantour G (1981) Production of tetraploid hybrids between *Ornithopus pinnatus* and *O. sativus* using embryo culture. N Z J Bot 19:23–30

I.2 Transformation in Legumes

G.S. NISBET[1] and K.J. WEBB[2]

1 Introduction

Plant breeders aim to improve the performance of cultivated species by developing new varieties for specific conditions. With traditional breeding methods the available gene pool is restricted by the sexual incompatibility of many interspecific and intergeneric crosses. One solution is to manipulate (transform) existing species genetically, introducing foreign genes from diverse sources which would normally be inaccessible. The resulting transgenic plants may be used directly, or more probably, as the starting material in a conventional breeding programme. In this way, the gene pool may be substantially broadened.

In simple terms, genetic transformation may be defined as "the uptake and expression of foreign DNA by a cell" (Cocking 1983). This may range from transient expression of non-integrated DNA to the regeneration of fertile transgenic plants. In the past decade, progress in molecular genetics and tissue culture has provided the impetus for dramatic advances in plant transformation (see Bajaj 1989). Until recently, these advances have been achieved mainly with model species such as tobacco and tomato (Horsch et al. 1985; McCormick et al. 1986), both of which are highly responsive in culture. Comparatively little success has been reported on the legumes, mainly because of their recalcitrance in culture. This is especially true of the grain legumes and cereal crops, where protoplast regeneration systems have yet to be perfected (Dale 1983; Webb 1988).

There are, broadly speaking, three ways of transforming plants. The method selected depends to a large extent on the nature of the DNA to be introduced and the intended recipient. The most commonly used method is *Agrobacterium*-mediated transformation, whereby foreign DNA is introduced into plant cells utilizing the natural gene-transfer system of the bacterium itself (the *Agrobacterium* transformation process has been extensively documented e.g. Matthyse 1986). This approach is technically simple, but is limited by the host range of *Agrobacterium* (De Cleene and De Ley 1977). By contrast, direct gene transfer permits the introduction of foreign DNA into cells without an intermediate vector. Both electroporation and PEG-induced uptake of small fragments of DNA into proto-

---

[1] ICI Seeds, Jealott's Hill Research Station, Bracknell, Berkshire RG12 6EY, United Kingdom
[2] AFRC Institute for Grassland and Animal Production, Welsh Plant Breeding Station, Aberystwyth, Wales, SY23 3EB, United Kingdom

Biotechnology in Agriculture and Forestry, Vol. 10
Legumes and Oilseed Crops I (ed. by Y.P.S. Bajaj)

plasts have been used extensively, but are dependent on reliable plant regeneration from protoplasts. Larger segments of DNA may be introduced by microinjection, while macroinjection of developing florets is an attractive option, following recent success with the cereals (de la Pena et al. 1987).

Finally, a third option, fusion of protoplasts (somatic hybridization) is available to combine two complete genomes. This technique has shown some promise in the legumes (Wright et al. 1986), but will not be discussed further here.

The susceptibility of most legumes to *Agrobacterium* infection and the difficulties of protoplast regeneration in many species makes *Agrobacterium*-mediated transformation the obvious first choice. Consequently, the majority of published material on legume transformation deals with this subject and it will be the main consideration of this review.

2 Targets for Transformation

There are many potential transformation targets. One obvious aim is to introduce genes which confer resistance to broad-spectrum herbicides, which cannot normally be used with susceptible broad-leaved, dicotyledonous crops. Such resistance has been achieved in tobacco, using a gene from *Salmonella typhimurium* which confers resistance to the widely used herbicide, glyphosate (Comai et al. 1985). Transformation was performed in this case with *A. rhizogenes*, so the technology may be directly applicable to many legumes. The introduction of disease- or pest-resistance genes from other plant species or from bacteria such as *Bacillus thuringiensis* (Fischhoff et al. 1987) has been achieved in a similar fashion.

Less obvious targets for transformation include the possibility of eliminating bloat in animals grazing on clover, by introducing tannin genes such as those found in *Lotus corniculatus* (Webb et al. 1987). Improving the protein content of grain and forage legumes is a subject which has received much attention but little practical work. However, there is one example of improved protein content being achieved via transformation. A genetically engineered alfalfa variety containing a pea albumin gene is presently being tested in Australia (Ford 1988). Because wool growth is limited by the availability of sulphur amino acids, sheep grazing on this variety show increased productivity. This gene may now be transferred to other forage crops.

Transformation techniques have also been used in more fundamental gene studies, without an immediate application to crop improvement. The highly conserved genes involved in the nodulation process of the legume-*Rhizobium* symbiosis are of particular interest. It is hoped that the study of these genes and the nitrogen-fixing (*nif*) genes may help to extend the benefits of nitrogen fixation to other crops which are normally not susceptible to *Rhizobium* infection. Transformation studies may also help in the understanding of gene regulation, especially of those genes which appear to be temporally regulated or "tissue-specific". For example, the seed storage protein genes vicilin and legumin are known to be differentially regulated during legume seed development (Boulter et al. 1987). In order to identify the sequences involved in this regulation, a series of truncated

Table 1. Summary of major reports on genetic transformation of legumes

Plant species	Explant used	Technique/Vector	Gene	Comment	Reference
Glycine max	Plants	*A. tumefaciens* – wild-type			Pederson et al. (1983)
		A. tumefaciens – oncogenic	Rubisco-*nptII*	Inducible expression	Facciotti et al. (1985)
	Protoplasts	*A. tumefaciens* – oncogenic			Baldes et al. (1987)
	Protoplasts	Electroporation	*nos-nptII*		Christou et al. (1987)
Lotus corniculatus	Plants	*A. tumefaciens* – wild-type			Webb (1986)
	Seedlings	*A. rhizogenes* – oncogenic	Soybean leghaem-*cat*	Transgenic plants	Jensen et al. (1986)
Medicago sativa	Plants	*A. tumefaciens* – wild-type			Mariotti et al. (1984)
	Stem	*A. tumefaciens* – disarmed binary	*nos-nptII*	Transgenic plants	Shahin et al. (1986)
	Protoplasts	Microinjection – Ti plasmid			Reich et al. (1986)
		A. rhizogenes – wild-type		Transgenic plants	Spano et al. (1987)
		A. rhizogenes – oncogenic binary		Transgenic plants	Sukhapinda et al. (1987)
		A. tumefaciens	Pea albumin	Transgenic plants	Ford (1988)
M. varia	Stem	*A. tumefaciens* – disarmed	*nos-nptII*	Transgenic plants/cells	Deak et al. (1986)
Onobrychis vicifolia	Plants	*A. tumefaciens* – wild-type			Webb (1986)
Sesbania rostrata	Plants	*A. tumefaciens* – wild-type			Vlachova et al. (1987)
	Plants	*A. rhizogenes* – wild-type			Vlachova et al. (1987)
Stylosanthes spp.	Leaves	*A. tumefaciens* – wild-type			Manners (1987)
		A. tumefaciens – oncogenic binary	*nptII*		Manners (1987)
Trifolium hybridum	Plants	*A. tumefaciens* – wild-type			Webb (1986)
T. pratense	Plants	*A. tumefaciens* – wild-type			Webb (1986)
T. repens	Plants	*A. tumefaciens* – wild-type			Webb (1986)
	Stolon	*A. tumefaciens* – disarmed binary	*nos-nptII*	Transgenic shoots	White and Greenwood (1987)
Vigna aconitifolia	Protoplasts	PEG-uptake	*nptII*		Kohler et al. (1987)
V. unguiculata	Leaves	*A. tumefaciens*	CPMV mRNA		Garcia et al. (1987)

genes will be used with overlapping deletions in the suspected controlling regions. These genes would ideally be re-introduced into *P. sativum*, but as yet no transformation system is available for this species. Model species have to be used instead. The Legumin A (*LegA*) gene has been successfully introduced into tobacco (Ellis et al. 1988) and the protein identified in the seeds, but the study of regulation in such a heterologous species is obviously limited. One alternative is to transform a more closely related species such as *L. corniculatus* (Nisbet 1987).

With most legume species, transformation work has not yet reached this level, but is instead still concerned with the development of transformation systems. This point can be illustrated by considering some of the published work on this topic (Table 1).

3 Approaches for Transformation

3.1 *Agrobacterium-Mediated Transformation*

3.1.1 Wild-Type Vectors

Much of the early work on the development of legume transformation involved the inoculation of whole plants (or seedlings) with wild-type, oncogenic strains of both *A. tumefaciens* and *A. rhizogenes*. Susceptible species and varieties were easily identified by the typical neoplastic growths (galls or hairy roots) which developed at the infection sites. During the development of these transformation systems, the importance of the interaction between the plant genome and the bacterial strain became evident. In a study of *A. tumefaciens* transformation of several *Medicago* varieties, Mariotti et al. (1984) detected distinct variety-strain interactions. Some of the varieties were generally more susceptible than others, but much of the specificity seemed to be determined by the type of plasmid (octopine or nopaline encoding) carried by the bacteria.

Specific interactions which determine the response to infection and tumor phenotype were noted by Webb (1986) in several legume species. Both Mariotti et al. (1984) and Pedersen et al. (1983) reported that galls formed on *Medicago sativa* and *Glycine max*, respectively, remained disorganized in situ and in culture. Webb (1986) confirmed this in *Medicago* and *Trifolium* species, but found that the galls formed on *Lotus corniculatus* plants were frequently teratomatous (shooty).

The use of wild-type bacteria in whole-plant inoculation experiments has helped to identify the best plant region or explant to inoculate (for example, hypocotyl, leaf, stem) and also the optimum age of the tissue (Armstead and Webb 1987). Even *Agrobacterium*-protoplast co-cultivation techniques have been developed (Baldes et al. 1987, using *Glycine max*; Nisbet 1987, using *L. corniculatus*), although these are of limited use. The simplicity of the oncogenic system has facilitated the spread of transformation techniques to a wide range of legume species, including tropical legumes such as *Sesbania rostrata* (Vlachova et al. 1987) and *Stylosanthes* spp. (Manners 1987).

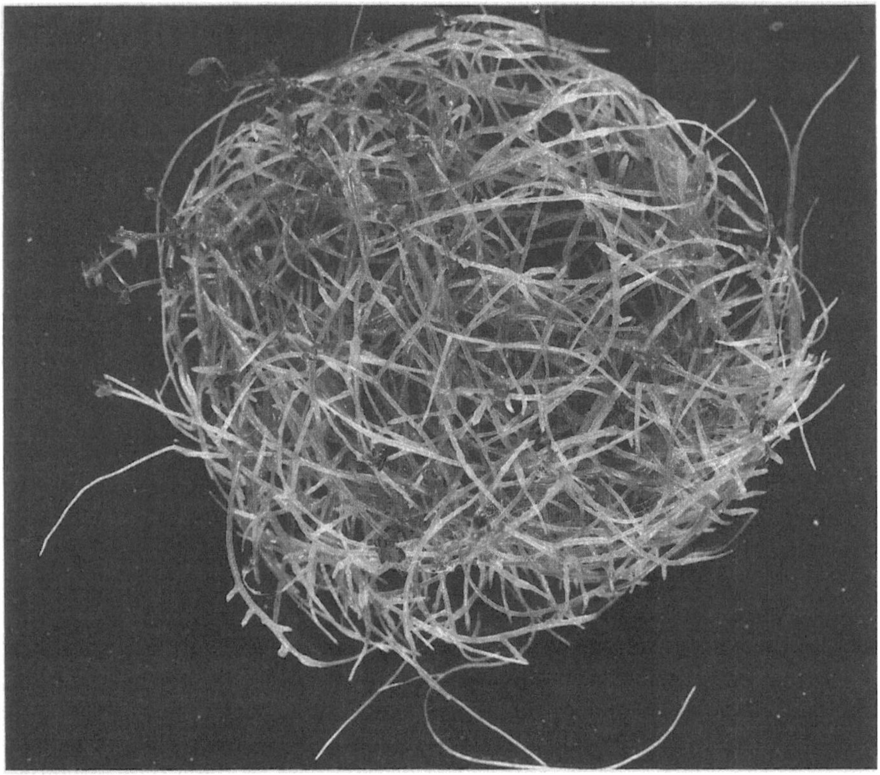

Fig. 1. Transformed root culture of *Lotus corniculatus* showing spontaneous shoot regeneration. ×1.12

Although transformation studies with oncogenic *Agrobacterium* strains have contributed greatly to our understanding of the transformation process (Petit et al. 1986), the transformed material generated is of little practical use. Disorganized galls yield no regenerants, and the teratomatous shoots produced by some species-strain interactions are phenotypically abnormal, non-rooting and infertile if grafted.

With some legumes these problems may be avoided by using oncogenic *A. rhizogenes* strains. Transgenic shoots frequently regenerate spontaneously from the hairy roots formed after infection (Fig. 1). This is the case for both *L. corniculatus* (Jensen et al. 1986; Webb pers. obs.) and *Medicago sativa* (Spano et al. 1987; Sukhapinda et al. 1987). These shoots can root when excised, and hence a reliable transformation/regeneration system is available. However, in other species, including *T. repens* and *T. pratense*, no such spontaneous regeneration has yet been found (Webb pers. obs.). Also, where field-release of the transformed plants is intended, it is desirable to use non-oncogenic strains to achieve transformation. This can be achieved by the use of genetically modified strains of bacteria.

3.1.2 Binary Vectors and Marker Genes

The recent trend in transformation work has been to use "disarmed" strains of *A. tumefaciens*, where the oncogenes (*onc*) which give the tumorous phenotype have been deleted from the plasmid. The early vectors such as A208 (Matzke and Chilton 1981) were based on the large Ti-plasmid carrying the T-DNA. Most modern vectors are based on a binary system (Bevan 1984), where the T-DNA borders are located on a small, wide host range replicon such as Bin 19. These small vector plasmids may be cloned in *E. coli* and are easily manipulated in vitro before being introduced into *Agrobacterium*. The virulence (*vir*) genes of the Ti-plasmid are located on an independent replicon and act in trans to effect the excision of the T-DNA from the vector plasmid. Disarmed binary vectors permit transformation of plants without the formation of neoplasms, but require a reliable regeneration and selection system in order to generate transgenic plants.

Chimaeric gene constructs, formed by the fusion of the coding region of antibiotic resistance genes to constitutively expressed promoters, can be introduced between the T-DNA borders. These act as dominant selectable markers to enable the selective regeneration of transformed cells in culture (Herrera-Estrella et al. 1983). The most common examples of this are fusions of the nopaline synthase (*nos*) or Cauliflower Mosaic Virus (CaMV) promoters to the neomycin phosphotransferase (*neo* or *nptII*) coding region. Transformed cells show resistance to the antibiotics kanamycin, neomycin and G-418.

There are numerous examples within the Leguminosae of successful transformation systems being developed on this basis. For example, Shahin et al. (1986) developed an *A. tumefaciens*-mediated transformation system for alfalfa stem explants, using a *nos-nptII* fusion in the disarmed binary vector pARC8. Later White and Greenwood (1987) transformed stolon internodes of a highly regenerative genotype of *T. repens* with a similar vector, generating the first reported transgenic white clover shoots. Disarmed vector systems are frequently developed after first determining the susceptibility of different species and genotypes with wild-type bacterial strains (for example Manners 1987 on *Stylosanthes*).

Selectable marker genes have proven to be of particular benefit in the study of the integration, expression and inheritance of foreign genes in plants (De Block et al. 1984; Hain et al. 1985). However, these genes have a much wider role to play in plant transformation. By close linkage of constitutively expressed marker genes to useful (but non-selectable) genes in vector constructs, transgenic plants containing both genes may be recovered. Whereas the selection is based on constitutive expression of the marker genes, the other genes can be under the control of developmentally regulated plant promoters.

Tissue-specific and temporally regulated genes can be introduced in this way. A marker gene and a reporter gene, such as the *gus* gene which codes for β-glucuronidase production, are often linked on either site of the gene of interest (e.g. a seed protein gene). To ensure the reliable selection of plants carrying the seed protein gene, only those regenerants which show expression of both of the marker/reporter genes are used.

3.1.3 Successful Transformations

As was outlined earlier, there are many potential targets for transformation in the legumes. To date, very few of these targets have been achieved. With the exception of the high sulphur alfalfa strain, no agronomically useful transformations have been reported. However, some very interesting gene regulation studies have been attempted in the legumes, using transformation techniques.

One illustration of this is the study of the plant-bacterial interactions involved during the nodulation of legume species by *Rhizobium* symbionts. Research into the regulation of the nodulin (*nod*) genes has been hampered by the inability to develop a reliable transformation/regeneration system for many of the species involved. Jensen et al. (1986) were able to introduce a chimaeric leghaemoglobin-chloramphenicol acetyl transferase (*cat*) gene construct into *L. corniculatus*. Shoots regenerated spontaneously from the hairy roots which developed and their transformed status was confirmed by expression of the *cat* reporter gene. Following infection with *Rhizobium loti*, efficient expression of the *leg-cat* gene was found. This method may be used to examine the regulation of the leghaemoglobin and other nodulin genes. Similarly, Vlachova et al. (1987) have developed a reliable regeneration system for *Sesbania rostrata* and have shown its susceptibility to *A. tumefaciens* infection. It is intended that disarmed bacterial vectors will be used to study the nodulation process in *S. rostrata-Azorhizobium sesbaniae* infection.

Other examples of gene studies include the work of Facciotti et al. (1985) and Garcia et al. (1987). Working with soybean seedlings, Facciotti et al. used an *A. tumefaciens* strain containing an oncogenic Ti-plasmid pTiCGN609, which has the coding region for *npt II* linked to the 5′ portion of the soybean small subunit (SSU) of ribulose 1,5-bisphosphate carboxylase (Rubisco) gene. Expression of the SSU is normally light-regulated in the plant (Smith and Ellis 1981). Facciotti et al. were able to demonstrate that expression of the chimaeric Rubisco-*npt II* gene was light-inducible in transformed soybean callus.

3.2 Direct Gene Transfer (DGT)

There are now numerous reports and reviews of direct gene transfer techniques (see Potrykus et al. 1985; Negrutio et al. 1987). These methods are widely employed where *Agrobacterium*-mediated transfer is impossible or unreliable, which includes most monocotyledonous species. Within the legumes, *Agrobacterium* is almost always the preferred method, and there are consequently very few reports of DGT on leguminous species. However, there are a few examples which can be considered here.

3.2.1 Electroporation

Successful transformation of soybean protoplasts has been reported by two groups (Hauptmann et al. 1987; Christou et al. 1987). Hauptmann's group reported transient expression of chimaeric 35S-*cat* constructs of soybean cultures, but failed to achieve stable integration. Having first determined the optimal conditions for transient expression, Christou et al. stably transformed soybean protoplasts with a plasmid containing a *nos-npt II* construct. Transformed cultures were selected on kanamycin-containing media. As well as attempting transformation of several *G.max* cultivars, Christou et al. also worked with *Glycine canescens*, a species which has been reportedly regenerated from protoplasts (Newell and Luu 1985). This technique may therefore provide another way of generating transgenic plants.

3.2.2 Microinjection

Microinjection into protoplasts and cells is a useful technique for the introduction of macromolecules. Reich et al. (1986) transformed protoplasts of alfalfa with a number of Ti-plasmids. The plasmids were injected directly into the nuclei which were visualized under UV light after Hoechst 33258 staining. The transformation frequency was found to be high (15% – 26%) and integration of the DNA was apparently not controlled by the T-DNA borders.

Problem Species

Several legume species present particular problems with regard to transformation, either because of their recalcitrance in culture, their resistance to *Agrobacterium*, or both. A prime example is *Pisum sativum*. Despite repeated attempts at *A. tumefaciens* transformation (Koshiba and Theologis 1987), no stable transformants have yet been obtained. Direct gene transfer is not possible as no regeneration system has been developed. Most of the problem may be solved by ascertaining the best *Agrobacterium* strains (whether *A. tumefaciens* or *A. rhizogenes*) for the task and the best tissues for inoculation.

4 Concluding Comments

Progress in legume transformation has been slow, but the recent advances outlined in this review reflect the high level of research currently targeted in this area. This intensive effort should help to overcome many of the inherent problems of legume transformation, especially genotype-dependent regeneration and transformation.

The trend of future transformation work is hard to predict. The development of broad host range "supervirulent" vectors such as *A. tumefaciens* pTiBo542 (Hood et al. 1984) may help to overcome problems of strain and cultivar specificity. However, explant regeneration systems will still be required. Genetically engineered *A. rhizogenes* vectors can be used to avoid this problem, but release of material transformed by oncogenic vectors into the environment may be restricted. One possible solution would be to use disarmed strains to transform embryos, embryogenic cells, or shoot meristems. However, somatic embryoids have so far proved immune to *Agrobacterium* infection.

Of the possible alternatives to *Agrobacterium* transformation, direct gene transfer to protoplasts is unlikely to be pursued due to the high genotype dependence of regeneration systems. Microinjection of pre-embryoid structures or possibly even laser-induced transformation, such as has been demonstrated in animal and yeast cells (Tao et al. 1987) may be a possible solution.

As well as the successful development of transformation systems, it is of paramount importance that physiological and biochemical studies be carried out into plant metabolic processes such as disease resistance. At present, our limited understanding of these processes and the genes involved restricts the agronomic success which we can hope to achieve through plant transformation. However, as our knowledge increases and transformation techniques are perfected, a powerful new tool in plant improvement will result.

Acknowledgments. GSN received a post-graduate studentship from DAFS. We wish to thank J.R. Ellis for his critical reading of the manuscript.

References

Armstead IP, Webb KJ (1987) Effect of age and genotype of tissue on genetic transformation of *Lotus corniculatus* by *Agrobacterium tumefaciens*. Plant Cell Tissue Organ Cult 9:95–101

Bajaj YPS (ed) (1989) Biotechnology in agriculture and forestry. 9. Plant protoplasts and genetic engineering II. Springer, Berlin Heidelberg New York Tokyo

Baldes R, Moos M, Geider K (1987) Transformation of soybean protoplasts from permanent suspension cultures by cultivation with cells of *Agrobacterium tumefaciens*. Plant Mol Biol 9(2):135–146

Bevan M (1984) Binary *Agrobacterium* vectors for plant transformation. Nucl Acids Res 12(22):8711–8721

Boulter D, Evans IM, Ellis JR, Shirsat AH, Gatehouse JA, Croy RRD (1987) Differential gene expression in the development of *Pisum sativum*. Plant Physiol Biochem 25(3):283–289

Christou P, Murphy JE, Swain WF (1987) Stable transformation of soybean by electroporation and root formation from transformed callus. Proc Natl Acad Sci USA 84:3962–3966

Cocking EC (1983) Genetic transformation through somatic hybridisation. SEB Seminar Series. In: Mantell SH, Smith H (eds) Plant Biotechnology. Cambridge Univ Press, pp 241–250

Comai L, Facciotti D, Hiatt WR, Thompson G, Rose RG, Stalker DM (1985) Expression in plants of a mutant AroA gene from *Salmonella typhimurium* confers tolerance to glyphosate. Nature (Lond) 317:741–744

Dale PJ (1983) Protoplast culture and plant regeneration of cereals and other recalcitrant crops. 6th Int Protoplast Symp, Lec Proc. Birkhäuser-Verlag, Basel, pp 31–41

Deak M, Kiss GB, Koncz C, Dudits D (1986) Transformation of *Medicago* by *Agrobacterium*-mediated gene transfer. Plant Cell Rep 5:97–100

De Block M, Herrerra-Estrella L, Van Montagu M, Schell J, Zambryski P (1984) Expression of foreign genes in regenerated plants and their progeny. EMBO J 3:1681–1689

De Cleene M, De Ley J (1977) Host range of crown gall. Bot Rev 42:389–466

de la Pena A, Lorz H, Schell J (1987) Transgenic rye plants obtained by injecting DNA into young floral tillers. Nature (Lond) 325:274–276

Ellis JR, Shirsat AH, Hepher A, Yarwood JN, Gatehouse JA, Croy RRD, Boulter D (1988) Tissue-specific expression of a pea legumin gene in seeds of *Nicotiana plumbaginifolia*. Plant Mol Biol 10:203–214
Facciotti D, O'Neal JK, Lee S, Shewmaker CK (1985) Light-inducible expression of a chimeric gene in soybean tissue transformed with *Agrobacterium*. Bio/technology 3:241–246
Fischoff DA, Bowdish KS, Perlak FJ, Marrone PG, McCormick SM, Niedermeyer JG, Dean AD, Kusano-Kretzmer K, Mayer EK, Rochester DE, Rogers SG, Fraley RT (1987) Insect-tolerant transgenic tomato plants. Bio/technology 5:807–813
Ford J (1988) Sheep will grow woollier on a bioengineered diet. New Sci 117(10 March):24
Garcia JA, Hille J, Vos P, Goldbach R (1987) Transformation of cowpea *Vigna unguiculata* with a full-length DNA copy of cowpea mosaic virus m-RNA. Plant Sci 48:89–98
Hain R, Stabel P, Czernilofsky AP, Steinbiss HH, Herrerra-Estrella L, Schell J (1985) Uptake, integration, expression and genetic transmission of a selectable chimaeric gene by plant protoplasts. Mol Gen Genet 199:161–168
Hauptmann RM, Ozias-Akins P, Vasil V, Tabaeizadeh Z, Rogers SG, Horsch RB, Vasil IK, Fraley RT (1987) Transient expression of electroporated DNA in monocotyledonous and dicotyledonous species. Plant Cell Rep 6:265–270
Herrerra-Estrella L, De Block M, Messens E, Hernalsteens JP, Van Montagu M, Schell J (1983) Chimaeric genes as dominant selectable markers in plant cells. EMBO J 2(60):985–987
Hood EE, Jen G, Kayes L, Kroner J, Fraley RT, Chilton M-D (1984) Restriction endonuclease map of pTiBO542, a potential Ti-plasmid vector for genetic engineering of plants. J Bacteriol 166:88–94
Horsch RB, Fry JE, Hoffman NL, Eichholz D, Rogers SG, Fraley RT (1985) A simple and general method for transferring genes into plants. Science 227:1229–1231
Jensen JS, Marcker KA, Otten L, Schell J (1986) Nodule-specific expression of a chimaeric soybean leghaemoglobin gene in transgenic *Lotus corniculatus*. Nature (Lond) 321:669–674
Kohler F, Golz C, Eapen S, Kohn H, Schieder O (1987) Stable transformation of moth bean *Vigna aconitifolium* via direct gene transfer. Plant Cell Rep 6:313–317
Koshiba T, Theologis A (1987) Expression of auxin regulated genes during early stages of *Agrobacterium tumefaciens*-mediated oncogenesis in pea tissue. Plant Physiol 4 Suppl 83:153
Manners JM (1987) Transformation of *Stylosanthes* spp. using *Agrobacterium tumefaciens*. Plant Cell Rep 6:204–207
Mariotti D, Davey MR, Draper J, Freeman JP, Cocking EC (1984) Crown gall tumorigenesis in the forage legume *Medicago sativa* L. Plant Cell Physiol 25(3):473–482
Matthysse AG (1986) Initial interactions of *Agrobacterium tumefaciens* with plant host cells. CRC Crit Rev Microbiol 13:281–307
Matze AJM, Chilton M-D (1981) Site-specific insertion of genes into T-DNA of the *Agrobacterium* tumour-inducing plasmid: an approach to genetic engineering of higher plant cells. J Mol Appl Genet 1:39–49
McCormick S, Niedermeyer J, Fry J, Barnason A, Horsch R, Fraley R (1986) Leaf disc transformation of cultivated tomato (*L. esculentum*) using *Agrobacterium tumefaciens*. Plant Cell Rep 5:81–84
Negrutiu I, Mouras A, Horth M, Jacobs M (1987) Direct gene transfer to plants: present development and some future prospectives. Plant Physiol Biochem 25(4):493–503
Newell CA, Luu HT (1985) Protoplast culture and plant regeneration in *Glycine canescens* F.J Herm. Plant Cell Tissue Organ Cult 4:145–149
Nisbet GS (1987) Transformation of *Lotus corniculatus* and regeneration of transgenic plants. MSc Thesis, Univ Wales
Pedersen HC, Christiansen J, Wyndaele R (1983) Induction and in vitro culture of soybean crown gall tumours. Plant Cell Rep 2:201–204
Petit A, Berkaloff A, Tempe J (1986) Multiple transformations of plant cells by *Agrobacterium* may be responsible for the complex organisation of T-DNA in crown gall and hairy root. Mol Gen Genet 202:388–393
Potrykus I, Shillito RD, Saul MW, Paszkowski J (1985) Direct gene transfer – state of the art and future potential. Plant Mol Biol Rep 3:117–128
Reich TJ, Iyer VN, Miki BL (1986) Efficient transformation of alfalfa protoplasts by the intranuclear microinjection of Ti Plasmids. Bio/technology 4:1001–1004
Shahin EA, Spielmann A, Sukhapinda K, Simpson RB, Yashar M (1986) Transformation of cultivated alfalfa using disarmed *Agrobacterium tumefaciens*. Crop Sci 26:1235–1239

Smith SM, Ellis RJ (1981) Light-stimulated accumulation of transcripts of nuclear and chloroplast genes for ribulosebisphosphate carboxylase. J Mol Appl Genet 1:127–137

Spano L, Mariotti D, Pezzotti M, Damiani F, Arcioni S (1987) Hairy root transformation in alfalfa (*Medicago sativa* L.). Theor Appl Genet 73:523–530

Sukhapinda K, Spivey R, Shahin EA (1987) Ri-plasmid as a helper for introducing vector DNA into alfalfa plants. Plant Mol Biol 8:209–216

Tao W, Wilkinson J, Stanbridge EJ, Berns MW (1987) Direct gene transfer into human cultured cells facilitated by laser micropuncture of the cell membrane. Proc Natl Acad Sci USA 84:4180–4184

Vlachova M, Metz BA, Schell J, de Bruijn FJ (1987) The tropical legume *Sesbania rostrata*: tissue culture, plant regeneration and infection with *Agrobacterium tumefaciens* and *rhizogenes* strains. Plant Sci 50:213–223

Webb KJ (1986) Transformation of forage legumes using *Agrobacterium tumefaciens*. Theor Appl Genet 72:53–58

Webb KJ (1988) Recent developments in the regeneration of agronomically important crops from protoplasts. Plant Cell Tissue Organ Cult 12:127–131

Webb KJ, Woodcock S, Chamberlain DA (1987) Plant regeneration from protoplasts of *Trifolium repens* and *Lotus corniculatus*. Plant Breed 98:111–118

White DWR, Greenwood D (1987) Transformation of the forage legume *Trifolium repens* L. using binary *Agrobacterium* vectors. Plant Mol Biol 8:461–469

Wright RL, Somers DA, McGraw RL (1986) Somatic hybridisation between birdsfoot trefoil (*Lotus corniculatus*) and *L. coimbriensis*. 6th Int Congr Plant Tissue Cell Cult, abstr Univ Minnesota, Minneapolis, p 82

I.3 Cryopreservation of Germplasm of Legumes and Oilseed Crops

Y. P. S. BAJAJ [1]

1 Introduction

The seeds of legumes and most of the oilseed crops are "orthodox" and present no problem for long-term storage; however (1) "recalcitrant" species, in which seed is short-lived due to degeneration of the embryo, and where the germplasm cannot be conserved, require cryopreservation, (2) for the long-term preservation of elite and useful variants obtained by in vitro cell culture, innovative methods need to be evolved. Cryostorage of in vitro cultures has been recommended as one of the unconventional approaches for the long-term conservation of genetic stocks and for international exchange of germplasm (Bajaj 1976a, 1979a, b, 1983a, 1984a, b, 1985, 1986b). In this chapter, literature on the cryopreservation of tissue cultures of legumes and oilseed crops is reviewed, and prospects for this technology in the long-term conservation of germplasm are discussed.

2 Cryopreservation of Cultures

Cryopreservation brings cultures to a state of zero metabolism. This is achieved by freezing the cultures in liquid nitrogen. During the past 10 years complete plants have been regenerated from frozen cultures of a number of plant species (see Bajaj 1986b), the literature on the legumes and oilseed crops is summarized in Table 1, and discussed below.

2.1 Cell Suspensions and Callus Cultures

Callus cultures on periodical subcultures undergo genetic erosions and result in somaclonal variation and genetic instability, causing polyploidy, aneuploidy, mutations, chromosome translocation, gene amplification, etc. At present, no method is known by which the genetic stability of the long-term cultures can be ensured; cryopreservation offers this possibility. With a view to avoiding periodical subculturing of cell suspensions, Bajaj for the first time attempted the freezing

[1] Formerly Professor of Tissue Culture, Punjab Agricultural University, Ludhiana, India. Present address: A-137 New Friends Colony, New Delhi 110065, India

Biotechnology in Agriculture and Forestry, Vol. 10
Legumes and Oilseed Crops I (ed. by Y. P. S. Bajaj)

Table 1. Summary of the cryopreservation (−196 °C) studies on cells, tissues, and organ cultures of legumes and oilseed crops

Crop	Culture	Response	Reference
Arachis hypogaea (peanut)	Meristem	Plants from cultures cryopreserved for 20 months	Bajaj (1979c, 1983b)
	Pollen embryo/anther segment	29% Survival, shoots	Bajaj (1983c)
A. villosa	Pollen embryo/anther segments	38% Embryos survived, plants	Bajaj (1983c)
Brassica campestris (mustard)	Pollen embryos	31% Survival	Bajaj (1983c)
B. napus (rape)	Pollen embryos	44% Survival, callus, plants	Bajaj (1983c)
Cicer arietinum (chickpea)	Meristems	Plants from cultures cryopreserved for 20 months	Bajaj (1979c, 1983b)
Cocos nucifera (coconut)	Young embryos	17.6% Survival	Bajaj (1984c)
	Transversely cut halves of embryo	25% Callus	Bajaj (1984c)
Elaeis guineensis (oil palm)	Somatic embryos	Up to 95% survival, plants	Engelmann and Duval (1986)
Glycine max (soybean)	Cell suspension	50% Cell survival	Bajaj (1976b)
	Cell suspension	19% Plating efficiency	Weber et al. (1983)
	Protoplasts	60% Survival, callus	Takeuchi et al. (1982)
Gossypium arboreum (cotton)	Proliferating anthers	34% Survival	Bajaj (1982a)
G. hirsutum	Cultured ovules	42% Survival	Bajaj (1982a)
Onobrychis viciifolia (sainfoin)	Cell suspension	Survival	Withers (1983)
Pisum sativum (pea)	Meristem	Plant regeneration	Haskins and Kartha (1980)
	Meristems	Evaluation of ATP	Sugawara and Takeuchi (1986)
	Fused protoplasts		
	(pea × wheat)	12−18% Survival	Bajaj (1983d)
	(pea × rice)	11−14% Survival	Bajaj (1983d)
Medicago sativa (alfalfa)	Callus	Plant regeneration	Finkle et al. (1979)
Vigna mungo × *Vigna radiata*	Hybrid embryo	34% Survival, plant	Bajaj (unpubl.)

of legume cultures (Bajaj 1976b). The actively growing soybean cell suspensions were frozen to −196 °C at the rate of 1−2 °C in the presence of a mixture of cryoprotectants (DMSO 5% and sucrose 4%). The cultures showed a viability of 50%, which underwent a lag period of about 2 weeks to 3 months, and then resumed activity. The effect of various concentrations of sorbitol on the viability of soybean cell cultures showed that 1 M gave the highest viability (46%), followed by 0.9 M (36%) (Weber et al. 1983).

Table 2. Effect of various cryoprotectants on survival of tissue cultures of *Gossypium arboreum* frozen in liquid nitrogen. Data based on the resumption of growth. (Bajaj 1982a)

Cryoprotectant	Percent survival of control cultures	
	Anther callus	Ovule callus
Sucrose 8%	7	9
Glycerol 10%	15	19
DMSO 10%	29	27
Sucrose + glycerol + DMSO (5% each)	34	42

The proliferating anthers and ovules of two species of cotton (*Gossypium hirsutum* and *G. arboreum*) resumed growth after exposure to LN (Bajaj 1982a). The retrieved proliferating anthers and the ovules turned brown and showed no sign of growth for the first 3–4 weeks. The callus then underwent localized activity, and at places the cells, especially at the periphery, resumed growth and formed islets of meristems. A mass of callus was formed in 6–7 weeks. The cultures with little callus were observed to revive better than those with masses of callus. In the latter, the callus became spongy and died. The cell suspensions obtained from the retrieved callus resumed active growth, and contained highly cytoplasmic aggregates of cells. The effect of various cryoprotectants is shown in Table 2. A mixture of 5% each of sucrose, glycerol, and DMSO resulted in higher viability than that obtained from chemicals used singly. The anther-derived callus was more sensitive than that obtained from the ovules, as evidenced by 34% and 42% survival respectively.

2.2 Meristem Culture (Fig. 1)

The preservation of meristem cultures has a number of advantages over the callus cell cultures. (1) They are genetically more stable, while the callus cultures on subculturing over an extended period show chromosomal aberrations, and change in nuclear and ploidy level. (2) In some tissues it is difficult to regenerate entire plants from the callus, whereas meristematic tips are relatively easy and quick to grow, and present an easy method of vegetative propagation. (3) Excised meristems yield pathogen-free plants. (4) Meristems taken from haploid plants ensure their maintenance. (5) While cell cultures require a regulated rate of freezing, meristems can withstand sudden freezing, thus eliminating the need for elaborate and expensive cryostats.

Meristems of peanut (*Arachis hypogaea*) and chickpea (*Cicer arietinum*) freeze-preserved in LN, and cultured on MS medium (Murashige and Skoog 1962) + IAA (2 mg/l) + kin (0.5 mg/l), MS + IAA (2 mg/l) + BAP (0.2 mg/l), or on MS + 2,4-D (2 mg/l), showed signs of growth within 2 weeks (Bajaj 1979c). Survival was judged by increase in size, capacity to form callus, and roots or shoots, or both. Complete plants were eventually regenerated from cultures freeze-preserved for 20 months in LN (Bajaj 1983b). These plants set normal seeds, and no

Fig. 1A–H. Various stages in the regeneration of plants of *Arachis hypogaea* from frozen (–196 °C) meristems. **A–C** Frozen-thawed meristems after 6, 9, and 11 weeks of culture respectively, note the initial all-round callusing (perhaps due to cryoinjury) of the meristem in **A**. **D, E** Normal growth of the retrieved meristem (preserved for 20 months), and regeneration of a complete plant 4 and 12 weeks respectively after culture on MS+IAA+BAP. **F, G** Test-tube plants 3 and 7 weeks respectively after transferring to soil. **H** Normal seeds obtained from such plants (total time 24 weeks from the culture of excised meristems to seed setting). (Bajaj 1983b)

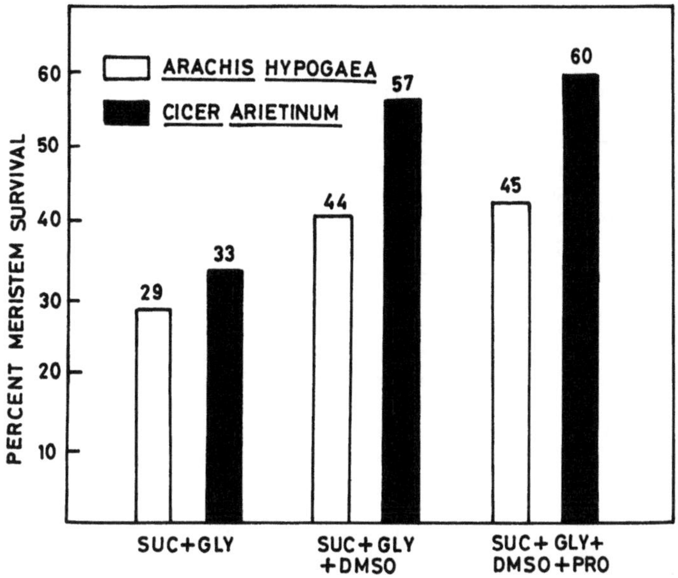

Fig. 2. Histogram showing the effect of mixture of various cryoprotectants on the survival of frozen meristems of *Arachis hypogaea* and *Cicer aeritinum* (*SUC* sucrose; *GLY* glycerol; *DMSO* dimethyl sulfoxide; *PRO* proline)

change in their chromosome number was observed. Three was maximum survival of 42% and 57% respectively in *Arachis* and *Cicer.* Growth in the case of peanut was faster than that in chickpea. On MS+IAA+kin, there was initially a tendency to form callus at the basal end. However, later the leaves unfolded and the shoots developed. The addition of 2,4-D to MS caused an all-round proliferation to form a mass of callus, while the development of shoots was inhibited.

The extent of survival of the meristems and their eventual regeneration into plants dependent on a number of factors, such as preculture conditions, method of freezing, cryoprotectant, and method of thawing. The data on some of these are shown in Tables 3, 4, and Fig. 2.

2.2.1 Effect of Preculture

Before freezing, meristems of *Cicer arietinum* were precultured on a basal medium containing DMSO (Table 3). The control showed 52% survival, the addition of 3% DMSO increasing it to 60%; however, DMSO at 10% was inhibitory.

2.2.2 Effect of Cryoprotectants

Figure 2 shows the effect of various cryoprotectants on meristems of *Arachis* and *Cicer.* The best response (60% survival) was obtained with a mixture of four cryoprotectants, i.e., sucrose, glycerol, DMSO, and proline. It is advisable to em-

Table 3. Effect of various preculture treatments on survival of excised meristems of *Cicer arietinum* freeze-preserved for 3 months at −196 °C. (Bajaj 1983b)

Preculture treatment	No. of meristems cultured	No. of meristems growing	Percent showing growth
1. Basal medium (BM)	29	15	52
2. BM + DMSO (3%)	23	14	60
3. BM + DMSO (5%)	25	14	56
4. BM + DMSO (10%)	17	8	47

Table 4. Effect of various thawing temperature on the survival of unfrozen and frozen (stored at −196 °C for 3 months) excised meristems of *Cicer arietinum*. (Bajaj 1983b)

Thawing temperature (°C)	Treatment	No. of meristems cultured	No. of meristems showing growth	Percent showing growth	Percent of control
25	Unfrozen	29	23	80	–
	Frozen	32	10	31	39
35	Unfrozen	21	16	76	–
	Frozen	23	9	39	51
40	Unfrozen	24	17	70	–
	Frozen	26	9	34	48

For control purposes, before culture the unfrozen meristems were treated with sterile distilled water at the appropriate temperature (25°, 35°, 40 °C).

ploy a combination of different cryoprotectants at low concentrations rather than one single cryoprotectant at a high concentration.

2.2.3 *Effect of Thawing*

The frozen cultures were thawed at 25, 35 and 40 °C (Table 4), the best results were obtained with cultures thawed at 35 °C (51%), followed by 40 °C (48%) and 25 °C (39%).

The observations on the beneficial influence of preculture, and the combination of cryoprotectants thus emphasize the role of these factors for further studies on cryopreservation. The regeneration of complete plants and setting of normal seeds further establish the use of this technology for long-term storage of genetic stocks.

2.3 Embryos

There are three areas where cryopreservation of embryos (Bajaj 1985) can be applied: (1) haploid cell cultures are known to be genetically highly unstable and revert to a diploid state within a short time. In such cases, crypreservation of pollen embryos or their callus would be rewarding; (2) in the incompatible crosses in which the hybrid embryos generally abort at early stages, the young embryos could be excised, cryopreserved, and cultured as the need arises (especially in cases where the flowering period is short and all the embryos cannot be cultured); (3) the "recalcitrant" species where seed cannot be stored.

The large-sized embryos are particularly vulnerable to cryodamage and methods for reviving them after freezing still need to be refined. The main task is the prevention of recrystallization and intracellular freezing, which is invariably lethal. Thus, the water content of these embryos needs to be controlled (by desiccation) to bring it to a level at which intracellular freezing is minimal. The cryodamage can be further prevented by adopting a method of thawing which discourages growth of ice crystals. The technology of the cryopreservation of embryos is thus a multiple event, involving dehydration-freezing-storage-thawing-culture, and faulty step at any stage may cause irreparable damage.

2.3.1 Pollen Embryos

The excised anthers of *Arachis villosa* and *A. hypogaea* cultured on MS+IAA (4 mg/l)+BAP (2 mg/l) for 5–6 weeks (Bajaj et al. 1980, 1981) were removed, cut into two to four segments, and used for cryopreservation studies (Bajaj 1983c). The segments from 50 anthers were either pooled, treated with the cryoprotectant solution (5% each of DMSO, sucrose, glycerol) and wrapped in an aluminum foil, or were put in 1 ml of the cryoprotectant solution in an ampule. From some anthers a suspension of pollen embryos was prepared. The materials were either: (1) subjected to the vapors of LN and immersed gradually (like prefreezing); or (2) were quickly immersed in LN, and stored for 1 year. They were thawed at 35–40 °C, washed with the liquid medium and cultured.

The survival was judged by (1) increase in size of the embryos and the segments of the anthers, (2) callusing, (3) greening, and formation of roots and shoots. The pollen embryos underwent a lag phase and showed no sign of growth for the first 4 weeks in cultures; however, thereafter they started to elongate. The embryos resumed growth and proliferated to form callus (Fig. 3A–D). *Arachis villosa* showed higher survival as compared to *A. hypogaea* (Table 5). Viability in the two species varied between 14 and 31% and 21 and 46% in anther segments and pollen embryos respectively. The callus occasionally differentiated malformed shoots and plants.

The survival of anther segments was observed in both methods employed for freezing. However, anther segments wrapped in the aluminum foil (dry method) yielded better results compared to those frozen in ampules. In the latter case, the anthers had a tendency to become spongy.

An almost similar response was observed in frozen anther segments and pollen embryos of *Brassica* (Bajaj 1983c). *Brassica napus* was more tolerant to freezing, compared to *B. campestris* (Table 5), and 44% of the pollen embryos survived, resumed growth, callused or regenerated plants (Bajaj 1983c).

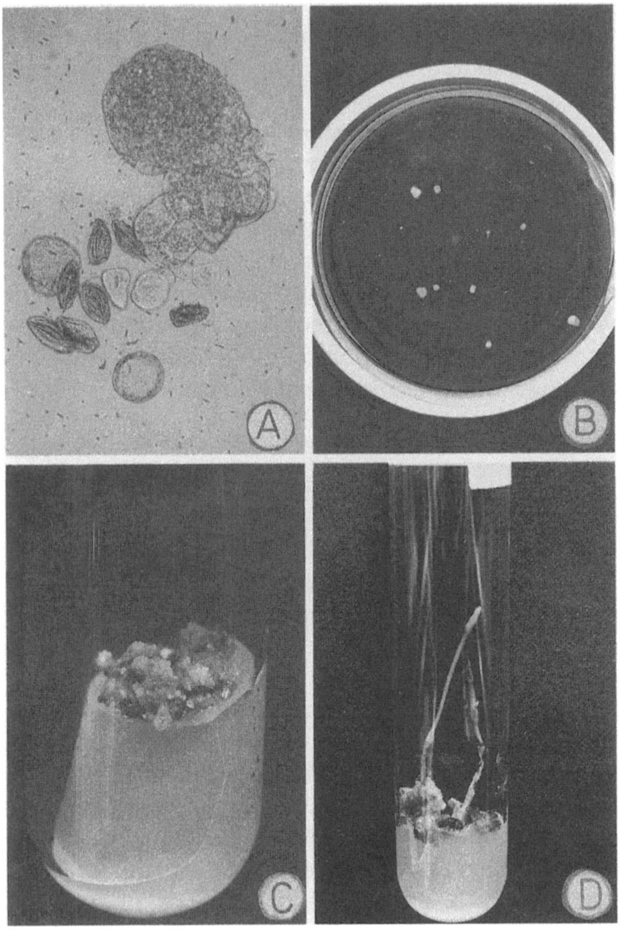

Fig. 3 A–D. Survival of pollen embryos and anther segments of *Arachis hypogaea* and *A. villosa* stored in liquid nitrogen for 1 year. **A, B** A retrieved pollen embryo immediately after thawing (**A**), and subsequent growth into callus colonies (*A. hypogaea*). **C, D** A mass of callus obtained from frozen anther segments of *A. villosa*, and its differentiation into shoots. (Bajaj 1983 c)

Table 5. Survival of pollen-embryos and segments of the androgenic anthers of *Arachis* and *Brassica* cryopreserved for 1 year. (Bajaj 1983 c)

Pant species	No. of pollen embryos frozen	No. of pollen embryos survived	Percent survival	No. of anther segments frozen	No. of anther segments resumed growth	Percent survival
Arachis hypogaea cv. M 13	73	21	29	107	15	14
A. villosa	121	46	38	120	37	31
Brassica campestris cv. TL 15	94	29	31	111	12	11
B. napus cv. Tower	81	35	44	114	21	19

2.3.2 *Zygotic and Hybrid Embryos*

The legume seeds are "orthodox", can be easily stored on a long-term basis, and there seems to be no deterioration. Thus the need for the cryopreservation of zygotic embryos does not arise. However, in wide hybridization programs, especially in incompatible crosses in which hybrid embryos abort and the flowering period is short, such embryos can be cryopreserved until required. In the "recalcitrant" group of plants, such as coconut, where the zygotic embryos abort and there is no method of vegetative propagation, cryopreservation assumes great significance in the conservation of germplasm.

The immature embryos of the West Coast Tall cv. of *Cocos nucifera*, removed from nuts and stored for 1 month, were partially dehydrated, then cut into transverse halves. They were treated with a cryoprotectant solution (7% DMSO and 7% sucrose in MS liquid medium), blotted dry, then wrapped in a single layer of sterile aluminum foil and frozen (Bajaj 1984c). The retrieved embryos and their segments in cultures showed a lag period of up to 4 months, and up to 25% of them survived, elongated and callused (Table 6).

In an incompatible cross involving *Vigna mungo* × *V. radiata* (blackgram and greengram), the hybrid embryos normally abort; however, if excised and cultured, hybrid plants can be obtained (Gosal and Bajaj 1983). An attempt has been made to freeze the immature embryos of *Vigna mungo, V. radiata* and their hybrids (Table 7). The excised immature embryos (11 – 15 days after pollination) were treated with a cryoprotectant mixture (containing 5% each of DMSO, sucrose, and glycerol) for 1 h, then frozen in LN, and thawed at 35 – 40 °C. After washing with the liquid medium, the retrieved embryos cultured on MS supplemented with IAA

Table 6. Survival of embryos of coconut subjected to freezing in the presence of DMSO (7%) + sucrose (7%). (Bajaj 1984c)

Explant	No. of cultures frozen	No. of cultures showing signs of growth	Percent survival	Remarks
1. Embryo (1 – 1.5 cm)	17	3	17.6	Embryo elongation
2. Transverse halves of embryo	28	7	25.0	Proliferation at the cut end

Table 7. Survival of seeds and immature embryos of *Vigna mungo*, *V. radiata*, and their hybrids at – 196 °C

Plant species	Material	Response (percent survival)
Vigna mungo	Seeds	87, normal germination
	Embryos	53, callus and plants
Vigna radiata	Seeds	91, normal germination
	Embryos	59, callus and plants
Vigna mungo × *Vigna radiata*	Hybrid Embryos	34, callus and plants

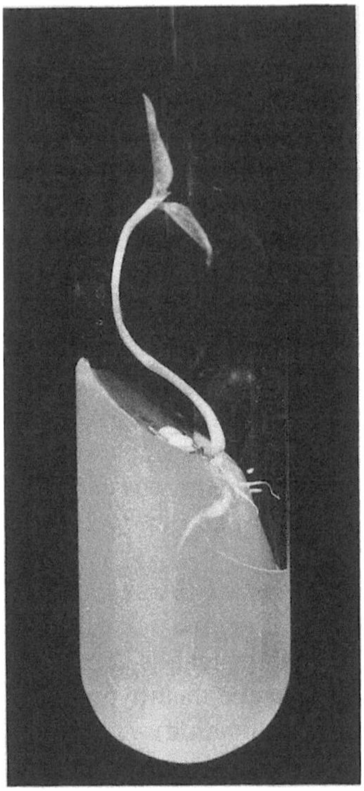

Fig. 4. A plantlet obtained from a frozen hybrid embryo of a cross *Vigna mungo* × *Vigna radiata* 4 weeks after cultures on MS + IAA (1 mg/l) + kin (0.2 mg/l) + casein hydrolysate (500 mg/l)

(1 mg/l) + kin (0.2 mg/l) + casein hydrolysate (500 mg/l) underwent a lag period of about 1 week. Thereafter they elongated, turned greenish, and in some cases proliferated to form callus. Eventually plants were formed (Fig. 4), which on transfer to the field plots matured and set seed.

2.3.3 Somatic Embryos

The cryopreservation of somatic embryos has already been described (Bajaj 1985). Somatic embryos are regarded as "seeds" in plants which do not set seed. In such cases somatic embryos can be cryostored and used. Since somatic embryos can be produced in large numbers from cell suspensions, or by in vitro culture of segments and tissues, it would be rewarding to use them for large-scale multiplication of desirable and elite plants in which seeds are produced in small quantities, or in recalcitrant species where the seeds cannot be stored. In oilpalm, an attempt has been made to cryopreserve somatic embryos. The embryos showed a survival rate of 50–80%, and adventive embryogenesis occurred in embryoid clumps stored up to 15 months in LN (Engelmann and Duval 1986).

2.4 Protoplasts

The cryopreservation of protoplasts has been reviewed recently (Bajaj 1989). The frozen protoplasts, like the cells, retain their morphogenetic potential and

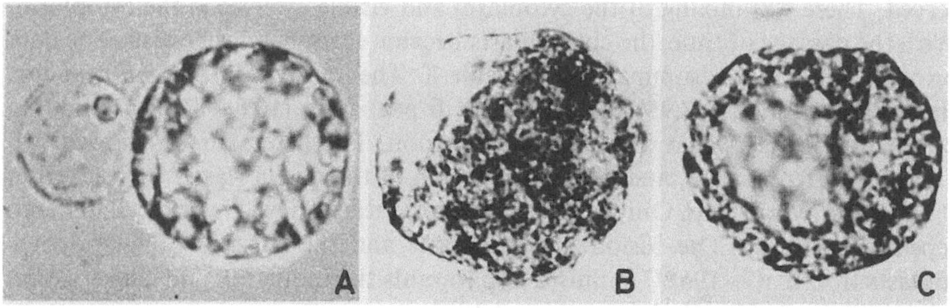

Fig. 5 A–C. Survival of somatic hybrid cells (*Triticum aestivum* × *Pisum sativum* subjected to freezing in liquid nitrogen. **A** Freshly isolated protoplast from cell suspension of wheat (*small, colorless*) and a mesophyll protoplast of pea (*large, dark*) showing adhesion after treatment with polyethylene glycol. **B** A fused product immediately after thawing at 35 °C. **C** Same, 21 days after culture in a microdrop; note the mixing of the protoplasts and the healthy nature of the retrieved somatic hybrid cell. (Bajaj 1983 d)

Table 8. Survival of somatic hybrid protoplasts of wheat × pea and rice × pea frozen in liquid nitrogen in the presence of cryoprotectants, and thawed at 35 °C. (Bajaj 1983 d)

Hybrid protoplast	Experiment number	No. of frozen hybrid protoplasts observed	No. showing signs of growth	Percent survival
Pisum sativum × *Triticum aestivum*	1	75	11	15.0
	2	41	15	12.2
	3	67	12	17.9
Pisum sativum × *Oryza sativa*	1	37	4	10.8
	2	59	7	11.8
	3	43	6	13.9

regenerate complete plants (Bajaj 1988). They differ, however, from the cells in their lack of a cell wall and plasmodesmata, and are more prone to bursting. Thus they require a critical concentration of an appropriate osmoticum cum cryoprotectant to prevent them from bursting or causing undue damage to the plasma membrane. Protoplats devoid of plasmodesmata are better to study the mechanism of cryopreservation at a cellular level, the isolated protoplasts providing an ideal system for quantitative cryomicroscopic studies (Steponkus et al. 1983).

Information on freeze preservation of protoplasts of legumes and oilseed crops is scanty. Soybean protoplasts (in 0.7 M sorbitol) frozen in LN, when cultured, develop clones from individual protoplasts with an efficiency of 10^{-3} as compared to nonfrozen control (Weber et al. 1983). Takeuchi et al. (1982) reported a survival of 60%.

The protoplasts of pea, *Pisum sativum* (Bajaj 1983 d) isolated from the in vitro-grown seedlings and fused with those of wheat and rice, using polyethylene glycol, were suspended in a mixture of 0.5 M mannitol and dimethyl sulfoxide (5%) for 1 h, and frozen (Fig. 5). The retrieved protoplasts (Fig. 5 B) lay quiescent for the first 2 weeks in culture. Thereafter, changes in their morphology were observed. There was mixing of the cytoplasm and visible cyclosis in the cytoplasm. With the passage of time, the chloroplasts became scanty and yellowish. The data on their survival are summarized in Table 8. The survival in pea x wheat was slighly higher (12.2–17.9%) as compared to pea x rice (10.8–13.9%). Budding of the protoplasts was a common phenomenon. The heterokaryon showed two nuclei of different sizes, and occasional fusion of the nuclei was also observed in the somatic hybrid cells. Only in one instance did the retrieved hybrid cell undergo repeated divisions. The fusion of protoplasts and the survival of hybrid protoplasts frozen at −196 °C is thus a step towards the induction and conservation of genetic variability (Bajaj and Gosal 1982).

3 Germplasm Banks

Since entire plants can be regenerated from freeze-preserved cells, tissues, and organs, this phenomenon may be applicable for the long-term storage of rare germplasm, and for the establishment of germplasm banks (Fig. 6). This system would be analogous to the cryopreservation of semen. For this purpose meristems are better, since they are genetically more stable than callus cultures, which undergo genetic aberrations, making them usually unfit for the maintenance of clonal stocks. However, the main objective of the conservation of induced genetic variability can be achieved through freeze preservation of novel, mutant, elite and rare cell lines which are of limited current use, but might be required in future. The material in this form can be exchanged and transferred to different locations. The establishment of repositories and banks of cell cultures based on the technology of cryopreservation is therefore very desirable. Before cryopreservation is recommended as a general tool, two points need to be considered: (1) the viability of the retrieved cultures should be high, so that there is no selection pressure. This can be achieved by optimizing factors such as age and physiological state of the cultures, effect of cryoprotectants, and the method of freezing, storage, and thaw-

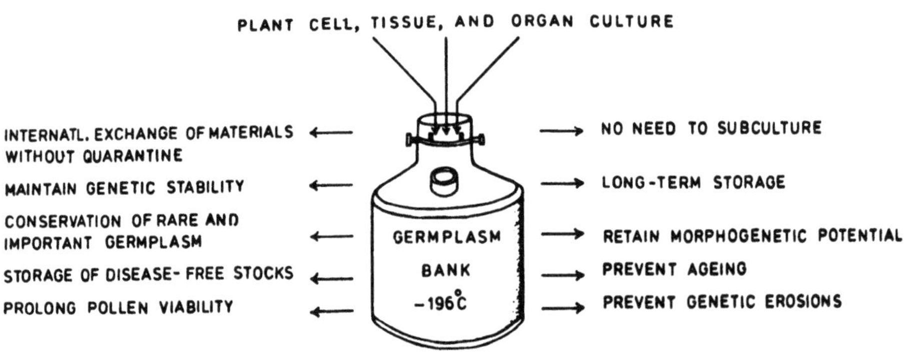

Fig. 6. Potentials of the cryopreservation of plant cell, tissue, and organ culture, and the establishment of a germplasm bank. (Bajaj 1982b)

ing; (2) it must also be ensured that during long-term storage there is no genetical change or deterioration of germplasm.

4 Summary and Conclusions

Cells and tissue cultures are a rich source of genetic variability, but they undergo genetic erosion, and at present no method is known to conserve the in vitro-induced variability on a long-term basis. The storage of cultures in liquid nitrogen has been recommended for this purpose. During the last decade, considerable progress has been made in the area of cryopreservation, and complete plants have been regenerated from tissues, meristems, pollen embryos, and hybrid embryos freeze-preserved for various lengths of time. The freeze preservation of cultures of "recalcitrant" species should enable the conservation of germplasm from these plants.

References

Bajaj YPS (1976a) Gene preservation through freeze storage of plant cell, tissue and organ cultures. Acta Hortic 63:75–84

Bajaj YPS (1976b) Regeneration of plants from cell suspensions frozen at −20°, −70° and −196 °C. Physiol Plant 37:263–268

Bajaj YPS (1979a) Technology and prospects of cryopreservation of germplasm. Euphytica 28:267–285

Bajaj YPS (1979b) Establishment of germplasm banks through freeze-storage of plant tissue culture and their implications in agriculture. In: Sharp WR, Larsen PO, Paddock EF, Raghavan V (eds) Plant cell and tissue culture – principles and applications. Ohio State Univ Press, Columbus, pp 745–774

Bajaj YPS (1979c) Freeze preservation of meristems of *Arachis hypogaea* and *Cicer arietinum*. Indian J Exp Biol 17:1405–1407

Bajaj YPS (1982a) Survival of anther-, and ovule-derived cotton callus frozen in liquid nitrogen. Curr Sci 51:139–140

Bajaj YPS (1982b) Cryopreservation of germplasm of potato, cassava, peanut and chickpea. In: Fujiwara A (ed) Plant Tissue Culture 1982. Maruzen, Tokyo, pp 799–800

Bajaj YPS (1983a) Cryopreservation and international exchange of germplasm. In: Sen SK, Giles KL (eds) Plant Cell Culture in Crop Improvement. Plenum Press, New York, pp 19–41

Bajaj YPS (1983b) Production of normal seeds from plants regenerated from meristems of *Arachis hypogaea* and *Cicer arietinum* cryopreserved for 20 months. Euphytica 32:425–430

Bajaj YPS (1983c) Regeneration of plants from pollen-embryos of *Arachis, Brassica* and *Triticum* spp. cryopreserved for one year. Curr Sci 52:484–485

Bajaj YPS (1983d). Survival of somatic hybrid protoplasts of wheat x pea, and rice x pea subjected to −196 °C. Indian J Exp Biol 21:120–122

Bajaj YPS (1984a) Cryopreservation of germplasm of crops – Present status and prospects. In: Novak FJ, Havel L, Dolezel J (eds). Int Symp Plant tissue and cell culture – application to crop improvement; Czech Acad Sci, Prague, pp 607–616

Bajaj YPS (1984b) Biotechnology of the conservation of germplasm and its implications in agriculture and forestry. Proc Int Symp Genetic manipulations of crops. Acad Sin, Beijing

Bajaj YPS (1984c) Induction of growth in frozen embryos of coconut and ovules of citrus. Curr Sci 53:1215–1216

Bajaj YPS (1985) Cryopreservation of embryos. In: Kartha KK (ed) Cryopreservation of plant cells and organs. CRC Boca Raton, pp 227–242

Bajaj YPS (1986a) Biotechnology in agriculture and forestry. Vols I, 2, Springer, Berlin Heidelberg New York Tokyo

Bajaj YPS (1986b) In vitro preservation of genetic resouces – techniques and problems. Proc Int Symp Nuclear techniques and in-vitro culture for plant improvement, IAEA/FAO, Vienna, pp 43–57

Bajaj YPS (1988) Regeneration of plants from frozen (–196 °C) protoplasts of *Atropa belladonna* L., *Datura innoxia* Mill and *Nicotiana tabacum* L. Indian J Exp Biol 26:289–292

Bajaj YPS (1989) Cryopreservation of plant protoplasts. In: Bajaj YPS (ed) Biotechnology in agriculture and forestry 8. Plant protoplasts and genetic engineering I. Springer, Berlin Heidelberg New York Tokyo, pp 97–106

Bajaj YPS, Gosal SS (1982) Induction of genetic variability in grain-legumes through tissue culture. In: Rao AN (ed) Proc Int. Symp Tissue culture of economically important crops. COSTED, Singapore, pp 25–41

Bajaj YPS, Labana KS, Dhanju MS (1980) Induction of pollen-embryos and pollen-callus in anther cultures of *Arachis hypogaea* and *A. glabrata.* Protoplasma 103:397–399

Bajaj YPS, Ram AK, Labana KS, Singh H (1981) Regeneration of genetically variable plants from the anther-derived callus of *Arachis hypogaea* and *A. villosa.* Plant Sci Lett 23:35–39

Engelmann F, Duval Y (1986) Cryoconservation d'embryons somatiques de palmier à huile (*Elaeis guineensis* Jacq.): Résultats et perspectives d'application. Oleagineux 41:169–172

Finkle BJ, Ulrich JM, Rains DW, Tisserat BB, Schaeffer GW (1979) Survival of alfalfa, rice and date-palm callus after liquid nitrogen freezing (Abstr). Cryobiology 16:583

Gosal SS, Bajaj YPS (1983) Interspecific hybridization between *Vigna mungo* and *Vigna radiata* through embryo culture. Euphytica 32:129–137

Haskins RH, Kartha KK (1980) Freeze-preservation of pea meristems: cell survival. Can J Bot 58:833–840

Murashige T, Skoog F (1962) A revised medium for rapid growth and bioassays with tobacco tissue culture. Physiol Plant 15:473–497

Steponkus PL, Dowgert MF, Ferguson JH, Levin RL (1983) Cryomicroscopy of isolated plant protoplasts. Cryobiology 21:209

Sugawara Y, Takeuchi M (1986) A rapid method for the determination of cell survival in shoot apex after freezing and thawing. Int Congr Plant tissue cell cult, abstr. Univ Minnesota, Minneapolis, p 233

Takeuchi M, Matsushima H, Sugawara Y (1982) Totipotency of protoplasts after long-term freeze preservation. In: Fujiwara A (ed) Plant tissue culture 1982. Maruzen, Tokyo, pp 797–798

Weber G, Roth JE, Schweiger HG (1983) Storage of cell suspensions and protoplasts of *Glycine max* (L.) Merr., *Brassica napus* (L.). *Datura innoxia* (Mill.), and *Daucus carota* (L.) by freezing. Z Pflanzenphysiol 109:29–39

Withers LA (1983) Germplasm storage in plant biotechnology. In: Mantell SH, Smith H (ed) Plant biotechnology. Cambridge Univ Press, Cambridge, pp 187–218

Section II Food Legumes

II.1 Pea (*Pisum sativum* L.)

M. GRIGA[1] and F. J. NOVÁK[2]

1 Introduction

The genus *Pisum* is a member of the order Fabales, family Fabaceae, tribe Viciae. It consists of a broad range of morphologically distinct types spread worldwide, many of which are described as separate species. *Pisum* is indigenous to Central Asia, Eurasia, and, as a secondary place of origin, the eastern Mediterranean (Hawkes 1983). Recent cytological and genetic research has proved that all species have a chromosome number of $2n = 14$, and *no* natural polyploid forms have been discovered (Jaranowski and Micke 1985). All wild types are divided into the following five species: *Pisum formosum Stev., P. fulvum* Sibth. et Sm., *P. abyssinicum* Braun., *P. humile* Boiss. et Noe, and *P. elatius* Stev. *Pisum sativum* L. is divided into two subspecies: ssp. *hortense* A. et Gr. – the garden pea, and *ssp. arvense* (L.) A. et Gr. – the field pea (syn. *P. arvense* L.).

This traditional classification system has recently been revised (Lamprecht 1966; Blixt 1972, 1975; Jaranowski and Micke 1985). Lamprecht (1966) regards the genus as monospecific and in *Pisum arvense* L. s.l. includes all the existing forms formerly considered to be separate species. Since there is no crossing barrier between these forms, and complete genetic recombination of all traits, these views appear valid. He thus makes a distinction within the species *P. arvense* between the ecotypes *abyssinicum*, *arvense*, *fulvum*, and *humile* – all occurring in nature as wild-growing populations. The ecotype *sativum* comprises the domesticated pea forms with no known wild types.

Since tissue culture experiments with pea have thus far been conducted only with the cultural forms, only *P. sativum* and *P. arvense* will be referred to as pea in this chapter.

Garden pea is an annual plant and consists of both summer and winter forms. In temperate climates, pea is autogamous (cleistogamous). Fertilization takes place approximately 24 h before opening of the flowers. Allogamy sometimes occurs under tropical and subtropical conditions (Govorov 1928; Harland 1948). The flowering time is rather long, extending over a period of 12–20 days, and each flower blossoms for about 3 days. Early genotypes begin bearing flowers at

[1] OSEVA-Plant Breeding and Seed Corporation, Research and Breeding Institute of Technical Crops and Legumes, Sumperk-Temenice, Czechoslovakia
[2] Plant Breeding Unit, Joint FAO/IAEA Programme, International Atomic Energy Agency's Laboratories, Seibersdorf, Austria

Biotechnology in Agriculture and Forestry, Vol. 10
Legumes and Oilseed Crops I (ed. by Y. P. S. Bajaj)

the leaf axils of the lower leaves, while late-maturing varieties begin at the higher leaves. The vegetation period of the spring pea varies from between 70 and 140 days.

Flower initiation begins at the embryo stage, when the plant has only two leaves. The number of sterile nodes varies considerably, from 2 to 34 (Marx 1969). Flower initiation seems to be under very strict genetic control, with the environment playing only a small part (Haupt 1952).

There are great differences among varieties in their reaction to day length. Day-neutral, short and long photoperiod genotypes all exist. Central and West European varieties are day-neutral.

The importance of grain legumes lies in their high content and quality of protein. Pea represents both a first-class component of fodder for livestock, and food for human consumption. Seed storage protein content varies from 15.5% to 39.7% (Monti and Grillo 1983). This wide range is due to both genetic differences among varieties and interactions of genotype and environment (Bajaj et al. 1971 a, b). Peas are limited in the sulfur-containing amino acids, cysteine and methionine, but are rich in lysine and other essential amino acids.

Peas contain lecithin, vitamins A, B_1, B_2, C in seeds, and vitamin E in the radicle. Peas also contain 46% – 58% glycides, 5% – 6% fiber and 2% – 4% ash. Large amounts of minerals such as potassium, phosphorus, magnesium, sulfur, calcium, and iron are also present (Wright 1985).

However, peas also contain negative or antinutritional factors including trypsin inhibitors, hemagglutinins, and toxic amino acids. Fortunately, most of these are inactivated by cooking. The tannin content may be twice as high in genotypes carrying the dominant alleles for colored flowers and seeds compared to those with recessive alleles (Jaranowski and Micke 1985). The pea is grown worldwide (Table 1). The variety greatly influences the yield of both green and dry peas. Goals for variety improvement are higher yield, disease resistance, and better quality (Table 2). One serious problem is the amount of variation in agronomic performance from location to location and from season to season. Snoad (1985) listed the following characters that should be improved:

1. Uniform germination and rapid seedling establishment, especially at low temperatures and in other stress environments.
2. Improved winter hardiness.
3. Thicker, stronger stems, especially the hypocotyl.
4. Greater uniformity of crop-plant development. This may be aided by reduction of the haulm and by precision drilling.
5. Increased synchronous seed development combined with a significant reduction in maturity rate just prior to harvesting of vining peas.
6. Decreased synchronous seed development for dried peas.
7. Improved standing ability.
8. Reduced seed losses prior to and during harvesting of dried peas.

Breeding for disease and insect resistance is one of the major objectives for pea improvement (Hagedorn 1985). However, this may require a minimum of 20% of the plant breeder's time, effort, and budget (Anonymous 1985).

Table 1. World production of dry peas and green peas. (Taken from FAO 1988)

Continent/ country	Dry peas			Green peas		
	Area (1000 ha)	Yield (kg/ha)	Production (1000 tons)	Area (1000 ha)	Yield (kg/ha)	Production (1000 tons)
World	9770	1488	14534	772	6326	4882
Africa	423	699	295	25	6448	163
Algeria	11F	377	4F	–	–	–
Burundi	62F	613	38F	–	–	–
Congo	5F	700	4F	–	–	–
Egypt	3F	1667	5F	9F	9425	82F
Ethiopia	146F	822	120F	–	–	–
Lesotho	7F	479	4F	–	–	–
Libya	–	5060	2F	1F	2316	2F
Madagascar	–	–	–	1F	962	1F
Mauritius	–	–	–	–	2308	–
Morocco	50F	700	35F	8F	4416	34F
Rwanda	25F	360	9	–		–
Sierra Leone	2F	750	2F	–	–	–
South Africa	4F	1750	7F	4F	6250	25F
Tanzania	35F	250	9F	–	–	–
Tunisia	–	–	–	3F	5676	19F
Uganda	20F	1000	20F	–	–	–
Zaire	52F	731	38F	–	–	–
North and Central America	344	1893	651	143	7621	1090
Canada	255	1825	465	13F	3881	50F
Dominican Rep.	2F	900	2F	–	–	–
Jamaica	4F	1034	5F	–	–	–
Mexico	4F	500	2F	12F	3333	40F
USA	79	2255	177	118	8468	1000
South America	131	728	96	52	2329	121
Argentina	7F	1106	7F	10F	2500	25F
Bolivia	5F	1133	5F	14	1400	20
Chile	9	808	7	4F	5000	20F
Colombia	57F	622	36F	–	–	–
Ecuador	13F	305	4F	8F	813	7F
Guyana	2F	522	1F	–	–	–
Paraguay	1F	833	1F	–	–	–
Peru	33F	909	30F	16F	3125	50F
Uruguay	1F	2273	3F	–	–	–
Venezuela	4	531	2	–	–	–
Asia	2031	1202	2442	179	4032	720
Burma	26	839	22	–	–	–
China	1300F	1431	1860F	56F	5393	302F
Cyprus	–	–	–	–	9346	2
Gaza Strip	–	–	–	–	5000	1F
India	483	891	430	94F	2777	261F
Iran	70	726	51	–	–	–
Israel	1F	500	–	1F	4000	6F
Japan	1F	2159	2F	10F	6854	66F
Jordan	–	–	–	–	3500	–
Korea Rep.	2F	900	1F	–	–	–
Lebanon	–	1091	–	2F	5455	9F
Pakistan	145F	476	69F	–	–	–
Philippines	–	–	–	8F	3584	28F

Table 1 (continued)

Continent/ country	Dry peas			Green peas		
	Area (1000 ha)	Yield (kg/ha)	Production (1000 tons)	Area (1000 ha)	Yield (kg/ha)	Production (1000 tons)
Syria	1	912	1	2	5458	11
Turkey	3	1964	6	6F	6034	35
Europe	990	3278	3245	277	8304	2303
Austria	–	1286	–	1	8571	12
Belgium-Lux.	2F	4471	8F	12F	16414	190F
Bulgaria	21F	1171	25F	10F	1041	10F
Czechoslovakia	75	2345	175	4	3829	15
Denmark	145F	2759	400F	13F	7692	100F
Finland	2	1682	4	2	4500	7
France	428	4164	1782	40F	10000	400
German DR	23F	2304	53F	7F	3138	20F
Germany FR	–	–	–	2	10588	18
Greece	–	1167	–	3	5987	17
Hungary	67	2700	182	32F	9277	295
Ireland	–	3500	1F	–	–	14F
Italy	3F	1667	5F	31	6742	209
Netherlands	36F	4250	153F	7F	11942	82F
Norway	–	–	–	1F	5807	5F
Romania	45F	2133	96F	30F	6000	180F
Spain	5F	778	4F	10	5500	55
Sweden	5F	2020	10F	12F	4466	52F
Switzerland	–	–	–	1F	4429	6F
UK	119	2748	327	47F	12262	580F
Yugoslavia	14F	1500	21F	14F	2465	35F
Oceania	351	2862	1005	20	10275	210
Australia	310	2861	887	10F	15538	162F
New Zealand	41	2868	118	10F	4800	48F
USSR	5500F	1236	6800F	75F	3667	275F

F – FAO estimate

Table 2. Main objectives for breeding dry peas. (Based on Poulsen and Sebro, 1983, after Gantotti and Kartha, 1986)

A. *Canopy*
- Early development
- Many short nodes before flowering
- Small leaflets (semi-leafless)
- Strong stem (short, erect, and stiff)
- High harvest index

B. *Flowering*
- Early flowering
- Short flowering period
- Early ripening
- Uniform ripening

C. *Seed yield*
- Many pods per plant
- Many seeds per pod
- Two pods per node
- Long nonshattering pods
- Intermediate seed weight
- Large embryo

D. *Seed quality*
- High protein content
- Rich in essential amino acids
- Tannin free
- Uniform seed size
- High volume weight
- Thin and strong testa
- Good color
- Good processing qualities

The pea is host to a whole series of viruses shared by other pulses. Of greatest economic importance are the pea common mosaic virus (PCMV), pea enation mosaic virus (PEMV), bean yellow mosaic virus (BYMV), pea leaf-roll mosaic virus (PLRMV), syn. pea seed-borne mosaic virus (PSBMV), and the broad bean common mosaic virus (BBCMV). These viruses cause (apart from colored necrotic lesions on the leaves) morphological changes and secondary yield losses due to pod deformation, fewer seeds in pods, shoot deformed apices, twisted and necrotized leaves, rolling along the main vein, phloem necrotization, dwarf plants, poor flower setting, flower shedding, and rod-like growth. Most viruses are transmitted by insects, such as greenfly and thrips, and fewer mechanically or through seeds (PLRMV).

High economic damage is also caused by fungal and bacterial diseases. The most important is *Ascochyta* blight – a term designating three diseases manifested by similar symptoms but caused by three different fungus species: *Mycosphaerella pinodes*, *Ascochyta pisi*, and *Ascochyta pinodella*. Symptoms are lesions on leaves, which gradually spread to pods and seeds. The yield may be reduced by as much as 50%. This disease is most frequently transmitted through the seeds (primary infection) and is evident from the characteristic symptoms on the first leaves of the affected plants. The frequency with which secondary infections occur depends upon weather, especially frequent rain. Soil-borne fungi of the genus *Fusarium*, primarily *F. solani* f. sp. *pisi* causing root rot (rotting of stem bases and roots), and *F. oxysporum* f. sp. *pisi*, responsible for *Fusarium* wilt (withering and drying of plants due to the parasite hyphae vein bundles, enhanced by the effect of toxins excreted by the fungus) are also dangerous pathogens. Other diseases include: pea rust (*Uromyces pisi*), powdery mildew (*Erysiphe pisi*, syn. *E. polygoni*), downy mildew of pea (*Perenospora viciae*) and bacterial blight (*Pseudomonas syringae* cv. *pisi*).

The above is only a brief list (Hagedorn 1984) but represents the principal economic diseases in the main production areas (North America, Europe, Asia).

Wild *Pisum* species are a rich germplasm source of resistance genes. However, remarkably, this variation has been very little explored and most modern cultivars have a very narrow genetic base. There are several problems in pea improvement that cannot be solved by conventional breeding methods. Plant biotechnology (see Bajaj 1986), however, stimulates hope for new dimensions in pea breeding, in particular opening up the possibility of genetic variation induction, early screening, and single cell selection, directed to agronomically important characters.

2 In Vitro Approach

Cell and tissue culture can serve as an extension or supplement to conventional methods of pea improvement. Shoot meristem culture is an effective method for mass propagation of genetically uniform pea plants and for the production of pathogen-free materials. Other tissue culture techniques, e.g., callus and cell culture, have proven useful to induce genetic variability by somaclonal variation, which has a great potential in plant breeding (Larkin and Scowcroft 1981). More recently, Mroginski and Kartha (1984) and Gantotti and Kartha (1986) published comprehensive reviews of in vitro techniques and their applications to pea improvement. In this chapter, several aspects of in vitro cultures are discussed with regard to the development of a new pea breeding system.

2.1 Meristem Culture (Shoot-Tip Culture)

Meristem culture is the most highly developed in vitro technique for pea and is the closest to practical application. The first successful regeneration from shoot apical meristems was published by Kartha et al. (1974). Their research was directed along two lines: (1) elimination of virus diseases, and (2) freeze preservation of meristems (Kartha and Gamborg 1978; Kartha et al. 1979; Kartha 1979, 1981; Haskins and Kartha 1980).

Pea meristem culture is a simple technique (Fig. 1). Seeds are surface sterilized and germinated under aseptic conditions in the dark at 25 °C. From 5 to 21 days after seed germination, meristems, shoot apices, or axillary buds are isolated with the aid of a stereomicroscope (Bajaj and Dhanju 1979; Kartha 1984; Griga et al. 1984a, b, 1986) and placed on a medium-inducing shoot proliferation. The most commonly used media are B5 (Gamborg et al. 1968) or MS (Murashige and Skoog 1962) with BAP and NAA. The growing shoots are excised and transferred onto a root-inducing medium containing auxin. Complete plants are then placed in either hydroponics or soil. The regenerated plants flower early to give seed progenies. Individual steps of the procedure are described in Part 4 – Protocols (see also Kartha 1984; Griga et al. 1986).

Isolation of apical meristems 0.2–0.3 mm in diameter and their subsequent culture on B5 medium made it possible to obtain virus-free regenerants. Kartha and Gamborg (1978) freed more than 100 breeding lines from PSBMV. This virus is also transmitted through seeds, at a frequency sometimes as high as 80% to 90%.

Advantage for in vitro clonal propagation and freeze preservation are physiological homogeneity and genetic uniformity of meristematic cells. The probability of genetic changes by direct regeneration from meristem cells is low compared to that from callus and cell suspension cultures, thus guaranteeing the genetic stability of progenies during long-term germplasm conservation and in vitro clonal propagation.

Freeze preservation of pea meristems in liquid nitrogen has been described (Kartha et al. 1979; Kartha and Leung 1979). The meristems precultured in the presence of 5% dimethylsulfoxide (DMSO) were subjected to a slow cooling gradient up to a terminal freezing temperature of −40 °C. The ampules containing the meristems were then transferred to liquid nitrogen (−196 °C). In the reverse procedure, the meristems were thawed rapidly to 37 °C in 90 s, washed and freed of the freeze-preservation agents, and then placed on the culture medium. The viable meristems showed cell division and growth within 4–5 days and normal plants were obtained within 3–4 weeks. Sixty percent of the meristems preserved for more than 2 years were viable.

Fig. 1A–G. A High proliferation of pea shoots from axillary bud of the first primary scale on modified MS medium with 0.1 μM NAA and 20 μM BAP. Second subculture. **B** Shoot proliferation from pea meristem explants of different origin on modified MS medium with 0.1 μM NAA and 20 μM BAP. From *left to right* shoot apex, axillary bud of the first normal leaf, axillary bud of the second primary scale, and axillary bud of the first primary scale. *Bar* = 1 cm. **C** A meristem explant of pea derived from the bud of the first primary scale axil. *Bar* = 1 cm. **D** A meristem explant of pea derived from the bud of cotyledonary axil. *Bar* = 1 cm. **E** Root formation on regenerated pea shoots on root-inducing medium with 5 μM NAA. **F** Pea plants derived from shoot tip transplanted into hydroponic pots filled with perlite and half-strength MS medium. **G** A regenerated pea plant with ripening pods. *Bar* = 10 cm

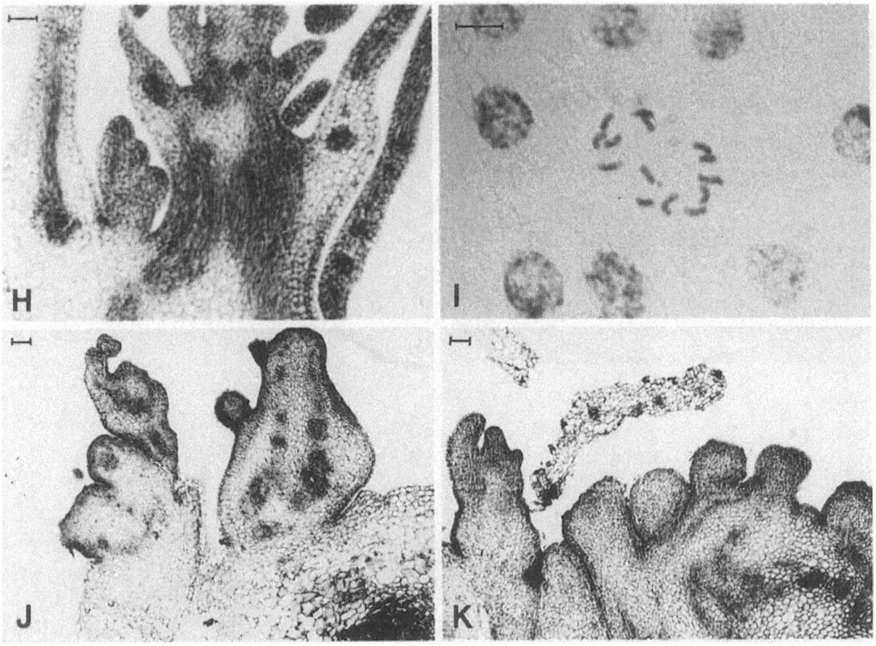

Fig. 1H–K. H Proliferation of axillary bud from pea shoot apex region in the initial culture. *Bar* = 100 μM. **I, J** Diploid chromosome number (2 n = 14) in the root tip of a regenerated pea plant. *Bar* = 10 μM. **K** Differentiation of adventitious buds on meristem explants during successive subculture. *Bar* = 100 μM. (**A–K** after Griga et al. 1986)

Griga et al. (1984a, b, 1986) diagramed in vitro meristem propagation of pea (Fig. 2). The method was also used to test more than 15 varieties or lines of *P. sativum* and *P. arvense*. They studied the influence of axillary meristems from various locations for the production of multiple shoots, the propagation coefficient for in vitro culture, and the effect of various factors on the induction of rhizogenesis.

The effect of various doses of gamma radiation of seeds on the growth of a long-term pea meristem culture was studied by Novák et al. (1984a). A dose of 10 Gy was found to stimulate growth in vitro when compared with the untreated control. Doses higher than 10 Gy inhibited growth of the meristem culture. A new phenomenon observed with pea meristem culture was the flowering induction and pod development in vitro (Novák et al. 1984b; Figs. 3, 4). In vitro flowering was dependent on variety and was frequently observed in the cv. Kleine Rheinländerin. The references on meristem pea cultures are summarized in Table 3.

2.2 Embryo Culture

Most studies with pea, including the oldest (Bonner 1938; Netien and Beauchesne 1951; Kruyt 1952), have dealt predominantly with the nutritional requirements of isolated embryos in in vitro culture, or the influence of various exogenously

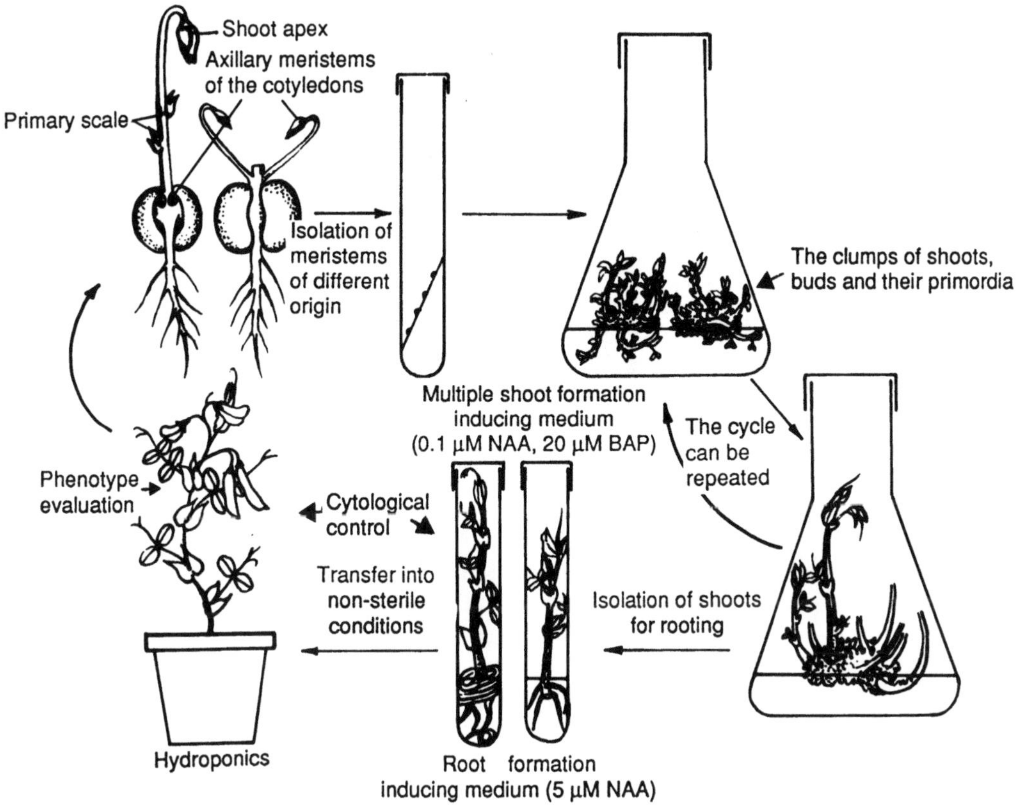

Fig. 2. Schematic representation of in vitro vegetative propagation cycle of pea (Griga, original, unpublished).

applied substances on embryo growth and development. These have included studies on the effects of nicotinic acid (Bonner 1938), growth regulators (Netien and Beauchesne 1951; Kruyt 1952), fluorinated pyrimidines (Paranjothy and Raghavan 1970), and various sugars (Davis 1983a, b).

In vitro culture of immature cotyledons and embryos was described by Millerd et al. (1975) and Thompson et al. (1977). Stafford and Davies (1979) cultured immature pea embryos in various developmental stages in a liquid medium containing a broad range of amino acids. Growth was not substantially improved by the exogenously applied growth regulators kin, GA3, NAA, BAP or ethylene. Growth improved if the medium was renewed every 2 or 3 days. Adequate aeration of the medium was also important. However, the critical factor was the osmotic potential. A very high potential (equivalent to 18% of the sucrose in the medium) was required for young embryos isolated at a developmental stage in which the endosperm was the major component of seeds. To the contrary, 4% sucrose was enough for embryos excised from seeds in which the endosperm was nearly resorbed. Optimal osmotic potential of the medium is obviously closely related to the development stage of the embryos. When embryos of seeds of various sizes

Fig. 3. Flower formation in shoot-tip culture of pea (Novák, unpublished)

are compared, those cultured in vitro are found to grow at about the same rate as those in vivo (Millerd et al. 1975; Thompson et al. 1977; Stafford and Davies 1979).

The influence of various carbon sources (sucrose, glucose, fructose, and maltose) on the growth of pea embryonic axes (excised from ripe seeds) and their α-amylase activity was studied by Davis (1983 a, b). The optimum sucrose concentration for fresh weight accumulation was 5%, although the growth was not statistically different from that obtained with 2%. The optimum sucrose concentration for dry weight accumulation was 10%–20%. Higher concentrations (30%, i.e., about 0.9 M) inhibited both the accumulation processes. Sucrose concentration also affected growth of the various plant organs; e.g., epicotyl growth was inhibited with 5%, while 10% was required to inhibit root growth. Excised embryonic axes grew equally well on 5% concentrations of fructose, glucose or maltose. The osmotic effects on growth were similar to that observed for sucrose. No effects of exogenous sugars on α-amylase activity was demonstrated.

Fig. 4. Detail of regular flowers formed in shoot-tip culture of pea (Novák, unpublished)

Table 3. Summary of meristem culture in pea

Explant source	Experimental approach, results or application	Reference
Apical meristem	Plant regeneration (first published report)	Kartha et al. (1974)
Apical meristem	Elimination of PSBMV from over 100 breeding lines	Kartha and Gamborg (1978)
Apical meristem	Cryopreservation of pea meristems in liquid nitrogen plant regeneration	Kartha et al. (1979), Haskins and Kartha (1980)
Apical meristem	Plant regeneration	Bajaj and Dhanju (1979)
Apical meristem	Optimization of the medium for clonal propagation in vitro	Griga et al. (1984a)
Apical meristem, axillary meristems of cotyledons, primary scales and normal leaves	Clonal propagation in vitro	Griga et al. (1984b), Griga et al. (1986)
Apical meristem, axillary meristems	Effect of gamma-irradiation on meristem culture growth	Novák et al. (1984a)
Apical meristem, axillary meristems	Flowering induction and pod development in vitro	Novák et al. (1984b)
Nodal segments with axillary buds	Comparison of shoot proliferation from axillary buds of various position on seedling; in vitro flowering	Gould et al. (1987)
Nodal segments with axillary buds; shoot apices	Optimization of rhizogenesis on isolated shoots in vitro	Kubaláková et al. (1988)

The effects of glutamine and asparagine on proteosynthesis in maturing pea cotyledons cultured in vitro was studied by Lea et al. (1979). Legumin, the principle pea embryo storage protein, was analyzed in immature embryos both in vivo and in vitro by Domoney et al. (1980) utilizing a highly sensitive immunotest. The amount of legumin and the total protein quantity per unit of fresh weight were much higher in embryos growing in vitro than in the corresponding embryos in vivo.

The influence of abscisic acid (ABA) on the accumulation of storage proteins in pea embryos growing in vitro was studied by Davies and Bedford (1982). Embryos of different developmental stages were cultured in media containing 0.29 or 0.53 M sucrose in the presence or absence of ABA. ABA concentrations of 10^{-6} or 10^{-4} M in medium with 0.53 M sucrose resulted in a slight reduction of growth in smaller, though not in larger embryos. No effect of ABA on the accumulation of total storage proteins (especially legumin) in embryos growing in vitro was demonstrated.

The morphogenetic capacity of early heart stage embryos in vitro was reported by Paal et al. (1982). These embryos yielded plants and the growth of roots and shoots was supported by transferring plants with four to five internodes to liquid B5 medium supplemented with GA3 and NAA.

Closely related to the problems of pea embryoculture are also studies dealing with the in vitro cultures of isolated ovules (Srivastava et al. 1980), seeds (Krechting et al. 1978), and pods (de Capite 1955; Baldev et al. 1965; Hahn et al. 1974; Frydman and MacMillan 1976; van Staden and Button 1978; Srivastava et al. 1980; Garcia-Martinez and Carbonell 1985).

2.3 Anther and Pollen Culture

Little attention has so far been paid to the induction of haploid plants through anther or pollen cultures in pea (Gupta et al. 1972; Gupta 1975; Gosal and Bajaj 1988).

Haploid callus was obtained from pollen of *Pisum sativum* cv. B22 cultured on basic White's medium (White 1943) with the addition of 15% v/v coconut milk and 0.5 mg/l 2,4-D. The callus was subcultured on the same medium with the 2,4-D replaced by 0.5 mg/l NAA. The callus was whitish brown and friable, containing parenchymatous, mostly mononuclear, cells of various shapes and sizes. Callus growth continued through eight subcultures but there was no differentiation. Root differentiation began with the 9th subculture at a frequency of 3% and increased to about 80% in 36th subculture. After 14 subcultures, shoot differentiation was observed and from the 21st subculture on, embryo-like structures began to appear. Step-by-step transfer of embryo-like structure, to media of different composition, was followed by their development from the globular, through to the heart, and on to the torpedo stages; however, no haploid plants were obtained. Histological characteristics of the initiation zones have been presented by the authors. They consist of groups of meristem cells found in the peripheral and inner layers of the callus. The embryo-like structures originated predominantly from the epidermal areas of the callus and were made up of very small cells (20 to 30 μm in diameter) with a dense cytoplasm and large nuclei.

Recently, Gosal and Bajaj (1988) reported pollen embryos, and various levels of chromosome number ranging from haploid to highly polyploid and aneuploid. Diploid cells occurred at highest frequency (60%), followed by tetraploid (15%) and triploid (8%). Factors such as addition of sucrose, charcoal, and cold treatment (4 °C) of cultures for 3 days enhanced embryogenesis.

2.4 Morphogenesis in Callus and Suspension Cultures

A summary of experiments dealing with the induction of in vitro morphogenesis in pea is presented in Table 4. In vitro regeneration was first reported by Hildebrandt et al. (1963). The callus, derived from stem explants, was placed on medium supplemented with coconut milk and the development of shoots was induced.

Table 4. Induction of in vitro morphogenesis in *Pisum sativum* L.

Original explant	Regeneration type	Regeneration stage reached	Reference
Stem segment	Organogenesis in callus	Shoot regeneration	Hildebrandt et al. (1963)
Macerated apical meristems	Organogenesis in callus	Complete plants[a]	Gamborg et al. (1974)
Immature embryos	Organogenesis in callus	Complete plants[a]	Atanassov and Mehandjiev (1979); Natali and Cavallini (1987a)
Epicotyl segments	Organogenesis in callus	Complete plants[a]	Malmberg (1979, 1982)
Epicotyl segments	Organogenesis in callus	Root regeneration	Jacobsen et al. (1980)
Immature and mature leaves	Organogenesis in callus	Complete plants[a]	Mroginski and Kartha (1981); Rubluo et al. (1982, 1984)
Macerated apical meristems	Organogenesis in callus	Shoot regeneration	Preobrazenskaja (1983)
Epicotyl base	Organogenesis in callus	Complete plants[a]	Hussey and Gunn (1984)
Leaf, epicotyl and root segments	Organogenesis in callus	Shoot regeneration	Kunakh et al. (1984a, b, c)
Epicotyl	Organogenesis in callus	Complete plants[a]	Sladky and Jandova (1984)
Leaf segments	Somatic embryogenesis in cell suspension	Torpedo embryonic stage with partial root proliferation	Jacobsen and Kysely (1984); Kysely and Jacobsen (1984)
Apical meristem, epicotyl, internodal and leaf segments	Organogenesis in callus	Shoot regeneration	Ezhova et al. (1985); Gostimskij et al. (1985)
Immature embryos, shoot apex segments	Somatic embryogenesis in callus	Complete plants	Kysely et al. (1987)

[a] In case of complete plant regeneration it is a two-stage process:
(1) induction of buds and shoots, their isolation, and
(2) induction of roots on these shoots

Isolated shoot tips (2 – 3 mm) from aseptically germinated seeds were macerated and placed on basic B5 and MS media with NAA and BAP (Gamborg et al. 1974). The explants were cultivated under two different environmental conditions: (1) 18-h photoperiod, 2000 lx, 26° ± 1 °C, (2) 18-h photoperiod, 4000 lx, in 25 °C temperature gradually dropping (4 °C h^{-1}) to 15 °C in the dark. Toward the end of the light period, the temperature was gradually rising at the same rate. Shoot proliferation was observed after 4 – 6 weeks in culture and was stimulated by BAP (0.2, 0.4, 1, 2, or 5 μM) or in combination with NAA (0, 1, or 5 μM). Shoot formation was more frequent on media with higher NH_4NO_3 concentration. There was also sporadic root formation and on the evidence of histological sections the authors demonstrated the de novo regeneration of stem primordia. The plants were grown to maturity in a greenhouse and were phenotypically similar to the original material (the karyotype was not analyzed). The system of shoot production is regarded by Gamborg et al. (1974) as reproducible, the critical factor for complete regeneration being root formation.

Leroux (1968, 1973) studied the induction of roots on pea stem segments in vitro. Glucose (20 g/l) and NAA (3 mg/l) were found to be important at optimum temperature 25 °C.

Atanassov and Mehandjiev (1979) described shoot morphogenesis in the culture of young embryos of various ages: 15 – 17 days, 20 – 22 days, and 25 – 27 days after the completion of flowering. The cotyledons and vegetative apices of the oldest embryos were isolated and cultured separately. The highest production of adventitious buds was obtained at the vegetative apices from the oldest embryos on MS-medium supplemented with 2 mg/l BAP and 0.2 mg/l NAA. Organogenesis was associated with minimum callus formation, or the callus was not formed at all. Younger embryos produced more callus, but shoot regeneration was reduced. The most intensive callus formation was observed in isolated cotyledons of the oldest embryos. Calli with differentiated buds were subcultured on a medium which contained a lower concentration of growth regulators (0.5 mg/l BAP + 0.05 mg/l NAA). A high proliferation of new buds and continued shoot growth was observed. A minimum of 1500 buds could be obtained from one embryonic axis (from 25-day-old pea embryos) over a 150-day culture period. The regenerated shoots were transferred to White's medium supplemented with 1% sucrose and 1 mg/l IAA or 1 mg/l NAA. The frequency of rhizogenesis was, however, low. The attempt to transfer the rooted shoots into soil failed. Callus induction and shoot morphogenesis in pea was also studied by Gosal and Bajaj (1979) and Czosnowski and Domka (1981), but regular plant regeneration was not achieved.

Malmberg (1979) found in vitro genetic differences in the morphogenic capacity of pea and discussed the existence of factors determining regeneration of whole plants from callus cultures. Sixteen genetic lines of pea were chosen for the study. Epicotyl segments from proliferating embryos were placed on a callus-inducing medium (MS + 2 mg/l NAA + 1 mg/l BAP) and incubated in the dark at 26 °C. The callus was subcultured monthly onto the same medium and then transferred onto a regeneration medium (MS + 0.2 mg/l IAA + 5 mg/l BAP) at a 16-h photoperiod. At the early stage of culture, six lines regenerated plants directly from epicotyl segments; however, after 2 months the plants were regenerating from the callus. Two lines maintained the regeneration ability after 6 months of callus growth. Well-developed shoots were placed on a rooting medium containing 1 mg/l IAA, and 75% formed roots. The complete plants were then transferred to soil. They morphologically resembled plants grown from seeds, and had a diploid chromosome number (2 n = 14).

In his next study, Malmberg (1982) provided a genetic analysis of the regeneration capacity of *P. sativum* in vitro. By standard crossing he obtained F_1, F_2 and backcross generations of two lines, line B 77-259 (regenerating well) and line A 1078-234-0 (nonregenerating). Both lines were characterized by a series of genetic markers. Calli were derived from germinated seeds of the F_1 and F_2 generations, and backcrosses. Shoot regeneration ability was also tested. Calli derived from F_1 seeds did not regenerate shoots, while 23% of the cultures from F_2 seeds did. These results indicated the existence of a single recessive gene for plant regeneration. The backcross data were consistent with these conclusions. Backcrosses with the parent line (B 77-259) sometimes regenerated, while those with the parent line (A 1078-234-0) did not. However, not all the regenerated shoots could be successfully rooted. There was no problem with the original line in this respect. The author suggested a possible second genetic factor responsible for transplantation and hardiness. Analysis of the F_2 generation showed segregation for genes st, b, le, cp, tl, suggesting that they are unlinked to the regeneration factors. Because of the small sample size (32 regenerated pea plants), a linkage could not be assigned to the regeneration genetic factors.

Murashko (1974) studied interline and intervariety differences in the ability to form callus in vitro. Significant differences were found among both genotypes and explants from various parts of the plant. Bulygina and Erkejev (1979) investigated nine pea varieties for their capacity to form callus in relation to the yield parameters and did not find direct correlation.

Effective systems of in vitro pea regeneration were developed by Mroginski and Kartha (1981) and Rubluo et al. (1984). Immature leaflets (0.9–1.8 mm) were the best explants for the regeneration of shoots. Of the broad spectrum of growth regulators (BAP, kin, 2iP, ZEA, NAA, IAA, ABA, 2,4-D, picloram) tested, the combination of BAP (10 μM) and NAA (10 μM) was the best for callus induction and subsequent regeneration. Maximum callus proliferation was induced by the ZEA-picloram combination, but this culture lacked the ability for shoot regeneration. Explant age was a crucial factor influencing shoot regeneration. The best environment was an alternating temperature of 20°/15 °C with a 16-h photoperiod at 2000 lx. The genetic stability of regenerants was appraised karyologically, by isoenzyme spectra, and by selected morphological characters.

Preobrazenskaja (1983) studied the organogenetic ability of eight pea genotypes. The callus cultures were obtained from macerated shoot tips of 3- to 4-day-old pea seedlings, according to the method of Gamborg et al. (1974). Bud and shoot proliferation were observed within 9 weeks. Three genotypes of high regeneration capacity (lines 1072 and 83 and the cv. Ranyj zelenyj) were recommended for the study of regeneration from protoplasts.

Kunakh et al. (1984 a, b, c) observed the positive influence of GA3 and ABA upon organogenesis. Calli were derived from leaf, stem, and root segments (about 5 mm long) of 2-week-old pea seedlings cultured on medium with 2,4-D (5 mg/l). The growth rate of subcultured tissue was dependent on both genotypes and the original explant. Callus growth rate during subculture decreased in the order root-stem-leaf and it was very difficult to induce morphogenesis. Of the six genotypes tested, only two responded positively (cv. Uladovskij jubilejnyj and line L 1288). Six medium variants and 300 transfer combinations were tested. Organogenesis was induced on B5 medium supplemented with 5 mg/l kin, 5 mg/l BAP, and 0.2 mg/l GA3. For regeneration from calli cultured over a long period, GA3 had to be replaced by 1 mg/l of ABA. Gibberellin also promoted the initiation of lateral shoots. Auxins inhibited organogenesis. Organogenesis could be induced in a cultured callus over a long period (2.5 years), but the frequency decreased distinctly with the age of the culture. Abnormalities not seen in younger cultures, e.g., disturbed geotropism in plantlets, formation of albinos, whorl formation, etc., were also observed. Root formation could not be induced either in callus or on regenerated shoots. However, results obtained by Jacobsen et al. (1980) indicated that the induction of rhizogenesis in callus was not difficult, but shoot regeneration proved to be a failure.

Sladky and Jandova (1984) derived the callus from pea epicotyl on MS medium with 2,4-D, IAA, and kin. The 5-μM 2,4-D concentration initiated callus proliferation; shoot morphogenesis was stimulated by BAP (5 – 10 μM).

Shoot regeneration from callus cultures of five varieties (Puget, Upton, Maro, Melton and Vedetta) was investigated by Hussey and Gunn (1984). Well-growing calli could be obtained from cv. Puget and Upton on MS medium with 1 mg/l BAP and 4 or 8 mg/l IBA. Seedling epicotyl bases served as the initial explant. Shoots were regenerated 2 to 4 weeks after transfer to the same medium, except that IBA was lowered to 0.25 mg/l. Regenerated shoots were then propagated as stem cultures (apical meristem cultures) and rooted on half-strength MS medium with 1.5% sucrose and 2 mg/l IAA.

Ezhova et al. (1985) and Gostimskij et al. (1985) induced shoot organogenesis in calli derived from shoot apices, epicotyls, internodes, and leaves. The study included 25 cultivars and genetic lines (cultured on B5 medium with 30 g/l sucrose, 2 g/l caseine hydrolysate, 0.8 g/l NH_4NO_3 and NAA and BAP of various concentrations). Shoots were regenerated from 71% to 85% of the calli derived from shoot apices. The ability to regenerate shoots in callus derived from other explants decreased in the following order: epicotyl (0% – 31.3%), internodes (0% – 11.1%), and leaves (0% – 5.1%). Histology proved that shoot bud and shoot production was de novo. Shoots were rooted using Malmberg's (1979) method or were grafted onto pea rootstocks of the same variety (cv. Ranyj zelenyj).

Somatic embryogenesis in vitro, elaborated in great detail for the forage legumes, *Medicago* and *Trifolium*, was not reported for grain legumes until after 1980, when somatic embryos were induced in soybean (Phillips and Collins 1981; Christianson et al. 1983; Gamborg et al. 1983; Lippmann and Lippmann 1984; Kysely and Jacobsen 1984).

Jacobsen and Kysely (1984) used two genotypes of *Pisum sativum* (cv. Dippes Gelbe Victoria and fasciated mutant 489 C) and one genotype of *Pisum arvense* for induction of somatic embryogenesis. Stem (8 mm long) and leaf (6 mm in diameter) segments were excised from 13-day-old seedlings. The most suitable medium for callus induction was a modified MS (mineral part according to MS, 2 mg/l thiamin and 250 mg/l inositol) with 0.06 mg/l picloram (4 amino-3,5,6-trichlor-

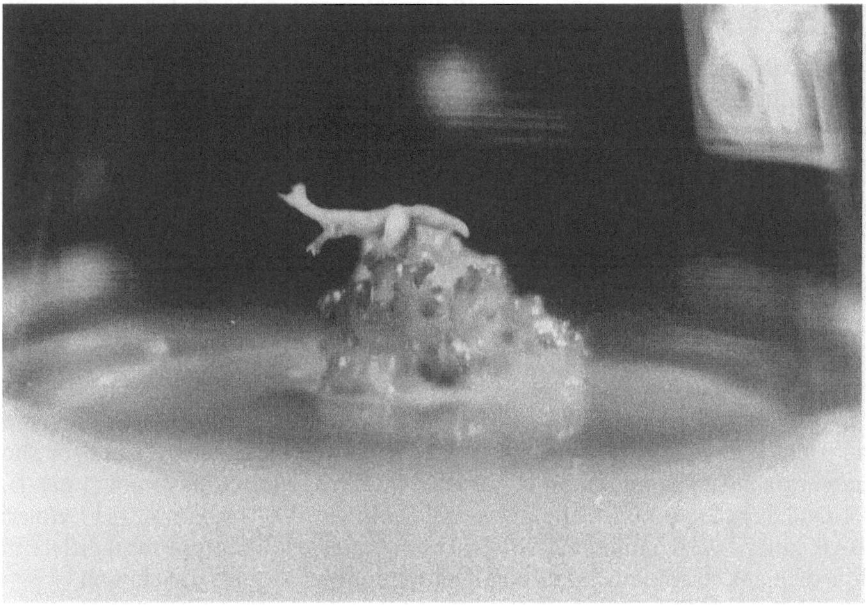

Fig. 5. Differentiated pea somatic embryo on surface of the callus (Griga and Novák, unpublished)

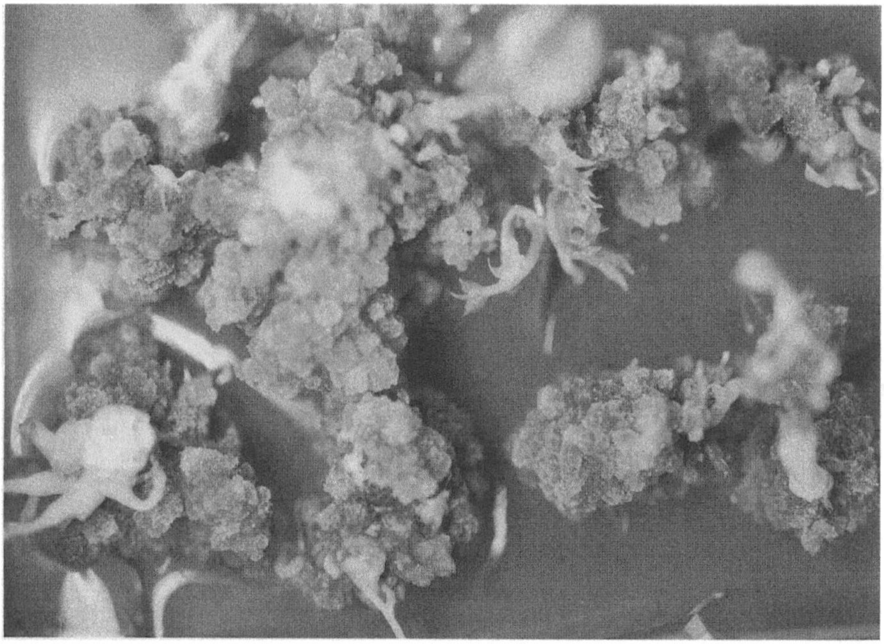

Fig. 6. Shoot differentiation from embryo-like structure on callus culture derived from axis of zygotic embryo (Pevreal and Novák, unpublished)

opicolinic acid) and 0.1 mg/l BAP. *Pisum sativum* produced embryogenic callus only from leaf explants, while *P. arvense* did so from stem segments as well. Calli formed on agar medium were transferred to liquid medium of the same composition and the white cell clusters gave rise to globular embryo-like structures. Somatic embryos spontaneously separated from calli in liquid culture. Development to the torpedo stage took place on the agar MS medium with 1.5% glucose, 1.5% sucrose, 0.006 mg/l picloram, and 0.1 mg/l BAP. After 2 days, the embryos began germinating, but only roots were formed.

Embryogenic callus was induced from zygotic embryo axes and cotyledons of unripe seed embryos on MS medium with 1 μM picloram and yeast extract (Griga and Novák, unpubl.; Fig. 5). The embryo-like structures developed into shoots on differentiation medium supplemented by zeatin, GA3, and NAA (Pevreal and Novák, unpubl.; Fig. 6).

More recently, Kysely et al. (1987) obtained whole plant regeneration in pea via somatic embryogenesis using explants from immature embryos and/or shoot apex segments. The induction of somatic embryos required picloram (1 mg/l) or 2,4-D (0.5 mg/l). Transfer of embryogenic cultures to MS medium with cytokinin (BAP, kin, ZEA) stimulated development of young embryo stages and differentiation of plantlets. Regenerated plantlets were grown to maturity in soil.

2.5 Isolation, Culture, and Fusion of Protoplasts

Constabel et al. (1973) isolated protoplasts enzymatically from the mesophyll of *P. sativum* L., cv. Century and cultured them in 0.2-ml droplets of B5 medium with 1 mg/l 2,4-D or 1 mg/l NAA and 2 mg/l kin. Within 2–3 days the protoplasts regenerated cell walls and began to divide. Calli were formed after 19 days of culture. Protoplast viability depended upon incubation of the donor plants in the dark for at least 30 h before protoplast isolation. Landgren and Torrey (1973) reported on the isolation and culture of protoplasts from pea roots.

Gamborg et al. (1975) isolated protoplasts from stem apices of 3-day-old pea seedlings. The protoplasts originated from cells of the meristematic domes and parenchymatous tissues of 2–3-mm tip sections. The isolation was made in two consecutive steps of enzyme treatment in the presence of hexitols and calcium salts. Cell wall regeneration occurred after 24 h of culture, and cell division was observed within 3 to 5 days. Supplementing the culture medium with 2–5 mM L-glutamine, 5–10 μM kin, 1–5 μM 2,4-D and 0.1–0.5 μM NAA resulted in sustained cell divisions and callus formation in 5 weeks. Glucose at 0.3 M appeared to be more suitable than sorbitol and mannitol as osmotic stabilizers. Regeneration of plant from calli, however, failed.

The physiological condition of donor plants had a significant influence upon the number of isolated protoplasts, and the survival and division of protoplasts (Arnold and Eriksson 1976, 1977). The best results were achieved with plants possessing three to five leaves cultured in a culture box under controlled conditions.

Protoplasts from the variety Rondo exhibited the highest cell division capacity (Jia 1982; Jia et al. 1982). The greatest division rate was observed in protoplasts isolated from the highest, fully developed leaf of young plants at the three- to five-leaf stage. Maximum division frequency was obtained at a protoplast density of 5×10^4 to $1 \times 10^5\ ml^{-1}$ at pH 5.6, optimum Ca^{2+} concentration of 850 mg/l $CaCl_2 \times 2H_2O$ and supplementing the culture medium with 2,4-D and zeatin-riboside. The continual and gradual decrease in osmotic pressure during culture was necessary to achieve a high cell division rate.

Landgren (1976a) analyzed the influence of culture conditions on mitotic activity of protoplasts derived from the cortical part of *Pisum sativum* roots. A series of culture regimes were tested and protoplast division observed. A marked inhibition of mitoses was recorded after exposure to high light intensities. Glucose as a carbon source inhibited division, but the protoplasts divided actively in the presence of 2% sucrose (Landgren 1976b, 1981; Crowder et al. 1979).

Studying fresh protoplasts isolated from pea leaves stained with acetocarmine, Lee and Min (1981) observed 91% mononuclear, 8% binuclear and 0.7% multinuclear forms. After 9 days of culture, the percent of bi- and multinuclear protoplasts rose to 25.6% and 8.1%, respectively. The authors attributed this phenomenon to mitotic division rather than fusion.

Fusion of protoplasts from soybean cell cultures with pea mesophyll protoplasts was induced by Constabel et al. (1975). The heterokaryon rate was 20% and differential staining proved most of them to contain one or two nuclei from each of the parents. In a following study, Constabel et al. (1976) observed cell division of the products of intergeneric protoplast fusion between soybean (cv. Mandarin) and pea (cv. Century and Laxton's Progress). Based on this work, Fowke et al. (1977) made a structural study of fusion products between soybean and pea protoplasts using an electron microscope. Interphase pea and soybean nuclei fused in some cells. The detection of heterochromatin typical for pea in the synkaryon (even after division was over) indicated that the cells were hybrid.

Somatic hybridization of pea and soybean protoplasts was also studied by Praznovsky et al. (1981, 1982). After the application of polyethylene glycol (PEG), there was a 2% – 10% fusion between protoplasts of the two species. Delayed division took place in the fused cells and reached 33% of the parental frequency. Mixing of cytoplasm and asynchronic mitoses of parental nuclei in heterokaryons were observed. The chromosomes of both parents were also easily identified in the fused cells.

Experimental fusion of pea and barley mesophyll protoplasts was performed by Lee et al. (1980). The fusion was induced after 60 min in a PEG medium enriched with calcium salts. Bajaj (1983) explored the possibilities of somatic hybridization between *Pisum sativum*, *Triticum aestivum*, and *Oryza sativa*, and the survival rates of these somatic hybrid cells following freezing in liquid nitrogen (–196 °C). The heterokaryons displayed two nuclei of different sizes; occasionally, these nuclei fused at a low frequency. Only one hybrid cell underwent repeated division.

Protoplast technology has also been used to study the symbiotic relationship between nitrogen-fixing bacteria and cells of legume root nodules. Direct transfer of *Rhizobium* to pea protoplasts was documented using an electron microscope (Davey and Cocking 1972).

Another alternative procedure for introducing bacteria is to fuse protoplasts from root nodules with those from a different portion of the plant. Protoplasts are first removed from the root nodules. Next, the nitrogen-fixing ability of the protoplasts is determined and they are fused with protoplasts not containing bacteria (Syono et al. 1979; Kajita et al. 1980). Protoplasts best suited for isolation were those from nodules of flowering pea plants. The protoplasts were then classified into four types ranging from the small spheres of the early stage (type 1) to the large elongated protoplasts of the late infection stage of the host cells (type 4). The absence of cell wall components on the plasmalemma was shown by electron microscopy.

The removed nodule protoplasts (particularly the spheroid types) and tobacco mesophyll protoplasts were agglutinated by PEG and then fused. Fusion of protoplasts from legume root nodules with those from other higher plants is a method used to overcome difficulties associated with infection barriers when new symbiotic relationships are being established in vitro. This method protects bacteria and their nitrogenase system, since membranes of the original host plants surround them in the cells of the root nodules. Furthermore, the bacteria are introduced into the nonleguminous cells in their active, nitrogen-fixing form.

Up to date only one report of successful regeneration from pea protoplasts was published (Puonti-Kaerlas and Eriksson 1988). The authors induced division of protoplasts (2,4-D and BAP) and growth of organogenic calli (BAP, NAA). In two tested varieties (Petra and Stivo) they were able to obtain bud proliferation and shoot development on regeneration medium with BAP, kin, and ABA. However, no roots formed on these shoots.

Protoplast cultures can also be used to study physiological and biochemical processes in plants, e.g., the intake, transport, and metabolism of endogenous substances. A method to follow sugars and amino acids transport through mesophyll protoplast membranes of *Pisum sativum* was developed by Guy et al. (1978). Using two isomers (the first designated as ^{3}H and the other as ^{14}C), it was demonstrated that sugars are not taken in by passive diffusion through damaged membranes, because the protoplast membranes allowed the reception of only one isomer. To further characterize transport, the authors made use of nonmetabolized glucose analogs (3-0-methylglucose) and an amino acid analog (α-aminoisobutyric acid). Comparing the characteristics of a leaf tissue segment and isolated protoplasts, the latter were found to preserve about 40%–50% of the active intake ability of whole cells. A report on the effect of ascochytine on pea protoplasts can be found in Lepoivre (1981). The light-dependent intake of rubidium by isolated pea mesophyll protoplasts was studied by Leurs et al. (1982) and Rahat and Reinhold (1983), and the intake of ^{14}C-labeled glucose protoplasts of the pea by Plemenkova and Lozovaja (1983). Using vacuoles removed from protoplasts, Guy and Kende (1984) investigated the compartmentation of ethylene synthesis in the pea leaves. Nandi and Eriksson (1977) studied the nuclear behavior of pea protoplasts.

2.6 Cytogenetics of in Vitro Cultures and Regenerated Pea Plants

The cytology of pea callus cultures has been intensively studied in the Laboratory of Genetics and Cytology at the University of Tartu (Estonia) (Kallak and

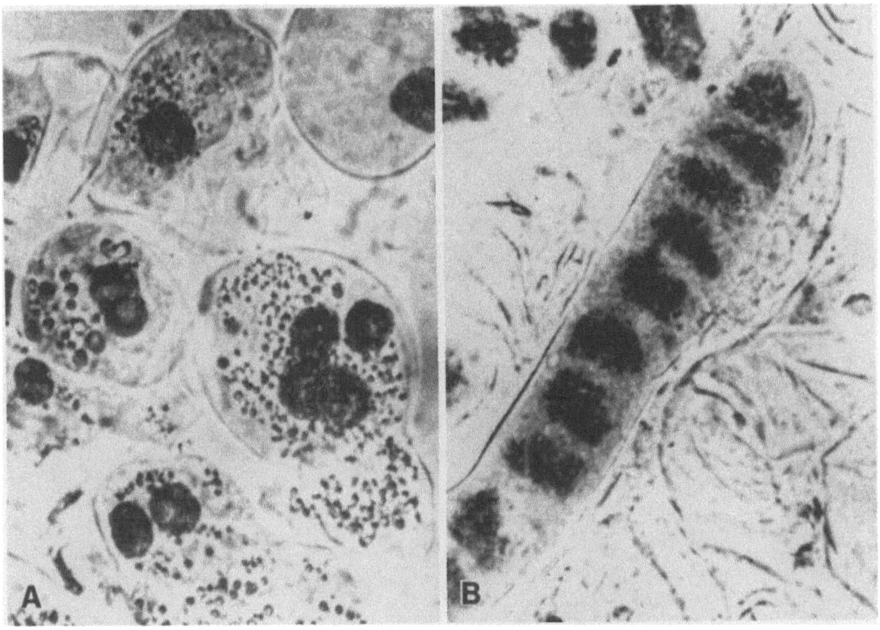

Fig. 7 A, B. Multinuclear cells in long-term callus tissue of pea (Kallak and Yarvekylg 1976)

Yarvekylg 1968, 1970, 1971, 1972, 1976, 1977 a, b; Kallak and Vapper 1979; Kallak 1983). Calli derived from the cotyledons of cv. Kiir were analyzed for their highly variable morphology and tissue structure, mitotic activity, and varying number of chromosomes. The compact callus contained xylem elements and many cells were of higher ploidy levels (e.g., in 90-day-old cultures 15.5% of the cells were diploid, 20% triploid, 17.3% tetraploid, and 47.2% were higher than 4 n). A typical feature of pea calli is very slow growth. They are usually subcultured at 2–3-month intervals and the mitotic index is less than 2% (Kallak and Yarvekylg 1968, 1977 a). Cytogenetic effects of 2,4-D on pea calli were investigated by the same authors (Kallak and Yarvekylg 1971). Globular callus was resistant to high 2,4-D concentrations. There was little variation in mitotic activity with concentrations up to 20 mg/l. Age of subculture also influenced ploidy level, i.e., due to the high 2,4-D concentration, the number of dividing diploid cells increased and the number of polyploid cells decreased as the age of the culture increased. Irregular mitoses, including C-mitoses, were more frequent than regular mitoses in callus cells. Other types of cell division (amitosis, nuclear fragmentation – Kallak and Yarvekylg 1976) also existed resulting in multi-nuclear cells and other karyological abnormalities (Figs. 7–10).

Van't Hof and McMillan (1969) analyzed the population kinetics of diploid and polyploid cells in pea callus cultures derived from root segments. A rapid increase in polyploid mitoses was observed within 6–7 days after stimulation with kinetin (3.2 μM).

Gupta et al. (1972) and Gupta (1975) reported on cytological characteristics of long-term cultured pollen callus of *Pisum sativum* (Table 5). During the

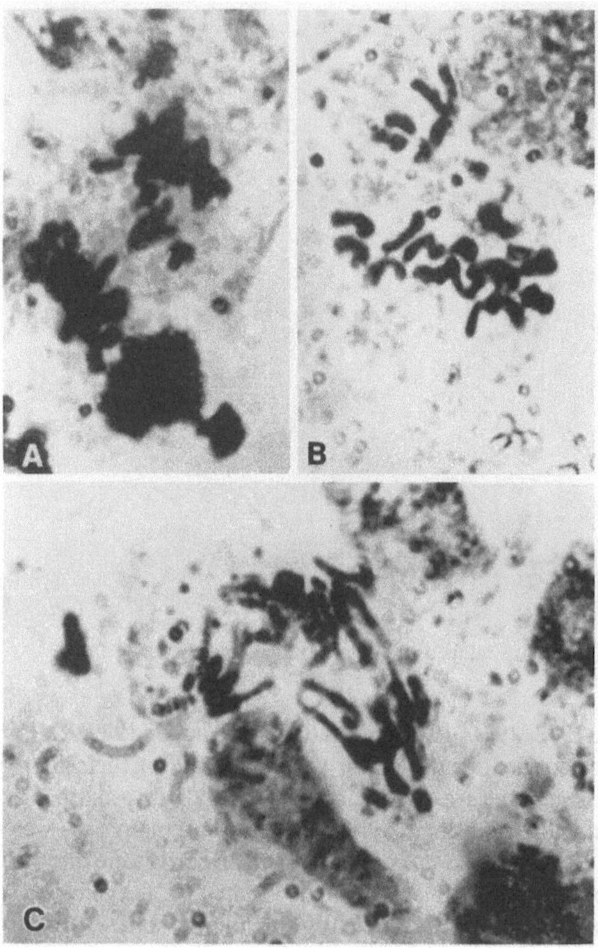

Fig. 8 A–C. The mitotic irregularities in the pea callus cells. Somatic reduction (**A**), nonequal chromosomal distribution (**B**), tripolar mitosis (**C**) (Kallak and Yarvekylg 1976)

36 months of culture there was a marked drop in the occurrence of haploid, and a rise in tetraploid cells. Comparative analyses of the distribution of ploidy showed a predominance of aneuploid and haploid cells in an undifferentiated callus.

Cytological changes in a callus culture of *Pisum sativum* were also demonstrated by Frolova and Shamina (1974). During subculture there was a tendency for the cells to polyploidize. In the first subculture diploid cells prevailed but by the fourth, the chromosome number varied from 7 to 84; only 18% of the analyzed metaphases were diploid. Diploid cells did, however, prevail in zones of secondary callus meristems.

Michajlov and Bessonova (1975) performed cytogenetic analyses of callus from immature pea cotyledons (cv. Ramon 77). The number of the polyploid cells (up to 8 n and more) increased with culture age, as did the frequency of chromosomal and mitotic aberrations (including primarily chromatid bridges, acentric fragments, lagging chromosomes, and tripolar mitoses).

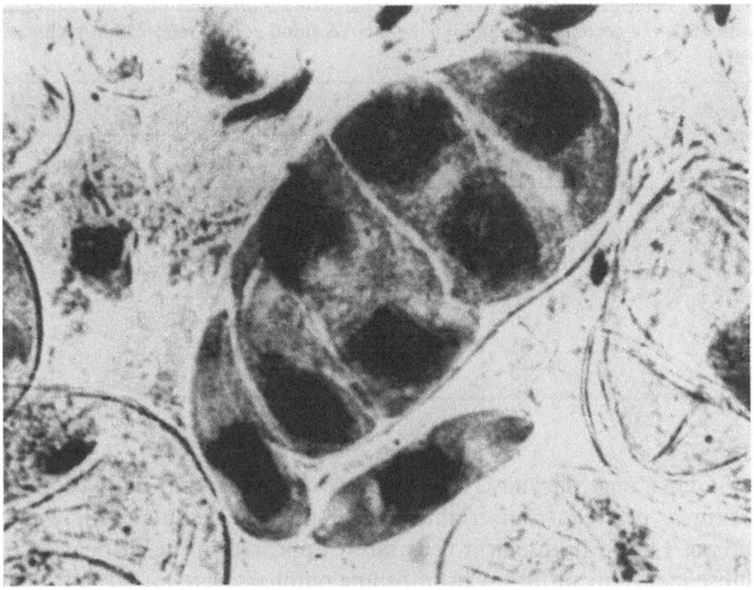

Fig. 9. Nuclear fragmentation in a cell in pea callus (Kallak and Yarvekylg 1976)

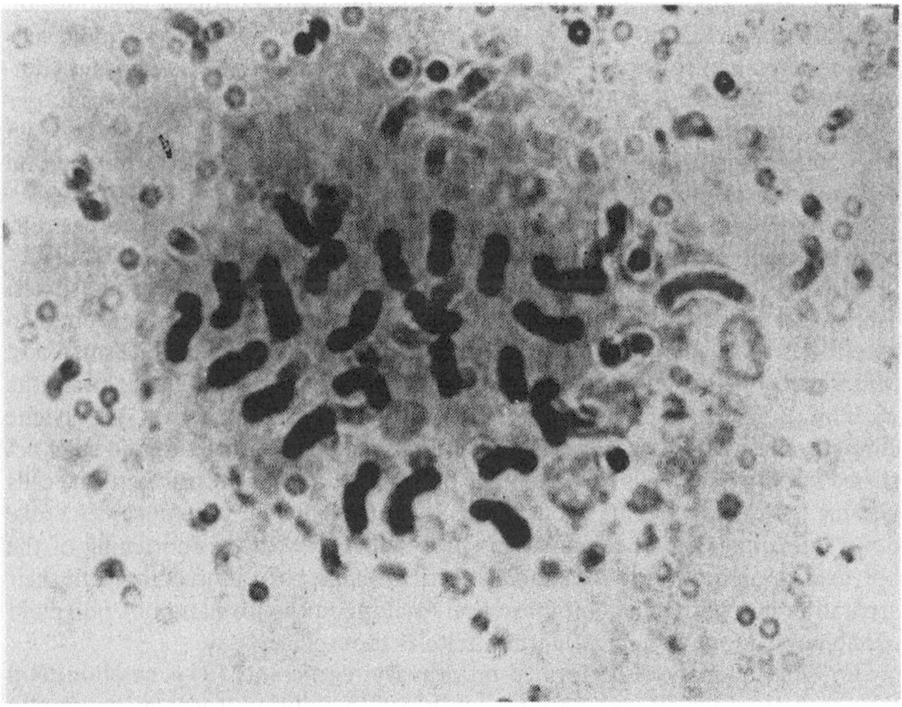

Fig. 10. Hypertriploid metaphase in pea callus cell (Kallak and Yarvekylg 1976)

Table 5. Percentage of ploidy cells with the age of *Pisum* callus on White's medium supplemented with 15% v/v coconut milk and 0.5 mg/l NAA (total counts were 250 in each case). (After Gupta 1975)

Age in months	Percentage of ploidy cells						Cytological irregularities
	n	2n	3n	4n	>4n	Aneuploids	
1	81	7	–	4	–	8	10
6	67	15	1	8	2	7	11
12	61	20	3	4	2	10	15
18	42	15	2	18	7	16	22
24	40	20	5	11	6	18	
30	32	21	7	26	2	12	17
36	25	12	2	41	8	12	21

Ghosh and Sharma (1979) explored chromosomal changes in cell suspension cultures, derived from friable callus, in MS medium with 4 mg/l 2,4-D and 1.5 mg/l kin. Multinuclear cells rarely appeared in the suspension (0.4%) during subsequent subcultures. Chromosome numbers were found to be within the range of 7 to 14. The frequency of aneuploidy fluctuated during the time in culture. The occurrence of asynchronic mitoses and mitotic irregularities was noted. Suspension cultures appear to offer a potential means of attaining cytogenetic variability in this species.

Spontaneous polyploidization in pea tissue cultures is a frequent phenomenon (von Knösche and Günther 1979, 1980; von Knösche 1981). Highly polyploid cells develop often and consist of 32 n or more chromosomes (von Knösche and Günther 1980; Therman and Murashige 1984). The frequency of restitution mitoses in callus cultures was reported by von Knösche and Günther (1980) to be from 0% – 50% (mean 25%). Thus, restitution cycles exercise considerable influence on the spectrum of ploidy.

Endoreduplication, induced by growth regulators in in vitro cultured pea cells leads not only to polyploidization and an increase in nuclear DNA content (Libbenga and Torrey 1973), but also to the formation of polytene chromosomes. Marks and Davies (1979) induced the formation of polytene chromosomes in cotyledons cultured on MS medium containing kin and IAA, and Therman and Murashige (1984) found polytene chromosomes in nuclei of root cells using the medium of Torrey and Shigemura (1957) with 2,4-D and kin (Fig. 11). Polytene chromosomes were often present in a diploid form, while the tetraploid state occurred in only a few cells (in these cells polytenization was present until the cells became tetraploid). During prolonged culture, the polytene chromosomes broke down into simple chromosomes. This process began with the shortening of the polytene chromosomes to an almost spherical shape, then proceeding with their separation into individual chromosomes, resulting in the formation of polyploid metaphases. From this stage, normal mitosis took place.

Cytological analyses performed on primary regenerants (R_1-generation) and their seed progenies (R_2-generation), confirmed that during regeneration, even in highly mixoploid calli, normal (mostly diploid) cells are preferred (Malmberg

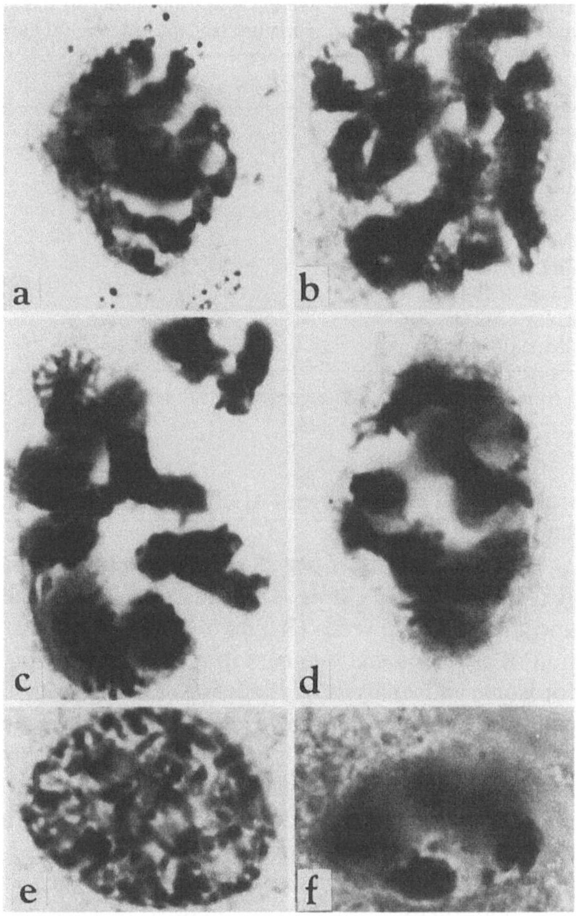

Fig. 11a–f. The behavior of polytene chromosomes in cultured pea roots. Intermediate stage between chromosomes with multiple chromatids and polytene chromosomes (**a**). Polytene chromosome (**b**). Section of nucleus with polytene chromosome; telomere attached to nuclear membrane (**c**). Polytene chromosome (**d**). Prophase which probably has arisen through the falling apart of polytene chromosomes (**e**). Condensed polytene chromosomes (**f**) (Therman and Murashige 1984)

1979; Rubluo et al. 1984). In some cases, where no cytological assessment was made, the phenotype of the regenerants corresponded favorably to the original diploid parents (Gamborg et al. 1974).

Kunakh et al. (1984a, b, c) confirmed the high genetic variability of pea calli cultured over long time periods. Shoot regeneration was still achieved after 2.5 years. Cytological analyses of differentiated meristems, however, proved that regenerated shoots have both diploid and tetraploid. The cv. Uladovskij jubilejnyj had an even higher frequency of tetraploidy than diploidy in regenerated shoots.

Hussey and Gunn (1984) discovered that most plants regenerated within the first year of culture were diploid and morphologically normal. Regenerants obtained after 2 years of culture displayed variable shoot morphology and difficulty in rooting.

Ahmed et al. (1987) performed a detailed chromosomal study of pea plant propagated through shoot-tip culture. The regenerated plants showed a narrow range of variation in chromosome number. In addition to numerical variations, chromosome aberrations were observed in root tips of regenerants and of con-

Table 6. Morphological characters in seed plants (control) and in in vitro regenerated plants of *Pisum sativum* (regeneration from immature leaflets). (After Rubluo et al. 1984)

Characters	Control plants	Regenerated plants	
Flowering node	16.97 ± 0.68	17.01 ± 0.74	NS[a]
Flowering day	50.16 ± 3.28	51.22 ± 3.03	NS
Length of internode	6.32 ± 0.66	6.39 ± 0.76	NS
Length of pod	5.45 ± 0.41	5.36 ± 0.44	NS
Weight of seeds (100 seeds)	22.51 ± 0.92	23.11 ± 1.40	NS
Fertility (mature ovules)	89.52%	90.60%	NS
Color of flowers	White	White	
Color of seeds	Yellowish-green	Yellowish-green	
Shape of seeds	Smooth	Smooth	

[a] Nonsignificant ($P < 0.05$) according to Student's t-test

trol plants. The karyotypes of culture and seed-derived plants were basically the same.

Genetic analysis of pea in vitro regenerants was made by Rubluo et al. (1984). A total of 128 R_1 plants obtained from leaflet cultures were grown to maturity and the harvested seeds (R_2) divided into three parts: (1) 60 seeds for determination of chromosome numbers in root-tip meristems, (2) 50 seeds grown to plants for isoenzyme analysis of the leaves, (3) 180 seeds for plants in which the genetically defined morphological characters were evaluated. Karyotype analyses of both regenerants and control plants clearly showed the diploid number of $2n = 14$. Analyses of four isoenzymes (esterase, glutamate-dehydrogenase, 6-phospho-gluconate-dehydrogenase, and leucylamino-peptidase) from R_2 plants showed no genetic changes. The assessment of morphological characters also demonstrated high genetic stability for all regenerants (Table 6).

Contradictory results were obtained by Gostimskij et al. (1985) from analyses of R_1-regenerants derived from callus cultures of macerated apical meristems and their progeny (R_2-, R_3-generations). The changed morphology of the R_1 plants was attributed to their development as grafts on rootstocks of the initial genotype under greenhouse conditions. Among 65 lines of the R_2-generation, the "chlorotica" (chi) mutation was recorded in one line, and plants without wax layers in another line. The most interesting results, however, were obtained from studying the physiological and quantitative characteristics. The following changes were discovered for 70% – 80% of the R_2-generation lines:

- a more robust habitus when compared with the control variety (cv. Ranyj zelenyj)
- dark green instead of light green leaves of the control
- elongated compared to oval-shaped leaves
- earlier or later onset of flowering (beginning of the flower period).

The same changes were found for the R_3-generation, and the altered characteristics were observed for all plants and all progenies of the same line. Five hundred cytological analyses of calli yielding regenerants showed 85% of the metaphase cells to have 2 n = 14, 11% 4 n = 28, and 4% an aneuploid set of chromosomes (8 – 15). Only 1.3% of anaphase cells had deviations (chromosomal fragments). The cytological data indicated that the phenotypic variability of regenerants was not connected with large reconstructions of the karyotype. Also, it is unlikely that stability for a trait such as flowering period (controlled by four basis genes) to be the result of single gene mutation. In the opinion of Gostimskij et al. (1985), these are some of the mechanisms of hereditary variability, characteristic for cells of in vitro cultured plants (amplification of some genome segments and their transposition).

A callus formation prior to plant regeneration was a potential source of cytogenetic instability. Pea callus induced from shoot-tip meristem contained cells with different chromosome number (hypodiploid, diploid, aneuploid, and tetraploid). Aneuploid cells originated from a process of nuclear fragmentation and abberant mitoses in the first days of culture (Natali and Cavallini 1987b). Aneusomaty was found in regenerated plants; however, this phenomenon was sharply reduced during plantlet development, which may indicate the advantage of diploid over aneuploid cells (diplontic selection). Kysely et al. (1987) examined nine pea plants regenerated via somatic embryogenesis. Three of them were tetraploid and six diploid. In conclusion, plants regenerated from pea callus either via organogenesis or via somatic embryogenesis can carry chromosomal variation. However, more extensive studies are needed in order to analyze genetic variability in tissue culture systems of *Pisum*.

2.7 In Vitro Induction of Salt Tolerance

Recent developments in in vitro technology offer a meaningful tool for screening and developing salt-tolerant genotypes. Gosal and Bajaj (1984) isolated salt-tolerant cell lines of *Pisum sativum* cv. Pb 88 by exposing cultures to increasing levels of NaCl (0.5%, 1%, 2%, 3%). Actively growing cell suspensions treated with EMS (0.25%) for 2 – 4 h increased the efficiency of salt tolerance. With an increase in salt concentration from 0% – 3%, growth was decreased, whereas the compactness and browning increased. Nevertheless, small sectors of well-growing tolerant cells were observed at 2%. Continuous challenging of the cells for 3 months to the elevated levels of NaCl resulted in the development of tolerant cell lines; however, no plants were regenerated from these calli.

3 Conclusions and Prospects

In vitro technology has developed rapidly and extensively in recent years. Despite such development, plant regeneration from various tissues and culture systems has been achieved to only a limited extent. The processes of shoot differentiation, somatic embryogenesis, rooting of plantlets and their regular transplantation into soil are crucial factors for further implementation of in vitro technology for pea-crop improvement.

Without new genetic variation there is only a limited chance for developing improved pea variability. Systematic studies are needed to observe somaclonal variation at both the tissue-culture and whole-plant levels where agronomically important traits must be expressed.

The pea is a classical object for genetic and induced mutagenesis study. The development of a cellular system of mutagenesis can open new avenues, not only for the theoretical study of mutagenesis, but also for an effective breeding system including the possibilities of single cell selection and early screening of the desired characters.

In vitro techniques are appropriate for improvement of the pea crop and current research should be applied to solve the basic problems of in vitro morphogenesis, in order to integrate biotechnology with conventional pea breeding programs.

4 Protocols

In Vitro Vegetative Propagation

1. a) Seeds are immersed in 70% ethanol for 2 min followed by surface sterilization in 2.6% solution of sodium hypochlorite (e.g., 50% commercial bleach Clorox) with a few drops of detergent (Tween 20). The seeds in hypochlorite solutions are placed on gyratory shaker for 20 min and afterwards rinsed 3–5 times in sterile distilled water. Sterilized seeds are germinated in flasks or in test tubes on agar medium or on paper wicks immersed in liquid medium or on moistened cotton with half-strength major and minor MS salts.
   b) The terminal and/or axillary buds are dissected from field-growing plants and surface sterilized in 1.3% solution of sodium hypochlorite (e.g., 25% commercial bleach Clorox) with a few drops of detergent (Tween 20) for 20 min. Subsequently, the buds are rinsed 3–5 times in sterile distilled water.
2. The shoot tips (meristematic dome with 1–2 pairs of leaf primordia) are aseptically excised under the stereomicroscope in the air-flow cabinet. The shoot tips are transferred to test tubes with slant agar medium. The composition of the basal medium is: MS minerals + B5 vitamins + 100 mg/l myo-inositol + 30 g/l sucrose solidified with 0.8% Difco Bacto-agar. Growth regulators supplemented into the medium control the morphogenetic pattern:
   i. Single shoot formation affected by 1 μM BAP alone or in combination with 1 μM NAA.
   ii. Multiple shoot formation affected by 20 μM BAP with 0.1 μM NAA.
   iii. Shooting of nodal cuttings affected by 0.1μM BAP.
   iv. Rooting on 1 μM IAA with 2 μM NAA in combination with half-strength MS, full B5 vitamins and 40 g/l sucrose.
3. The cultures are incubated at 20 °C in 16 h white fluorescent light (40 W/m^2) and at 15 °C in 8 h dark.
4. After 3–4 weeks the isolated shoot tips and/or nodal cuttings with one axillary bud are recultured on shooting medium or rooted in vitro.
5. Rooted plants are grown in hydroponic pots with perlite saturated with Hoagland solution till flowering and seed maturity. The aeration of the hydroponic solution has positive effects on root growth, flowering, and pod formation.

Plant Regeneration in the Culture of Immature Leaflets

The following protocol was developed by Mroginski and Kartha (1981).

1. Dissect aseptically immature leaflets from the second and third apical leaves from the apex of 2–3-day-old seedlings.
2. Culture medium consists of major and minor mineral salts of MS, vitamins of B5, 3% sucrose, 0.8% Difco Bacto-agar, supplemented with 0.1 μM NAA and 10 μM BAP.
3. Incubate culture at 20 °C with 70% relative humidity and 16 h light (combined intensity of 40 W/m^2 supplied by fluorescent and incandescent lamps).
4. The explants form callus and regenerate shoots in 45–90 days of culture.
5. Excise individual shoots and transfer on half-strength B5 medium with 1 μM NAA for rooting. Callus with primordial structures can be subcultured on the shooting medium for further regeneration.
   Note: For rooting of shoots half-strength MS medium with 40 g/l sucrose supplemented with 1 μM IAA and 2 μM NAA can also be used.

Somatic Embryogenesis in Callus Culture

1. The immature pods are surface sterilized by immersion for 2 min in 70% ethanol and 20 min in 1.3% solution of sodium hypochlorite (e.g., 25% commercial bleach Clorox) with few drops of detergent (Tween 20).
2. Embryos of size 0.6 to 12 mm are aseptically excised and placed on MS medium with B5 vitamins, 30 g/l sucrose, 100 mg/l myo-inositol, 8 g/l Difco Bacto-agar supplemented with 1 μM picloram (4-amino-3,5,6-trichloropicolinic acid).
3. After 20 days culture on an induction medium the embryogenic callus is transferred to MS medium with B5 vitamins, 30 g/l sucrose, 100 mg/l myo-inositol, 8 g/l Difco Bacto-agar supplemented with 5 μM zeatin, 1 μM GA3 and 1 μM NAA.
4. Embryogenic shoots and/or small plantlets are placed on MS medium with 40 g/l sucrose supplemented with 1 μM IAA and 2 μM NAA for rooting. Rooted plants are transferred to hydroponic pots and grown to maturity.

References

Ahmed R, Gupta SD, Ghosh PD (1987) The cytological status of plants regenerated from shoot-meristem culture of *Pisum sativum*. Plant Breed 98:306–311

Anonymous (1985) Mutation breeding for disease resistance using in vitro culture techniques. IAEA-TECDOC 342, Int Atomic Energy Agency, Vienna

Arnold SV, Eriksson T (1976) Factors influencing the growth and division of pea mesophyll protoplasts. Physiol Plant 36:193–196

Arnold SV, Eriksson T (1977) A revised medium for growth of pea mesophyll protoplasts. Physiol Plant 39:257–260

Atanassov AI, Mehandjiev AD (1979) In vitro induced morphogenesis in pea embryos. CR Acad Bulg Sci 32:115–118

Bajaj S, Mickelson O, Lillivek A, Baker LR, Bergen WG, Gill JL (1971 a) Prediction of protein efficiency ratio of peas from their albumin contents. Crop Sci 11:813–815

Bajaj S, Mickelson O, Baker LR, Markarian D (1971 b) The quality of protein in various lines of peas. Br J Nutr 25:207–212

Bajaj YPS (1983) Survival of somatic hybrid protoplasts of wheat×pea and rice×pea subjected to −196 °C. Indian J Exp Biol 21:120–122

Bajaj YPS (1986) Biotechnology in agriculture and forestry. 2. Crops I. Springer, Berlin Heidelberg New York Tokyo

Bajaj YPS, Dhanju MS (1979) Regeneration of plants from apical meristem tips of some legumes. Curr Sci 48:906–907
Baldev B, Lang A, Agatep AO (1965) Gibberellin production in pea seeds developing in excised pods: Effect of growth retardant AMO-1618. Science 147:155–156
Blixt S (1972) Mutation genetics in *Pisum*. Agr Hort Genet 30:1–293
Blixt S (1975) The pea. In: King CR (ed) Handbook of genetics, vol 2. Plenum Press, New York, pp 181–221
Bonner J (1938) Nicotinic acid and the growth of isolated pea embryos. Plant Physiol 13:865–868
Bulygina NG, Erkejev MI (1979) Rate of callus formation in varieties of pea with different genotypes. 3-ja Vses Konf Kult Kletok Rast, Abovyan, pp 182–183 (in Russian)
Christianson ML, Warnick DA, Carlson PS (1983) A morphogenetically competent soybean suspension culture. Science 222:632–634
Constabel F, Kirkpatrick JW, Gamborg OL (1973) Callus formation from mesophyll protoplasts of *Pisum sativum*. Can J Bot 51:2105–2106
Constabel F, Dudits D, Gamborg OL, Kao KN (1975) Nuclear fusion in intergeneric heterokaryons: A note. Can J Bot 53:2092–2095
Constabel F, Weber G, Kirkpatrick JW, Pahl K (1976) Cell division of intergeneric protoplast fusion products. Z Pflanzenphysiol 79:1–7
Crowder AJ, Landgren CR, Rockwood LL (1979) Cultivar differences in starch content and protoplasts yields from root cortical explants of *Pisum sativum*. Physiol Plant 46:85–88
Czosnowski E, Domka L (1981) In vitro callus formation from organs of some Papilionaceae plants. Bull Soc Am Sci Lett Poznan, Ser D 21:39–46
Davey MR, Cocking EC (1972) Uptake of bacteria by isolated higher plant protoplasts. Nature (Lond) 239:455–456
Davies DR, Bedford ID (1982) Abscisic acid and storage protein accumulation in *Pisum sativum* embryos grown in vitro. Plant Sci Lett 27:337–343
Davis BD (1983a) Growth of excised pea embryonic axes on different sugars. Am J Bot 70:816–820
Davis BD (1983b) Effects of sugars on alpha-amylase activity in pea embryonic axes. Am J Bot 70:821–826
de Capite L (1955) La coltura dei frutti in vitro da fiori recisi di *Fragaria chiloensis* Ehrh.×*F. virginia* Duch. var. Marshali e di *Pisum sativum* L. var. Zekka. Ric Sci Ital 25:532–538
Domoney C, Davies DR, Casey R (1980) The initiation of legumin synthesis in immature embryos of *Pisum sativum L.* grown in vivo and in vitro. Planta 149:454–460
Ezhova TA, Bagrova AM, Gostimskij SA (1985) Shoot formation in callus tissues derived from stem apices, epicotyls, stem internodes, and leaves of various pea genotypes. Fiziol Rast 32:513–520 (in Russian)
FAO (1988) Production Yearbook 1987, vol. 41. FAO, Rome, pp 151, 196
Feldman JL, Torey JG (1977) Nuclear changes associated with cellular dedifferentiation in pea root cortical cells cultured in vitro. J Cell Sci 28:87–105
Fowke LC, Constabel F, Gamborg OL (1977) Fine structure of fusion products from soybean cell culture and pea protoplasts. Planta 135:257–266
Frolova LV, Shamina ZB (1974) Cytogenetic characterization of tissue culture in plants from the Fabaceae family. Citol Genet (Kiev) 8:413–418 (in Russian)
Frydman VM, MacMillan J (1976) Gibberellins in developing seed of *Pisum sativum* cv. Progress No 9. Acta Univ N Copernici Biol 18:133–136
Gamborg OL, Miller RA, Ojima K (1968) Nutrient requirements for suspension cultures of soybean root cells. Exp Cell Res 50:151–158
Gamborg OL, Constabel F, Shyluk JP (1974) Organogenesis in callus from shoot apices of *Pisum sativum*. Physiol Plant 30:125–128
Gamborg OL, Shyluk J, Kartha KK (1975) Factors affecting the isolation and callus formation in protoplasts from the shoot apices of *Pisum sativum* L. Plant Sci Lett 4:285–292
Gamborg OL, Davies BP, Stahlhut RW (1983) Somatic embryogenesis in cell cultures of *Glycine* species. Plant Cell Rep 2:209–212
Gantotti BV, Kartha KK (1986) Pea. In: Evans DA, Sharp WA, Ammirato PV (eds) Handbook of plant cell culture, vol 4. Techniques and applications. Macmillan, New York, pp 370–418
Garcia-Martinez JL, Carbonell J (1985) Induction of fruit set and development in pea ovary explants by gibberellic acid. J Plant Growth Reg 4:19–28

Ghosh P, Sharma AK (1979) Chromosome analysis in suspension culture of *Vigna sinensis* var. black and *Pisum sativum* L. Caryologia 32:419–424

Gosal SS, Bajaj YPS (1979) Establishment of callus tissue cultures and the induction of organogenesis in some grain legumes. Crop Improv 6:154–160

Gosal SS, Bajaj YPS (1984) Isolation of sodium chloride resistant cell lines in some grain legumes. Indian J Exp Biol 22:209–214

Gosal SS, Bajaj YPS (1988) Pollen embryogenesis and chromosomal variation in anther culture of three food legumes – *Cicer arietinum, Pisum sativum* and *Vigna mungo.* Sabrao J 20

Gostimskij SA, Bagrova AM, Ezhova TA (1985) Discovery and cytogenetic analysis of variation originated during plant regeneration in pea tissue culture. Dokl Akad Nauk (USSR) 283:1003–1011 (in Russian)

Gould KS, Cutter EG, Young JPW, Charlton WA (1987) Positional differences in size, morphology, and in vitro performance of pea axillary buds. Can J Bot 65:406–411

Govorov LI (1928) The peas of Afghanistan. Trudy Prikl Bot Genet Sel 19:517–522

Griga M, Tejklová E, Novák FJ (1984a) Hormonal regulation of growth of pea (*Pisum sativum* L.) shoot apices in in vitro culture. Rostl Vyr 30:523–530 (in Czech)

Griga M, Tejklová E, Novák FJ (1984b) In vitro propagation of pea by axillary and adventitious bud technique. In: Novák FJ, Havel L, Dolezel S (eds) Proc Int Symp Plant tissue cell cult – appl crop improv, Czech Acad Sci, Prague, pp 507–508

Griga M, Tejklová E, Novák FJ, Kubaláková M (1986) In vitro clonal propagation of *Pisum sativum* L. Plant Cell Tissue Organ Cult 6:95–104

Gupta S (1975) Morphogenetic response of haploid callus tissue of *Pisum sativum* (var. B22). Indian Agric 19:11–21

Gupta S, Ghosal KK, Gadgil VN (1972) Haploid tissue culture of *Triticum aestivum* var. Sonalika and *Pisum sativum* var. B22. Indian Agric 16:277–278

Guy M, Kende H (1984) Ethylene formation in *Pisum sativum* and *Vicia faba* protoplasts. Planta 160:276–280

Guy M, Reinhold L, Laties GG (1978) Membrane transport of sugars and amino acids in isolated protoplasts. Plant Physiol 61:593–596

Hagedorn DJ (ed) (1984) Compendium of pea diseases. Am Phytopathol Soc, St Paul, Minn

Hagedorn DJ (1985) Diseases of peas: Their importance and opportunities for breeding for disease resistance. In: Hebblethwaite PD, Heath MC, Dawkins TCK (eds) The Pea Crop. A Basis for Improvement. Butterworths, London, pp 205–213

Hahn H, De Zacks R, Kende H (1974) Cytokinin formation in pea seeds. Naturwissenschaften 61:170

Harland SC (1948) Inheritance of immunity to mildew in Peruvian forms of *Pisum sativum.* Heredity 2:263–269

Haskins RH, Kartha KK (1980) Freeze preservation of pea (*Pisum sativum* cv. Century) meristems: Cell survival. Can J Bot 58:833–840

Haupt W (1952) Untersuchungen über den Determinationsvorgang der Blütenbildung bei *Pisum sativum.* Z Bot 40:1–32

Hawkes JG (1983) The diversity of crop plants. Harvard Univ Press, Cambridge, Massachusetts

Hildebrandt AC, Wilmar JC, Johns H, Riker AJ (1963) Growth of edible chlorophyllous plant tissue in vitro. Am J Bot 50:248–254

Hussey G, Gunn HV (1984) Plant production in pea (*Pisum sativum* L. cvs. Puget and Upton) from long-term callus with superficial meristems. Plant Sci Lett 37:143–148

Jacobsen HJ, Kysely W (1984) Induction of somatic embryos in pea, *Pisum sativum* L. Plant Cell Tissue Organ Cult 3:319–324

Jacobsen HJ, Ingensiep HW, Herlt M, Kaul MLH (1980) Tissue culture studies in *Pisum sativum.* In: Sala F, Parisi B, Cella R, Ciferri O (eds) Plant cell cultures: results and perspectives. Developments in plant biology, vol 5. Elsevier, Amsterdam, pp 319–324

Jaranowski J, Micke A (1985) Mutation breeding in peas. Mut Breed Rev 2:1–23

Jia SR (1982) Factors affecting the division frequency of pea mesophyll protoplasts. Can J Bot 60:2192–2196

Jia SR, Kao KN, Knott DR (1982) Factors affecting the division frequency of pea mesophyll protoplasts. Sci Agr Sin 4:20–25

Kajita S, Matsui C, Syono K, Suzuki M, Nagata T (1980) Fine structure of fusion bodies formed between pea root nodule and tobacco mesophyll protoplasts. Z Pflanzenphysiol 97:233–240
Kallak HI (1983) On karyotypic differentiation of callus cells in long-term culture. Acta Comment Univ Tart 583:25–36 (in Russian)
Kallak HI, Vapper MA (1979) Tendencies to karyotypic variability in prolonged tissue culture. 3-ja Vses Konf Kult Kletok Rast Abovyan, pp 188–189 (in Russian)
Kallak H, Yarvekylg L (1968) On morphological and cytological differences of pea callus. Citol Genet (Kiev) 2:408–414 (in Russian)
Kallak H, Yarvekylg L (1970) On cytological characterization of pea callus culture. In: Butenko RG (ed) The culture of isolated plant organs, tissues and cells. Nauka, Moscow, pp 140–144 (in Russian)
Kallak H, Yarvekylg L (1971) On the cytogenetic effects of 2,4-D on pea callus in culture. Acta Biol Acad Sci Hung 22:67–73
Kallak H, Yarvekylg L (1972) Cytological characterization of some callus lines of pea. In: Genetics of grain legumes. All-Union Res Breed Inst of Legumes, Orel, pp 7–16 (in Russian)
Kallak H, Yarvekylg L (1976) On nuclear form and division in plant tissue culture. Acta Comment Univ Tart 383:52–63 (in Russian)
Kallak K, Yarvekylg L (1977a) Changes in chromosome complement in long-term pea callus cultures. Acta Biol Acad Sci Hung 28:183–189
Kallak H, Yarvekylg L (1977b) Nuclear behaviour in callus cells: morphology and division. Biol Plant 19:48–52
Kartha KK (1979) Freeze preservation of plant meristems as a means of germplasm preservation. Agrologist 8:29
Kartha KK (1981) Meristem culture and cryopreservation-methods and applications. In: Thorpe TA (ed) Plant Tissue Culture Methods and Application in Agriculture. Academic Press, New York, pp 181–211
Kartha KK (1984) Culture of shoot meristems: pea. In: Vasil IK (ed) Cell Culture and Somatic Cell Genetics of Plants, vol 1. Academic Press, Orlando, pp 106–110
Kartha KK, Gamborg OL (1978) Meristem culture techniques in the production of disease-free plants and freeze-preservation of germplasm of tropical tuber crops and grain legumes. In: Maraite H, Meyer JA (eds) Diseases of Tropical Food Crops. Univ Catholique de Louvain, Belgium, pp 267–283
Kartha KK, Leung NL (1979) Cryopreservation of plant meristems as a means of germplasm storage. Cryobiology 16:582–583
Kartha KK, Gamborg OL, Constabel F (1974) Regeneration of pea (*Pisum sativum* L.) plants from shoot apical meristems. Z Pflanzenphysiol 72:172–176
Kartha KK, Leung NL, Gamborg OL (1979) Freeze-preservation of pea meristems in liquid nitrogen and subsequent plant regeneration. Plant Sci Lett 15:7–15
Knösche von R (1981) Untersuchungen zur Polyploidisierung einer Gewebekultur von *Pisum sativum* L. II. Die Beziehung zwischen Mitoseindex und der Häufigkeit der Restitutionsmitosen. Biol Zbl 100:55–65
Knösche von R, Günther G (1979) Restitution cell cycles important processes in the tissue culture of *Pisum sativum* L. leading to polyploidy. 3-ja Vses Konf Kult Kletok Rast Abovyan, p 188 (in Russian)
Knösche von R, Günther G (1980) Untersuchungen zur spontanen Polyploidisierung einer Gewebekultur von *Pisum sativum* L. I. Der Nachweis von Restitutionszyklen. Biol Zbl 99:311–323
Krechting HCJM, Varga A, Bruinsma J (1978) Absence of cytokinin biosynthesis in pea seeds developing in vitro. Z Pflanzenphysiol 87:91–94
Kruyt W (1952) Effects of some plant growth substances on early growth of pea embryos in sterile culture; a study in connection with the problem of hormonisation of seeds. Konikl Nederl Akad Wetenschap Proc C 55:503–514
Kubaláková M, Tejklová E, Griga M (1988) Some factors affecting root formation on in vitro regenerated pea shoots. Biol Plant 30:179–184
Kunakh VA, Alkhimova EG, Voitjuk LI (1984a) Variation in chromosome number in callus tissues and regenerants of pea. Citol Genet (Kiev) 18:20–26 (in Russian)

Kunakh VA, Alkhimova EG, Voitjuk LI, Alpatova LK (1984b) Obtaining of tissue cultures and organogenesis induction in *Pisum sativum* L. In: Novák FJ, Havel L, Dolezel J (eds) Proc Int Symp Plant tissue cell cult – appl crop improv, Olomouc, pp 135–136

Kunakh VA, Voitjuk LI, Alkhimova EG, Alpatova LK (1984c) Callus tissue formation and induction of organogenesis in *Pisum sativum* L. Fiziol Rast (Moscow) 31:542–548 (in Russian)

Kysely W, Jacobsen HJ (1984) Induction of somatic embryos in pea and soybean. In: Novák FJ, Havel L, Dolezel J (eds) Proc Int Symp Plant tissue cell cult – appl crop improv, Olomouc, pp 131–132

Kysely W, Myers JR, Lazzeri PA, Collins GB, Jacobsen HJ (1987) Plant regeneration via somatic embryogenesis in pea (*Pisum sativum* L.). Plant Cell Rep 6:305–308

Lamprecht H (1966) Die Entstehung der Arten und höheren Kategorien. Springer, Berlin Heidelberg New York

Landgren CR (1976a) The influence of culture conditions on mitotic activity in protoplasts derived from *Pisum* root cortical explants. Protoplasma 87:49–69

Landgren CR (1976b) Patterns of mitosis and differentiation in cells from pea root protoplasts. Am J Bot 63:473–480

Landgren CR (1981) Gibberellin enhancement of the enzymic release of *Pisum* root cell protoplasts. Physiol Plant 52:349–352

Landgren CR, Torrey JG (1973) The culture of protoplasts derived from explants of seedling pea roots. Coll Intern CNRS 212:281–285

Larkin PJ, Scowcroft WR (1981) Somaclonal variation – a novel source of variability from cell cultures for plant improvement. Theor Appl Genet 60:197–214

Lea PJ, Hughes JS, Miflin BJ (1979) Glutamine- and asparagine-dependent protein synthesis in maturing legume cotyledons cultured in vitro. J Exp Bot 30:529–537

Lee KW, Min CK (1981) Cell wall regeneration of pea mesophyll protoplasts. Korean J Bot 24: 73–86

Lee KW, Cho SH, Cha HC (1980) The isolation and fusion of pea and barley mesophyll protoplasts. Korean J Bot 23:49–54

Lepoivre P (1981) Effect de l'ascochitine sur les protoplasts de *Pisum sativum* L. Bull Rech Agron Gembloux 16:27–33

Leroux R (1968) Action de l'acide gibbérellique sur la rhizogenèze de fragments de tiges de Pois (*Pisum sativum* L.) cultivés in vitro en présence d'auxine à l'obscurité ou à la lumière. CR Acad Sci Paris 266:106–108

Leroux R (1973) Contribution a l'étude de la rhizogenèze de fragments de tiges de Pois (*Pisum sativum* L.) cultivés in vitro. Rev Cytol Biol Veg 36:1–132

Leurs CJ, Winter H, Wiersema PK, Helder RJ (1982) Light-dependent rubidium uptake into isolated mesophyll protoplasts from leaves of *Pisum sativum*. Physiol Plant 56:339–342

Libbenga KR, Torrey JG (1973) Hormone-induced endoreduplication prior to mitosis in cultured pea root cortex cells. Am J Bot 60:293–299

Lippmann B, Lippmann G (1984) Induction of somatic embryos in cotyledonary tissue of soybean, *Glycine max* (L.) Merr. Plant Cell Rep 3:215–218

Malmberg RL (1979) Regeneration of whole plants from callus culture of diverse genetic lines of *Pisum sativum* L. Planta 146:243–244

Malmberg RL (1982) The inheritance of the ability to regenerate plants from cell cultures of *Pisum sativum* L. A preliminary analysis. Pisum Newsl 14:39–40

Marks GE, Davies DR (1979) The cytology of cotyledon cells and the induction of giant polytene chromosomes in *Pisum sativum*. Protoplasma 101:73–80

Marx GA (1969) Two additional genes conditioning wax formation. Pisum Newsl 1:10–11

Michajlov OF, Bessonova VP (1975) Some data on the cytogenetic analysis of callus obtained from regenerating cotyledons of pea. Probl Onkol Teratol Rastenij, Nauka, Leningrad, pp 52–54 (in Russian)

Millerd A, Spencer D, Dudman WF, Stiller M (1975) Growth of immature pea cotyledons in culture. Austr J Plant Physiol 2:51–60

Monti LM, Grillo S (1983) Legume seed improvement for protein content and quality. Qual Plant Plant Foods Hum Nutr 32:253–266

Mroginski LA, Kartha KK (1981) Regeneration of pea (*Pisum sativum* L. cv. Century) Plants by in vitro culture of immature leaflets. Plant Cell Rep 1:64–66

Mroginski LA, Kartha KK (1984) Tissue culture of legumes for crop improvement. In: Janick J (ed) Plant breeding reviews, vol 2. AVI, Westport, pp 215–264
Murashige T, Skoog F (1962) A revised medium for rapid growth and bioassays with tobacco tissue cultures. Physiol Plant 15:473–497
Murashko LN (1974) Interline and intervarietal differences in callus formation in stem explants of pea. Nauc Dokl Vyss SK Biol Nauk 10:111–113 (in Russian)
Nandi S, Eriksson T (1977) Nuclear behaviour of pea leaf protoplasts. Hereditas 85:49–55
Natali L, Cavallini A (1987a) Regeneration of pea (*Pisum sativum* L.) plantlets by in vitro culture of immature embryos. Plant Breed 99:172–176
Natali L, Cavallini A (1987b) Nuclear cytology of callus and plantlets regenerated from pea (*Pisum sativum* L.) meristems. Protoplasma 141:121–125
Netien G, Beauchesne G (1951) Action des sels de potassium des acides alpha-naphtylacétique et alpha-naphtylpyruvique sur la croissance des embryos excisés de pois "Alaska". Bull Mens Soc Linn Lyon 20:56–59
Novák FJ, Lucretti S, Hermelin T, Donini B, Afza R, Daskalov S, Kubaláková M, Griga M (1984a) Gamma ray irradiation effects on multiple shoot cultures of pea (*Pisum sativum* L.). In: Novák FJ, Havel L, Dolezel J (eds) Proc Int Symp Plant Tissue cell cult – appl crop improv, Olomouc, pp 453–454
Novák FJ, Lucretti S, Donini B, Afza R, Hermelin T (1984b) Flower and pod development in shoot tip and axillary bud culture of pea (*Pisum sativum* L.). In: Novák FJ, Havel L, Dolezel J (eds) Proc Int Symp Plant Tissue cell cult – appl crop improv, Olomouc, pp 133–134
Paal H, Banyai D, Balatincz Z, Kurnik E (1982) In vitro morphogenetic capacity of pea embryos at the early heart stage. Agr Közlem 41:249
Paranjothy K, Raghavan V (1970) Effects of fluorinated pyrimidines on the growth of excised pea embryos. Plant Cell Physiol 11:259–271
Phillips GC, Collins GB (1981) Induction and development of somatic embryos from cell suspension cultures of soybean. Plant Cell Tissue Organ Cult 1:123–129
Plemenkova SF, Lozovaja VV (1983) Uptake of ^{14}C-glucose by pea isolated protoplast. 3-ja Vses Konf Transport Assimiljatov V Rast I Probl Sacharonakoplenija, Frunze, p 69 (in Russian)
Poulsen MH, Sebro C (1983) Breeding of dried peas for Danish conditions. In: Thompson R, Casey R (eds) Perspectives for peas and lupins as protein crops. Nijhoff, The Hague, pp 169–172
Praznovsky T, Kurnik E, Paal H, Dudits D (1981) Fusion between soybean (*Glycine max*) and pea (*Pisum sativum*): Protoplast division in heterokaryons. Biologia (Budapest) 29:213–218
Praznovsky T, Dudits D, Kurnik E, Paal H (1982) Use of soybean and pea protoplast fusion in cell hybridization experiments. Agr Közlem 41:212
Preobrazenskaja EV (1983) The comparison of organogenetic ability of different pea (*Pisum sativum* L.) lines and cultivars in callus culture. Tr 14 Konf Mol Ucenych Biol Fak MGU, Moscow, pp 134–137 (in Russian)
Puonti-Kaerlas J, Eriksson T (1988) Improved protoplast culture and regeneration of shoots in pea (*Pisum sativum* L.). Plant Cell Rep 7:242–245
Rahat M, Reinhold L (1983) Rb^+ uptake by isolated pea mesophyll protoplasts in light and darkness. Physiol Plant 59:83–90
Rubluo A, Mroginski L, Kartha K (1982) Morphogenetic responses of pea leaflets cultured in vitro. In: Fujiwara A (ed) Plant Tissue cell cult 1982. Maruzen, Tokyo, pp 151–152
Rubluo A, Kartha KK, Mroginski LA, Dyck J (1984) Plant regeneration from pea leaflets cultured in vitro and genetic stability of regenerants. J Plant Physiol 117:119–130
Sladky Z, Jandova B (1984) Micropropagation of pea, cucumber and potato. In: Novák FJ, Havel L, Dolezel J (eds) Proc Int Symp Plant Tissue cell cult appl crop Improv, Olomouc, pp 509–510
Snoad B (1985) The need for improved pea-crop plant ideotypes. In: Hebblethwaite PD, Heath MC, Dawkins TCK (eds) The pea crop. A basis for improvement. Butterworths, London, pp 31–41
Srivastava PS, Varga A, Bruinsma J (1980) Growth in vitro of fertilized ovules of pea, *Pisum sativum* L., with and without pods. Z Pflanzenphysiol 98:347–354
Stafford A, Davies DR (1979) The culture of immature pea embryos. Ann Bot 44:315–321
Syono K, Nagata T, Suzuki M, Kajita S, Matsui C (1979) Fusion of pea root nodule protoplasts with tobacco mesophyll protoplasts. Z Pflanzenphysiol 95:449–458

Therman E, Murashige T (1984) Polytene chromosomes in cultured pea roots (*Pisum*, *Fabaceae*). Plant Syst Evol 148:25–34

Thompson JF, Madison JT, Muenster AE (1977) In vitro culture of immature cotyledons of soya bean (*Glycine max* L. Merr.). Ann Bot 41:29–39

Torrey JG, Shigemura Y (1957) Growth and controlled morphogenesis in pea root callus tissue grown in liquid media. Am J Bot 44:333–344

van Staden J, Button J (1978) The cytokinin content of aseptically cultured pea fruits. Z Pflanzenphysiol 87:129–135

van't Hof J, McMillan B (1969) Cell population kinetics in callus tissue of cultured pea root segments. Am J Bot 56:42–51

White PR (1943) A handbook of plant tissue culture. J Cattel, Lancaster

Wright DJ (1985) Combining peas for human consumption. In: Hebblethwaite PD, Heath MC, Dawkins TCK (eds) The pea crop. A basis for improvement. Butterworths, London, pp 441–451

II.2 Chickpea (*Cicer arietinum* L.)

N. ALTAF and M.S. AHMAD[1]

1 Introduction

Chickpea (*Cicer arietinum* L.) is a small annual plant with green pods and toothed leaflets. It has the status of an important winter seed legume crop in the Indo-Pakistan subcontinent, where it is commonly known as gram, Channa, or Chhola. Young plants and green pods are eaten like spinach. The green seeds are cooked as a vegetable; mature seeds are used as dry pulse, parched, boiled, fried, or in various dishes. Dhal is the split chickpea seed, without the coat, and is eaten cooked in a thick soup. The flour is used in many bakery products, sweets, and ceremonial dishes. It is a good source of carbohydrates, vitamins, and minerals (Duke 1980). It is rich in protein, 22% (Jambunathan and Singh 1979) and is fed to animals to obtain animal protein.

The crop is cultivated in rain-fed areas of various provinces, especially Punjab, where no other crop can be grown on such a large scale without artificial irrigation. As rains provide the only source of soil moisture and it fixes atmospheric nitrogen, there is little need of nitrogen fertilizers; therefore the chickpea crop has the greatest economic importance for the bulk of the farming populations of these areas.

Chickpea is important not only for the Indo-Pakistan subcontinent, but it is also cultivated in other countries of Asia, Africa, Europe, and the Americas (Table 1). In 1985, the total world area harvested was 9512000 ha, the yield was 675 kg/ha, and the production was 6416×1000 mt (40 kg = 1 mt). The major producing countries are India, Pakistan, Turkey, and Burma in Asia; Ethiopia, Tunisia, and Malawi in Africa; Spain, Portugal and Italy in Europe; Mexico and Colombia in the Americas (FAO 1985).

The genus *Cicer* contains some 40 species of annual or perennial herbs or small perennial shrubs. Nine annual species are recognized: *C. arietinum, C. bijugum, C. chorassanicum, C. cuneatum, C. echinospermum, C. judaicum, C. pinnitifidum, C. reticulatum*, and *C. yamashitae*. All have the same somatic chromosome number (2 n = 16).

Two races are recognized within the cultivated chickpea, referred to as the microsperma and macrosperma group, which are equivalent to the desi and kabuli types (Moreno and Cubero 1978; Hawtin and Singh 1980). The genetic diversity

[1] Nuclear Institute for Agriculture and Biology, PAEC, P.O. Box 128, Jhang Road, Faisalabad, Pakistan

Biotechnology in Agriculture and Forestry, Vol. 10
Legumes and Oilseed Crops I (ed. by Y.P.S. Bajaj)

Table 1. World chickpea production (FAO 1985)

Location	Area under cultivation (1000 ha)	Yield (kg/ha)	Production (1000 MT7)
World	9512	675	6416
Asia	8763	669	5866
Africa	422	673	284
North Central America	150	1000	150
South America	41	636	26
Europe	137	660	90
Burma	250	862	215
India	6857	663	4547
Pakistan	1001	518	519
Turkey	350	1086	380
Ethiopia	180	750	135
Tunisia	65	774	50
Sudan	3	1000	3
Egypt	10	1737	17
Mexico	150	1000	150
Italy	9	1265	12
Portugal	29	340	10
Spain	88	659	58

within each group is considerable and introgression is used as an important strategy in crop improvement (Byth et al. 1980). Presently chickpea crop improvement research at the international level is going on at the International Crops Research Institute for the Semi-Arid Tropics (ICRISAT), India, and the International Centre for Agricultural Research in the Dry Areas (ICARDA) at Tel Hadia in Syria. These two institutes co-ordinate their efforts on chickpea improvement (Byth et al. 1980; Allen 1983).

This important crop is beset with three very destructive diseases, namely chickpea or gram blight, chickpea wilt, and smalling of leaves (Sattar and Hafiz 1952). Gram blight is the most serious disease and is caused by *Ascochyta rabiei*. The fungus causes brown spots of varying size on the entire surface of the plant, and from February onward, the diseased plants show partial or total drying. If the weather is wet because of rains, the whole field may be completely destroyed and the plants appear as if scorched by fire. The second important disease of gram is wilt, and in this case the roots are attacked by fungus, mostly *Fusarium oxysporum*. As a result, the vascular system is clogged up, the plants start wilting, and ultimately dry up. Wilting of plants takes place at two stages of growth, seedling and flowering, because at both these stages the soil moisture is optimum for the growth of fungus. The third disease is smalling of leaves, in which plants show symptoms of arrested growth and reduced size of various organs such as leaves, leaflets, branches, and stem. Plants remain stunted in growth and leaves ultimately change color and are shed. In severe cases, there is no pod formation. The cause of this disease is unknown; however, genetic resistance is most effective in controlling it. Chickpea is strictly a self-pollinating species (Malik et al. 1982; Malhotra and Singh 1986). The high degree of self-pollination in chickpea is attributed to the fact that flowers are autogamous and that only certain papillate cells of the stigma are specialized to receive pollen when it is shed prior to full flower bloom. The low success rate of artificial hybridization in chickpea may be related to the timing of stigma receptivity and to the limited number of cells on the stigmatic surface which appear to be receptive to pollen. Functional specialization of stigmatic cells may thus be responsible for its self-pollinating nature (Turnao et al. 1983).

For improvement of the chickpea crop, conventional breeding and induced mutation methods are used. Conventional breeding includes the methods of self-pollinated species such as the pedigree method, bulk method, backcross method and multiple crosses. However, large populations are required in the case of more complex inheritance and to study the segregating generations and obtain a desirable combination of characters when two factors are closely linked. Their recombination is infrequently obtained in F_3 by growing progeny of F_2 plants that contain one of the two desired characters. Induced mutation breeding is most suitable, especially when one or two characters are to be incorporated in a well-adapted high-yielding local variety. Mutation frequency can be enhanced by chemical mutagens and radiation.

In Pakistan, by introduction and selection, C44, a high-yielding and *Ascochyta rabiei*-resistant variety, developed by the Ayub Agricultural Research Institute, Faisalabad, has been released for general cultivation, as has another high-yielding and *A. rabiei*-resistant variety, CM72, developed by the Nuclear Institute for Agriculture and Biology. Variety CM72 is an induced mutant of a local, well-adapted and high-yielding variety 6153 that is susceptible to *A. rabiei*. The mutant was a result of 15 krad exposure to γ-radiation (Haq and Hassan 1980). With the development of these two blight-tolerant varieties, CM72 and C44, the situation of the chickpea crop in Pakistan has much improved. The yield level is almost normal and the areas has increased from 642000 ha in 1980–81 to almost 809000 ha in 1985–86.

The crop suffers damage from *Heliothis armigera* at the podding stage. Insecticides like Endosulfan and Quinalphos are used for crop protection against *Heliothis* attack (Vyas et al. 1983).

In the last decade, in vitro techniques have been explored as a potential tool for plant improvement (see Bajaj 1986). However, it is essential to understand the growth characteristics of a plant before utilizing these techniques. Ultimately these unconventional methods will supplement conventional breeding.

Cicer reticulatum hybridizes readily with cultivated species and is considered as the wild progenitor of chickpea. Another species that has cross-compatibility with *C. arietinum* is *C. echinospermum* (Ladizinsky and Adler 1976a, b). *Cicer reticulatum* has *Ascochyta* blight resistance, and crosses were made to transfer the resistance to the cultigen (van der Maesen et al. 1980). Blight resistance among certain accessions of other wild chickpea, *C. bijugum*, *C. pinnatifidum*, and *C. judaicum* (Nene and Haware 1980; van der Maesen et al. 1980; Singh et al. 1981), and wilt resistance in *C. judaicum* (Nene and Haware 1980), are known but these have no cross-compatibility with *C. arietinum*. A further radical technique such as somatic hybridization is needed to transfer blight/wilt resistance in the cultigen.

2 In Vitro Culture

Tissue culture techniques have their own specific purpose. Plant propagation by means of shoot apex culture ensures genetic stability, efficient control of quality production of stocks, and continuous production throughout the year. Endosperm cultures are important for raising triploid lines for genetic mapping and for the production of seedless fruits. Embryogenic callus provides an ideal and stable system for the study of embryogenesis and the factors influencing this process, which directly form an intact plant. Protoplast cultures are important for somatic hybridization and DNA uptake experiments for genetic engineering. Isolated protoplasts can be handled like microbes and can be treated with various mutagenic chemicals and radiation. Haploids rapidly achieve homozygosity, especially in self-sterile plants, and facilitate interspecific hybridization. Induction and selection of mutants at the haploid level is important. Haploids also give an easy recovery of recessives. The technique of embryo culture helps to procure hybrid plants which are otherwise difficult to obtain. The literature relating to the in vitro culture of chickpea is summarized in Table 2.

Table 2. Work on growth responses of explants of various genotypes of *Cicer arietinum*

Variety	Salt medium	Growth regulators (mg/l)	Explants	Morphogenetic responses	Reference
–	PRL-4C	2,4-D	Stem, root	Unlimited cell division	Hosel et al. (1972)
–	PRL-4C	2,4-D	Stem, root	Unlimited cell division	Muhle et al. (1976)
–	PRL-4C	2,4-D	Stem, root	Unlimited cell division	Hosel et al. (1977)
–	PRL-4C	2,4-D	Stem, root	Unlimited cell division	Hosel et al. (1978)
–	PRL-4C	2,4-D	Stem, root	Unlimited cell division	Burmeister and Hosel (1980)
–	–	–	Leaf (protoplast)	Cell divisions	Gamborg et al. (1974)
G543	MS	IAA(2)+K(0.5)	Shoot apices freeze preserved	Complete plants	Bajaj (1979b)
G543	MS	2,4-D(2)	Shoot meristem freeze preserved	Callus	Bajaj (1979b)
G543	MS	NAA(0.1)+BA(0.02)	Apical meristem	Rooting	Bajaj and Dhanju (1979)
G130	MS	IAA(2)+kin(0.5)	Apical meristem	Complete plants	Bajaj and Dhanju (1979)
G543, L550, Hare chhole No. 1	MS	2,4-D(2)	Cotyledons	Callus	Gosal and Bajaj (1979)
–	MS	BA(0.2)	Shoot meristem	Shoot development and profuse axillary branching	Kartha et al. (1981)

Table 2 (continued)

Variety	Salt medium	Growth regulators (mg/l)	Explants	Morphogenetic responses	Reference
T3	B5	2,4-D(0.5)+kin(0.1)	Hypocotyl	Callus	Singh et al. (1982)
T3	B5	2,4-D(0.1)	Hypocotyl	Rooting	Singh et al. (1982)
UC5	Modified	Nil	Apical and subapical stem cutting	Rooting	Davis and Foster (1982)
B108	MS	2,4-D(2)+Coconut milk (10% v/v)	Anther	Multicellular pollen	Khan and Ghosh (1983)
G543	MS	IAA(2)+BA(0.2)	Shoot meristem	Complete plants	Bajaj (1983)
B17, BR77, BR65, Type 1, K5	MS	2,4-D(0.5)+pCPA(2) +Picloram(0.12) +BA(0.1)	Stem	Cell suspension growth	Sinha et al. (1983)
B108	B5	2,4-D(2)+kin(1)	Cotyledonary node	Callusing	Khan and Ghosh (1984)
B108	B5	NAA(0.5)	Shoot bud	Rooting	Khan and Ghosh (1984)
B108	B5	NAA(0.5)	Cotyledonary node	Shoot bud and root	Khan and Ghosh (1984)
6153	MS	BA(1-2)+IAA (1-2)	Shoot apices, cotyledonary nodes	Complete plants	Altaf and Ahmad (1986a, b)
6153	MS	2,4-D(1-2)+BA (1-2)+NAA(0.5)	Seed, shoot apices, young leaves, stem, cotyledonary nodes, root, anther	Callusing and rooting	Altaf and Ahmad (1986a, b)
ICCC-4, Annigeri, H208	B5	BA(0.5)+NAA(2)	Cotyledons	Callusing and rooting	Neelam et al. (1986a)
ICCC-4, Annigeri, H208	B5	BA(1)+kin(1) +IAA(0.5)	Hypocotyl and shoot apex	Good shoot growth	Neelam et al. (1986a)
ICCC-4	B5	BA(0.5)+IAA(0.1)	Hypocotyl	Shoot development	Neelam et al. (1986b)
ICCC-4	B5	BA(1)+kin(1) +IAA(0.5)	Shoot apices and hypocotyls	Multiple branching	Neelam et al. (1986c)
G534, L550, Hare chhole 1	MS	IAA(4)+kin(2)	Anther	Multicellular pollen, embryoids	Gosal and Bajaj (1988); Bajaj and Gosal (1987)

2.1 Sterilization of Explants/Organs, Media Composition, Callus, and Meristem

Sterilization of seeds is mostly done with $HgCl_2$, ranging from 0.1% to 0.5% for 5 to 10 min, depending on the specific genotype and seed hardiness, especially the

Table 3. Effect of growth regulators and additives on the callus subculture growth

Treatment	Callus responses[a]
1. Control MS basal medium	+ +
2. Callus maintenance medium MS + BA (5 mg/l) + IAA (0.1 mg/l) + GA (0.5 mg/l)	+ + +
3. Treatment 2 + zeatin (5 mg/l)	+ + + +
4. Treatment 2 + ADP (3 mg/l)	+ + +
5. Treatment 2 + maleic acid (10 – 50 mg/l) + oxalic acid (5 mg/l)	+ + + +
6. N6 + BA (1 mg/l) + IAA (0.5 mg/l)	+ + +
7. Treatment 2 + chickpea seedling extract (70%)	+ + + +
8. Treatment 1 + green seed extract (50%)	+ + + +
9. Treatment 2 + chickpea seedling root extract (70%)	+ + + +
10. Treatment 2 + chickpea cotyledons extract (70%)	+ + + +
11. Treatment 2 + coconut milk (10% – 15%)	+ + + +
12. Treatment 2 + honey (2 g/l)	+ + + +
13. Treatment 2 + potato extract (15% – 20%) (Fig. 1G)	+ + + +

[a] Maximum response was graded as + + + +.

seed coat texture. Since chickpea is self-pollinated and pure lines of different genotypes are available, varietal seed is good material for raising seedlings.

For explant and organ sterilization, NaOCl is commonly used in a range 0.5% to 1.2% for 5 to 20 min. Sometimes dual treatment with ethanol and NaOCl is applied. Freshly prepared chlorine water for 12 – 15 min has been also used successfully. For anther culture studies, small floral buds can be sterilized by dipping in 90% ethanol for 1 min. The sterilizing chemical may be shaken or standing left.

MS and B5 media are frequently used (Table 2). At NIAB (Faisalabad) MS, B5, Yu-Pei, PC-L2, Blaydes, soybean, T&L media, and N6 were tried. MS, N6, and B5 showed good responses. MS and N6 were slightly better than B5, but the two media were equally good for growth responses of explants and calli. The addition of different amounts of NH_4NO_3 and KH_2PO_4 in MS medium was studied for callus growth responses. Amounts of NH_4NO_3 and KH_2PO_4 were doubled for MS medium and also reduced to $\frac{1}{2}$ and $\frac{1}{4}$th in strength. The callus growth indicated that MS was the most suitable medium for chickpea. Sinha et al. (1983) used many media, such as SS-K-8, SS-B-8, PC-L2, B5, MS, and SH; and found that fresh weight increase of suspension culture derived from stem tissue was better on MS medium for cultivars B17, BR65, BR77, Type 1, and K5.

The growth regulators used were K, Z, BA, 2ip, GA, IAA, IPA, NAA, IBA, 2,4-D, adenine-sulfate, and 2,3 – 5 TIBA. The additives studied were coconut water, honey, potato extract, chickpea whole seedling extract, seedling shoot extract, seedling root extract, cotyledon extract, germinating seed extract, casamino acids, malt extract, and yeast extract. The effects of growth regulators and additives on callus growth are shown in Table 3. Casamino acids, malt extract, and yeast extract at concentration 0.1% to 0.3% had slightly better effects on callus growth responses. Of the γ-radiation doses (1, 5, 7.5 krad), 5 krad increased callus mass as compared to the control (Fig. 1I). Among glycine, L-alanine, L-valine, L-isoleucine, L-leucine, L-serine, L-threonine, L-aspartic acid, L-glutamic acid, L-lysine, L-arginine, L-asparagine, L-glutamine, L-cystine, L-cysteine, L-methionine, L-

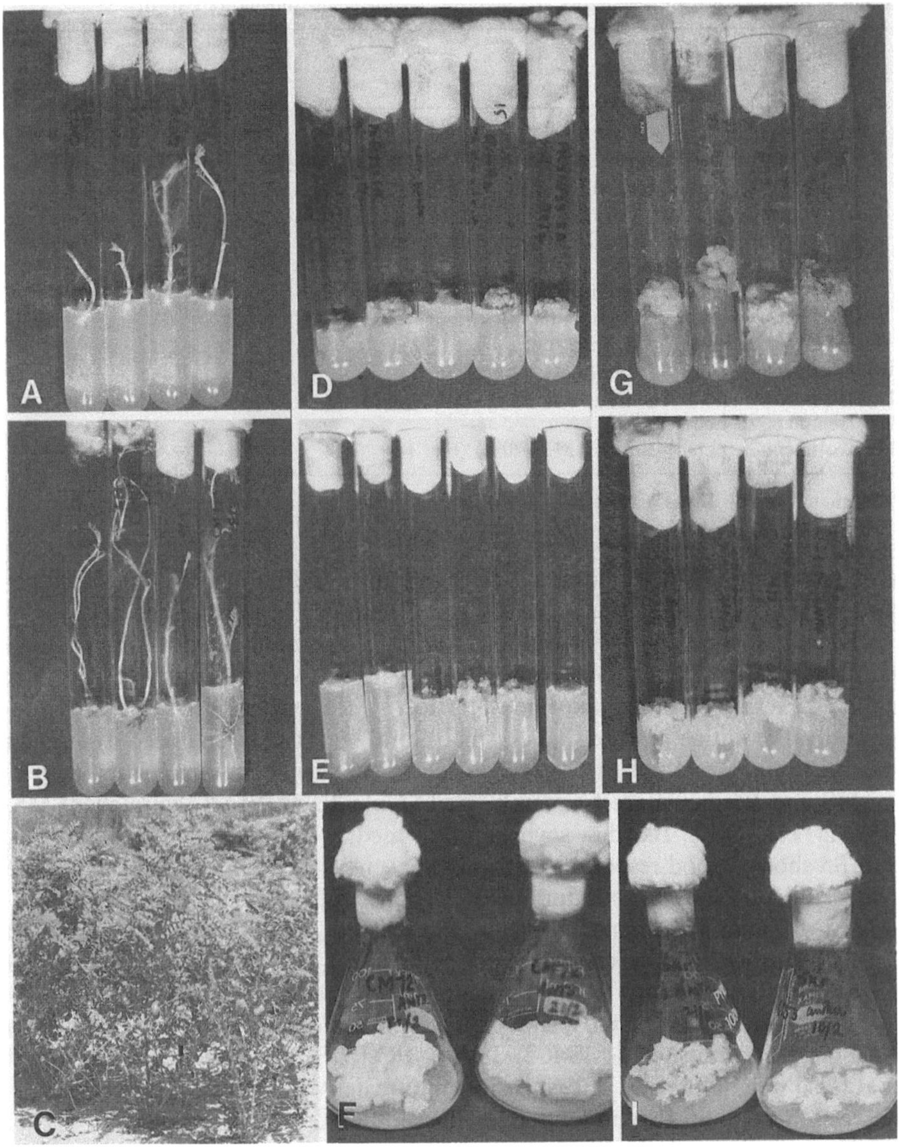

Fig. 1A–I. In vitro culture studies on *Cicer arietinum*. **A** Growth of shoots, explants:apical meristems, medium: MS+BA (1 mg/l). **B** Plantlets from cotyledonary nodes, two tubes at *left*; plantlets from shoot apices, two tubes at *right*. Medium: MS+BA (0.5 mg/l)+IAA (1.5 mg/l). **C** Plantlets derived from shoot apices and nodes are grown to maturity in the field. Plants 3 months after transplantation. **D** Genotypes *from left to right*, C727, K850, C141, CM72, C44. Medium: MS+2,4-D (1 mg/l)+BA (0.5 mg/l)+NAA (0.5 mg/l). Age: 1-month-old cultures. **E** Globular structures from cotyledonary node callus, genotype CM72. Medium: MS+BA (1 mg/l)+2,4-D (0.5 mg/l)+IAA (0.1 mg/l). **F** Anther callus from variety CM72, 6th subculture. Medium: 2,4-D (1.5 mg/l)+BA (1 mg/l)+IAA (0.1 mg/l). Age: 2 months. **G** Callus growth with 15% potato extract as additive, two tubes at *left*; callus growth with 20% potato extract as additive, two tubes at *right*. **H** Callus growth in medium containing 2,4-D (1 mg/l), two tubes at *left*; callus growth in medium containing 2,4-D (10 mg/l), two tubes at *right*. **I** Anther callus derived from explant given 5 krad γ-radiation dose. Variety 6153. Age: 2 months tryptophan, L-phenylalanine, L-tyrosine, L-histidine, L-proline, folic acid, biotin and hydroxy proline, when added each as 1 mg/l to MS medium (containing MS vitamins and amino acids), proline and glutamine induced vigorous callus growth (each 2–4 mg/l), the next being biotin, cysteine, and glycine (1 mg/l each).

Among the cytokinins, zeatin proved best. Calli remained healthy and kept growing in zeatin (2) at a concentration ranging from 2–5 mg/l, for 3 months without any visible adverse effects. Second best was BA in optimal amounts of 5–6 mg/l. BA can be used as a substitute for Z, and subcultures can be kept growing till 2 months. BA gives more vigorous and healthy calli than K and 2 ip; adenine sulfate causes swelling of explants. For callus growth the cytokinin to auxin ratio was studied. The two auxins NAA and 2,4-D were used to initiate callus (2,4-D 2 mg/l + NAA 0.5 mg/l) and then reduced each to 0.1 mg/l and kept constant and the BA concentration was varied as 1, 2, 3, 4, 5, 6, 7, 8, 9, and 10 mg/l, making the cytokinin to auxin ratio 5×, 10×, 15×, 20×, 25×, 30×, 35×, 40×, 45×, 50× (Altaf and Ahmad 1986b). It was found that 25× to 30× was the best for growth of calli. The selected combination was BA (5 mg/l) + NAA (0.1 mg/l) + 2,4-D (0.1 mg/l). With the overall growth regulators in this 25× ratio, when reduced to $\frac{1}{2}$ or $\frac{1}{4}$th there were no adverse effects, but the original amounts in the 25× ratio were slightly better.

Since genotype has a strong influence on the culturability of the tissue, we also studied some genotypic influences on culture. A number of genotypes were studied (Altaf and Ahmad 1986a, b) and finally K850, C44, CM72, 6153, C141, and ILC195 were selected on the basis of better culture response (Fig. 1D). Most of the experiments were conducted with 6153 or its mutant CM72. It was also concluded that genotypes with high yielding potential give better growth response. White-seeded genotypes give a comparatively poor culture response, and ILC195 was the best among the genotypes studied.

Almost every part of the plant produces callus. Roots, cotyledonary nodes, hypocotyl, shoot apices, and small floral buds show good cell proliferation. Mature seeds give more callus than immature seeds. The growth responses of a few explants in callusing media and the influence of genotype is shown in Table 4. Cotyledons are modified leaves, and adventitious bud formation has been reported from the basal region of the leaves in many species. Cotyledons of chickpea in callusing medium produce globular structures as well as callus and when these globular structures are put on to the same medium, they revert to a callus mass (Fig. 1E). However, on rare occasions shoot primordia develop but the exact origin of these primordia is uncertain.

Regeneration of complete plants from shoot tips of chickpea has been demonstrated (Bajaj and Dhanju 1979; Kartha et al. 1981; Altaf and Ahmad 1986a; Neelam et al. 1986c. Regeneration of plantlets from cotyledonary nodes was also observed (Table 2). Complete plants have been obtained via organogenesis (Figs. 1A, B) from apical meristem and cotyledonary node calli (Altaf and Ahmad 1986a; Neelam et al. 1986b). These calli may have meristems already present, and possibly shoots originate from these pre-existing shoot primordia. However, regeneration of actively dividing cells of callus has still not been reported.

Table 4. Genotype responses in the callusing medium

Explant and response[a]	Genotypes					
	C141	6153	K850	C44	CM72	ILC195
Cotyledons (globular structure formation)	5	4	4	3	3	3
Seeds (callusing response)	4	4	5	4	4	3
Shoot apices (callusing response)	4.5	5	5	4.5	4.5	3
Cotyledonary nodes (callusing response)	4.5	5	5	4.5	4.5	4

[a] Maximum response was graded as 5.

Table 5. Effect of growth regulators on cultured shoot apices and cotyledon nodes, var. 6153. Morphogenetic responses. Basal medium was MS in all treatments

Treatment (mg/l)	Shoot apices	Cotyledonary nodes
1. BA(2) + IAA(2)	Shoot growth + callus + roots	Shoot buds + callus + roots
2. BA(1) + IAA(2)	Shoot growth + rooting	Shoot growth + rooting
3. BA(2) + IAA(1)	Shoot growth + rooting	Shoot growth + rooting + swelling of lower portion
4. IAA(2)	Slow shoot growth + rooting	Small shoots + rooting
5. BA(2)	Shoot growth	Shoot growth
6. BA(2) + IAA(0.1) + IBA(0.1) + NAA(0.1)	Multiple branches and swelling of embedded portion	Multiple branches + few roots
7. BA(2) + NAA(0.5)	Complete plants	Complete plants
8. kin(1) + 2ip(2) + BA(1) + 2,4-D(1.5)	Cell proliferation + callusing	Cell proliferation + callusing
9. BA(1.5) + 2,4-D(2) + NAA(0.5)	Callusing	Callusing
10. Adenine sulfate (10 – 25)	Swelling of explants with slow growth	Swelling of explants with slow growth

Root regeneration has been induced in cuttings (Davis and Foster 1982; Rupela and Dart 1982). Rooting is also observed in cultures (Singh et al. 1982; Altaf and Ahmad 1986a; Neelam et al. 1986c).

BA is suitable for shoot growth and IAA, NAA, and IBA are suitable for root growth. 2,4-D is good for continued cell proliferation (Table 5).

2.2 Embryo, Anther, and Protoplast Culture

Immature, mature, and germinating embryos of variety CM72 were studied for their callusing response. More cell proliferation occurred in germinating embryos. Gosal and Bajaj (1979) cultured chickpea ovules and embryos at three stages of development: (1) young (globular to heart-shaped), (2) medium (heart-shaped to differentiating), and (3) late (maturing) embryos. They observed that very young embryos invariably proliferated to form callus; the younger and medium ones produced callus and plantlets, and maturing embryos developed directly into plantlets. However, the rate of proliferation differed among cultivars, Hare chhole 1 being better than L550 and G543.

Interspecific hybridizations between *Cicer arietinum*, *C. cunetum*, and *C. bijugum* have been tried and as these crosses are not compatible. Embryo culture techniques are being used to raise hybrid plants (Chowdhury et al. 1984).

Anthers of chickpea undergo healthy callus formation in BA (1.5 mg/l) + 2,4-D (2 mg/l) + IAA or NAA (0.1 mg/l). All those treatments, which are good for explant-derived calli, also work equally well with anther calli. However, anther calli of 6153 kept growing in 2,4-D even up to 10 mg/l (Fig. 1 H). The anther callus of 6153 grew best and good growth was also observed in variety CM72 (Fig. 1 F). Mature anthers gave no or only very slight response in some cases. However, greenish white anthers having meiotic cells or pollen mitosis showed maximum cellular proliferation. Callus was creamy to greenish in color and became loose and friable after three to four subcultures. Anther-derived calli undergo rhizogenesis (Khan and Ghosh 1983; Altaf and Ahmad 1986a).

Recently pollen embryogenesis and chromosomal variation in cultured anthers of 3 Indian cvs. (G543, L550 and Hare chhole 1) of chickpea have been reported (Gosal and Bajaj 1988; Bajaj and Gosal 1987). The anthers cultured at the uninucleate pollen stage showed various types of development, ranging from multinucleate, bicellular, multicellular and the formation of embryoids. The anther-derived callus of cv. G543 and Hare chhole 1 showed exuberant growth. Varietal differences were observed, cv. Hare chhole 1 being the best (61.7% anthers callused and 0.52% showed pollen embryos). The chromosome numbers in the callus varied from haploid ($n = 8$) to octoploid. The haploid callus was unstable, the older callus after 6–8 weeks subculturing completely lacked haploid cells.

Gamborg et al. (1974) have reported cell divisions in protoplasts obtained from leaves. Similarly, calli from protoplasts, derived from leaves, were obtained by Tayyab et al. (1985). Regeneration of complete plants from protoplast-derived calli has not been demonstrated.

2.3 Transfer of Regenerated Plants to Soil

Regenerated plantlets should preferably be of small size (2–3 inches), green and healthy for transfer to pots. Roots should be clearly developed. Vermiculite and soil mixtures have been used for establishment. Plantlets can also be transferred to sand or simple soil. Soil should be kept moist with frequent watering. Plantlets may be covered for 1–2 days to maintain humidity. Temperatures should preferably be 20°–22 °C. After 10–15 days hardening in pots, the plants can be transferred to the field (Fig. 1 C).

On the other hand, plantlets with soft, weak stems, and elongated internode and callus between shoot and root are more difficult to handle and do not survive transfer into the soil. Sometimes there are no or poor vascular connections between shoot and root, and these hinder plant survival.

2.4 Cryopreservation

Bajaj (1979a, b) has done pioneering work on cryopreservation of germplasm of plant species, including chickpea (Bajaj 1982, 1983). He has successfully shown the survival of shoot apices and meristems in chickpea and has produced normal seeds from plants derived from meristems which were cryopreserved (−196 °C) for 20 months.

2.5 Physiological Studies on Cell Cultures

Salt-tolerance studies have been made by Gosal and Bajaj (1984) and Pandey and Ganapathy (1984). They successfully induced salt resistance in calli of cultivars G543 and BG203. However, regeneration of complete plants from these calli was still lacking. Gupta et al. (1984a), working with root callus cultures of chickpea cultivar G543, have shown that nitrogenase activity in callus cultures infected with *Rhizobium* strains in vitro is comparable with that in vivo. The root exudate (68 mg/l) obtained from young chickpea seedlings promoted the nitrogenase activity of the *Cicer-Rhizobium* association in vitro (Gupta et al. 1984b).

Metabolism of flavonoids and isoflavonoids in plant cell suspension cultures has been extensively studied by Hosel and co-workers. In chickpea plants, flavonols occur as 3-glycosides, and isoflavones as 7-0-glycosides, and it has been demonstrated that flavonol aglycones and flavonol 7-glycosides are readily degraded by peroxidases in vitro as well as in plant cell culture (Hosel et al. 1972). The naturally occurring flavonol 3-glycoside (kaempferol 3-glucoside) is rapidly metabolised by cell cultures of mung bean, soybean, and chickpea; the liberated aglycone is oxidized by the activity of peroxidases; the hydrolytic activity of flavonol 3-glucosides is three to four times higher than of flavonol 7-glucosides. Parsley cell cultures are an exception, as these have moderate ability to split kaempferol 3-glucoside; in contrast, kaempferol 7-glucoside is attacked by peroxidase while the sugar portion is still attached to the aglycone (Muhle et al. 1976; Hosel et al. 1977).

Isoflavone 7-0-glucosides, formononetin and biochanin A 7-0-glucosides, applied to plant cell suspension cultures of soybean, mung bean, chickpea, and parsley, have been studied by Hosel et al. (1977). All four cell suspension cultures rapidly deglucosilated the isoflavone 7-glucosides. The aglycone portion and free glucose are metabolized further, but the exact sequence of degradation reaction and enzymes involved is not characterized. A specific β-glucosidase for isoflavone 7-glucosidases is demonstrated in chickpea plants. The β-glycosidase activity of proteins extracted from chickpea plants is nearly the same when measured at pH 7.5 or pH 5, due to the high content of the isoflavone 7-glucoside specific β-glucosidase, which has a pH optimum at 7.5. In contrast, the β-glycosidase activity of chickpea cell cultures at pH 5 is five times higher than at pH 7.5, indicating that large amounts of β-glycosidase with acidic pH optimum were present. Furthermore the β-glycosidase pattern of cell suspension cultures, separated by DEAE and CM chromatography, differed strikingly from that of original plant tissue.

Comparison of chickpea β-glycosidases from cell cultures and whole seedlings has been carried out (Burmeister and Hosel 1980). Chickpea cell lines initiated either from stem or root develop a β-glycosidase pattern which is different from either plant organ. Invariably, 1-year-old culture has a β-glycosidase pattern identical to that of 8-year-old cultures. The isoflavone 7-glucoside-specific β-glucosidase, which is present in large amounts in plant organs, is only a minor component in cell cultures. The isoflavone 7-apiosylglucoside-specific β-glycosidase, which was present in stem and leaves of chickpea, is entirely absent from cultures. However, several other β-glycosidases accumulate in cell cultures in large amounts, whereas they are only detected with great difficulty in whole plants. The isoenzyme pattern of the isoflavone 7-glucoside-specific β-glucosidase of chickpea cell culture is quantitatively different from that of leaves, stems, and roots.

Although the physiological role of most of these β-glucosidases is uncertain, one of them has been shown to have specificity for coniferin, and it is suggested that it may play a role in lignification (Hosel et al. 1978).

3 Conclusion

There is a need to improve the yield potential of chickpea, and to provide disease-resistant varieties. As conventional breeding methods take longer to tailor a variety for a particular purpose, nonconventional methods such as micropropagation and tissue culture technique should be developed to speed up the evolution of a new variety. It has been shown that vegetative propagation through apical shoots or nodal meristems or rooting in plant cuttings is feasible in chickpea. It has also been established that callusing can be induced in explants taken from any organ of the plant, but that regeneration from truely dividing callus cultures is lacking. It appears that regeneration in callus cultures reported by some workers was from pre-existing meristematic cells. Regeneration of plants from truely dividing callus cultures of chickpea is thus eagerly awaited.

References

Allen DJ (1983) The pathology of tropical food legumes. Disease in crop improvement. Wiley, New York

Altaf N, Ahmad MS (1986a) Plant regeneration and propagation of chickpea (*Cicer arietinum* L.) through tissue culture techniques. IAEA, Vienna, pp 407–417

Altaf N, Ahmad MS (1986b) Callusing and regeneration response of various chickpea genotypes. PAEC-KfK Symposium/Workshop on biotechnology in agriculture and energy. March 3–7, 1986, NIAB, Faisalabad

Bajaj YPS (1979a) Technology and prospects of cryopreservation of germplasm. Euphytica 28:267–285

Bajaj YPS (1979b) Freeze preservation of meristems of *Arachis hypogaea* and *Cicer arietinum*. Indian J Exp Biol 17:1405–1407

Bajaj YPS (1982) Cryopreservation of germplasm of potato, cassava, peanut and chickpea. In: Fujiwara A (ed) Plant tissue culture 1982. Maruzen, Tokyo, pp 799–800

Bajaj YPS (1983) Production of normal seeds from plants regenerated from the meristems of *Arachis hypogaea* and *Cicer arietinum* cryopreserved for 20 months. Euphytica 32:425–430

Bajaj YPS (1986) Biotechnology in agriculture and forestry 2. Crops I. Springer, Berlin Heidelberg New York Tokyo

Bajaj YPS, Dhanju MS (1979) Regeneration of plants from apical meristem tips of some legumes. Curr Sci 48:906–907
Bajaj YPS, Gosal SS (1987) Pollen embryogenesis and chromosomal variation in cultured anthers of chickpea. Chickpea Newslett 17:12–13
Burmeister G, Hosel W (1980) Comparison of β-glycosidases from *Cicer arietinum* L. cell culture and whole seedlings. Planta Med 4:40–48
Byth DE, Green JM, Hawtin GC (1980) ICRISAT/ICARDA chickpea breeding strategies. Proc Int Worshop Chickpea Improv, Hyderabad, India 28 Feb–2 March, 1979, ICRISAT, pp 11–22
Chowdhury JB, Sharma DR, Chowdhury VK, Jain RK, Dhawan RS (1984) IAPTC Newslett 42(March): 10
Davis TM, Foster KW (1982) A method for rooting chickpea cuttings. Int Chickpea Newslett 7:6–8
Duke JA (1980) Handbook of legumes of world economic importance. Plenum Press, New York
FAO (1985) Production yearbook. FAO, Rome
Gamborg OL, Constabel F, Fowke L, Kao KN, Ohyama K, Kartha K, Pelcher L (1974) Protoplast and cell culture methods in somatic hybridization in higher plants. Can J Genet Cytol 16:737–750
Gosal SS, Bajaj YPS (1979) Establishment of callus tissue cultures, and the induction of organogenesis in some grain legumes. Crop Improv 6(2):154–160
Gosal SS, Bajaj YPS (1984) Isolation of sodium chloride resistant cell lines in some grain legumes. Indian J Expl Biol 22(4):209–214
Gosal SS, Bajaj YPS (1988) Pollen embryogenesis and chromosomal variation in anther culture of three food legumes – *Cicer arietinum, Pisum sativum* and *Vigna mungo.* Sabrao J 20
Gupta RP, Kalra MS, Bajaj YPS (1984a) Nitrogen fixation in cell cultures of some legumes and non-legumes. Indian J Expl Biol 22:560–563
Gupta RP, Kalra MS, Bajaj YPS (1984b) Role of root exudates on in vitro establishment of symbiosis between *Cicer* cells and *Rhizobium*. In: Ghai BS (ed) Symbiotic nitrogen fixation I. USG, Ludhiana
Haq MA, Hassan M (1980) Induced mutations for resistance to *Ascochyta* blight in chickpea. Pak J Agric Sci XVII (3–4):57–64
Hawtin GC, Singh KB (1979) Kabuli-Desi introgression. Problems and prospects. Proc Int Workshop Chickpea Improv, Hyderabad, India 28 Feb–2 March, 1979, ICRISAT, pp 51–60
Hosel W, Shaw PD, Barz W (1972) Über den Abbau von Flavonolen in pflanzlichen Zellsuspensionskulturen. Z Naturforsch 278:946–954
Hosel W, Burmeister G, Kreysing P, Surholt E (1977) Enzymological aspects of flavonoid catabolism in plant cell cultures. In: Barz W et al. (eds) Plant tissue culture and its biotechnological application. Springer, Berlin Heidelberg New York, pp 172–177
Hosel W, Surholt E, Borgmann E (1978) Characterization of β-glucosidase isoenzymes possibly involved in lignification from chickpea (*Cicer arietinum* L.) cell suspension cultures. Eur J Biochem 84:487–492
Jambunathan R, Singh U (1979) Studies on Desi and Kabuli chickpea (*Cicer arietinum* L.) cultivars, chemical composition. Proc Int Workshop Chickpea Improv, Hyderabad, India 28 Feb–2 March, 1979, ICRISAT, pp 61–65
Kartha KK, Pahl K, Leung NL, Mroginski LA (1981) Plant regeneration from meristems of grain legumes: soybean, cowpea, peanut, chickpea and bean. Can J Bot 59:1671–1679
Khan SK, Ghosh PD (1983) In vitro induction of androgenesis and organogenesis in *Cicer arietinum* L. Curr Sci 52(18):891–893
Khan SK, Ghosh PD (1984) Plantlet regeneration from cotyledonary nodes of chickpea. Int Chickpea Newslett 11:22–24
Ladizinsky G, Adler A (1976a) The origin of chickpea *Cicer arietinum* L. Euphytica 25:211–217
Ladizinsky G, Adler A (1976b) Genetic relationships among the annual species of *Cicer arietinum* L. Theor Appl Genet 48:197–203
Malhotra RS, Singh KB (1986) Natural cross pollination in chickpea. Int Chickpea Newslett 14:4–5
Malik BA, Syed AH, Haqqani AM, Chaudhury AH (1982) Floral biology and yield in chickpea (*Cicer arietinum* L.). Pak J Agric Res 3(1):1–5
Moreno MT, Cubero JI (1978) Variation in *Cicer arietinum* L. Euphytica 27:465–485
Muhle E, Hosel W, Barz W (1676) Catabolism of flavonol glucosides in plant cell suspension cultures. Phytochemistry 15:1669–1972
Neelam A, Reddy CS, Reddy GM (1986a) Growth and differentiation in tissue cultures of *Cicer arietinum* L. Int Chickpea Newslett 14:9–12

Neelam A, Reddy CS, Reddy GM (1986b) Plantlet regeneration from callus cultures of *Cicer arietinum* L. Int Chickpea Newslett 14:12–13

Neelam A, Reddy CS, Reddy GM (1986c) Multiple shoot and plantlet regeneration from shoot apex and hypocotyl explants of *C. arietinum* L. Int Chickpea Newslett 14:13–16

Nene YL, Haware MP (1980) Screening chickpea for resistance to wilt. Plant Dis 64:379–380

Pandey R, Ganapathy PS (1984) Effect of NaCl stress on callus cultures of *Cicer arietinum* L. cv. BG-203: Growth and ion accumulation. J Exp Bot 35(157):1194–1199

Rupela OP, Dart PJ (1982) Rooting chickpea cuttings from field grown plants. Int Chickpea Newslett 7:9–10

Sattar A, Hafiz A (1952) Researches on plant diseases of the Punjab. Pak Assoc Adv Sci Univ Inst Chem, Lahore

Singh KB, Hawtin GC, Nene YL, Reddy MV (1981) Resistance in chickpea to *Ascochyta rabiei*. Plant Dis 65:586–587

Singh RP, Singh BD, Jaiswal HK, Singh RM, Singh RB (1982) Organogenesis in callus cultures of chickpea. Indian J Agric Sci 52(2):86–90

Sinha RR, Das K, Sen SK (1983) Nutritional requirement of tissue cultures of some tropical legume crops. Indian J Exp Biol 21:113–119

Tayyab H, Parveen T, Riazuddin S (1985) Isolation and growth of mesophyll protoplasts from *Cicer arietinum* L. Proc 2nd Natl Meet Plant Tissue Culture, July 15–19, 1985, Baragali-Abbottabad, Pakistan, p 9

Turano MJ, Baird LM, Webster BD (1983) Characteristics of the stigma of chickpea. Crop Sci 23:1033–1036

van der Maesen LJG, Rundir RPS, Remanandan P (1980) The current status of chickpea germplasm work at ICRISAT. Proc Int Workshop Chickpea Improv, Hyderabad, India 28 Feb–2 March, 1979 CRISAT, pp 28–32

Vyas HN, Ghari MS, Beria NN (1983) Bioefficacy of quinalphos against gram pod borer, *Heliothis armigera* hubner attaching chickpea in Gujrat through aerial spraying. Int Chickpea Newslett 9:22–24

II.3 Soybean: Plant Regeneration and Somaclonal Variation

U. B. BARWALE and J. M. WIDHOLM[1]

1 Introduction

The soybean, a legume of Eastern Asian origin, is a major source of edible vegetable oils and of high protein feed supplements for livestock. The demand for soybean, being the most produced, traded, and utilized meal and oil in the world, has resulted in increased soybean production, and research on this crop has also increased proportionally (Smith and Huyser 1987).

The composition of soybean seed is unique, with 20% oil and 40% protein. Despite this quality composition, there are quality problems which could be solved. Eight percent of the total oil is an 18:3 fatty acid, linolenic acid, which is unstable in air and light due to its high degree of unsaturation. Because of this, soybean oil becomes rancid and deteriorates in flavor when exposed to air and light. Soybean flour also contains a number of antinutritional factors including trypsin inhibitors, a hemagglutinin complex, and a goitrogenic factor which require special processing procedures, thus increasing the cost of soybean products (Cowan 1973). The protein is well balanced for animal consumption as far as amino acid composition is concerned, except that it is somewhat deficient in methionine.

Soybeans are self-pollinated, so are highly inbred. Due to the high cost of producing hybrids and to no clear heterotic response of the F_1 hybrids, no hybrid soybeans are commercially produced. Pedigree selection and single-seed descend methods are most often used for soybean breeding. Mutation breeding (Hammond et al. 1972) and interspecific hybridization (Hymowitz et al. 1970) have been used to select for specific traits and to increase variation.

With the possiblity of tissue culture-induced variation (Larkin and Scowcroft 1981), desired changes in many diverse traits may be seen in regenerated plant progeny. In vitro techniques thus may provide a complementary approach for crop improvement.

2 History of Soybean Tissue Culture

Regeneration of whole plants from undifferentiated cells in culture is necessary before any of the in vitro techniques can be applied to plant breeding. Gamborg et al. (1968) developed a medium (B5) which induced and maintained soybean cal-

[1] Department of Agronomy, University of Illinois at Urbana-Champaign, Urbana, IL 61801, USA

Biotechnology in Agriculture and Forestry, Vol. 10
Legumes and Oilseed Crops I (ed. by Y. P. S. Bajaj)

lus and suspension cultures. This medium was characterized by a high nitrate to ammonium ratio. In the late 1960's and early 1970's several reports emphasized the importance of nitrogen source and proper nutritional balance for growth and maintenance of soybean cultures (Gamborg and Shyluk 1970; Baley et al. 1972a, b; Ohira et al. 1975; Ojima and Ohira 1977). Mott et al. (1984) reported that ammonium stunted the growth of shoots unless relatively high concentrations of nitrate were also supplied to soybeans in vitro.

With the information available on the nutritional requirements for soybean callus cultures, plantlet formation was still not attained. Media modifications by Beversdorf and Bingham (1977) and Oswald et al. (1977) resulted in some progress, since structures resembling somatic embryos were obtained, but these did not form plantlets. Root formation from hypocotyl sections was noted by Kimball and Bingham (1973). No shoots formed on sections which already had roots, however. Adventitious bud development was also demonstrated.

Somatic embryo-like structures were obtained from suspension cultures which did not germinate and produce whole plants (Phillips and Collins 1981). In this system, *G. soja* produced one shoot structure which did not grow to maturity. Gamborg et al. (1983) developed procedures which produced somatic embryos in cell cultures of *Glycine* species. Genotype-specific responses were also noted. In another suspension culture system the number and stages of somatic embryos formed varied from genotype to genotype (Kerns et al. 1986). In all of these cases the embryoids did not develop further to form whole plants.

Plant regeneration in certain wild perennial relatives of soybean has been seen. Kameya and Widholm (1981) were able to obtain shoots directly on the explants of *G. canescens* when the young seedling explants were used. Further, shoots were produced at a high frequency from *G. canescens* callus and suspension cultures, but rooting was not seen (Widholm and Rick 1983). Grant (1984) regenerated whole plants of *G. canescens* callus via embryogenesis. Plant regeneration protocols from protoplast-derived calli were also reported for *G. canescens* (Newell and Luu 1985; Hammatt et al. 1987) and for *G. clandestina* (Hammatt et al. 1987). Use of leaf explants of *G. clandestina* also resulted in plant regeneration (Hymowitz et al. 1986). Infrequent regeneration was also seen from seedling tissues of *G. tomentella* (Kameya and Widholm 1981). Hildebrand et al. (1986) have reviewed plant regeneration protocols for several *Glycine* species.

Soybean anthers have also been cultured in an attempt to obtain haploid plants unsuccessfully by Ivers et al. (1974), but Yuyu et al. (1984) induced 19 plantlets and numerous embryoids from anther cultures of soybean. In this case, uninucleate pollen produced the embryogenic callus. Guangchu et al. (1984) also reported haploid callus formation from pollen which later produced plantlets. The plant regeneration frequency was low in these studies, however.

The ease of soybean suspension culture allowed the early development of protoplast isolation and culture techniques. Kao et al. (1970) first reported continued cell division in protoplast-derived cultures. Several other reports, using different explants or culture conditions, obtained callus regeneration from isolated protoplasts; however, no plant regeneration was obtained from this callus (Kartha et al. 1974; Zieg and Outka 1980; Schwenk 1981; Schwenk et al. 1981; Xu et al. 1982; Gamborg et al. 1983; Lu et al. 1983; Tricoli et al. 1986).

A different approach was taken by Cheng et al. (1980) and Saka et al. (1980) to attempt plant regeneration. Shoot formation at the cotyledonary node was induced and up to 11 shoots could be obtained from one seedling. In this instance, a complex explant with preexisting meristems was utilized. Genotype-specific responses were noted for the number of shoots produced at the cotyledonary node when this system was used as a screening procedure to identify genotypes which are more responsive under in vitro conditions (Barwale et al. 1986a).

Christianson et al. (1983) first described a morphogenically competent soybean suspension culture. The explant (immature embryo) and culture medium used were critical for plant regeneration. Lippmann and Lippmann (1984) induced somatic embryos, which did not develop further, on cotyledons of immature embryos. Ranch et al. (1985, 1986), Lazzeri et al. (1985) and Barwale et al. (1986b) were, however, able to produce somatic embryos from many different genotypes which reproducibly were able to develop into whole plants. Li et al. (1985) described a complex, multistep procedure for plant regeneration, involving at one stage single somatic cells. Ghazi et al. (1986) induced both a smooth and a rough callus, from which embryogenic cultures were produced from the smooth type, and whole plantlets were produced.

Plant regeneration in soybean has also been reported via organogenesis using immature embryos and cotyledonary nodes as the explants (Barwale et al. 1986b; Wright et al. 1986). While immature embryos have been the usual source of regenerable calli, Wright et al. (1987) showed that primary leaf tissue and epicotyls produced cultures which were regenerable as well.

In the remainder of this review, we will discuss the plant regeneration procedures established in this laboratory, the variation found in regenerated plants resulting from the tissue culture process, the possible mechanisms causing this variation, and the possible application of this variation to soybean breeding.

3 Establishment of Cultures

Initially, the genotypes used in this study were selected based on a multiple shoot screening assay, where shoot numbers were used to pick out genotypes responding to the tissue culture environment (Barwale et al. 1986a). Other genotypes used were commercial cultivars and genotypes with specific reactions to different diseases. A total of 64 genotypes were evaluated for plant regeneration.

To initiate embryogenic cultures, 0.5- to 10-mm-sized immature embryos were removed from surface sterilized pods and were placed on ENT medium, which is a modification of standard MS medium (Murashige and Skoog 1962), supplemented with the optimum concentration of 43 μM naphthaleneacetic acid (NAA) and 5 μM thiamine, as described by Barwale et al. (1986b). The pH of the medium was adjusted to 5.8 before autoclaving. Of the different embryo sizes used to initiate cultures, 4 to 6 mm was determined to be most efficient in producing the maximum number of somatic embryos and, in turn, plants (Barwale et al. 1986b). The high concentration of NAA in the medium stimulated normal embryogenesis and development.

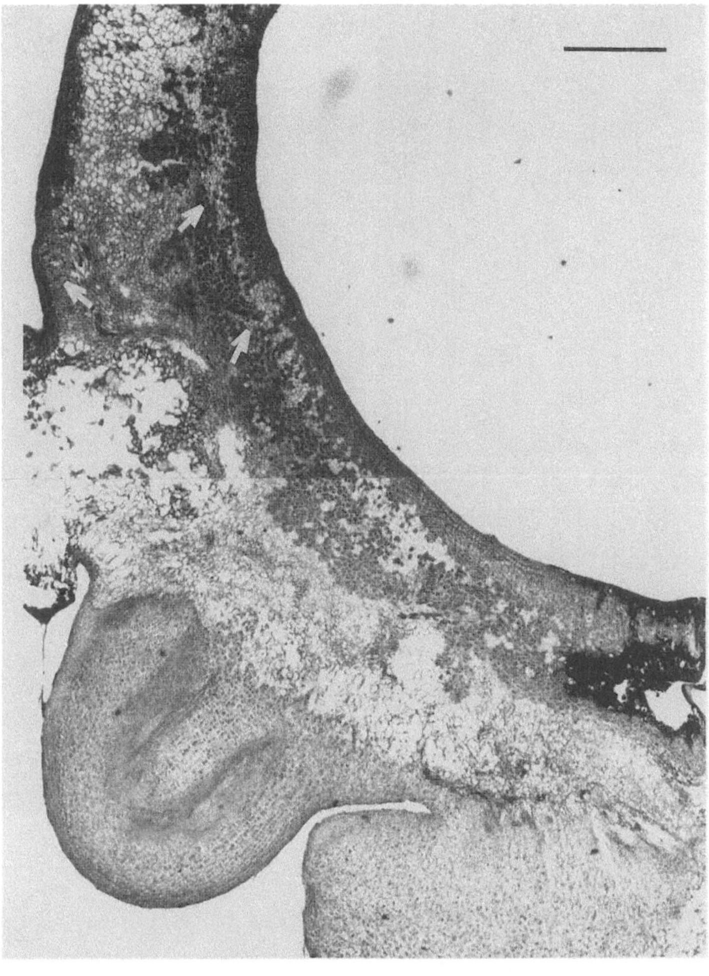

Fig. 1. Meristematic activity seen 5 days after initiating cultures, beneath the epidermis of the immature cotyledons on ENT medium. *Arrows* indicate regions with actively dividing cells. (Bar = 0.5 mm)

After 5 days in culture, the zygotic embryos exhibited periclinal cell divisions in the subepidermal cell layers of the cotyledons (Fig. 1). Seven to ten days after initiation, globular structures were seen directly on the cotyledon surface. These further developed into heart-stage embryos, at which time these somatic embryos were transferred to a germination medium, MSR or R5. Both MSR and R5 are MS basal media with 1.7 μM benzylaminopurine (BAP), 0.2 μM indolebutyric acid (IBA), and 9.8 μM IBA+5 nM BAP, 5 μM gibberellic acid (GA_3), respectively (Fig. 2A–D). Germination of these somatic embryos occurred at the late torpedo stage, where the presence of a shoot apex could be shown by scanning electron microscopy (Fig. 3).

Germinating somatic embryos were then transferred to MS^- medium (no growth regulator) followed by transplanting to the greenhouse in a hydroponic

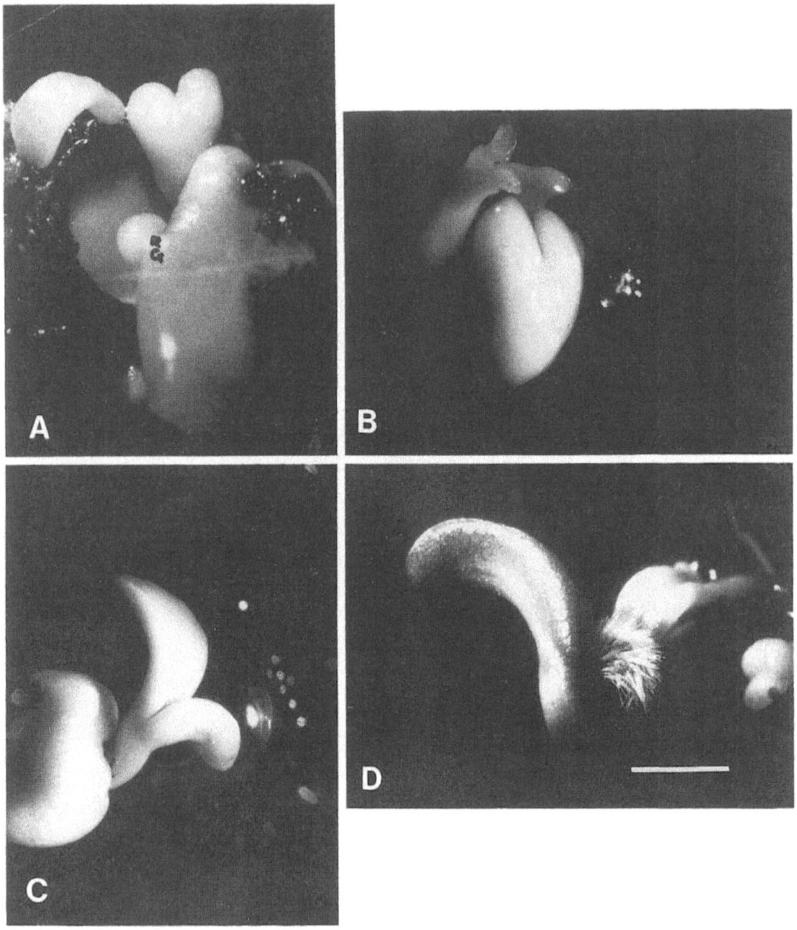

Fig. 2A–D. Different developmental stages of somatic embryo development. **A** Globular embryo (G_T). **B** Heart stage. **C** Torpedo stage. **D** Embryo germination (Bar = 3 mm). Barwale et al. (1986b)

system with quarter strength Hoagland's solution to stimulate root growth. The regenerated plants are then grown to maturity in a soil mix, as discussed by Barwale et al. (1986b). The regenerated plants are designated R_0, selfed seed of R_0 are R_1 seeds, etc. Seed from all plants were harvested individually and maintained separately.

Organogenic cultures were obtained when embryos were placed on 4MSII medium, which contains the major elements of MS basal medium + 4× concentrations of MS minor elements +B5 (Gamborg et al. 1968) vitamins +13.3 μM BAP and 0.2 μM NAA. Three to four weeks after embryos (optimally 4 to 6 mm) are initially placed on 4MSII medium, shoot formation is seen at the cotyledonary node and less frequently on the cotyledonary surface. Once shoot formation is initiated, the need for high concentrations of BAP and minor elements is elimi-

Fig. 3. Scanning electron micrograph showing the presence of meristematic dome with the shoot apex. (×37)

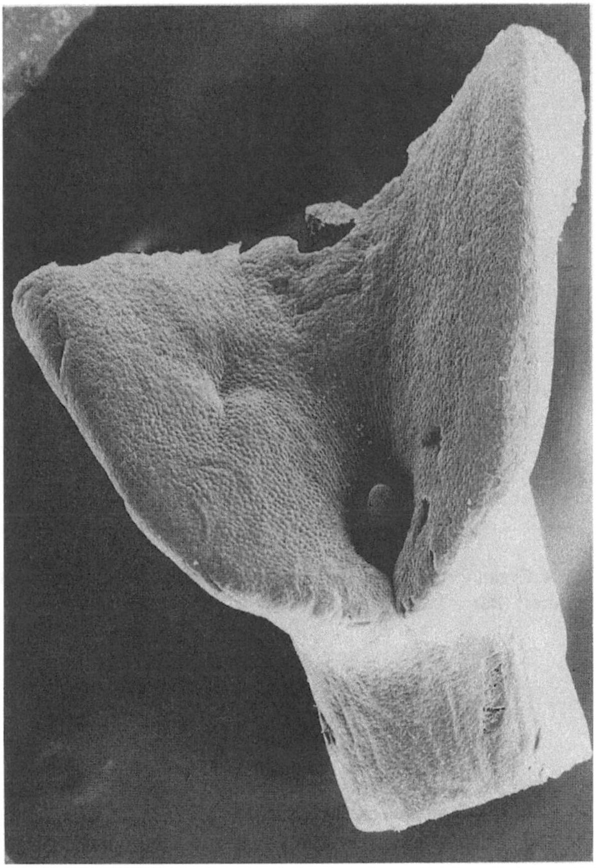

nated, and the cultures can be transferred to MSR or R5 medium to maintain them as continuously shoot-producing (Fig. 4). Plants are then obtained from these cultures as described above for the embryogenic cultures.

Thidiazuron (N-phenyl-N′-1,2,3-thiadiazol-5-ylurea), an experimental cotton defoliant, has been shown to have cytokinin-like activity on plant cell cultures (Mok et al. 1982; Capelle et al. 1983) and when organogenic soybean cultures were placed on medium containing thidiazuron in place of BAP, an increase in shoot growth was noted. The shoots have a greener appearance, as has been noted with *Phaseolus* (Mok et al. 1982), and more shoot elongation also occurs. In comparison to MSR and R5 medium, the thidiazuron medium appears to induce healthier shoots; however, the number of shoots that are formed is approximately 20% less than that on MSR and R5 medium. A greater increase in fresh weight is noted with the thidiazuron medium as compared to MSR and R5 medium, with some of this being nonregenerable, friable callus. A concentration of 0.03 mg/l of thidiazuron was found to produce the best response in combination with 0.3 mg/l IBA in a standard MS medium (Fig. 5). Use of this new medium has produced faster-growing cultures for many genotypes. Cultures, however, are routine-

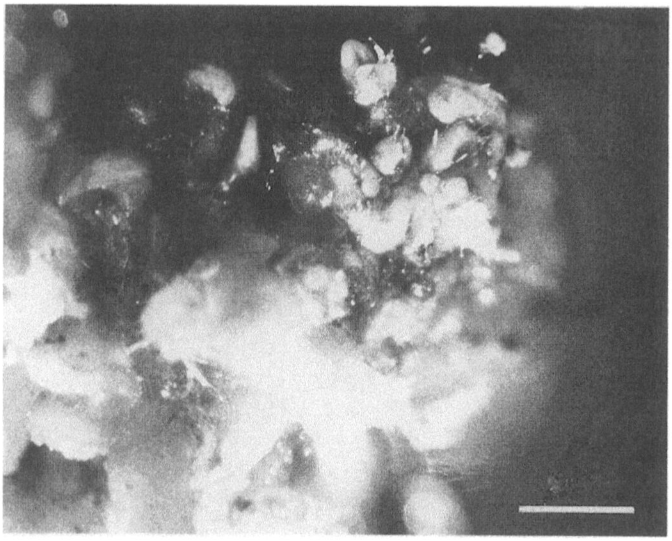

Fig. 4. Organogenic cultures maintained on MSR medium for over seven transfers as continuous shoot cultures. (Bar = 1 mm)

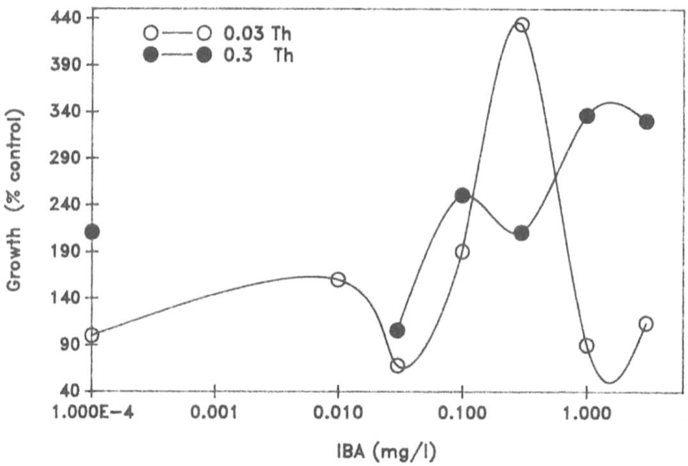

Fig. 5. Stimulation of organogenic Williams 82 callus growth by Th in the presence of varying concentrations of IBA. Each *point* represents the % increase in fresh weight of 20 calli pieces after 21-day incubation period

ly maintained on MSR medium (a sequence of alternate cycles with thidiazuron medium) to increase the number of new meristematic areas, followed by transfer to the thidiazuron medium to obtain vigorous shoots.

Cultures have been maintained for over 3 years using the methods described above, with the embryogenic culture system being less efficient in producing numbers of plants as compared to the organogenic culture system.

4 Somaclonal Variation

With the present ability to attain soybean plant regeneration, it is now possible to regenerate many plants to be used to study tissue culture-derived variation (somaclonal variation, Larkin and Scowcroft 1981) in soybeans. Studies with different legumes have shown that somaclonal variation occurs as a result of tissue culture conditions. Mathews et al. (1986) showed that 14% of the R_1 progeny of the regenerants from short-term mung bean cultures segregated for chlorophyll deficiency and morphological mutations. Variants in alfalfa cell and protoplast cultures have also been studied extensively (Bingham and McCoy 1986). Regenerated plants of pea derived from cultured leaflets showed no variation for four isozyme and nine genetically defined morphological characters (Rubluo et al. 1984), however.

Recently, studies with regenerated soybean progeny of self-fertilized plants have also demonstrated the presence of new variations. Polyploidy (tetraploidy) was found among tissue culture-derived plant progeny, derived from embryogenic cultures and was associated with a sterile phenotype (Ranch and Palmer 1987). Four families of independent origin showed the sterile phenotype. Of the R_2 and R_3 families, 27% showed other morphological variation. Variation sometimes was limited to a few plants in the row, and in other cases, the entire row showed the variant phenotype (Ranch and Palmer 1987).

Graybosch et al. (1987) discovered decreased yield in two of the 19 regenerated families studied. This yield loss might be expected, since mutations which are deleterious probably occur in tissue culture. One family of the 89 studied was also taller than the control, while two showed putative sterility.

We have also found both quantitative and qualitative variation in regenerated soybean plant progeny (Barwale 1987; Barwale and Widholm 1987), and these results are summarized below.

4.1 Qualitative Variation

A total of 263 plants were regenerated from both embryogenic and organogenic cultures of the genotypes Asgrow A3127, Adams, Capitol, PI 36653, PI 404155A and Williams 82, and R_0, R_1, R_2, R_3, and R_4 generations (R_0 being the regenerated plant) have been extensively evaluated. Variant phenotypes seen include chlorophyll deficiency, complete or partial sterility, wrinkled leaves, twin seeds, abnormal leaf morphology, abnormal leaflet number, dwarf growth habit, and multiple shoots. These variants can be divided into three categories:

1. Traits inherited in a Mendelian fashion.
2. Stably inherited traits with aberrant segregation ratios.
3. Possible epigenetic traits.

Complete sterility was seen in an R_2 family of CN210, in which 15.6% of the plants were completely sterile. When R_1 and R_3 families of this R_2 family were also studied, 40% of the plants in the R_1 generation were sterile and 10 of the

Fig. 6. Wrinkled leaf phenotype seen in an R_3 family of A3127. The row segregated for wrinkled leaf type and all plants on a given wrinkled plant show the same phenotype (Bar = 1 cm). (Barwale and Widholm 1987)

16 R_3 families segregated for sterility. These data suggest a stable inheritance of sterility from R_1 to the R_3 generation, and a Chi-square analysis shows that the numbers of sterile and fertile plants obtained fits a 3:1 model for a single gene, recessive trait. Complete sterility was seen in more than seven independent families derived from different genotypes.

Wrinkled leaf type was noted in an R_3 family of A3127 and was associated with small flower size and partial sterility (Fig. 6). In the R_3 generation, 32% of

Table 1. Probability levels and Chi-square values for a single gene model for wrinkled leaf type seen in the R_3 and R_4 generations of A3127

Line	Normal		Wrinkled		X^2 (3:1)	df	p
	Observed	Expected	Observed	Expected			
A3127 R_3[a]	17	18.75	8.00	6.25	0.33	1	0.56[c]
A3127 R_4[a]	37	38.25	14.00	12.75	0.06	1	0.81[c]
A3127 R_4[b]	5	5.33	3.00	2.67	0.16	1	0.90[c]

[a] Within family segregation.
[b] Row segregation determined by counting the number of rows showing segregation for wrinkled leaf phenotype and number of rows with all normal plants using a 2:1 model.
[c] Segregation ratios fit the expected model at 95% probability level

Fig. 7. Sectors of normal leaf phenotype seen on an otherwise wrinkled leaf plant of A3127, R_4 somaclone. *N* normal leaf type; *W* wrinkled leaf type

the plants segregated for wrinkled leaf phenotype and R_2, R_3 and R_4 progeny of this family also segregated for this phenotype. A total of four of the eight R_3 families studied showed segregation. Also, two-thirds of the R_3 families planted from normal R_2 plants showed segregation, which is what would be expected from a heterozygous R_2 source (Table 1). Progeny of the wrinkled leaf plants are all wrinkled. Chi-square values indicate that wrinkled leaf phenotype is controlled by a single, recessive nuclear gene. The wrinkled leaf phenotype was noted in two independent families, one derived from A3127 and the other from PI 36653.

An interesting phenomenon was noted with four plants in rows segregating for wrinkled leaf morphology. A single branch with normal leaf phenotype was found on two otherwise wrinkled-leaved plants. Also two other plants had a single branch with wrinkled leaf morphology on an otherwise normal plant (Fig. 7). The presence of a chimera could be explained in R_0 plants, due to the possible multicellular origin of the regenerants with more than one cell type. However, the sectors seen here are in the R_4 generation (four self-pollination cycles away from R_0) and could possibly be due to the movement of transposable elements. No movement of the putative soybean transposable elements (Tgm1) has ever been demonstrated (Vodkin et al. 1983), but activity of transposable elements could result in morphological instability, which could produce chimeric plants.

Complete sterility and wrinkled leaf morphology have been evaluated in three or more generations and both traits are inherited as single, recessive nuclear gene mutations.

Partial sterile phenotypes were noted in several R_3 families. In one family, 14% of the R_3 plants were partially sterile, while in the R_4 generation two of the eight plants derived from the normal plants in the R_3 family show partial sterility. In other families the segregation ratios are much lower than one-fourth, indicating that the trait is not simply inherited. The conclusions, however, are limited by a small sample size.

Multiple branching, i.e., more than one shoot at the cotyledonary node, was noted in both the R_2 and R_3 generations in one family. In the R_2 generation, 20 of 24 plants had two or three shoots arising from the node. Progeny of these also show two shoots.

One of the three plants in the R_3 generation of PI 36653 was dwarf. In the R_4 generation, plants produced by the R_3 dwarf plants retained the dwarf growth habit with plant heights of 20.3, 19.1, 20.3 and 14 cm, while the two normal plants from the same family produced normal progeny. The control plants grew to an average height of 71.1 cm. The small sample size does not allow us to draw any conclusions regarding the genetic nature of this trait; however, the trait is stably inherited in more than one generation.

The partial sterility, multiple branching, and dwarf growth habit are due to apparent genetic changes, as discussed above, since the traits are heritable. However, these variants do not fit a simple genetic model, but more studies are needed. The persistence of the phenotypes over several generations indicates that these changes are not environmental.

While nonlethal chlorophyll deficiency was seen in a few R_2 and R_3 plants, the R_4 progeny from these chlorophyll-deficient plants and other plants of the same family did not show the chlorophyll deficiency. Thus, due to the lack of expression of the chlorophyll-deficient phenotype in the R_4 generation, the trait is apparently not due to genetic change.

Other phenotypes such as twin seeds, abnormal leaf morphology, and abnormal leaflet number have been noted as random events. However, when more than one generation was examined, the variant phenotype was not seen, indicating a random or physiological nature of this variation rather than genetic change.

Plants derived from both embryogenic and organogenic cultures were evaluated in this study, but the number of plants derived from embryogenic cultures was very small as compared to organogenic culture-derived plants. While none of the inherited variation was derived from the embryogenic cultures, three chimeral albinos were noted in R_0 plants only of embryogenic culture origin (Fig. 8). These variants could not be grown to maturity as the plants died upon transfer to the greenhouse. With the limited number of plants studied from embryogenic cultures, it is not possible to conclude if more variation is seen in plants derived from one system or the other. However, we have shown that organogenic cultures do produce somaclonal variation, and Ranch and Palmer (1987) have shown that embryogenic cultures can do likewise.

Frequency of qualitative variation was calculated in progeny up to the R_4 generation per R_0 plant. Only the established mutations of wrinkled leaf type, sterility, partial sterility, dwarf growth habit, and multiple branching were included in calculating the frequency. In a given line, the total number of mutant phenotypes seen in all R_0 progeny are divided by the total number of R_0 families. Identical mutants in two or more R_1 families of the same genotype are counted as a single mutational event. For cv. A3127 (76 initial R_0 plants), the frequency of mutation

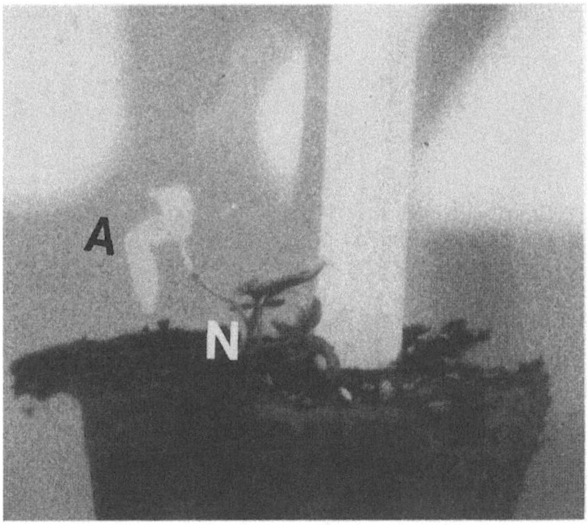

Fig. 8. Sectoral albino seen in the R_0 plant of Williams 82, derived from the embryogenic cultures only. Three such events were seen. *A* albino sector; *N* normal sectors. (Barwale et al. 1986b)

was 0.05 (5%) per R_0 plant while for cv. CN210 (4 R_0 plants), the frequency of mutation per R_0 plant was 1.00 (100%). The low frequency for A3127 is somewhat misleading, since one phenotype which occurred more than once was counted as a single event and a larger number of R_0 plant progeny were studied.

The above results show the presence of morphological variation in soybean plants derived from tissue culture. None of the variations seen is of economic use, as one might expect, since mutations are in large part deleterious. Directed selection for desired traits can now be applied since spontaneous variation is occurring in tissue culture-derived material of soybeans.

Screening for Brown Stem Rot (BSR) Tolerance Among Soybean Somaclones. BSR is a vascular disease of soybeans caused by the fungus *Phialophora gregata* W. Gams (Allington and Chamberlain 1948). Yield losses in heavily infested fields could be as high as 40%. Studies have indicated the involvement of a toxin in the induction of the BSR disease symptoms (Gray and Chamberlain 1975).

Recently, Gray et al. (1986) demonstrated that an apparently host-specific toxin found in *P. gregata* culture filtrate could be used to differentiate between callus from BSR-sensitive and resistant genotypes. The results indicate that insensitivity to the toxin may be one of the mechanisms of BSR resistance in PI 437833, a resistant soybean genotype.

In this study, progeny of regenerated plants derived from resistant and susceptible genotypes were evaluated for sensitivity to culture filtrate, using friable callus initiated from seedlings. These seedlings were then grown for seed. Friable calli from susceptible soybean genotypes exposed to 20% culture filtrate levels were inhibited severely, while calli from resistant genotypes were not inhibited under similar conditions (Fig. 9).

Calli derived from 579 R_1 and R_2 seedlings of the susceptible (Corsoy) and resistant (PI 437833 and BSR201) genotypes were tested on culture filtrate (20%–25%) containing medium. These concentrations permitted the resistant calli to grow, and inhibited the susceptible calli (Fig. 9). The calli (nine 1-mg

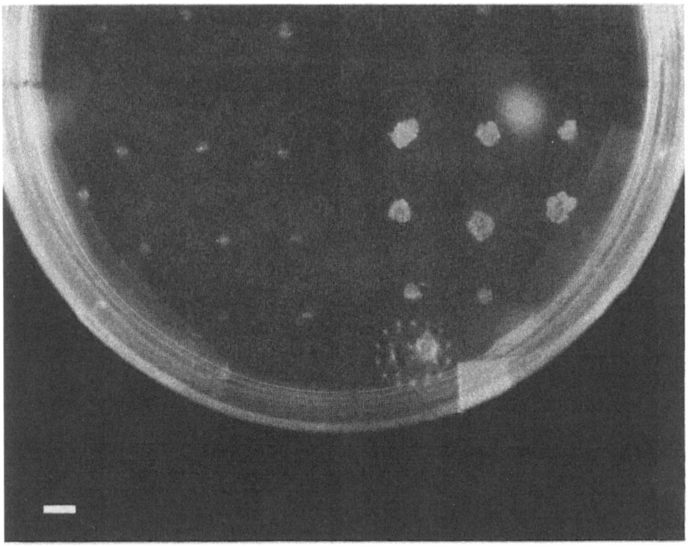

Fig. 9. Effect of *Philaphora gregata* culture filtrate on growth and browning of calli derived from Corsoy (susceptible genotype) and PI 437833 (resistant genotype). The observations are taken 5 days after the calli are placed on the 20% culture filtrate medium. (Bar = 2 mm). (Barwale and Widholm 1987)

pieces) were evaluated for percent browning and increase in fresh weight after 5 days on the culture filtrate-containing medium.

Corsoy (susceptible genotype) callus growth was 102 mg on culture filtrate-free medium and 69 mg on 20% culture filtrate medium from the initial weight of 9 mg with four or more pieces turning brown. PI 437833 (resistant genotype) grows to 71.4 mg on culture filtrate-free medium and 247 mg on the 20% culture filtrate with no browning. A stimulatory effect is seen in growth with PI 437833 as compared to the inhibitory effect on Corsoy due to the culture filtrate. Calli of BSR201 (resistant genotype) show complete inhibition of growth and total browning at the 20% culture filtrate concentration. While BSR201 is a resistant genotype under field conditions, in tissue culture the response is more like that of a susceptible genotype. When 118 R_1 and R_2 somaclones with 539 plants derived from eight BSR201 R_0 plants were tested, family U, a R_2 family, showed a greater tolerance, with callus growth to more than 500 mg fresh weight and no browning. The control in this experiment shows complete inhibition of growth and total browning by the culture filtrate. Families B4, AX9, and BD5 show susceptibility similar to or greater than the control. The majority of the somaclones tested show sensitivity similar to that of the control, with the few exceptions mentioned above.

The screening done here was based on the system developed by Gray et al. (1986), which had not been used before for screening callus from individual plants for BSR sensitivity. Thus we need to confirm the results by testing the whole plants for BSR sensitivity. Stem injections on the seedlings (from which the calli were derived) with fungal mycelial fragments and root dip inoculations are presently being carried out to determine the in vivo responses. Screening for resistance to BSR is also being carried out by growing organogenic cultures on inhibitory culture filtrate medium to select for resistant calli. Plants regenerated from these calli would hopefully carry the resistance also.

The preliminary results presented do indicate that variation for level of tolerance to the *P. gregata* culture filtrate is seen in regenerated plant progeny. If this variation is stable and is expressed at the whole plant level as BSR resistance, it could be exploited to obtain plants with greater tolerance to BSR.

4.2 Quantitative Somaclonal Variation

Somaclonal variation in soybeans is not restricted to qualitative variation only. Initial observations with R_3 generation plants show maturity variation with later and earlier maturity dates in evidence. When these selected families were grown in the R_4 generation, two families which matured earlier in the R_3 generation also showed this trait in the R_4 generation, and one family remained green for several more days than the control.

In the summer of 1987, R_4 plant progeny of A3127 regenerants, which showed no apparent variation in the R_3 generation, were grown on a large scale in the field (over 85000 plants from ca. 1200 families). Earlier and later maturing families were seen, but no other obvious changes were evident. Further evaluation of the selected plants may produce lines with earlier or later maturity. This study also demonstrates that, although the plant genome may be reorganized to some extent during tissue culture, the genome does stabilize, as was seen with lines which have no apparent variation in the R_4 generation.

Oil and protein contents were also determined in seed of ca. 700 R_3 families of A3127, Adams, Capitol, PI 36653, and PI 404155A using wide-line nuclear magnetic resonance (NMR) and near-infrared reflectance spectroscopy, respectively (Alexander et al. 1967; Hymowitz et al. 1974). Most somaclones show oil and protein contents similar to that of the control. However, a few exceptions were seen where oil contents in somaclones were increased by ca. 25%. Changes in protein contents were also noted, where four families showed levels at least 5% greater than the respective controls and three families showed protein levels 3% to 5% lower than the control.

The oil and protein analyses were carried out on only one generation, so need to be repeated on successive generations grown under different conditions to determine if indeed the changes noted are valid, stable and heritable. If these differences are verified in the somaclone families, this may provide a source of new variation for desirable changes in oil and protein composition.

4.3 Mechanism of Somaclonal Variation

In order to attempt to determine the mechanisms causing somaclonal variation in soybeans, two studies were carried out: (a) cytological evaluations for changes in chromosome numbers, and (b) Ouchterlony double diffusion test for reversion of lectin null (le^-) lines to lectin positive (Le). In le^- lines, the absence of this protein is due to an insertion of a transposable element like DNA segment in the coding region of the lectin gene (Vodkin et al. 1983), designated Tgml (transposable element, *Glycine max*). If tissue culture conditions cause a genomic shock, Tgml may transpose, which would result in the Le phenotype (lectin production).

Chromosome numbers were determined on root-tip squashes of variant and normal somaclones, as described by Palmer and Heer (1973), and Ouchterlony double diffusion tests were performed using chips of dried R_1 seeds of lectin null plants (Columbia and Sooty) and Le plants (A3127) and purified antiserum to lectin (Orf et al. 1978).

The chromosome numbers of control preparations are normal ($2n = 40$). In the case of the sterile, partial sterile, and wrinkled leaf mutants which have lower seed set, mixoploidy is seen. The chromosome numbers ranged from 10 to 60 in root tips of these plants. This mixoploidy might help explain the partial or complete sterility seen with these mutants.

Nuclear fragmentation followed by mitosis, occurring at an early stage of culture, can give rise to a wide variation in ploidy levels in cultured cells (D'Amato et al. 1980). However, this variation in ploidy level should be seen in the regenerated plants directly, unlike the mixaploidy seen here in the R_3 and R_4 generations (three or four self-pollinations away from the regenerated plants). Plants with chromosome mosaicism in the R_0 generation could survive, since some of the pollen produced might have the normal chromosome complement. In the later generations, however, stabilization of the plant genome would be expected, since the seed is formed from a single cell. However, the ploidy level in barley has been shown to be regulated by a single gene (Takahashi et al. 1964). If mutation(s) occur in this gene, ploidy changes including mixaploidy can be transmitted over many generations. The preliminary indications from the present information indicate that such mechanisms may also be present in soybeans and a possible mutation in the gene regulating ploidy level results in mixaploidy.

Measurements of lectin levels using the Ouchterlony double diffusion test showed that none of the 168 R_1 seeds of Sooty and Columbia reverted to the Le phenotype. All 78 seeds of A3127 (Le, tissue cultured positive controls) show the white precipition band indicating presence of lectin while none of the Sooty or Columbia regenerants do. In order to have lectin production from a null line, the excision of the transposable element insert has to be precise, without changing the reading frame of the gene. The host DNA duplications created at the site of insertion or the transposable element inserted repeats which can remain behind after the excision may change the reading frame. If this does occur, even in the absence of the insert, the gene product, lectin, will not be produced. Secondary mutations created by the insertion element may also prevent expression of the gene even in the absence of the insertion element.

The results obtained with the Ouchterlony technique do not rule out the possibility that changes in the transposable element or in its activity may have occurred, which were not detectable as lectin production. Likewise, larger numbers of samples need to be analyzed. It is also possible that Tgml may be a defective element or is activated at very low frequencies (Vodkin et al. 1983), so that lectin production will never be recovered in these nulls.

Based on the chromosome numbers, we might conclude that chromosomal abnormalities contribute toward some of the variation seen here.

5 Concluding Remarks

In the last few years, many plant regeneration protocols for soybeans have become available, allowing further studies with such systems. Plant regeneration is possible via either embryogenesis or organogenesis, with the most efficient systems using immature embryos as the starting material. Other tissues, e.g., leaf, hypocotyls, have also been used in some instances, but the rate of success in plant regeneration in these systems appears to be much lower. As seen with our culture systems, the organogenic culture system is more productive in regenerating plants in a shorter time as compared to the embryogenic culture system.

Using one or more of the culture systems, selection for desirable traits could be carried out at the callus stage. Cell lines resistant to amino acid analogs could result in improved plants with better chemical composition (Widholm 1983). Using a tobacco suspension culture, Gonzales et al. (1984) obtained a ethionine-resistant line which overproduced methionine by more than a 100-fold. However, when a soybean suspension culture with resistance to ethionine was selected, the methionine level was unchanged and the mechanism of resistance was determined to be the lack of ethionine uptake (data not shown).

Screening procedures using in vitro techniques would be beneficial, since these would be carried out under controlled conditions. Plants with putative resistance to brown stem rot were obtained when organogenic calli were grown on inhibitory levels of culture filtrate. Calli surviving such treatment were maintained on the culture filtrate medium until plant regeneration. These regenerants are being tested presently to determine if the plants and their progeny express the resistance.

Screening for chemical composition of seeds of regenerated plant progeny was also carried out with various somaclones showing some variation. Like some other results presented here, these changes need to be studied further to verify the stability. However, without directed selection, apparent variation is still noted in plant progeny of regenerated plants, as has been reported with many other species.

Chromosomal abnormalities may also contribute to somaclonal variation. The presence of mixaploidy is a unique phenomenon not reported before in soybeans. No variation of agronomic value has been noted for soybeans as yet. This is what one would expect, since changes in the plant genome during the culture process are probably random, usually deleterious, events. A longer period in tissue culture before plant regeneration may saturate the plant genome with many mutations, resulting in poor growth habit. However, the study here is a preliminary evaluation and specific directed selection may produce novel plants or phenotypes.

Plant tissue culture may provide new items like the wrinkled leaf type and mixaploidy, which have not been reported before in soybean. Also the activation of transposable elements would be very useful in producing new variation. The chimeric plants with both normal and the wrinkled leaf phenotype may be an example of how transposable elements may be activated due to the tissue culture process in the regenerated plant progeny. Tissue cultures can provide variation in a very homozygous genetic source of soybeans which may otherwise never be found. Such variation may enlarge the existing diversity in the germplasm and be of use to plant breeders in improvement of existing soybean.

6 Protocol for Soybean Plant Regeneration Through Embryogenesis or Organogenesis

1. Immature embryos ranging in size from 1 to 10 mm were initially used to start the cultures. 4 to 6 mm was determined to be the optimum size for highest efficiency of both embryogenic and organogenic cultures.
2. Immature embryos were placed on 4MSII medium (OR medium lacking proline and thiamine, Barwale et al. 1986b) to initiate organogenic cultures and on ENT (EB medium without the nicotinic acid, Barwale et al. 1986b) for embryogenesis. Cultures were incubated in the dark at 25 °C for up to 4 weeks.
3. Organogenic cultures were divided in smaller pieces and transferred to MSR (Barwale et al. 1986b) or Th medium as described in this chapter. Embryogenic cultures were transferred to the ENT medium or to MSR when the embryos were in later developmental stages. Cultures were placed under a 16-h photoperiod with temperature being 27 °C during the day and 18 °C at night. Cultures were routinely transferred to fresh medium every 3 to 4 weeks. The original tissue was subdivided into smaller pieces. Cultures could be maintained in this manner for more than 2 years.
4. To regenerate plants, calli with small shoots or embryos were transferred to MS medium without any growth regulator. Shoots were allowed to grow in this medium for over 2 weeks, at which time they were transferred to the greenhouse. Initial stages of growth in the greenhouse were in hydroponics, after which the plants were planted in soil. R_1 seeds were harvested from these plants.

Acknowledgment. The unpublished research presented here was supported by funds from the Illinois Agricultural Experimentation Station and from Agrigenetics Inc./Lubrisol.

References

Alexander DE, Silvera L, Collins FI, Rodgers RC (1967) Analysis of oil content of maize by wide-line NMR. J Am Oil Chem Soc 44:555–558

Allington WB, Chamberlain DW (1948) Brown stem rot of soybean. Phytopathol 38:793–802

Baley JM, King J, Gamborg OL (1972a) The effect of the source of inorganic nitrogen on growth and enzymes of nitrogen assimilation in soybean and wheat cells in suspension cultures. Planta 105:15–24

Baley JM, King J, Gamborg OL (1972b) The ability of amino compounds and conditioned medium to alleviate the reduced nitrogen requirement of soybean cells grown in suspension cultures. Planta 105:25–32

Barwale UB (1987) Somaclonal variation in soybeans. Ph D Thesis, Univ Illinois Urbana-Champaign

Barwale UB, Widholm JM (1987) Somaclonal variation in plants regenerated from cultures of soybean. Plant Cell Rep 6:365–368

Barwale UB, Meyer MM, Widholm JM (1986a) Screening of *Glycine max* (L.) Merr. and *G. soja* Sieb. and Zucc. genotypes for multiple shoot formation at the cotyledonary node. Theor Appl Genet 72:423–428

Barwale UB, Kerns HA, Widholm JM (1986b) Plant regeneration from callus cultures of several soybean genotypes via embryogenesis and organogenesis. Planta 167:473–480

Beversdorf WD, Bingham ET (1977) Degrees of differentiation obtained in tissue cultures of *Glycine* species. Crop Sci 17:307–311
Bingham ET, McCoy TJ (1986) Somaclonal variation in alfalfa. Plant Breed Rev 4:123–152
Capelle SC, Mok DWS, Kirchner SC, Mok MC (1983) Effects of thidiazuron on cytokinin autonomy and the metabolism of N^6-(delta^2-isopentyl) [8-^{14}C] adenosine in callus tissues of *Phaseolus lunatus* L. Plant Physiol 73:796–802
Cheng TY, Saka H, Voqui-Dinh TH (1980) Plant regeneration from soybean cotyledonary node segments in culture. Plant Sci Lett 19:91–99
Christianson ML, Warnick DA, Carlson PS (1983) A morphogenetically competent soybean suspension culture. Science 222:632–634
Cowan JC (1973) Processing and products. In: Caldwell BE (ed) Soybeans: improvement, production, and uses. Am Soc Agron Madison, WI, pp 619–664
D'Amato F, Bennici A, Cionini PG, Baroncelli S, Lupi MC (1980) Nuclear fragmentation followed by mitosis as mechanisms for wide chromosome number variation in tissue culture: its implications for plantlet regeneration. In: Sala F, Parisi R, Cella R, Ciferri D (eds) Plant cell cultures: results and perspective. Elsevier Amsterdam, pp 67–72
Gamborg OL, Shyluk JP (1970) The culture of plant cells with ammonium salts as the sole nitrogen source. Plant Physiol 45:598–600
Gamborg OL, Miller RA, Ojima K (1968) Nutrient requirements of suspension cultures of soybean root cells. Exp Cell Res 50:151–158
Gamborg OL, Davies BP, Stahlhut RW (1983) Cell division and differentiation in protoplasts from cell cultures of *Glycine* species and leaf tissue of soybean. Plant Cell Rep 2:213–215
Ghazi TD, Cheema HV, Nabors MW (1986) Somatic embryogenesis and plant regeneration from embryogenic callus of soybean, *Glycine max* L. Plant Cell Rep 5:452–456
Gonzales RA, Das PK, Widholm JM (1984) Characterization of cultured tobacco cell lines resistant to ethionine, a methionine analog. Plant Physiol 74:640–644
Grant J (1984) Plant regeneration from cotyledonary tissue of *Glycine canescens*, a perennial wild relative of soybean. Plant Cell Tissue Organ Cult 3:169–173
Gray LE, Chamberlain DW (1975) Evidence for toxin production by a strain of *Cephalosporium gregata*. Phytopathol 65:89–90
Gray LE, Guan YQ, Widholm JM (1986) Reaction of soybean callus to culture filtrates of *Phielophora gregata*. Plant Sci 47:45–55
Graybosch RA, Edge ME, Delannay X (1987) Somaclonal variation in soybean plants regenerated from the cotyledonary node tissue culture system. Crop Sci 27:803–806
Guangchu Y, Zhiyin Z, Zhen X, Li C, Xuezhan L, Fengyun B (1984) Studies on the anther culture and its androgenesis in vitro of soybean (*Glycine max* L. Merr.). Int Symp Genetic Manipulation in Crops. Oct 22–26, Beijing, China
Hammatt N, Kim H-I, Davey MR, Nelson RS, Cocking EC (1987) Plant regeneration from cotyledon protoplasts of *Glycine canescens* and *G. clandestina*. Plant Sci 48:129–135
Hammond EG, Fehr WR, Synder HE (1972) Improving soybean quality by plant breeding. IBID 49:33–35
Hildebrandt DF, Phillips GC, Collins GB (1986) Soybean (*Glycine* max (L.) Merr.). In: Bajaj YPS (ed) Biotechnology in agriculture and forestry 2. Crops I. Springer, Berlin Heidelberg New York, pp 283–308
Hymowitz T, Collins FI, Sedgwick VE, Clark RW (1970) The oil and fatty acid content of some accessions of *Glycine wightii*. Trop Agric (Trinidad) 48:265
Hymowitz T, Dudley JW, Collins FI, Brown CM (1974) Estimation of protein and oil concentration in corn, soybean and oat seed by near-infrared light reflectance. Crop Sci 14:713–715
Hymowitz T, Chalmers NL, Costanza SH, Saam MM (1986) Plant regeneration from leaf explants of *Glycine clandestina* Wendl. Plant Cell Rep 3:192–194
Ivers DR, Palmer RG, Fehr WR (1974) Anther culture in soybeans. Crop Sci 14:891–893
Kameya T, Widholm JM (1981) Plant regeneration from hypocotyl sections of *Glycine* species. Plant Sci Lett 21:289–294
Kao KN, Keller WA, Miller RA (1970) Cell division in newly formed cells from protoplasts of soybean. Exp Cell Res 62:338–340
Kartha KK, Gamborg OL, Constabel F (1974) Regeneration of pea (*Pisum sativum* L.) plants from shoot apical meristems. Z Pflanzenphysiol 72:172–176

Kerns HR, Barwale UB, Meyer MM Jr, Widholm JM (1986) Correlation of cotyledonary node shoot proliferation and somatic embryoid development in suspension cultures of soybean (*Glycine max* L. Merr.). Plant Cell Rep 5:140–143
Kimball SL, Bingham ET (1973) Adventitious bud development of soybean hypocotyl sections in culture. Crop Sci 13:758–760
Larkin PJ, Scowcroft WR (1981) Somaclonal variation – A novel source of variability from cell cultures for plant improvement. Theor Appl Genet 60:197–214
Lazzeri PA, Hildebrand DF, Collins GB (1985) A procedure for plant regeneration from immature cotyledon tissue of soybean. Plant Mol Biol Rep 3:160–167
Li BJ, Langridge WHR, Szalay AA (1985) Somatic embryogenesis and plantlet regeneration in soybean *Glycine max*. Plant Cell Rep 4:344–347
Lippmann B, Lippmann G (1984) Induction of somatic embryos in cotyledonary tissue of soybean, *Glycine max* L. Merr. Plant Cell Rep 3:215–218
Lu DY, Cooper-Bland S, Pental D, Cocking EC, Davey MR (1983) Isolation and sustained division of protoplasts from cotyledons of seedlings and immature seeds of *Glycine max* L. Z Pflanzenphysiol 111:389–394
Mathews VHM, Rao PS, Bhatia CR (1986) Somaclonal variation in cotyledonary plants of mung bean. Z Pflanzenzucht 96:169–173
Mok MC, Mok DWS, Armstrong DJ, Shudo K, Isogai Y, Okamato T (1982) Cytokinin activity of N-phenyl-N′-1,2,3-thiadiazol-5-ylurea (Thidiazuron). Phytochem 21:1509–1511
Mott RL, Cordts JM, Larson AM (1984) Nitrogen and growth regulator effects on shoot and root growth of soybean in vitro. Plant Tissue Cult Meet, Knoxville, Tenn
Murashige T, Skoog F (1962) A revised medium for rapid growth and bioassays with tobacco tissue cultures. Physiol Plant 15:473–497
Newell CA, Luu HT (1985) Protoplast culture and plant regeneration in *Glycine canescens* FJ Herm. Plant Cell Tissue Org Cult 4:145–149
Ohira K, Yamaya T, Ojima K (1975) Studies on the greening of cultured soybean and ruta cells. 1. Pigmentation as influenced by the composition of the medium. Tohoku J Agric Res 25:136–178
Ojima K, Ohira K (1977) Solubilization and utilization of Fe in the medium by suspension cultured cells of rice and soybean. J Sci Soil Manure 48:96–100
Orf JH, Hymowitz T, Pull SP, Pueppke SG (1978) Inheritance of a soybean seed lectin. Crop Sci 18:899–900
Oswald TH, Smith AE, Phillips DV (1977) Callus and plantlet regeneration from cell culture of ladino clover and soybean. Physiol Plant 39:129–134
Palmer RG, Heer HE (1973) A root tip squash technique for soybean chromosomes. Crop Sci 13:389–391
Phillips GC, Collins GB (1981) Induction and development of somatic embryos from suspension cultures of soybean. Plant Cell Tissue Organ Cult 1:123–129
Ranch JP, Palmer RG (1987) A ploidy variant regenerated from embryogenic tissue cultures of soybean. Soybean Genet Newl 14:116–163
Ranch JP, Oglesby L, Zielinski AC (1985) Plant regeneration from embryo-derived tissue cultures of soybeans. In Vitro 21:653–658
Ranch JP, Oglesby L, Zielinski AC (1986) Plant regeneration from tissue cultures of soybean by somatic embryogenesis. In: Vasil I (ed) Plant regeneration and genetic variability. Cell Culture and Somatic Cell Genetics of Plants, vol 3. Academic Press, Orlando, pp 97–110
Rubluo A, Kartha KK, Mroginski LA, Dyck J (1984) Plant regeneration from pea leaflets cultured in vitro and genetic stability of regenerants. J Plant Physiol 117:119–130
Saka H, Voqui-Dinh TH, Cheng TY (1980) Stimulation of multiple shoot formation on soybean stem nodes in culture. Plant Sci Lett 19:193–201
Schwenk FW (1981) Callus formation from mechanically isolated mesophyll cells of soybean and sweet potato. Plant Sci Lett 23:147–151
Schwenk FW, Pearson CA, Roth MR (1981) Soybean mesophyll protoplasts. Plant Sci Lett 23:153–155
Smith KJ, Huyser W (1987) World distribution and significance of soybean. In: Wilcox JR (ed) Soybeans: improvement, production, and uses, 2nd edn. American Soc Agron, Madison, WI, USA, pp 1–21

Takahashi R, Tsuchiya T, Moriya I (1964) Heritable mixaploidy in barley. III. On a dwarf mutant from a cultivar, Taisho-mugi. Nogaku Kenkyu 50:123–129
Tricoli DM, Hein MB, Carnes MG (1986) Culture of soybean mesophyll protoplasts in alginate beads. Plant Cell Rep 5:334–337
Vodkin LO, Rhodes PR, Goldberg RB (1983) A lectin gene insertion has the structural features of a transposable element. Cell 34:1023–1031
Widholm JM (1983) Isolation and characterization of mutant plant cell cultures. In: Sen SK, Giles KL (eds) Plant cell culture in crop improvement. Plenum Press, New York, pp 71–86
Widholm JM, Rick S (1983) Shoot regeneration from *Glycine canescens* tissue cultures. Plant Cell Rep 2:19–20
Wright MS, Koehler SM, Hinchee MA, Carnes MG (1986) Plant regeneration by organogenesis in *Glycine max*. Plant Cell Rep 5:150–154
Wright MS, Ward DV, Hinchee MA, Carnes MG, Kaufman RJ (1987) Regeneration of soybean (*Glycine max* L. Merr.) from cultured primary leaf tissue. Plant Cell Rep 6:83–89
Xu ZH, Davey MR, Cocking EC (1982) Callus formation from root protoplasts of *Glycine max* (soybean). Plant Sci Lett 24:111–115
Yuyu J, Dupu L, Ximing L, Guilan Z (1984) Studies on induction of pollen plants in soybean (*Glycine max* [L.] Merr.). Int Symp Genetic Manipulation in Crops, Oct 22–26, Beijing China, pp 22–26
Zieg RG, Outka DE (1980) The isolation, culture and callus formation of soybean pod protoplasts. Plant Sci Lett 18:105–114

II.4 Soybean: Wide Hybridization Through Embryo Culture

J. E. GRANT[1]

1 Introduction

Soybean, *Glycine max* (L.) Merr. is one of the world's most economically important crops. Its seed contains about 40% protein and 20% oil, both of which are used for human consumption, livestock feed, and industrial purposes (Fehr 1980). Soybean is the world's most important grain legume, and is grown in many countries, with the USA, China and Brazil accounting for over 90% of the world production (Weiss 1983).

The USA is the largest exporter of soybeans, yet the current varieties in use there rely on a narrow genetic resource base. Investigation into the pedigrees of these varieties reveals a limited number of plant introductions. Additionally, many of the predominant introductions originated from the same geographic area (Delannay et al. 1983). In the past, a narrow genetic resource base has led to major economic loss in crops such as potato, tobacco, sugar, and maize (Harlan 1976). Such examples serve to highlight the need for sound knowledge of the prospects and limitations of the methods used for broadening the genetic base.

Techniques such as mutagenesis (for examples, see Green and Marshall 1984; Carroll et al. 1985) and somaclonal variation (Larkin and Scowcroft 1981; Evans et al. 1984) have been and will continue to be important methods in increasing the genetic variability in an existing crop. Somatic hybridization and DNA manipulation are technologies that hold promise for the future. Interspecific hybridization with wild species is currently a method that has the ability to open a potentially vast resource of variability for exploitation by plant breeders. For the genus *Glycine* interspecific hybridization is a feasible method.

2 The Genus *Glycine*

The genus *Glycine* Willd. has been divided into two subgenera (Table 1). The subgenus *Soja* (Moench) F.J. Herm comprises annuals, including soybean and its immediate progenitor *G. soja*. The subgenus *Glycine* Willd. contains 15 formally described perennial species (Table 1). In addition, there are at least two newly discovered taxa awaiting formal description, and several morphologically distinct en-

[1] APPM, Forestry Research Unit, P.O. Box 63, Ridgley, TAS 7321, Australia

Biotechnology in Agriculture and Forestry, Vol. 10
Legumes and Oilseed Crops I (ed. by Y.P.S. Bajaj)

Table 1. The Genus *Glycine*

Species	Chromosome no.	Distribution
A. Subgenus *Glycine* (perennials)		
G. clandestina Wendl.	40	NE, SE, S and SW Australia
G. canescens F.J. Herm.	40	Dryland Australia
G. latifolia (Benth.) Newell & Hymowitz	40	NE Australia
G. latrobeana (Meissn.) Benth.	40	SE Australia
G. falcata Benth.	40	Queensland, Northern Territory
G. cyrtoloba Tind.	40	NE Australia
G. argyrea Tind.	40	S Queensland coast
G. microphylla (Benth.) Tind.	40	NE, SE and S Australia, Norfolk Island
G. curvata Tind.	40	N Queensland
G. arenaria Tind.	40	Northern Territory NE West Australia
G. tabacina (Labill.) Benth.	40, 80	NE and SE Australia, Taiwan, S China, S Pacific Islands
G. tomentella Hayata	38, 40, 78, 80	N Australia, Papua New Guinea, Taiwan, S China, Philippines
G. lactovirens Tind.&Craven	40	Kimberley WA
G. albicans Tind.&Craven	40	Kimberley WA
G. hirticaulis Tind.&Craven	80	Northern Australia
B. Subgenus *Soja* (annuals)		
G. soja Sieb.&Zucc.	40	China, Taiwan, Japan, Korea, USSR
G. max (L.) Merr.	40	Cultigen

tities within currently designated species that will be described as separate species. All these wild species are native to Australia and only three occur naturally outside Australia. *Glycine microphylla* has been found on Norfolk Island (Tindale 1986) and the diploid ($2n = 40$) form of *G. tomentella* occurs in Papua New Guinea. The tetraploid forms of *G. tabacina* and *G. tomentella* ($2n = 80$ and $2n = 78$, 80, respectively) occur in the Western Pacific Basin and their distribution extends north to China.

Relatively recently, significant steps have been taken to collect, study, and exploit the genetic resources of the more distant wild relatives of soybean. As might be expected in wild relatives, the perennial *Glycine* species encompass a broad range of morphological, ecological, and cytological variability (Grant 1986b). Screening studies of agronomically important traits in the subgenus *Glycine* are limited, and speculation as to their importance for soybean breeding is based largely on characters related to the wide ecological amplitude of the wild species, e.g., tolerance to drought and cold, daylength insensitivity, and disease resistance.

Among the diseases that could prove to be highly destructive to the soybean industry, especially in the USA, is soybean leaf rust caused by the pathogen *Phakopsora pachyrhizi* Syd. Presently it is the most serious disease of soybean in Asia. It frequently causes yield losses of 20%–30% and occasionally causes complete crop failures (Weiss 1983; Burdon and Speer 1984). Recently, the pathogen was reported, for the first time, in Central America, and thus the disease is now in a position to threaten soybean production in the USA (Burdon and Speer

1984). Essentially there is a lack of resistance to this pathogen in *Glycine max*, but Burdon and Speer (1984) have identified a large number of genes in the wild perennial species, some of which show resistance to all the races of the pathogen currently identified.

The wild species have been surveyed for resistance to other diseases that affect soybean, e.g. cyst nematode (W.J. Kenworthy, pers. commun.; Riggs 1986) and brown spot (*Septoria*) tolerance (Hymowitz 1986). So far these surveys have only been preliminary but reflect the increasing interest in the agronomic characters of the wild species that may have important implications for soybean breeding.

3 Hybridization Between the Subgenera

A key step in the exploitation of the wild species is to determine the nature of the biological relationship between the cultigen and its relatives. Reproductive isolation may be achieved by a great variety of mechanisms acting at different stages in the life cycle.

Early hybridization attempts between the two subgenera showed that up to 15% of pollinations resulted in pod initiation (Ladizinsky et al. 1979; Hood and Allen 1980; Sedova 1982; Broué et al. 1982) but no mature seed formed. The percent of pod initiation was both species- and genotype-dependent. Hood and Allen (1980) confirmed, by histological analysis, that fertilization did take place. It therefore seemed likely that one of the first visible isolating mechanisms between the subgenus *Soja* and the subgenus *Glycine* was endosperm failure leading to starvation of the embryo in situ. Thus embryo rescue has been an important technique in obtaining interspecific hybrids. In crosses between soybean and the wild species, embryo rescue is essential (Broué et al. 1982; Newell and Hymowitz 1982; Brown et al. 1985; Singh and Hymowitz 1985; Grant et al. 1986; Grant 1986a, b; Newell et al. 1987; Singh et al. 1987). Embryo rescue is optional, depending on the genotype and the cross, in crosses within the wild species (Grant et al. 1984a; Doyle et al. 1986). The successes have been achieved using chemically defined media for the growth of a single embryo through to an early seedling stage before being transferred to soil. The media defined in these reports are not infallible recipes for the growth of embryos. The transfer from tissue culture to soil is also a critical step where losses do occur.

3.2 Tissue Culture of Hybrids Between the Subgenera

The path from embryo to plant when the wild perennial relatives of soybean are crossed with the soybean is by no means assured. Progress in bridging the gap between the subgenera by producing hybrid plants has been slow and limited as far as the number of hybrids and the species involved is concerned (Table 2). Additionally, Newell et al. (1987) obtained hybrids between *G. tabacina* and *G. max*, *G. clandestina* and *G. soja*, and *G. max* and *G. falcata*, but these could not be maintained to flowering, indicating the existence of hybrid barriers. Although pod, hence ovule, production is highly genotype-dependent, results would indicate that only 1%–4% of embryos result in plants (Newell 1986). Stebbins (1958) considered that the time of degeneration of the embryo most often coincides with some critical or maximal period of differentiation of its tissue.

Development of media suitable for rescuing embryos of hybrids is difficult because hybrids do not necessarily respond in the same way as either parent, and as noted above, the number of embryos available that are capable of developing into plants is very small.

The tissue culture requirements for the development of wild species embryos (Grant 1984, 1986b) and soybean embryos (Lazzeri et al. 1985; Ranch et al. 1985; Ghazi et al. 1986; Barwale et al. 1986) have not, so far, been directly applicable to the hybrids between the wild species and soybean. However, the recent successes with soybean tissue culture may have more influence on the hybrid tissue culture.

From 10 days after emasculation and pollination, putative hybrid pods are removed from the maternal parent and surface sterilized (Broué et al. 1982; Newell and Hymowitz 1982; Grant 1986b). Ovules are then either cultured intact (Broué et al. 1982; Newell and Hymowitz 1982; Grant et al. 1986) or dissected for culture as immature embryos (Broué et al. 1982; Newell et al. 1987). After initial ovule culture, Grant (1986b) dissected the embryos from the ovules in a solution containing 10% polyethylene glycol and 2% sucrose, as this was found to lessen transfer shock and prevent possible loss of electrolytes (Woodstock and Tao 1981; Woodstock and Taylorson 1981a, b).

Subsequent growth of embryos has been achieved using several media. These include media based on Gamborg et al. (1968) B5 medium (Newell and Hymowitz 1982; Singh and Hymowitz 1985) and Murashige and Skoog's (1962) medium (Broué et al. 1982; Grant et al. 1986; Grant 1986b). Newell et al. (1987) used a Schenk and Hildebrandt-based medium and a medium of Shepard. Most of the hybrids that developed were by the pathway where a single embryo gave rise to a single plant.

Grant (1986b) induced multiple shoots from the cotyledonary node in the hybrid (*G. argyrea*×*G. canescens*)×*G. max* when using two media based on Murashige and Skoog (1962; MS) medium, one containing five times the concentration of minor salts, 0.1 µM NAA and 10 µM zeatin while the other contained 0.1 µM NAA, 5 µM benzylaminopurine, 3.4 mM glutamine, and 56.8 µM asparagine. The latter medium was also used in the initiation of multiple buds on the hybrid (*G. latifolia*×*G. canescens*)^2×*G. max* (Grant 1986b). Singh et al. (1987) were able to produce 21 plants from one seed of the cross *G. max*×*G. clandestina* through organogenesis on a modified B5-based medium.

Because the chance of obtaining a hybrid plant between soybean and the wild species is small, the development of media to give multiple plants and somatic embryos (Fig. 1) not only ensures against loss on transfer from tissue culture to soil, but also offers the possibility of tissue culture-induced variation. In a study of tissue culture-induced variation in an F_2 hybrid embryo, from a hybrid between *G. argyrea* and *G. canescens* induced to produce multiple plants, Grant (1986b) found some variation in pollen fertility and morphology, but no gross chromosomal changes or differences in 14 isozymes analyzed on starch gel electrophore-

Fig. 1. Plant regeneration from the hypocotyl and cotyledons of the interspecific hybrid *G. argyrea* × *G. clandestina*. (Grant 1986b)

sis. This study was carried out using 35 primary regenerants. Barwale et al. (1986) noted a few small plants which produced only two or three seeds amongst primary regenerants from soybean, but more than 400 progeny of regenerated plants appeared normal morphologically. The 21 plants from a *G. max* × *G. clandestina* embryo were found to be identical morphologically, cytologically, and by their banding pattern for the isozyme phosphoglucomutase (PGM; Singh et al. 1987).

3.2.1 The Critical Stage of Differentiation in Regeneration

It is notable that although there have been many reports of bud formation and early growth of somatic embryos in *Glycine* tissue cultures (Kimball and Bingham 1973; Beversdorf and Bingham 1977; Cheng et al. 1980; Saka et al. 1980; Kartha et al. 1981; Kameya and Widholm 1981; Phillips and Collins 1981; Widholm and Rick 1983; Gamborg et al. 1983; Christianson et al. 1983; Grant 1984; Lippman and Lippman 1984; Li et al. 1985; Grant 1986b) and reports of early growth in embryos between the subgenera (Newell and Hymowitz 1982; Broué et al. 1982; Grant 1986b; Newell et al. 1987), most of these would not proceed to the stage of shoot elongation. Also in the reports where regeneration to whole plants was achieved, many more buds were produced than could be developed (Lazzeri et al. 1985; Newell and Luu 1985; Ranch et al. 1985; Wright et al. 1986; Ghazi et al. 1986; Hammatt et al. 1987; Wright et al. 1987; Grant and Kenworthy, unpubl.). Such reports confirm that bud elongation is a critical stage in *Glycine* differentiation and hence a critical stage in regeneration pathways. Lazzeri et al. (1985) reported a "lag phase" for soybean embryos of up to 4 months. Finkelstein and Crouch (1984) monitored the transition from embryogeny to germination in rapeseed, using morphological and biochemical markers. They found that in immature embryos the processes of embryogeny and germination can occur concurrently and the capacity to develop is acquired gradually during the maturation

Table 2. Hybrids obtained between soybean and its wild perennial relatives

Cross	Chromosome no.	Reference
1. [a] (*G. tomentella* × *G. canescens*)^2 × *G. max*	2n = 59	Broué et al. (1982)
2. [b] *G. max* × *G. tomentella*	2n = 59	Newell and Hymowitz (1982)
3. *G. max* × *G. tomentella*	2n = 60	Newell and Hymowitz (1982)
4. *G. tomentella* × *G. max*	2n = 59	Singh and Hymowitz (1985)
5. (*G. argyrea* × *G. canescens*) × *G. max*	2n = 40	Grant (1986a)
6. *G. canescens* × *G. max*	2n = 40	Newell et al. (1987)
7. *G. max* × *G. clandestina*	2n = 40	Singh et al. (1987)
8. (*G. latifolia* × *G. canescens*)^2 × *G. max*	2n = 60[c]	Grant (1986b)

[a] (*G. tomentella* × *G. canescens*)^2 implies that the F_1 hybrid *G. tomentella* × *G. canescens* had its chromosome number doubled with colchicine.
[b] Hybrids are given with the female parent designated first.
[c] Expected chromosome number

process, a period of 6 weeks in rapeseed embryos. Finkelstein and Crouch (pers. commun.) found that so far the most effective way to confer maturity on rapeseed embryos was exposure to high osmotic concentration. The success of using polyethylene glycol (Grant 1986b) and the addition of mannitol to the culture medium (Newell et al. 1987) indicate that regulation of the process of embryogeny to germination may be governed by factors similar to those found for rapeseed embryos, i.e., some maturation process has to occur to switch the embryogenic process properly to the process of germination and plant development. The interaction of chemical, physical, and temporal factors would seem to be extremely important in allowing the maturation process to occur.

3.3 Hybrid Plants Obtained

So far, all the hybrids between the subgenera have been produced by embryo or ovule culture (Broué et al. 1982; Newell and Hymowitz 1982; Singh and Hymowitz 1985; Newell et al. 1987; Grant et al. 1986; Grant 1986a, b; Singh et al. 1987). Table 2 shows the species involved in the hybrids produced.

3.3.1 Cytology of the Hybrids

In the crosses between soybean and the wild relatives at the diploid (2 n = 40) level, the wild species involved are all in one evolutionary group (Grant et al. 1984a, 1986; Singh and Hymowitz 1985). The small number of bivalents found at pollen mother cell meiosis in the hybrid (*G. argyrea* × *G. canescens*) × *G. max* (an average of 2.61, Fig. 2A; Grant et al. 1986), the hybrid *G. canescens* × *G. max* hybrids (an average of three bivalents, Newell et al. 1987) and the hybrid *G. max* × *G. clandestina* (70% of pollen mother cells having one or no bivalents;

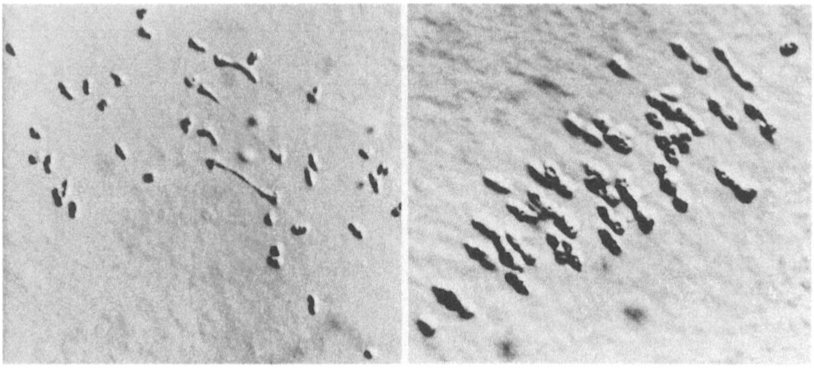

Fig. 2. Pollen mother cell meiosis in **A** (*G. argyrea*×*G. canescens*)×*G. max* showing 3 bivalents and 34 univalents, and **B** {(*G. argyrea*×*G. canescens*)×*G. max*}^{2} showing 38 bivalents and 4 univalents. (Grant et al. 1986)

Singh et al. 1987) would indicate little homology between soybean and the wild species in the *G. canescens*-related group (Grant et al. 1984a, 1986).

In the polyploid complex *G. tomentella*, isozyme analysis coupled with cytological analysis allowed a comprehensive overview of relationships in the group (Grant et al. 1984b; Doyle and Brown 1985; Doyle et al. 1986). These studies recognized five tetraploid and seven diploid subspecific groups based on isozyme variation. The cytological analyses showed that hybrids within each group were fertile, while those between groups were sterile. The *G. tomentella* (2n = 78) accessions used in hybrids with soybeans all belong to one subspecific group, while the *G. tomentella* (2n = 80) accessions used in the hybrids belong to a second subspecific group. All the hybrids between soybean and the tetraploid *G. tomentella* and the hybrid between *G. max* and the synthetic amphiploid (*G. tomentella*×*G. canescens*)^2 showed an average of between three and seven bivalents at pollen mother cell meiosis (Broué et al. 1982; Newell and Hymowitz 1982; Singh and Hymowitz 1985; Newell et al. 1987).

3.3.2 Morphology of the Hybrids

Morphologically, the hybrids obtained between the subgenus *Soja* and the subgenus *Glycine* have a number of features in common whether they are triploid with chromosome number 2n = 59, 60 (Broué et al. 1982; Newell and Hymowitz 1982; Singh and Hymowitz 1985) or diploid with 2n = 40 (Brown et al. 1985; Grant 1986a, b; Newell et al. 1987; Singh et al. 1987). These features include the very vigorous twining growth habit, the intermediate leaflet shape, flower color, number of flowers per raceme, and the indeterminate and continuous flowering. However, Newell et al. (1987) found two exceptions amongst their *G. max*×*G. tomentella* hybrids. They were both slow-growing and had distinctive leaf morphology. Table 3 gives a morphological comparison of the hybrid (*G. argyrea*×*G. canescens*)×*G. max* with each of its parents. The hybrid shows intermediate char-

Table 3. Some morphological comparisons of the hybrid between diploid perennial *Glycine* and soybean and its parents. (Grant 1986a)

Character	*G. arg*[a] × *G. can*[b] G1420×G1232	(*G. arg* × *G. can*) × *G. max* (G1420×G1232)×Improved Pelican	*G. max* cv. Improved Pelican
Growth habit	Twining, perennial	Twining, perennial	Erect, annual
Terminal leaflet shape	Narrowly oblong	Lanceolate	Ovate
Stipules			
– shape	Triangular	Triangular	Triangular
– length mm	2.0	3.5	6.0
– indumentum	Hairs white, appressed and ascendant	Hairs white, ascendant	Hairs white, ascendant
Indumentum			
– lower leaf	Sericeous, hairs white, appressed, moderately dense, ca. 0.5 mm long	Hirsute, hairs white, appressed, moderately dense, ca. 0.8 mm long	Hirsute, hairs white, appressed, moderately dense, ca. 1.0 mm long
– midrib of lower leaf	Hairs white and pale brown, appressed	Hairs white and pale brown, ascendant	Hairs white, erect
– margins	Hairs white and pale brown, appressed	Hairs pale brown, appressed, and ascendant	Hairs pale brown, appressed and ascendant
– stem	Retrorse, hairs white and pale brown, appressed	Retrorse, hairs white, ascendant	Retrorse, hairs pale brown, suberect
– calyx	Hairs white and brown, ascendant	Hairs white and pale brown, ascendant	Hairs white, suberect

[a] *G. arg* = *Glycine argyrea.*
[b] *G. can* = *Glycine canescens*

acters, e.g., leaf shape, stipule and hair length; some distinctly wild species characters, e.g., the twining growth habit, and some soybean characters, e.g., indumentum type (see also Fig. 3).

3.3.3 Colchicine Doubling of Hybrids

As might be expected, all the hybrids so far analyzed have been sterile, and the trend has been to double these hybrids with colchicine. One method has been to take cuttings of the parent material, root them and tie cotton wool at the axillary nodes. A solution of 0.1% colchicine was applied four times in a 24-h period into the cotton wool (Grant et al. 1986). Newell et al. (1987) found that hybrids with wild *Glycine* as the female parent show less vigor and are more difficult to propagate by cuttings and to double with colchicine than those with *G. max* as the female parent.

For the most part, doubling with colchicine has not been an effective means of restoring fertility to these intersubgeneric hybrids. None of the doubled hybrids has produced mature seed when backcrossed to the parental species. Grant (1986a) noted some pod initiation in the doubled (*G. argyrea* × *G. canescens*)

Fig. 3. The hybrid (center) between the fertile F_1 hybrid *G. argyrea* × *G. canescens* (*right*) and the soybean cultivar Improved Pelican (*left*). (Grant 1986b)

× *G. max* hybrid when backcrossed to either the male or female parent, but the ovules were too small to be rescued by tissue culture. For the doubled *G. max* × *G. tomentella* hybrids, both Hymowitz and Singh (1984) and Newell et al. (1987) reported obtaining a very small number of mature selfed seed. Hymowitz and Singh (1984) attempted to backcross some F_2 hybrid plants, but progeny are yet to be reported.

Analysis of pollen mother cell meiosis in the doubled hybrid (*G. argyrea* × *G. canescens*) × *G. max* showed an average of 38.08 bivalents and 3.84 univalents. Although the chiasmata appeared to terminalize (Fig. 2B), no stainable pollen was observed. The common abnormality was that of lagging chromosomes resulting in micronuclei (Grant 1986a).

3.4 The Nature of the Interaction Between the Wild Species and Soybean

All the triploid hybrids produced have two genomes derived from the wild parent and one genome derived from the soybean parent, and are highly sterile. Those with *G. tomentella* as the wild parent show an average of 7 bivalents with a range of three to 15 (Newell and Hymowitz 1982; Singh and Hymowitz 1985; Newell et al. 1987). However, Newell et al. (1987) found that when *G. max* was the female parent in their *G. max*×*G. tomentella* hybrids the range of bivalents was 0–13 with an average close to four. For the triploid hybrids produced by Broué et al. (1982), the range of bivalents was similar, i.e., three to nine bivalents, but the average was slightly lower at 3.9 bivalents. All triploids were completely pollen-sterile. The diploid hybrids (*G. argyrea*×*G. canescens*)×*G. max* and *G. canescens*×*G. max* have a ratio of one genome of wild chromosomes to one genome of soybean chromosomes. They are also completely pollen-sterile but the range of bivalents, 0–10, and their average of approximately three bivalents was somewhat lower than in triploid hybrids.

Chromosome pairing in hybrids may be due to homoeologous associations within the soybean or the wild species parent to associations between the soybean and the wild species chromosomes, or to a combination of both. In haploid soybeans, at least one bivalent has usually been observed and is attributed to homoeologous chromosome association (Sorrells and Bingham 1979; Crane et al. 1982).

Newell and Hymowitz (1983) cite an example of low chromosome pairing in a *Glycine* hybrid between *G. tomentella* ($2n = 80$) and *G. tabacina* ($2n = 80$). This hybrid had an average of three bivalents, which suggested that little homoeologous association was taking place within either the 40-chromosome *G. tomentella* or the 40-chromosome *G. tabacina* genomes when they were isolated in a hybrid in the absence of their true homologs. Further examples of low levels of pairing in diploids can be found for all the species used as wild parents in the reported hybrids between soybean and the wild species. For example: (1) in a *G. canescens*×*G. falcata* hybrid the bivalent average was 2.89 (Grant et al. 1984a); (2) in a *G. argyrea*×*G. cyrtoloba* hybrid the bivalent average was 2.25 (Grant et al. 1986). This also supports the suggestion that little homoeologous pairing takes place when the genomes are isolated in a hybrid. Singh and Hymowitz (1985) favored the option that the chromosome pairing in their *G. tomentella*×*G. max* hybrid was between the two *G. tomentella* genomes, as *G. tomentella* has an allopolyploid origin.

The complete lack of pollen fertility in spite of regular pairing, as in the hybrid [(*G. argyrea*×*G. canescens*)×*G. max*]^2, is unusual in the genus. In hybrids between *G. tomentella*, accessions with similar pollen mother cell meiosis figures, i.e., an average of 38 bivalents, there was very high fertility (Doyle et al. 1986). From a triploid hybrid, *G. argyrea*×*G. tomentella*, with $2n = 59$ and an average of nine bivalents at pollen mother cell meiosis several F_2 seeds have been produced. This was unexpected, although in the same *G. tomentella* chromosomal race crossed with a different chromosomal race of *G. tomentella* to give $2n = 79$ hybrids, plants were obtained that produced occasional seed. This could be attributed to restitution gametes being formed on both the male and the female side

(Doyle et al. 1986). Sorrells and Bingham (1979) found a bias toward diploid progeny in soybean haploids crossed with diploids, indicating restitution gametes.

The progeny from the doubled *G. max* × *G. tomentella* hybrids (2 n = 118) reported by Newell et al. (1987) and Hymowitz and Singh (1984) had chromosome numbers similar to (2 n = 117) or the same as (2 n = 118) the doubled parent. They have been produced in one- or two-seeded cleistogamous pods. The mechanism that gave rise to these seeds is assumed to be selfing.

3.5 Possible Reasons for Infertility

The reasons for infertility in these hybrids could be genetic factors, environmental factors, or a combination of both. With regard to genetic factors there seems to be a large group of genes, in the genomes of higher plants, that lead to the breakdown of microsporogenesis when they are present in their mutant condition. In a review of genetic control of microsporogenesis, Gottschalk and Kaul (1974) found that the majority of the male-sterile genes influenced the final stages of meiosis between interphase II and pollen formation, i.e., in all these genotypes there was a completely normal course of microsporogenesis until the stage at which the degeneration set in.

This may be due to a mutation or interaction in the wild species. In a cross between *G. argyrea*, accession G1420, and a *G. tomentella* accession, with 2 n = 38, microsporogenesis was severely disturbed at an early stage, so that pollen mother cell number was reduced and almost no pollen grains could be seen at the time of anthesis (Grant et al. 1986). In hybrids between *G. argyrea* and *G. clandestina*, fertile F_1 plants gave rise to progeny that showed a number of disturbances both morphologically and cytologically; included in these are plants that show normal microsporogenesis but very low pollen fertility. On such plants no seed was set until hand cross-pollinated with parental lines.

In soybean at least four nuclear male sterility genes have been documented. The action of such male sterility genes or modifiers of them in a different genetic background, such as an interspecific hybrid, is unpredictable. The causes of female fertility or sterility would also have a major effect on the production of progeny from these intersubgeneric hybrids. As Stebbins (1958) pointed out, it is difficult to establish a new balanced condition when such a condition is upset, as in the case of certain interspecific pollinations.

This balance may be not only genetically but also environmentally conditioned. Hybrid microsporogenesis in the wild species is environmentally conditioned either by daylength or temperature or both. In adverse conditions, all stages of pollen mother cell meiosis can be detected in the anthers of one flower at any one time, while in congenial conditions all the anthers in one flower are at a similar stage of meiosis. In *G. max* × *G. soja* hybrids, Ahmad et al. (1984) found that cooler temperatures reduced the number of aberrant tetrads in all cases, although the genotype of the hybrids influenced the absolute number. A dramatic example of the influence of the environment on an interspecific hybrid concerns the triploid hybrid between *Arachis hypogaea* and the diploid species of section *Arachis nom. nud.* (Singh and Moss 1984; Singh 1985). These hybrids were observed to be sterile in Reading, England. However, cuttings from these plants grown at the International Crops Research Institute for the Semi-Arid Tropics (ICRISAT) in India produced pods and seed in all triploid combinations.

4 Conclusions

Hybrid barriers do exist between the wild species and soybean. However there is a gradually increasing number of wild species and genotypes within those species that can be crossed with soybean as well as a greater understanding of the wild species per se. This along with the recent advances in regeneration techniques (see Chap. II.3, this Vol.) and with the advances in culturing hybrids between soybean and its wild perennial relatives, should lead to a new phase in the exploitation of the wild germplasm for soybean breeding.

5 Protocol

Following is one possible protocol for the culture of wild species × soybean hybrids to give multiple plants.

1. Emasculate flowers before anthesis and pollinate with soybean pollen.
2. Cover plant with a plastic bag for 24 h to prevent dessication.
3. After 10–15 days remove the pod.
4. Surface sterilize the pod in, e.g., 1.2% Zephiran (Winthrop Laboratories, Australia) with 10% ethanol.
5. Dissect ovules from the pod.
6. Place ovules for 3–4 weeks on an ovule culture medium [Bourgin and Nitsch (1967) major salts + 12.8 mM KNO_3, 9 mM NH_4NO_3, 1.5 mM $(NH_4)_2SO_4$; 1/10 MS minor salts; B5 vitamins; 0.05 mM NaFeEDTA; 0.0876 M sucrose; 0.3 μM IAA; 0.05 μM BAP; 25 g/l yeast extract; 60 μM asparagine; 15 mM glutamine; 5.6 mM inositol; 0.7% agar; pH 6.0].
7. Dissect embryo from ovule in 10% polyethylene glycol + 2% sucrose.
8. Place embryos on MS salts, KM vitamins (Kao and Michayluk 1975), 0.05 mM NaFeEDTA, 0.1 μM NAA, 5 μM BAP or 2.3 μM zeatin, 56.8 μM asparagine, 3.4 mM glutamine, 0.7% agar, pH 5.8.
9. Subculture at 3–4-week intervals.
10. Root on 1/2 B5 with 2 μM NAA and transfer to soil in the normal manner.

Acknowledgments. To Drs. M.J. Doyle, D.W. de Little and A.H.D. Brown for helpful comments on the manuscript. To CSIRO Division of Plant Industry, the Australian Rural Credits Development Fund, and the International Board of Plant Genetic Resources for supporting the work on the wild perennial relatives of soybean in Australia.

References

Ahmad QN, Britten EJ, Byth DI (1984) Effects of interacting genetic factors and temperature on meiosis and fertility in soybean×*Glycine soja* hybrids. Can J Genet Cytol 26:50–56

Barwale, UB, Kerns HR, Widholm JM (1986) Plant regeneration from callus cultures of several soybean genotypes via embryogenesis and organogenesis. Planta 167:473–481

Beversdorf WD, Bingham ET (1977) Degrees of differentiation obtained in tissue cultures of *Glycine* species. Crop Sci 17:307–311

Blaydes DF (1966) Interaction of kinetin and various inhibitors in growth of soybean tissue. Physiol Plant 19:748–753

Bourgin JP, Nitsch JP (1967) Obtention de *Nicotiana* haploide à partir d'etamines cultivées in vitro. Ann Physiol Veg 9:377–382

Broué P, Douglass J, Grace JP, Marshall DR (1982) Interspecific hybridization of soybeans and perennial *Glycine* species indigenous to Australia via embryo culture. Euphytica 31:715–724

Brown, AHD, Grant JE, Burdon JJ, Grace JP, Pullen R (1985) Collection and utilization of wild perennial *Glycine*. Proc World Soybean Res Conf III. Shibles R (ed) Westview. Boulder, Colorado, pp 345–353

Burdon JJ, Speer SS (1984) A set of differential *Glycine* hosts for the identification of races of *Phakopsora pachyrhizi* Syd. Euphytica 33:891–896

Carroll BJ, McNeil DL, Gresshoff PM (1985) Isolation and properties of soybean [*Glycine max* (L.) Merr.] mutants that nodulate in the presence of high nitrate concentrations. Proc Natl Acad Sci 82:4162–4166

Cheng T-Y, Saka H, Voqui-Dinh TH (1980) Plant regeneration from soybean cotyledonary node segments in culture. Plant Sci Lett 19:91–99

Christianson ML, Warnick DA, Carlson PS (1983) A morphogenetically competent soybean suspension culture. Science 222:632–634

Crane CR, Beversdorf WD, Bingham ET (1982) Chromosome pairing and associations at meiosis in haploid soybean (*Glycine max*). Can J Genet Cytol 24:293–300

Delannay X, Rodgers DM, Palmer RG (1983) Relative genetic contributions among ancestral lines to North American soybean cultivars. Crop Sci 23:944–949

Doyle MJ, Brown AHD (1985) Numerical analysis of isozyme variation in *Glycine tomentella*. Biochem Syst Ecol 13(4):413–419

Doyle MJ, Grant JE, Brown AHD (1986) Reproductive isolation between isozyme groups of *Glycine tomentella* (Leguminosae), and spontaneous doubling in their hybrids. Aust J Bot 34:523–535

Evans DA, Sharp WR, Medina-Filho HP (1984) Somaclonal and gametoclonal variation. Am J Bot 71(6):759–774

Fehr WR (1980) Soybean. In: Fehr WR, Hadley HH (eds) Hybridization of crop plants. Am Soc Agron, Crop Sci Soc Am, Madison WL, pp 589

Finkelstein RR, Crouch ML (1984) Precociously germinating rapeseed embryos retain characteristics of embryogeny. Planta 162:125–131

Gamborg OL, Miller R, Ojima K (1968) Nutrient requirements of suspension cultures of soybean root cells. Exp Cell Res 50:151–158

Gamborg OL, Davis BP, Stahlhut RW (1983) Somatic embryogenesis in cell cultures of *Glycine* species. Plant Cell Rep 2:209–212

Ghazi TD, Cheema HV, Nabors MW (1986) Somatic embryogenesis and plant regeneration from embryogenic callus of soybean *Glycine max* (L.). Plant Cell Rep 5:452–456

Gottschalk W, Kaul MLH (1974) The genetic control of microsporogenesis in higher plants. Nucleus 17:133–166

Grant JE (1984) Plant regeneration from cotyledonary tissue of *Glycine canescens*, a wild perennial relative of soybean. Plant Cell Tissue Organ Cult 3:169–173

Grant JE (1986a) Hybridization of soybean with its diploid wild perennial relatives. In: Williams TA, Wratt GS (eds) Proc Symp Plant Breeding, DSIR. Agron Soc NZ Spec Publ 5:27–29

Grant JE (1986b) Interspecific hybridization in the genus *Glycine*. Ph D Thesis, Australian National Univ, Canberra

Grant JE, Grace JP, Brown AHD, Putievsky E (1984a) Interspecific hybridization in *Glycine* Willd. subgenus *Glycine* (Leguminosae). Aust J Bot 32:655–663

Grant JE, Brown AHD, Grace JP (1984b) Cytological and isozyme diversity in *Glycine tomentella* Hayata (Leguminosae). Aust J Bot 32:665–677

Grant JE, Pullen R, Brown AHD, Grace JP, Gresshoff PM (1986) Cytogenetic affinity between the new species *Glycine argyrea* and its congeners. J Hered 77:423–426

Green AG, Marshall DR (1984) Isolation of induced mutants in linseed (*Linum usitatissimum*) having reduced linolenic acid content. Euphytica 33:321–328

Hammatt N, Kim H-I, Davey MR, Nelson RS, Cocking EC (1987) Plant regeneration from cotyledon protoplasts of *Glycine canescens* and *G. clandestina*. Plant Sci 48:129–135

Harlan JR (1976) Genetic resources in wild relatives of crops. Crop Sci 16:329–332

Hood MJ, Allen FL (1980) Interspecific hybridization studies between cultivated soybean, *Glycine max*, and a perennial wild relative *G. falcata*. Agron Abstr, p58

Hymowitz T (1986) Screening studies with perennial soybeans. Soybean Breed Pathol Workshop, St Louis, MO

Hymowitz T, Singh RJ (1984) A soybean×*Glycine tomentella* hybrid: progress and problems. Soybean Genet Newsl 11:90

Kameya T, Widholm J (1981) Plant regeneration from hypocotyl sections of *Glycine* species. Plant Sci Lett 21:289–294

Kao KN, Michayluk MR (1975) Nutritional requirements for growth of *Vicia hajastana* cells and protoplasts at a very low population density in liquid media. Planta 126:105–110

Kartha KK, Pahl K, Leung NL, Mroginski LA (1981) Plant regeneration from meristems of grain legumes: soybean, cowpea, peanut, chickpea and bean. Can J Bot 59(9):1671–1679

Kimball SL, Bingham ET (1973) Adventitious bud development of soybean hypocotyl sections in culture. Crop Sci 13:758–760

Ladizinsky G, Newell CA, Hymowitz T (1979) Wide crosses in soybean: prospects and limitations, Euphytica 28:421–423

Larkin PJ, Scowcroft WR (1981) Somaclonal variation – a novel source of variability from cell cultures for plant improvement. Theor Appl Genet 60:197–214

Lazzeri PA, Hildebrandt DF, Collins GB (1985) A procedure for plant regeneration from immature cotyledon tissue of soybean. Plant Mol Biol Rep 3(4):160–167

Li BJ, Langridge WHR, Szalay AA (1985) Somatic embryogenesis and plantlet regeneration in the soybean *Glycine max*. Plant Cell Rep 4:344–347

Lippman B, Lippman G (1984) Induction of somatic embryos in cotyledonary tissue of soybean, *Glycine max* L. Merr. Plant Cell Rep 3:215–218

Murashige T, Skoog F (1962) A revised medium for rapid growth and bioassays with tobacco tissue cultures. Physiol Plant 15:473–497

Newell CA (1986) Description of the genus *Glycine*. Soybean Breed Pathol Workshop, St Louis, MO

Newell CA, Hymowitz T (1982) Successful wide hybridization between the soybean and a wild perennial relative, *G. tomentella* Hayata. Crop Sci 22:1062–1065

Newell CA, Hymowitz T (1983) Hybridization in the genus *Glycine* subgenus *Glycine* Willd. (Leguminosae, Papilioniodeae). Am J Bot 70(3):334–348

Newell CA, Luu HT (1985) Protoplast culture and plant regeneration in *Glycine canescens* F.J. Herm. Plant Cell Tissue Organ Cult 4:145–149

Newell CA, Delannay X, Edge ME (1987) Interspecific hybrids between the soybean *Glycine max* (L.) Merr. and wild perennial relatives. J Hered 78:301–306

Phillips GC, Collins GB (1981) Induction and development of somatic embryos from cell suspension cultures of soybean. Plant Cell Tissue Organ Cult 1:123–129

Ranch JP, Oglesby L, Zielinski AC (1985) Plant regeneration from embryo-derived tissue cultures of soybeans. In vitro Cell Dev Biol 21(11):653–658

Riggs RD (1986) Screening for cyst nematode resistance. Soybean Breed Pathol Workshop, St Louis, MO

Saka H, Voqui-Dinh TH, Cheng T-Y (1980) Stimulation of multiple shoot formation on soybean stem nodes in culture. Plant Sci Lett 19:193–201

Sedova TS (1982) Interspecific hybridization among the cultivated and wild species of soybean belonging to the subgenera *Glycine* and *Soja*. Soviet Genet 18:1142–1146

Singh AK (1985) Genetic introgression from compatible wild species into cultivated groundnut. In: Moss JP (ed) Cytogenetics of *Arachis*. Proc Int Workshop, 1983, ICRISAT, India, pp 107–117

Singh RJ, Hymowitz T (1985) An intersubgeneric hybrid between *Glycine tomentella* Hayata and the soybean *G. max* (L.) Merr. Euphytica 34:187–192
Singh AK, Moss JP (1984) Utilization of wild relatives in genetic improvement of *Arachis hypogaea* L. VI. Fertility in triploids: Cytological basis and breeding implications. Peanut Sci 11:17–21
Singh RJ, Kollipara KP, Hymowitz T (1987) Intersubgeneric hybridization of soybeans with a wild perennial species, *Glycine clandestina* Wendl. Theor Appl Genet 74:391–396
Sorrells ME, Bingham ET (1979) Reproductive behaviour of soybean haploids carrying the *ms 1* allelle. Can J Genet Cytol 21:449–455
Stebbins GL (1958) The inviability, weakness, and sterility of interspecific hybrids. Adv Genet 9:147–215
Tindale MD (1986) Taxonomic notes on three Australian and Norfolk Island species of *Glycine* Willd. (Fabaceae: Phaseolae) including the choice of a neotype for *G. clandestina* Wendl. Brunonia 9:179–191
Weiss EA (1983) Soyabean. In: Weiss EA (ed) Oilseed crops, Chap 8. Trop Agric Ser Longman, London, pp 341–401
Widholm JM, Rick S (1983) Shoot regeneration from *Glycine canescens* tissue cultures. Plant Cell Rep 2:19–20
Woodstock LW, Tao K-LJ (1981) Prevention of imbibitional injury in low vigor soybean embryonic axes by osmotic control of water uptake. Physiol Plant 51:133–139
Woodstock LW, Taylorson RB (1981 a) Soaking injury and its reversal with polyethylene glycol in relation to respiratory metabolism in high and low vigor soybean seeds. Physiol Plant 53:263–268
Woodstock LW, Taylorson RB (1981 b) Ethanol and acetaldehyde in imbibing soybean seeds in relation to deterioration. Plant Physiol 67:424–428
Wright MS, Koehler SM, Hinchee MA, Carnes MG (1986) Plant regeneration by organogenesis in *Glycine max*. Plant Cell Rep 5:150–154
Wright MS, Williams MH, Pierson PE, Carnes MG (1987) Initiation and propagation of *Glycine max* (L.) Merr: Plants from tissue-cultured epicotyls. Plant Cell Tissue Organ Cult 8:83–90

II.5 Soybean: Isolation, Culture, and Fusion of Protoplasts

N. HAMMATT[1] and M.R. DAVEY[2]

1 Introduction

The genus *Glycine* contains the annual cultivated soybean (*G. max*), its wild annual progenitor *G. soja*, and several wild perennial species, some of which are twiners (Fig. 1a), while others are prostrate in their growth habit (Fig. 1b). The perennial species represent a germplasm resource of potential for improving the cultivated crop. Useful traits so far identified in perennial *Glycine* include tolerance to cold, drought, heat and salinity, daylength neutrality (Marshall and Broué 1981; Newell and Hymowitz 1982), and resistance to soybean cyst nematode (Riggs and Hamblen 1962, 1966), yellow mosaic virus (Singh et al. 1974), powdery mildew (Mignucci and Chamberlain 1978), and rust (Singh et al. 1974; Burdon and Marshall 1981a, b; Burdon and Speer 1984).

Hybrids between soybean and perennial *Glycine* species have been difficult to produce by conventional breeding, although some success has been obtained following embryo rescue in crosses between soybean and *G. tomentella* (Newell and Hymowitz 1982; Singh and Hymowitz 1985; Newell et al. 1987), *G. canescens*, *G. falcata* and *G. tabacina* (Newell et al. 1987), a synthetic amphiploid of *G. canescens* × *G. tomentella* (Broué et al. 1982), and the sexual hybrid, *G. argyrea* × *G. canescens* (Grant 1986).

Somatic hybridization, involving protoplast fusion, should constitute a method for generating interspecific hybrids in the genus *Glycine*. Such technology demands prior understanding of procedures to isolate, fuse and culture protoplasts from soybean and perennial *Glycine* species, and to regenerate plants from protoplast-derived tissues. Few details have been published on the isolation and culture of soybean protoplasts, and even less regarding perennial *Glycine* species.

Reports of the successful culture of *Glycine* protoplasts are summarised in Table 1 (see also Newell and Luu 1988). In brief, early studies with soybean protoplasts involved those isolated from cell suspensions. Indeed, some of the first reports of division in isolated plant protoplasts were observed in those obtained from soybean suspension cultures (Kao et al. 1970; Miller et al. 1971).

[1] Present address: Institute of Horticultural Research, East Malling, Maidstone, Kent, ME19 6BJ, United Kingdom
[2] Plant Genetic Manipulation Group, Department of Botany, University of Nottingham, University Park, Nottingham NG7 2RD, United Kingdom

Biotechnology in Agriculture and Forestry, Vol. 10
Legumes and Oilseed Crops I (ed. by Y.P.S. Bajaj)

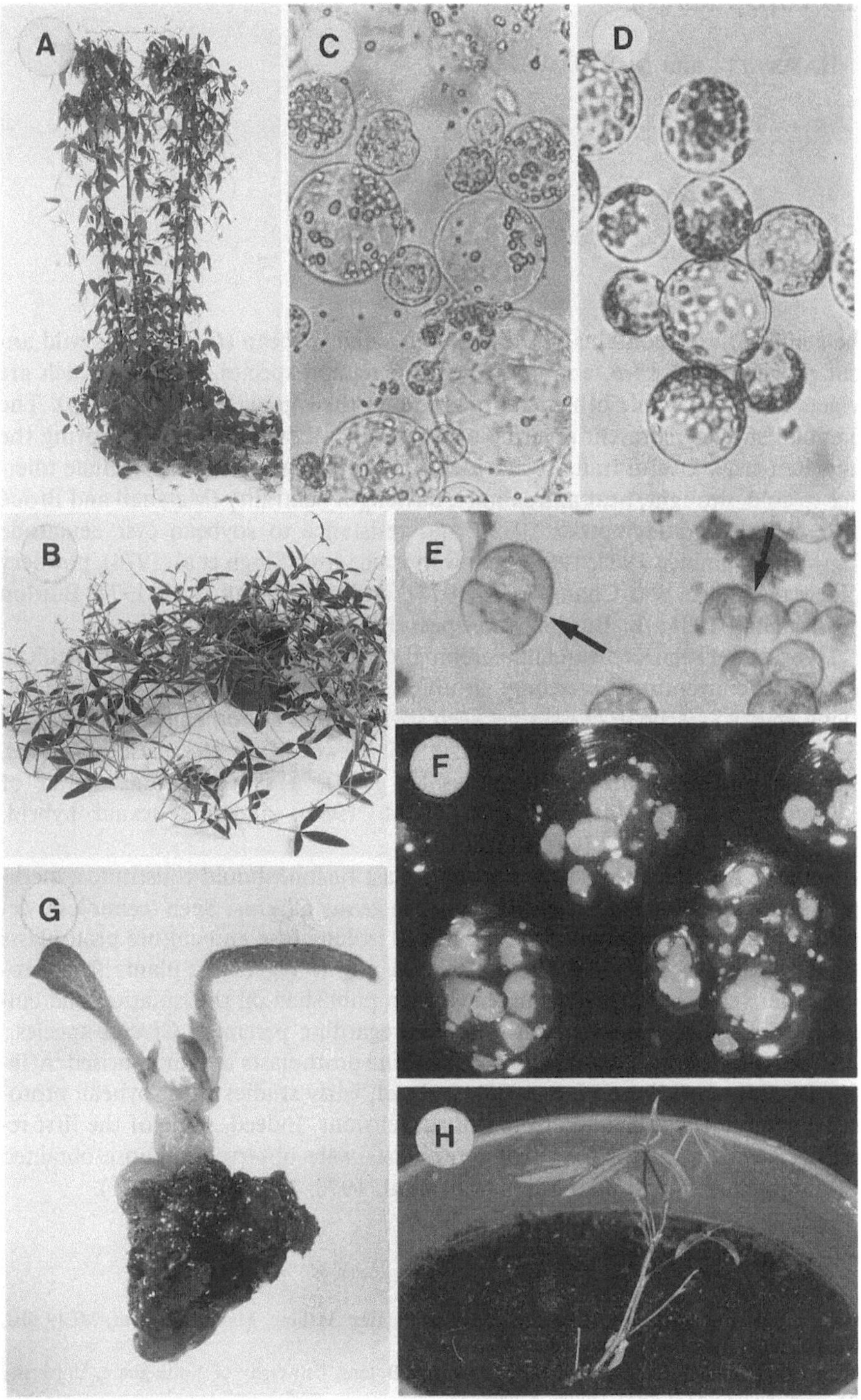

Fig. 1A–H

Table 1. *Glycine* species in which protoplasts have been successfully cultured

Species	Protoplast source	Growth response	Reference
G. argyrea	Cotyledon	Plants	Hammatt et al. (1989)
G. canescens	Hypocotyl	Plants	Newell and Luu (1985)
G. canescens	Cell suspension	Callus	Hammatt (1986)
G. canescens	Cell suspension	Plants	Myers et al. (1986)
G. canescens	Cotyledon	Plants	Hammatt et al. (1987a)
G. clandestina	Cell suspension	Callus	Myers et al. (1986)
G. clandestina	Cotyledon	Plants	Hammatt et al. (1987a)
G. falcata	Cotyledon	Callus	Hammatt et al. (1987a)
G. latifolia	Cotyledon	Callus	Hammatt et al. (1987a)
G. max	Cell suspension	Cell clusters	Kao et al. (1970)
G. max	Cell suspension	Cell clusters	Kao et al. (1971)
G. max	Cell suspension	Cell colonies	Michayluk and Kao (1975)
G. max	Pod tissue	Root formation	Zieg and Outka (1980)
G. max	Root	Callus	Xu et al. (1982)
G. max	Leaf	Cell cultures	Gamborg et al. (1983)
G. max	Immature cotyledon	Callus	Lu et al. (1983)
G. max	Seedling cotyledon	Callus	Lu et al. (1983)
G. max	Leaf	Callus	Oelck et al. (1983)
G. max	Crown gall	Callus	Pedersen et al. (1983)
G. max	Cell suspension	Cell suspension following freezing of protoplasts	Weber et al. (1983)
G. max	Callus	Cell regeneration	Bojsen and Wyndaele (1985)
G. max	Photoautotrophic cell suspension	Callus	Chowdhury and Widholm (1985)
G. max	Leaf	Callus	Tricoli et al. (1986)
G. max	Hypocotyl	Callus	Hammatt and Davey (1988)
G. max	Cotyledon	Plants	Wei and Xu (1988)
G. soja	Cell suspension	Embryo-like structures	Gamborg et al. (1983)
G. tabacina	Cell suspension	Embryo-like structures	Gamborg et al. (1983)
G. tabacina	Cotyledon	Callus	Hammatt et al. (1987a)
G. tomentella	Cell suspension	Callus	Hammatt (1986)
G. tomentella	Cotyledon	Callus	Hammatt et al. (1987a)

Protoplasts derived from soybean seedling root tips (Xu et al. 1982) and cotyledons (Lu et al. 1983) have been cultured to callus. There have also been reports of the isolation of soybean leaf mesophyll protoplasts (Schwenk et al. 1981; Lin 1983; Franceschi et al. 1984; Rees et al. 1985), but only Gamborg et al. (1983), Oelck et al. (1983) and Tricoli et al. (1986) have succeeded in culturing them to

Fig. 1. A Mature plant of *G. tabacina* (×0.04). **B** Mature plant of *G. falcata* (×0.13). **C** Freshly isolated hypocotyl protoplasts from dark-grown seedlings of soybean cv. HP-20-20 (×415). **D** Freshly isolated protoplasts of *G. canescens* G1171 (×415). **E** Divisions (*arrows*) in soybean hypocotyl protoplasts after 8 days of culture in liquid K8P medium. Phase contrast (×415). **F** Cell colonies of *G. canescens* G1171 after culture of cotyledon protoplasts for 3 weeks in agarose droplets (×5). **G** Shoot emerging from dark-coloured cotyledon protoplast-derived callus of *G. canescens* G1171 after four 3-weekly subcultures on agar solidified SC2 medium (×4.4). **H** Plantlet regenerated from cotyledon protoplast-derived tissue of *G. canescens* G1171 2 weeks after transfer to soil (×1)

Table 2. Fusions of soybean protoplasts that have resulted in successful culture of heterokaryons

Origin of soybean protoplasts	Fusion partner		Culture response of heterokaryons	Reference
	Species	Origin of protoplasts		
Suspension culture	*Hordeum vulgare*	Leaf	Divisions	Kao et al. (1974)
Suspension culture	*Zea mays*	Leaf	Divisions	Kao et al. (1974)
Suspension culture	*Pisum sativum*	Leaf	Divisions	Kao et al. (1974)
Suspension culture	*Vicia hajastana*	Suspension culture	Divisions	Kao et al. (1974)
Suspension culture	*H. vulgare*	Leaf	Cell clusters	Kao and Michayluk (1974)
Suspension culture	*Brassica napus*	Leaf	Cell clusters	Kartha et al. (1974)
Suspension culture	*Pisum sativum*	Leaf	Divisions	Constabel et al. (1975a)
Suspension culture	*Melilotus* sp.	Leaf	Cell clusters	Constabel et al. (1975b)
Suspension culture	*Medicago sativa*	Leaf	Cell clusters	Constabel et al. (1975b)
Suspension culture	*Caragana aborescens*	Leaf	Cell clusters	Constabel et al. (1975b)
Suspension culture	*Pisum sativum*	Leaf	Divisions	Constabel et al. (1976)
Suspension culture	*Nicotiana glauca*	Leaf	Divisions	Constabel et al. (1976)
Suspension culture	*N. langsdorffii*	Leaf	Divisions	Constabel et al. (1976)
Suspension culture	*N. tabacum*	Leaf	Divisions	Constabel et al. (1976)
Suspension culture	*N. rustica*	Leaf	Divisions	Constabel et al. (1976)
Suspension culture	*Colchicum autumnale*	Leaf	Divisions	Constabel et al. (1976)
Suspension culture	*Melilotus officinalis*	Leaf	Cell clusters	Fowke et al. (1976)
Suspension culture	*N. glauca*	Leaf	Asymmetric somatic hybrid callus	Kao (1977) Wetter (1977)
G. max(+) *N. glauca* somatic hybrid callus (Kao 1977)	*N. glauca*	Leaf	Stable somatic hybrid callus following two "back fusions" to *N. glauca*	Wetter and Kao (1980)
Suspension culture	*N. tabacum*	Leaf	Somatic hybrid callus	Chien et al. (1982)
Callus	*Oryza sativa*	Callus	Cybrid callus	Niizeki et al. (1985, 1986)

callus. Other protoplasts that have undergone sustained division include those from pod tissue (Zieg and Outka 1980), callus (Bojsen and Wyndaele 1985), crown gall suspension cells (Pederson et al. 1983), and a photoautotrophic soybean cell culture (Chowhury and Widholm 1985).

In the case of wild *Glycine* species, protoplasts have been successfully isolated from cell suspensions and cultured to callus in *G. canescens* and *G. tomentella* (Hammatt 1986). Those obtained from suspensions of *G. soja* and *G. tabacina*, were cultured to embryoid-like structures (Gamborg et al. 1983). To date, plants have been regenerated from seedling hypocotyl (Newell and Luu 1985), suspension culture (Myers et al. 1986) and cotyledon (Hammatt et al. 1987a) protoplasts of

G. canescens, and cotyledon protoplasts of *G. clandestina* (Hammatt et al. 1987a).

Soybean protoplasts have been used in many studies of intergeneric protoplast fusion, as summarized in Table 2. There are, however, no reports at present, on the fusion of soybean protoplasts with those obtained from perennial *Glycine* species. Baldes et al. (1987) have recently reported the production of transformed soybean callus following cocultivation of protoplast-derived cells with *Agrobacterium tumefaciens*.

This chapter summarizes recent advances from this laboratory in the manipulation of *Glycine* protoplasts, specifically:

a) The isolation and culture of protoplasts from soybean and perennial *Glycine* species.
b) Plant regeneration from protoplasts of *G. argyrea, G. canescens* and *G. clandestina*.
c) Fusion of protoplasts from soybean and *G. canescens*.

2 Isolation of Protoplasts

Seedlings and cell suspension cultures have been used as sources of protoplasts.

2.1 Seedlings

Soybean seeds were provided by Dr. W. Ellingson, AgriPro, Nickersons American Plant Breeders, Ames, Iowa, USA. Seeds of *G. canescens* 9944, *G. soja* 3080, *G. tabacina* 9940 and *G. tomentella* 9941 were supplied by Dr. T.C. Seng, AVRDC, P.O. Box 42, Shanua, Tainan 741, Taiwan, while those of *G. latifolia* PI 253.238 and *G. falcata* PI 246.591 were obtained from Dr. R.L. Bernard, USDA, Department of Agronomy, University of Illinois, IL 61801, USA. All other accessions were provided by Dr. A.H.D. Brown, CSIRO Division of Plant Industry, P.O. Box 1600, Canberra, A.C.T. Australia.

Seeds were surface sterilized in 10% (v/v) Domestos bleach solution (Lever Bros., London) for 20 min, followed by thorough rinsing with sterile tap water. Those of *G. max* were sown in 175-ml screw-capped jars, each containing 40 ml agar solidified (0.8% w/v; Sigma) hormone-free MSO medium (Table 3). Seeds of *G. soja* were scarified before sowing by removal of a piece of the testa, while those of perennial *Glycine* species were scarified and soaked in sterile tap water (48 h with a change after 24 h) prior to sowing on MSO medium. Culture vessels were incubated in the light (1.6 W/m^2, daylight fluorescent tubes; continuous illumination) at 27 °C. Soybean seeds were also germinated in the dark.

2.2 Cell Suspensions

Callus of *G. canescens* 9944, *G. soja* 3080 and *G. tomentella* G1300 was initiated by incubating hypocotyl sections (5 mm in length), excised from 8–9-day-old axenic seedlings, on agar UM medium (Uchimiya and Murashige 1974; Table 3) contained in 9-cm diameter Petri dishes (30 ml medium/dish). Callus was maintained by subculture every 3 weeks to fresh UM medium. Cell suspensions were established by transferring 3 g of callus to 30 ml aliquots of liquid UM medium contained in 250-ml Erlenmeyer flasks. The latter were incubated on a horizontal shaker (80 rpm) in the light (1.6 W/m^2) at 25 °C. Suspensions were maintained by regular transfer, every 8 days, of 5 ml of cells to 45 ml of fresh liquid medium. Those of *G. soja* 3080 were cream in colour, those of *G. canescens* 9944 white-green, and those of *G. tomentella* G1300 olive-green.

Table 3. Composition of media used in studies on *Glycine* species

Medium code	Salts and organics	Sucrose (g/l)	Additions (mg/l)	
MSO	MS	30	–	–
SC2	B5	30	1.1	BAP
			0.005	IBA
SC4	B5	30	5.0	BAP
			0.005	IBA
SC6	B5	30	0.2	BAP
			0.005	IBA
UM	MS	30	2.0	2,4-D
			0.25	Kinetin
			9.9	Thiamine·HCl
			9.5	Pyridoxine·HCl
			4.5	Nicotinic acid
			2% (w/v)	Casein hydrolysate
HB50	Half-strength B5	15	–	–
CPW9M	CPW Salts	–	9% (w/v)	Mannitol
CPW13M	CPW Salts	–	13% (w/v)	Mannitol
K8P	Modified after Kao (1977)	0.25	0.2	2,4-D
			0.5	Zeatin
			1.0	NAA
			31600	Glucose
K8	Modified after Kao (1977)	20	0.1	2,4-D
			0.2	Zeatin
			1.0	NAA

MS, Murashige and Skoog (1962); B5, Gamborg et al. (1968); CPW salts, Frearson et al. (1973). In both K8P and K8 media the salts and organics were modified from Kao (1977) by the inclusion of 1 mg/l instead of 10 mg/l thiamine HCl. For K8 (cell) medium, casamino acids were reduced from 250 mg/l, in the original formulation, to 125 mg/l, and coconut milk from 2% to 1% (v/v). MSO was solidified with 0.8% (w/v) agar and the other media, as required, with 0.6% agar. K8P and K8 were filter sterilized (0.2 µm pore size), while the other media were autoclaved (121 °C; 20 min)

2.3 Enzyme Treatment

The sources of the enzymes and the mixtures used for protoplast isolation are summarized (Tables 4 and 5). All solutions were filter sterilized (0.2-μm pore size) and frozen at −20 °C until required.

Cells from suspension cultures were separated from the medium by pouring the cultures onto a 64-μm mesh nylon sieve. Three ml packed volume of cells were incubated in 20 ml of ES1 enzyme solution in a 250-ml Erlenmeyer flask on a shaker (80 rpm) at 25 °C in the light (1.6 W/m^2).

Hypocotyls from soybean seedlings were cut transversely into slices 2−3 mm thick, and the latter preplasmolyzed (30 min) in CPW13M solution (Table 3). The preplasmolysis solution was replaced with enzyme solution ES2, 1−2 g of hypocotyl tissue (three to six seedling hypocotyls) being incubated with 6 ml of enzyme contained in a 5 cm diameter plastic Petri dish. Enzyme incubation was carried out on a slow rotary shaker (20 rpm, 16 h, 25 °C). Dishes containing material from dark-grown seedlings were wrapped in aluminium foil during the incubation period. At the end of this time, hypocotyl sections were squeezed gently with fine forceps to enhance protoplast release.

Table 4. Suppliers of plant cell wall-degrading enzymes

Enzyme	Manufacturer/supplier
Cellulase R10	Yakult Honsha Co., Nishinomiya, Japan
Meicelase	Meiji Seika Kaisha, Tokyo, Japan
Rhozyme HP150	Röhm and Haas Co., Philadelphia, USA
Pectolyase Y23	Seishin Pharmaceutical, Tokyo, Japan
Macerozyme R10	Kinki Yakult, Nishinomiya, Japan
Driselase	Kyowa Hakko Kogya, Tokyo, Japan
Pectinase (ex. *Rhizopus*)	Sigma Chemical Co., Poole, UK

Table 5. Enzyme solutions used for the isolation of *Glycine* protoplasts

Enzyme solution		
ES1	ES2	ES3
0.8% Rhozyme	2% Rhozyme HP150	1% Rhozyme HP150
0.8% Cellulase R10	2% Meicelase	0.2% Cellulase R10
0.4% Driselase	0.03% Macerozyme R10	0.1% Pectolyase Y23
0.4% Pectinase	CPW salts[a]	CPW salts
0.1% $CaCl_2 \cdot 2H_2O$		
0.01% $NaH_2PO_4 \cdot H_2O$		
0.059% MES[b]		
6.4% Sorbitol		
6.4% Mannitol	13% Mannitol	9% Mannitol
pH 5.6	pH 5.8	pH 5.8

[a] CPW salts, Frearson et al. (1973).
[b] MES, 2-(N-morpholino)ethane sulfonic acid.
All percentages are w/v

Cotyledons were removed from 6- to 14-day-old seedlings of perennial *Glycine* species, cut into slices 1–2 mm thick, and preplasmolyzed in CPW9M (Table 3; 30 min). 100–300 mg of tissue were incubated in 10 ml ES3 enzyme solution in 9-cm Petri dishes (20 rpm, dark, 25 °C, 16 h).

2.4 Protoplast Purification

Following enzyme incubation, crude digests were passed through a 45-µm mesh nylon sieve to remove large undigested material. Protoplasts were pelleted by centrifugation (55 g, 7 min), and the enzyme supernatant removed. A proportion of the cotyledon protoplast preparations was purified further by resuspending in 1 ml of CPW9M, layering onto a solution of 30% (v/v) Percoll (Sigma) made up in K8P medium (Table 3) and centrifuging (25 g, 5 min). All preparations were washed twice by resuspending and centrifuging in either CPW13M (hypocotyl and cell suspension protoplasts) or CPW9M (cotyledon protoplasts). Protoplasts were counted using a haemocytometer.

2.5 Protoplast Yields

Protoplasts were released from exponentially growing cell suspensions of *G. soja* 3080, *G. canescens* 9944 and *G. tomentella* G1300, 4–8 days after subculture, using enzyme ES1. Protoplast yields were 1.3×10^6, 4.5×10^5 and 5.7×10^5/ml packed volume of cells for the three species. Suspensions of *G. canescens* and *G. tomentella* in the lag or stationary growth phases did not release protoplasts.

Detailed studies were carried out on seedling hypocotyl protoplasts of *G. max* cv. HP-20-20. The ES2 enzyme mixture gave reproducible protoplast release (Fig. 1c), a proportion of the protoplasts being purple in colour due to the presence of anthocyanins. Despite the use of a purification procedure involving CPW13M, all preparations were contaminated with subcellular debris. When the protoplasts were suspended in CPW13M solution, the diameters of protoplasts from hypocotyls of 5- and 6-day-old seedlings ranged from 10 to 60 µm, with a mean of 25µm. The highest yield of protoplasts from HP-20-20 (approx. 2.3×10^6/g fr. wt.) was obtained from 5-day-old seedlings (Fig. 2). The same isolation procedure was also applicable to cvs. Essex, Fiskeby, Mitchell, ACCO 101, and AP 120. Yields from hypocotyls of dark-grown soybean seedlings were less than those from material grown in the light; the yields from 4-, 5- and 6-day-old hypocotyls were 4.7, 5.2 and 3.2×10^5/g fr. wt., respectively.

ES3 enzyme solution released protoplasts from seedling cotyledons in all the perennial species studied (Fig. 1d), the yields varying from species to species and between genotypes of *G. canescens* (Table 6). Although protoplast release was enhanced by carefully drawing digested cotyledon material into and out of a Pasteur pipette once or twice at the end of the enzyme incubation period, this physical agitation usually led to increased contamination by undigested cells and debris.

Fig. 2. Effect of the age of soybean seedling hypocotyls on protoplast yield

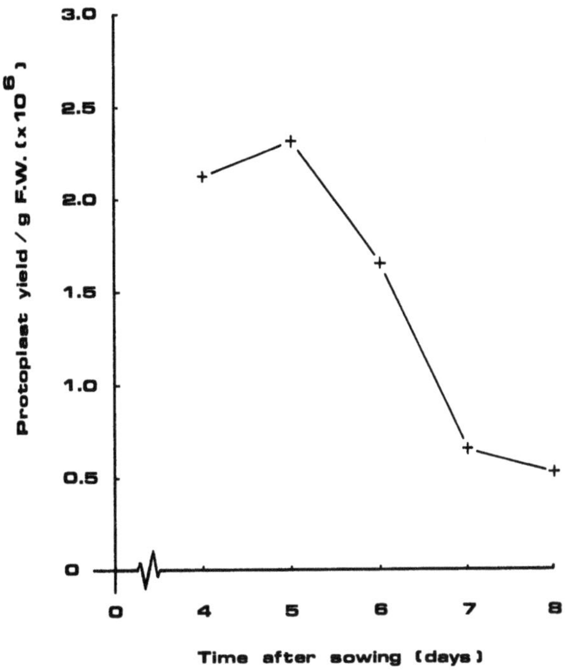

Table 6. Yields and division frequencies obtained for cotyledon protoplasts of wild *Glycine* species using enzyme mixture ES3

Species	Accession	Protoplast yield ($\times 10^6$/g fr.wt.)	Division frequency (%)
G. argyrea	G1626	3.6	65
G. canescens	G1171	2.5	45
	G1301	2.6	40
	G1340	4.1	26
	9944	3.5	32
G. clandestina	G1145	1.9	Not tested
	G1231	4.1	35
G. falcata	PI 246.591	1.4	31
G. latifolia	PI 253.238	2.2	29
G. latrobeana	G1387	1.4	0
G. tabacina	9940	2.7	44
G. tomentella	9941	2.5	43

Flotation on a 30% Percoll solution greatly reduced contamination, but protoplasts were lost, the final yield being reduced by 30% in *G. canescens* 9944, 89% in *G. clandestina* G1231, 52% in *G. latifolia*, 37% in *G. tabacina* and 57% in *G. tomentella*. A size distribution of cotyledon protoplasts from *G. canescens* G1171 showed protoplast diameters to range from 15 to 54 µm, with a mean of 37.5 µm.

3 Methods of Culturing Glycine Protoplasts

Protoplasts were cultured at a range of densities in:

a) Nunclon 56×14 mm plastic Petri dishes (A/S Nunc, Kamstrup, DK-4000 Roskilde, Denmark) each containing 3 ml of liquid K8P medium (Table 3).
b) Sterilin 96-well microtitre plates, each well containing 25–30 µl of liquid K8P medium, or
c) Sterilin 9-cm diameter Petri dishes containing 3 ml 0.6% w/v agarose-solidified K8P medium.

To prepare the agarose medium used in (c), a molten autoclaved solution of 1.2% w/v agarose (Sigma type VII) was added to an equal volume of filter sterilized, double strength K8P (Table 3) medium. The agarose medium was cooled to 40 °C before addition of the protoplasts and, after mixing with the latter, dispensed in the bottom of the dishes as small droplets (each approximately 30 µl in volume), or as a thin layer. After the medium had gelled, 10 ml of liquid K8P was added to each culture dish. All cultures were maintained for the first 7 days in the dark (27°) before transfer to the light (30 cm below Thorn Pluslux cool fluorescent tubes; continuous illumination, 1.6 W/m^2). The division frequency was recorded as the percentage of surviving protoplasts that had divided at least once by day 7 of culture.

During liquid culture in Petri dishes, the osmoticum was reduced gradually by replacing, every 7 days, 1 ml of culture medium with a similar volume of a 2:1, 1:1 or 1:2 mixture of K8P:K8 medium (Table 3), and finally with K8 medium (Table 3) alone. Similar media mixtures were used to dilute microtitre plates, one drop (30 µl) of fresh medium being added to each well at weekly intervals. For culture in agarose droplets, all of the liquid medium was removed weekly and replaced by 5 ml of a mixture of (a) K8P:K8, (b) K8P:SC2 or (c) K8P:SC4, again mixed in the proportions of 2:1, 1:1 and 1:2 (v/v). Subsequently, tissues in liquid or agarose media were maintained with regular, weekly subculture using liquid K8, SC2 or SC4 media as appropriate. If necessary, cell colonies could be released from the agarose matrix by application of gentle pressure to each agarose bead, using a spatula. Colonies released into the surrounding liquid medium were subcultured every 7 days with liquid SC2 or SC4 media. The plating efficiency was calculated as the percentage of the protoplasts originally plated which developed into cell colonies.

4 Behaviour of Isolated Protoplasts in Culture

4.1 Soybean Seedling Hypocotyl Protoplasts

Hypocotyl protoplasts from 5-day-old seedlings of *G. max* cv. HP-20-20 divided within 2 to 3 days of culture in liquid medium (Fig. 1e). Following the testing of a range of plating densities, hypocotyl protoplasts of HP-20-20 achieved a division frequency of 54% when plated at 5×10^4/ml (4.7 on $\log_{10}$ scale; Fig. 3). At

Fig. 3. Effect of plating density (plotted on a $\log_{10}$ scale) on division frequency in liquid medium for hypocotyl protoplasts isolated from 5-day-old soybean seedlings

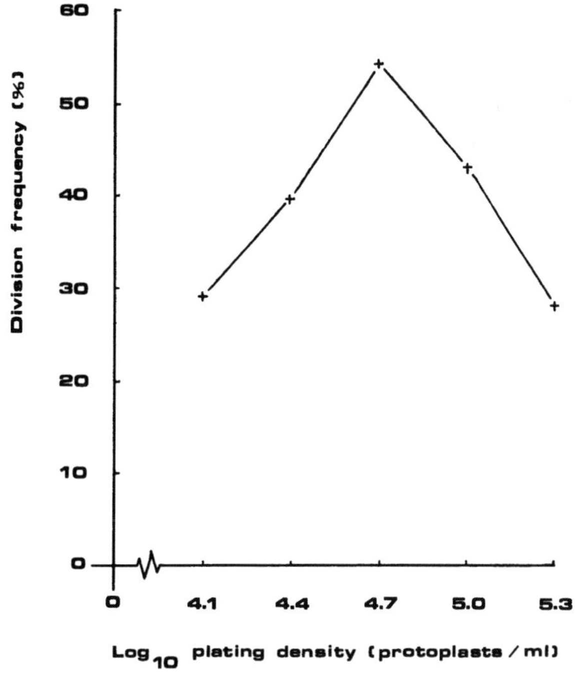

this density, the division frequency of protoplasts from 5-day-old plants was higher than that of protoplasts from younger or older seedlings (Fig. 4). Cell colonies were obtained from protoplasts cultured at 2.5×10^4/ml, 5×10^4/ml and 1×10^5/ml, but not at densities of, or below, 1.2×10^4/ml. Such tissues developed into callus following transfer to agar-solidified K8 medium. Only those hypocotyl protoplasts from soybean cv. HP-20-20 cultured at 5×10^4/ml in microtitre plates with flat-bottomed wells formed cell colonies. Protoplasts cultured in wells with U- or V-shaped profiles failed to divide. These results indicate that relatively small numbers of soybean protoplasts can be cultured using small droplets, with as few as 1250 to 1666 protoplasts forming callus using this microtitre system.

After 6 weeks of culture, hypocotyl protoplasts had a plating efficiency of 10% in droplets of agarose K8P medium coupled with the use of K8P mixed with liquid K8, SC2 or SC4 media to reduce the osmoticum. Cell colonies produced callus following transfer to either semi-solid K8 or SC2 media as appropriate. Callus produced on agar K8 was green and friable; that produced on SC2 was glossy, green and nodular, and similar to callus obtained from explants of perennial species cultured on this medium (Hammatt et al. 1986, 1987b).

4.2 Cell Suspension Protoplasts of Wild *Glycine* Species

Protoplasts released from exponentially growing suspensions of *G. soja* 3080 plated in liquid medium at densities of 2.5×10^4/ml and 5×10^4/ml exhibited division frequencies of 29% and 23% respectively. Although suspensions of this species

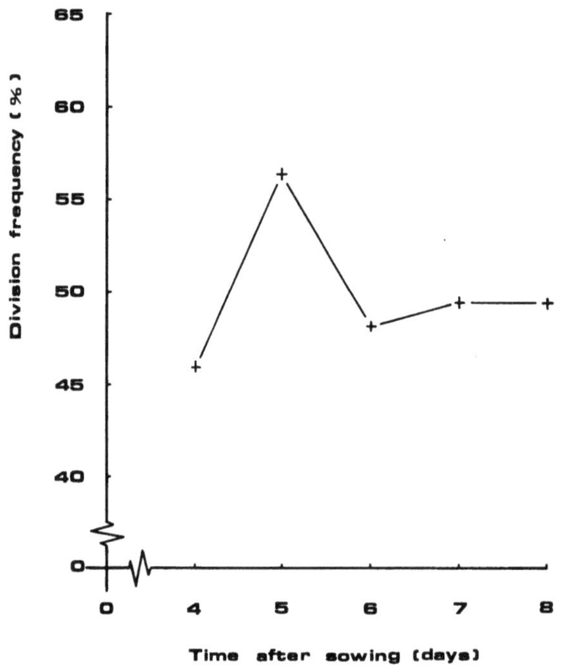

Fig. 4. Effect of seedling age on division frequency of soybean hypocotyl protoplasts cultured in liquid medium at 5.0×10^4/ml

yielded 1.1×10^5 protoplasts/ml packed volume of cells 3 or 9 days after subculture (lag and stationary phases respectively), these protoplasts failed to divide in liquid culture medium. Protoplasts of *G. canescens* 9944 and *G. tomentella* G1300 plated in liquid medium at 5×10^4/ml exhibited division frequencies of 20% and 14% respectively.

4.3 Seedling Cotyledon Protoplasts of Wild *Glycine* Species

Seedling cotyledon protoplasts plated in K8P liquid medium at 2.5×10^4/ml underwent sustained division in all species except *G. latrobeana* (G1387), the division frequencies being summarized in Table 6. Purification of protoplasts of *G. canescens* 9944 with Percoll increased the frequency from 32% to 39%. Although reduction of the culture osmolality using K8 medium led to browning of the developing cell colonies, regular replacement of spent medium with fresh liquid resulted in the production of callus in all species except *G. latrobeana*. However, the plating efficiency of cotyledon protoplasts from perennial *Glycine* species never exceeded 0.1% using liquid culture.

As in the case of soybean hypocotyl protoplasts, those isolated from *G. canescens* G1171 cotyledons and cultured in microtitre plates at 2.5×10^4/ml developed into cell colonies only in flat-bottomed wells.

With the exception of cotyledon protoplasts of *G. latrobeana* G1387, which failed to divide, protoplasts of all other accessions plated at 6×10^3/ml, 1.25×10^4/ml or 2.5×10^4/ml in agarose droplets underwent sustained division when subcultured using mixtures of K8P and K8 media, and then maintained in K8 alone. In most genotypes, the plating efficiencies reached 5%, although this was reduced to 1% or 2% at 6×10^3/ml when the initial plating medium was diluted with SC2 or SC4 instead of K8. The browning encountered in liquid medium was circumvented by culturing protoplasts in agarose.

Cell colonies of *G. argyrea* G1626, *G. canescens* G1171 and *G. clandestina* G1231 produced following osmoticum reduction with mixtures of K8P, SC2, or SC4 media, were glossy, green and nodular, and approximately 1–2 mm in diameter after 5 to 6 weeks of culture (Fig. 1 f), while those of other species were yellow in colour and friable in texture. The plating density for *G. canescens* G1171 was critical; the plating efficiency was approx. 1% at 6×10^3/ml, but protoplasts failed to undergo sustained division when plated at densities in excess of 1.0×10^4/ml or less than 2.0×10^3/ml. The plating efficiency for *G. clandestina* G1231 at 6.0×10^3/ml was approx. 5%, and 43% for *G. argyrea* G1626 plated at the same density.

5 Regeneration of Plants from Protoplasts of Wild *Glycine* Species

Green nodular tissues of *G. argyrea* G1626, *G. canescens* G1171 and *G. clandestina* G1231, obtained by culturing protoplasts in agarose medium with osmoticum reduction using SC2 or SC4 media, survived transfer to 30-ml aliquots of agar SC2 medium overlaid with 1 ml of liquid SC2 medium in 9-cm diameter Petri dishes. In contrast, friable protoplast-derived tissues from other *Glycine* species died on subculture to SC2.

Colonies of *G. canescens* G1171 were transferred to agar-solidified SC2 medium when 1–2 mm in diameter. Most responded by turning black. However, during subsequent subcultures, each of 3 weeks duration, as many as 90% of the colonies developed into green nodular callus from which shoot buds developed (Fig. 1 g). The percentage of callus tissues producing buds increased after each subculture to about 90% after six culture periods on SC2. Following transfer and a 3-week period on SC6 agar medium, buds developed into shoots, of which about 58% could be rooted on hormone-free HB50 medium. As more than one shoot developed from each morphogenic tissue, rooted plants were recovered from each protoplast-derived callus. These plants were transferred to soil (Fig. 1 h).

The regeneration procedure for *G. argyrea* G1626 and *G. clandestina* G1231 was more complex. After 4–5 weeks in culture, protoplast-derived cell colonies were released from the agarose droplets into the surrounding liquid medium using gentle pressure from a spatula and subcultured regularly with liquid SC2 or SC4 media. It was essential, following a further 3 months in liquid, to reduce the cell density in each dish. Consequently, the colonies from each original 9-cm dish were redistributed between four 9-cm dishes. Colonies larger than 4 mm in diameter were transferred to 30-ml agar solidified SC2 medium overlaid with 1 ml of liquid SC2, contained in 9-cm diameter Petri dishes. Under these conditions, 90% of the colonies produced green, nodular callus. Shoot bud formation occurred in approximately 15% of the green tissues. Unlike *G. canescens* G1171 and

G. argyrea G1626, regular subculture of protoplast-derived colonies of *G. clandestina* to SC2 failed to have a major effect on the frequency of shoot formation. Shoots developed from buds following a 20–30-day period on SC6 medium, with about 70% of the shoots producing roots following excision from the callus and transfer to HB50 medium containing 0.2% w/v activated charcoal.

Overall, about 6.4% of protoplasts of *G. argyrea* G1626, 0.8% of those from *G. canescens* G1171 and 0.68% in *G. clandestina* G1231 produced shoots after six subcultures of protoplast-derived tissues on agar-solidified SC2 medium.

6 Fusion of Protoplasts

Electrofusion has been found to be the most convenient method of fusing soybean protoplasts. Protoplasts from hypocotyls of dark-grown soybean cv. HP-20-20 seedlings were labelled with fluorescein by the addition of 12 µl of a 1.1 mg/ml solution of fluorescein diacetate (FDA) in acetone to each 6.0 ml volume of the enzyme solution used for protoplast isolation. FDA-labelled protoplasts were mixed with those from green seedling cotyledons of *G. canescens* G1171 (1 : 1 ratio) in an autoclaved solution containing 11% w/v mannitol and 147 mg/l $CaCl_2 \cdot 2H_2O$. The final protoplast density was 5.0×10^4/ml. One ml volumes of the protoplast mixture were dispensed into 2.0 cm^2 wells of a Sterilin 10-cm square Petri plate (only the eight squares surrounding the centre square of each plate were used). The electrode system, similar to that described by Watts and King (1984), containing five flat brass electrodes (each 1.8×3.0 cm) separated by five perspex spacers (each 1.8×1.0×0.4 cm), was sterilized by immersion in absolute ethanol, allowed to dry in a laminar flow cabinet, and inserted into one of the square wells. After electrofusion, the electrode was moved to other protoplast-containing wells. The electrical parameters applied were an AC field of 500 KHz, 20–30 V/cm, and a DC pulse of 200–400 V/cm. Dielectrophoretic movement of protoplasts into dimers and "pearl chains" of tightly adhering protoplasts, was achieved during a 30- to 60-s exposure to the AC field. Fusion was induced using a 2-ms DC pulse (Fig. 5a). There was no noticeable difference in the effect of DC voltages between 200–400 V/cm upon the rate of fusion (Table 7). A typical field of protoplasts 1 h post fusion is shown using UV microscopy (Fig. 5b).

Fig. 5A–F. A Electrofusion of a soybean hypocotyl protoplast (*H*) with a green cotyledon protoplast (*C*) of *G. canescens* G1171 (Bright field ×540). **B** Protoplasts under UV illumination 1 h post-electrofusion. Unfused, FDA-labelled soybean hypocotyl protoplasts (*g*) show green fluorescence; unfused protoplasts of *G. canescens* G1171 (*r*) fluoresced red owing to chlorophyll autofluorescence. Heterokaryons (*arrowed*) exhibit both green and red fluorescence (×173). **C, D** Fluorescence profiles revealed by flow cytometric analysis of FDA-labelled, dark-grown soybean hypocotyl protoplasts (**C**) showing mainly green fluorescence (X-axis), and green cotyledon protoplasts of *G. canescens* G1171 (**D**) showing mainly red fluorescence (Y-axis). A preparation 1 h post-electrofusion (**E**) contains protoplasts exhibiting parental fluorescence profiles, and a separate population (*arrowed*) of heterokaryons with both green and red fluorescence. Vertical axis is the relative number of particles. **F** Bifluorescent heterokaryons formed between hypocotyl protoplasts of *G. max* cv. HP-20-20 and cotyledon protoplasts of *G. canescens* G1171 isolated by flow cytometry (×100)

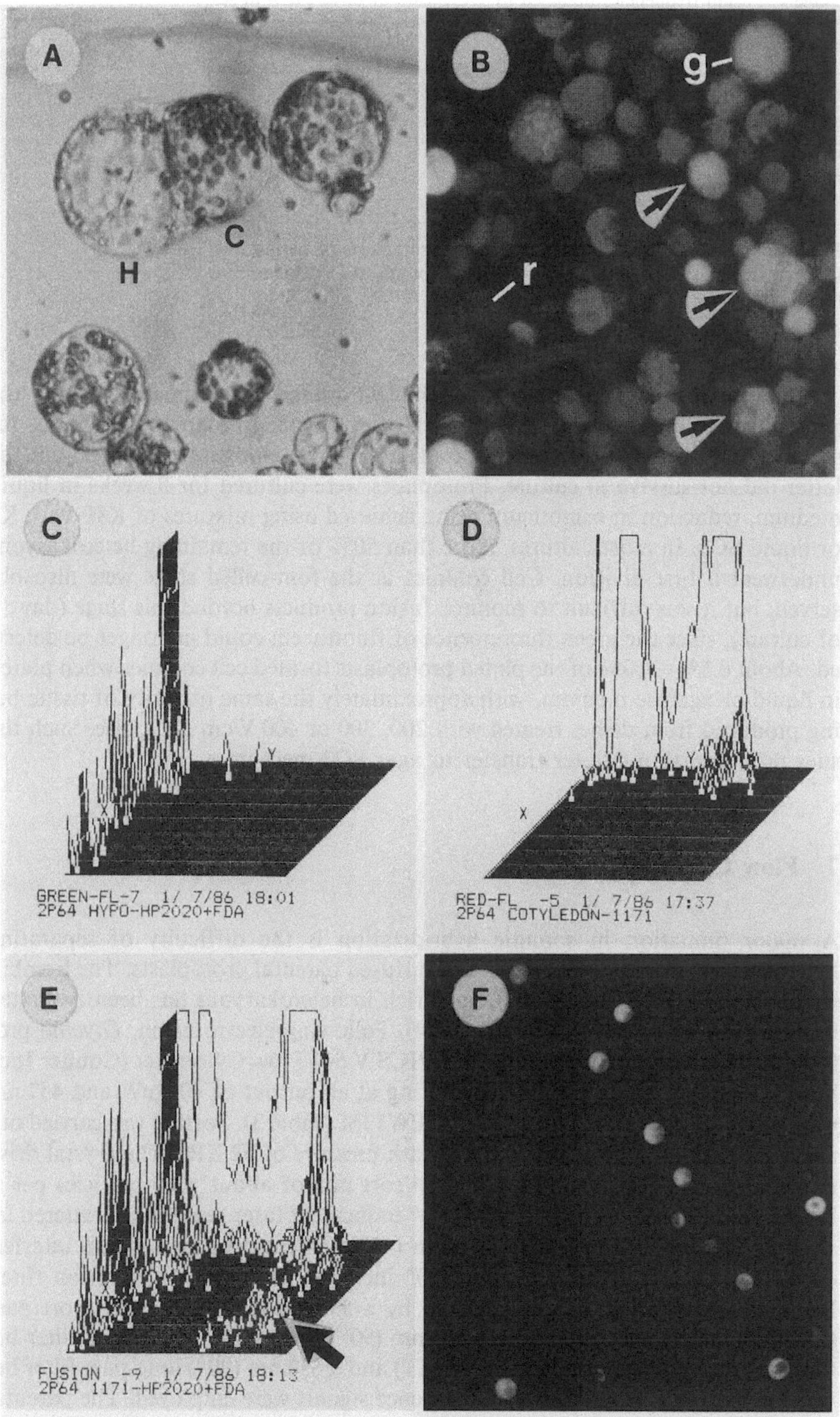
A
H
C
B
g
r
C
GREEN-FL-7 1/ 7/86 18:01
2P64 HYPO-HP2020+FDA
D
RED-FL -5 1/ 7/86 17:37
2P64 COTYLEDON-1171
E
FUSION -9 1/ 7/86 18:13
2P64 1171-HP2020+FDA
F

Table 7. Fusion frequencies obtained by DC pulses of different voltages

Voltage (V)	Fusion rate (%)
0	0
100	0
200	3.3 ± 1.7
300	3.4 ± 2.3
400	2.8 ± 2.2

Results of six fusion experiments. Hypocotyl protoplasts of dark-grown soybean seedlings fused with green cotyledon protoplasts of *G. canescens* G1171.

In most treatments, irrespective of the DC voltage, approximately 50% of the protoplasts collapsed during the first 4 h after fusion. In many pearl chains, fusion occurred between adjacent protoplasts to form elongate structures, but the latter did not survive in culture. Protoplasts were cultured for 8 weeks in liquid medium, reduction in osmoticum being achieved using mixtures of K8P with K8 or liquid SC2. In most cultures, more than 50% of the remaining heterokaryons underwent a first division. Cell colonies at the four-celled stage were also observed, but it was difficult to monitor fusion products beyond this stage (day 10 of culture), since the green fluorescence of fluorescein could no longer be detected. About 0.5% – 1.0% of the plated protoplasts formed cell colonies when plated in liquid or agarose medium, with approximately the same quantity of tissue being produced from dishes treated with 200, 300 or 400 V/cm DC pulse. Such tissues produced shoots after transfer to agar SC2 medium.

7 Flow Cytometry

A major limitation in somatic hybridization is the difficulty of separating heterokaryons from homokaryons and unfused parental protoplasts. The feasibility of selecting protoplast populations rich in heterokaryons has been evaluated using a flow cytometer (Galbraith 1984). Following electrofusion, *Glycine* protoplasts have been analyzed using an EPICS V 541 Flow Cytometer (Coulter Inc., Florida) with an argon ion laser operating at an output of 100 mW and 457 nm wavelength. Sheath fluid consisted of CPW13M (Table 3). Sorting was carried out using a 200-μm diameter tip with a sheath pressure of 52 KPa. The crystal drive frequency was 8 kHz, giving a potential sort rate of about 1000 particles per s. Fluorescence signals were detected at 90° to incident laser radiation. Scattered laser light was blocked by a 450–470-nm (45° incidence) band rejection interference filter, backed up by a 495-nm (90° incidence) long pass absorption filter. Red-green discrimination was achieved by a 590 nm (45° incidence) short pass (dichroic) interference filter with a 565 nm (90°) short pass interference filter before the green photomultiplier tube (PMT) and a 630 nm (90°) long-pass filter before the red PMT. Log integral fluorescence signals were employed. The parental protoplast populations possessed distinct fluorescence profiles (Fig. 5c, d). In preparations containing FDA-labelled soybean hypocotyl protoplasts fused with green cotyledon protoplasts of *G. canescens*, the former emitted green fluorescence, while the latter showed red chlorophyll autofluorescence. In a 3-D plot of a protoplast preparation 3 h after fusion, a scattered population of heterokaryons was present, showing both green and red fluorescence (Fig. 5e). Heterokaryons were sorted from the parental protoplasts at a rate of 1000 per flat-bottomed well of a Sterilin 96-well microtitre plate (Fig. 5f). Each well contained 30 µl of liquid K8P medium. Following sorting, the heterokaryons were allowed to settle, after which, while observing the protoplasts using an inverted microscope, the K8P medium was carefully removed in 20-µl volumes using a micropipette in order to minimize disturbance of the heterokaryons. Forty to 60 µl of molten 0.6% (w/v) agarose-solidified K8P medium (Table 3), previously cooled to 30 °C, were added to each well to achieve a plating density of 200 heterokaryons/30 µl medium. The resulting protoplast suspension was dispensed as 20-µl drops in the bottom of a 9-cm diameter Petri dish. The position of each droplet was marked, prior to filling the rest of the dish with 30-µl droplets of agarose K8P containing cotyledon protoplasts of *G. canescens* G1171 at a density of 6×10^3/ml. Subsequent to gelling of the agarose, 10 ml of liquid K8P medium was added to each dish. Protoplasts were cultured as described for cotyledon protoplasts of *G. canescens* G1171. Osmoticum reduction was achieved using mixtures of K8P and SC2 media, and later, SC2 alone (see Sect. 3).

Cell colonies that developed from the droplets that had initially contained heterokaryons were transferred to agar-solidified SC2 medium, on which, following regular, fortnightly subculture, they developed into green, nodular tissue. Approximately 10% of these callus tissues became morphogenetic and gave rise to shoots.

8 Conclusion and Prospects

Recent work with *Glycine* species has shown that it is now possible to:

a) isolate protoplasts from hypocotyls of soybean and from cell suspensions and seedling cotyledons of perennial *Glycine* species,
b) culture protoplasts from *Glycine* species to cell colonies,
c) regenerate whole plants from protoplast-derived tissues of *G. argyrea* G1626, *G. canescens* G1171 and *G. clandestina* G1231,
d) fuse protoplasts of soybean with those from *G. canescens* G1171,
e) select heterokaryons from unfused parental protoplasts using flow cytometry, and
f) culture flow sorted heterokaryons to shoots.

The ability to regenerate plants from cotyledon protoplasts of *G. argyrea*, *G. canescens* and *G. clandestina* now enables these perennial *Glycine* species to be incorporated into programmes employing somatic hybridization to improve the commercial soybean, *G. max*.

Since heterokaryons isolated using a flow cytometer have already been cultured to morphogenetic callus, it is possible that heterokaryon-derived tissues will give rise to novel plants. Extension of the regeneration procedures described in this chapter, either with or without modification, to other perennial *Glycine* accessions, should enable a range of somatic hybrid and cybrid plants to be produced following protoplast fusion. In addition, reproducible plant regeneration will enable perennial *Glycine* species to be subjected to other genetic manipulation procedures, including transformation by isolated DNA, and *Agrobacterium*-mediated delivery of genes.

9 Protocol for the Isolation, Culture, and Electrofusion of Glycine Protoplasts

Isolation

1. Incubate sliced tissue from three 5-day-old, dark-grown hypocotyls of soybean cv. HP-20-20 for 16 h at 27 °C, in 10 ml ES2 enzyme, to which has been added 50 µl of a 1 mg ml/l solution of FDA in acetone.
2. Incubate 20 sliced cotyledons of *G. canescens* G1171 in 10 ml ES3 enzyme (16 h; 27 °C).
3. Separate protoplasts from enzyme by sieving (45 µm pore), followed by centrifugation (55 *g*).
4. Wash protoplasts by centrifugation twice through CPW13M (soybean) or CPW9M (*G. canescens*).

For electrofusion proceed to step 8.

Culture

5. Suspend soybean protoplasts in K8P medium (Table 3) at a density of 5×10^4/ml, and dispense as 3-ml aliquots into 5-cm diameter Petri dishes.
6. Suspend cotyledon protoplasts of *G. canescens* in 0.6% agarose-solidified K8P medium (30 °C at a density of 6×10^3/ml, and dispense as 30-µl droplets into a 9-cm diameter Petri dish (3 ml total per dish). Following agarose solidification, add 10 ml liquid K8P to each dish.
7. Incubate protoplasts in the dark (27° ± 2 °C).

Electrofusion

8. Wash protoplasts twice by centrifugation, through electrofusion solution (11% w/v mannitol, 147 mg/l $CaCl_2\cdot2H_2O$, pH 5.8).
9. Resuspend protoplast pellets in electrofusion solution and adjust their densities to 5×10^4/ml.
10. Mix parental protoplasts in equal volumes.
11. Transfer separate 1-ml aliquots of the suspension of mixed parental protoplasts to each of the eight wells surrounding the central compartment of a 100 mm × 100 mm Petri dish (Sterilin).
12. Align protoplasts by inserting an electrode similar to that described by Watts and King (1984) into each well sequentially, and applying an AC field of 20–30 V/cm (500 KHz, 30 s).
13. Fuse protoplasts in each well by application of a DC pulse (100–400 V/cm, 0.2 ms).
14. Isolate heterokaryons from parental protoplasts using an EPICS V flow cytometer (see text).
15. Suspend 200 heterokaryons in 30-µl volumes of 0.6% agarose-solidified K8P medium.
16. Dispense as droplets in the bottom of a 9-cm diameter plastic Petri dish, and mark the position of these droplets.

17. Surround beads containing heterokaryons with 3 ml of *G. canescens* G1171 cotyledon protoplasts, suspended in agarose-solidified K8P medium at a density of 6×10^3/ml, dispensed as 30-μl droplets.
18. After agarose solidification, add 10 ml liquid K8P medium and culture as described in step 7.
19. For details on plantlet recovery from parental *Glycine* protoplasts and heterokaryons, see text.

Acknowledgments. The work described was carried out while N.H. was in receipt of an SERC CASE Studentship with Shell Research Ltd, Sittingbourne Research Centre, Sittingbourne, UK. The authors thank B.V. Case (photographic assistance), A. J. C. Parrish (graphical illustrations) and Dr. A. Lister (cell sorting), and express their gratitude to Professor E. C. Cocking (University of Nottingham) and Dr. R. S. Nelson (Shell Research Ltd.) for stimulating discussions.

References

Baldes R, Moss M, Geider K (1987) Transformation of soybean protoplasts from permanent suspension cultures with cells of *Agrobacterium tumefaciens*. Plant Molec Biol 9:135–145

Bojsen KM, Wyndaele R (1985) Influence of lowered temperature storage on growth and protoplasts isolation of soybean callus. J Plant Physiol 118:95–103

Broué P, Douglass J, Grace JP, Marshall DR (1982) Interspecific hybridisation of soybean and perennial *Glycine* species indigenous to Australia via embryo culture. Euphytica 31:715–724

Burdon JJ, Marshall DR (1981 a) Evaluation of Australian native species of *Glycine* for resistance to soybean rust. Plant Dis 65:44–45

Burdon JJ, Marshall DR (1981 b) Inter- and intra-specific diversity in the disease response of *Glycine* species to the leaf rust fungus *Phakopsora pachyrhizi*. J Ecol 69:381–390

Burdon JJ, Speer SS (1984) A set of differential *Glycine* hosts for the identification of races of *Phakopsora pachyrhizi* Syd. Euphytica 33:891–896

Chien YC, Kao KN, Wetter LR (1982) Chromosome and isozyme studies of *Nicotiana tabacum-Glycine max* hybrid cell lines. Theor Appl Genet 62:301–304

Chowhury VK, Widholm JM (1985) Callus production from photoautotrophic soybean cell culture protoplasts. Plant Cell Rep 4:289–292

Constabel F, Dudits D, Gamborg OL, Kao KN (1975 a) Nuclear fusion in intergeneric heterokaryons. Can J Bot 53:2092–2095

Constabel F, Kirkpatrick JW, Kao KN, Kartha KK (1975 b) The effect of canavanine on the growth of cells from suspension cultures and on intergeneric heterokaryocytes of canavanine-sensitive and tolerant plants. Biochem Physiol Pflanz 168:319–325

Constabel F, Weber G, Kirkpatrick JW, Pahl K (1976) Cell division of intergeneric protoplast fusion products. Z Pflanzenphysiol 79:1–7

Fowke LC, Rennie PJ, Kirkpatrick JW, Constabel F (1976) Ultrastructure of fusion products from soybean cell culture and sweet clover leaf protoplasts. Planta 130:39–45

Franceschi VR, Ku MSB, Wittenbach VA (1984) Isolation of mesophyll and paraveinal mesophyll protoplasts from soybean leaves. Plant Sci Lett 36:181–186

Frearson EM, Power JB, Cocking EC (1973) The isolation, culture and regeneration of *Petunia* leaf protoplasts. Dev Biol 33:130–137

Galbraith DW (1984) Selection of somatic hybrid cells by fluorescence-activated cell sorting. In: Vasil IK (ed) Cell culture and somatic cell genetics of plants, vol I. Academic Press, Orlando, pp 433–448

Gamborg OL, Miller RA, Ojima K (1968) Nutrient requirements of suspension cultures of soybean roots. Exp Cell Res 50:151–158

Gamborg OL, Davis BP, Stahlhut RW (1983) Cell division and differentiation in protoplasts from cell cultures of *Glycine* species and leaf tissue of soybean. Plant Cell Rep 2:213–215

Grant JE (1986) Hybridisation of soybean with its diploid wild perennial relatives. Proc Symp Plant Breed, Lincoln, N Z Pap 35

Hammatt N (1986) Plant regeneration and protoplast fusion in the genus *Glycine*. PhD Thesis, Univ of Nottingham, UK

Hammatt N, Davey MR (1988) Isolation and culture of soybean hypocotyl protoplasts. In Vitro 24:601–604
Hammatt N, Davey MR, Nelson RS (1986) Plant regeneration from seedling cotyledons, petioles and leaves of *Glycine clandestina* Wendl. Physiol Plant 68:125–128
Hammatt N, Kim H-I, Davey MR, Nelson RS, Cocking EC (1987a) Plant regeneration from cotyledon protoplasts of *Glycine canescens* and *G. clandestina*. Plant Sci 48:129–135
Hammatt N, Nelson RS, Davey MR (1987b) Plant regeneration from seedling explants of perennial *Glycine* species. Plant Cell Tissue Organ Cult 11:3–11
Hammatt N, Jones B, Davey MR (1989) Plant regeneration from seedling explants and cotyledon protoplasts of *Glycine argyrea* Tind. In Vitro 25:669–672
Kao KN (1977) Chromosomal behaviour of somatic hybrids of soybean-*Nicotiana glauca*. Molec Gen Genet 50:225–230
Kao KN, Michayluk MR (1974) A method for high-frequency intergeneric fusion of plant protoplasts. Planta 115:355–367
Kao KN, Keller WA, Miller RA (1970) Cell divisions in newly formed protoplasts of soybean. Exp Cell Res 62:338–340
Kao KN, Gamborg OL, Miller RA (1971) Cell divisions in cells regenerated from protoplasts of soybean and *Happlopappus gracilis*. Nature New Biol 232:124
Kao KN, Constabel F, Michayluk MR, Gamborg OL (1974) Plant protoplast fusion and growth of intergeneric hybrid cells. Planta 120:215–227
Kartha KK, Gamborg OL, Constabel F, Kao KN (1974) Fusion of rapeseed and soybean protoplasts and subsequent division of heterokaryocytes. Can J Bot 52:2435–2436
Lin W (1983) Isolation of mesophyll protoplasts from mature leaves of soybeans. Plant Physiol 73:1067–1069
Lu DY, Cooper-Bland S, Pental D, Cocking EC, Davey MR (1983) Isolation and sustained division of protoplasts from cotyledons of seedlings and immature seeds of *Glycine max* L. Z Pflanzenphysiol 111:389–394
Marshall DR, Broué P (1981) The wild relatives of crop plants indigenous to Australia and their use in plant breeding. J Aust Inst Agric Sci 47:149–154
Michayluk MR, Kao KN (1975) A comparative study of sugars and sugar alcohols on cell regeneration and sustained cell division in plant protoplasts. Z Pflanzenphysiol 75:181–185
Mignucci JS, Chamberlain DW (1978) Interactions of *Microsphaera diffusa* with soybeans and other legumes. Phytopathology 68:169–173
Miller RA, Gamborg OL, Keller WA, Kao KN (1971) Fusion and division of nuclei in multinucleated soybean protoplasts. Can J Genet Cytol 13:347–353
Mroginski LA, Kartha KK (1985) Tissue culture of legumes for crop improvement. Plant Breed Rev 2:215–264
Murashige T, Skoog F (1962) A revised medium for rapid growth and bioassays with tobacco tissue cultures. Physiol Plant 15:473–497
Myers JR, Kysely W, Lazzeri PA, Hildebrandt DF, Collins GB (1986) Protoplast isolation and culture of *Glycine* species with plant regeneration of *G. canescens*. In: Somers DA, Gengenbach BG, Biesboer DD, Hackett WP, Green CE (eds). 6th Int Congr plant tissue and cell culture. Abstr, Univ Minnesota, Minneapolis, p 269
Newell CA, Hymowitz T (1982) Successful wide hybridisation between the soybean and a wild perennial relative, *G. tomentella* Hayata. Crop Sci 22:1062–1065
Newell CA, Luu HT (1985) Protoplast culture and plant regeneration in *Glycine canescens* F. J. Herm. Plant Cell Tissue Organ Cult 4:145–149
Newell CA, Luu HT (1988) Isolation and culture of *Glycine* protoplasts. In: Bajaj YPS (ed) Biotechnology in agriculture and forestry, 8. Plant protoplasts and genetic engineering. Springer, Berlin Heidelberg New York Tokyo, pp 173–192
Newell CA, Delannay X, Edge ME (1987) Interspecific hybrids between the soybean and wild perennial relatives. J Hered 78:301–306
Niizeki M, Tanaka M, Akada S, Hirai A, Saito K (1985) Callus formation of somatic hybrid of rice and soybean and characteristics of the hybrid callus. Jpn J Genet 60:81–92
Niizeki M, Tanaka M, Saito K (1986) Response of somatic hybrid callus between rice and soybean to streptomycin. Jpn J Breed 36:75–79

Oelck MM, Rao PS, Ozias-Akins P, Schieder O (1983) Protoplast regeneration from some legume species. Experientia Suppl 45:50–51
Pederson HC, Christiansen J, Wyndaele R (1983) Induction and in vitro culture of soybean crown gall tumors. Plant Cell Rep 2:201–204
Rees RT, Cobb AH, Pallett KE (1985) A developmental study of *Glycine max* cell and protoplast isolation in relation to leaf age and photosynthetic competence. Physiol Plant 65:252–256
Riggs RD, Hamblen ML (1962) Soybean cyst-nematode host studies in the family Leguminosae. Agric Exp Sta, Div Agric, Univ Arkansas Rep Ser 110:20
Riggs RD, Hamblen ML (1966) Further studies on the host range of the soybean cyst nematode. Agric Exp Sta, Div Agric, Univ Arkansas Bull 718
Schwenk FW, Pearson CA, Roth MR (1981) Soybean mesophyll protoplasts. Plant Sci Lett 23:153–155
Singh RJ, Hymowitz T (1985) An intersubgeneric hybrid between *Glycine tomentella* Hayata and the soybean, *G. max* (L.) Merr. Euphytica 34:187–192
Singh BB, Gupta SC, Singh BD (1974) Sources of field resistance to rust and yellow mosaic diseases of soybean. Indian J Genet Plant Breed 34:400–404
Tricoli DM, Fein MB, Carnes MG (1986) Culture of soybean mesophyll protoplasts in alginate beads. Plant Cell Rep 5:334–337
Uchimiya H, Murashige T (1974) Evaluation of parameters in the isolation of viable protoplasts from cultured tobacco cells. Plant Physiol 54:936–944
Watts JW, King JM (1984) A simple method for large-scale electrofusion and culture of plant protoplasts. Biosci Rep 4:335–342
Weber G, Roth EJ, Schweiger HG (1983) Storage of cell suspensions and protoplasts of *Glycine max* (L.) Merr., *Brassica napus* (L.)., *Datura innoxia* (Mill.), and *Daucus carota* (L.) by freezing. J Plant Physiol 62:29–39
Wei Z, Xu Z (1988) Plant regeneration from protoplasts of soybean (*Glycine max* L.). Plant Cell Rep 7:348–351
Wetter LR (1977) Isoenzyme patterns in soybean-*Nicotiana* somatic hybrid lines. Mol Gen Genet 150:231–235
Wetter LR, Kao KN (1980) Chromosome and isoenzyme studies on cells derived from protoplast fusion of *Nicotiana glauca* with *Glycine max-Nicotiana glauca* cell hybrids. Theor Appl Genet 57:273–276
Xu Z-H, Davey MR, Cocking EC (1982) Callus formation from root protoplasts of *Glycine max* (Soybean). Plant Sci Lett 24:111–115
Zieg RG, Outka DE (1980) The isolation, culture and callus formation of soybean pod protoplasts. Plant Sci Lett 18:105–114

II.6 Winged Bean [*Psophocarpus tetragonolobus* (L.) D.C.]

S. VENKETESWARAN[1]

1 Introduction

The winged bean, *Psophocarpus tetragonolobus* (L.) D.C., is a member of the Leguminoseae familiy and is a tropical plant found chiefly in rural areas of Papua New Guinea and Southeast Asia. It grows abundantly in hot, humid equatorial countries such as Indonesia, Malaysia, Thailand, the Philippines, India, Bangladesh, Burma, and Sri Lanka. Although at one time considered as a "poor man's food", the potential economic importance of the plant has attracted worldwide attention and it is now recognized as "A High Protein Crop for the Tropics" (NAS 1981). Because of this, several important advances in growing winged bean in the fields and research in various laboratories are being made all over the world so that it will become a significant source of food and proteins, particularly to developing countries. The advancements and breakthroughs in biotechnology and its applications to agricultural plants like the winged bean have opened new avenues for crop improvement and significant progress in achieving these goals are anticipated in order to exploit the full potential of this important plant.

1.1 Description of the Plant

The winged bean plant grows as a vine with climbing stems and leaves, 3–4 m high. It is a herbaceous perennial; but can be grown generally as an annual. Although there are several varieties, with wide differences in morphology and physiology, the plant produces an abundance of leaves and inflorescences of white, blue, deep purple, or pink flowers which quickly develop into pods. The pods are four-sided with fringed wings and can be 6–30 cm in length, consisting of 5–20 seeds per pod. The seeds, which are rich in protein, are comparable to soybean in composition and nutritional value (Kailasapathy 1986), and contain similar proportions of protein, carbohydrates, oil, minerals, vitamins, essential amino acids, and other constituents. The plant also produces underground tubers of varying sizes, rich in carbohydrates (25%–30%) and proteins (10%–15%). The plant is one of the best nitrogen fixers, with nodulation accomplished by the soil bacterium, *Rhizobium*. Because of its ability to fix nitrogen from the atmo-

[1] Department of Biology, University of Houston, 4800 Calhoun, Houston, TX 77004, USA

Biotechnology in Agriculture and Forestry, Vol. 10
Legumes and Oilseed Crops I (ed. by Y.P.S. Bajaj)

Table 1. Proximate chemical composition of winged bean and soybean seeds (g/100 g dry weight)[a]. (Kailasapathy 1986)

Analysis	Winged bean cultivars		Soybean (PB-1)
	SLS-40	TPT-2	
Protein[b]	32.65	36.76	40.3
Fiber	6.45	7.53	3.20
Fat (ether extract)	15.84	16.85	21.00
Ash	3.51	4.32	3.20
Carbohydrates[c]	41.55	34.54	32.3

[a] Average of three analyses.
[b] N×6.25.
[c] Calculated by difference

sphere, the plant requires very little or no fertilizers and the seeds have a relatively high protein content (30%–40%) and edible oil (15%–20%). Table 1 gives the approximate composition of winged bean and soybean seeds in g/100 g dry weight (Kailasapathy 1986). Besides the various economical and industrial uses of seeds for commercial exploitation, they are also useful as food when steamed, boiled, fried, roasted, fermented, made into milk, and cooked by other methods. The young tender immature pods are consumed as a vegetable, the young sprouts, shoots, and leaves are cooked and the tubers are boiled, steamed, fried, or baked; therefore, the entire plant is useful for human consumption as food. Lastly, the mature leaves and other parts of the plant can be useful as animal feedstock and for other purposes. The winged bean can therefore be considered as a multipurpose plant. The botanical and economical value of the winged bean and the related research activities on this plant have been described in detail in several publications. (*The Winged Bean* 1978; NAS 1981; Khan 1982; The Winged Bean Flyer(s) 1977–1978; 1980–1985; Venketeswaran 1985; Gunasena 1986; Tran Thanh et al. 1986).

1.2 Distribution and Area Under Cultivation

The taxonomy and distribution of the genus *Psophocarpus* has been described and discussed in detail (Khan 1982; Eagleton et al. 1982). Of the nine species recorded, only *P. tetragonolobus* and *P. palustris*, which is indigenous to Africa, is widely used as food with *P. palustris* remaining as a semi-wild plant in West Africa (NAS 1981), and *P. tetragonolobus* being the most popular food species. A suggested classification of the species of *Psophocarpus* is documented (Verdcourt and Halliday 1979) and the taxonomy of the genus and key to the species has been described (Khan 1982). The origin of the genus has been the subject of much speculation, and three widely separated areas have been suggested as centers of origin, i.e. Africa, Indo-Malaya (Asia), and Papua New Guinea (Burkill 1906; Vavilov 1949/50; Hymowitz and Boyd 1977). Eagleton et al. (1982) examined the historical, anthropological, and taxonomic evidence and the available germplasm

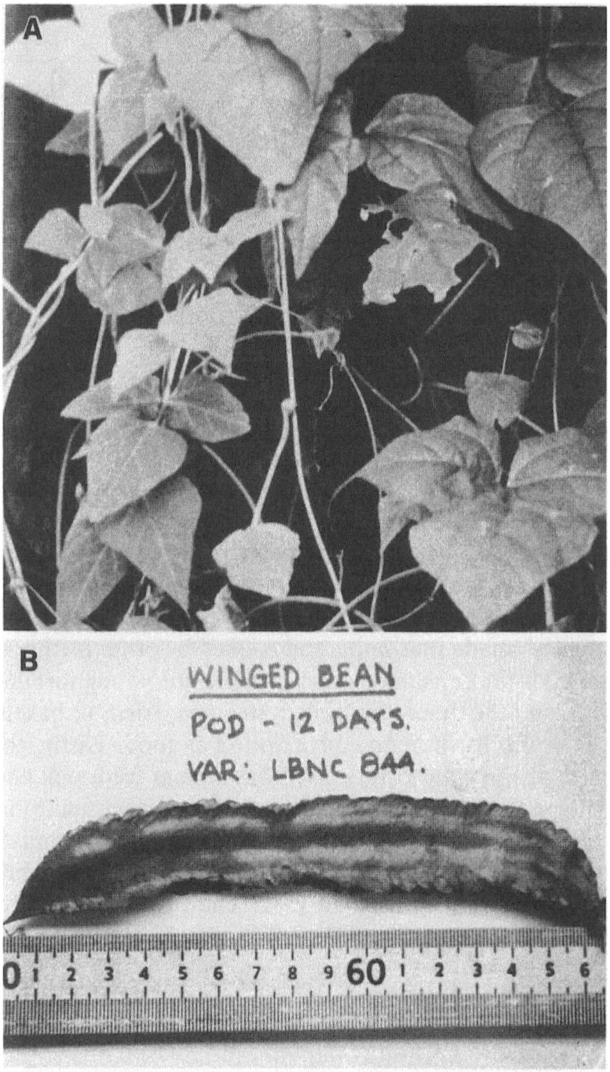

Fig. 1. A Winged bean plant growing in the backyard in 10911 Valley Hills, Houston, Texas, USA. **B** Pod of the winged bean variety LBNC844 at 12 days of age

studies, and support the possibility of the Asian origin. Khan (1982) indicates that more taxonomic and germplasm studies comparing and covering a wider gene pool will be needed for understanding the origin of the winged bean. In recent years, cultivation of winged bean, particularly for experimental purposes, has been attempted in several tropical, semi-tropical, and subtropical areas of the world with varying success. A winged bean plant growing in the backyard in Houston, Texas, USA, is illustrated in Fig. 1.

1.3 Relevance of the Winged Bean Plant to Developing Countries

The protein-rich winged bean can be considered as playing a dominant role in short- and long-range development and improvement of food production in developing countries of Asia and Africa. In Sri Lanka, although at present, the winged bean is grown as a vegetable crop by individuals (locally called "dambala") for their daily cooking, – an International Winged Bean Institute has been established to demonstrate its vital role in expanding its resources for commercial production and various industrial uses. In the Philippines, a national program on the winged bean has been targeted. The Government of Indonesia has taken winged bean as a priority research crop. In Ghana, the bean has been adapted as a crop for its green pod, and extensive research on agronomy, dietary use, etc. is being undertaken. In India, the green pod is grown for use as a vegetable in certain areas and some research is carried out to expand its germplasm resources. These are just a few examples of the tremendous impact of and interest in the winged bean as a potential high-protein crop for the tropics and developing nations. Expansion of winged bean cultivation for its pods and tubers for improving the standard of nutrition and large-scale production of seeds for other purposes can lead to the creation of many village-scale industries, like the production of winged bean milk and milk-based products like yogurt, winged bean coffee, and other beverages. Further, large-scale industrial projects for the manufacture of winged bean seed oil, leaf protein concentrates and pellets for animal feed, etc. can make winged bean as significant as the soybean in world agriculture.

1.4 Diseases and Pests

The winged bean is susceptible to a variety of diseases and pests depending on how, where, and in which climate it is grown. Besides certain nonpathogenic disorders which may appear when grown in undesirable locations, diseases caused by viruses, bacteria, fungi and mycoplasm, nematode disorders, and a variety of insect pests have been recorded. Although only few detailed studies are available, descriptions of some of these diseases are reported and a table of insect pests of winged bean has been compiled (*The Winged Bean* 1978; NAS 1981; Khan 1982). A number of chemicals which have been used against these diseases and insect pests under experimental conditions with varying success have also been documented (Khan 1982).

1.5 Conventional Practices for Propagation and Improvement

The conventional cultivation practices of winged bean depend largely on whether they are grown in small plots or in large fields. These have been investigated in great detail (Rachie and Luse 1978; *The Winged Bean* 1978; NAS 1981; Khan 1982). The plant tolerates a wide range of soil conditions and can withstand high temperatures, but never survives frost. It grows at sea level and altitudes up to 2000 m high. They are sensitive to daylength and set flowers only when days are short. The plants are capable of producing exceptional amounts of seeds (2–3 tons/ha). Seeds are generally sown in rows and after 3–4 weeks, when the plant attains a height of 2–3 ft, support systems for climbing are generally made in the form of stakes, trellises, or mixed cropping in association with support crops, which may be hardy herbs, shrubs, annuals, or even perennials. Since the requirement for plant support is a major limitation for growing this crop plant on a commercial and industrial level, due to agricultural labor costs, etc., attempts are being carried out in different regions to produce dwarf and/or other mutants which will not require a support system, and which will allow mechanization of seed collection (Anonymous 1977; Shivashankar 1984).

1.6 Improved Methods of Propagation and Rationale

The advent of biotechnology and its applications to agricultural plants has opened up new avenues for crop improvement (Bajaj 1986, 1989). "Biotechnology" in plants means "tissue culture", which Bollinger (1980) calls the "botanical equivalent of laser", suggesting the vast area of potential applications of tissue culture to a variety of agricultural crops toward crop improvement. The development of in vitro culture techniques on the winged bean thus offers an alternative method to conventional plant breeding programs. The basic development goals of the world for increased food production and the development of high protein-rich food crops will need large quantities of seeds. Every effort which can lead to increased seed production or development of plants in large numbers for mass propagation will relieve the shortage of plant stock materials. In vitro methods offer this advantage in that it will be possible to mass produce potential plants which can be transferred to the greenhouse and eventually transported worldwide to different regions. Further, genetic improvement through selection (somaclonal variation methods) and development of plants which can be grown on various saline/alkaline soils or under stress conditions, etc. can be achieved.

Conventional plant breeding techniques and propagation by seed is limited by the long time required for obtaining the desired results. Seed propagation also has the disadvantage of heterogeneity, as some cross-pollination does occur, as in the case of winged bean, which is basically a self-pollinating plant. Therefore, the development of a self-supporting, high-yielding variety with good quality seed by conventional breeding methods will involve heavy expenditure and considerable time. In vitro techniques offer the best solution, as they will enable development of desirable varieties and their multiplication with less time.

2 In Vitro Approaches

Although, traditionally, the winged bean has been grown as a garden agricultural crop, in vitro studies, including regeneration of plants from single cells, protoplasts, etc., and development of new cultivars can be advantageously used to

Table 2. Summary of tissue culture studies on winged bean

Inoculum	Medium (mg/l)	Growth response	Reference
Hypocotyl explants	MS + kin(1) + NAA(0.1)	Roots Embryos	Venketeswaran and Huhtinen (1978)
Hypocotyl explants	B5 + 2,4-D(0.5) B5 + kin(0.3) + IAA(5)	Callus Roots	Bottino et al. (1979)
Leaf explants	MS + NAA(0.2) + BAP(2.5)	Callus	Gregory et al. (1980)
	MS + NAA(2) + BAP(0.02) MS + BAP(2,10) + IAA(0.2)	Roots Shoot buds	
Mature leaves	MS + NAA(0.2) + BAP(2) 3% Sucrose	Callus	Evans et al. (1981)
	MS + IAA(0.2) + BAP(2) MS + NAA(0.4) + BAP(2) 0.5% Sucrose	Buds Shoots, plantlets	
Seedlings	B5 + BAP(1.125 – 6.75) B5 + NAA(0.09 – 0.9)	Shoots Roots	Mehta and Mohan Ram (1981)
Epidermal cells, cotyledon, leaf	MS + IAA(0.175) + BAP(2.25) →no growth regulators	Buds	Trinh et al. (1981) Brunel et al. (1982)
Suspension	Modified B5 + 2,4-D(0.1) + BAP(0.5) + CW(2%) + N-Z amine type as (250 mg/l)	Protoplasts and calli	Cuddihy and Bottino (1982)
Embryo, cotyledons, stem explants, leaf	MS + 2,4-D(1) + kin(1) MS + BAP(1) + kin(1) MS + IAA(1) + IBA(1)	Calli Shoots Roots and plants	Venketeswaran et al. (1985) Venketeswaran (1986)
Leaf, protoplasts	MS + 2,4-D(2) + kin(0.25) MS + NAA(0.2) + BAP(2)	Callus	Wilson et al. (1985)
	MS + BAP(1) + NAA(0.2)	Shoot buds and plants	
	Modified B5 MS + NAA + BAP (various combinations)	Callus Plants	Zakri (1983)
Seed	MS + BAP(1 – 5) Liquid cultures	Plantlets	Dias et al. (1986)
Anthers	B5 + 2,4-D(10) 2iP(5) + kin(5) + BAP(1)	Nodulated callus	Mohan Ram et al. (1982)
Anthers	B5 + 2,4-D,IAA, kin, BAP, GA_3 (all 0.5 – 50) (various concentrations)	Callus, multicellular masses, root	Mehta et al. (1983)
Anther wall	Nitsch and Nitsch	Callus	Brunel et al. (1982)
Pollen	Not mentioned	Embryos	Trinh et al. (1985)

Abbreviations: MS: Murashige and Skoog (1962); B5: Gamborg et al. (1968).
Supplements: IAA: Indoleacetic acid; NAA: α-Naphthaleneacetic acid; kin: Kinetin; BAP: Benzylaminopurine; 2,4-D: 2,4-dichlorophenoxyacetic acid; GA_3: Gibberellic acid; 2iP: 2-isopentyladenine; CW: Coconut Water.

develop this as a commercially valuable crop. A summary of the work done so far on the tissue culture of winged bean from several laboratories is shown in Table 2.

2.1 Methods of Sterilization of Explant Sources

Surface sterilization of seeds of different varieties of winged bean with commercial laundry bleach (Chlorox), chlorine water, aqueous Domestos (Lever Bros., London, UK) and calcium hypochlorite at various concentrations (15%–25%, v/v) for 10–20 min, subsequent washing in ethyl alcohol and several times in distilled water have been standard routine procedures. In some cases, surface sterilization of other plant parts, like 1-cm stem segments, punched out circular leaf segments, buds, etc., was carried out with lower concentration of Chlorox (5%–10%, v/v). Since the seed coat of winged bean is thin and permeable and not hard, no other procedures of scarification and sterilization have been employed.

2.2 Media Composition

The culture medium which was generally employed for the winged bean was the modified Murashige and Skoog (1962) medium (MS). In certain experiments, a modified mineral salt solution (Gamborg et al. 1968), referred to as B5 medium, was employed. These two media have been routinely used by several workers in experiments with addition of different growth hormones at appropriate concentration. Certain modifications to this medium, like half-strength MS, or replacement of certain salts with others, omission of agar (for liquid suspension cultures), addition and/or deletion of certain chemicals according to a particular experiment, have been employed. The composition of the two culture media, with appropriate additions, are given in Table 3.

2.3 Anther Culture

Genetic variability in anther-derived callus of legume plants has been recorded (Bajaj et al. 1980). Mohan Ram et al. (1982) reported anther culture of the winged bean plant up to the callus stage. Callus was obtained from the anther wall (Brunel et al. 1982) and embryo-like structures from anther cultures were obtained, which eventually developed into plants (Trinh et al. 1985). Anther culture of winged bean (Thailand var.) was initiated in our laboratory on an MS medium containing NAA (0.5 mg/l) and BAP (2 mg/l). Haploid callus was established in 3 weeks and maintained for several months, but attempts to regenerate plants were not successful (Venketeswaran 1984).

2.4 Culture of Thin Cell Layer Explants

Tran et al. (1986) have reported a very detailed account of regeneration of plantlets and development of buds from thin cell layer explants of the winged bean variety TPT 6. In vitro roots, formation of buds and eventual plant regeneration were successful either through a callus stage or direct from explants. Con-

Table 3. Compositions of plant culture media (Murashige and Skoog 1962; Gamborg et al. 1968)

	MS (mg/l)	B5 (mg/l)
NH_4NO_3	1650	
KNO_3	1900	2500
$CaCl_2$-$2H_2O$	440	150
$MgSO_4$-$7H_2O$	370	250
KH_2PO_4	170	
$(NH_4)_2SO_4$		134
$NaH_2PO_4-H_2O$		150
KI	0.83	0.75
H_3BO_3	6.2	3
$MnSO_4-4H_2O$	22.3	
$MnSO_4-H_2O$		10
$ZnSO_4-7H_2O$	8.6	2
$Na_2MoI_4-2H_2O$	0.25	0.25
$CuSO_4-5H_2O$	0.025	0.025
$CoCl_2-6H_2O$	0.025	0.025
Na_2EDTA	37.3	37.3
$FeSO_4-7H_2O$	27.8	27.8
Supplements:		
Inositol	100	100
Nicotinic acid	0.5	1
Pyridoxine HCl	0.5	1
Thiamine HCl	0.1	10
Glycine	2	
Auxin[a]	0.1 – 5	1 – 5
Cytokinins[b]	0 – 10	0.1 – 5
Sucrose (g)	30	20

[a] Auxins = ∝-naphthaleneactic acid (NAA); 2,4-dichlorophenoxyacetic acid (2,4-D); Indole acetic acid (IAA); Indole butyric acid (IBA).
[b] Benzylaminopurine (BAP); Kinetin (K).
pH 5.5
For solid cultures: 8000 mg/l agar. For liquid cultures agar is omitted.
Depending on the nature of the tissue culture and material, appropriate growth hormones are added at concentrations required for optimum growth, etc.
For medium for protoplasts: In addition, glutamine 300, serine 100, glycine 2, yeast extract 25, soluble starch 50, folic acid 0.5, biotin 0.05, mannitol 36430, sorbitol 36430 (all in mg/l).
For protoplast isolation: Pectinase, Cellulase, Hemicellulase, Drieselase, and other enzymes are used in various combinations and concentrations from callus and other plant parts.
For medium for anther culture: Nitsch (1977) medium and in addition Glutamine 730, Serine 105, Inositol 4505 (all in mg/l).

trolled organ formation was obtained by varying the ratio of IAA and BAP with 1% sucrose. One hundred percent explant forming direct vegetative buds was obtained with BAP (1×10^{-5} M) and IAA (1×10^{-6} M), and direct root formation was achieved at the reverse ratio, with only callus formation with BAP (1×10^{-5} M). The possible influence of the physiological age of the donor plant, photoperiod treatment, and light quality was also demonstrated by these studies (Tran et al. 1986). These authors conclude that lectins, which are phytohemagglutinins possessing the property of binding saccharide and saccharide-containing proteins in a highly specific way, may possibly play a role in plant differentiation.

2.5 Establishment of Callus and Suspension Cultures

Most of the tissue culture work on winged bean thus far reported in the literature has been directed toward callus establishment and attempts to regenerate plants from callus, including protoplast isolation, callus establishment, and regeneration. Generally, seeds are used as the starting material.

1. Materials. Seeds of several varieties of winged bean, *Psophocarpus tetragonolobus* (L.) D.C., were obtained from the International Winged Bean Institute and the University of Peradeniya, Sri Lanka. These seed varieties represented material from different geographical locations in the world, particularly from Papua New Guinea, Nigeria, Sri Lanka, Thailand, and Indonesia.

Experiments were carried out on many varieties (depending on the particular nature of the experiment and the response of the variety). Special emphasis was placed on those varieties which are important to Sri Lanka, Papua New Guinea, and Nigeria and those six varieties in which certain experiments have been carried out already (Venketeswaran 1985). Plants from all these varieties were also grown in the greenhouse to obtain research material (stems, leaves, flowers, etc.) and new seeds for various experimental procedures.

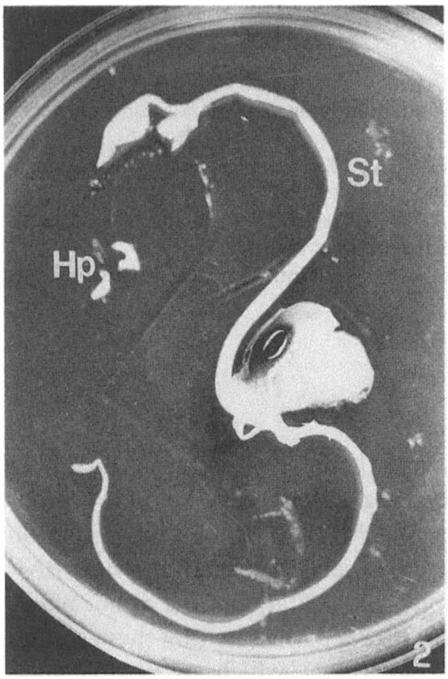

Fig. 2. Explants used in tissue culture of winged bean. *Hp* hypocotyl segments; *St* stem of the seedling germinated aseptically

2. Methods of Culture. The seeds were surface-sterilized for 10 min in 12% commercial chlorox and rinsed several times with sterile distilled water. They were germinated on the salt-sucrose culture medium referred to as basal medium (BM) with no growth regulators. The three explants initially used for tissue culture were: (a) hypocotyl axis from embryos of soaked seeds, (b) 1-cm segments from elongated regions of aseptic germinated seedlings, and (c) 1-cm segments from stem of plants (Fig. 2). In addition, other plant parts, like thin slices of cotyledons, punched out circular leaf segments, buds, etc. were used. These explants were placed on agar media for callus induction and the cultures were incubated at 25° ± 1 °C in the incubator illuminated with fluorescent lights (ca. 7000 lx) to give a 12-h day. At periodic intervals, the callus masses were transferred to shooting media (SM) with different hormonal combinations and concentrations, to rooting media (RM), and the callus maintenance media (CMM) (see Protocol). Other manipulations for controlled initiation of organ primordia and eventually regeneration of plantlets were followed, since each tissue had its own requirement and the

Table 4. Responses of explants of six varieties of winged bean after 4 weeks in culture. (Venketeswaran 1984)

Variety	Explant	Media composition	Response
1. Thailand	Stem segments from mature plant (old)	MS or B5 (BM)	Enlargement of segments
		BM + 2,4-D (0.5 mg/l)	Formation of friable, white callus
		BM + 2,4-D (0.5 mg/l) + kin (2 mg/l) MS + NAA (0.5 mg/l) + 6-BAP(2 mg/l)	Formation of fast-growing, green callus at the end of 4 weeks
2. LBNG 1 3. LBNG 3	Embryo segments from seeds sown on MS + 2,4-D (0.5 mg/l) for 5 days	BM	Formation of callus, leafy shoots, roots and plantlets differentiation at the end of 4 weeks
		BM + 2,4-D (0.5 mg/l) + kin(2)	Formation of white callus
4. SL 50	Epicotyl segments and sections of cotyledon from 15–25-day-old seedling	BM	Slight proliferation from the cut ends
		MS + NAA (0.5 mg/l) + 6-BAP (2 mg/l)	Formation of fast-growing, compact green callus at the end of 4 weeks
LBNG 3	Epicotyl segments and sections of cotyledon from 15–25-day-old seedling	MS + NAA (0.5 mg/l) + 6-BAP (2 mg/l)	Formation of fast-growing, compact green callus at the end of 4 weeks
5. SLS 23	Epicotyl segments and sections of cotyledon from 15–25-day-old seedling	MS + NAA (0.5 mg/l) + 6-BAP (2 mg/l)	Formation of fast-growing, compact green callus at the end of 4 weeks
6. PNG ups 102	Epicotyl segments and sections of cotyledon from 15–25-day-old seedling	MS + NAA (0.5 mg/l) + 6-BAP (2 mg/l)	Formation of fast-growing, compact green callus at the end of 4 weeks

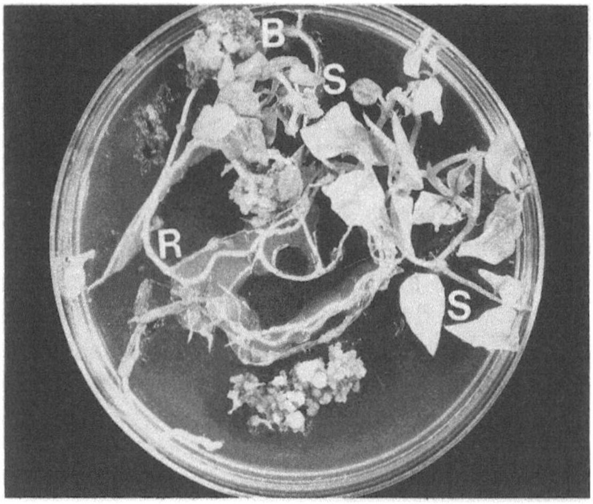

Fig. 3. Leafy shoots (*S*), roots (*R*), and buds (*B*) formed on embryo explants after 5 weeks, on basal medium with no hormones (Table 4, var. LBNG 1, LBNG 3)

phenomenon was dependent on a number of factors such as genotype, varietal differences, age, endogenous growth hormone levels and physical factors.

3. Callus and Organogenesis Response from Explants. Segments of embryo, epicotyl portion of seedling and stem segments of fully grown plants were used as explants. When growth regulators (2,4-D, NAA, BAP) were used alone or in combination, it resulted in callus formation (Table 4). The callus formed on 2,4-D (0.5 mg/l) medium was friable and white, whereas NAA (0.5 mg/l) and BAP (2 mg/l) in combination produced hard green callus in all varieties; 1-cm embryo segments from seeds which were initially sown on MS + 2,4-D (0.5 mg/l) for 5 days and then transferred to BM alone responded differently: leafy shoots, roots, shoot buds, and even fully differentiated plantlets were produced (Figs. 3, 4; Table 4). Some varieties, like SLS 23, SLS 50, and Thailand, showed no response. These observations indicate that explants at all stages of development respond to the growth regulator treatment in a similar fashion, although some varietal differences were evident. Other observations have demonstrated that induction of organs and regeneration of plantlets can be obtained on the winged bean, using explants with different media composition. Roots and shoots developed and subsequently plantlet formation was possible in several varieties.

4. General Callus Growth and Suspension Culture. The embryo explants or cotyledonary slices growing in agar solid medium containing 2,4-D, kinetin, or NAA in various combinations or concentrations showed evidence of callus initiation and growth within 10 days after inoculation. Further growth produced callus proliferation in large quantities within 3 weeks, so that several subcultures could be made from one initial culture. The nature of callus growth and texture varied considerably (hard, compact, or loose), depending on the type of culture medium containing the growth substances (Table 5). Subcultures made from the original callus maintained in light appeared green-pigmented in certain media and were white, yellow, or yellowish brown in other media and/or under dark conditions.

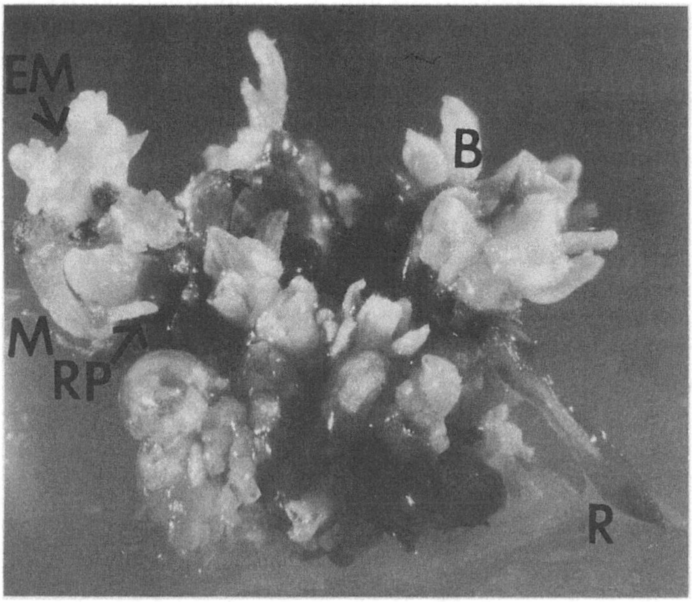

Fig. 4. Development of bud-like primordia (*B*), root (*R*), root primordia (*RP*) from mid-hypocotylar region (*M*)

Table 5. Growth characteristics of callus of winged bean in media with varying concentrations of growth factors after 3 weeks in culture. (Venketeswaran 1985)

Medium[a] + growth factors (mg/l)	Texture	Pigmentation[b]
2,4-D (1)	Friable	White
2,4-D (5) + kin (0.1)	Very friable	White-yellow
2,4-D (1) + kin (1)	Hard	Yellow
2,4-D (0.5) + kin (1)	Hard, compact	Yellow-green
2,4-D (1) + NAA (1.5)	Friable	Yellow
2,4-D (1) + NAA (1.5) (liquid)	Very friable	White-yellow

[a] Modified Murashige and Skoog (1962) medium.
[b] Pigmentation varied from yellowish, yellow brown, brown as cultures were grown for longer periods

Callus tissue growing in a 2,4-D medium and particularly without kin generally were friable, loose, and less compact.

Suspension cultures were initiated by placing 75–100 mg of callus tissue into 50 ml of MS or B5 liquid supplemented with 2,4-D (1 mg/l) in Erlenmeyer flasks. The cultures were incubated on an orbital shaker maintained at 60 rpm. Filtered suspension of cells from these cultures was used for plating on agar to obtain clones and strains. They dispersed into a fine suspension of cells and cell aggregates when transferred to liquid medium on the shaker. The liquid suspension cultures were generally white, brown, or yellowish brown in color. Microscopic observations of the cells generally showed a uniform morphology with active protoplasmic streaming, distinct nucleus, and various other cell organelles. Differentiated cells or cell aggregates with lignified walls and tracheid-like elements were often visible when squash preparations were stained with feulgen or other dyes.

Cultures transferred to the liquid medium produced actively growing suspension cultures. After two to three subcultures, 5 – 10 ml inoculum of the suspensions containing cells and cell aggregates could initiate new cultures in fresh liquid medium. The tissues were highly friable and maintained such an active growth rate that subculturing had to be effected very frequently (every 2 weeks). Occasionally, root primordia and differentiation occurred in tissue clumps (large aggregates) growing in the liquid media. Filtration through cheesecloth or wire meshes removed big tissue pieces and 1 ml of this filtered inoculum, which generally contained about 600 – 700 cell groups (of which approximately 70% were single cells), was plated onto agar Petri plates. Colonies derived from single cells and cell groups grew into callus masses within 2 weeks and eventually covered the entire surface of the Petri plate. Isolated small clumps or cell masses (0.1 – 0.5-cm diameter) when transferred to Petri plates with agar media developed callus tissue from either one or several regions. The new callus growing from these centers was generally white in color. It appeared very loose and friable. The small callus masses, tissue pieces, or colonies were routinely used for regeneration experiments.

5. Root and Bud Differentiation. Tissues which were growing in solid cultures for several months (with routine subcultures) were used. Small pieces of tissues (approximately 30 mg) or small cell aggregates from suspension cultures were inoculated on Petri plates to solid agar media supplemented with 0.1, 0.5 or 1 mg/l NAA and 0.01 mg/l kin or without kin to induce root initiation and further development. 2,4-D was omitted in these cultures. Callus was also transferred to agar-solid medium without growth hormones (i.e., 2,4-D, NAA or kin) and maintained generally in such media for two to three subcultures.

Attempts to obtain root initiation and eventual sequence of development of well-developed roots were successful in cultures transferred to a medium containing NAA (Table 6).

Root primordia initiated either singly from one callus mass, or several roots arose from a single callus piece. 2,4-D was omitted totally in all attempts to induce root organogenesis, and the occasional appearance of callus on the root initiation surface was possibly due to the presence of residual growth substances in the callus or of kin in the medium. Low concentration of NAA (0.1 mg/l) induced several root organogenesis, usually from cultures maintained in dark. In several cases, many nodules (roots ?) appeared on callus surfaces. Cultures transferred to agar medium in growth jars or Erlenmeyer flasks produced such an intense number and amount of root growth within 3 weeks that the containers were filled with roots growing into the medium.

Tissues growing in callus medium (1 mg/l 2,4-D + 1 mg/l kin), when transferred to medium without growth factors or different concentrations of BAP, produced several bud-like primordia. Eventually, these bud-like primordia became heavily green-pigmented and produced green leaf-like structures. In many cultures, several embryoid or embryoid-like structures with regions of mid-

Table 6. Growth and root organogenesis of winged bean callus in solid media with varying concentrations of growth factors after 3 weeks in culture. (Venketeswaran 1985)

Medium[a] + growth factors (mg/l)	Callus	Roots	
		Dark	Light
kin (0.01) + NAA (0.1)	+ + + +	72 (6)	12 (4)
kin (0.01)	+ +[b]	4 (3)	–
NAA (1)	+ +	–	–
NAA (1) + BAP (0.1)	+ + +	–	12 (4)

[a] Modified Murashige and Skoog (1962) medium.
[b] Many nodules (roots?) appear in callus.
() = Number of Petri plates.
+ + to + + + + = Amount of growth (visual).
Dark vs. Light = Observations only

hypocotyl root primordia and leafy structures were formed. Several of the bud-like primordia and/or embryoid-like structures (with a small amount of attached callus mass) continued to grow and elongate in the medium without growth substances.

6. Controlled Organogenesis and Regeneration of Plantlets from Established Tissue Cultures. After a series of experiments and trials, it was possible to obtain routinely plantlet regeneration of winged bean from established "old" (8–10 months) callus cultures, by controlled organogenesis, for eventual transfer to the growth chamber and greenhouse. After transfer to soil (outside environment), the plants flowered normally.

Protocol for Regeneration of Plantlets

1. Callus masses growing as stock cultures in culture medium (CMM), viz. BM + 1 mg/l each of 2,4-D + Kin or BM + 1 mg/l each of NAA and 2,4-D.
2. Transfer to shoot-inducing medium (SM), BM + 1 mg/l or 2 mg/l BAP: shoots formed after two to three subcultures.
3. Transfer shoots to root-inducing medium (RM), i.e., BM + 1 mg/l of IAA or IBA or both IAA + IBA: root formed in 3–4 weeks.
4. Transfer plantlets with shoots and roots from agar to liquid medium on cheese cloth or cotton base: 2–3 weeks.
5. Remove plantlets, wash in water several times non-aseptically and let stand 3–4 h or overnight. Transfer to vermiculite-perlite soil mixture (nonaseptic). To minimize bacterial growth and contamination, dipping roots in $CuSO_4$ + antibiotic solution before planting is desirable.
   Maintain plant in incubator maintained at 25° ± 1 °C for production of new healthy green leaves: 2–3 weeks.
6. Transfer to growth chamber in laboratory at room temperature and maintain 4–5 weeks before transfer to greenhouse.

Figures 5–7 illustrate these different steps for regeneration of winged bean.

7. Regeneration of Isolated Cell Strains from Suspension Cultures or Direct. The callus obtained from stem and epicotyl segments of seedlings growing on BM + 2,4-D (1 mg/l) medium dissociated to single cells and cell clumps in suspension culture at the end of a 2–3-week culture period. Techniques were perfected

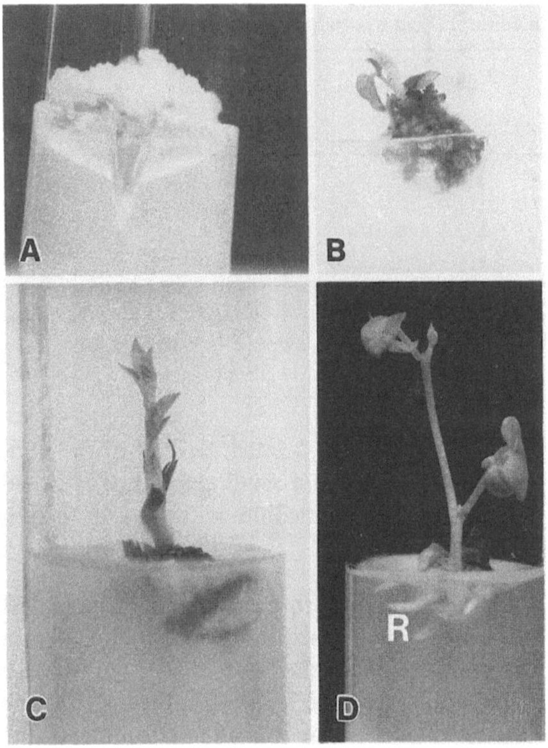

Fig. 5 A– D. Regeneration of plants. **A** After 4 weeks of culture. **B** Shoot aged 3 weeks, initiated on callus following subculture to agar medium containing MS salts and 1 mg/l BAP. **C** Shoot, ready to transfer to rooting media on MS + 2 mg/l BAP at the age 6 weeks. **D** Root formation after 2 weeks, on shoot obtained through tissue culture on MS + 1 mg/l IAA

Fig. 6. Fully developed plantlets being conditioned to nonaseptic environment, prior to transfer to potting media

Fig. 7A,B. Plantlets growing **A** on potting media – perlite: soil, 1:1 mixture, **B** in the green house

to grow liquid cultures of winged bean (several varieties) routinely which form suspensions of cell aggregates, small clumps of cells and single cells, etc., and isolation of "clones" was possible in some varieties which were attempted. The single cells or small cell aggregates, when plated to agar on MS medium alone, formed callus masses. Organogenesis and eventually regeneration of plantlets were obtained by varying the growth hormone levels, as described in the protocol. Several plantlets were obtained in this manner (Dias 1988).

Organogenesis, particularly the development of roots directly from cell clumps, was observed in suspension cultures of varieties SLS-1, UPS 122, and TPT 2 (Fig. 8). These rooted clumps, when transferred to a shoot-inducing medi-

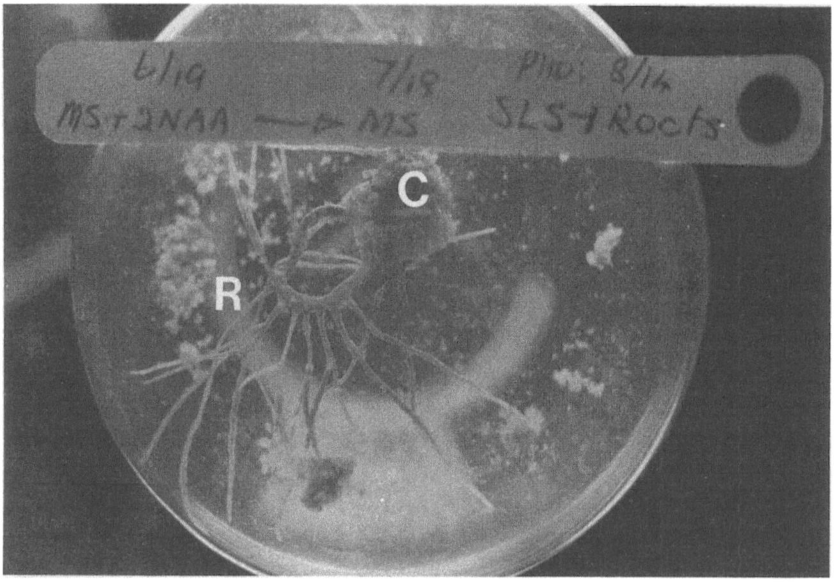

Fig. 8. Roots (*R*) formed directly on callus (*C*) in suspension culture following subculture from MS+2 mg/l NAA to medium supplemented with only MS salts

um (SM), produced shoots and eventually plantlets. Occasionally, shoot buds or shoots were also evident from the cell clumps growing in suspension cultures and they could be transferred to appropriate media to produce plantlets.

8. Somatic Embryogenesis in Suspension Cultures. Recent attempts to induce somatic embryogenesis in winged bean suspension cultures on some newly initiated calli of two varieties, TPT-2 and SLS-23, have had some success. Reports in the literature indicate that callus cultures exposed only to a 2,4-D medium, when subsequently transferred to a hormone-free medium, will induce embryogenesis. Liquid cultures growing in a medium containing 1 mg/l of each 2,4-D and NAA were first transferred to a medium containing only 1 mg/l of 2,4-D. After 3–4 weeks,

Flow chart for induction of somatic embryogenesis in winged bean:

Seeds —4 wks→ Calli initiation in MS+NAA(1)+2,4-D(1) Agar medium —4 wks→ MS+2,4-D(1) Agar media —4 wks→ 1/2 MS or MS liquid culture media —3–4 wks→ Embryoid

Calli initiation —4 wks→ Subculture in MS+NAA(1)+2,4-D(1) Agar media (Callus maintenance media) → MS+2,4-D(1) Agar media

Note: Concentrations of growth regulators are given in mg/l in parenthesis.

Table 7. Media which induced embryoid or embryoid structures. (Dias 1988)

Variety	Initial agar callus medium or callus maintenance medium (CMM) (mg/l)	First or subsequent subculture liquid culture medium
TPT 1	MS+NAA (5)	MS or 1/2 MS only or 2,4-D (0.1)
	MS+NAA+2,4-D (both 2)	MS or 1/2 MS only or 2,4-D (0.1)
UPS122	MS+2,4-D (0.1)	MS or 1/2 MS only
	MS+2,4-D (5)	MS or 1/2 MS only
	MS+NAA (5)	MS+1/2 MS+2,4-D (0.1)
TPT 2	MS NAA (0.12)+2,4-D (2)	MS+1/2 MS only

1/2 MS = half-strength concentration of Murashige and Skoog salts

the cultures were transferred to a hormone-free medium (minus 2,4-D). After 2–3 weeks, several embryo-like growths or "embryos" were observed. Histological and cytological studies, and scanning electron microscopy of such structures confirmed this observation. Somatic embryogenesis is a superior and excellent method to produce large quantities (in millions) of plantlets, and eventually plants in a quicker, cheaper way within a smaller area (space) than conventional plant propagation methods.

A flow chart depicting the procedure to induce somatic embryogenesis is presented above (p. 186). Appropriate changes and/or modifications were made as and when necessary.

Table 7 shows the response of three varieties, viz. TPT-1, TPT-2 and UPS-122, to culture medium combination to form somatic embryoid or embryoid-like structures in liquid suspension cultures.

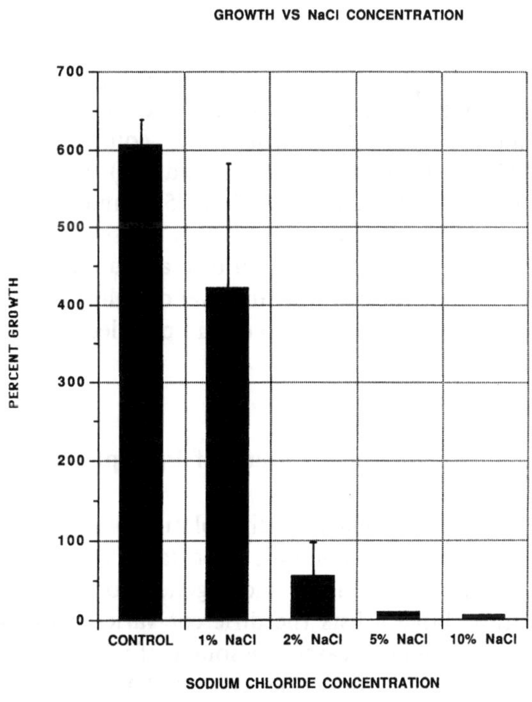

Fig. 9. Histogram showing the effect of various concentrations of sodium chloride on callus cultures

9. Selection of Stress Tolerance, e.g., Salt. Callus masses (200–500 mg) each derived from established tissue cultures (from stem and epicotyl segments of seedlings var. LBNG 3) were transferred to MS medium + 0.5 mg/l NAA + 2 mg/l BAP containing NaCl concentrations of 0%, 1%, 2%, 5%, and 10% w/v. Appropriate quantities of NaCl were weighed and dissolved in the culture medium before autoclaving. The above medium was poured into Petri plates and allowed to gel. The experimental set-up consisted of four Petri plates with four individual callus pieces in each Petri plate, sealed with parafilm. Cultures were allowed to grow in a growth chamber maintained at 24°±1 °C with 12 h/day illumination. After 4 weeks, the weight of each individual callus mass was determined for each concentration. The results were tabulated and analyzed statistically.

Figure 9 shows the quantitative data and the growth of the callus pieces of winged bean in control medium and with different concentrations of NaCl incorporated into the culture medium. The quantitative data were obtained by recording the initial and final fresh weight of each of the individual callus mass and determining the amount of growth in 4 weeks. The callus pieces appeared healthy and green, and grew extremely well in the control medium after 4 weeks of growth. At a concentration of 1% NaCl, several of the cultures grew reasonably well and appeared comparatively healthy, although growth was reduced. In comparison, at a concentration of 2% NaCl, the growth of the callus declined very considerably, to a minimum. At 5% and 10% concentrations of NaCl, the callus masses turned brown and died.

These experiments suggest that callus cultures of winged bean can tolerate NaCl concentration levels up to 2%, and it may be possible to obtain salt-tolerant callus tissue within the concentration level of 1%–2%.

Once it was shown that winged bean tissue cultures can tolerate NaCl concentrations of up to 2% (w/v) and at 1% NaCl approximately only 30% inhibition of growth was observed in the growth of the callus as against the control, another set of experiments was carried out to examine whether callus which generally shows excellent growth in a control mineral salt-sucrose agar medium (MS+1 mg/l each of 2,4-D and kin) will tolerate NaCl concentrations at 0%, 0.05%, 0.10%, 0.25%, 0.5%, 0.75%, and 1% (w/v). At the end of a 6-week culture period, the growth of the callus masses at various concentrations of NaCl compared with control was uniform and uninhibited even at 1% w/v. Experiments to determine the salt-tolerant level of callus in suspension cultures and to select the salt-tolerant lines and to obtain plantlet regeneration from these lines, will be the ultimate goals.

2.6 Isolation and Culture of Protoplasts

Legume protoplasts, their isolation and culture have been of significant interest because of the economic importance of the family. Protoplast isolation, culture, callus establishment, and regeneration have been very successful with winged bean; Table 8 lists the different varieties on which these studies have been reported. In most cases, the source of protoplasts was suspension cultures from callus obtained from either epicotyl (Wilson et al. 1985), or leaf explants (Cuddihy

Table 8. Summary of protoplast isolation and culture of winged bean varieties

var.	Origin of protoplast	Enzyme mixture	Culture medium	Growth response	Reference
TPT-2	Suspension cultures growing in MS + 2,4-D (0.5 mg/l) + kin (0.3 mg/l)	Cellulysin (6%) + Macerase (1%) + Rhozyme (1%) + Sorbitol (0.4 M) + $CaCl_2$ (0.1 M)	Mod. B_5 − 2,4-D (0.1 mg/l) + BAP (0.5 mg/l) + Coconut water (2%) + N-Z amine type AS (250 mg/l)	Callus	Cuddihy and Bottino (1982)
Su 623	Suspension cultures growing in MS + 2,4-D (1 mg/l) + kin (0.1 mg/l)	Cellulase Onozuka R10 (2%) + Macerozyme R10 (0.4%) + 2-mercapto-ethanol (1 mM)	UM + 2,4-D (2 mg/l) + kin (0.25 mg/l) + mannitol (3%)	Callus	Wilson et al. (1985)
Su 623	Suspension cultures growing in MS + 2,4-D (1 mg/l) + kin (0.1 mg/l)	Cellulase Onozuka R10 (2%) + Macerozyme R10 (0.4%) + 2-mercapto-ethanol (1 mM)	UM + IAA (0.2 mg/l) BAP (1 mg/l)	Plants	Wilson et al. (1985)
TPT-2 SLS-1 UPS-53	Suspension cultures growing in MS + 2,4-D (1 mg/l) + kin (0.1 mg/l) mesophyll cells	Drieselase (3%) Hemicellulase (1%) Pectinase (1%)	MS + 2,4-D (1 mg/l) + kin (1 mg/l)	Callus	Dias (1988)
	Young leaves	Cellulysin (2%) Macerease (0.5%) Mannitol (0.5 M)	Modified B5	Callus	Zakri (1983)
	Young leaves	Cellulysin (2%) Macerase (0.5%)	MS + NAA + BAP (various combinations)	Plants	Zakri (1983)

MS: Murashige and Skoog (1962); UM: Uchimiya and Murashige (1974); B5: Gamborg et al. (1968)

and Bottino 1982; Zakri 1983). Although leaf mesophyll cells are generally the primary source of protoplasts in most of the plants studied, protoplast isolation from leaves of winged bean have been less successful and difficult. Wilson et al. (1985) attributes this to the thick-walled nature of the mesophyll cells, which essentially confirms earlier observations in our laboratory of protoplast isolation from leaves, that they tend to stick with each other with no success in culture.

Wilson et al. (1985) obtained regeneration of plants from the callus derived from protoplast culture by manipulation of auxin/kinin combinations. These plants matured and flowered normally. Further experiments will determine whether it is possible to obtain regeneration from protoplast culture of other varieties of winged bean, because differences in response to in vitro conditions of several varieties have been observed. Experiments with winged bean protoplast culture, callus production, and regeneration is an important step, since genetic manipulations in legumes can now be successfully explored.

2.7 Genetic Stability in Callus and Regenerated Plants

The normal chromosome number of *P. tetragonolobus* has been reported to be 2 n = 18 (Haq and Smartt 1978; Pickersgill 1980; Khan 1982), consisting of 12 long and 6 short chromosomes. Chow and Subha (1986) indicated that they observed an additional pair of dot-type chromosomes, but mentioned the 2 n = 18. The cytology of *P. tetragonolobus* has been described in detail (Khan 1982).

Routine squash preparation of callus has shown generally a diploid number of 2 n = 18 for the majority of cells (over 80%), with polyploid or aneuploid cells being occasionally evident (Venketeswaran 1986). Bottino et al. (1979) reported chromosome analysis of suspension cultures over a period of 1 year to be predominantly diploid (2 n = 18), with a frequency of polyploid cells being less than 10% and not increasing at any appreciable rate.

Regenerated plants (over 100 examined) do not exhibit any variation in leaf shape or other morphological characteristics. These plants, transferred to field conditions in Houston, have grown normally and produced flowers and seeds. Squash preparation of root tip cells also revealed normal diploid chromosomes numbers. Continued observation of the callus tissue after several subcultures (1–2 years) needs to be monitored to determine the cytological stability.

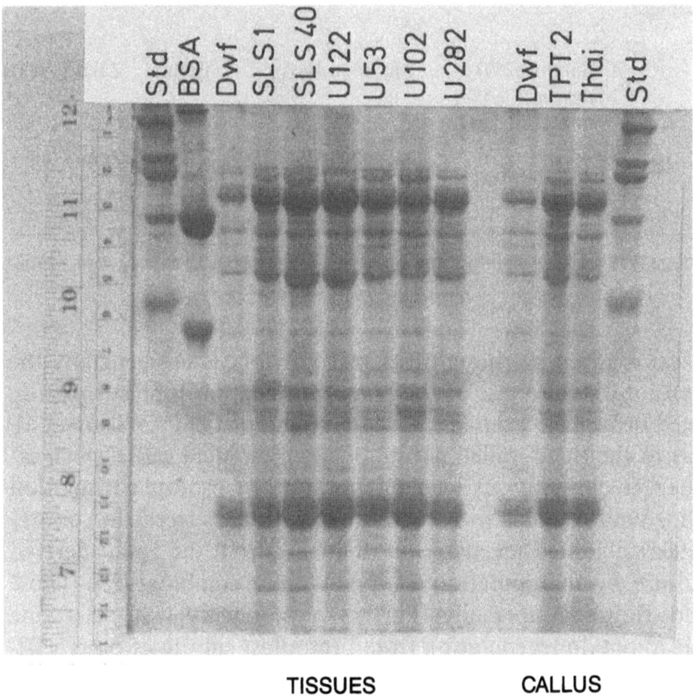

Fig. 10. SDS/PAGE seed protein profiles of different varieties of winged bean. Lanes *from left to right*: Lane *1* protein standard with molecular weights ranging from 97400 to 14400; lane *2* bovine serum albumin; lane *3* dwarf winged bean; lane *4* variety SLS1; lane *5* variety SLS40; lane *6* variety UPS122; lane *7* variety UPM53; lane *8* variety UPM 102; lane *9* variety UPM282; lane *10* blank; lane *11* dwarf variety; lane *12* variety TPT2; lane *13* variety Thailand; lane *14* protein standard

2.8 Protein Analyses and Electrophoresis of Callus

Protein analyses and gel electrophoretic studies of proteins can reveal information about protein metabolism in tissue cultures and plant parts. Comparison of protein profiles of callus vs. tissues of plants can detect differences in protein, since changes in protein profiles often reflect genetic changes and therefore can be used as markers to detect genetic stability of the callus. Protein profiles can show differences among different varieties also (Fig. 10), thereby serving as a useful taxonomic tool.

Callus tissues of three winged bean varieties, initiated in MS medium + 2,4-D (2 mg/l), were subcultured after 4 weeks to liquid media. After 30–35 days in culture, callus masses from the two sets were homogenized using standard methods (Gillespie and Blagrove 1978) to obtain protein extracts. Acrylamide gel electrophoresis of the protein extracts were carried out according to the method of Laemmili (1970). After appropriate staining and destaining, 22 bands on the gel were observed.

On comparison of proteins from calli grown in liquid media against solid media, more protein bands were visible in callus grown in liquid media. The most probable reason is that callus grown on solid media accumulates brown pigments (inhibitory substances, i.e., phenols) which are known to precipitate proteins. In the callus grown in the liquid media, these inhibitory substances are leached out of the callus into the liquid media. The different varieties of winged bean also showed differences in protein profile in terms of number of bands.

Electrophoretic studies are useful for comparing the original plant vs. regenerated plants from tissue cultures. Such comparisons can demonstrate the genetic stability of plants obtained through in vitro procedures, and further experiments on regenerated plants will provide such information.

3 Conclusions and Prospects

The applications of biotechnology, i.e., tissue culture of plants in general, are directed toward plant regeneration in large numbers for mass propagation. In the case of winged bean, this approach will provide large numbers of plants, particularly because seeds are not available in large quantities and, if available, can be employed for other useful purposes.

Cell culture techniques also provide a method for selection of plant types with desirable characteristics. The study of stability and variation, and the possibility of selecting variants from plants regenerated from tissue culture of winged bean is significant.

Results on NaCl tolerance of winged bean show that cell cultures can tolerate, survive, and grow in culture medium containing up to 2% NaCl (Venketeswaran 1985). Based on these data, it is possible to obtain salt-tolerant callus from several varieties and cultivars of winged bean on solid media, liquid suspension cultures and on platings of cell types and aggregates. After isolation of tolerant cell lines, regeneration studies will have to be evaluated with the ultimate aim of producing salt-tolerant winged bean plant types.

Expansion of winged bean cultivation is of vital importance, because of the immense potential the crop has in improving the standard of nutrition. It is unique in its potential for village-scale development in view of its numerous possible uses as the crop, for "food, feed, fodder, and fertilizer". This is particularly true in less developed countries, where protein malnutrition is a serious problem. Increasing animal protein is long-term and expensive, and land is not available for livestock expansion. Besides, religious inhibitions increase the problem. Therefore, the winged bean, which is adaptable to different forms of cropping/farming systems, could provide higher outputs of protein per hectare per year much cheaper and sooner. Further, the technology of processing the seed is known through the work on soybean. The availability of high-yielding, short-age varieties could alleviate protein malnutrition problems. In addition, due to high protein levels in the tuber, a nutritious root crop could also be developed where root tubers are consumed as a staple food. The development of winged bean and the expansion of production thus would not only increase primary productivity of vegetable proteins, but would also contribute to the development of appropriate agro-industries and new exports.

Winged bean has potential in the rehabilitation and development of denuded or degraded lands, as it can be grown as a live mulch or cover crop, either singly or in combination with commercial perennial crops. The marginal lands can be rehabilitated by this means, provided adequate quantities of desirable planting material are available. The experiments on salt resistance will help develop plants that could be grown on saline/alkaline soils. Thus the availability of plants that could grow in such soils would help reclaim areas with salinity problems and thereby increase land availability and productivity. In vitro culture techniques thus offer a promising potential toward the development of this high protein-rich food crop for less developed countries.

Acknowledgments. Acknowledgments are made for support of this research to the Agency for International Development (AID), Washington, DC, Grant # DPE-5542-G-SS-3056-00, and in part to the University of Houston Coastal Center. I am grateful to Dr. Ossi Huhtinen (Finland) and Dr. R. Nagmani (Weyerhauser Co., Madison, WI, USA) for their help during the initial investigations of this research. I feel indebted to Mr. Dilip Dias for the excellent assistance in the laboratory, photography, etc., without which this task would have been very difficult. Other technical assistance from F. Sultanbawa, Mrs. Kalyani Dias, Ursula V. Weyers, Sunita, and a number of other students in the laboratory are appreciated. I thank Dr. Glenn D. Aumann and Dr. Joe R. Cowles, University of Houston, for their support and encouragement. Grateful appreciation is extended to Profs. Walter Herath, Don Senanayake, H.M. Gunasena (University of Peradeniya, Sri Lanka), and Mr. Louis Lazaroff, ICDUP, California for their collaboration, cooperation, and support. Lastly, I thank Prof. T. Hymowitz, University of Illinois, for sending me the first batch of four seeds which initiated my tissue culture studies in 1977.

References

Bajaj YPS (ed) (1986) Biotechnology in agriculture and forestry. 2. Crops I. Springer, Berlin Heidelberg New York Tokyo

Bajaj YPS (ed) (1989) Biotechnology in agriculture and forestry. 6. Crops II. Springer, Berlin Heidelberg New York Tokyo

Bajaj YPS, Singh H, Gosal SS (1980) Haploid embryogenesis in anther cultures of pigeon pea (*Cajanus cajan*). Theor Appl Genet 58:157–159

Bollinger WH (1980) Sustaining renewable resources. Techniques from applied botany. Academic Press, New York

Bottino PJ, Maire CE, Goff LM (1979) Tissue culture and organogenesis in the Winged Bean. Can J Bot 57:1733–1776

Brunel A, Landre C, Chardard R, Kovoor A (1982) Studies in the tissue culture of the Winged Bean. In: Rao AN (ed) Tissue culture of economically important plants. COSTED, Singapore, pp 63–65

Burkill IH (1906) Goa bean in India. Agric Ledger (Calcutta) 4:51–64

Chow KH, Subha N (1986) Effects of gamma rays on Winged Bean tissue cultured in vitro. Proc Symp Nucl tech in vitro cult plant improv, IAEA, Vienna, pp 175–179

Cuddihy AE, Bottino PJ (1982) Winged Bean protoplast: Isolation and culture to callus. Plant Cell Tissue Organ Cult 1:201–209

Dias, MADL (1988) Plantlet regeneration of Winged Bean (*Psophocarpus tetragonolobus* (L.) D.C.) from single cells and tissues. MS Diss, Univ Houston, TX

Dias, MADL, Weyers UV, Venketeswaran S (1986) Plant regeneration in the Winged Bean, *Psophocarpus tetragonolobus* (L.) D.C. In: 6th Int Congr plant tissue and cell cult, Univ Minnesota, Minneapolis, Abstr, p 104

Eagleton GE, Khan TN, Chai NF (1982) Winged Bean origins. In: Khan TN (ed) Winged Bean production in the tropic, Plant Production and Production Paper 38. FAO, Rome, p 8

Evans PK, Haq N, Cooper-Bland S (1981) Tissue culture of the Winged Bean. In: Islam AS (ed) Proc Int Workshop on Improvement of Tropical Crops through Tissue Culture. Dacca University, Bangladesh, pp 107–111

Gamborg OL, Miller RA, Ojima K (1968) Plant cell cultures. I. Nutrient requirements of suspension cultures of soybean root cells. Exp Cell Res 50:151–158

Gillespie JM, Blagrove RJ (1978) The proteins of Winged Bean seed. In: The Winged Bean (ed) 1st Int Symp on developing the potentials of the Winged Bean. Los Baños, Philippines, pp 358–362

Gregory HM, Haq N, Evans PK (1980) Regeneration of plantlets from leaf callus of the Winged Bean, *Psophocarpus tetragonolobus* (L.) D.C. Plant Sci Lett 18:395–400

Gunasena HPM (1986) In: Gunasena HPM (ed) Special Issue on Winged Bean. Sri Lankan J Agric Sci Nattl Agric Soc Sri Lanka, pp 1–249

Haq N, Smartt J (1978) Cytology of *Psophocarpus* spp. In: The Winged Bean (ed) 1st Int Symp on developing the potentials of the Winged Bean. Los Baños, Philippines, pp 96–100

Hymowitz T, Boyd J (1977) Origin, ethnobotany and agricultural potential of the Winged Bean (*Psophocarpus tetragonolobus*). Econ Bot 31:180–186

Kailasapathy K (1986) Comparative studies on chemical, nutritional properties and biological evaluation of soybean (*Glycine max.* (L.) Merr.) and Winged Bean (*Psophocarpus tetragonolobus* (L.) D.C.) seeds from selected cultivars grown in Sri Lanka. Sri Lankan J Agric Sci 23:105–128

Khan TN (1982) Winged Bean production in the tropics. Plant Production and Protection Paper 38, FAO, Rome, pp 1–217

Laemmli UK (1970) Cleavage of structural proteins during the assembly of the head of bacteriophage T4. Nature (Lond) 227:680–685

Mehta U, Mohan Ram HY (1981) Tissue culture and whole plantlet regeneration in the Winged Bean (*Psophocarpus tetragnolobus*) Ann Bot 47:163–166

Mehta U, Ramajuja Rao IV, Marasimham M (1983) Haploid plant production in Winged Bean (*Psophocarpus tetragonolobus*): I. Induction of haploid and diploid calli from pollen. 15th Int Congr Genet, Abstr, Part I, New Dehli, pp 415

MohanRam HY, Mehta U, Ramanuja Rao IV, Narasimham M (1982) Haploid induction in legumes. In: 5th Int Congr plant tissue and cell culture. Tokyo, pp 541–542

Murashige T, Skoog F (1962) A revised medium for rapid growth and bioassays with tobacco tissue cultures. Physiol Plant 15:473–497

NAS (1981) The Winged Bean: high-protein crop for the tropics. By Ad hoc panel of the Advisory Committee on Technology Innovation. Board on Science and Technology for International Development (2nd edn) Natl Acad Sci, Washington, DC, pp 1–48

Nitsch C (1977) Culture of isolated microspores. In: Reinert J, Bajaj YPS (eds). Applied and fundamental aspects of plant cell, tissue, and organ culture. Springer, Berlin Heidelberg New York, pp 268–278

Pickersgill B (1980) Cytology of two species of Winged Bean, *Psophocarpus tetragonolobus* (L.) D.C. and *P. scandens* (Endl.) Verd. (Leguminosae). Bot J Linn Soc 80:279–291

Rachie KO, Luse RA (1978) Support system for climbing food crops. In: The Winged Bean (ed) 1st Int Symp on developing potentials of the Winged Bean. Los Baños, Philippines, pp 183–190

Shivashankar G (1984) Dr. Shivashankar develops bush type mutant in India. The Winged Bean Flyer 4:32–33

The Winged Bean (1978) Papers presented in the 1st Int Symp on developing the potentials of the Winged Bean. Manila, Philippines, pp 1–448

The Winged Bean Flyer (1977–1978). Vol I–II. Dep Agron, Univ Illinois, Urbana, IL (USA)

The Winged Bean Flyer (1980–1985) Vol 3–5. In: Stephensen RA (ed) Int Documentation Center for Winged Bean. College, Laguna 3720, Philippines

Tran Thanh van K, Lie-Schricks H, Marotte JL, Trinh TH (1986) Winged Bean (*Psophocarpus tetragonolobus* (L.) D.C.). In: Bajaj YPS (ed) Biotechnology in Agriculture and Forestry 2. Crops I. Springer, Berlin Heidelberg New York Tokyo, pp 556–567

Trinh TH, Lie-Schricks H, Tran Thanh van K (1981) Formation directe de bourgeon a partir des fragments et des courches cellulaires minces de differents organes chez le *Psophocarpus tetragonolobus* (L) D.C. Z Pflanzenphysiol 102:127–139

Trinh TH, Lie-Schricks H, Tran Thanh van K (1985) Régénération des plantes à partir de cultures d'anthéres et d'ovules non fécondes du haricot ailé (*Psophocarpus tetragonolobus* L.). VI colloque IAPTC Section Française Obtention d'haploide in vitro: Etat actuel et perspectives

Uchimiya H, Murashige T (1974) Evaluation of parameters in the isolation of viable protoplasts from cultured tobacco cells. Plant Physiol 54:936–944

Vavilov NI (1949/50) The origin, immunity and breeding of cultivated plants. Chronica Botanica 13:364, Waltham, MA

Venketeswaran S (1984) Isolation of strains, clones and regeneration of plants from single cells of winged bean. Report I. Agency Int Dev, Washington, D.C., pp 1–15

Venketeswaran S (1985) Isolation of strains, clones and regeneration of plants from single cells of winged bean. Report II and III. Agency Int Dev, Washington, D.C., pp 1–15

Venketeswaran S (1986) Isolation of strains, clones and regeneration of plants from single cells of winged bean. Final report. Agency Int Dev, Washington, D.C., pp 1–64

Venketeswaran S, Huhtinen O (1978) In vitro root and shoot differentiation from callus cultures of a legume. The Winged Bean, *Psophocarpus tetragnolobus* (L.) D.C. In Vitro 14:355

Venketeswaran S, Nagmani R, Weyers UV (1985) Plantlet regeneration from callus tissue of *Psophocarpus tetragonolobus* (L.) D.C. In Vitro 21:36A

Verdocourt B, Halliday P (1979) A revision of *Psophocarpus* (Leguminosae, Papilionoideae, Phaseoleae). Kew Bull, 33:191–227

Wilson VM, Haq N, Evans PK (1985) Protoplast isolation, culture and plant regeneration in the winged bean, *Psophocarpus tetragonolobus* (L.) D.C. Plant Sci Lett 41:61–68

Zakri AH (1983) Isolation of mesophyll protoplasts and plant regeneration in the winged bean. 15th Int Cong Genet, Abstr, Part I, New Delhi, p 434

Section III Forage Legumes

III.1 Alfalfa, Lucerne (*Medicago* spp.)

S. ARCIONI[1], F. DAMIANI[1], M. PEZZOTTI[1], E. LUPOTTO[2]

1 Introduction

The genus *Medicago* comprises more than 60 species. Lesins and Gilles (1972) report 62 species, but some of them are interfertile and would be better defined as varieties. The native area of the genus is the Mediterranean Basin and the Middle East, but today it is dispersed all over the world. The genus can be divided into annuals and perennials, diploid and polyploid species. The basic chromosome number (x) is 7 or 8 so that the diploid species have 14 or 16 chromosomes. There are some autotetraploid species ($2n = 32$), and two species are hexaploid ($2n = 48$). Phylogenetic investigations have been carried out with chromosomal analysis (Mariani 1968) and more recently with isoenzymatic (Mariani et al. 1978; Quiros 1983) and restriction fragment length (Johnson and Palmer 1986) polymorphism.

All but one species are herbaceous, and most of them are utilized for animal feeding. *Medicago sativa* is the world's most important forage crop and hence the most important species of this genus from the agronomical point of view. *M. sativa* is cultivated in all the continents, but mainly in the two Americas and in Europe; the world surface area is about 32 million ha (Michaud et al. 1987). Alfalfa cultivation has decreased in the last 15 years in Europe and in Oceania, but has increased in Asia, Africa, and slightly in North America. Worldwide a decrease of almost 1 million ha in the cultivation of this species has been observed (Table 1). Forage yield is highly variable as a consequence of environmental and agronomical influences and in general the production of hay ranges from 3–4 to 10–15 metric tons per ha per year (Bolton et al. 1972). The extensiveness of this species is witnessed by the number of different folk names for the plant of which the American and Spanish word "alfalfa" and the North European "lucerne" are the most common. Alfalfa is an autotetraploid species ($2n = 32$), prevalently allogamous, and insect-pollinated. High levels of selfing were also observed (Lesins 1961), depending upon the environment and the weather conditions. Alfalfa is greatly affected by a large number of diseases of different origin: Graham et al. (1972) reported losses of 24% for the forage and 9% for the seed yield in the United States.

[1] Centro di Studio per il Miglioramento Genetico delle Piante Foraggere, Facoltà di Agraria, Borgo XX Giugno, 06100 Perugia, Italy
[2] Istituto Sperimentale per la Cerealicoltura, Sezione Maiscoltura, Via Stezzano 24, 24100 Bergamo, Italy

Biotechnology in Agriculture and Forestry, Vol. 10
Legumes and Oilseed Crops I (ed. by Y.P.S. Bajaj)

Table 1. Alfalfa world distribution in the late 1960's (Bolton et al. 1972) and in the early 1980's (Michaud et al. 1987). The cultivated areas in million ha are reported

Continent	1960's	1980's
Europe	9.36	7.99
North America	13.14	13.35
South America	7.80	7.80
Asia	1.32	2.50
Africa	0.17	0.43
Oceania	1.20	0.22
World total	33.02	32.27

The most widespread diseases are: wilt, that can be caused by the bacteria *Corineobacterium insidiosus* and fungi (*Verticillium* spp. and *Fusarium* spp.), powdery mildew (*Peronospera trifoleorum, Erisiphe polygony, Oidiopsis* spp.), anthracnose (*Colletotricum trifolii*) and leafspot (*Xantomonas alfalfae, Phoma medicaginis*). Considerable damage can be caused by pests, insects and nematodes that are sometimes also vectors of viruses, bacteria, and fungi (App and Manglitz 1972). Reviews concerning the principal alfalfa diseases in Europe (Raynal 1986) and the USA (Graham et al. 1979) are available.

As lucerne has been cultivated for centuries, it has been unknowingly selected by the farmers for very long time. Their work greatly reduced the genetic variability and led to the development of local high-yielding populations which are still difficult to improve with traditional breeding work (Bocsa 1981). Nevertheless, several problems affect the breeding of this crop: an increase in forage yield can be brought about by introducing resistance to biotic and abiotic stresses; improvement in forage quality must aim at obtaining bloat-free forage, and varieties appropriate for pastures must be developed. Problems related to the optimization of the symbiosis rhizobium-plant and to interspecific competition are also currently faced by the alfalfa breeders.

Another outstanding problem is the seed production, which is related to many variables: flower fertility, plant health, environmental features, and pollinating insects. The concomitance of these factors makes seed production very unpredictable. Apart from reduced variability within the species, a major impediment in genetic improvement of alfalfa is its autotetraploid status, that implies a slow introgression of recessive genes and discourages the use of backcross breeding methods.

On the basis of the assumption that the highest productivity is associated with the tetraallelic status (Demarly 1963; Busbice and Wilsie 1966; Dunbier and Bingham 1975), an interesting breeding strategy has been developed. It consists of obtaining tetraallelic lines through the intercross of monoallelic ones. Because monoallelic lines cannot be easily obtained in *M. sativa* because of its high inbreeding depression, recourse to the unreduced gametes has been proposed to reduce locus variability (Bingham 1979). Biotechnology could offer a valid support and alternative routes to traditional plant breeding in alfalfa. Somaclonal variation and in vitro selection could solve the problems of introducing resistance to pathogens (Johnson et al. 1984a) and in some cases single or a few loci controlled traits such as for ethionine-resistant cell lines (Reisch et al. 1981). The protoplast fusion (Damiani et al. 1988), and the embryo (McCoy and Smith 1986) and pod culture (Wang et al. 1984) offer a valid system to bypass the incompatibility barriers among species and thereby enlarge the genetic background of *M. sativa*. Alien genes could be introduced recurring to bacterial vectors (Deak et al. 1986; Spanò et al. 1987) and/or via direct introduction of naked DNA into protoplasts. However, the approach that could offer the most conspicuous contribution to the improvement of alfalfa is the induction of haploid plants through microspore, anther, or ovule culture. The haploids are of fundamental importance for the study of the genetic structure, and represent the basic material for the constitution of tetraallelic lines, avoiding recourse to the use of lines with high production of unreduced gametes; in addition they could be used for obtaining somatic hybrids between sexual incompatible species without increasing the DNA content.

Medicago species of limited agronomical interest are the shrub *M. arborea* utilized for pasture in dry and marginal lands of the Mediterranean Basin, *M. falcata* characterized by winter hardiness that allows the cultivation of alfalfa in the cold areas of the Northern Hemisphere, *M. glutinosa* characterized by grazing resistance due to its creeping habit, *M. coerula* drought- and frost-resistant, and tolerant to saline soil (Lubenec 1967). There are also a number of annual species which could be interesting crops in hot and dry environments. Somaroo and Witcombe (1982) report promising performances in Syria of *M. rigidula, M. blancheana, M. aculeata, M. constricta*, and *M. noeana*. Other species with a limited area of cultivation are *M. arabica, M. orbicularis*, and *M. polymorpha* (Duke 1981). All these species are mainly interesting as donor lines of agronomically important traits in cross-breeding programs with alfalfa.

2 In Vitro Approaches

Alfalfa represents the forage legume most studied at the in vitro level (Tables 2, 3). Because of its economic importance and relative responsiveness to in vitro approaches, it represents something like a model system among the crop species for modern biotechnologies. Since the pioneer work of Saunders and Bingham (1972), alfalfa tissue culture has come of age very rapidly. The following is an account of history and present status of the techniques involved in the establishment of tissue culture of alfalfa.

2.1 Influence of the Genotype

Alfalfa germplasm, bred all over the world, varies greatly in its composition, but most of the in vitro studies deal with North American varieties and some with European varieties. Landraces, which might also be interesting, have not been intensively studied so far. Although present technology is not adequate for the regeneration of the full range of alfalfa germplasm, many efforts have been made

Table 2. Plant regeneration from callus culture in *Medicago* species

Species and cultivars	Explant	Callus induction and/or growth medium (mg/l)	Pre-regeneration medium (mg/l)	Regeneration medium (mg/l)	Type of regeneration	Reference
M. sativa						
cv. Saranac	Immature anthers ovaries hypocotyls internodes	BII 2 2,4-D 2 NAA 2 kin		BOi2Y	Shoot and embryos	Saunders and Bingham (1972)
cv. Regen S	Hypocotyls cotyledons	BII 2 2,4-D 2 NAA 2 kin		BOi2Y LS 1 IAA 6 kin or SH 1 2,4-D 21.5 kin	Shoot and embryos Shoots	Bingham et al. (1975) Stavarek et al. (1980)
cv. Regen clone RA-3	Ovaries and petioles	SH 5 NAA 2 kin	SH 11 2,4-D 1 kin	BOi2Y	Embryos	Walker et al. (1979)
cv. Europe	Leaves	UM 2 2,4-D 0.25 kin		MS 0.05 NAA 0.5 6-BAP	Shoots and embryos	Dos Santos et al. (1980)
clone A-15	Leaves hypocotyls cotyledons stems petioles	BOi2Y 22 2,4-D 1 kin	BOi2Y 0.2 NAA 2 kin	BOi2Y	Embryos	Novák and Konecna (1982)
cv. Adriana	Roots leaves hypocotyls	B5 2 2,4-D 0.1 kin		MS 1 2-iP 0.1 IAA	Embryos	Pezzotti et al. (1984)
cv. Rangelander cv. Vista cv. Rambler (1985) cv. Regen Y cv. Regen S	Hypocotyls Cotyledons	B5h 1 2,4-D 0.2 kin	SHb 11.1 2,4-D 1.1 kin	BOi2Y	Embryos	Brown and Atanassov (1985)

cv. CMF3713						
cv. Rambler						
cv. Vernal						
cv. Heinricks						
cv. Spreader						
cvs. Roamer, Hardistan						
cvs. Saranac, Anchor						
cvs. DuPuits, Drylander						
cvs. Trek, Beaver						
cv. Norseman						
cvs. Ladak, Chino						
cvs. Pickstar, Answer						
cv. Algonquin						
cvs. Nomad, Ladak cv. Norseman cv. Turkistan	Hypocotyls	SH 4.65 NAA 2.1 kin	SH 1.1 2,4-D 1.05 kin	BOi2Y	Embryos	Mitten et al. (1987)
M. sativa (diploid) clone HG2	Immature ovaries	BII 2 2,4-D 2 kin		BOi2Y	Shoots	McCoy and Bingham (1977)
M. arborea	Leaves roots hypocotyls cotyledons	UM 2 2,4-D 0.25 kin or MS 1 NAA 1 BAP		MS 0.5 6-BAP 0.05 NAA	Shoots	Mariotti et al. (1984a)
M. coerulea	Leaves	UM 2 2,4-D 0.25 kin UM 2 2,4-D		MS 0.05 NAA 0.5 6-BAP or MS 1.0 2,4-D 0.1 6-BAP	Embryos	Arcioni et al. (1982)
	cotyledons	 0.25 kin		MS 1.0 2,4-D 0.1 6-BAP	Embryos	Meijer and Brown (1985)

Table 2 (continued)

Species and cultivars	Explant	Callus induction and/or growth medium (mg/l)	Pre-regeneration medium (mg/l)	Regeneration medium (mg/l)	Type of regeneration	Reference
M. glutinosa	Leaves	UM 2 2,4-D 0.25 kin		MS 0.1 – 1.0 2,4-D 0.1 zeatin	Embryos	Arcioni et al. (1982)
M. varia *M. hemicycla* *M. falcata* Accession F2	Hypocotyls	B5 1 2,4-D 0.2 kin	SH 11 2,4-D 1 kin	BOi2Y	Embryos	Meijer and Brown (1985)
Accessions: 172, 202, SL 195 1886A, 1886B 2296, SL 194	Cotyledons hypocotyls roots, leaves	SH 4.65 NAA 2.1 kin	SH 11.05 2,4-D 1.05 kin	BOi2Y	Embryos	Chen and Marowitch (1987)
170, 198	Hypocotyls roots, leaves					
SL 196	Cotyledons hypocotyls, leaves					

References of media: MS, Murashige and Skoog (1962); UM, Uchimiya and Murashige (1974); SH, Schenk and Hildebrandt (1972); BII, modified Blaydes' medium, Bingham et al. (1975); BOi2Y, modified Blaydes' medium, Bingham et al. (1975); B5, Gamborg et al. (1968); B5h, modified B5 medium, Brown and Atanassov (1985); LS, Linsmaier and Skoog (1965).

to understand the role of the genetic background in somatic embryogenesis of *Medicago* species. Brown and Atanassov (1985) tested 76 cultivars of *Medicago* (*M. sativa* L., *M. falcata* L., and *M. varia* Martyn) with different background for their capacity to produce callus and somatic embryos. Of the 11 best regenerating cultivars all possessed genetic background from the Ladak and *M. varia* sources and 10 had *M. falcata* background. In particular, nine of the cultivars had their major contribution of germplasm from both *M. falcata* and Ladak. The choice of protocol medium may affect the number of embryos produced by a given genotype but high regenerating cultivars compared to low regenerating ones show a superior response regardless of medium protocol used. Work by Mitten et al. (1984) showed that among 35 landrace cultivars of various origin some displayed regenerative capability, namely cultivars Lakak, Norseman, Turkestan, and Nomad. These genotypes were originally from the region of the Turko-Iranian border. Other cultivars were not or very poorly regenerative (1% – 10%), and at low frequency. Italian landraces, gathered in specific regions, namely Lombardy, Venetia, Emilia, Toscany, have been screened for their responsiveness in vitro. Although all of them developed callus cultures under test conditions, only few showed morphogenesis, at very low frequency (Lupotto unpubl.). Meijer and Brown (1985) examined 19 accessions of diploid *M. sativa* L., *M. coerulea* L., *M. falcata* L., and *M. varia* for their ability to produce somatic embryos from hypocotyl-derived callus. All the accessions showed poor regenerability, and no correlation was found between high regeneration frequency and germplasm sources. However, some regenerating genotypes were identified in each species and some *M. falcata* accessions contained a high percentage of in vitro regenerating genotypes. Chen and Marowitch (1986) screened 17 accessions of *M. falcata* for the ability to produce somatic embryos on callus derived from different explants. Regenerative genotypes were identified in ten accessions and the efficiency of somatic embryo formation varied among genotypes and accessions. The work carried out by Brown and Atanassov (1985), Meijer and Brown (1985) and Chen and Marowitch (1986) has shown that *M. falcata* germplasm is characterized by a high potentiality for plant regeneration and this species has contributed significantly to the in vitro regeneration response of *M. sativa*.

Genotypic variation in embryogenesis is a widespread phenomenon in alfalfa and genotypes with a good regeneration capability can be recovered in all the cultivars if enough plants are screened. Highly regenerable cultivars show superior embryogenic response regardless of the medium protocol or explant source used. However, even in a less regenerable cultivar a superior embryogenic genotype could be identified which produced more embryos per callus than the best genotypes from the highly regenerable cultivars. Therefore, the highly regenerable cultivars are not necessarily the best source for obtaining superior embryogenic genotypes (Chen et al. 1987). Because in vitro regeneration is highly heritable (Bingham et al. 1975), the regeneration potential of a given population can be increased by selecting regenerable and fertile plants, intercrossing them, and then performing at least two cycles of recurrent selection. Following this procedure, Bingham and his colleagues increased the frequency of regenerant genotypes from 12 to 67% and released two tetraploid (Regen S and Regen Y) and one diploid (HG-2; McCoy and Bingham 1977) strains with improved regeneration capacity.

Table 3. *Medicago* species in which protoplasts have been successfully cultured

Plant species	Origin of protoplasts	Enzyme mixture*	Incubation condition	Plating density (protoplast/ml)	Division frequency (%)	Culture medium	Growth response	Reference
M. sativa								
cv. Canadian	Leaves	1	24 °C; 5 – 6 h; occasional agitation		60 – 80	1 : 3 of cell and protoplast culture medium, Kao (1977)	Embryos, plants	Kao and Michayluk (1980)
cv. Adriana	Leaves	1	28 °C; 4 – 5 h; occasional agitation	$5 - 10^4$	30 – 40	KM8P (Kao and Michayluk 1975)	Embryos, plants	Pezzotti et al. (1984)
cv. Europe	Leaves	2	28 °C; 4 h; occasional agitation	1×10^5	60	KM8P	Shoots, embryos, plants	Dos Santos et al. (1980)
cv. Regen S	Leaves	3	27 °C; 16 – 18 h; gentle agitation	$0.2 - 1 \times 10^5$		[a]	Embryos, plants	Johnson et al. (1981)
cv. (several)	Leaves	4	25 °C; 5 – 7 h; gentle agitation	5×10^3		[b]	Embryos, plants	Atanassov and Brown (1984)
cv. Europe	Cotyledons	5	25 °C; 16 h; agitation (10 – 20 rpm)	5×10^4	66 – 76	Kao (1977)	Embryos, plants	Lu et al. (1982)
cv. Europe	Roots	5	25 °C; 16 h; agitation (60 rpm)	$2.5 - 3 \times 10^4$	9 – 12	Kao (1977)	Embryos, plants	Xu et al. (1982)
cv. Rangelander	Leaves	7	25 °C; 18 h; agitation	1×10^5	22	[c]		Dijak and Brown (1987)
cv. Rambler	Leaves	7	25 °C; 18 h; agitation	1×10^5	6	[c]		Dijak and Brown (1987)
cv. Adriana	Cell suspension	5	23 °C; 16 – 18 h; agitation (40 rpm)	2×10^5	70 – 90	KM8P	Embryos, plants	Arcioni et al. (unpubl.)
cv. Regen S Answer	Cell suspension	4	25 °C; 16 – 18 h; agitation (40 rpm)	1×10^5	30 – 80	[b]	Embryos, plants	Atanassov and Brown (1984)
M. coerulea	Leaves	2	28 °C; 4 h	1×10^5	40 – 60	KM8P	Embryos, plants	Arcioni et al. (1982)
and *M. glutinosa*	Cell suspension	5	25 °C; 16 h; agitation (60 rpm)	1×10^5	70 – 80	KM8P	Embryos, plants	Arcioni et al. (1982)

M. glutinosa	Cotyledons	8	25 °C; 16 h; agitation (30 rpm)	5×10^4	65	KM8P	Embryos	Gilmour et al. (1987)
M. arborea	Leaves	3	28 °C; 6 h; gentle agitation	5×10^4	75	KM8P	Shoots, plants	Mariotti et al. (1984)
M. arborea	Roots	5	23 °C; 8 h; agitation (50 rpm)	3×10^4	20	KM8P	Shoots, plants	Mariotti et al. (1984)
M. falcata	Leaves	1	26 °C; 16 h	5×10^4		Modified Kao (1977): − coconut water + extra glucose (42 g/l)	Embryos, plants	Teoule (1983)
M. falcata	Cotyledons	8	25 °C; 16 h; agitation (30 rpm)	5×10^4	61	KM8P	Embryos, plants	Gilmour et al. (1987b)
M. difalcata	Cotyledons	8	25 °C; 16 h; agitation (30 rpm)	5×10^4	77	KM8P	Embryos, plants	Gilmour et al. (1987b)
M. hemicycla	Cotyledons	8	25 °C; 16 h; agitation (30 rpm)	5×10^4	62	KM8P	Embryos, plants	Gilmour et al. (1987b)
M. varia	Cotyledons	8	25 °C; 16 h; agitation (30 rpm)	5×10^4	81	KM8P	Embryos, plants	Gilmour et al. (1987b)
M. lupulina and *M. disciformis*	Leaves	6	27 °C; 16 – 18 h; gentle agitation	$0.5 - 1 \times 10^5$		[a]	Embryo-like structure in *M. lupulina* only	Johnson et al. (1982)
M. scutellata	Leaves	3	27 °C; 16 – 18 h; gentle agitation	$0.5 - 1 \times 10^5$		[a]	Few calli	Johnson et al. (1982)

* The composition of the enzymatic mixtures is reported in Table 5, p. 212.
[a] Modified Kao and Wetter (1977) for: NH_4NO_3 (250 mg/l); KNO_3 (2500 mg/l); $CaCl_2 \cdot 2H_2O$ (300 mg/l); NAA (1 mg/l); 2,4-D (0.2 mg/l); Zeatin (0.5 mg/l).
[b] Modified Kao (1977): + extra glucose (12.6 g/l); penicillin G (75 mg/l); streptomycin sulfate (5 mg/l).
[c] Modified Kao (1977) to contain glucose (0.6 M), NAA (1 mg/l); 2,4-D (0.2 mg/l); 6-BAP (0.5mg/l).

An overview of the genotypes actually screened in literature is given in Table 2, with respect to their specific responsiveness in vitro for callus induction and plant regeneration, and in Table 3 for protoplast isolation growth and regeneration.

Most of the work so far done in *M. sativa* is based on tissue culture of the line Regen S and its clones improved for regenerative capability, released by Bingham's laboratory. This line appears to be suitable for callus induction or suspension culture (Stavarek et al. 1980) either for protoplast isolation and culture (Johnson et al. 1984b; Atanassov and Brown 1984). Derivatives of Regen S have been successfully used in isolating cell lines and regenerating plants tolerant to NaCl (line W75RS: Smith and McComb 1983; line W74RS: Croughan et al. 1978) and the diploid derivative HG2, for isolating and regenerating plants selected for ethionine resistance (Reisch and Bingham 1981; Reisch et al. 1981). Some genotypes, although responding positively to callus induction and plant regeneration, are not always suitable for protoplast isolation and culture (Orr et al. 1985). As previously reported for plant regeneration from callus induced from different explants, the development pattern of protoplasts in culture varies among genotypes. Protoplasts from Regen S and Rambler grow into calli which could be induced to form embryos, while protoplasts from Rangelander undergo direct embryogenesis (Dijak and Brown 1987). In diploid alfalfa the qualitative inheritance of in vitro regeneration is controlled by at least two dominant genes. Reisch and Bingham (1980) and Phillips (1983) concluded that the genes controlling in vitro plant regeneration appear at the same relative frequency in southwestern US populations of alfalfa. However, Brown and Atanassov (1985) suggested that the genes controlling in vitro regeneration are probably present at higher frequency in populations of creeping-rooted alfalfa containing the germplasm of *M. falcata* and Ladak.

2.2 Tissue Culture Initiation: The Explant and Sterilization

Callus can be successfully induced from both immature and mature explants. So far, tissue culture has been initiated from immature ovaries and anthers (Saunders and Bingham 1972; Walker et al. 1978, 1979), stem sections and petiol tissues (Walker et al. 1979; Walker and Sato 1981), hypocotyl sections (Bingham et al. 1975), shoot tips (Kao and Michayluk 1981), leaf, hypocotyl, and cotyledon explants of 2–3-week-old seedlings (Atanassov and Brown 1984), hypocotyl and cotyledon tissues of 2–3-day germinating seedlings (Lupotto 1983, 1986). When seedling explants are used, these come from axenically germinated seeds on basic hormone-free medium (Bingham et al. 1975; Orr et al. 1985) or simply on filter paper discs (Lupotto 1983). Sterilization is done by immersion in ethanol and commercial bleach or bleach alone, supplemented with a surfactant, generally Tween 80 (Merck).

Alternatively, a treatment of 10 min in 0.2% (w/v) $HgCl_2$, containing a drop of Tween 80, can be used (Atanassov and Brown 1984). When shoot tips of outdoor-grown plants are being used, a rapid sterilization in 60% ethanol can be done as described by Kao and Michayluk (1981).

2.3 Callus Induction and Culture Media

As clearly indicated in a review (Stuart et al. 1985a), alfalfa needs a special sequence of media and hormonal stimuli to be cultured and regenerated in vitro. Explants are induced to callus in different basal medium compositions such as BOi2Y, modified Blaydes' medium (Blaydes 1966) as reported in Saunders and Bingham (1972), B5 medium (Gamborg et al. 1968) as described in Atanassov and Brown (1984), MS medium (Murashige and Skoog 1962) as in Lupotto (1983), SH medium (Schenk and Hildebrandt 1972) as described in Walker and Sato (1981), or in more complex media reported in Kao and Michayluk (1981). In Table 4 the basal medium compositions reported in various cases are listed. In some cases the addition of organic complexes such as yeast extract or casein hydrolysate to the basal medium are reported (Kao and Michayluk 1981; Novák and Konecna 1982; Lupotto 1983). Callus initiation needs the presence of a synthetic auxin, NAA or 2,4-D, in combination with kin or 6-BAP in the range between 0.5–2 mg/l auxin and 0.1–0.5 mg/l cytokinin, depending on the culture. This step induces the proliferation of a callus type generally friable and rapidly growing. Callus induction can be obtained either in the light with a photoperiod of 16/8 h (Walker et al. 1979; Lupotto 1983), in continuous light (Saunders and Bingham 1975) or in very dim light (Kao and Michayluk 1981): light seems not to have a great effect on the final result. Careful examinations of the influence of the hormones at the callus induction level, for the further support of callus growth and plant regeneration, have been done by Saunders and Bingham (1975) and by Walker and co-workers (Walker et al. 1978; 1979; Walker and Sato 1981). In the first case, callus cultures derived from immature ovaries of the tetraploid clone S-4 (cv. Saranac) on Blaydes' basal medium containing various combinations of kin, 2,4-D, and NAA, were studied in their response for budding on Blaydes' hormone-free medium. No budding occurred when 2,4-D was absent from the starting medium, whereas the frequency of bud formation was directly correlated with the 2,4-D concentration in the range 2.3–54 μM; 2,4-D appeared not only to influence callus budding but also to increase callus growth as final fresh weight. Kin was shown to be not strictly necessary in the first medium but in the second medium its presence in the range 2.3–36 μM increased bud formation. NAA did not seem to play an essential role in bud induction: 2,4-D was the sole hormone responsible for the concomitant callus induction and subsequent organogenesis. The effect of the duration of the induction period on 2,4-D-containing medium has been shown by Walker et al. (1979). Tissues grown in the presence of 2,4-D, and therefore "induced", were capable of further organogenesis when transferred to hormone-free medium, and the induction was attributable to 2,4-D alone. Furthermore, the hormonal balance between 2,4-D and kin in the induction medium was essential to shift the organogenesis toward shoot formation (high 2,4-D, low kin) or toward root formation (low 2,4-D, high kin).

Callus propagation in the presence of 2,4-D seems, in some cases, particularly sensitive to the medium composition. In callus cultures of cv. Robot, MS medium proved more suitable than B5 in maintaining the regenerative capability of the calli: propagation on B5 medium resulted in gradual loss of embryogenic potential, with gradual deviation of the callus morphogenesis toward rhizogenesis

Table 4. Tissue culture medium compositions for *Medicago* species

Tissue culture medium components (mg/l)	MS[1]	LS[2]	SHb[3]	BOi2Y[4]	B5h[5]	A medium[6]
NH_4NO_3	1650	1650	–	1000	–	600
KNO_3	1900	1900	2500	1000	3000	2200
$CaCl_2 \cdot 2H_2O$	440	440	200	–	895	295
$MgSO_4 \cdot 7H_2O$	370	370	400	35	500	310
$(NH_4)_2SO_4$	–	–	–	–	134	67
KH_2PO_4	170	170	–	300	–	170
$Ca(NO_3)_2 \cdot 4H_2O$	–	–	–	347	–	–
$NH_4H_2PO_4$	–	–	300	–	–	–
$NaH_2PO_4 \cdot H_2O$	–	–	–	–	150	75
KCl	–	–	–	65	–	–
H_3BO_3	6.2	6.2	5	1.6	3	3
$FeSO_4$	36.7	27.8	15	23.6	40	–
Na_2-EDTA		37.3	20	32		
Sequestrene 330	–	–	–	–	–	28
$MnSO_4 \cdot 4H_2O$	22.3	22.3	10	4.4	10	10
$ZnSO_4 \cdot 7H_2O$	8.6	8.6	1	1.5	2	2
KJ	0.82	0.83	1	0.8	0.8	0.75
$Na_2MoO_4 \cdot 2H_2O$	0.25	0.25	0.1	–	0.25	0.25
$CoCl_2 \cdot 6H_2O$	0.025	0.025	0.1	–	0.025	0.025
$CuSO_4 \cdot 5H_2O$	0.025	0.025	0.2	–	0.025	0.025
Nicotinic acid	0.5	–	5	0.5	1	1
Thiamine·HCl	0.1	–	5	0.1	10	10
Pyridoxine·HCl	0.5	–	0.5	0.1	1	1
Glycine	2	–	–	2	–	–
M-inositol	100	100	1000	100	100	100
Yeast extract	–	–	–	2000	–	–
N-Z-amine	–	–	–	–	–	500
L-glutamine	–	–	–	–	800	–
L-serine	–	–	–	–	100	–
Adenine	–	–	–	–	1	–
Glucose	–	–	–	–	–	g 5
Xylose	–	–	–	–	–	250
Organic acids[a]						
Sucrose	g 30	g 30	g 30	g 30	g 30	g 25

References:

[1] Murashige and Skoog (1962).

[2] Linsmaier and Skoog (1965).

[3] Schenk and Hildebrandt medium (1972) as described in Atanassov and Brown (1984).

[4] Blaydes' medium (1966) as reported by Bingham et al. (1975).

[5] Gamborg et al. (1968) as reported by Atanassov and Brown (1984).

[6] As indicated in Kao and Michayluk (1980).

[a] Na-pyruvate 5 mg/l; citric acid, malic acid, fumaric acid, each in 10 mg/l.

(Lupotto 1983). Callus cultures can maintain the regenerative potential for a long period in culture or a relatively short period, mostly depending on the genotype. Sometimes long-term cultures can be reinduced to regenerate by adding high cytokinin levels to the medium (kin 27.9 – 100 μM), as reported by Stavarek et al. (1980).

2.4 Suspension Culture

In the cases in which the callus is particularly friable, suspension cultures can be established and maintained in liquid medium. Suspension cultures, depending on the genotype used and the explant source for callus induction, can be grown in relatively simple medium (McCoy and Bingham 1977) or in more complex medium (Kao and Michayluk 1981; Atanassov and Brown 1984). The suspensions are generally established from young primary callus originated from different explants obtained on agar medium (Novák and Konecna 1982; Binarova and Novák 1984), fragmented in liquid medium, and maintained on a gyrotory shaker in direct light (Atanassov and Brown 1984) or in dim light (Kao and Michayluk 1981; Pezzotti et al. 1984). Subsequent continuous sieving of the material allows the recovery of fine suspensions of cell clumps. The cell clusters are formed by small, round, cytoplasmic dividing cells which represent the actively growing populations, plus several single, bean-shaped vacuolated cells not showing any mitotic activity (Pezzott et al. 1984). In other cases, also somatic embryos at various stages of development are present in the cell suspension (Atanassov and Brown 1984). Suspension cultures represent particularly suitable material for experiments of selection and recovery of resistant lines. Alfalfa cell cultures subjected to NaCl selection released morphogenetically competent cell lines which regenerated plants (Smith and McComb 1983). Mutagenized suspension cultures obtained from the diploid alfalfa strain HG2, cultured in the presence of 0.02 mM ethionine, allowed the isolation of clones resistant to up to 1 mM ethionine. Some of these clones were also resistant to the addition of lysine+threonine to the medium, and one of the cell lines isolated showed an increase in soluble methionine content compared to the parental HG2 callus (Reisch et al. 1981). Some of the resistant clones maintained their regenerative ability (Reisch and Bingham 1981). Cell suspensions may serve also as optimal source of protoplasts in *M. sativa* (Atanassov and Brown 1984; Pezzotti et al. 1984), *M. coerulea*, and *M. glutinosa* (Arcioni et al. 1982). In all these cases plant regeneration from protoplast-derived colonies was achieved. Plant regeneration is obtained by plating the suspension cultures onto an appropriate medium, directly by embryoid formation and plant regeneration (Kao and Michayluk 1981) or by embryoid budding on an intermediate callus (Novák and Konecna 1982; Atanassov and Brown 1984).

2.5 Initiation, Propagation and Regeneration of Alfalfa in Vitro: A Protocol

a) *Seed Sterilization and Germination*

1. Surface sterilize the seeds for a few seconds, washing in 70% (v/v) ethanol, followed by 15 min in Domestos bleach (Lever Bros. UK, final concentration active Cl^- 3%) plus 1 drop of Tween 80 (Merck).
2. Wash five times with sterile deionized water.
3. Germinate seeds 48 h on moistened filter paper at 24° ±2 °C, 2000 lx daylight fluorescent tubes.

b) *Callus Induction and Propagation*

1. Cut roots and cotyledons and save hypocotyls of each seedling.

2. Culture hypocotyls on MS medium containing 30 g/l sucrose, 2 mg/l 2,4-D, 0.25 mg/l BAP, gelled with 0.8% Difco Bacto Agar; incubate at 26° ± 2 °C, 16/8 h day/night cycle at about 1000 lx.
3. Within 3 weeks callus appears from the explants. Subculture this callus explant-free on MS medium with 1 mg/l 2,4-D in the same conditions. Callus can be propagated in this way for up to 1 year, with transfers into fresh medium every 3 weeks.

c) *Initiation and Propagation of Somatic Embryogenesis*

1. Transfer callus cultures onto MS hormone-free medium into 50×14-mm Petri dishes, in direct light at 3000 lx, 16/8 h day/night cycle.
2. Within 40 days, large, compact, green bodies appear from the calli: remove them and subculture on Boi2Y medium (Blaydes' medium; Blaydes 1966, as described by Bingham et al. 1975) containing 100 mg/l m-inositol, 2 g/l yeast extract Difco. Incubate in the same condition as step 1.
3. Within 20 days profuse somatic embryogenesis arises from these tissues on Boi2Y. Subculture every 20 days in the same conditions. The embryogenic tissue can be propagated for up to $1\frac{1}{2}$ years under these conditions.
4. Recurrent somatic embryogenesis can be obtained directly on Boi2Y medium from the hypocotyls of the cotyledonary embryos left in the same conditions or in some lines on MS hormone-free medium. Complete embryos are recovered every 20 days.

d) *Plant Regeneration*

1. Detach and explant single somatic embryos on MS hormone-free medium. Embryos suitable for plantlet development germinate within 1 or 2 weeks. A complete branched plantlet with well-developed root system can be obtained in 2–3 weeks.
2. For the establishment of several clones out of one regenerated plantlet, cuttings can be derived from the branches and rooted directly on MS hormone-free medium. The rooting process takes about 2 weeks.
3. After careful washing of the roots in sterile water, complete plantlets and rooted clones are transplanted in 80×120-mm plastic pots containing sterile potting mixture. Cover with a glass beaker for the first 3 days. Acclimatize the plants 15 days in a growing chamber or greenhouse, then transfer outdoor.

3 Isolation and Culture of Protoplasts

A variety of tissues, including cotyledons (Lu et al. 1982; Gilmour et al. 1987), leaves (Kao and Michayluk 1980; Dos Santos et al. 1980; Johnson et al. 1981; Arcioni et al. 1982; Teoule 1983; Atanassov and Brown 1984; Mariotti et al. 1984a; Dijak and Brown 1987); roots (Xu et al. 1982; Mariotti et al. 1984a; Pezzotti et al. 1984), cell suspensions (Mezentsen 1981; Atanassov and Brown 1984; Pezzotti et al. 1984), and leaf calli (Mezentsen 1981) have been used for protoplast isolation. In general, attention must be paid to numerous factors such as the growth conditions and status of the source plant, the composition of the enzymatic mixture, the culture medium, and the culture conditions.

3.1 Protoplast Sources

Leaves. Leaves are a convenient source of material, but the conditions in which the plants are grown and their age is reflected in the subsequent behavior of the isolated protoplasts in culture.

Plants are generally maintained in a growth cabinet with standardized light and temperature regimes (12-h photoperiod, 20 °C, 70% relative humidity), to ensure uniform development of seedlings. Best results have been obtained using leaves from fast growing plants under low light intensity (4000 – 6000 lx; Kao and Michayluk 1980; Johnson et al. 1981). Sometimes good protoplasts have also been obtained by growing the plants in glasshouse under partly shaded conditions (Kao and Michayluk 1980) or from young shoots (8 – 10 cm length, with no axillary bud development) excised from a plant, placed in a beaker filled with distilled water and kept in the growth chamber for 3 days in the dark (Dijak and Brown 1987). Plant age is critical for protoplast division and plants older than 30 – 35 days yield protoplasts which fail to divide (Dos Santos et al. 1980). Moreover, protoplast division differs from plant to plant within the same variety, as is expected, since *Medicago* species are open-pollinated.

Before enzymatic incubation, sterilized leaves are peeled off (abaxial epidermal layer removed) and floated on the surface of a plasmolyzing solution and then the solution is replaced with an enzyme mixture. Alternatively leaves are "feathered" with a razor blade (or chopped into approximately 1-mm^2 sections for *M. lupulina*) and vacuum infiltrated with protoplast isolation medium (Johnson et al. 1982).

Cotyledons and Roots. Cotyledons and roots are useful sources of protoplast in *M. sativa* and *M. arborea, M. difalcata, M. falcata, M. glutinosa, M. hemicicla* and *M. varia.* Young seedlings need reduced space to grow and allow saving of time on growth of plant material. Moreover, a short growth period (5 – 7 days for cotyledons, 36 – 60 h for roots) avoids any environmental influence which may affect the reproducibility of isolation and planting efficiency of protoplasts. Root age influences protoplast yield, and good results have been achieved from 36- and 60-h-old rootlets in *M. sativa* (Pezzotti et al. 1984) and *M. arborea* (Mariotti et al. 1984a) respectively.

Two major disadvantages are associated with the use of root protoplasts, such as the loss of material during washing procedures after enzymatic digestion, and the difficulty in sterilizing the seeds.

For roots the latter problem can be partially overcome by adding antibiotics (ampicillin 400 mg/l, gentamycin 40 mg/l, tetracyclin 10 mg/l) in the enzyme mixture, and for cotyledons by tissue resterilization (Lu et al. 1982).

For protoplast isolation, root tips (1 – 1.5 cm long) are excised, cut transversally into 0.5 – 1-mm-thick segments and plasmolyzed for 1 h in a salt solution (CPW; Frearson et al. 1973) with 13% w/v mannitol. Root segments are vacuum infiltrated and incubated on an enzymatic mixture. Cotyledons do not require vacuum infiltration (Lu et al. 1982).

Cell Suspension. Cell suspension from different organs has to be considered as the best source of high-yielding and viable protoplasts. Suitable cell suspension

Table 5. Enzymatic solutions for protoplast isolation from different plant organs in *Medicago* species

Enzyme	Enzyme mixtures (%)								
	Leaves						Cotyledons, roots and cell suspensions		
	1	2	3	4	5	6	4	7	8
Cellulase Onozuka R-10	1[a]	5[a]	2	0.1	2	0.5	0.1		
Driselase				1			1		
Rhozyme HP 150	1[a]		2[b]	1	2	0.5	1	2	2
Macerozyme R-10		1.25[a]	1[b]		1	0.25		0.3	0.03
Pectolyase Y23					0.5				
Pectinase	0.5[a]								
Meicelase								4	2
	Sorbitol (0.3 M) Mannitol (0.3 M) NaH_2PO_4 (0.7 mM) MES (3 mM) pH = 5.5 dilute 1:1 with Kao's (1977) culture medium	Mannitol (0.6 M) in CPW salts pH = 5.6	Sorbitol (0.3 M) Mannitol (0.3 M) NaH_2PO_4 (0.7 mM) MES (3 mM) pH = 5.7 dilute 1:1 with modified Kao and Wetter's (1977) culture medium	Mannitol (0.4 M) pH = 5.8	Sorbitol (0.3 M) Mannitol (0.3 M) NaH_2PO_4 (0.7 mM) MES (3 mM) pH = 5.7 dilute 1:1 with modified Kao and Wetter's (1977) culture medium	$NaH_2PO_4 \cdot H_2O$ (1.1 mM) MES (1.4 mM) Mannitol (0.6 M)	Mannitol (0.4 M) pH = 5.8	Mannitol (0.6 M) in CPW salts	Mannitol (0.7 M) in CPW salts

References: 1. Kao and Michayluk (1980); 2. Dos Santos et al. (1980); 3. Johnson et al. (1981); 4. Atanassov and Brown (1984); 5. Johnson et al. (1982); 6. Dijak and Brown (1987); 7. Arcioni et al. (1982), Xu et al. (1982), Lu et al. (1982); 8. Gilmour et al. (1987b)

[a] The enzymes were desalted. [b] The enzymes were dialyzed.

cultures for protoplasting are characterized by a high amount of clusters of small, round, cytoplasmatic cells and very few single, long, vacuolated cells. Only the former give rise to dividing protoplasts, while the latter produce very large-sized protoplasts that usually do not regenerate cell wall. A strict subculture/sampling protocol has to be found in order to maintain the cell suspension in a state which is most responsive to protoplast isolation. In *M. sativa* a cell suspension yielding a large amount of dividing protoplasts can be obtained inoculating cell clusters (diameter between 125 and 500 μm) in B5 medium at a density of 5×10^4/ml (22 °C, 2000 lx, 150 rpm). Protoplasts can be isolated after 1 week of culture (unpubl.).

Enzymatic Solution. Some protoplast isolation solutions (Table 5) have been utilized to release mesophyll protoplasts in *Medicago* and they differ with respect to the quality and concentration of the digestion enzymes, osmotic components, and incubation conditions. Cellulase Onozuka R-10, Macerozyme and/or Rhozyme, mixed at different concentrations, are the fundamental enzymes for the isolation of viable protoplasts from young leaves of *Medicago* spp., with the exception of *M. disciformis* and *M. lupulina*, which require also the presence of Pectolyase Y-23 (0.1% final concentration; Johnson et al. 1982). This last enzyme is more efficient than Macerozyme in the digestion of pectic polysaccharides present in the cell wall of dicots because it contains both endo-polygalatturonase and endo-pectin-lyase plus a factor stimulating maceration not found in other enzymatic mixtures (Nagata and Ishii 1979). The use of desalted enzymes (Dos Santos et al. 1980; Kao and Michayluk 1980) reduces the incubation period, preventing discoloration of the enzyme solution and leaf tissue, and enhances protoplast viability. However, sustained divisions of mesophyll protoplasts have been reported by using crude enzyme and a longer incubation period (5–18 h).

An enzymatic mixture consisting of hemicellulases (2% Rhozyme and 2%–4% Meicelase) and pectinase (0.03%–0.3% Macerozyme) has been used successfully for protoplast isolation from cell suspension, roots, and cotyledons. Roots require a shorter (6 h; Xu et al. 1982) and cotyledons a longer (16 h; Xu et al. 1982; Gilmour et al. 1987b) incubation time. Cell suspension protoplasts have also been isolated with the same enzyme mixture used to digest leaves from shoot cultures (Atanassov and Brown 1984). In all protoplast isolation solutions, mannitol alone (0.6 M), or in combination with sorbitol (0.3 M mannitol, 0.3 M sorbitol) is used as osmotic component. The enzyme treatments have generally been carried out in the dark at temperatures ranging from 23° to 28 °C with gentle or occasional agitation for mesophyll protoplasts and at 30–60 rpm for cotyledon, cell suspension and root protoplasts. The incubation time varies from 4 to 18 h, depending on the concentration of the enzymatic mixture.

3.2 Protoplast Isolation and Culture

Following enzymatic treatment, a mixture of undigested cells, components of broken or burst cells, and protoplasts is obtained. The most frequently used purification technique for mesophyll protoplasts is filtration-centrifugation. The protoplast mixture is passed through a filter, 63 μm or less pore size, to retain undigested cells, cell clumps and vascular tissue and the suspension of protoplasts and cell fragments is centrifuged. Speed of centrifugation (100 g 5 min) should be sufficient to pellet protoplasts resuspended in a washing solution (CPW 13 M; Frearson et al. 1973, or in the protoplast culture medium, etc.) and centrifuged again. The washing is repeated three times and the purified protoplast preparation is resuspended and counted with hemocytometer. For root and cell suspension protoplasts, flotation is required for achieving a clean preparation. After the first working step, the pelleted protoplasts are resuspended in a sucrose solution (21% w/v), centrifuged (150 g 10 min) and recovered at the top of the centrifuge tube. Protoplasts are then washed twice and counted.

Mesophyll protoplasts of *M. sativa* (Dos Santos et al. 1980), cell suspension and mesophyll protoplasts of *M. coerulea* and *M. glutinosa* (Arcioni et al. 1982) have been successfully cultured (dark) at 1×10^5/ml in Kao and Michayluk's (1975) KM8P liquid medium (0.2 mg/l 2,4-D, 1 mg/l NAA, 0.5 mg/l zeatin) over the same agar solidified medium separated by a sterile filter paper (Whatman No. 1). Even though the role of filter paper is unknown, it can stimulate gaseous exchanges and absorb brown substances which otherwise accumulate and result in cell necrosis. By this procedure cell colonies are visible to the naked eye 2–3 weeks after culture. The plasmolyticum level in the cultures was reduced weekly by transferring the filter paper to a medium consisting of a mixture of KM8P and Kao and Michayluk's (1975) cell culture medium KM8 (0.1 mg/l 2,4-D; 1 mg/l NAA and 0.2 mg/l zeatin) in the ratio of 3:1, 2:1, 1:1, and 1:2. Cell colonies were transferred to KM8 and then to UM (Uchimiya and Murashige 1974) medium for further growth.

Johnson et al. (1981, 1982) successfully cultured mesophyll protoplasts of *M. sativa* and *M. lupulina* (Table 2) at $0.2-0.1 \times 10^5$/ml in droplets of KWM medium (derived from Kao and Wetter 1977) at 25 °C, dark the first 2–3 days, 16-h photoperiod 1800–3200 lx successively.

After 8–10 days the protoplasts and small colonies were pipetted onto the surface of unsolidified warm agar (0.35%) KW medium (three parts KWM and one part of cell culture medium of Kao and Michayluk 1980).

Kao and Michayluk (1980) cultured mesophyll protoplasts (23°–25 °C, 8-h photoperiod, 10–20 lx during the first 2 weeks followed by gradual increase to 1500 lx) in flattened droplets using Kao's medium (1977). After 1 week, fresh medium with a lower osmolarity (cell culture medium 1:3 parts of the protoplast culture medium; Kao 1977) was added. When protoplasts divided and developed to globular structures, the hormone-free medium "A" (Kao and Michayluk 1980) was gradually added to the cultures (two to three times at 4–5-days interval). The presence of starch (10 g/l) in the medium "A" inhibited the growth of individual cells and small clusters, but not that of embryoids. Embryo formation was observed after transfer to fresh medium. Mesophyll protoplasts of *M. sativa* (Pezotti et al. 1984) and *M. arborea* (Mariotti et al. 1984a) were cultured in KM8P liquid medium at $3-5 \times 10^4$/ml (dark and 28 °C for the first 4 weeks and then 1500 lx and 23 °C). Starting from the 2nd mitotic division the cultures were diluted with a mixture of KM8P and KM8 (2:1, 1:1, 1:2, 0:1 v/v). Following the last dilution, minicolonies were transferred (for 1 month) to KM8 solid medium and then to media for further callus growth. Plating efficiency was satisfactory in both species.

Atanassov and Brown (1984) cultured in diffused fluorescent light mesophyll (droplets at 5×10^3/ml) and cell suspension (liquid culture, 1×10^5/ml) protoplasts in Kao medium modified to contain 81 g/l glucose, 75 mg/l penicillin G and 5 mg/l streptomycin sulfate.

Osmotic potential was reduced (after 21 days) by adding Kao medium modified to contain 36 g/l glucose and 5 g/l sucrose. At 45 days, cell colonies were diluted further and the Kao medium containing cell colonies was solidified with an equal volume of warm B5 medium with 12 g/l of Difco Bacto agar. Protoplast division was stimulated by the combination of 2,4-D (0.2 mg/l), NAA (1 mg/l) and BAP (0.5 mg/l).

It has been observed that BAP in combination with 2,4-D and NAA was superior to zeatin and kin for cell colony formation. BAP inhibited embryogenesis, while kin stimulated the phenomenon. Cotyledon protoplasts of *M. sativa* (Lu et al. 1982) have been cultured at 5×10^4/ml in liquid medium (Kao 1977; 23 °C and dark for the first 10 days and then 600 lx). At intervals of 7 days, protoplast medium was diluted with a mixture of protoplast medium and cell culture medium (Kao 1977) in the proportion of 2:1, 1:1, and 0:1 (v/v). Plating efficiency was around 70%. Protoplast colonies (diameter of 2–3 mm) were transferred to plant regneration media. Cotyledon protoplasts of some wild *Medicago* species (Gilmour et al. 1987b) have been cultured (dark, 5×10^4/ml) using three different procedures: (1) liquid culture in KM8P; (2) drops of a mixture of equal volumes of 1.2% agarose and double-strength KM8P medium, surrounded by liquid medium; (3) cotyledon protoplasts were added to dividing albino cell suspension protoplasts (1–2 days old) of *M. sativa* (ratio of $\frac{1}{2}$; v/v) in KM8P liquid medium. The highest division frequency (60%–80%) was reached with a culture method that involved co-cultivation with albino nurse protoplasts. However, agarose was also beneficial to protoplast division in comparison to culture liquid medium.

Root protoplasts (Xu et al. 1982) and cell suspension protoplasts (Pezzotti et al. 1984) of *M. sativa* have been cultured in KM8P : KM8 (2 : 1 v/v) liquid medium at $3-5 \times 10^4$/ml and 2×10^5/ml respectively (28 °C, dark for the first 2 weeks and subsequently 25 °C and 700–1500-lx fluorescent tubes). Cultures were diluted weekly adding a mixture of KM8P : KM8 (ratio of 1 : 1, 1 : 3, 0 : 1 v/v). Plating efficiency was around 10% for root and 20%–30% for cell suspension protoplasts.

Root protoplasts of *M. arborea* required the same cultural conditions previously reported for *M. sativa* with the exception that the initial culture medium was KM8P. Plating efficiency was low and around 2%. Examples of protoplast isolated from various sources, culture and regeneration in *M. sativa* and *M. arborea*, are given in Fig. 1.

3.3 Protoplast Processing

Protoplasts can be isolated from different organ explants (cotyledons, roots, leaf mesophyll, cell suspensions; the following protocol is one of several which may be used in protoplast isolation and culture.

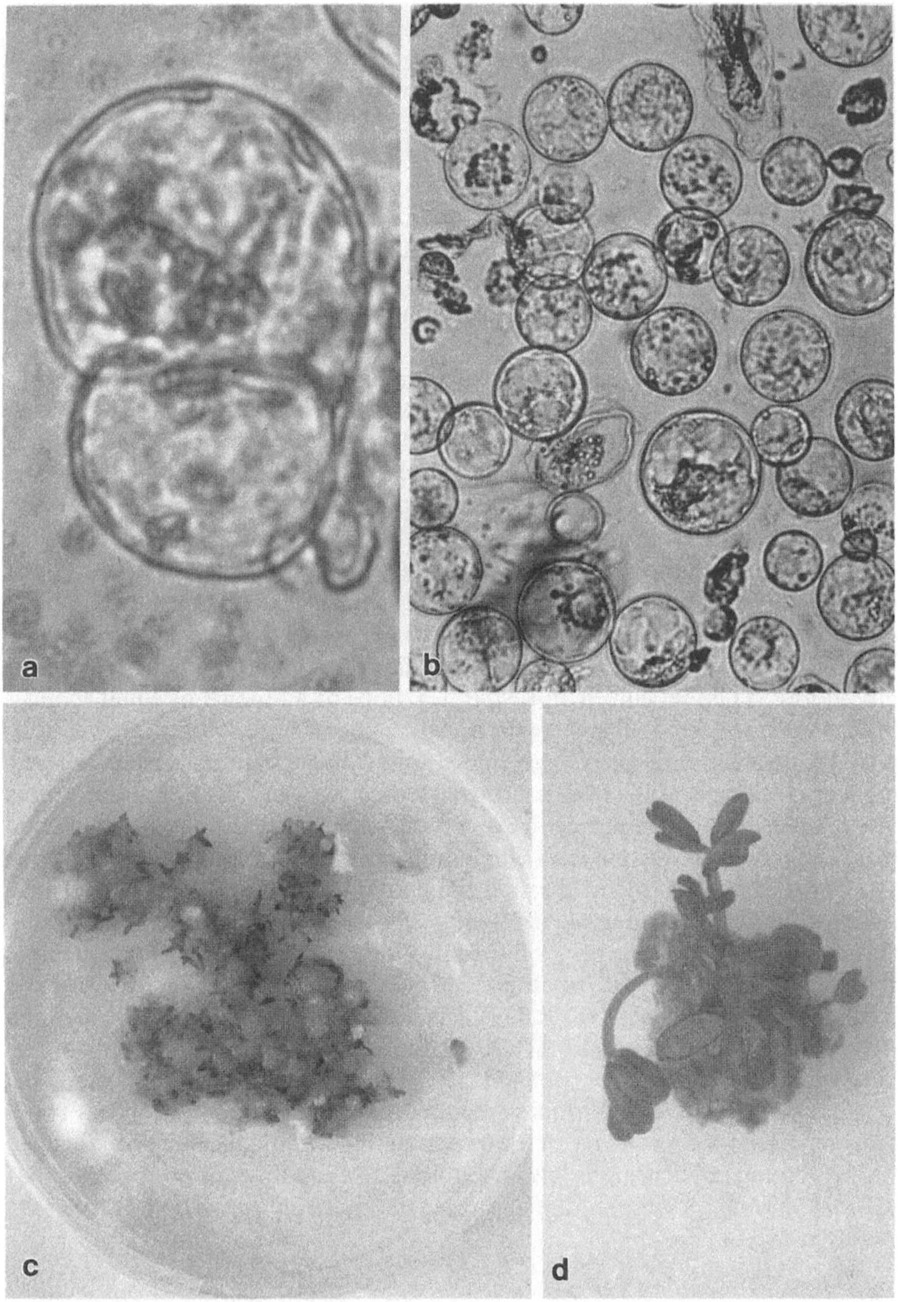

Fig. 1. **a** First mitotic division of mesophyll protoplasts of *M. sativa* after 3–4 days of culture in KM8P liquid medium. **b** Freshly isolated root protoplasts suspended in CPW 13 M. **c** Embryogenetic callus of *M. sativa*. **d** Meristematic shoot from mesophyll protoplast-derived callus of *M. arborea*

a) *Roots and Cotyledons*

1. Surface sterilize seeds with 0.1% w/v of both mercuric chloride and sodium lauryl sulfate (20 min) then in 20% v/v Domestos bleach (Lever Bros. UK) (20 min) and wash with sterilized tap water (five changes).
2. Germinate seeds in hormone-free MS ($\frac{1}{4}$ strength) solid medium (23 °C, 2000 lx daylight fluorescent tubes).
3. Cut roots and cotyledons from seedlings after 36 h and 4–6 days, respectively, into approximately 1-mm-wide strips and plasmolyze for 1 h in a salt solution (CPW: Frearson et al. 1973), with 13% w/v mannitol (CPW13M).
4. Replace the CPW13M solution with the enzymatic mixture of 2% w/v Rhozyme HP 150 (Genencor), 4% w/v Meicelase (Meiji Seika Kaisha) and 0.3% w/v Macerozyme R 10 (Yakult Biochemicals) in CPW13M, pH 5.8. For roots only vacuum infiltrate for 10 min. Incubate on a rotary shaker under the following conditions: roots: 6 h, dark, 23 °C, 50 rpm; cotyledons: 16 h, dark, 25 °C, 10–20 rpm
5. Gently squeeze plant material for protoplast release with a Pasteur pipette; filter the protoplast suspension through a sterile sieve (pore size: 64 µm).
6. Centrifuge the protoplast suspension at 80 g for 5 min.
7. For cotyledons, resuspend the pellet in CPW13M and centrifuge at 80 g for 5 min. Repeat, washing two more times and go to step 9.
8. For roots, resuspend the pellet, coming from step 6, in CPW21S (the same salt solution as CPW13M, but with 21% w/v sucrose replacing mannitol) and centrifuge at 100 g for 10 min. Collect the floating protoplasts, resuspend them in CPW13M and centrifuge at 80 g for 5 min. Repeat the washing twice.
9. For cotyledons, suspend the final pellet in liquid protoplast medium (Kao 1977). For roots, suspend the final pellet in a mixture of protoplast culture medium KM8P and cell culture medium KM8 in the ratio of 2:1 v/v.
10. Culture both types of protoplasts at a density of 5×10^4/ml in 54×14 mm (4 ml/dish) in the dark at 23°–25 °C for the first 2 weeks and then under continuous illumination (600–1500 lx) at 25 °C.
11. Decrease osmotic concentration of the culture medium at regular intervals of 7 days by adding 0.5 ml of the following culture media: cotyledons: a mixture of protoplast culture medium and cell culture medium (Kao 1977) in the ratio of 2:1, 1:1, 1:0 v/v; roots: a mixture of KM8P and KM8 in the ratio of 1:1, 1:3, 0:1 v/v. Finally, maintain minicolonies in the respective cell culture media.
12. Grow minicolonies to the calli stage on B5 solid medium for 4 weeks.
13. Transfer calli to MS agarized medium with 2-iP (1 mg/l) and IAA (0.1 mg/l) for inducing somatic embryogenesis.

b) *Cell Suspension*

1. Aseptically germinate seeds, as reported above for roots and cotyledons.
2. Culture hypocotyl sections (0.5–1 cm length) from 1-week-old seedlings on B5 solid medium (23 °C, 16 h light, 1500 lx).
3. Subculture the callus to fresh medium every 2 weeks for 2 months.
4. Initiate cell suspension culture by transferring 5 g of callus to a 250-ml Erlenmeyer flask containing 35 ml of B5 liquid medium. Place the flasks on a rotary shaker (150 rpm, 23 °C, 1500 lx).

5. After 1 week, filter the suspension through 500-μm and 125-μm pore size sieves. Collect the cells present on the 125-μm sieve and plate this fraction at a density of 5×10^4 cell clumps/ml. Culture for 1 week (150 rpm, 23 °C, 1500 lx).
6. Centrifuge the cell suspension at 150 g for 15 min.
7. Resuspend the pellet in the enzymatic mixture of 1% w/v Driselase (Kyowa Hakko Kogyo), 1% w/v Rhozyme HP 150 (Genencor), 0.1% w/v Cellulase Y-C (Seishin) in CPW salt solution with 10% w/v mannitol, pH 5.8 (8 ml of enzymatic solution for 1 ml of Package Cell Volume) and incubate on a rotary shaker (6–7 h, dark, 28 °C, 20 r.p.m.).
8. Follow the above procedure for root protoplasts starting from step 5, but plate protoplasts at a density of 2×10^5/ml.

c) *Mesophyll*
1. Surface-sterilize young, pale green, fully expanded leaves (from 2-month-old plants grown in a greenhouse) with 5% v/v Domestos bleach (10 min) and wash with sterilized tap water (six changes).
2. Remove lower epidermis and float leaves upside down on the surface of the CPW13M solution. Plasmolyze for 1 h.
3. Replace the CPW13M solution with the enzymatic solution of 1% w/v Cellulase R-10 (Yakult Biochemical), 1% w/v Rhozyme HP 150 (Genencor), 0.1% w/v Pectinase (Serva), 0.3 M sorbitol, 0.3 M mannitol, 0.7 mM Na_2HPO_4, 3 mM MES pH 5.6 diluted 1 : 1 with the protoplast culture medium KM8P. Incubate in the dark 28 °C, for 4–5 h with occasional agitation.
4. Follow the above procedure for cotyledon protoplasts, starting from step 5, but use KM8P culture medium instead of CPW13M as the washing solution. Culture on KM8P.

3.4 Protoplast Fusion

The difficulty of applying conventional techniques to alfalfa breeding, combined with recent success in regenerating plants from protoplasts of *M. sativa* and other leguminous species, has led to considerable interest in somatic hybridization as an additional means of improving alfalfa cultivars. Many problems are involved with the use of this technique: the choice of the species and of the protoplast source, the selection of the most appropriate fusing technique and the setting of the optimal parameters, the identification and isolation of the heterokaryons. Few papers about protoplast fusion in the genus *Medicago* have been published yet, and each faces these problems differently.

Chemical fusion of protoplasts has been attempted in some laboratories, Teoule and Dattèe (1987) observed that PEG 1540 does not manage to fuse agglutinated protoplasts, on the contrary, PEG 6000 is toxic in concentration over 20%. The fusion in the presence of high pH and calcium ions resulted in efficient, but low heterokaryon yield. Deionized PEG can improve the efficiency of heterokaryon formation and their viability as reported by Kao and Saleem (1986) who fused alfalfa mesophyll protoplasts with cotyledon protoplasts of soybean.

Gilmour et al. (1987a) fused *M. sativa* protoplasts with *M. falcata* and *M. quasifalcata* using a short application of PEG 6000 (30%) followed by 10-min incubation in high pH solution with Ca^{2+}. The heterokaryon yield was low (<2%), but they managed to pick up hybrids, distinguishable by color, and to grow them in an albino protoplast nurse culture. Isoenzymatic analysis confirmed the hybrid nature of the six green colonies recovered. An alternative method of culturing protoplasts at low density, as in the case of manually selected heterokaryons, has been proposed by Gilmour et al. (1987c). It consists of a well, where protoplasts at low density must be grown, protected by a membrane filter in the middle of an agarose plate containing protoplasts in active division. Gilmour et al. (1986) attempted electrofusion between *M. sativa* and *M. borealis*; even though no further results have been published as yet, the same authors observed an advantage of electrofusion with respect to the chemical method for increasing somatic hybrid colonies (Gilmour et al. 1987a). Damiani et al. (1987) set up the optimal conditions for the electrical fusion between mesophyll protoplasts of *M. arborea* and cell suspension and root protoplasts of *M. sativa* (Fig. 2). For each type of protoplast they established the optimal conditions in order to maximize the double unit chains, the most suitable for protoplast fusion. Cell suspension-derived protoplasts were the most sensitive to the electric field, thus requiring the lowest current values, root protoplasts were less sensitive. The authors attributed the different degrees of reaction to the fusion procedure mainly to the protoplasts' size.

For electrofusion, equal amounts of parental protoplasts at the respective optimal densities (mesophyll and root 3×10^5, cell suspension 2×10^5 protoplasts/ml) were suspended in a fusogen medium (0.6 M mannitol, 0.6 mM histidine HCl, 0.2 mM $CaCl_2$). The electrical parameters for mesophyll × root protoplast fusion were: an AC field of 250 V/cm, pulse voltage of 1750 V/cm and three to six pulses, each 16 s long, and for mesophyll × cell suspension were: AC field of 150 V/cm, pulse voltage 1500 V/cm and three pulses, each 15 μs long. Electrofusion has been carried out in a helical chamber of 200 μl and the fused and non-fused protoplasts were embedded in KM8P 0.3% low gelling agarose medium. Protoplast cultures were scored under the inverted microscope and the coordinates of the heterokaryons (half green and half colorless) were recorded. The somatic hybrids underwent the first division a few days after culture and some developed into hybrid minicolonies (Fig. 1A–H).

The combination mesophyll × cell suspension yielded higher percentages of heterokaryons (9%) than the combination mesophyll × root (5%). When fusion is between cell suspension and mesophyll protoplasts, the somatic hybrids can sometimes be mistaken for mesophyll protoplasts with all chloroplasts concentrated in one hemisphere (Fig. 2E). In this case heterokaryon identification is greatly improved by labelling cell suspension protoplasts by adding fluorescein isothiocyanate (FITC, 1 μg/ml) in the enzymatic mixture. Under fluorescent light, mesophyll protoplasts appear red, cell suspension protoplasts yellowish, and the heterokaryons yellow and red. Moreover the labelling remains for a few days after fusion, facilitating the manual isolation of dividing heterokaryons only (Arcioni et al. 1987).

Walton and Brown (1988), aiming for protoplast fusion within the genus *Medicago*, tested the effect of the duration and amplitude of the electric field on

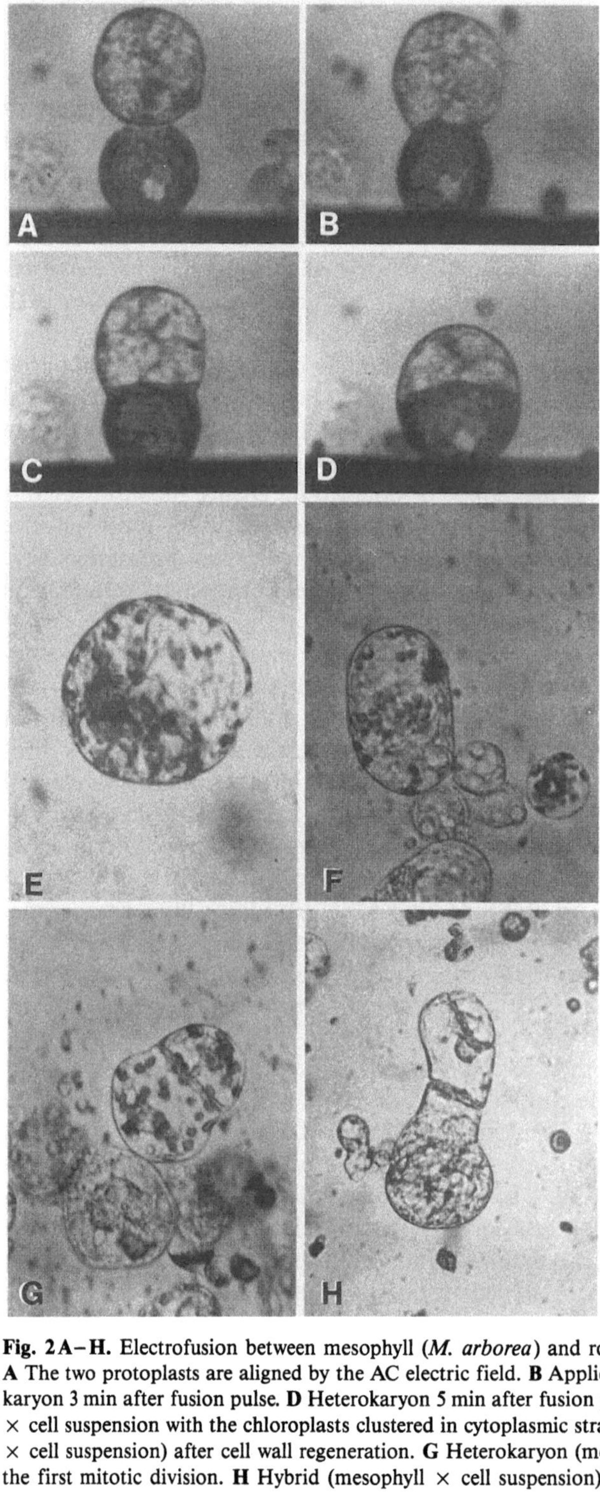

Fig. 2 A–H. Electrofusion between mesophyll (*M. arborea*) and root (*M. sativa*) protoplasts: **A** The two protoplasts are aligned by the AC electric field. **B** Application of fusion pulse. **C** Heterokaryon 3 min after fusion pulse. **D** Heterokaryon 5 min after fusion pulse. **E** Heterokaryon mesophyll × cell suspension with the chloroplasts clustered in cytoplasmic strands. **F** Heterokaryon (mesophyll × cell suspension) after cell wall regeneration. **G** Heterokaryon (mesophyll × cell suspension) after the first mitotic division. **H** Hybrid (mesophyll × cell suspension) minicolony

the viability of protoplasts isolated from different accessions of three species. The problem of increasing heterokaryon recovery has been faced utilizing parental protoplasts isolated from different explants and for this reason of different color. The use of fluorescent dyes, already utilized for manual recovery of heterokaryons, were more suitable for automatic recovery using a fluorescent activator cell sorter (FACS) (Lucretti et al. 1986; Gilmour et al. 1988). An in vitro early screening of heterokaryons has been proposed by Davis (1988) who fused kanamycin-resistant protoplasts of *M. sativa* treated with iodoacetic acid with *Lotus corniculatus* and *Onobrychis viciifolia* protoplasts. Only hybrid products should be able to survive in a medium containing kanamycin; in fact, *M. sativa* protoplasts cannot divide as a consequence of iodoacetic acid treatment and the protoplasts of the other partner are not kanamycin-resistant. The main problem in somatic hybridization in *Medicago* sp. is the regeneration of hybrid calli. Up to now regeneration has been observed only once by Teoule (1983). She chemically fused mesophyll protoplasts of *M. sativa* with mesophyll protoplasts of *M. falcata* (the two species are sexually compatible), and without any selection procedure identified hybrid plants by morphological and isoenzymatic evidence. Further demonstration of the hybrid nature of these plants was given by the occurrence of recombinant bands in their mitochondrial DNA (D'Hont et al. 1987). Arcioni et al. (1987), in their study of the chromosomal structure of *M. sativa* × *M. arborea* somatic hybrid calli, observed some structural changes and an increase of the chromosome number up to 120–150 that cannot be accounted for in the in vitro treatment. These observations led to the idea that regeneration is prevented by the high chromosome numbers that are obtained fusing autotetraploid species such as *sativa* and *arborea*. In this case the asymmetric fusion could be more efficient in regenerating hybrid plants.

4 Somatic Embryogenesis and Plant Regeneration

Regeneration in alfalfa has been interpreted as occurring via organogenesis and embryogenesis based on the morphology of regenerating structures. Stuart et al. (1985 b) found somatic embryos of alfalfa to contain 11S seed protein that was used as evidence of embryogenesis to distinguish adventitious shoots from somatic embryos in *Brassica* (Crouch 1982). The level of this storage protein expression may also be useful in verifying the similarity of somatic embryos with their zygotic embryo counterparts. Stuart et al. (1988) and Stuart and Nelsen (1988) showed that alfalfa somatic embryos accumulated storage proteins similar to those found in seed, but the level in somatic embryos was ten-fold less than in seed. Furthermore, these authors found that the level of 2,4-D, used in the embryo induction process, affected the accumulation of the storage protein. In particular the culture exposed to 50 μM 2,4-D germinated at lower levels and expressed less 11S storage protein than embryos from culture exposed to 10 μM 2,4-D. Moreover, it has been reported that maltose, when added to regeneration media, yielded embryos which were larger and germinated better than sucrose-treated cultures. Embryos produced on maltose had a higher 11S protein level than sucrose-treated cultures.

Almost all research on the role of reduced nitrogen, amino acids and carbohydrate sources in somatic embryo formation in *M. sativa* has been carried out with the clone RA3 of the cultivar Regen S. In this material very good regeneration via somatic embryogenesis occurs when calli are exposed for a short period (3–4 days) to a high concentration of 2,4-D and then transferred to a regeneration medium without hormones and with a level of NH_4^+ between 25 and 100 mM (Walker and Sato 1981). Moreover, the concentration of NH_4^+ is responsible for root or embryo formation. When the callus growth medium contains kin (50 μM) and 2,4-D (5 μM), roots appear if the level of ammonium ion is less than 12.5 mM, while somatic embryos occur when the NH_4^+ concentration is above 25 mM (Walker and Sato 1981). Similar results were obtained by Meijer and Brown (1987) who investigated the effect of reduced nitrogen on somatic embryogenesis from petiole-derived calli of diploid and tetraploid regenerable clones of *M. sativa* spp. *falcata*. From this study the presence of 5 mM NH_4^+ was essential for embryo induction and the level of 10–20 mM was optimum for differentiation of somatic embryos. Stuart and Strickland (1984a, b) have investigated the effect of adding amino acids to cultures of alfalfa undergoing in vitro somatic embryogenesis. By supplementing the regeneration medium with L-proline, somatic embryogenesis was stimulated three-fold when compared with the normal medium. Several other proline analogs have also been effective. Alanine, glutamine, arginine, lysine, serine, asparagine and ornithine also stimulate the numbers of somatic embryos formed in alfalfa cultures. The structural quality of embryos has been enhanced by glutamine, arginine, or alanine in the regeneration medium. Both embryo size and conversion to plantlets have been increased by these amino acids. Stuart and Strickland (1984a, b) have suggested that adding proline with other amino acids which stimulate embryo conversion may be a method for achieving high frequency and high quality of somatic embryogenesis. They have also reported that the proline-enhanced regeneration of alfalfa depends upon the ammonium ion. Optimization of regeneration occurs when 100 mM proline and 25 mM ammonium are added to the regeneration medium. On the contrary, glutamine is the only amino acid capable of stimulating somatic embryogenesis which is not dependent on ammonium for its effects on alfalfa. Glutamine could substitute for ammonium in proline-stimulated embryogenesis. Maltose, maltotriose and soluble starch all contribute to improve the morphology and eventual conversion to plantlets of somatic embryos produced on callus derived from RA3 clone. In particular maltose gave the highest embryo yield and improved embryogenesis of alfalfa genotypes which do not readily undergo high regeneration. The effect of maltose on embryogenesis is dependent on the presence of NH_4^+. An effective regeneration medium for high frequency of embryogenesis can be achieved by the combination of 3% maltose plus 15 mM NH_4^+ plus 30 or 100 mM proline. A complex mixture of organic substances such as yeast extract and myo-inositol has been reported to stimulate morphogenesis (Saunders and Bingham 1975; Walker et al. 1979) and embryogenesis (Lupotto 1983). In *M. falcata*, high concentrations of amino acids, particularly of proline, usually did not increase embryo yield or embryo quality (and frequently had detrimental effects) unlike the clone RA3. Recently, interesting results have been reported on the strong effect of the donor plant genotype on somatic embryogenesis in *M. sativa*,

Fig. 3a–d. Callus culture and plant regeneration in *M. sativa* cv. Robot. **a** Start of green globular structures on undifferentiated callus obtained in hormone-free medium 2 weeks after transfer. **b** Budding of somatic embryos on the green structure of **a** developed in 2–3 weeks after transfer on yeast extract containing medium. **c** Germination of somatic embryos 4 weeks after transfer of embryogenic calli in hormone-free medium; plantlets are ready to be explanted singularly. **d** Somatic embryos at the cotyledonary stage: cotyledons, hypocotyl and root meristem are clearly distinguishable

M. media, M. varia and *M. falcata.* In the case of three cultivars, Rangelander, Rambler, and Heinrichs, Chen et al. (1987) observed strong genotype × medium interactions on the resulting somatic embryogenesis and plant regeneration: better-performing genotypes give good response irrespective of the culture medium of the explant source. Besides, the genotypic variability present in a cultivar allows the identification of a better-performing embryogenic genotype even if the cultivar itself is of less regenerable material. Evidence of strong interactions in genotype × medium, genotype × explant and genotype × explant × medium is

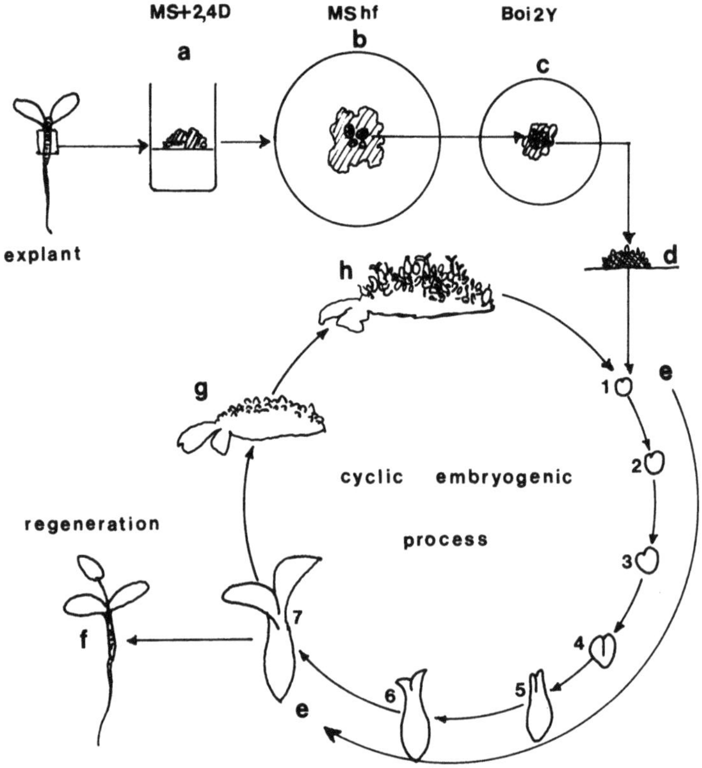

Fig. 4a–h. Origin, development, propagation, and regeneration via somatic embryogenesis in *M. sativa* cv. Robot. **a** Callus induction from hypocotyls on 2,4-D-containing medium; **b** formation of the proembryogenic tissues on MS hormone-free medium; **c** budding of the first embryos; **d** somatic embryo development as in **e**; **e** development of the embryos in culture from globular (1) to the cotyledonary stage (7); **f** development of the complete embryos into a new plantlet; **g** deviation toward recurrent somatic embryogenesis on the hypocotyl and new generation of complete somatic embryos (**h**)

also reported in the case of *Medicago media* and *M. sativa* cultivars (Nagarajan et al. 1986); it is also pointed out that better-performing genotypes in producing profuse embryogenesis are not always also the best producers of regenerated plants, because several embryos do not develop into whole plantlets capable of growing in the soil. These results are in agreement with those reported previously by Brown and Atanassov (1985) on 76 cultivars. Anyway, optimal conditions for regeneration with respect to these factors vary from genotype to genotype. In *M. sativa* cv. Robot, somatic embryogenesis could be induced in callus cultures derived from hypocotyls of germinating seedlings by transferring the callus tissue from 2,4-D containing medium to hormone-free medium in the light. Under these conditions, most of the callus died, except some embryogenic cell clusters, which developed globular green structures (Fig. 3 A). Subculture of these green bodies on yeast extract-containing medium produced profuse somatic embryogenesis (Fig. 3 B). Embryogenic cultures became stabilized and could be subcultured for

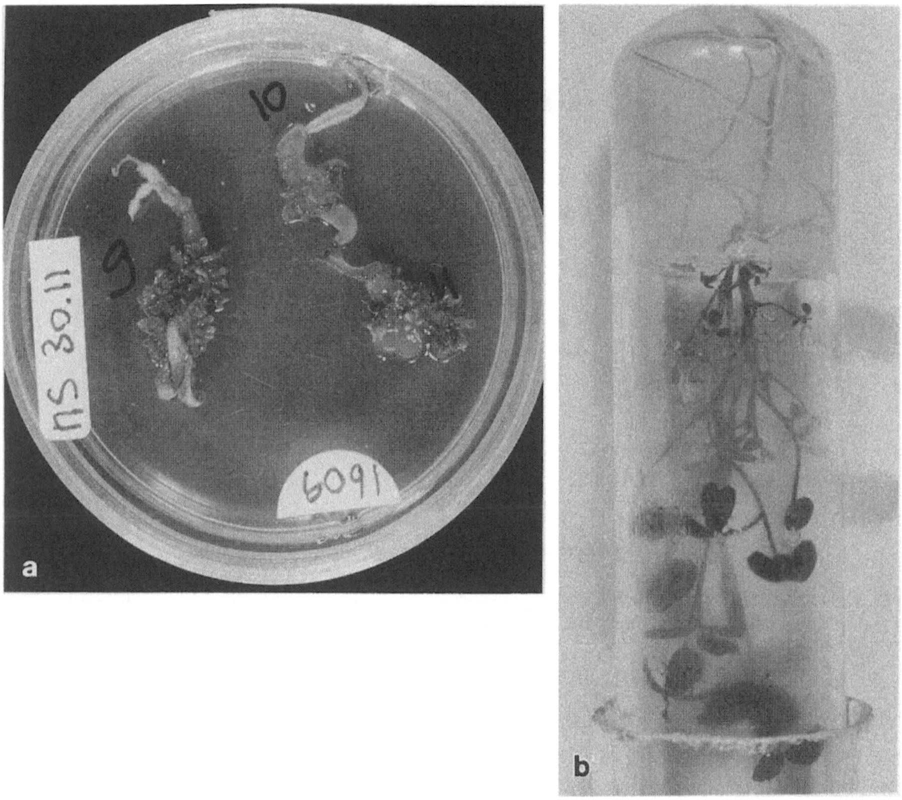

Fig. 5a, b. In vitro establishment of clones from regenerants. **a** Recurrent somatic embryogenesis on the hypocotyl of germinating somatic embryos on MS hormone-free medium, picture taken 2 weeks after the primary embryo explant. **b** Rooted clone ready to be transferred into pots in a growing chamber; this stage is reached 15–20 days after the cuttings are planted

several months (Fig. 3C): somatic embryos generated at the surface of the callus tissue underwent typical pattern of development from globular to torpedo and cotyledonary stages; embryos at this latter stage detached and germinated (Fig. 3D). One peculiar behavior of these cultures was that sometimes, when somatic embryos were maintained in regenerative conditions, cells of the hypocotyl budded into new embryos and gave rise to the cyclic process of somatic embryogenesis illustrated in Fig. 4. This characteristic is particularly suitable for propagating and regenerating alfalfa in long-term schemes of plant recovery from tissue culture and evaluation of the regenerants. In cv. Robot, the propagation of the cultures through the recurrent process of embryogenesis on the hypocotyls of mature somatic embryos allowed reproduction of each particular regenerant in several individuals before transplantation into soil as illustrated in Fig. 5 (Lupotto 1983, 1986). When embryogenic cultures are propagated in this way, the increase in fresh weight of the tissues is due only to the proliferation of new embryoids, not to callus production. There is a linear correlation between the growth of the em-

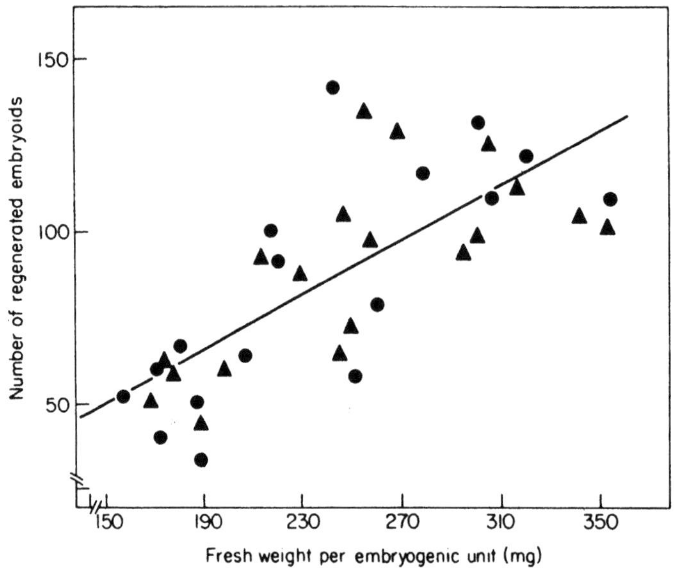

Fig. 6. Correlation between the number of regenerated somatic embryos on the hypocotyl and fresh weight of the tissue after 20 days in culture, in two embryogenic lines (● line 1425, ▲ line 2416)

bryogenic tissue and the number of complete embryoids recovered (Fig. 6). Histological examination of the origin and development of the embryoids showed that embryogenic cell clusters are located in the external layer of meristematic cells, surrounding a compact nodule of callus tissue. These cells are clearly distinguishable because of their dense staining reaction, and give rise to the embryoids (Dos Santos et al. 1983).

Embryos and shoots induced on regeneration medium are transferred to hormone-free agar medium. Plantlets are moved to soil-less compost under clear plastic to prevent dessication (23 °C, 8000 lx fluorescent tube, 12-h photoperiod). After 3 to 4 weeks the plastic bags are removed and plants are grown in the greenhouse. Figure 7 illustrates the various passages from in vitro to field conditions, from the first transfer of each plant into pots outdoor, through the selfing and field evaluation of the progenies of each regenerant.

5 Genetic Stability and Somaclonal Variation

Somaclonal variation commonly occurs in many plant species following regeneration from tissue culture, even in the absence of deliberate mutagens (Larkin and Scowcroft 1981). An extensive review on somaclonal variation in alfalfa has recently been published by Bingham and McCoy (1986). Several mechanisms for such variation are proposed, which include changes in gene expression as a result of chromosome structural modifications (heterochromatin and position effects)

Fig. 7a, b. Steps from in vitro to the field. **a** Plantlets in acclimatization in a growth chamber 1 week after transfer from in vitro conditions. **b** Field evaluation of progenies of single regenerante or movement of transposable elements, chromatin loss, DNA amplification, somatic crossing-over, somatic reduction, and changes in the DNA of cytoplasmic organelles (Larkin and Scowcroft 1981; Meins 1983; Orton 1984). These alterations can occur at different stages of the tissue culture. As a matter of fact, ploidy levels are altered during the tissue culture. In 16-month-old colonies, derived from cell suspension protoplasts, 50% of the calli was tetraploid, 40% octoploid, and 10% decaploid. Moreover, at each ploidy level, chromosome loss was observed (Atanassov and Brown 1984). The increase in ploidy level was related to the rapid loss of regeneration capacity: genetic instability due to tissue culture is better observed and analyzed on the regenerants. Binarova and Dolezel (1988) observed that the insurgence of genetic instability in cell suspension culture could be avoided with a short subculture interval (7 days); following this strategy they managed to maintain a stable and embryogenic cell suspension for more than 1 year.

Somaclonal variation in alfalfa was first reported by Saunders and Bingham (1972): in the regenerants they observed few octoploid and several off-type plants, including albinos. Alfalfa regenerants from callus culture frequently show chromosome doubling (McCoy and Bingham 1977) and chromosome loss. Groose and Bingham (1984) observed that of 116 regenerated plants, at least 11% lost one or more chromosomes and the chromosome number of the aneuploid plants ranged from 2n = 28 to 31.

Even if the first papers on alfalfa protoplasts did not mention phenotypic and karyotypic alterations of regenerants, complete and partial chromosome doubling has been observed in plants regenerated from mesophyll protoplasts (Johnson et al. 1984a, b; Latunde-Dada and Lucas 1983). Recent work on *M. media* cultivars revealed that the cultivar and therefore the genetic background, may play an important role in the cytological stability or instability of the regenerants from tissue culture (Nagarajan and Walton 1987).

Tissue culture can provide chromosomal structural changes (Johnson et al. 1984a) and qualitative genetic changes such as single gene recessive mutations or deficiency. Changes in gene expression were observed by Groose and Bingham (1984): when a white-flowered mutant, regenerated from a purple-flowered plant with one functional allele for anthocyanin synthesis was recultured, about 40% of the regenerants reverted to purple flowers. Genetic analysis showed that a functional allele of a locus for anthocyanin pigmentation was in the simplex condition in the original purple-flowered plant, and this allele mutated to an unstable recessive allele in the white-flowered mutant and then reverted at high frequency in vitro. Reversion also occurred in the plant, but at much lower frequency than in vitro. Smith et al. (1986) found changes in chlorophyll pigmentation arising in the tissue culture of the progeny of the white-flowered variant. A variant with chlorophyll-deficient sectors regenerated along with 25 normal green plants. Chlorophyll-deficient sectors were transmitted biparentally as a cytoplasmic trait and the transmission was higher through the pollen than through the seed. Baertlein and McDaniel (1987) observed differences among regenerated plants in protein content and the appearance of a growth habit compressed and planar. However, it has not been demonstrated whether these alterations are due to genetic changes or to environmental factors (e.g., lateral illumination of the in vitro grown plants). No changes in the restriction patterns of mitochondrial DNA was observed in 20 alfalfa protoclones when digested with five restriction endonucleases (Rose et al. 1984; Johnson et al. 1984b). Variability among regenerated plants for quantitative traits, such as forage yield, has also been noted. Reisch and Bingham (1981) isolated a somaclonal variant yielding three times more herbage than the parental material in field experiments. However, the sexual method produced a clone yielding double the best in vitro variant.

In order to verify the practical use of somaclonal variation, 75 plants were regenerated via somatic embryogenesis from calli induced from leaves of two alfalfa plants (Arcioni et al. 1988). A field evaluation of regenerated plants showed the absence of genetic alterations exploitable in plant breeding for all the morphological and agronomic traits evaluated. Moreover, no differences were detected for the electrophoretic banding patterns of several enzymes. The reasons for these results could be due to: the short time of callus phase (40 days from explant culture to embryo formation), the plant regeneration system (embryogenesis reduces the range of genetic variability), the genotypes used, and the possible presence of only recessive mutations that are displayed in the progenies.

Magnification of the somaclonal variation could be attempted with the use of mutagens. Experiments in many species have been made to test the relative influence of mutagens on tissue culture in the induction of novel variability (Novák et al. 1986). In the *Medicago* species there is limited information about this topic. Bingham and co-workers treated callus culture with EMS and ethionine separately and also sequentially. Ethionine treatments, also when combined with EMS, succeeded in amplifying the somaclonal variation; in contrast, EMS treatments showed no significative effect (Bingham and McCoy 1986). However, the very limited number of experiments carried out on this topic allows no definitive conclusion. Somaclonal variation is magnified if tissue culture is supported by selection. Arcioni et al. (1987) induced resistance to *Fusarium oxysporum* by selecting calli in the presence of fungus culture filtrate, while unselected calli from the same genotypes did not display variation at any level. Other workers succeeded in exploiting the somaclonal variation to select for particular variants. Tolerance or resistance to toxic agents is the simplest way to drive somaclonal variation in a particular direction because the addition of toxic compounds to the culture medium allows screening the advantageous mutations. In *Medicago sativa*, Hartman et al. (1984) selected plants resistant to *Fusarium oxysporum*, Latunde-Dada and Lucas (1983) obtained plants resistant to *Verticillium albo-atrum*; cell lines tolerant to high concentrations of NaCl were selected by Croughan et al. (1978) and Smith and McComb (1983); ethionine-resistant cell lines were obtained by Reisch and Bingham (1981). In *Medicago* species, particularly in *M. coerulea*, cell lines resistant to sodium chloride and methyolammonium have been reported (Indiogine et al. 1986).

6 Anther Culture

The interest in haploids stems largely from their considerable potential use in plant breeding (Melchers and Labib 1970; Kasha 1974; Hlasmikova 1977; Bajaj 1983). Haploids may be utilized to detect mutations and to facilitate the recovery of unique recombinants. Furthermore, doubling of the chromosome number of haploids offers a method for rapid production of homozygous plants, from which inbred lines to be utilized for hybrid production may be obtained.

In *M. sativa* results in haploid plant production are very scarce. For the first time (Bingham 1969, 1971) several haploids were isolated by crossing tetraploid alfalfa with diploid *M. sativa* and *M. falcata*. The 4x×2x cross method proved to be effective and to have the advantage of obtaining haploids from specific parents. Nevertheless, no more than one or two small pods per raceme were obtained, and each pod often contained only one, not necessarily haploid, seed.

Because this method requires a lot of work to emasculate the mother plants, to cross the parentals, and to identify the haploids, an in vitro approach was attempted. Zagorska et al. (1984) obtained alfalfa plants with a reduced chromosome number, via anther culture. For inducing androgenesis, buds (0.2–0.3 mm long) were used which were collected at the mononuclear stage of the pollen from ten alfalfa lines and clones. For callus induction, anthers were cultured on Blaydes' (1966) solidified medium with 2 mg/l of both 2,4-D and NAA, and kin or BAP (2 mg/l). The anthers bloated 2 weeks after isolation and white-grained callus appeared 20 days later. Callus was induced in all the lines tested and its growth was strongly affected by the genotype. Organogenesis was induced by transferring the calli to Blaydes' solid medium supplemented with 1 mg/l of 2iP. Rhizogenesis was a common phenomenon and stem morphogenesis was observed only in three lines and was enhanced by the addition of mesoinositol (500 mg/l) and yeast extract (2 g/l). Shoots were rooted on B5 medium without hormones; plantlets were grown in perlite and 3 weeks later were planted in soil. Analysis of the chromosome number of the regenerants revealed a high degree of variability. Some of the plants were mixoploid, with the chromosome number at the haploid level (2n = 16–20), while others were normally diploid. Similarly, anther culture of *M. sativa* lines, selected for high regeneration ability through somatic embryogenesis, has been carried out in Perugia (Piccirilli unpubl.) and at CSIRO in Canberra (Tanner pers. commun.), but all the regenerated plants showed the normal somatic chromosome number (2n = 4x = 32). Anther culture induced nonphysiological divisions of the nucleus in uninucleate pollen grains. The division frequency was affected by the genotype, age of the donor plant as well as the type and concentration of auxins, cytokinins and osmoticum (Tanner et al. 1988).

Haploids of tetraploid alfalfa are of great importance in genetic, physiological, and breeding research. Their potential use in research would be similar to that of potato haploids (Hougas and Peloquin 1958; Peloquin and Hougas 1958) and the elaboration of efficient methods of inducing haploids in alfalfa would be of enormous significance to further work on the implementation of the genetic selection program involving this species.

7 Ovule, Embryo and Pod Culture

In *Medicago* species, the recovery of interspecific hybrids is limited by the presence of post-fertilization barriers (Oldemeyer 1956; Fridriksson and Bolton

Table 6. Number of flowers pollinated, number of ovules cultured, number of embryos cultured and number of hybrid plants obtained for various interspecific combinations

Cross	No. of flowers pollinated	No. of ovules cultured[a]	No. of embryo cultured[b]	No. of hybrid plants obtained
M. sativa (2×)×*M. rupestris* (2×)	2.645	451	317	66
F_1 (*M. sativa*×*M. rupestris*) (2×)×*M. sativa* (2×)	2.084	547	450	323
M. sativa (2×)×*M. rhodopea* (2×)	2.494	329	256	157
M. sativa (2×)×*M. papillosa* (2×)	2.160	83	30	13
M. sativa (2×)×*M. marina* (2×)	2.016	39	4	2
M. sativa (2×)×*M. hybrida* (2×)	2.933	28	12	2
F_1 [*M. sativa* (2×)×*M. dzhawakhetica* (4×)](3×) ×*M. cancellata* (6×)	1.560	75	57	40
F_1 [*M. sativa* (2×)×*M. dzhawakhetica* (4×)](3×) ×*M. saxatilis* (6×)	2.057	91	73	52

[a] For *M. sativa*×*M. rupestris* F_1 and BC_1 the NH_4NO_3 concentration in L_2 medium was 25 mM, and for all other combinations the NH_4NO_3 concentration was 12.5 mM.
[b] Indicates the number of embryos that appear to be viable when dissected from the ovule and placed on fresh medium.

1963a; Sangduen et al. 1983; McCoy 1985). Attempts to produce *Medicago* interspecific hybrids, via embryo rescue, have been unsuccessful (Fridriksson and Bolton 1963b; Elgin et al. 1977). McCoy (1985) proposed a modification of the embryo rescue technique, designated the ovule-embryo culture method, which was successful for recovering the first interspecific hybrids, diploid *M. sativa* × diploid *M. rupestris*. Basically, the technique involves the culture of fertilized ovules 14–20 days after pollination; ovules were precultured for 6–9 days and then embryos were removed and cultured in fresh medium, where they germinated directly. Removal of the embryo prior to 4 days or after 10 days of culture in the ovule was unsuccessful. The optimum medium for both embryo and culture phases was L_2 (Phillips and Collins 1979) without growth hormones and containing 12.5 mM NH_4NO_3 and sucrose 25 g/l (original L_2 medium). Shoots were maintained on L_2 medium until five trifoliate leaves emerged; plantlets were transferred to Jiffy-7 peat pellets and placed in a high humidity chamber (16-h day, 105 $\mu E/m^2/s$ from cool white fluorescent lamps) until roots emerged from the peat pellet. After 14 days of hardening, plants were transferred to a glasshouse. The interspecific hybrids reported in Table 6 were obtained by applying this technique (McCoy 1985; McCoy and Smith 1986).

The interspecific hybrid plants, whose hybridity was confirmed by isozyme analysis (perioxidase or leucine amino-peptidase), resulting from diploid parents, mostly showed the diploid chromosome number (2n = 16) while the majority of the trispecies hybrids [*M. sativa* (2×) × *M. dzawhakhetica* (4×) (3×)] × *M. cancellata* (6×) or *M. saxatilis* (6×) were at or near the hexaploid level.

The morphological traits of the hybrid plants were intermediate to those of the parents. Pod culture has been successfully employed for growing embryos produced through interspecific crosses in *Medicago* species. There are few reports on this topic, whose results, however, stress the particular significance of this technique in breeding program. Pods excised 3 days after pollination and cultured in vitro produced viable seeds in annual (diploids *M. blancheana* Boiss, *M. disciformis* D.C.; tetraploid *M. scutellata* L.) and perennial (diploid *M. jaleata* L., tetraploid *M. sativa*) *Medicago* species. Mature seeds were rescued in all the species, though to different degrees, the most responsive being the annual diploids: over 80% of *M. blancheana* and *M. disciformis* pods produced viable seeds (Wang et al. 1984). Rescue of embryos deriving from interspecific crosses has also been studied for incorporating specific agronomic traits brought by *Medicago* species into *M. sativa*. The annual species *M. scutellata* is characterized by dense glandular stem and leaf hairs which provide a natural defensive system against alfalfa weevil (*Hypera postica* Gyllenhal) and potato leafhopper (*Empoasca fabae* Harris). Limited success has been obtained in trials for transferring the glandular haired trait from *M. scutellata* to *M. sativa*: one cause for this is embryo abortion. A study conduced on in vitro zygotic embryogenesis was undertaken in *M. sativa* and *M. scutellata* in order to set up the conditions to rescue and develop early embryos derived from the cross, and grow the hybrids (Bauchan 1987).

These new techniques, such as ovule-embryo, fertilized pod cultures and somatic hybridization are very useful for the incorporation of important agronomic traits from wild *Medicago* species into cultivated alfalfa.

8 Transformation

Plant biotechnology today is broadening its potentiality to introduce exogenous isolated genes into a plant genome and have them stably integrated, expressed, and transmitted to the progeny of the transformed plant. The transformation process, primarily discovered and extensively studied in the natural system of *Agrobacterium tumefaciens*, is now broadened to many species, most of them of great agronomical value (for recent review see Hooykaas and Schilperoort 1984; Bajaj 1989).

Legumes fall into the host range of *A. tumefaciens* and therefore the natural system of transformation, in which a part of the bacterial plasmid DNA (T-DNA) is inserted into the plant genome and integrated and expressed, can be efficiently used. Examples have recently been given in soybean (*Glycine max*, *G. soja*, and *G. canescens*), in which the effect of cultivar specificity in the interaction bacterium-plant, when in vivo inoculation is used, is outlined (Byrne et al. 1987). Forage legumes can be transformed via *A. tumefaciens*, and their response depends mostly on the plant species treated. *M. sativa* develops small, green, disorganized and slow-growing tumors (Mariotti et al. 1984b; Webb 1986).

Medicago species are sensitive to both *A. tumefaciens* and *A. rhizogenes*. *M. varia* has been effectively transformed via *Agrobacterium*-mediated transfer of the neomycin phosphotransferase gene and kanamycin-resistant plants have been regenerated at very high frequency, in which the transferred gene was detected as stably integrated in the nuclear genome and expressed (Deak et al. 1986). *M. sativa* has also been transformed using a disarmed vector residing in the bacterium (pARC8); kanamycin-resistant calli could undergo somatic embryogenesis and regenerated transformed plants. The system was particularly efficient, and transformation rates of 12.5% to 51% on the number of inoculated stems have been obtained (Shahin et al. 1986). Recently, alfalfa plants have been transformed with an *A. tumefaciens* derived binary vector, containing the neomycin phosphotransferase gene under the control of the mannopine biosynthesis promoter (Pezzotti et al. 1988). While *A. tumefaciens*-mediated transformation gives rise to problems once plant regeneration has to be obtained as final result, *A. rhizogenes* appears more helpful in this context, since *Rhizogenes*-induced hairy roots are capable of regeneration and whole fertile plants containing full-length T-DNA can be obtained (Costantino et al. 1984; Tepfer 1984). Indeed, the plasmid of *A. rhizogenes* (Ri-plasmid) can be used effectively to transform alfalfa and transgenic plants could be recovered, by somatic embryogenesis, from the hairy root derived calli (Sukhapinda et al. 1987). In this case, regenerated plants recovered from hairy root tissue culture were phenotypically normal compared to those regenerated from nontransformed calli, although displaying an extensive shallow-root system, which results in a common trait among hairy root-derived plants. In another report, Spanò and coworkers (Spanò et al. 1987), describe morphological changes in the transgenic plants, ascribed to the transformation system itself. Indeed, modifications affecting both the root system, which was fasciculated in the transformants, and the areal parts, internode length and number, leaf size and stem numbers, were also observed. Transformation with Ri-plasmid did not affect the sexual apparatus of the regenerants, which set seeds. Moreover, Ri T-DNA was also detected in the progeny of each transformant, thus proving to be transmitted through meiosis (Costantino et al. 1984).

Recently, a novel technique of transformation by using microinjection of cell suspension-derived protoplast of alfalfa has been developed (Reich et al. 1986). The reported transformation frequencies were quite high: 15%−26% over the total of injected nuclei and transformed cell lines confirmed by Southern blot analyses; transgenic lines remained stable in culture for over 1 year. Moreover, direct injection of DNA into the cells broadens the application of this technique to species which are not sensitive to *Agrobacterium* infection. However, in this case, no transgenic plants could be regenerated (Reich et al. 1986). The transformation system represents one of the most advanced areas in alfalfa genetic manipulation studied at the in vitro level. Effective transformation of alfalfa with specific genes might represent the most dramatic impact of modern biotechnologies on plant improvement.

9 Cryopreservation

The maintenance and preservation of germplasm collections, representing a wide array of genetic diversity of cultivated plants and their wild relatives, are essential for plant breeding programs. Plant genetic improvement is directed toward high-yielding varieties of crop plants with a narrow genetic base. As a consequence, problems can arise, such as particular diseases, pest infections, or environmental perturbation. Thus, the germplasm, which may be of less importance but also holds future potential, has to be conserved (see Chap. I.3, this Vol.).

For *Medicago* species, whose germplasm is maintained through seeds, there are particular problems with its preservation. In fact, seeds of *M. sativa* are orthodox and their viability is routinely retained for long periods by storing them in sealed containers at about −18 °C with a moisture content of 5%−6% (Lorenzetti, Plant Breeding Institute of Perugia, Italy; pers. commun.). Although seeds effectively store genetic material, plants obtained from them sometimes show traits different from those of the mother plants (Bajaj 1979a). Deterioration of stored seeds is reduced at temperatures below −20 °C. Cryopreservation techniques (e.g., liquid nitrogen at −196 °C) provide the potential for indefinite preservation by reducing all biochemical processes to such a low level that biological deterioration is virtually stopped (Bajaj 1979b). The utilization of liquid nitrogen (LN) as a storage technique is based on the ability of seeds to survive LN exposure without significant reduction of viability.

Seeds of *Medicago* spp. (Bequerel 1950) and some cultivars of *M. sativa* (Busse 1927; Lipman and Lewis 1934; Sakai and Noshiro 1975) are tolerant to exposure to LN (rate of cooling from 50 °C/s to 200 °C/min) and can be rewarmed to 25°−30 °C (rewarming rate from 25 °C/s to 30 °C/min) without loss of viability.

Alfalfa pollen can be preserved for several months or years by vacuum drying (Hanson 1961; Hanson and Campbell 1972). For this purpose pollen was collected in glass tubes and predried for 48 h in a desiccator (20 °C) with 20% relative humidity and then vacuum dried (1 h and 20 min). The tubes were sealed under vacuum and stored in a freezer at −21 °C. Following this procedure, alfalfa pollen was successfully stored for breeding studies for 11 years, although some deterioration in viability (reduction of pod set and seeds per pod) occurred.

Storage of alfalfa germplasm has also been achieved by maintaining plantlets from cultured shoot tips at low temperatures (Cheyne and Dale 1980). Shoot tips, 0.2−4.3 mm long, were cultured on B5 culture medium with 0.2 mg/l of NAA for 6−8 weeks and the regenerated plantlets (2−5 cm tall) were transferred to low temperature storage conditions (dark, 4°−6 °C; fluorescent light at 300 lx and 8 h of photoperiod, 2°−4 °C) for 15−18 months, then removed for subculturing. The survival of lucerne plants under in vitro storage conditions was very high (94%).

Very little information is available on alfalfa germplasm preservation through cell culture freezing. Cell survival of alfalfa callus during freezing was greatly improved with the addition of a cryoprotective mixture of polyethylene glycol (PEG) : glucose : dimethylsulfoxide (DMSO) in the proportion of 10% : 8% : 10% (w/v) (Finkle et al. 1985). The addition of PEG to the glucose-DMSO mixture increased the relative viability index of cryoprotectant-treated unfrozen cells (decreased toxicity), and with frozen cells it increased their viability index.

10 Summary and Prospects

The genus *Medicago*, and in particular the species *sativa*, has greatly improved its response to tissue culture and in vitro methods in the past years. Plant regeneration through somatic embryogenesis and organogenesis has been achieved and many regenerative lines have been produced. Protoplast systems are available from different plant organs and more sophisticated techniques, such as protoplast fusion and transformation, are in progress. Transgenic plants have been obtained through *Agrobacterium* infection of both *tumefaciens* and *rhizogenes* species. For these reasons *M. sativa* can now be considered a model plant for in vitro studies.

However, there are many problems in this species which slow its genetic improvement through conventional and unconventional techniques. One critical point is the lack of knowledge of the genetic background. Only sparse and unreliable data are available on gene-enzyme relationships, on intra- and inter-allelic interactions and on gene linkage and chromosomal structure. The lack of this basic knowledge is probably due to a number of factors, the most important of which are its autotetraploid status and its high sensitivity to inbreeding depression. These characteristics hamper the achievement of homozygous genotypes, the most useful for carrying on genetic studies. Tissue culture could be useful for solving these problems if it can succeed in regenerating haploid plants through pollen or ovule culture, but up-to-date studies on this topic have not yet produced well-documented success.

Furthermore, the introduction in cultivated varieties of single useful traits, such as resistance to biotic and abiotic stresses, could be achieved more quickly through in vitro selection than with the backcross method.

Other practical problems which could take advantage of tissue culture and molecular biology methods are those of forage quality and seed production. Alfalfa produces the most protein-rich forage, but its feeding value is strongly reduced by the occurrence of bloating, and by the elimination of most of the proteins as nitrogen compounds. These problems could be overcome by transferring genes for the production of tannins into the leaves of lucerne. Seed production is affected by pollination, which is limited by the absence of insects and by climatic factors. The production of apomictic plants could allow stable seed production and the maintenance of useful gene combinations. These goals could be achieved through interspecific hybridization and the introduction of alien genes, which are the new tools available to modern plant breeding.

References

App BA, Manglitz GR (1972) Insects and related pests. In: Hanson CH (ed) Alfalfa science and technology. Am Soc Agron, Madison, WI, pp 527–554

Arcioni S, Davey MR, Dos Santos AVP, Cocking EC (1982) Somatic embryogenesis in tissues from mesophyll and cell suspension protoplasts of *Medicago coerulea* and *M. glutinosa*. Z Pflanzenphysiol 106:105–110

Arcioni S, Pezzotti M, Damiani F (1987) In vitro selection of alfalfa plants resistant to *Fusarium oxysporum* f. spp. *medicaginis*. Theor Appl Genet 74:700–705

Arcioni S, Pupilli F, Pezzotti M, Falistocco E, Damiani F (1987) Interspecific hybrid lines of *M. sativa* and *M. arborea* by protoplast electrofusion. In: Puite KJ, Dons JJM, Huizing HJ, Kool AJ, Koorneef M, Krens FA (eds) Progress in plant protoplast research. Kluwer, Dordrecht Boston London, pp 259–260

Arcioni S, Pezzotti M, Damiani F, Pupilli F (1988) Plant regeneration study and somaclonal variation in *Medicago sativa*. In: Rep 31st North American alfalfa improvement conference, June 19–23, 1988, Greenbelt, Maryland, p 83

Atanassov A, Brown DCW (1984) Plant regeneration from suspension culture and mesophyll protoplasts of *Medicago sativa* L. Plant Cell Tissue Organ Cult 3:149–162
Baertlein DA, McDaniel RG (1987) Molecular divergence of alfalfa somaclones. Theor Appl Genet 73:575–580
Bajaj YPS (1979a) Technology and prospects of cryopreservation of germplasm. Euphytica 28:267–285
Bajaj YPS (1979b) Establishment of germplasm banks through freeze-storage of plant tissue culture and their implications in agriculture. In: Sharp WR, Larsen PO, Paddock EF, Raghvan V (eds) Plant cell and tissue culture: principles and applications. Ohio State Univ Press, Columbus, pp 745–774
Bajaj YPS (1983) In vitro production of haploids. In: Ammirato PV, Evans DA, Sharp WR, Yamada Y (eds) Handbook of plant cell culture 1. McMillan, New York, pp 228–287
Bajaj YPS (ed) (1989) Biotechnology in agriculture and forestry. 9 protoplasts and genetic engineering II. Springer, Berlin Heidelberg New York Tokyo
Bauchan GR (1987) Embryo culture of *Medicago scutellata* and *M. sativa*. Plant Cell Tissue Organ Cult 10:21–29
Bequerel PM (1950) La via latente des graines aux confins du zéro absolute. CR Hebd Sci 231:1274
Bingham ET (1969) Haploids of cultivated alfalfa (*Medicago sativa* L.). Nature (Lond) 22:865–866
Bingham ET (1971) Isolation of haploids of tetraploid alfalfa. Crop Sci 11:433–435
Bingham ET (1979) Maximizing heterozygosity in autopolyploids. In: Lewis WH (ed) Polyploidy, biological relevance. Plenum Press, New York, pp 471–489
Bingham ET, McCoy TJ (1986) Somaclonal variation in alfalfa. In: Janick J (ed) Plant Breeding Reviews, vol 4. AVI, Westport, Connecticut, pp 123–152
Bingham ET, Hurley LV, Kaatz DM, Saunders JW (1975) Breeding alfalfa which regenerates from callus tissue in culture. Crop Sci 15:719–721
Binarova P, Novák FJ (1984) Regulation and somatic embryo development in cell culture of alfalfa (*Medicago sativa* L.). In: Novák FJ, Havel L, Dolezel J (eds) Plant tissue and cell culture application to crop improvement. Proc Int Symp Olomouc, 24–29 September 1984. Czech Acad Sci Prague, pp 139–141
Binarova P, Dolezel J (1988) Alfalfa embryogenic cell suspension culture: growth and ploidy level stability. J Plant Physiol 133:561–566
Blaydes DF (1966) Interaction of kinetin and various inhibitors in the growth of soybean tissue. Physiol Plant 19:748–753
Bocsa I (1981) Amélioration du rendement en matière sèche chez la luzerne. In: Bosca I (ed) Proc EUCARPIA Fodder Crop Section. July 1–3, 1980, Agricultural Research Institute, Kompolt, Hungary, pp 19–28
Bolton JL, Goplen BP, Baezinger H (1972) World distribution and historical developments. In: Hanson CM (ed) Alfalfa science and technology. Am Soc Agron, Madison, WI, pp 1–34
Brown DCW, Atanassov A (1985) Role of genetic background in somatic embryogenesis in *Medicago*. Plant Cell Tissue Organ Cult 4:111–122
Busbice JH, Wilsie CP (1966) Inbreeding depression and heterosis in autotetraploids with application to *Medicago sativa* L. Euphytica 15:52–57
Busse WF (1927) Effects of low temperatures on germination of impermeable seeds. Bot Gaz 89:169
Byrne MC, McDonnell RE, Wright MS, Carnes MG (1987) Strain and cultivar specificity in the *Agrobacterium*-soybean interaction. Plant Cell Tissue Organ Cult 8:3–15
Chen THH, Marowitch J (1987) Screening of *Medicago falcata* germplasm for in vitro regeneration. J Plant Physiol 128:271–277
Chen THH, Marowitch J, Thompson BG (1987) Genotypic effects on somatic embryogenesis and plant regeneration from callus cultures of alfalfa. Plant Cell Tissue Organ Cult 8:73–81
Cheyne VA, Dale PJ (1980) Shoot tip culture in forage legumes. Plant Sci Lett 19:303–309
Costantino P, Spanò L, Pomponi M, Benvenuto E, Ancora G (1984) The T-DNA of *Agrobacterium rhizogenes* is transmitted through meiosis to the progeny of hairy root plants. J Mol Appl Genet 2:465–470
Crouch ML (1982) Non-zygotic embryos of *Brassica napus* L. contain embryo specific storage proteins. Planta 156:520–524
Croughan TP, Stavarek SJ, Rains DW (1978) Selection of a NaCl tolerant line of cultured alfalfa cells. Crop Sci 18:959–963

Damiani F, Pezzotti M, Arcioni S (1988) Electric field mediated fusion of protoplasts of *Medicago sativa* L. and *Medicago arborea* L. J Plant Physiol 132:474–479

Davis PA (1988) Cell culture of forage legumes for the production of nutritionally improved, bloat-safe lucerne. Ph D Thesis (submitted to the Australian National University)

Deak M, Kiss GB, Koncs C, Dudits D (1986) Transformation of *Medicago* by *Agrobacterium*-mediated gene transfer. Plant Cell Rep 5:97–100

Demarly Y (1963) Genétique de tetraploides et amélioration des plantes. Ann Amelior Plant 13:307–400

D'Hont A, Quétier F, Teoule E, Dattèe Y (1987) Mitochondrial and chloroplast DNA analysis of interspecific somatic hybrids of a *Leguminosae: Medicago* (alfalfa). Plant Science 53:237–242

Dijak M, Brown DCW (1987) Patterns of direct and indirect embryogenesis from mesophyll protoplasts of *Medicago sativa*. Plant Cell Tissue Organ Cult 9:121–130

Dos Santos AVP, Outka DE, Cocking EC, Davey MR (1980) Organogenesis and somatic embryogenesis in tissue derived from leaf protoplast and leaf explants of *Medicago sativa*. Z Pflanzenphysiol 99:261–270

Dos Santos AVP, Cutter EG, Davey MR (1983) Origin and development of somatic embryos in *Medicago sativa* L. (alfalfa). Protoplasma 117:107–115

Duke JA (ed) (1981) Handbook of legumes of world economic importance. Plenum Press, New York

Dunbier MW, Bingham ET (1975) Maximum heterozygosity in alfalfa: results using haploid-derived autotetraploids. Crop Sci 15:527–531

Elgin JH Jr, McMurtrey JE III, Schaeffer GW (1977) Attempted interspecific hybrydization of *Medicago scutellata* and *M. sativa*. Agron Abstr 69:54

Finkle BJ, Zuval ME, Ulrich JM (1985) Cryoprotective compounds in the viable freezing of plant tissues. In: Kartha KK (ed) Cryopreservation of plant cells and organs. CRC, Boca Raton, Florida, pp 75–113

Frearson EM, Power JB, Cocking EC (1973) Isolation, culture and regeneration of *Petunia* leaf protoplasts. Dev Biol 33:130–137

Fridriksson S, Bolton JL (1963a) Development of the embryo of *Medicago sativa* L. after normal fertilization and after pollination by other species of *Medicago*. Can J Bot 41:23–33

Fridriksson S, Bolton JL (1963b) Preliminary report on the culture of alfalfa embryos. Can J Bot 4:439–440

Gamborg OL, Miller RA, Ojima K (1968) Nutrient requirements of suspension culture of soybean root cells. Exp Cell Res 50:151–158

Gilmour DM, Davey MR, Cocking EC (1986) Somatic hybridization in *Medicago*. In: 6th Int Congr plant tissue and cell cult, Abstr, Aug 3–8, Univ Minnesota, Minneapolis, p 435

Gilmour DM, Davey MR, Cocking EC (1987a) Isolation and culture of heterokaryons following fusion of protoplasts from sexually compatible and sexually incompatible *Medicago* species. Plant Sci 53:263–270

Gilmour DM, Davey MR, Cocking EC (1987b) Plant regeneration from cotyledon protoplasts of wild *Medicago* species. Plant Sci 48:107–112

Gilmour DM, Davey MR, Cocking EC, Pental D (1987c) Culture of low numbers of forage legume protoplasts in membrane chambers. J Plant Physiol 126:457–465

Gilmour DM, Golds TJ, Davey MR (1988) *Medicago* protoplasts: isolation, fusion, culture and plant regeneration. In: Bajaj YPS (ed) Biotechnology in agriculture and forestry. 8 protoplasts and genetic engineering I. Springer, Berlin Heidelberg New York Tokyo

Graham JK, Kritlow KW, Faulkner LR (1972) Disease. In: Hanson CH (ed) Alfalfa science and technology. Am Soc Agron, Madison, WI, pp 497–526

Graham JH, Frosheiser FI, Stuteville DL, Erwin DC (1979) A compendium of alfalfa diseases. Am Phytopathol Soc, St Paul, Minnesota

Groose RE, Bingham ET (1984) Variation in plants regenerated from tissue culture of tetraploid alfalfa heterozygous for several traits. Crop Sci 24:655–658

Hanson CH (1961) Longevity of pollen and ovaries of alfalfa. Crop Sci 1:114–116

Hanson CH, Campbell TA (1972) Vacuum-dried pollen of alfalfa (*Medicago sativa* L.) viable after eleven years. Crop Sci 12:874

Hartman CL, McCoy TJ, Knous TR (1984) Selection of alfalfa (*Medicago sativa*) cell lines and regeneration of plants resistant to the toxin (s) produced by *Fusarium oxysporum* f. spp. *medicaginis*. Plant Sci Lett 34:183–194

Hlasmikova A (1977) Androgenesis in vitro evaluated from the aspect of genetics. Z Pflanzenzuecht 78:44–56
Hooykaas PJJ, Schilperoort RA (1984) The molecular genetics of crown-gall tumorigenesis. Adv Genet 22:209–283
Hougas RW, Peloquin SJ (1958) The potential of potato haploids in breeding and genetic research. Am Pot J 35:701–707
Indiogine SEP, Mita G, Terzi M (1986) Isolation of variants resistant to sodium chloride and methylammonium through somaclonal variation. In: 6th Int Congr plant tissue and cell culture, Abstr, Aug 3–8, Univ Minnesota, Minneapolis, p 377
Johnson LB, Palmer JD (1986) Some phylogenetic relationships within the genus *Medicago* as determined by chloroplast DNA restriction mapping. Communication at Joint Meeting of 30th NAAIC, July 27–31, 1986, St Paul, Minnesota
Johnson LB, Stuteville DL, Higgins RK, Skinner DZ (1981) Regeneration of alfalfa plants from protoplasts of selected Regen S clones. Plant Sci Lett 20:297–304
Johnson LB, Stuteville DL, Higgins RK, Douglas HL (1982) Pectolyase Y-23 for isolating mesophyll protoplasts from several *Medicago* species. Plant Sci Lett 26:133–137
Johnson LB, Stuteville DL, Schlarbaum SE, Skinner DZ (1984a) Variation in phenotype and chromosome number in alfalfa protoclones regenerated from nonmutagenized calli. Crop Sci 24:948–952
Johnson LB, Rose RJ, Schlarbaum SE, Stuteville PL (1984b) Effects of protocloning in "Regen S" alfalfa on somatic chromosomes and on mitochondrial and chloroplast DNA. Rep 29th Alfalfa Improv Conf Lethridge, Alberta, Canada, p 74
Kao KN (1977) Chromosomal behaviour in somatic hybrids of soybean *Nicotiana glauca*. Mol Gen Genet 150:225–230
Kao KN, Michayluk MR (1975) Nutritional requirement for growth of *Vicia hajastana* cells and protoplasts at very low population density in liquid media. Planta 126:105–110
Kao KN, Michayluk MR (1980) Plant regeneration from mesophyll protoplasts of alfalfa. Z Pflanzenphysiol 96:135–141
Kao KN, Michayluk MR (1981) Embryoid formation in alfalfa cell suspension cultures from different plants. In Vitro 17:645–648
Kao KN, Saleem M (1986) Improved fusion of mesophyll and cotyledon protoplasts with PEG and high pH-Ca^{++} solution. J Plant Physiol 122:217–225
Kao KN, Wetter LR (1977) Advances in techniques of plant protoplast fusion and culture of heterokaryocytes. In: Brinkley BR, Porte K (eds) International cell biology. Rockefeller Univ Press, New York, p 216
Kasha KJ (1974) Haploids in higher plants: advances and potential. Univ Press, Guelph
Larkin PJ, Scowcroft WR (1981) Somaclonal variation: a novel source of variability from cell cultures for plant improvement. Theor Appl Genet 60:197–214
Latunde-Dada AO, Lucas JA (1983) Somaclonal variation and reaction to *Verticillium wilt* in *Medicago sativa* L. plants regenerated from protoplasts. Plant Sci Lett 32:205–211
Lesins K (1961) Mode of fertilization in relation to breeding methods in alfalfa. Z Pflanzenzuecht 43:31–54
Lesins K, Gillies CB (1972) Taxonomy and cytogenetics of *Medicago*. In: Hanson CH (ed) Alfalfa science and technology. Am Soc Agron, Madison, WI, pp 53–86
Linsmaier EM, Skoog F (1965) Organic growth factor requirements of tobacco tissue cultures. Physiol Plant 18:100–127
Lipman CB, Lewis GN (1934) Tolerance of liquid-air temperatures by seeds of higher plants for sixty days. Plant Physiol 9:392
Lu DJ, Pental D, Cocking EC (1982) Plant regeneration for seedling cotyledon protoplasts. Z Pflanzenphysiol 107:59–63
Lubenec PA (1967) Utilizing wild perennial species. Proc All Univ Lenin Acad Agric Sci 5:19–22
Lucretti S, Dolezel J, Spanò M, Moretti F (1986) Legume protoplast culture and conditions for fluorescent labelling and cell sorting. 6th Int Congr plant tissue and cell cult, Abstr, Aug 3–8, Univ Minnesota, Minneapolis, p 452
Lupotto E (1983) Propagation of an embryogenic culture of *Medicago sativa* L. Z Pflanzenphysiol 111:95–104
Lupotto E (1986) The use of single somatic embryo culture in propagating and regenerating lucerne (*Medicago sativa* L.). Ann Bot 57:19–24

Mariani A (1968) Impiego dell'incrocio interspecifico nel miglioramento genetico dell'erba medica. Genet Agr 22:35–51

Mariani A, Arcioni S, Veronesi F (1978) Cytological analysis and electrophoretic patterns of seed proteins in *Medicago sativa*, *Medicago glutinosa* and their hybrids. Genet Agr 32:21–39

Mariotti D, Arcioni S, Pezzotti M (1984a) Regeneration of *Medicago arborea* L. plants from tissue and protoplasts culture of different organ origin. Plant Sci Lett 37:149–156

Mariotti D, Davey MR, Draper J, Freeman JP, Cocking EC (1984b) Crown-gall tumorigenesis in the forage legume *Medicago sativa* L. Plant Cell Physiol 25:473–482

McCoy TJ (1985) Interspecific hybridization of *Medicago sativa* L. and *M. rupestris* M.B. using ovule-embryo culture. Can J Genet Cytol 27:238–245

McCoy TJ, Bingham ET (1977) Regeneration of diploid alfalfa plants from cells grown in suspension culture. Plant Sci Lett 10:59–66

McCoy TJ, Smith LY (1984) Uneven ploidy levels and a reproductive mutant required for interspecific hybridization of *Medicago sativa* L. and *Medicago dzhawakhetica* Bordz. Can J Genet Cytol 26:511–518

McCoy TJ, Smith LY (1986) Interspecific hybridization of perennial *Medicago* species using ovule-embryo culture. Theor Appl Genet 71:772–783

Meijer EGM, Brown DCW (1985) Screening of diploid *Medicago sativa* germplasm for somatic embryogenesis. Plant Cell Rep 4:285–288

Meijer EGM, Brown DCW (1987) Role of exogenous reduced nitrogen and sucrose in rapid high frequency somatic embryogenesis in *Medicago sativa*. Plant Cell Tissue Organ Cult 10:11–19

Meins F Jr (1983) Heritable variation in plant cell culture. Annu Rev Plant Physiol 34:327–346

Melchers G, Labib G (1970) Die Bedeutung haploider höherer Pflanzen für Pflanzenphysiologie und Pflanzenzüchtung. Durch Antheres Kultur erzeugte Haploide, ein neuer Durchbruch für die Pflanzenzüchtung. Ber Dtsch Bot Ges 83:129–150

Mezentsen AV (1981) Mass regeneration of plants from the cells and protoplasts of alfalfa. Dokl Vaskuil 4:22–23

Michaud E, Leman WF, Rumbaugh MD (1987) World distribution and historical development. In: American Society of Agronomy (ed) Alfalfa and alfalfa improvement (in preparation)

Mitten DH, Sato SJ, Skokut TA (1984) In vitro regenerative potential of alfalfa germplasm sources. Crop Sci 24:943–945

Murashige T, Skoog F (1962) A revised medium for rapid growth and bioassays with tobacco tissue cultures. Physiol Plant 15:473–497

Nagarajan P, McKenzie JS, Walton PD (1986) Embryogenesis and plant regeneration of *Medicago* spp. in tissue culture. Plant Cell Rep 5:77–80

Nagarajan P, Walton PD (1987) A comparison of somatic chromosomal instability in tissue culture regenerants from *Medicago media* Pers. Plant Cell Rep 6:109–113

Nagata T, Ishii S (1979) A rapid method for isolation of mesophyll protoplasts. Can J Bot 57:1820–1823

Novák FJ, Konecna D (1982) Somatic embryogenesis in callus and cell suspension cultures of alfalfa (*Medicago sativa* L.). Z Pflanzenphysiol 105:279–284

Novák FJ, Hermelim T, Daskav S, Nesticky M (1986) In vitro mutagenesis in maize. In: Horn W, Jensen CJ, Odenbach W, Schieder O (eds) Genetic manipulation in plant breeding. De Gruyter, Berlin, pp 563–576

Oldemeyer RD (1956) Interspecific hybridization in *Medicago*. Agron J 48:584–585

Orr W, Singh J, Brown DCW (1985) Induction of freezing tolerance in alfalfa cell suspension cultures. Plant Cell Rep 4:15–18

Orton TJ (1984) Somaclonal variation: theoretical and practical consideration. In: Gustafson JP (ed) Gene manipulations in plant improvement. Plenum Press, New York, pp 427–468

Peloquin SJ, Hougas RW (1958) Fertility in two haploids of *Solanum tuberosum*. Science 128:1340–1341

Pezzotti M, Arcioni S, Mariotti D (1984) Plant regeneraton from mesophyll, root and cell suspension protoplasts of *Medicago sativa* cv. Adriana. Genet Agr 38:195–208

Pezzotti M, Pupilli F, Damiani F, Arcioni S (1988) Genetic transformation of *M. sativa*. Agr 42:477

Phillips GC (1983) Screening alfalfas adapted to the south western United States for regenerator genotypes. In Vitro 19(3):265

Phillips GC, Collins GB (1979) In vitro tissue culture of selected legumes and plant regeneration from callus cultures of red clover. Crop Sci 19:59–64
Quiros CF (1983) Alfalfa, Lucerne (*Medicago sativa* L.). In: Tanksley SD, Orton TJ (eds) Isozymes in plant genetics and breeding, Part B. Elsevier, Amsterdam, pp 253–294
Raynal G (1986) Les maladies de la luzerne en Europe. In: Guy P, Massenot M (eds) Compte-rendu des groupes de travail *Medicago sativa*. EUCARPIA, INRA, Paris, pp 7–14
Reich TJ, Iyer VN, Miki BL (1986) Efficient transformation of alfalfa protoplasts by the intranuclear microinjection of Ti plasmid. Bio/technology 4:1001–1004
Reisch B, Bingham ET (1980) The genetic control of bud formation from callus cultures of diploid alfalfa. Plant Sci Lett 20:71–77
Reisch B, Bingham ET (1981) Plants from ethionine-resistant alfalfa tissue cultures: variation in growth and morphological characteristics. Crop Sci 21:783–788
Reisch B, Duke SH, Bingham ET (1981) Selection and characterization of ethionine-resistant alfalfa (*Medicago sativa* L.) cell lines. Theor Appl Genet 59:89–94
Rose RJ, Johnson LB, Kemble RJ (1984) Restriction endonuclease studies on the chloroplast and mitochondrial DNA, of alfalfa (*Medicago sativa* L.) protoclones. Curr Top Plant Biochem Physiol 3:178
Sakai A, Noshiro M (1975) Some factors contributing to the survival of crop seeds cooled to the temperature of liquid nitrogen. In: Frankel OH, Hawkes JG (eds) International Biology Program: Crop genetic resources for today and tomorrow. Cambridge Univ Press, Cambridge, pp 317–326
Sańgduen N, Kreitner GL, Sorensen EL (1983) Light and electron microscopy of embryo development in an annual × perennial *Medicago* species cross. Can J Bot 61:1241–1257
Saunders JW, Bingham ET (1972) Production of alfalfa plants from callus tissue. Crop Sci 12:804–806
Saunders JW, Bingham ET (1975) Growth regulator effects on bud initiation in callus cultures of *Medicago sativa*. Am J Bot 62:850–855
Schenk RU, Hildebrandt AC (1972) Medium and techniques for induction and growth of monocotyledonous and dicotyledonous plant cell cultures. Can J Bot 50:199–204
Shahin EA, Spielmann A, Sukhapinda K, Simpson RB, Yashar M (1986) Transformation of cultivated alfalfa using disarmed *Agrobacterium tumefaciens*. Crop Sci 26:1235–1239
Smith MK, McComb JA (1983) Selection for NaCl tolerance in cell cultures of *Medicago sativa* and recovery of plants from a NaCl-tolerant cell line. Plant Cell Rep 2:126–128
Smith SE, Bingham ET, Fulton RW (1986) Transmission of chlorophyll deficiencies provides evidence for biparental inheritance of plastids in *Medicago sativa*. Heredity 77:35–39
Somaroo BH, Witcombe JR (1982) The evaluation and utilization of annual medics at ICARDA. In: Hayward MD (ed) The utilization of genetic resources in fodder crop breeding. Rep EUCARPIA Fodder Crops Section, Aberystwyth, Wales, UK, 13–16 Sept 1982, pp 198–207
Spanò L, Mariotti D, Pezzotti M, Damiani F, Arcioni S (1987) Hairy root transformation in alfalfa (*Medicago sativa* L.). Theor Appl Genet 73:523–530
Stavarek SJ, Croughan TP, Rains DW (1980) Regeneration of plants from long-term cultures of alfalfa cells. Plant Sci Lett 19:2253–2261
Strickland GS, Nichol JW, Stuart DA (1987) Effect of carbohydrate source on somatic embryogenesis. Plant Sci 48:113–121
Stuart DA, Strickland SG (1984a) Somatic embryogenesis from cell cultures of *Medicago sativa* L. I. The role of the amino acid addition to the regeneration medium. Plant Sci Lett 34:165–174
Stuart DA, Strickland SG (1984b) Somatic embryogenesis from cell cultures of *Medicago sativa* L. II. The interaction of amino acids with ammonium. Plant Sci Lett 34:175–181
Stuart DA, Nelsen J, McCall CM, Strickland SG, Walker KA (1985a) Physiology of the development of somatic embryos in cell cultures of alfalfa and celery. In: Zaitlin M, Day P, Hollaender A (eds) Biotechnology in plant science: relevance to agriculture in the eighties. Academic Press, New York, p 35
Stuart DA, Nelsen J, Strickland SG, Nichol JW (1985b) Factors affecting developmental processes in alfalfa cell cultures. In: Proc Tennessee Symp on Propagation of Higher Plants. Knoxville, p 373
Stuart DA, Nelsen J (1988) Isolation and characterization of alfalfa 7S and 11S seed storage proteins. J Plant Physiol 132:129–133
Stuart DA, Nelsen J, Nichol JW (1988) Expression of 7S and 11S alfalfa seed storage proteins in somatic embryos. J Plant Physiol 132:134–139

Sukhapinda K, Spivey R, Shahin EA (1987) Ri-plasmid as a helper for introducing vector DNA into alfalfa plants. Plant Mol Biol 8:209–216

Tanner GY, Moore AE, Arcioni S, Larkin PJ (1988) Initiation of non-physiological division in cultured lucerne microspores. In: International Congress on genetic manipulation in plant breeding; organized by EUCARPIA, 11–12 Sept, Helsingor, Denmark, p 74

Teoule E (1983) Hybridization somatique entre *Medicago sativa* L. et *Medicago falcata* L. CR Acad Sci Paris 297:13–16

Teoule E, Dattee Y (1987) Recherche d'une mèthode fiable de culture de protoplastes, d'hybridation somatique et de régénération chez *Medicago*. Agronomie 7:575–584

Tepfer D (1984) Transformation of several species of higher plants by *Agrobacterium rhizogenes*: sexual transmission of the transformed genotype and phenotype. Cell 37:955–967

Uchimiya H, Murashige T (1974) Evaluation of parameters in the isolation of viable protoplasts from cultured tobacco cells. Plant Physiol 54:936–944

Walker KA, Sato SJ (1981) Morphogenesis in callus tissues of *Medicago sativa*. The role of ammonium ion in somatic embryogenesis. Plant Cell Tissue Organ Cult 1:109–121

Walker KA, Yu PC, Sato SJ, Jaworski EG (1978) The hormonal control of organ formation in callus of *Medicago sativa* L. cultured in vitro. Ann J Bot 65:654–659

Walker KA, Wendeln ML, Jaworski EG (1979) Organogenesis in callus tissue of *Medicago sativa*. The temporal separation of induction processes from differentiation processes. Plant Sci Lett 16:23–30

Walton PD, Brown DCW (1988) Electrofusion of protoplasts and heterokaryon survival in the Genus *Medicago*. Plant Breeding 101:137–142

Wang JW, Sorensen EL, Liang GH (1984) In vitro culture of pods from annual and perennial *Medicago* species. Plant Cell Rep 3:146–148

Webb KJ (1986) Transformation of forage legumes using *Agrobacterium tumefaciens*. Theor Appl Genet 72:53–58

Xu ZH, Davey MR, Cocking EC (1982) Organogenesis from root protoplasts of the forage legumes *Medicago sativa* and *Trigonella foenum-graecum*. Z Pflanzenphysiol 107:231–235

Zagorska N, Robeva P, Dimitrov B, Shtereva R, Ganacheva V (1984) Induction of regeneration in anther cultures in *Medicago sativa* L. CRA Bulgare Sci 37(8):1099–1102

III.2 Clovers (*Trifolium* spp.)

E. G. WILLIAMS[1,2], G. B. COLLINS[1], and J. R. MYERS[1]

1 Introduction

Trifolium is a genus of some 250 species indigenous to Europe, the Middle East, Africa, and the Americas, with a major center of diversity in the Eurasian region. About 20 species have been considered agriculturally significant for provision of food to domesticated herbivores in cool-temperate to subtropical regions (Table 1). These are generally grown with companion grasses as grazed pasture or rangeland, but mown feed and hay crops are also important functions of certain species. Annual species are generally self-pollinating, with simple tap roots, low chromosome numbers (e.g., $2n = 14$, 16), and adaptations for Mediterranean-type climates. Perennial species are mostly cross-pollinating, often with higher chromosome numbers (e.g., $2n = 48$, 64, 80, 130), and are frequently spread vegetatively by means of stolons or rhizomes (Taylor et al. 1979). For most species a wide range of ecotypic variability is available, and particularly in outcrossing species, substantial interplant genetic variability exists even within ecotypes and bred varieties.

Evans (1976) listed ten *Trifolium* species as having major agricultural importance. These were the annuals *T. resupinatum*, *T. incarnatum*, *T. alexandrinum*, *T. subterraneum*, and *T. dubium*, and the perennials *T. repens*, *T. pratense*, *T. hybridum*, *T. fragiferum*, and *T. ambiguum*. White (*T. repens*), red (*T. pratense*), and alsike (*T. hybridum*) clovers are grown across a broad spectrum of cool- to mild-temperate climates. The remaining species are more characteristic of mild- to warm-temperate climates, with the exception of *T. ambiguum*, which is one of the few perennial *Trifolium* species adapted to cold-temperate regions.

The major agronomic significance of the clovers, apart from their production of high-quality herbage, is their ability to fix atmospheric nitrogen in symbiotic association with soil bacteria of the genus *Rhizobium*. This function will become increasingly important as the cost of synthetic nitrogenous fertilizer escalates, rendering its application to marginal rangelands progressively less economic (Isely 1982). Clovers are also significant as a source of honey and as seed crops in many regions.

Production and usefulness of clovers is limited by many factors which can be grouped into four general categories: diseases and pests, forage quality problems, environmental limitations, and limitations to symbiotic nitrogen fixation. Clovers

[1] Department of Agronomy, University of Kentucky, Lexington, Kentucky 40546, USA
[2] Present address: Department of Botany, University of Georgia, Athens GA 30602, USA

Biotechnology in Agriculture and Forestry, Vol. 10
Legumes and Oilseed Crops I (ed. by Y. P. S. Bajaj)

are susceptible to a range of pathogens which reduce herbage and seed production (Leath 1985; Barnett and Diachun 1985). These include viruses (e.g., clover yellow mosaic, clover yellow vein, white clover mosaic, alfalfa mosaic, bean yellow mosaic, peanut stunt), fungi (e.g., *Sclerotinia trifoliorum* the clover rot, northern and southern anthracnose, *Fusarium* wilt), and mycoplasmas (e.g., clover phyllody which converts inflorescences to leafy structures, and red clover leaf mycoplasma). The plants are eaten, both above and below ground level, by a number of insect larvae and molluscs (slugs and snails) (Manglitz 1985). Forage quality problems are related to the physiology of grazing herbivores, and include the bloating effect of high-protein, low-flavonol herbage (Jones et al. 1973), the estrogenic activity of foliar coumarins (Essig 1985) and the toxic and goitrogenic effects of cyanogenic glucosides (Coop and Blakley 1950; Flux et al. 1960; Collinge and Hughes 1982). Symbiotic nitrogen fixation is inhibited by nitrate (Carroll and Gresshoff 1983) so that in fertile soils or after fertilizer application, the nitrogen-fixing potential of clovers is not realized.

These limitations define a number of important breeding objectives for clovers, i.e., disease and pest resistance, improvement of forage quality and optimization of nitrogen fixation in fertile as well as infertile soils. Other objectives are associated with extending the use of major species into more extreme climates, more marginal lands, and new or lower input management practices. This involves improvement of factors such as establishment, persistence, soil nutrient uptake, winter hardiness, salt tolerance, and drought tolerance. Among the more commonly grown species some typical breeding objectives are:

- *T. pratense*: yield, disease resistance, winter hardiness, nonbloating foliage and improved persistence to extend the life of the stand and to obtain higher yields the second and subsequent years after sowing (Taylor and Smith 1979).
- *T. repens*: virus tolerance, a more extensive, deeper root system, resistance to drought, and nonbloating foliage (Williams and Verry 1981; White and Bhojwani 1981; Pandey et al. 1987).
- *T. ambiguum*: more rapid establishment and greater regrowth vigor (Speer and Allinson 1985).
- *T. hybridum*: greater longevity, virus resistance and a more prostrate habit for resistance to grazing (Williams 1980).
- *T. subterraneum*: alteration of flowering date to provide regionally adapted cultivars (Tan 1985), and greater winter hardiness.
- *T. resupinatum*: greater self-compatibility for increased seed production (Morgner 1976), and increased winter hardiness.

Breeding of clovers has traditionally been based on controlled mating of plants selected from the available natural variation within single species. For a number of objectives, however, particularly disease resistance, forage quality and extended geographic adaptations, the generation of novel variation beyond natural species limits will be necessary (Razdan and Cocking 1981). Thus, procedures such as interspecific sexual and somatic hybridization, induction and screening of somaclonal variants, and genetic transformation are of interest for application to the clovers no less than to other agriculturally important plants. For these procedures to succeed, techniques are required for efficient in vitro manipulation

Table 1. Selected *Trifolium* species of agronomic interest (after Phillips and Collins 1984; Taylor 1985)

Species	Common name	Habit: annual (A) or perennial (P)	Breeding system: inbreeding (I) or outcrossing (O)	Common chromosome numbers (2n)	Features of interest for breeding
Section *Lotoidea* Zoh.					
T. alpinum L.		P	O (?)	16	Resistance to sclerotinia
T. ambiguum M.B.	Kura, Caucasian	P	O	16, 32, 48	Virus resistant, winter hardy
T. hybridum L.	Alsike	P	O	16	Resists waterlogging, cold hardy
T. nigrescens Viv.	Ball	A	O	16	Tolerant of heavy soils
T. repens L.	White, ladino	P	O	32	High production, rapid establishment
T. semipilosum Fresen.	Kenya	P	O	16	Drought tolerance
T. suffocatum L.		A	I	16	Resistance to sclerotinia
T. uniflorum L.		P	O	32	Deep tap roots
Section *Trifolium* Zoh.					
T. affine C. Presl		A	I	16	Foliar tannins
T. alexandrinum L.	Berseem, Egyptian	A	O	16	Provides high quality forage at times of the year when pasture production is low
T. alpestre L.		P	O	16	Perenniality
T. arvense L.	Rabbitfoot	A	I	14	Foliar tannins
T. diffusum Ehrh.		A	I	16	Source of interspecific variability for *T. pratense*
T. hirtum All.	Rose	A	I	10	Low maintenance winter annual
T. incarnatum L.	Crimson	A	O	14	Rapid cool season growth, high seed yields
T. lappaceum L.	Lappa, burdock	A	I	16	Tolerant of heavy soils
T. medium L.	Zigzag	P	O	64–80	Perenniality, rhizomes
T. pallidum Waldst. and Kit.		A	O	16	Source of interspecific variability for *T. pratense*

T. pannonicum Jacq.	Hungarian	P	O	130	Perenniality
T. pratense L.	Red	P	O	14	High production, rapid establishment
T. rubens L.		P	O	16	Efficient regeneration from culture
T. sarosiense Hazsl.		P	O	48	Perenniality, rhizomes
T. striatum L.	Striate, knotted	A	I	14	Tolerant of heavy soils
Section *Chronosemium* (Ser.) Zoh.					
T. aureum Poll. (= *T. agrarium* L.)	Hop	A	I	14	Tolerant of sandy soils
T. campestre Schreb.	Large hop	A	I	14	Provides early spring grazing
T. dubium Sibth.	Small hop, yellow suckling	A	I	14, 28	Tolerant of acid, sandy soils
Section *Tricocephalum* Zoh.					
T. subterraneum L.	Subterranean, sub	A	I	16	Tolerant of heavy grazing, acid soils, wet soils
Section *Vesicaria* Crantz					
T. fragiferum L.	Strawberry	P	O	16	Tolerant of wet and alkaline soils
T. resupinatum L.	Persian	A	O (I)	16	Tolerant of wet soils
Section *Mistyllus* (Presl) Zoh.					
T. vesiculosum Savi	Arrowleaf	A	O	16	Wide adaptation, drought tolerance

and regeneration from protoplast, cell, tissue, and organ cultures. Significant advances have been made, particularly in the last 10 years, so that currently, certain species of *Trifolium* are among the legumes best characterized and most responsive to tissue culture manipulations.

Phillips and Collins (1984), Mroginski and Kartha (1984), and Rupert and Collins (1985) have reviewed in detail the available procedures for tissue culture of clovers, with particular emphasis on techniques and the formulation of media. Our aim is to extend this theme, to present examples of a range of applications, and to highlight areas where tissue culture might be expected to contribute to the improvement of clovers for agricultural use.

2 Factors Important for Initiation of Cultures

2.1 Genotype

The most important factor in establishing successful in vitro cultures of clover species is the genotype selected. This has been repeatedly demonstrated in studies involving shoot cultures (Oelck and Schieder 1983; Campbell and Tomes 1982, 1984), callus, cell suspension and protoplast cultures (Phillips and Collins 1979a; Gresshoff 1980; Keyes et al. 1980; Mohapatra and Gresshoff 1982; Oelck et al. 1982; Bhojwani and White 1982; Ahuja et al. 1983; White 1984; Bhojwani et al. 1984) and embryo rescue (Williams and Verry 1981; Taylor et al. 1981; Phillips et al. 1982; Williams 1987). For most purposes, therefore, a prudent strategy is to select a genotype known to be responsive to the techniques intended, or to screen a range of cultivars, ecotypes or individual seedlings. By screening 50 seedlings each of four white clover lines of distinct geographic origin, White (1984) was able to select a genotype (WR 8) capable of sustaining long-term, high-frequency regeneration by organogenesis from cell cultures, and also capable of regeneration from protoplasts. For outbreeding species such as red and white clovers, considerable variation is available within cultivars. Thus, even where in vitro manipulations are to be applied to a specified cultivar, the screening of a number of different seedling genotypes is likely to improve the success rate of cultures.

2.2 Preculture Conditions

Although pretreatment of clover plants before tissue explanting has received little experimental attention, it is clear that in *Trifolium*, as in other genera, the physiological state of the plant influences the success of tissue culture manipulations. This is particularly true for protoplast studies and where efficient regeneration is required. Bhojwani and White (1982) found that a reproducible plating efficiency of about 37% could be achieved for *T. repens* mesophyll protoplasts only if plants were maintained in a controlled environment before protoplast isolation. Plants were grown in a soil-less potting mix in a growth cabinet at 20 °C with a 12/12 day/night photoperiod and light intensity of 15000 lx from tungsten bulbs and white fluorescent tubes. Only the second-to-top leaves were harvested. Similarly, Grosser and Collins (1984), in developing a protocol for regeneration from protoplasts of *T. rubens*, used plants grown axenically under controlled conditions in large jars with gas exchange through filter paper lids. Oelck et al. (1982) found preculture of materials important in protoplast studies of *T. resupinatum*. A high frequency of protoplast division was obtained only from in vitro shoot cultures selected for root initiation in hormone-free medium and maintained at least 6 months in culture after seed germination. In the design of streamlined culture protocols, the preculture conditions under which the plant is grown represent an important set of variables determining efficient culture establishment.

2.3 Age and Origin of the Explant

As in other groups of plants, the physiological age and origin of the explant also affect in vitro responses. Immature and meristematic tissues generally respond to induction more rapidly and regenerate more efficiently than mature and differentiated tissues, but it is normally possible to establish cultures from a range of immature, meristematic and maturing plant parts such as seedling stems, young roots, young leaves, petioles, and ovaries, if the medium has been optimized for the species concerned (Horvath Beach and Smith 1979; Phillips and Collins 1979a). Certain responses, however, may be developmentally limited. Maheswaran and Williams (1984, 1985, 1986a, b, c) found direct somatic embryogenesis limited to the torpedo and early cotyledonary stages of zygotic embryo development. It is likely that the stage of the life cycle from which the explant is taken, by means of the developmental gene program in operation within the cells, will influence the type of response most readily obtained in vitro. Thus, somatic embryogenesis appears to be most readily induced from pre-embryogenic determined cells (Evans et al. 1981; Sharp et al. 1980, 1982) which are characteristic of the

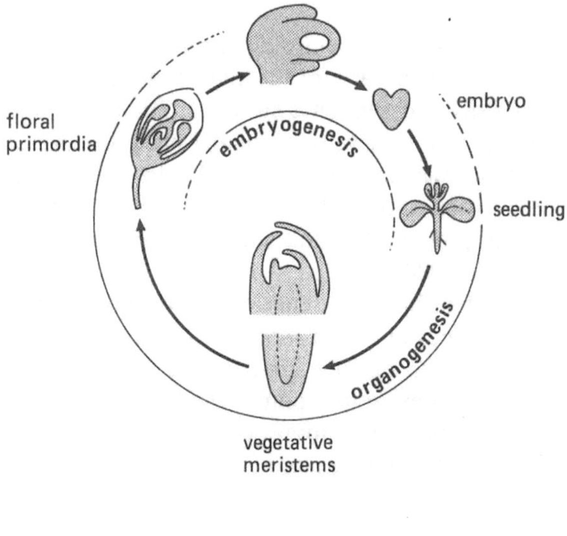

Fig. 1. Diagram illustrating the concept that the most readily expressed response in tissue culture is related to the stage of the life cycle from which the explant is taken. Cells from vegetative meristems are determined for organogenesis. Pre-embryogenic determined cells occur most commonly in certain reproductive tissues such as the nucellus and young embryo

ovule nucellus and young embryo (Fig. 1). Organogenesis is the typical response of tissues explanted in the vegetative phase of the life cycle. Certain genotypes of *T. rubens* and *T. pratense*, however, have been observed to undergo direct embryogenesis from mature somatic tissues (Fig. 8; Cui et al. 1988).

2.4 Surface Sterilization

The use of field- or greenhouse-grown material as an explant source for tissue culture requires that the tissue be freed of microorganisms to ensure aseptic cultures. In general, surface sterilization techniques for the clovers are similar to those used in many other plant species. One difference is that tissues of some clover species (red clover in particular) are sensitive to typical sterilization procedures and require reduced time in sterilizing solutions. Phillips and Collins (1979b) routinely sterilized meristems with a 5-sec dip in 70% ethanol, 8 min in 1% sodium hypochlorite (20% Clorox) and a rinse in sterile distilled water of at least 5 min duration. Reduction of time in sterilizing solutions is required when petioles and leaves are to be used directly as explants.

An effective method of sterilizing dry seed involves the use of chlorine gas. Seeds are placed in a desiccator (250 mm inside diameter) in a fume hood. A beaker containing 100 ml of 5% sodium hypochlorite (full-strength Clorox bleach) is placed in the center of the desiccator and 3.3 ml of concentrated hydrochloric acid are slowly dripped down the side of the beaker and allowed to mix with the sodium hypochlorite solution. The desiccator is sealed and left overnight. Seeds are removed, ventilated in a laminar flow hood, and stored in sealed sterile dishes until use.

3 Shoot Meristem Cultures

3.1 Clonal Propagation and Genotype Storage

Although clovers are normally propagated and distributed as seed, asexual multiplication is necessary for the maintenance of defined genotypes, including the nucleus stocks of synthetic cultivars. For perennial species such as white and red clovers, cuttings have traditionally been used for this purpose, with the disadvantage that losses or poor seed set due to accumulation of viruses can alter the seed composition of the cultivar. Multiple shoot cultures offer a reliable means for rapid micropropagation and storage of pathogen-free, defined genotypes (Cheyne and Dale 1980).

Proliferating shoot cultures are normally established from apical (sometimes axillary) buds of crown shoots or stolons. Differentiated leaf and stem tissues are removed after surface sterilization, and the apical dome with associated leaf primordia is explanted onto a medium with a relatively high cytokinin to auxin ratio. Campbell and Tomes (1984) found that the efficiency of shoot induction was increased when apices of red clover were placed horizontally rather than vertically on the medium. Shoots are usually multiplied and maintained on the induction medium, or a medium of similar composition, and rooted on a medium containing either no hormones or a low to moderate level of auxin (Cheyne and Dale 1980; Dale et al. 1980; Bhojwani 1981; Collins and Phillips 1982; Parrott and Collins 1983; Churova 1984; Campbell and Tomes 1984). Basal media have included MS (Bhojwani 1981), B5 (Cheyne and Dale 1980; Campbell and Tomes 1984), Blaydes (Cheyne and Dale 1980) and L2 (Table 2), a medium optimized for red clover (Phillips and Collins 1979a; Collins and Phillips 1982; Parrott and Collins 1983). Generally, BAP has been used as the cytokinin at concentrations of 0.2–2 mg/l, although 2iP also shows shoot stimulatory activity for clovers (Cheyne and Dale 1980). Bhojwani (1981) and Campbell and Tomes (1984) found no auxin necessary for shoot cultures, but where a low level of auxin has been required, this has been either IAA at 0.2 mg/l (Cheyne and Dale 1980) or 0.6 μM (Churova 1984), or PIC at 2–4 μg/l (LSP, ML8; Parrott and Collins 1983; Rupert and Collins 1985; Phillips and Collins 1984).

Currently, the most widely applicable shoot multiplication medium for clovers is ML8 (Table 2) in which the basal medium L2 is supplemented with 0.5 mg/l BAP and 3 μg/l PIC. In addition to *T. pratense* this medium has been found suitable for *T. alpestre*, *T. incarnatum*, *T. medium*, *T. repens*, *T. rubens*, and *T. subterraneum* (Parrott and Collins 1983). For rooting, shoots are transferred to RL medium with 0.2 mg/l IAA as the only growth regulator, or CR2 with 4.38 mg/l COU (Table 2). Where genotypic differences in shooting are of interest, these are likely to be visible early, after the first or second subculture (Campbell and Tomes 1984). Specific genotypes may require minor modification of the medium to attain optimal multiplication rates. Gas exchange in culture vessels may be important. Campbell and Tomes (1984) reported enhanced shoot multiplication of red clover at high levels of BAP (2 mg/l) when either the atmosphere inside culture vessels was enriched with CO_2 or the vials were capped with polypropylene closures which retain moisture but allow gas exchange.

Traditionally, clover genotypes have been stored as potted cuttings, a method which is both expensive and risky. Plants require frequent trimming, dividing, repotting, spraying and removal of flower heads. Individuals often die of disease or are replaced by volunteer seedlings. Maintenance in vitro as shoot cultures under minimal growth conditions at 4°–5 °C offers a space-saving and labor-efficient alternative which is also less complex and less subject to losses than cryogenic storage at ultra-low temperatures (Bhojwani 1981).

Cheyne and Dale (1980) and Dale et al. (1980) obtained 70%–92% survival after the first 15–18 months of minimal growth storage at 2°–6 °C of red and white clover shoot cultures on Blaydes', or B5 basal media with 0.2 mg/l 2iP and 0.2 mg/l IAA. There was little difference between light and dark treatments. Cultures could be removed from storage, grown under regenerative conditions for 4–8 weeks and then rooted, or returned to storage. Bhojwani (1981) developed a simple protocol for storage of *T. repens* genotypes on MS basal medium supplemented with 0.2 mg/l BAP. Proliferating shoot cultures could be stored in the dark at 5 °C for at least 10 months without affecting their survival or potential for rapid multiplication on return to light at 25 °C. To maintain red clover genotypes, Campbell and Tomes (1982) initiated multiple shoot cultures on B5 medi-

Table 2. Modifications of L2 culture medium for in vitro manipulations of *T. pratense*, *T. rubens* and other *Trifolium* species (Collins and Phillips 1982; Rupert and Collins 1985; Grosser and Collins 1984)

A. Basal media	mg/l (pH = 5.8)			
	L2 callus	SL suspension	RL rooting	SGL germination
NH_4NO_3	1000.0	600.0	500.0	100.0
KNO_3	2100.0	2100.0	1050.0	210.0
KH_2PO_4	325.0	250.0	325.0	32.5
$NaH_2PO_4 \cdot H_2O$	85.0	–	42.5	8.5
$MgSO_4 \cdot 7H_2O$	435.0	400.0	217.5	43.5
$CaCl_2 \cdot 2H_2O$	600.0	350.0	300.0	43.8
$FeSO_4 \cdot 7H_2O$ (EDTA)	25.0	25.0	25.0	2.8
$MnSO_4 \cdot H_2O$	15.0	13.5	7.5	–
$ZnSO_4 \cdot 7H_2O$	5.0	4.5	2.5	–
H_3BO_3	5.0	4.5	2.5	–
KI	1.0	0.9	0.5	–
$Na_2MoO_4 \cdot 2H_2O$	0.4	0.36	0.2	–
$CuSO_4 \cdot 5H_2O$	0.1	0.09	0.05	–
$CoCl_2 \cdot 6H_2O$	0.1	0.09	0.05	–
Thiamine·HCl	2.0	2.0	1.0	0.5
Pyridoxine·HCl	0.5	0.5	0.25	0.125
Myo-inositol	250.0	250.0	125.0	62.5
Nicotinic acid	–	–	1.0	–
3-Aminopyridine	–	–	2.5	–
Sucrose	25000.0	25000.0	10000.0	10000.0
Agar	8000.0	–	6500.0	6500.0

B. Growth regulator regimes

Medium code	Basal medium	Growth regulators (mg/l)		Use
		Auxin	Cytokinin	
L2	L2	PIC 0.06	BAP 0.1	Callus
SL2	SL	PIC 0.06	BAP 0.1	Suspension
SEL (LSE)	L2	2,4-D 0.01	ADE 2.0	Somatic embryogenesis
LSP	L2	PIC 0.002	BAP 0.2	Shoot promotion
ML8	L2	PIC 0.003	BAP 0.5	Meristem culture
ML12	L2	PIC 0.002	BAP 1.0	Shoot promotion
RL	RL	IAA 0.2	–	Rooting
CR2	RL[a]	COU 4.38	–	Rooting
SGL	SGL	–	–	Seed germination
LIH	L2[b]	PIC 0.06	ADE 3.0	Immature embryos

[a] Nicotinic acid and 3-aminopyridine omitted.
[b] 12.5% sucrose.

um with 2 mg/l BAP. Shoots were separated onto B5 maintenance medium without hormones for development of single plantlets. Once on maintenance medium, plants could be stored at 4 °C for at least 8 months before transfer to soil. Multiple shoot cultures could also be stored up to 8 months without subculture (Campbell and Tomes 1984).

Propagation and storage as multiple shoot cultures have given excellent preservation of clover genotypes. No abnormalities have been observed among many hundreds of red and white clover regenerants transferred to soil (Cheyne and Dale 1980; Campbell and Tomes 1982), although in a field study comparing regenerated plants with plants cloned as cuttings, Campbell and Tomes (1982) found red clover plants regenerated from shoot cultures to show greater winter hardiness and a higher second-year yield than conventionally propagated plants.

Avoidance of induced variation apparently depends on direct shoot induction from pre-formed meristematic centers without passage through a destabilizing callus phase. Where basal callus forms at the point of shoot initiation, somaclonal variants are more likely to occur. A compromise system in which shoots are cycled through a controlled basal callus phase without total dedifferentiation has been used by Tan and Stern (1985) to induce somaclonal variation in *T. subterraneum*, a species for which regeneration from fully dedifferentiated callus has proved difficult and unreliable. Shoot cultures were induced on MS basal medium with 0.2 mg/l BAP and 2 mg/l IAA. Small clusters including shoot buds were transferred for 2–3 weeks to MS with 0.1 mg/l BAP and 1 mg/l 2,4-D to induce basal callus, and were then returned to the former medium for 3–4 weeks to produce additional shoots. Shoots were excised and rooted on half-strength MS with 1 mg/l IBA. Two variant lines were obtained with significantly later flowering, and one with a seed coat color mutation.

Skucinska et al. (1984) devised a cotyledonary node multiplication system for micropropagation of red clover. Seedling hypocotyls with attached cotyledons were explanted vertically into L2 medium containing NAA (4.5 μM) and BAP (8.9 μM). The one to three shoots that developed were then subcultured to induce further shoot proliferation from axillary meristems. An average propagation coefficient of five to seven shoots was achieved. Nearly all shoots rooted on L2 medium without hormones.

A procedure related to shoot meristem culture was developed by Skucinska and Miszke (1980) for propagation of red clover from young inflorescences. Flowering heads were placed on MS medium with 5 mg/l BAP and 0.1 or 1 mg/l IAA, and direct adventitious shooting was observed in 2–3 weeks. Shoots could be subcultured on MS with 5 or 10 mg/l BAP and 0.1 mg/l IAA, or rooted on MS without hormones or with 5 mg/l IAA. After 8 months and a number of subcultures, 1000 true-to-type plantlets had been obtained, while the original cultures still maintained their morphogenic potential. Disinfection of flower heads can be difficult, but the method appears promising for rapid multiplication of mature plants when simultaneous virus elimination is not required.

Direct adventitious shooting can also be induced from the immature tissues of young seedlings. Webb et al. (1984) found approximately 10% of *T. repens* seedlings regenerated adventitious shoots from the hypocotyl, and at a somewhat lower frequency from the cotyledons. Responsive genotypes were polycrossed to test for increase in response frequency.

3.2 Pathogen Elimination

As in other plants, the apical meristems from rapidly growing clover shoots can be explanted free of invading viruses and regenerated in vitro to give virus-free plants with restored vigor and seed production. This method was first tested by Barnett et al. (1975) on the parent clones of Tillman white clover, and was found to be more effective than conventional stolon-tip cuttings from plants grown at 40° or 10 °C. Meristems with one or two leaf primordia were taken from stolon tips of plants in the 10 °C treatment and grown as single shoots on MS medium with 10^{-5} mg/l IAA to achieve both shoot and root growth. Survival was only 9%, but of these plants 57% were free of all detectable viruses. Viruses eliminated included white clover mosaic (WCMV), clover yellow vein (CYVV), alfalfa mosaic (AMV), and peanut stunt (PSV).

An improved survival rate was achieved by Phillips and Collins (1979b) using an ML medium (Table 2) for virus elimination from the ten parent clones of the synthetic red clover variety Kenstar. Survival of meristems was 65% for those with one leaf primordium and up to 80% for those with two leaf primordia. The smaller explants gave a greater frequency of virus-free plants. From clones infected with the insect-vectored viruses, bean yellow mosaic, peanut stunt, white clover mosaic, and alfalfa mosaic, two-thirds of the regenerants were apparently virus-free. Dale et al. (1981) also achieved a higher success rate with smaller meristem explants using the shoot culture method of Cheyne and Dale (1980) to eliminate white clover mosaic virus from red and white clovers.

Mycoplasmas causing clover phyllody and clover red leaf symptoms have also been apparently eliminated by meristem tip culture (Dale et al. 1980, 1981).

4 Embryo Culture and in Vitro Fertilization

4.1 Interspecific Hybridization by Embryo Rescue

Embryo culture studies in *Trifolium* were, until 1984, directed almost exclusively towards the rescue of hybrid embryos in interspecific crosses showing endosperm failure (Fig. 2). This application has been extensively reviewed in previous articles (Collins and Grosser 1984; Phillips and Collins 1984; Mroginski and Kartha 1984; Rupert and Collins 1985; Williams 1987; Williams et al. 1982, 1987).

Embryo rescue involves controlled pollination followed by explanting of the embryo onto a solid or liquid culture medium at the latest stage of development before degeneration occurs in situ. This stage is determined experimentally, and in *Trifolium* is normally about 7–14 days after pollination. The culture medium replaces the nutritional function of the endosperm to a limited extent. Success has been found, however, to depend on the embryo achieving at least the heart stage of development before explanting. This, in turn, depends to a large extent on the genotypes of the parent individuals (Taylor et al. 1981; Williams 1987), and also on factors such as maternal plant vigor and environmental conditions. The differentiation of cotyledonary primordia appears to be a natural break-point for embryo culture, such that most attempts to culture undifferentiated, globular proembryos have failed. Occasionally it has proved possible to induce proembryos to form callus capable

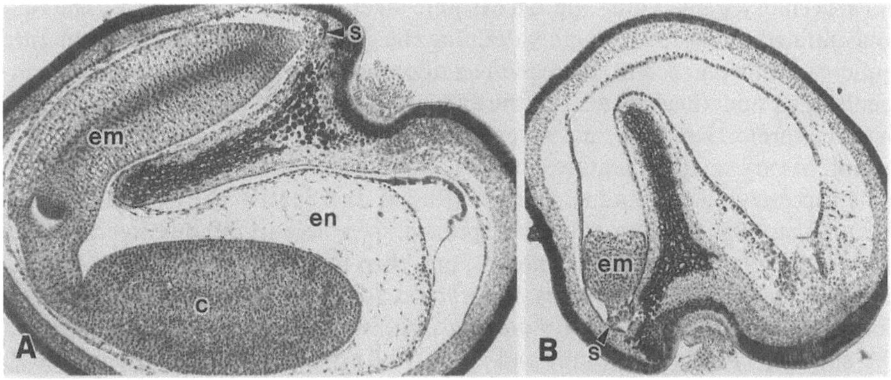

Fig. 2A, B. Sections of ovules of *Trifolium ambiguum*, kura clover, showing **A** normal development 10 days after intraspecific pollination, and **B** abortive development 10 days after interspecific pollination with *T. repens* pollen. (After Williams and White 1976). *em* embryo; *en* endosperm; *c* cotyledon; *s* suspensor. In **B**, endosperm development has failed, the hybrid embryo has not progressed beyond the heart stage, and the suspensor is becoming detached from the maternal tissues. Such embryos are rescued by explanting into culture at 7–10 days after pollination

of regeneration, but generally these extremely immature stages have proved unresponsive.

Keim (1953a, b) was the first to apply embryo rescue successfully in the production of clover hybrids. Using a simple agar-solidified Randolph and Cox mineral medium, he successfully cultured embryos of *T. repens*, *T. pratense* and *T. ambiguum* from as early as 7 days after pollination. A total of 22 plants was raised from crosses of *T. ambiguum* (6×)×*T. hybridum* (2×), and one vigorous hybrid of *T. ambiguum* (2×)×*T. nigrescens* was lost at transplanting. Hybrids were also obtained between *T. repens* and *T. nigrescens* at the diploid and tetraploid levels (Brewbaker and Keim 1953). Further hybrids between these two species were produced by Hovin (1962) using a modified version of Keim's technique and a liquid culture medium. An extensive series of interspecific crosses was made by Evans (1962) in the course of an evaluation of interspecies barriers in *Trifolium*. Crosses were made using diploids and induced autopolyploids of *T. pratense*, *T. repens*, *T. hybridum*, *T. incarnatum*, *T. alexandrinum*, *T. fragiferum*, *T. subterraneum*, *T. resupinatum*, and *T. semipilosum*. Hybrid embryos were observed in 21 crosses and their reciprocals, and a number of plantlets were obtained using simple mineral culture media. Altogether a total of 14 different hybrid types was raised in culture, of which the following were successfully grown in soil: *T. repens* (4×)×*T. nigrescens* (2×), *T. repens* (8×)×*T. nigrescens* (4×), *T. repens*×*T. uniflorum* (and reciprocal), *T. ambiguum* (2×)×*T. hybridum* (2×), *T. ambiguum* (6×)×*T. hybridum* (2×), *T. ambiguum* (6×)×*T. hybridum* (4×), and *T. hybridum* (4×)×*T. ambiguum* (2×). Further hybrids of *T. ambiguum* (6×)×*T. hybridum* (2×) were raised in culture by Firsova et al. (1980).

After the work of Evans (1962), numerous attempts were made to extend embryo rescue to other crosses characterized by smaller and less well differentiated hybrid embryos. Failures presumably resulted from immaturity of embryos and a general lack of knowledge of the nutritional and growth regulator requirements of early embryogeny. Three approaches were taken to solve this problem. The first concentrated on more thoroughly defining the growth requirements of immature embryos, while the second, more indirect approach sought to supply such requirements in an undefined form. The third involved culture of embryos in ovulo (i.e., ovule culture) to retain normal tissue interrelationships and to avoid difficult dissection of very small, immature embryos. All three approaches achieved some notable successes. Rupert and co-workers (Rupert and Seo 1977; Rupert and Evans 1980; Evans 1984), taking the first approach, were able to produce hybrid callus from heart stage and younger embryos, and then to achieve both indirect and direct regeneration of hybrid plants. Embryos excised 8–12 days after cross pollinations among species in section *Lotoidea* Zoh. were placed on MS medium with 3% sucrose, 100 mg/l i-inositol and 20 ml/l of a vitamin mixture (Staba 1969). For heart-stage or younger embryos this medium was supplemented with 0.5 mg/l 2iP and 2 mg/l 2,4-D for callus induction, while multiple shoots and roots were induced directly with 2–5 mg/l 2iP and 0.2 mg/l NAA from torpedo and older embryos. Altogether, ten hybrid types were raised, based on nine species and including the first hybrids of *T. ambiguum*×*T. montanum*, *T. ambiguum*×*T. occidentale*, and *T. isthmocarpum*×*T. repens*. Additional hybrids were also produced from the crosses *T. ambiguum*×*T. hybridum*, *T. ambiguum*×*T. repens*, *T. repens*×*T. isthmocarpum*, *T. repens*×*T. nigrescens* and *T. repens*×*T. uniflorum*.

Collins and co-workers (Collins et al. 1981; Collins and Phillips 1982; Phillips et al. 1982) also followed a direct approach in adapting the culture medium L2, which had been optimized for *T. pratense* (Phillips and Collins 1979a), to the rescue of hybrid embryos from crosses between *T. pratense* and the related perennial species, *T. sarosiense*, *T. alpestre* and *T. medium*. Embryos, predominantly at the heart stage, were explanted at 12–19 days after pollination and cultured 8–14 days on a modified L2 medium (LIH, Table 2) with high sucrose (12.5%), moderate auxin activity (25 nM PIC) and low cytokinin activity (15 μM adenine) for continued development of cotyledons. Survivors (19%–21%) were then transferred to a modified L2 medium (LSP, Table 2) with low auxin activity (4 nM PIC) and moderate cytokinin activity (0.66 μM BAP) to encourage shoot germination. On this medium 45% of embryos germinated shoots directly and 39% lost their organization to form callus. Calli were transferred to SEL (Table 2) for shoot regeneration and ML8 (Table 2) for shoot increase. Shoots were rooted on RL (Table 2) before transplanting to soil. Using this protocol, Phillips et al. (1982) produced hybrid clones from the cross *T. sarosiense* ($2n = 6\times = 48$)×*T. pratense* ($2n = 2\times = 14$) (Fig. 3), and Merker (1984) produced three hybrid clones from the cross *T. medium* ($2n = 10\times = 80$)×*T. pratense* ($2n = 4\times = 28$).

The indirect, second approach taken by Williams and co-workers depended on culture of hybrid embryos within or in contact with nurse endosperm dissected from ovules developing normally after intraspecific pollination (Fig. 4; Williams and DeLautour 1980; Williams et al. 1982; Williams 1987; Williams et al. 1987). The culture media (e.g., EC1, Table 3) contained vitamins and organic factors but no growth regulators. Nurse endosperm was presumed to provide nutritional factors, growth regulators and/or a more suitable physical environment for embryo development. It could not, however, fulfill all such requirements, since the composition of the culture medium remained important. Using this nurse-culture technique,

Fig. 3A–D. Interspecific hybrid between *Trifolium sarosiense* and *T. pratense* produced with the aid of embryo rescue. (After Phillips et al. 1982). **A** *left to right* flowers from *T. pratense*, hybrid, *T. sarosiense*. **B** *left to right* leaves of *T. pratense*, hybrid, *T. sarosiense*; **C** *left to right* rooting habit of *T. pratense*, hybrid, *T. sarosiense*; **D** root tip squash from the hybrid showing 31 chromosomes, 7 from *T. pratense* and 24 from *T. sarosiense*

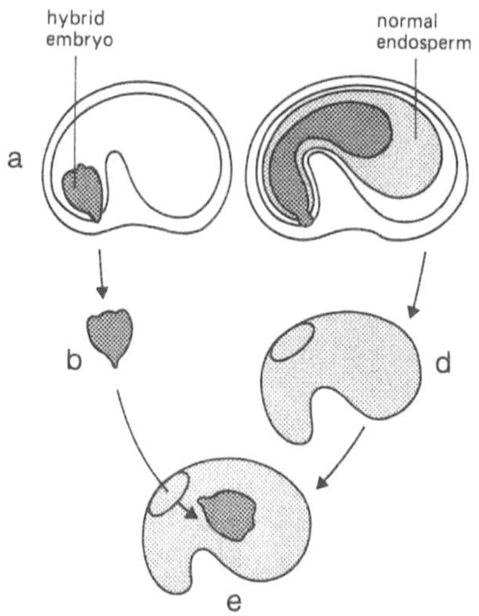

Fig. 4a–e. Endosperm transplant technique used for culture of hybrid embryos. **a, b** The hybrid embryo is removed from an ovule in which endosperm development has failed. **c** To provide nurse endosperm, an intraspecifically pollinated ovule is dissected to remove the gelatinous sac of cellular endosperm surrounding the embryo. **d** The normal embryo is pressed out of the endosperm, leaving an exit hole. **e** The hybrid embryo is inserted into the nurse endosperm through the exit hole, and the two are placed on the surface of a culture medium. (After Williams and DeLautour 1980)

first hybrids were obtained in the cross *T. ambiguum* (4×)×*T. repens* (Williams 1978; Williams and Verry 1981), and further hybrids were obtained from the crosses *T. ambiguum* (4×)×*T. hybridum* (2×) (Williams 1980) and *T. repens*×*T. uniflorum* (Pandey et al. 1987). Although nurse endosperm represents a useful expedient when embryo culture requirements are poorly defined, on the whole, results have been less satisfactory than those obtained using media optimized for the species concerned.

The third approach, ovule culture, was used successfully to raise hybrids in the cross *T. ambiguum* (6×)×*T. repens* by Yamada and Fukuoka (1985, 1986). Five days after interspecific pollination, ovules were explanted from the *T. ambiguum* female parent onto a mineral salts medium supplemented with 7% sucrose, 0.2% casein hydrolysate and 10% juice of either young cucumber fruits or immature clover seeds. These two treatments gave 10% and 15% germination of ovules respectively, on subculture after 15–20 days to basal medium with 1% sucrose and no juice. Four hybrid plantlets were transferred to soil, and one was raised to flowering. This hybrid ($2n = 5\times = 40$) showed a general resemblance to the *T. ambiguum* female parent but was sterile. The ovule culture approach allows the rescue of hybrid embryos too small for dissection, but does not obviate the requirement for complex growth factors in the culture medium.

It is clear that the two factors which will permit the rescue of progressively less differentiated embryos from a wider range of crosses are: (1) optimization of the basal medium for the species involved in the cross, and (2) better definition of the sequence of nutritional, osmotic, and growth regulatory factors required during development of very immature embryos.

Interspecific hybrids obtained to date in *Trifolium* have ranged from over 50% fertile to highly sterile, often with morphological and physiological abnormalities. These barriers to further breeding have been discussed in detail by Williams (1987). Sufficient numbers of partially fertile hybrids have been obtained, however, even in crosses where sterile isolates predominate, to suggest that with progressive improvements in embryo rescue technology, effective sexual gene exchange can be achieved between a number of *Trifolium* species. Further improvements in

Table 3. Media used for embryo rescue with nurse endosperm and embryo cloning in *Trifolium* (Williams and DeLautour 1980; Maheswaran and Williams 1984)

	mg/l (pH = 5.5 – 5.8)	
	EC1 Embryo rescue	EC6 Embryo cloning
$Ca(NO_3)_2$	200.0	–
Na_2SO_4	200.0	–
KCl	65.0	–
$NaH_2PO_4 \cdot 2H_2O$	21.5	–
KNO_3	80.0	950.0
$MgSO_4 \cdot 7H_2O$	720.0	185.0
NH_4NO_3	–	600.0
$CaCl_2 \cdot 2H_2O$	–	220.0
KH_2PO_4	–	170.0
$MnSO_2 \cdot 4H_2O$	7.0	2.23
$ZnSO_4 \cdot 7H_2O$	3.0	0.86
H_3BO_3	1.5	0.62
KI	0.83	0.083
$Na_2MoO_4 \cdot 2H_2O$	0.25	0.025
$CuSO_4 \cdot 5H_2O$	0.025	0.0025
$CoCl_2 \cdot 6H_2O$	0.025	0.0025
$FeSO_4 \cdot 7H_2O$	27.8	27.8
$Na_2EDTA \cdot 2H_2O$	37.3	37.3
Casein hydrolysate	500.0	–
Myo-inositol	100.0	–
Nicotinic acid	5.0	–
Pyridoxine·HCl	1.0	–
Thiamine·HCl	0.4	–
Ca pantothenate	0.1	–
Yeast extract	–	1000.0
Sucrose	20000.0[a]	20000.0[a]
Agar	7000.0	7000.0

[a] Or higher, e.g., 40000.0 as required.

embryo culture technology may require critical comparisons between in situ and in vitro embryogenesis. A single study of this type was made by Enyingi and Heszky (1973), who described arrested cotyledon development and premature shoot growth in embryos of *T. pratense* explanted into culture at the heart stage. Such comparisons will assist formulation of media and transfer schedules for optimal embryo maturation and germination.

4.2 In Vitro Fertilization

Intra-ovarian pollination has not to our knowledge been attempted in *Trifolium*, and only a single study has investigated in vitro pollination of intact, excised pistils. Richards and Rupert (1980) aseptically pollinated pistils of *T. repens* and placed them, with or without accessory flower parts (calyx and pedicel), on MS medium with 0.1 g/l CH, 0.1 g/l i-inositol, 20 ml/l vitamin mixture (Staba 1969), 3% sucrose and varying amounts of thiamine hydrochloride, BAP, GA, and IAA. Accessory flower parts were found necessary for fertilization to succeed, and 52%

of these pistils contained maturing ovules with embryos following compatible cross pollination. When *T. ambiguum* pistils were pollinated with *T. repens* or *T. hybridum* pollen, development closely paralleled the in vivo situations. Fertilization occurred, followed by partial embryo development and endosperm abortion. However, various combinations of growth regulators added to the medium showed some evidence of enhancing embryo growth. This technique, together with preculture of fertilized ovules before embryo excision (McCoy 1985; McCoy and Smith 1986), may assist in optimizing in situ embryo development in crosses where abortion occurs normally at an extremely early stage.

4.3 Embryo Cloning and Direct Somatic Embryogenesis

A recent new application of zygotic embryo culture among *Trifolium* species has been the cloning of immature embryos by direct somatic embryogenesis (Figs. 5, 6; Maheswaran and Williams 1984, 1985, 1986a, b, c; Williams and Maheswaran 1986). When immature (torpedo) embryos of *T. repens* were explanted at 7–8 days after pollination onto a simple basal medium, EC6 (Table 3), supplemented with 0.05 mg/l BAP and 1 g/l yeast extract (YE), somatic embryos were formed from superficial cells of the hypocotyl region (Maheswaran and Williams 1984). Initiation was direct, without preliminary callus formation, and predominantly multicellular (Fig. 7; Maheswaran and Williams 1985). Somatic embryos developed to form leafy shoots on the induction medium, and these could be separated and rooted on the basal medium to give a clone of plantlets in approximately the time taken to produce a single seedling after embryo maturation in situ. When somatic embryos were removed from the parent embryo at the torpedo to cotyledonary stage and subcultured on induction medium with a raised level of BAP (1–2 mg/l), secondary somatic embryos were induced directly, without callus formation, on tissue in contact with the medium (Maheswaran and Williams 1986a). These also could readily be rooted on basal medium to produce plantlets.

A total of 55 primary and secondary regenerants from three original zygotic embryos were assessed for uniformity with respect to morphological markers, chromosome number, breeding behavior, isoelectric focusing of leaf total proteins, and gel electrophoresis of four families of leaf isozymes. Major differences were observed between clones, but all primary and secondary regenerants from each original embryo were apparently identical (Maheswaran and Williams 1986c). It was concluded, therefore, that direct somatic embryogenesis on immature zygotic embryos is a conservative, clonal regeneration process. This is in accordance with the concept of embryonic cells as pre-embryogenic determined cells (PEDC) (Fig. 7; Evans et al. 1981; Sharp et al. 1980, 1982) which are already programmed for embryonic development and require no dedifferentiation. Only a mitotic stimulus and means of escaping from coordinated control of the parent embryo are needed for individual expression of their embryogenic potential (Williams and Maheswaran 1986). The cytokinin BAP provides the mitotic stimulus while one or more components of YE suppress growth of the parent embryo shoot-root axis to allow hypocotyl cells to escape from coordinated control (Pretova and Williams 1986).

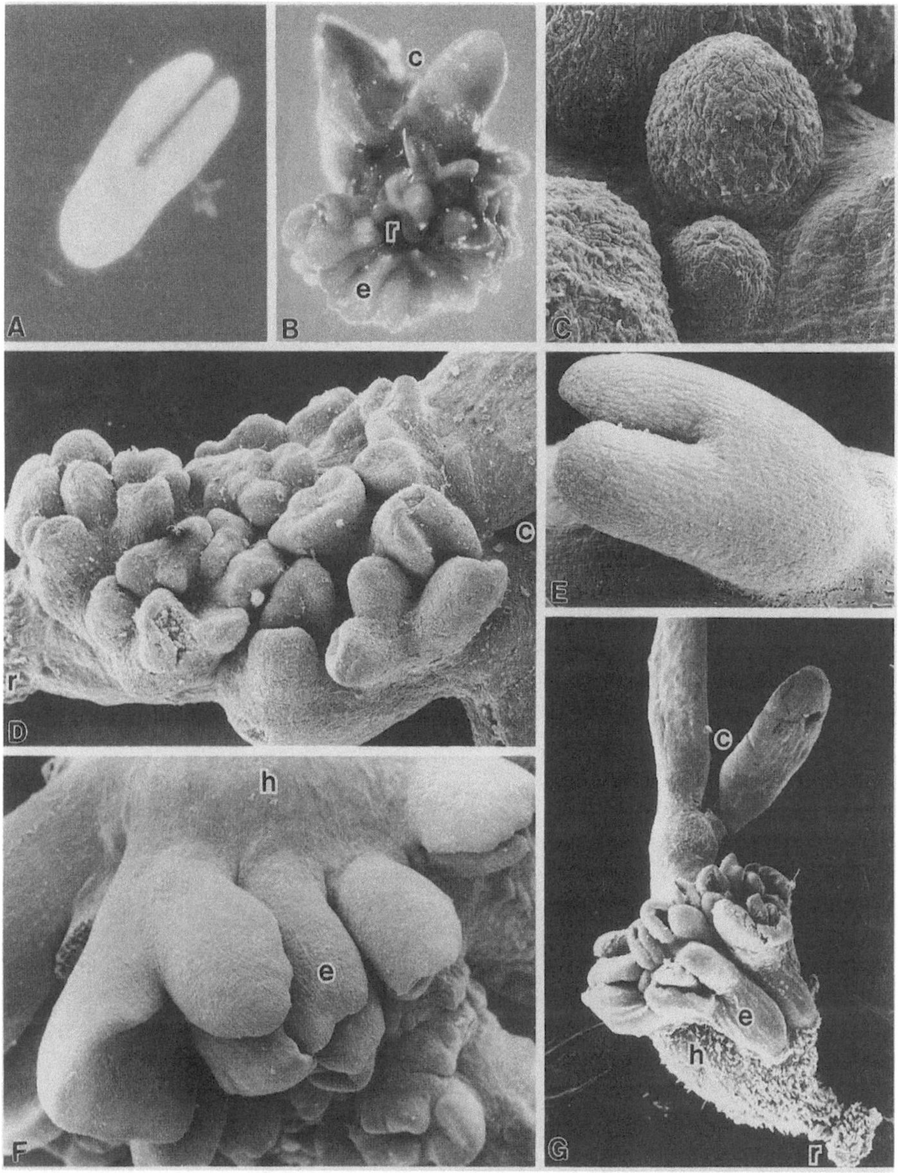

Fig. 5A–G. Direct somatic embryogenesis from immature zygotic embryos of *Trifolium* cultured on EC6 medium with 1 g/l YE and 0.05 mg/l BAP. **A, B** *T. pratense*. **C–G** *T. repens* (After Maheswaran and Williams 1985, 1986b). *c* Zygotic embryo cotyledons; *e* somatic embryos; *h* zygotic embryo hypocotyl; *r* root pole or root of zygotic embryo. **A** Zygotic embryo at the responsive torpedo stage. **B** Zygotic embryo after 20 days in culture. Growth of the shoot and root apices is inhibited, and somatic embryos have formed on the surface of the hypocotyl. **C–G** Scanning electron micrographs of somatic embryos. **C, D** Globular to early heart stages. **E** Heart stage. **F, G** Torpedo to cotyledonary stages

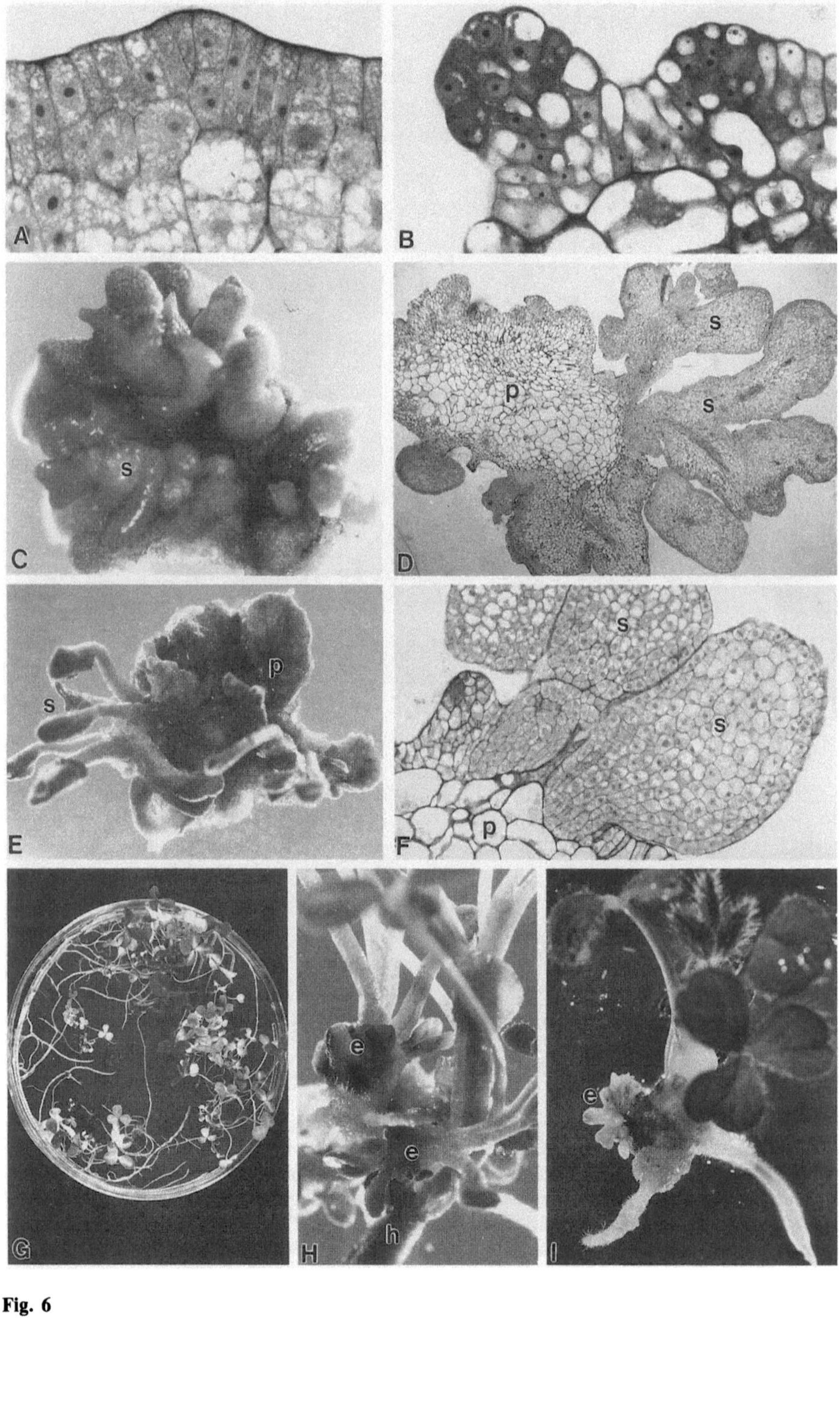

Fig. 6

Fig. 7. Probable relationship of the stage of zygotic embryo differentiation to the initiation of direct somatic embryogenesis from single cells or cell groups. *Shading* indicates pre-embryogenic determined cells (terminology of Sharp et al. 1980) IEDC's, induced embryogenic determined cells). (After Maheswaran and Williams 1985)

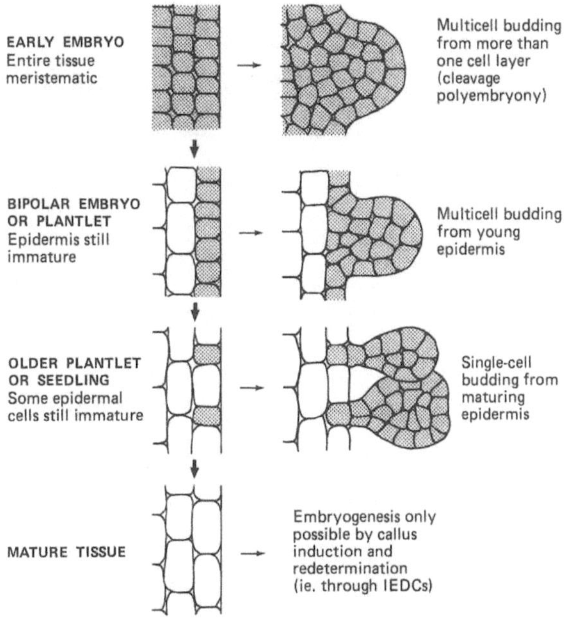

This direct embryo cloning procedure has been extended to *T. pratense* (Maheswaran and Williams 1984, 1986b), confirming an earlier observation for this species, of regeneration from immature embryos by somatic embryogenesis (Mezentsev et al. 1982). Embryos of *T. subterraneum* and *T. resupinatum* also responded similarly on appropriate basal media supplemented with 0.05 mg/l BAP and 1 g/l YE (Maheswaran and Williams 1986b). The basal medium/sucrose formulation, YE level and cytokinin level were all found to affect the response. In particular, for *T. pratense* and *T. subterraneum*, lowering YE suppressed initiation of somatic embryos, and raising BAP stimulated formation of nodular callus. In

Fig. 6A–I. Direct somatic embryogenesis from immature zygotic embryos of *Trifolium* cultured on EC6 medium with 1 g/l YE and 0.05 mg/l BAP (After Maheswaran and Williams 1985, 1986a, b). **A–G** *T. repens*. **H** *T. resupinatum*. **I** *T. subterraneum*. *e* Somatic embryos; *h* hypocotyl of zygotic embryo; *p* primary somatic embryo; *s* secondary somatic embryo. **A, B** Sections of the hypocotyl of zygotic embryos. **A** After 3 days in culture, meristematic activity, including irregular and periclinal divisions, is becoming localized as small swellings of the hypocotyl epidermis. **B** Developing globular somatic embryo after 6 days in culture. **C–G** Development of secondary somatic embryos after explanting primary somatic embryos on to EC6 medium with 1 g/l YE and 2.0 mg/l BAP. **C** Mass of secondary embryos forming on the lower surface of a primary somatic embryo 14 days after explanting on to the secondary induction medium. **D, F** Sections through primary somatic embryos similar to that shown in **C**, showing formation of secondary embryos from superficial cells. **E** Secondary embryos forming leafy shoots 21 days after transfer of the primary somatic embryo to secondary induction medium. **G** Plantlets obtained from a single primary somatic embryo, after rooting on EC6 basal medium. **H** Somatic embryos forming leafy shoots on the hypocotyl of a zygotic embryo of *T. resupinatum* after 49 days in culture. **I** Somatic embryos growing from compact callus on the hypocotyl of a germinating zygotic embryo of *T. subterraneum* after 21 days in culture. An accessory root has also formed from the hypocotyl

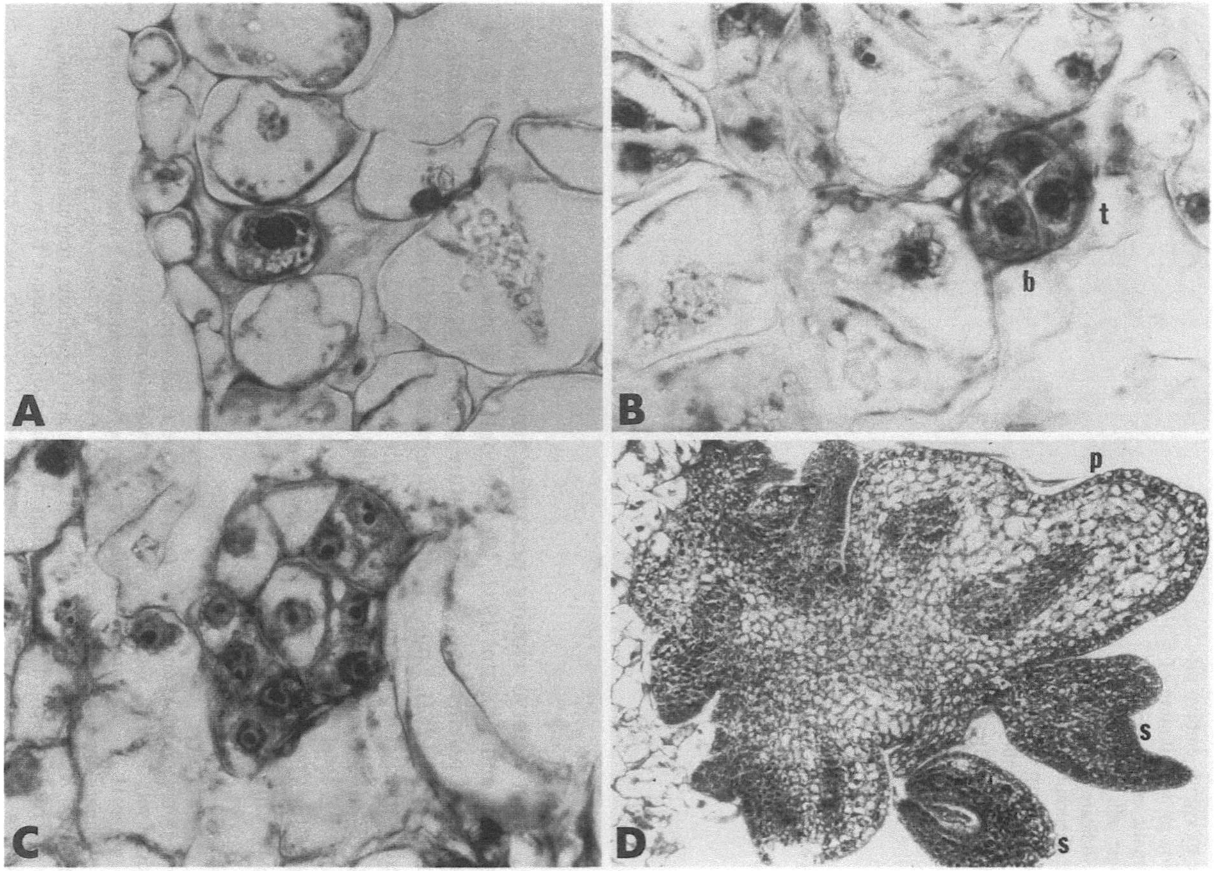

Fig. 8 A–D. Direct somatic embryogenesis from mature petioles of *Trifolium rubens* (After Cui et al. 1988). *t* Terminal cell; *b* basal cell; *s* somatic embryo; *p* primary somatic embryo. **A** Single outer cortical cell with dense cytoplasm and highly stained nucleus. **B** Three-celled proembryo found in the cortex showing differentiation into a single basal cell (*b*) and two terminal cells (*t*). **C** Multicelled globular embryo located in the cortex. **D** Secondary embryogenesis (*s*) from primary embryo (*p*)

T. resupinatum, the site of somatic embryo initiation could be altered from hypocotyl to cotyledons or both regions by alterations in the basal medium/sucrose formulation, suggesting effects on a physiological gradient within the parent embryo.

Direct somatic embryogenesis on zygotic embryos is probably best conceived of as embryo cleavage (Fig. 7). It has potential applications in cloning of valuable genotypes at the earliest possible stage of the life cycle, and in the provision of "seedling" clones in species where each seed normally represents a different genotype. Such an application would facilitate replication of individual genotypes for genetic testing and plant breeding. Currently, a year may be required for clones from crown cuttings to recover from the cloning process. Another major application appears to be in the experimental determination of factors controlling embryogeny.

Direct somatic embryogenesis from mature somatic tissues has also been observed in *Trifolium*. Unlike the analogous phenomenon from zygotic embryos, direct somatic embryogenesis from mature tissues is highly genotype-specific. This type of regeneration occurs readily in *T. rubens* (Fig. 8) and has been observed in *T. pratense*. In *T. rubens*, embryogenesis from single cells occurs in the cortex of petioles plated on L2 medium with 0.015 mg/l PIC and 0.1 mg/l BAP (Fig. 8; Cui et al. 1988). Plants from these embryos show increased chromosomal and morphological variation, unlike those plants derived by somatic embryogenesis from immature zygotic embryos.

5 Callus, Cell, and Protoplast Culture

5.1 Initiation and Maintenance of Callus and Cell Cultures

Callus, and in many cases suspension cultures, have been initiated from 20 species of *Trifolium* (Table 4). Explants have most frequently been taken from seedling tissues, particularly the hypocotyl, but young leaf, petiole, stem, root, meristem, and anther somatic tissues have also been used. In general, immature tissues have proved most responsive, but Phillips and Collins (1979a) have shown that with a medium optimized for the particular species (*T. pratense*), callus could be induced from a wide range of immature and mature plant parts.

Media have generally been based on the mineral salt formulations of MS, B5, L2, and occasionally SH, supplemented with both auxin and cytokinin. The most commonly used growth regulators have been 2,4-D (0.1–2 mg/l) in conjunction with kin (0.04–2 mg/l). NAA (0.5–2 mg/l) has also been used alone or together with 2,4-D, and 2iP (1–2 mg/l) or BAP (0.5 mg/l) have been used occasionally in place of kin. PIC (0.06 mg/l) has been used in conjunction with BAP (0.1 mg/l), particularly for induction of callus in *T. pratense* and closely related species. The diversity of media used to induce callus in *Trifolium* species indicates that a number of growth regulators are effective, and that the important factors are likely to be: (1) a basal medium optimized for the species concerned, and (2) a sufficiently high auxin : cytokinin balance in terms of biological activity, if not

Table 4. Cell and protoplast culture studies in *Trifolium*: initiation and regeneration (SE, somatic embryogenesis; SO, shoots only, whole plants not grown out)

Species[a]	Callus/cell culture initiated	Regeneration from callus/cell culture	Protoplast culture initiated	Regeneration from protoplasts
T. alexandrinum	Mokhtarzadeh and Constantin (1978)	Mokhtarzadeh and Constantin (1978)		
T. alpestre	Parrott and Collins (1983)			
T. ambiguum	Rupert et al. (1976); Jones et al. (1981b); Horak and Fojtik (1984); Pederson (1986)			
T. arvense	Bhojwani et al. (1984)	Bhojwani et al. (1984) (SO)	White and Bhojwani (1981)	
T. campestre	Parrott and Collins (1983)			
T. fragiferum	Smith and McComb (1981); Gustine and Moyer (1982); Rybczynski (1984)	Rybczynski (1984) (SE, SO)		
T. hirtum	Gustine and Moyer (1982)			
T. hybridum	Schenk and Hildebrandt (1972); Rupert et al. (1976); Jones et al. (1981b); Gustine and Moyer (1982)		Jones et al. (1981a); Dale et al. (1982); Webb et al. (1983, 1984)	Dale et al. (1982); Webb et al. (1983, 1984)
T. incarnatum	Horvath Beach and Smith (1979); Gustine and Moyer (1982); Parrott and Collins (1983); Rybczynski (1984); Pederson (1986)	Horvath Beach and Smith (1979); Rybczynski (1984) (SE, SO); Pederson (1986) (SE)		
T. medium	Gustine and Moyer (1982); Parrott and Collins (1983); Horak and Fojtik (1984)			
T. meduseum	Tan (1983)	Tan (1983) (SO)		
T. meneghinianum	Horak and Fojtik (1984)			
T. michelianum	Gustine and Moyer (1982)			
T. pannonicum	Horak and Fojtik (1984)	Horak and Fojtik (1984) (SE)		

T. pratense	Zakrzewski and Zakrzewska (1976); Horvath Beach and Smith (1979); Phillips and Collins (1979a); Keyes et al. (1980); Phillips and Collins (1980); Gustine and Moyer (1982); Snaselova (1982); Parrott and Collins (1983); Bhojwani et al. (1984); Horak and Fojtik (1984); Rybczynski (1984)	Horvath Beach and Smith (1979); Phillips and Collins (1979a); Keyes et al. (1980) (SE, SO); Phillips and Collins (1980) (SE); Snaselova (1982); Bhojwani et al. (1984) (SE); Horak and Fojtik (1984) (SE)	Jones et al. (1981a); Rakousky et al. (1984)	
T. repens	Hughes (1968); Pelletier and Pelletier (1971); Oswald et al. (1977a); Gresshoff (1980); Jones et al. (1981b); Gustine and Moyer (1982); Mohapatra and Gresshoff (1982); Parrott and Collins (1983); Bhojwani et al. (1984); White (1984); Pederson (1986)	Pelletier and Pelletier (1971); Oswald et al. (1977a) (SO); Gresshoff (1980); Mohapatra and Gresshoff (1982); Bhojwani et al. (1984); White (1984)	Gresshoff (1980); Bhojwani and White (1982); Ahuja et al. (1983); Jones et al. (1981a); Dale et al. (1982); Webb et al. (1983, 1984); Bhojwani et al. (1984); White (1984)	Gresshoff (1980); Ahuja et al. (1983); Webb et al. (1984); Bhojwani et al. (1984) (SO); White (1984)
T. resupinatum	Gustine and Moyer (1982); Oelck and Schieder (1983)	Oelck and Schieder (1983) (SO)	Oelck et al. (1982)	
T. rubens	Parrott and Collins (1983)	Parrott and Collins (1983) (SE)	Grosser and Collins (1984)	Grosser and Collins (1984) (SE)
T. subterraneum	Graham (1968); Parrott and Collins (1983); Tan (1983)	Tan (1983) (SE, SO)		
T. vesiculosum	Horak and Fojtik (1984); Pederson (1986)	Pederson (1986) (SE, SO)		

[a] See also Webb et al. (1987) for additional species producing callus, shoots and/or roots in culture.

actual concentration. Thus, Oswald et al. (1977a), using 2,4-D or 2,4,5-T and kin to induce callus from *T. repens* seedling tissues, found a 50:1 auxin:cytokinin balance by weight optimal for induction. Using the more active auxin PIC, however, callus induction is achieved with 0.06 mg/l PIC to 0.1 mg/l BAP (Phillips and Collins 1979a).

Complex additives have also been used occasionally in induction media, for example CH (Gresshoff 1980; Jones et al. 1981b; Mohapatra and Gresshoff 1982; White 1984), and corn milk replacing cytokinin (Zakrzewski and Zakrzewska 1976). Vitamin E at 2.12 μM was present in the culture medium used for *T. repens* by Oswald et al. (1977a), and appeared to enhance callus induction. Adjustment of the iron level may also be beneficial (Oswald et al. 1977a).

Although cultures were incubated in darkness in a number of early studies (e.g., Graham 1968; Oswald et al. 1977a; Mokhtarzadeh and Constantin 1978; Gresshoff 1980), in the studies since 1980, cultures have generally been initiated in continuous light or occasionally a 16-h photoperiod. The light regime appears to make relatively little difference to induction in *Trifolium* (Schenk and Hildebrandt 1972). Incubation temperatures have usually been in the range 20°–25 °C.

Once callus has been induced, it is subcultured regularly, either on the induction medium or a modified maintenance medium. In some studies the auxin: cytokinin balance has been either lowered (Pelletier and Pelletier 1971) or raised (Mokhtarzadeh and Constantin 1978; Mohapatra and Gresshoff 1982) for maintenance after induction.

Suspension cultures are normally initiated by inoculating a liquid maintenance medium with segments of rapidly growing, friable callus and agitating on a gyratory shaker at 50–150 rpm. Under these conditions, the callus dissociates into a mixture of single cells and cell aggregates which must be screened or filtered to provide a homogeneous suspension. For *T. pratense* the optimized suspension culture medium, SL (Table 2; Phillips and Collins 1980), has a 10% reduction of most mineral salts but the same concentrations of growth regulators as the callus induction medium.

Although callus and cell cultures of *Trifolium* species are often described as "undifferentiated", they may show varying morphological phenotypes and internal cellular nonhomogeneity. Callus may be soft and friable, hard and compact, wet and slimy, white, yellow or green, and may contain partially differentiated cells, particularly vascular elements. Phenotype and homogeneity can generally be controlled by adjusting growth regulators. For example, increasing the auxin level generally increases friability and decreases cellular differentiation. The property of friability, a tendency of cells to round up and separate early after division, presumably involves rupture of cell-cell contacts which might otherwise contribute to neighbor-mediated differentiation.

5.2 Applications for Screening and Studies of Physiology and Disease

Callus and cell cultures of *Trifolium* species have been used to study herbicide metabolism, response to salt (NaCl), interactions with *Rhizobium*, virus infection, phytoalexin production, and cyanogenesis. Red clover callus has also been used as a propagation medium for parasitic nematodes (Zakrzewski and Zakrzewska 1976).

Studies on the metabolism by *T. repens* cell suspensions of the phenoxyalkylcarboxylic acid herbicides (Oswald et al. 1977b; Smith and Oswald 1979; Smith 1979), showed that the relative insensitivity of certain legumes to the butyric acid homologs (2,4-DB and 2,4,5-TB) could be attributed to their rapid metabolism to nontoxic compounds rather than to the toxic acetic acid homologs 2,4-D and 2,4,5-T. It was found that up to a near eightfold increase in tolerance of cells to this group of herbicides could be obtained within 5 days of selection in culture. Cells selected with any of the homologs exhibited greater tolerance to all, and the highest tolerance to all was shown by cells selected with 2,4-DB. Tolerance was stable through mitotic divisions in selected lines.

Smith and McComb (1981) examined the comparative effects of increasing NaCl concentrations on intact plants and callus cultures of *T. repens* and *T. fragiferum*. In vitro responses closely paralleled those obtained in vivo, indicating that callus cultures could be used as a space-saving alternative to screening whole plants for salt tolerance. Although *T. fragiferum* has been considered moderately salt-tolerant, the cultivar Palestine used in this study was as sensitive as *T. repens* Ladino. Interestingly, in the same study, salt tolerance was detected in a line of *Medicago sativa* which had previously been selected for efficient regeneration from tissue culture. The point was made that selection for optimal performance on tissue culture media, which have comparatively high levels of mineral salts, may also inadvertantly select for NaCl tolerance. Salt-tolerant clover cell lines have not been selected using tissue culture, although this has been done for certain other legumes such as alfalfa (Croughan et al. 1978; Smith and McComb 1983).

The majority of in vitro studies of legume-*Rhizobium* interactions have involved soybean cells, often in association with rhizobia of the slower-growing "cowpea" type (e.g., Holsten et al. 1971; Phillips 1974; Child 1975; Reporter et al. 1980). Several investigators, however, have used *Trifolium* cultures and rhizobial strains of the faster-growing type. Such studies have shown that the nitrogenase genes are entirely bacterial, and that the host plant provides only an appropriate physical and nutritional environment. Co-culture of legume and rhizobium cells has resulted in both extracellular and intracellular nitrogen-fixing associations, the latter being associated with infection thread-like structures within host callus tissues (e.g., Holsten et al. 1971). Graham (1968) initiated co-cultures of *T. subterraneum* root callus with *R. trifolii* and *R. melilotti*. The rhizobial cells remained extracellular and did not interfere with callus growth. Similar co-cultures of *T. pratense* root, stem, and leaf cells with *R. trifolii* exhibited nitrogenase activity, although free rhizobial cells showed no such activity when host tissue was removed from the culture vessels (Ranga Rao 1976). Compared with uninoculated controls, infected tissue showed two characteristic growth responses. It continued to grow on nitrate-free medium, and it differentiated roots with curled root hairs. Release of kin and IAA by the bacterial cells was suggested to account for these responses. Child and Kurz (1978) observed nitrogenase activity in *T. pratense-Rhizobium* co-cultures when the bacterial and host cells were either in direct contact or separated in such a way that diffusion could occur between them. In the latter system, addition of a pentose sugar (arabinose) and a tricarboxylic acid (succinate) was necessary to achieve maximum activity.

Following up reports that plant cells in culture produced dialyzable components which induced nitrogenase activity in *Rhizobium* strains (e.g., Reporter et al. 1980), Gresshoff and co-workers investigated the effects of plant conditioned medium (PCM) in a *T. repens-R. trifolii* system (Gresshoff et al. 1981; Gresshoff 1982). Using a co-culture system with the host cell suspension and bacterial cells separated by a dialysis membrane or membrane filter, evidence was obtained that bacterial cells produce a dialyzable factor which influences plant cells to produce dialyzable compound(s) capable of inducing nitrogenase activity in free rhizobial cells. Results were too variable, however, and the system too oxygen-labile for routine use in characterizing host and rhizobial genotypes for nitrogen-fixation capability. A more suitable system for this purpose was the co-culture in vitro of axenic whole *T. repens* seedlings inoculated with specific rhizobial strains (Rolfe et al. 1980; Carroll and Gresshoff 1983; Bassam and Gresshoff 1986).

Jones et al. (1981 b) inoculated cell suspensions of *T. repens*, *T. ambiguum*, *T. hybridum*, and the hybrid *T. ambiguum* × *T. hybridum* with clover yellow mosaic virus (CYMV) or clover yellow vein virus (CYVV), and analyzed the cultures for infection and virus multiplication using latex serology and microscopy to detect intracellular virus particles. Results were compared with in vivo inoculation of whole plants. CYVV multiplied in *T. repens* and *T. hybridum* cells but not in cells of *T. ambiguum* or the hybrid. This result paralleled that obtained in vivo, i.e., CYVV infected *T. repens* and *T. hybridum* plants but not plants of *T. ambiguum* and the hybrid. CYMV multiplied in all cultures, but at a low rate in cells of *T. ambiguum* and the hybrid. After in vivo inoculation, however, CYMV infected *T. repens* and *T. hybridum*, did not infect *T. ambiguum*, and gave a low, symptomless infection of the hybrid. With reference to this close but not exact parallel, it was suggested that, of the two main components of virus disease, infection and replication, the first in particular may be modified in tissue culture. The in vitro and in vivo results were sufficiently similar to show that cell suspension cultures might be used as an effective screen for virus resistance.

For metabolic studies, cell cultures have a number of advantages (Ellis 1984), although care must be exercised in extrapolating results to the whole plant situation. Callus cultures have been used to study the regulation of phytoalexin production in nine *Trifolium* species: *T. fragiferum*, *T. hirtum*, *T. hybridum*, *T. incarnatum*, *T. medium*, *T. michelianum*, *T. pratense*, *T. repens*, and *T. resupinatum* (Gustine 1981; Gustine and Moyer 1982). Phytoalexin accumulation, elicited abiotically with $HgCl_2$, was found to depend on integrity of SH groups within the cell. None of the cultures produced the same complement of phytoalexins as reported for intact leaves inoculated with fungal pathogens. Calli of all species except *T. pratense* produced medicarpin. Maackiain was produced only by *T. pratense* and *T. medium* calli, while sativan and vestitol were not detected in calli, although they occur in leaves infected by fungi. From this work it was concluded that the medicarpin pathway is of major importance in this group of species, and that although the mechanism for eliciting phytoalexin production is not lost in callus culture, the biosynthesis of certain phytoalexins may be blocked.

Cell cultures have also been used to study the mechanism of cyanogenesis, the production of hydrocyanic acid (HCN) after tissue damage, which may be harmful to animals during grazing. In *T. repens* the two cyanogenic glucosides, linamarin and lotaustralin, produced by the *Ac* locus (Collinge and Hughes 1982) are hydrolyzed by a β-glucosidase, linamarase, produced by the *Li* locus, releasing glucose and an unstable aglycone. The aglycone dissociates to yield an alkanone and HCN. Hughes (1968) used *T. repens* suspension cultures with differing rates of glucoside production to show that two forms of β-glucosidase are involved, both controlled by the *Li* locus. Enzyme activity was found to vary among different subcultures from a single explant, even under identical environmental conditions.

Trifolium cell culture systems have so far been comparatively poorly exploited for selection and screening in breeding programs. From the preceding discussion it is clear that tissue cultures could provide compact systems for selection of herbicide-, virus- and salt-tolerant genotypes, independently of seasonal constraints and whole-plant growth facilities. Further work is needed to design routine protocols for such tests, and to extend this type of screening to fungal diseases as *Fusarium*, and the clover rot *Sclerotinia*, for which no effective test for resistance other than field screening is currently available (Evans and Williams 1984). For fungal diseases, both the incorporation of purified toxins into culture media (e.g., Hartman et al. 1984) and host-pathogen co-cultures deserve consideration.

5.3 Regeneration from Cell Culture

To permit a full spectrum of novel breeding methods to be applied to *Trifolium* species, reliable protocols must be available for regeneration from callus, cell and protoplast cultures. Only when whole plants can be recovered from single cells and disorganized tissues will we be able to exploit somaclonal variation, genetic transformation, single-cell selection, and somatic hybridization. Considerable progress has been made with the small-seeded pasture legumes including alfalfa and several *Trifolium* species. It is worth re-emphasizing at this point that the single most important factor in regeneration is genotype. For all species and cultivars investigated to date, it has proved necessary to screen a range of seedling genotypes to find those capable of regeneration under a particular set of cultural conditions (Phillips and Collins 1979a; Keyes et al. 1980; Gresshoff 1980; Phillips and Collins 1980; Mohapatra and Gresshoff 1982; Webb et al. 1983; Oelck and Schieder 1983; Bhojwani et al. 1984; White 1984). At the same time, however, species such as red and white clover show considerable plasticity in responses (Gresshoff 1980; Rupert and Collins 1985), so that regenerating genotypes can be selected on a range of basal media and growth-regulator regimes.

Regeneration of whole plants from callus or suspension cultures has been achieved for *T. repens*, *T. pratense*, *T. incarnatum*, *T. alexandrinum*, *T. panonicum*, *T. rubens*, and *T. vesiculosum*, and in addition shoot regeneration has been reported for *T. resupinatum*, *T. arvense*, *T. subterraneum*, *T. meduseum* and *T. fragiferum* (Table 4). Plants may originate either by organogenesis from structures resembling shoot and root meristems, or by somatic embryogenesis. The developmental pathway is determined by the growth-regulator regime in a fashion generally similar to that seen in other tissue culture systems. Subculture is usually necessary after bud or embryo initiation in order to obtain elongated shoots and subsequently roots. Shoot growth is promoted by high cytokinin to auxin ratios, rooting by high levels of auxins (other than 2,4-D), and somatic embryogenesis by 2,4-D (Phillips and Collins 1980; Bhojwani et al. 1984).

For shoot induction, various combinations of the cytokinins kin, 2iP, BAP, Z or ADE (0.1 – 10 mg/l) have been used in conjunction with low levels (0.01 – 1 mg/l) of the auxins NAA or IAA. For *T. pratense* best results were obtained with BAP (0 – 10 mg/l) in combination with PIC (0.006 mg/l) (Phillips and Collins 1979a). For *T. arvense*, shoot growth was observed only after transfer of callus to a medium supplemented with TIBA (2 mg/l) (Bhojwani et al. 1984). White (1984) found a high NH_4^+ level important in the differentiation of shoot meristemoids in suspension cultures of *T. repens*. Although a number of earlier studies included a low level of cytokinin in the rooting medium (e.g., Pelletier and Pelletier 1971; Mokhtarzadeh and Constantin 1978; Gresshoff 1980), shoots of several species have been rooted readily with IAA alone or on hormone-free basal medium, often with a reduced content of mineral salts (Collins and Phillips 1982; Snaselova 1982; White 1984). For *T. pratense* and *T. rubens*, somatic embryos were induced in callus cultures using L2 basal medium modified for somatic embryogenesis (SEL, Table 2) by addition of 2,4-D (0.01 mg/l) and ADE (2 mg/l) as a cytokinin (Phillips and Collins 1980; Parrott and Collins 1983). Embryos did not germinate on the induction medium. Transfer was required to a shoot-promoting medium (LSP, Table 2) containing BAP (0.2 mg/l) and PIC (0.002 mg/l), and finally to a rooting medium (RL, Table 2) containing only IAA (0.2 mg/l) and having reduced levels of most mineral salts. Somewhat similar observations were made by Bhojwani et al. (1984), who induced somatic embryos in callus cultures of *T. pratense* using an MS medium supplemented with 2,4-D (0.05 mg/l) and 2iP (0.5 mg/l). The resulting shoots were rooted on hormone-free basal medium. Embryo-like structures were reported in cell cultures of *T. ambiguum*, *T. hybridum*, and their interspecific hybrid by Rupert et al. (1976), but did not develop further.

Embryos in cultures initiated from differentiated, nonembryonic tissues are presumed to arise from cells which have undergone a redetermination event to reinstitute the embryogenic state. Such cells have been termed "induced embryogenic determined cells" (IEDC) by Evans et al. (1981) and Sharp et al. (1980, 1982), and once induced, with 2,4-D for example, can proliferate by mitotic division as callus showing a high frequency of somatic embryogenesis.

Further work is required to isolate a greater range of *Trifolium* materials capable of regeneration from cell cultures under the simplest possible conditions. A survey by Webb et al. (1987) showed that when seedling explants of 72 *Trifolium* species were cultured on a basal salts medium (MS or B5) supplemented with 3% sucrose, 0.5 mg/l NAA and 0.5 mg/l BAP, 32 species formed callus alone, but 40 formed shoots and/or roots. In vitro responses have not only been shown to be genotype-dependent, but properties such as callus growth rate, shoot growth, root growth, and somatic embryogenesis are all highly heritable and should respond to breeding (Keyes et al. 1980).

5.4 Somaclonal Variation

The many recent studies on somaclonal variation among regenerants from tissue culture have emphasized that disorganized callus and cell cultures are subject to a variety of heritable alterations, including chromosomal aberrations and changes in ploidy level. Such variants may be desirable or undesirable depending on the context of the experiment. Novel variation may be useful in breeding research, but for genotype conservation or physiological studies it must be avoided. Cultures may also "drift" with time in a less defined way from one phenotype or physiological state to another, possibly by inadvertant preferential selection of a certain cell type during successive subcultures. This possibility must be taken into account when performing series of experiments with long-term cultures. A common expression of such a drift is gradual loss of regenerative capacity with time (e.g., Oelck and Schieder 1983). Changes in cell cultures can be slowed or halted by storage under minimal growth or cryogenic conditions. Although such procedures have been investigated for other legumes, e.g., soybeans (Weber et al. 1983), they have so far received little attention for *Trifolium* species.

A number of morphological and ploidy variants have been recorded among regenerants from tissue cultures of *T. repens* (Pelletier and Pelletier 1971; White 1984), *T. pratense* (Phillips and Collins 1980), *T. incarnatum* (Horvath, Beach and Smith 1979), *T. hybridum* (Webb et al. 1984), and *T. rubens* (Grosser and Collins 1984). In addition, occasional variants were reported by Tan and Stern (1985) among regenerants from basal callus in partially dedifferentiated shoot cultures. Somaclonal variation is a random and opportunistic phenomenon with no control over the type of variant produced, other than that which can be applied after the fact as selective discrimination among variant cells. Tissue culture-induced variants in *Trifolium*, other than those selected for in vitro responses and regenerative potential, have so far been of little use. For example, Bagley and Taylor (1987) compared the response to low soil phosphorus levels of *T. pratense* somaclones regenerated from cell suspensions, cultured either at low (1.8 μM) or normal (1.8 mM) phosphorus levels. Although variation was observed among the somaclones, the two groups did not differ overall in response to soil phosphorus. In addition, many were morphologically abnormal and most showed poor flowering and reduced fertility. Possible potential is indicated, however, by studies such as that of Latunde-Dada and Lucas (1983), who found that higher ploidy regenerants from protoplasts of alfalfa were resistant to *Verticillium* wilt. Similarly, Smith (1985) found that tetraploid plants of red clover had greater resistance to northern anthracnose compared to the diploid parents.

5.5 Protoplast Culture, Uptake Studies and Somatic Hybridization

Protoplast cultures have been initiated for *T. arvense*, *T. hybridum*, *T. pratense*, *T. repens*, *T. resupinatum*, and *T. rubens*, with regeneration to whole plants being achieved for *T. hybridum*, *T. repens*, and *T. rubens* (Table 4; Fig. 9). As in other plants, *Trifolium* protoplast cultures are initiated by incubation of intact cells or tissues for several hours in an osmotically adjusted and buffered medium con-

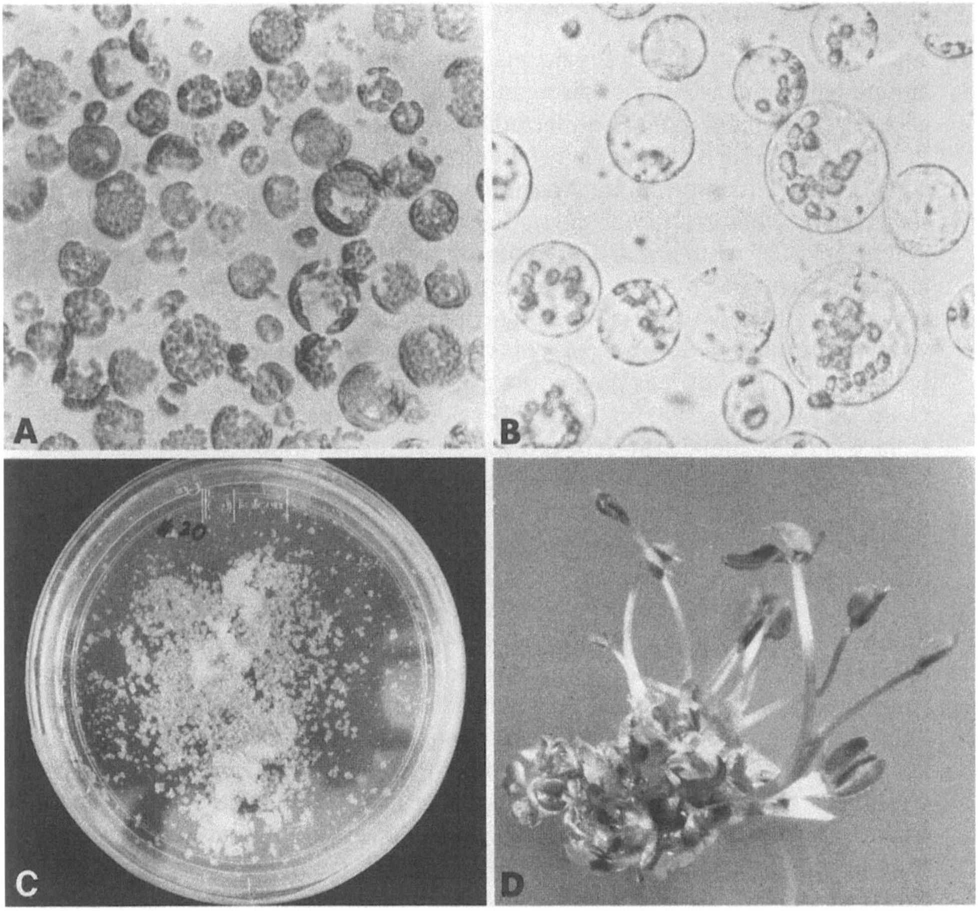

Fig. 9 A–D. Protoplast isolation, culture and regeneration in *Trifolium* (After Grosser and Collins 1984; unpubl.). **A** Mesophyll protoplasts of *T. rubens*. **B** Suspension culture-derived protoplasts of *T. pratense*. **C** Microcalli of *T. rubens* freshly plated onto solid medium. **D** Plant regeneration from mesophyll protoplast of *T. rubens*

taining hydrolytic enzymes (cellulases and hemicellulases) to remove the cell walls. Liberation of protoplasts is enhanced by slight plasmolysis of the cells before enzyme treatment. The isolation medium contains mineral salts, mannitol and/or sorbitol as an osmoticum, and is buffered to pH 5.5–6.0. Incubation is generally in the dark at 27°–28 °C with gentle shaking or swirling. After isolation, protoplasts are separated from cellular debris by centrifuging onto a pad of 20%–30% sucrose, washed in nutrient medium containing an osmoticum, and cultured in droplets or thin layers of the same medium at 27°–28 °C to reform walls and initiate divisions. The medium contains cytokinin (e.g., Z, BAP, kin or 2iP) and auxin (e.g., 2,4-D or NAA) at concentrations similar to those used for callus induction. Culture in the dark may be necessary to prevent browning of leaf mesophyll protoplasts (Bhojwani and White 1982). Inclusion of a filter paper as a support in the nutrient medium was found beneficial to division of *T. resupinatum* protoplasts by Oelck et al. (1982). Once small colonies of dividing cells have formed, usually after several days, the medium can be replaced with soft agar or agarose (e.g., 0.2%), and finally, once small calli are visible, these can be transferred to a standard nutrient agar medium without osmoticum and treated similarly to other callus cultures. Suitable detailed protocols for isolation and culture of *Trifolium* protoplasts have been described by Bhojwani and White (1982) and Grosser and Collins (1984). As pointed out in a previous section, the genotype, source tissue, and preculture conditions before protoplast isolation are critical to the success of cultures. Whole plants are preferably grown axenically (Grosser and Collins 1984) or under controlled conditions (Bhojwani and White 1982; Webb et al. 1983). Where existing in vitro cultures are used, their genotype, age, and physiological state can be much more important than details of media and growth regulators (Gresshoff 1980; Oelck et al. 1982).

To be useful for genetic manipulation and somatic hybridization studies, protoplast systems must show, firstly, a high frequency of colony and callus formation at a sufficiently low cell density to permit physical isolation of individual cells, and secondly, efficient regeneration to whole plants. Although Gresshoff (1980) obtained regenerants from *T. repens* protoplasts, the frequency of colony formation (plating efficiency) was only 5%–15% at a cell density of 2×10^3 – 4×10^4 per ml. Similarly, Oelck et al. (1982) found that a cell density greater than 10^5 per ml was required for division of *T. resupinatum* protoplasts. White and Bhojwani (1981) achieved a higher plating efficiency of 40%–60% for *T. arvense* protoplasts at a low cell density of 15 per ml, but without plant regeneration. Bhojwani and White (1982) increased the plating efficiency for *T. repens* to a reliable 37%. Grosser and Collins (1984), using *T. rubens*, reported plating efficiencies of 88% and 56% at cell densities of 5×10^4 – 5×10^5 per ml for suspension- and leaf-derived protoplasts, respectively, with regeneration to whole plants via somatic embryogenesis for both protoplast types. Further work is obviously required to standardize techniques for reliable high plating efficiency at low cell density, followed by efficient whole-plant regeneration. Plating efficiency, like the capacity for regeneration, is genotype dependent and presumably responsive to selection (Ahuja et al. 1983).

Protoplast cultures of *Trifolium* species have so far been little used other than for determination of the culture conditions required for cell division and plant regeneration. Viral infection has been studied using protoplasts of other plants such as tobacco and cowpea (Takebe and Otsuki 1969; Aoki and Takebe 1969; Rao and Hiruki 1978), and the effect of stage of the cell cycle on uptake of plasmid DNA has been examined using soybean protoplasts (Cress 1982). Graves and Beck (1984a) used protoplasts of *T. repens* to test a lytic bacteriophage for removal of extracellular *Rhizobium* bacteria after bacterial uptake by protoplasts. The phage was found to remove excess bacteria without adversely affecting protoplast viability. Internalized bacteria which had been previously labelled with fluorescein or rhodamine could be detected inside the protoplasts using fluorescence microscopy (Graves and Beck 1984b). Although not a culture procedure in the strict sense, isolated root nodule protoplasts of *T. repens* have been used to demonstrate viability and totipotency of bacteroids, and to show that, in most instances, only one *Rhizobium* genotype colonizes a single host cell (Gresshoff et al. 1977; Rolfe and Gresshoff 1980; Gresshoff 1982).

Somatic hybridization has not yet been achieved in *Trifolium*, although its use has been proposed for generation of extraspecific variation (Razdan and Cocking 1981). For example, fusion of *T. repens* protoplasts with protoplasts of *T. arvense*, a species with foliar tannins (Jones et al. 1973), may allow the development of a nonbloating clover (Jones et al. 1981a; White and Bhojwani 1981; Ahuja et al. 1983). Somatic hybridization of *T. pratense* (section *Trifolium*; x = 7) with *T. hybridum* (section Lotoidea; x = 8) has been attempted by Honkanen et al. (1986) and Honkanen and Ryoppy (1988). Green mesophyll protoplasts of one species were fused with colorless suspension protoplasts of the other, using polyethylene glycol (PEG). Heterokaryons were selected manually using a capillary micropipette. By using mutant cell lines of *T. pratense* or *Petunia* as nurse cultures, some microcalli were obtained from hybrid cells, but regenerants were not reported. A similar procedure has been used to obtain heterokaryons of the more closely related species *T. pratense* and *T. rubens* (both in section *Trifolium*), but although the hybrid cells are apparently capable of forming microcalli in nurse culture, no hybrid regenerants have been obtained to date (J.R. Myers, unpubl.). Among other forage legumes, somatic hybridization has been achieved in *Medicago* and *Lotus* using both chemical and electrofusion techniques, selective and manual isolation of heterokaryons, and albino cell lines as nurse cultures (Teoule 1983; Gilmour et al. 1986; Wright et al. 1986).

Apart from wide hybridization, somatic hybridization may be useful for maximizing heterozygosity in autotetraploid varieties. In particular, tetraploid cultivars of red clover are of increasing importance in Europe. Tetraploids have the potential for four different alleles at a single locus, compared to two for diploids. By optimizing allelic combinations, heterosis may be raised to its highest possible levels. Somatic hybridization of heterozygous diploids will incorporate all genetic variation from the parents, whereas in the two sexual methods for maximizing tetraploid heterozygosity, i.e., double crosses between tetraploids and fusion of 2n gametes from diploid parents, some heterozygosity will be lost owing to crossing-over (Bingham 1979).

6 Anther Culture and Haploid Production

The legumes have been generally recalcitrant to androgenesis in anther culture, the most common response being production of diploid callus from somatic tissues, for example in alfalfa (Saunders and Bingham 1972), soybean (Ivers et al. 1974), and *T. pratense* (Niizeki and Kita 1973; Niizeki 1977; Phillips and Collins 1979a; Bhojwani et al. 1984). Some haploid cells were reported in anther cultures of *Phaseolus*, together with diploid cells and cells of higher ploidies (Peters et al. 1977). In *Arachis* the production of rare pollen embryos and haploid callus was found to be genotype-dependent, and regenerants were mixaploid from haploid to octaploid (Bajaj et al. 1981). A single claim exists for the production of haploids from anther cultures of *Trifolium*. Mokhtarzadeh and Constantin (1978) reported haploid (n = 8) plantlets derived from anther callus of *T. alexandrinum*. However, since the single parent plant was a "dwarf" M_1 mutant grown from an irradiated seed, we might suspect that this plant may have been haploid to begin with. Androgenesis remains to be convincingly demonstrated in *Trifolium*.

Gynogenesis, the production of haploids from cells of the female gametophyte in cultured, unfertilized ovules (Yang and Zhou 1982), has not to our knowledge been attempted in *Trifolium*.

7 Transformation

Whereas transformation utilizing *Agrobacterium tumefaciens* is now routine for many solanaceous and some cruciferous species, the introduction of engineered genes into host plant cells has been achieved in relatively few legume species (see Chap. I.2, this Vol.).

Webb (1986) induced crown gall formation on stems, petioles, or stolons of six species of forage legumes including *T. repens*, *T. hybridum*, and *T. pratense*. The *Trifolium* species developed small galls which occasionally formed roots. These galls could be cultured axenically on hormone-free MS medium.

White and Greenwood (1987) were able to transform *T. repens* with disarmed strains of *Agrobacterium*. Stolon segments were inoculated with four binary *Agrobacterium*-plasmid combinations and incubated in a moist Petri dish for 1 to 3 days. One-mm slices were cut off the ends of the stolons and placed on callus induction medium with 100 µg/ml kanamycin and 500 µg/ml cefotaxime. Twenty-four to 44% of the explants produced kanamycin-resistant calli, from which kanamycin-resistant shoots were regenerated. Transformation was confirmed by a stable kanamycin-resistant phenotype, neomycin phosphotransferase II (NPT-II) activity, nopaline synthase activity, and integration of vector DNA sequences into plant DNA. Combinations of the *Agrobacterium* strain LBA4404 with the vectors pBin6 and pBin19 appeared to give greater frequencies of transformation than combinations of GV3850 with pGA470 and pGA472.

Among other forage legumes, Webb (1986) induced galls on *Medicago sativa*, *Onobrychis viciifolia*, and *Lotus corniculatus*. *M. sativa* formed small disorganized tumors, but teratomatous galls were formed on *O. viciifolia* and *L. corniculatus*. *L. corniculatus* has also been transformed with *A. rhizogenes*, and regenerated plants were used to study nodule-specific expression of a soybean leghemoglobin gene fused to a gene for chloramphenicol acetyltransferase (Jensen et al. 1986). *Medicago varia* and *M. sativa* were transformed with engineered *A. tumefaciens* strains, and regenerated plant tissues were used to demonstrate incorporation of the plasmid into host DNA with subsequent transcription and translation (Deak et al. 1986; Shahin et al. 1986).

As an alternative to *Agrobacterium*-mediated transformation, DNA has been incorporated into plant protoplasts by microinjection (Crossway et al. 1986) and direct uptake facilitated by electroporation (Fromm et al. 1986) or by chemical agents (Paszkowski et al. 1984; Hain et al. 1985). Acceleration of DNA-coated microprojectiles has also been used to carry DNA into intact cells and tissues

(Klein et al. 1987; McCabe et al. 1988; Christou et al. 1988). To our knowledge, these techniques have not so far been used to transform *Trifolium*, although work of this type is currently underway in a number of laboratories.

Introduction of agronomically useful genes into *Trifolium* species via *Agrobacterium* appears feasible, although additional efforts will be required to establish efficient inoculation and selection routines. In those *Trifolium* species where plant regeneration from protoplasts has been achieved, direct uptake and microinjection could be explored as alternatives to *Agrobacterium* transformation. The microprojectile technique has the advantage that it can be applied to intact meristems from which regeneration occurs readily for most genotypes. It therefore offers a means for direct gene transfer into elite cultivars without the need to backcross from a regenerating genotype.

8 Conclusion

Although the cycling of plant materials through animals can be regarded as an inefficient means of converting energy to human food, this is only true for high quality land which can be efficiently cropped. Harvesting by means of grazing animals will, in the forseeable future, remain the most effective means of utilizing steep, inaccessible, and marginally productive lands. This means a secure future for the clovers and other pasture legumes, even if the world's arable lands are some day given up almost entirely to direct human food production. In inaccessible and marginal areas forage legumes will also be particularly important for their ability to put nitrogen into the soil without the need for applied fertilizers.

A major objective in the future of farming is the reduction of costs. Plant materials must therefore be tailored to declining agricultural inputs. Clovers and other forage legumes will need to be regionally adapted for growth in a variety of high-stress, marginal environments, and will also need to be adapted for pest and disease resistance, rapid establishment, persistency, nutrient efficiency and low animal health care, all of which will help to reduce farming costs.

As a set of research and breeding tools, tissue culture techniques have the potential to contribute much to the improvement of clovers, particularly in the propagation and conservation of superior genotypes, and the generation and selection of new, adaptive variations.

Protocol

Initiation of Callus and Suspension Cultures

Protocol used for *T. pratense* and *T. rubens*, modified from Phillips and Collins (1979a, 1980).

Explant Sources

Petiole segments from greenhouse- or field-grown plants.
Seedlings germinated aseptically in vitro.

Note: It is more difficult to obtain clean callus cultures from field- or greenhouse-grown materials. Use aseptically grown seedlings if possible.

Media

Prepared beforehand and poured in 45-ml aliquots into 100×20 mm sterile disposable Petri dishes.

SGL for seed germination (Table 2).

Prepared beforehand and poured in 10–15-ml aliquots into 125-ml Erlenmeyer flasks.

SL 2 for cell suspension (Table 2).

Surface Sterilization of Petiole Segments

- Wash petioles gently with dishwashing detergent, and rinse thoroughly with tap water.
- Place petioles in sterile, wide-mouthed, capped jar and rinse with 95% ethanol or isopropanol for 30 s.
- Drain into waste beaker, and replace with 1.3% sodium hypochlorite solution (25% (v/v) Clorox household bleach) for 5 min.
- Drain hypochlorite solution and rinse petioles at least twice with sterile water. Each rinse should be for at least 1 min.
- Place petioles in sterile Petri dish for temporary storage during dissections.

Note: Sterilization fluid times and concentrations may be adjusted depending on petiole response.

Dry Sterilization and Aseptic Germination of Seeds

- If scarification is required, scarify seeds before sterilization.
- Spread seeds loosely in 60×20-mm plastic Petri dishes.
- Place dishes in a desiccator inside a fume hood, and tilt the lids of the dishes over on one side for gas exchange.
- Place a 150-ml beaker containing 100 ml of 5.25% sodium hypochlorite solution (Clorox household bleach) among the seed dishes in the desiccator.
- Slowly pipette 3.3 ml conc. HCl down the side of the beaker to form a layer over the bleach.
- Seal desiccator and post warning of the presence of chlorine gas.
- After 24 h carefully open desiccator and quickly replace lids on dishes. Remove dishes to laminar flow hood, re-open and allow to ventilate for several minutes.
- Dishes may be sealed and stored, or seeds may be germinated immediately.
- For germination, place up to ten seeds per dish on SGL medium.

Initiation of Callus Cultures

- Cut sterile petioles or seedling tissues (hypocotyl, or cotyledon tips) into 0.5-cm segments and place on L 2 medium.
- Culture in light at 25°C.
- Subculture monthly onto fresh L 2 medium, using actively growing light-colored callus as inoculum.

Initiation of Suspension Cultures

- Place small segments of the above callus in liquid SL 2 medium and incubate in light at 25 °C on a gyratory shaker at 100 rpm.
- Subculture ca. 20 ml of suspension into ca. 30 ml fresh medium at least once each week.

Regeneration from Callus Culture

Media

Prepared beforehand and poured in 45-ml aliquots into 100×20-mm Petri dishes.

LSE for shoot initiation (Table 2).
LSP for shoot proliferation (Table 2).
CR2 for rooting (Table 2).

Plant Regeneration

- Transfer pieces of actively growing and/or green, nodular callus to LSE medium.
- Subculture monthly on LSE until small shoots appear. If regeneration is by somatic embryogenesis, allow embryos to germinate.
- Subculture actively shooting tissue segments to LSP for maturation and proliferation of shoots. Do not crowd in the dishes or growth will be slow.
- Transfer established shoots to CR2 for rooting. Several subculture cycles may be necessary before roots appear.

Note: Shoot cultures and rooting shoots grow best when dead or brown tissue is discarded before subculturing. Select only shoots with a healthy meristem for subculturing. Regeneration is genotype dependent, even within *T. pratense* and *T. rubens*. The manner by which regeneration occurs (i.e., embryogenesis or apparent organogenesis), and the ability to regenerate shoots directly on L 2 without transfer to LSE, are also genotype dependent.

Isolation and Culture of Protoplasts

Protocol as used for *T. pratense* and *T. rubens*, modified from Grosser and Collins (1984).

Enzyme Solution

- 1% cellulase Onozuka RS (Yakult)
- 0.20% Pectolyase Y-23 (Seishim)
- 0.25% Driselase (Plenum Sci.)
- 0.25% Rhozyme HP150 (Corning)
- 0.35 M Sorbitol
- 0.35 M Mannitol
- 6 mM $CaCl_2.2H_2O$
- 3 mM MES
- 0.7 mM NaH_2PO_4
- pH 5.5
  Solution can be prepared in bulk and frozen in aliquots for up to 3 months (filter-sterilize).

Culture Solutions

- 8P (Kao and Michayluk 1975)
- C8P (Grosser and Collins 1984)
- SL2 (Table 2)
- L2 (Table 2).

Equipment

- Low speed centrifuge (0–500 g).
- Shaker (should have reliable settings at 60 rpm).
- Sterile 100×25-mm glass Petri dishes.
- 125-ml side-arm Erlenmeyer flask with rubber stopper. The vent to the side-arm is covered with 3–4 layers of Miracloth (Cal Biochem) held in place by autoclave tape. The stopper and side-arm are covered with aluminum foil and the entire assembly autoclaved. One flask is needed for each lot of leaf-derived protoplasts.
- 50-ml conical-bottom glass centrifuge tubes (capped with foil and autoclaved). One is needed for each lot of protoplasts to be isolated.
- Filters (one per 50-ml centrifuge tube). These are prepared by cutting a 50-c syringe barrel in half. The cut portion of the barrel is melted to 45-µm stainless steel mesh. The flanges on the syringe barrel support the filter when it is inserted into the 50-ml centrifuge tube. The filters are packed loosely with glass wool, wrapped in foil, and autoclaved. The glass wool catches the debris and undigested cells and prevents the filter from plugging.
- 15-ml conical-bottom centrifuge tubes with screw caps (one for each lot of protoplasts). The caps should be left loosely secured to the tube with foil when autoclaving.
- 60×20-mm sterile disposable plastic Petri dishes.
- Pasteur pipettes (18 cm length). The bulb end should be plugged with cotton and canisters of pipettes sterilized with dry heat (autoclaving causes water to collect in the narrow tip).
- 3- and 1-ml bulbs to attach to Pasteur pipettes.
- 5- to 10-ml sterile glass or plastic pipettes for transfer of solutions. Suspension cultures are transferred using a 5-ml plastic pipette with the tip removed. The tip is broken off while the pipette is still in its plastic wrapper.
- Test tube rack to support the 50- and 15-ml centrifuge tubes.

Isolation of Leaf-Derived Protoplasts

Start and maintain shoot cultures as described previously.

- With a sharp scalpel blade, cut trifoliate leaves from shoots and transfer to sterile 100×25-mm glass Petri dish containing 3 ml 8P medium. Shoot cultures should have been subcultured at most 2 weeks before use (on a 4-week subculture schedule).
- Using a sharp, sterile scalpel blade and replacing as it becomes dull, feather the leaflets by making cuts 1 mm apart perpendicular to the secondary veins of each leaflet.
- Transfer feathered leaflets to a 125-ml side-arm Erlenmeyer flask containing 2.5–3.5 ml enzyme solution and 6.5–7.5 ml 8P mixture (total volume of 10 ml), and vacuum infiltrate for 15 min.
- Enough leaflets should be prepared to completely cover the surface of the liquid for optimum isolation.
- Incubate for 6 h at 28 °C in the dark on a shaker rotating at 60 rpm.

Isolation of Suspension-Derived Protoplasts

Start and maintain suspension cultures as described previously.

- Pipette 2.5 ml liquid cell suspension into a sterile 60×20-mm plastic Petri dish.
- Add. 2.5–3.0 ml of enzyme solution and gently agitate to mix. (Amount of suspension culture and enzyme solution can be varied empirically to optimize protoplast release).
- Incubate for 3 h at 28 °C in the dark on a shaker rotating at 60 rpm.

Procedures for Both Protoplast Types

- Gently pour the crude protoplast preparation through the 45-µm mesh filter into a 50-ml centrifuge tube.

- Decant the filtered solution containing protoplasts into a 15-ml centrifuge tube.
- Centrifuge the filtered protoplast-enzyme mixture at 100 g for 8 min.
- Replace the supernatant with 5–10 ml 8P medium and resuspend the pellet by gentle bubbling with a Pasteur pipette.
- The suspension is again centrifuged at 100 g for 4 min, followed by removal and replacement of the supernatant.
- The above step is repeated an additional two to three times.
- After the final wash, protoplasts can be counted with a hemacytometer and resuspended in 8P medium at a density of 5×10^4 to 10^5 protoplasts per ml.
- Protoplasts are then plated in 1.5 ml 8P medium in 60×20-mm sterile plastic Petri dishes.
- Spread the suspension into a thin layer covering the bottom of the plate by gently shaking and tipping the Petri dish.
- For determination of division frequencies, protoplasts can be plated in 1.5 ml 8P medium partially solidified with 0.2% low melting point agarose (Bethesda Research Lab., Inc. or International Biotechnologies Inc.).
- Place in a sealed opaque box and incubate under low light levels at 25° ± 3 °C. (The less handling the protoplasts receive at this point, the higher the division frequencies).

Feeding and Recovery of Callus

- One week after isolation, cultures are fed with 5 to 10 drops (depending on colony vigor) of C8P+ medium (2:1 C8P:SL2).
- A second feeding of 10 drops of SL2+ medium (1:3 C8P:SL2) is made during the second week.
- Weekly feedings thereafter are made with the addition of 10 to 15 drops of SL2 medium.
- When colonies are 0.5 to 1.0 mm in diameter, transfer the cultures to L2 solid medium for callus growth.

Follow protocol for regeneration from callus culture.

Embryo rescue

For protocol see Williams (1987).

Abbreviations

Basal media: B5, Gamborg et al. 1968; Blaydes, Blaydes 1966; L2, Phillips and Collins 1979a; MS, Murashige and Skoog 1962; SH, Schenk and Hildebrandt 1972.

Growth regulators: ADE, 6-aminopurine (adenine); BAP (BA) 6-benzylaminopurine (6-benzyladenine); COU, coumarin; GA, gibberellic acid; IAA, indole-3-acetic acid; IBA, indole-3-butyric acid; kin, 6-furfurylaminopurine (kinetin); NAA, naphthalene acetic acid; pCPA, p-chlorophenoxyacetic acid; PIC, 4-amino-3,5,6-trichloropicolinic acid (picloram); TIBA, 2,3,5-triiodobenzoic acid; Z, zeatin; 2,4-D, 2,4-dichlorophenoxyacetic acid; 2,4-DB, 2,4-dichlorophenoxybutyric acid; 2,4,5-T, 2,4,5-trichlorophenoxyacetic acid; 2,4,5-TB, 2,4,5-trichlorophenoxybutyric acid; 2iP, 2-isopentenyladenine.

Complex Organic Additives: CH, casein hydrolysate; YE, yeast extract.

Acknowledgments. The authors gratefully acknowledge the helpful criticisms of Dr. Norman Taylor, Dr. Wayne Parrott, and Mrs. Elda Cunningham.

References

Ahuja PS, Lu DY, Cocking EC, Davey MR (1983) An assessment of the cultural capabilities of *Trifolium repens* L. (White clover) and *Onobrychis viciifolia* Scop. (Sainfoin) mesophyll protoplasts. Plant Cell Rep 2:269–272

Aoki S, Takebe I (1969) Infection of tobacco mesophyll protoplasts by tobacco mosaic virus ribonucleic acid. Virology 39:439–448

Bagley PC, Taylor NL (1987) Evaluation of phosphorus efficiency in somaclones of red clover. Iowa State J Res 61:459–480

Bajaj YPS, Ram AK, Labana KS, Singh H (1981) Regeneration of genetically variable plants from the anther-derived callus of *Arachis hypogaea* and *Arachis villosa*. Plant Sci Lett 23:35–39

Barnett OW, Diachun S (1985) Virus diseases of clovers. In: Taylor NL (ed) Clover science and technology. Agron Monogr No 25. Am Soc Agron, Crop Sci Soc Am, Soil Sci Soc Am, Madison, WI, pp 235–268

Barnett OW, Gibson PB, Seo A (1975) A comparison of heat treatment, cold treatment, and meristem tip-culture for obtaining virus-free plants of *Trifolium repens*. Plant Dis Rep 59:834–837

Bassam BJ, Gresshoff PM (1986) Use of neomycin for preferential selection against *Rhizobium trifolii* in symbiosis with white clover (*Trifolium repens*). Aust J Biol Sci 39:23–30

Bhojwani SS (1981) A tissue culture method for propagation and low temperature storage of *Trifolium repens* genotypes. Physiol Plant 52:187–190

Bhojwani SS, White DWR (1982) Mesophyll protoplasts of white clover: isolation, culture and organogenesis. Plant Sci Lett 26:265–271

Bhojwani SS, Mullins K, Cohen D (1984) Intra-varietal variation for in vitro plant regeneration in the genus *Trifolium*. Euphytica 33:915–921

Bingham ET (1979) Maximizing heterozygosity in autotetraploids. In: Lewis WH (ed) Polyploidy: biological relevance. Plenum Press, NY, pp 471–489

Blaydes DF (1966) Interaction of kinetin and various inhibitors in the growth of soybean tissue. Physiol Plant 19:748–753

Brewbaker JL, Keim WF (1973) A fertile interspecific hybrid in *Trifolium* (4n *T. repens* L.×4n *T. nigrescens* Viv.). Am Nat 87:323–326

Campbell CT, Tomes DT (1982) In vitro propagation and storage of red clover. Can J Plant Sci (Abstr) 62:253

Campbell CT, Tomes DT (1984) Establishment and multiplication of red clover plants by in vitro shoot tip culture. Plant Cell Tissue Organ Cult 3:49–57

Carroll BJ, Gresshoff PM (1983) Nitrate inhibition of nodulation and nitrogen fixation in white clover. Z Pflanzenphysiol 110:77–88

Cheyne VA, Dale PJ (1980) Shoot tip culture in forage legumes. Plant Sci Lett 19:303–309

Child JJ (1975) Nitrogen fixation by a *Rhizobium* sp. in association with non-leguminous plant cell cultures. Nature (Lond) 253:350–351

Child JJ, Kurz WGW (1978) Inducing effect of plant cells on nitrogenase activity by *Spirillum* and *Rhizobium* in vitro. Can J Microbiol 24:143–148

Christou P, McCabe DE, Swain WF (1988) Stable transformation of soybean callus by DNA-coated gold particles. Plant Physiol 87:671–674

Churova K (1984) Regeneration of plants in culture of red clover apical meristems. In: Proc Int Symp plant tissue and cell cult: Application to Crop Improvement, Olomouc, Czechoslovakia, pp 501–503

Collinge DB, Hughes MA (1982) In vitro characterization of the *Ac* locus in white clover (*Trifolium repens* L.). Arch Biochem Biophys 218:38–45

Collins GB, Grosser JW (1984) Culture of embryos. In: Vasil IK (ed) Cell culture and somatic cell genetics of plants, vol. 1. Academic Press, NY, pp 241–257

Collins GB, Phillips GC (1982) In vitro tissue culture and plant regeneration in *Trifolium pratense* L. In: Earle ED, Demarly Y (eds) Variability in plants regenerated from tissue culture. Praeger, NY, pp 22–34

Collins GB, Taylor NL, Phillips GC (1981) Successful hybridization of red clover with perennial *Trifolium* species via embryo rescue. Proc 4th Int Grassland Congr, Lexington, KY, pp 168–170

Coop IE, Blakley RL (1950) The metabolism and toxicity of cyanides and cyanogenetic glucosides in sheep. III. The toxicity of cyanides and cyanogenetic glucosides. NZ J Sci and Technol 31(5):44–58
Cress DE (1982) Uptake of plasmid DNA by protoplasts from synchronized soybean cell suspension cultures. Z Pflanzenphysiol 105:467–470
Crossway A, Oakes JV, Irvine JM, Ward B, Knauf VC, Shewmaker CK (1986) Integration of foreign DNA following microinjection of tobacco mesophyll protoplasts. Mol Gen Genet 202:179–185
Croughnan TP, Stavarek SJ, Rains DW (1978) Selection of a NaCl tolerant line of cultured alfalfa cells. Crop Sci 18:959–963
Cui D, Myers JR, Collins GB, Lazzeri PA (1988) In vitro regeneration in *Trifolium*. I. Direct somatic embryogenesis in *T. rubens* (L.). Plant Cell Tissue Organ Cult 15:33–45
Dale PJ, Cheyne VA, Dalton SJ (1980) Pathogen elimination and in vitro plant storage in forage grasses and legumes. In: Ingram DS, Helgeson JP (eds) Tissue culture methods for plant pathologists. Blackwell, Oxford, pp 119–124
Dale PJ, Dalton SJ, Cheyne VA (1981) Shoot tip culture in forage legumes. Welsh Plant Breed Stat Annu Rep, Univ College of Wales, Aberystwyth, p 38
Dale PJ, Arumuganathan K, Dalton SJ, Fay M, Jackson JA, Jones MGK, Pike LS (1982) Legume protoplasts. Welsh Plant Breed Stat Annu Rep, Univ College of Wales, Aberystwyth, pp 41–42
Deak M, Kiss GB, Koncz C, Dudits D (1986) Transformation of *Medicago* by *Agrobacterium* mediated gene transfer. Plant Cell Rep 5:97–100
Ellis BE (1984) Probing secondary metabolism in plant cell cultures. Can J Bot 62:2912–2917
Enyingi KH, Heszky L (1973) Raising of the heart-stage embryo of *Trifolium pratense* L. in embryo culture: the differences of embryogenesis in situ and in vitro. Bot Kozlemenyek 60:213–219 (in Hungarian; English summary)
Essig HW (1985) Quality and antiquality components. In: Taylor NL (ed) Clover science and technology. Agron Monogr No 25, Am Soc Agron, Crop Sci Soc Am, Soil Sci Soc Am, Madison, WI, Chap 11, pp 309–324
Evans AM (1962) Species hybridization in *Trifolium*. I. Methods of overcoming species incompatibility. Euphytica 11:164–176
Evans AM (1976) Clovers. In: Simmonds NW (ed) Evolution of Crop Plants. Longman, London, pp 175–179
Evans DA, Sharp WR, Flick CE (1981) Growth and behavior of cell cultures. Embryogenesis and organogenesis. In: Thorpe TA (ed) Plant tissue culture: methods and applications in agriculture. Proc UNESCO Symp, Sáo Paulo, Brazil, 1978. Academic Press, NY, pp 45–113
Evans DR, Williams TA (1984) Legume breeding: Disease resistance. Welsh Plant Breed Stat Annu Rep, Univ College of Wales, Aberystwyth, p 41
Evans PT (1984) Interspecific hybridization in *Trifolium*. Ph D Thesis, Clemson University, Clemson, South Carolina, USA. Plant Breed Abstr 54(4–5):3141
Firsova EK, Mezentsev AV, Rubstov MI (1980) Interspecific hybridization of clover using embryo culture. Selektsiya i Semenovodstvo, USSR 1:16–18 (in Russian)
Flux DS, Butler GW, Rae AL, Brougham RW (1960) Relationship between levels of iodine and cyanogenetic glucoside in pasture and the productive performance of sheep. J Agric Sci 55:191–196
Fromm ME, Taylor LP, Walbot V (1986) Stable transformation of maize after gene transfer by electroporation. Nature (Lond) 319:791–793
Gamborg OL, Miller RA, Ojima K (1968) Nutrient requirements of suspension cultures of soybean root cells. Exp Cell Res 50:151–158
Gilmour DM, Davey MR, Cocking EC (1986) Somatic hybridization in *Medicago*. In: 6th Int Congr plant tissue and cell cult, Abstr, Univ Minnesota, Minneapolis, p 386
Graham PH (1968) Growth of *Medicago sativa* L. and *Trifolium subterraneum* L. in callus and suspension culture. Phyton 25:159–162
Graves DA, Beck RW (1984a) The use of a lytic bacteriophage to remove *Rhizobium trifolii* from protoplast cultures of *Trifolium repens*. Plant Sci Lett 34:385–389
Graves DA, Beck RW (1984b) Detection of internalized *Rhizobium trifolii* in plant cell protoplasts using fluorescence microscopy. Plant Sci Lett 34:391–395
Gresshoff PM (1980) In vitro culture of white clover: callus, suspension, protoplast culture, and plant regeneration. Bot Gaz 141:157–164
Gresshoff PM (1982) Legume cell culture to study plant-bacterial interaction. In: Fujiwara A (ed) Plant tissue culture 1982. Maruzen, Tokyo, pp 661–662

Gresshoff PM, Skotnicki ML, Eadie JF, Rolfe BG (1977) Viability of *Rhizobium trifolii* bacteroids from clover root nodules. Plant Sci Lett 10:299–304

Gresshoff PM, Carroll B, Mohapatra SS, Reporter M, Shine J, Rolfe BG (1981) Host factor control of nitrogenase function. In: Gibson AH, Newton WE (eds) Current perspectives in nitrogen fixation. Austr Acad Sci, Canberra, pp 209–212

Grosser JW, Collins GB (1984) Isolation and culture of *Trifolium rubens* protoplast with whole plant regeneration. Plant Sci Lett 37:165–170

Gustine DL (1981) Evidence for sulfhydryl involvement in regulation of phytoalexin accumulation in *Trifolium repens* callus tissue cultures. Plant Physiol 68:1323–1326

Gustine DL, Moyer BG (1982) Retention of phytoalexin regulation in legume callus cultures. Plant Cell Tissue Organ Cult 1:255–263

Hain R, Stabel P, Czernilofsky AP, Steinbiss HH, Herrera-Estrella L, Schell J (1985) Uptake, integration, expression and genetic transmission of a selectable chimeric gene by plant protoplasts. Mol Gen Genet 199:161–168

Hartman CL, McCoy TJ, Knous TR (1984) Selection of alfalfa (*Medicago sativa*) cell lines and regeneration of plants resistant to the toxin(s) produced by *Fusarium oxysporum* f. sp. *medicaginis*. Plant Sci Lett 34:183–194

Holsten RD, Burns RC, Hardy RWF, Herbert RR (1971) Establishment of symbiosis between *Rhizobium* and plant cells in vitro. Nature (Lond) 232:173–176

Honkanen J, Ryoppy P (1988) Somatic hybridization in *Trifolium*. In: Bajaj YPS (ed) Biotechnology in agriculture and forestry 8. Plant protoplasts and genetic engineering I. Springer, Berlin Heidelberg, New York, Tokyo, pp 397–409

Honkanen J, Ryoppy P, Tigerstedt PMA (1986) Protoplast culture and fusion of red and alsike clover. In: Horn W, Jensen CJ, Odenbach W, Schieder O (eds) Genetic manipulation in plant breeding. Proc EUCARPIA Symp, 1985, Berlin. de Gruyter, Berlin, pp 685–687

Horak J, Fojtik A (1984) Callus formation and plant regeneration in tissue culture of genus *Trifolium*. In: Proc Int Symp Plant tissue and cell cult: Application to crop improvement, Olomouc, Czechoslovakia, pp 143–144

Horvath Beach K, Smith RR (1979) Plant regeneration from callus of red and crimson clover. Plant Sci Lett 16:231–237

Hovin AW (1962) Interspecific hybridization between *Trifolium repens* L. and *T. nigrescens* Viv. and analysis of hybrid meiosis. Crop Sci 2:251–254

Hughes MA (1968) Studies on β-glucosidase production in cultured tissue of *Trifolium repens* L. J Exp Bot 19:52–63

Isely D (1982) Leguminosae and *Homo sapiens*. Econ Bot 36:46–70

Ivers DR, Palmer RG, Fehr WR (1974) Anther culture in soybeans. Crop Sci 14:891–893

Jensen JS, Marcker KA, Otten L, Schell J (1986) Nodule-specific expression of a chimaeric soybean leghaemoglobin gene in transgenic *Lotus corniculatus*. Nature (Lond) 321:669–674

Jones MGK, Pike LS, Fay MF (1981a) Protoplast studies: forage legumes. Welsh Plant Breed Stat Annu Rep, Univ College of Wales, Aberystwyth, pp 39–40

Jones RA, Rupert EA, Barnett OW (1981b) Virus infection of *Trifolium* species in cell suspension cultures. Phytopathology 71:116–119

Jones WT, Anderson LB, Ross MD (1973) Bloat in cattle XXXIX. Detection of protein precipitants (flavolans) in legumes. NZ J Agric Res 16:441–446

Kao KN, Michayluk MR (1975) Nutritional requirements for growth of *Vicia hajastana* cells and protoplasts at a very low population density in liquid media. Planta 126:105–110

Keim WF (1953a) An embryo culture technique for forage legumes. Agron J 45:509–510

Keim WF (1953b) Interspecific hybridization in *Trifolium* utilizing embryo culture techniques. Agron J 45:601–606

Keyes GJ, Collins GB, Taylor NL (1980) Genetic variation in tissue cultures of red clover. Theor Appl Genet 58:265–271

Klein TM, Wolf ED, Wu R, Sanford JC (1987) High-velocity microprojectiles for delivering nucleic acids into living cells. Nature (Lond) 327:70–73

Latunde-Dada AO, Lucas JA (1983) Somaclonal variation and reaction to *Verticillium* wilt in *Medicago sativa* L. plants regenerated from protoplasts. Plant Sci Lett 32:205–211

Leath KT (1985) General diseases. In: Taylor NL (ed) Clover Science and Technology. Agron Monogr No 25. Am Soc Agron, Crop Sci Soc Am, Soil Sci Soc Am, Madison, WI, pp 205–233

Maheswaran G, Williams EG (1984) Direct somatic embryoid formation on immature embryos of *Trifolium repens*, *T. pratense* and *Medicago sativa*, and rapid clonal propagation of *T. repens*. Ann Bot 54:201–211

Maheswaran G, Williams EG (1985) Origin and development of somatic embryoids formed directly on immature embryos of *Trifolium repens* in vitro. Ann Bot 56:619–630

Maheswaran G, Williams EG (1986a) Direct secondary somatic embryogenesis from immature embryos of *Trifolium repens* cultured in vitro. Ann Bot 57:109–117

Maheswaran G, Williams EG (1986b) Clonal propagation of *Trifolium pratense*, *T. resupinatum* and *T. subterraneum* by direct somatic embryogenesis on cultured immature embryos. Plant Cell Rep 3:165–168

Maheswaran G, Williams EG (1986c) Uniformity of plants regenerated by direct somatic embryogenesis from zygotic embryos of *Trifolium repens*. Ann Bot 59:93–97

Manglitz GR (1985) Insects and related pests. In: Taylor NL (ed) Clover science and technology. Agron Monogr No 25. Am Soc Agron, Crop Sci Soc Am, Soil Sci Soc Am, Madison, WI, pp 269–294

McCabe DE, Swain WF, Martinell BJ, Christou P (1988) Stable transformation of soybean (*Glycine max*) by particle acceleration. Biotechnology 6:923–926

McCoy TJ (1985) Interspecific hybridization of *Medicago sativa* L. and *M. rupestris* M.B. using ovule-embryo culture. Can J Gen Cytol 27:238–245

McCoy TJ, Smith LY (1986) Interspecific hybridization of perennial *Medicago* species using ovule-embryo culture. Theor Appl Genet 71:772–783

Merker A (1984) Hybrids between *Trifolium medium* and *Trifolium pratense*. Hereditas 101:267–268

Mezentsev AV, Lyubavina LA, Karelina NA (1982) Cell cultures in the breeding of clover and alfalfa. Soviet Agric Sci 7:10–13

Mohapatra SS, Gresshoff PM (1982) Ecotypic variation of in vitro plantlet formation in white clover (*Trifolium repens*). Plant Cell Rep 1:189–192

Mokhtarzadeh A, Constantin MJ (1978) Plant regeneration from hypocotyl- and anther-derived callus of berseem clover. Crop Sci 18:567–572

Morgner F (1976) Using self-fertility in breeding of Persian clover. Rep Fodder Crops Sect, EUCARPIA Meet, Roskilde, pp 67–68

Mroginski LA, Kartha KK (1984) Tissue culture of legumes for crop improvement. Plant Breed Rev 2:215–264

Murashige T, Skoog F (1962) A revised medium for rapid growth and bioassays with tobacco tissue cultures. Physiol Plant 15:473–497

Niizeki M (1977) Haploid, polyploid and aneuploid plants from cultured anthers and calluses in species of *Nicotiana* and forage crops. J Faculty Agric, Hokkaido Univ, Japan 58:343–466

Niizeki M, Kita F (1973) Studies on plant cell and tissue culture III. In vitro induction of callus from anther culture of forage crops. J Faculty Agric, Hokkaido Univ, Japan 57:293–300

Oelck MM, Schieder O (1983) Genotypic differences in some legume species affecting the redifferentiation ability from callus to plants. Z Pflanzenzücht 91:312–321

Oelck MM, Bapat VA, Schieder O (1982) Protoplast culture of three legumes: *Arachis hypogaea*, *Melilotus officinalis*, *Trifolium resupinatum*. Z Pflanzenphysiol 106:173–177

Oswald TH, Smith AE, Phillips DV (1977a) Callus and plantlet regeneration from cell cultures of ladino clover and soybean. Physiol Plant 39:129–134

Oswald TH, Smith AE, Phillips DV (1977b) Herbicide tolerance developed in cell suspension cultures of perennial white clover. Can J Bot 55:1351–1358

Pandey KK, Grant JE, Williams EG (1987) Interspecific hybridization between *Trifolium repens* and *T. uniflorum*. Aust J Bot 35:171–182

Parrott WA, Collins GB (1983) Callus and shoot-tip culture of eight *Trifolium* species with regeneration via somatic embryogenesis of *T. rubens*. Plant Sci Lett 28:189–194

Paszkowski J, Shillito RD, Saul M, Mandak V, Hohn T, Hohn B, Potrykus I (1984) Direct gene transfer to plants. EMBO J 3:2717–2722

Pederson GA (1986) In vitro culture and somatic embryogenesis of four *Trifolium* species. Plant Sci 45:101–104

Pelletier G, Pelletier A (1971) Culture in vitro de tissus de trèfle blanc (*Trifolium repens*); variabilité des plantes regénérées. Ann Amélior Plant 21:221–233 (in French; English Summary)

Peters JE, Crocomo OJ, Sharp WR, Paddock EF, Tegenkamp I, Tegenkamp T (1977) Haploid callus cells from anthers of *Phaseolus vulgaris*. Phytomorphology 27:79–85

Phillips DA (1974) Promotion of acetylene reduction by *Rhizobium*-soybean cell associations in vitro. Plant Physiol 54:654–655
Phillips GC, Collins GB (1979a) In vitro tissue culture of selected legumes and plant regeneration from callus cultures of red clover. Crop Sci 19:59–64
Phillips GC, Collins GB (1979b) Virus symptom-free plants of red clover using meristem culture. Crop Sci 19:213–216
Phillips GC, Collins GB (1980) Somatic embryogenesis from cell suspension cultures of red clover. Crop Sci 20:323–326
Phillips GC, Collins GB (1984) Red clover and other forage legumes. In: Sharp WR, Evans DA, Ammirato PV, Yamada Y (eds) Handbook of plant cell culture, vol. 2. MacMillan, NY, pp 169–210
Phillips GC, Collins GB, Taylor NL (1982) Interspecific hybridization of red clover (*Trifolium pratense* L.) with *T. sarosiense* Hazsl. using in vitro embryo rescue. Theor Appl Genet 62:17–24
Pretova A, Williams EG (1986) Direct somatic embryogenesis from immature zygotic embryos of flax (*Linum usitatissimum* L.). J Plant Physiol 126:155–161
Rakousky S, Nasinec V, Nemcova B (1984) Protoplast cultures of *Trifolium pratense* L. Proc Int Symp Plant tissue cell cult: Application to crop improvement, Olomouc, Czechoslovakia, pp 403–404
Ranga Rao V (1976) Nitrogenase activity in *Rhizobium* associated with leguminous and non-leguminous tissue cultures. Plant Sci Lett 6:77–83
Rao DV, Hiruki C (1978) Infection of cowpea mesophyll protoplasts with clover yellow mosaic virus. J Gen Virol 38:303–311
Razdan MK, Cocking EC (1981) Improvement of legumes by exploring extra-specific genetic variation. Euphytica 30:819–833
Reporter M, Skotnicki ML, Rolfe BG (1980) Role of plant cell conditioned medium in the phenotypic expression of nitrogenase activity of *Rhizobium trifolii* strain T1. Aust J Biol Sci 33:613–621
Richards KW, Rupert EA (1980) In vitro fertilization and seed development in *Trifolium*. In Vitro 16:925–931
Rolfe BG, Gresshoff PM (1980) *Rhizobium trifolii* mutant interactions during the establishment of nodulation in white clover. Aust J Biol Sci 33:491–504
Rolfe BG, Gresshoff PM, Shine J (1980) Rapid screening for symbiotic mutants of *Rhizobium* and white clover. Plant Sci Lett 19:277–284
Rupert EA, Collins GB (1985) Tissue culture. In: Taylor NL (ed) Clover science and technology. Agron Monogr No 25. Am Soc Agron, Crop Sci Soc Am, Soil Sci Soc Am, Madison, WI, Chap. 16, pp 405–416
Rupert EA, Evans PT (1980) Embryo development after interspecific cross-pollinations among species of *Trifolium*, section Lotoidea. Agron Abstr, p 68
Rupert EA, Seo A (1977) Hybrid cell cultures from undifferentiated *Trifolium* embryos. Agron Abstr, p 69
Rupert EA, Seo A, Browne WC, Gibson PB (1976) Differentiation in callus and cell suspension cultures of *Trifolium* species and hybrids. Agron Abstr, p 60
Rybczynski JJ (1984) The in vitro culture of selected *Leguminosae* – plant regeneration and somatic embryogenesis. Proc Int Symp Plant tissue and cell cult: Application to crop improvement, Olomouc, Czechoslovakia, pp 129–130
Saunders JW, Bingham ET (1972) Production of alfalfa plants from callus tissue. Crop Sci 12:804–808
Schenk RU, Hildebrandt AC (1972) Medium and techniques for induction and growth of monocotyledonous and dicotyledonous plant cell cultures. Can J Bot 50:199–204
Shahin EA, Spielmann A, Sukhapinda K, Simpson R, Yashar M (1986) Transformation of cultivated alfalfa using disarmed *Agrobacterium tumefaciens*. Crop Sci 26:1235–1239
Sharp WR, Sondahl MR, Caldas LS, Maraffa SB (1980) The physiology of in vitro asexual embryogenesis. Hort Rev 2:268–310
Sharp WR, Evans DA, Sondahl MR (1982) Application of somatic embryogenesis to crop improvement. In: Fujiwara A (ed) Plant tissue culture 1982. Maruzen, Tokyo, pp 759–762
Skucinska B, Miszke W (1980) In vitro vegetative propagation of red clover. Z Pflanzenzücht 85:328–331
Skucinska B, Miszke W, Cebrat J (1984) Micropropagation from hypocotyl segments with cotyledons (fodder beet, red clover). Proc Int Symp Plant tissue and cell cult: Application to crop improvement, Olomouc, Czechoslovakia, pp 505–506

Smith AE (1979) Metabolism of 2,4-DB by white clover (*Trifolium repens*) cell suspension cultures. Weed Sci 27:392–396
Smith AE, Oswald TH (1979) Degradation of phenoxyalkylcarboxylic acids by white clover (*Trifolium repens*) cell suspensions. Weed Sci 27:389–391
Smith MK, McComb JA (1981) Use of callus cultures to detect NaCl tolerance in cultivars of three species of pasture legumes. Aust J Plant Physiol 8:437–442
Smith MK, McComb JA (1983) Selection for NaCl tolerance in cell cultures of *Medicago sativa* and recovery of plants from a NaCl-tolerant cell line. Plant Cell Rep 2:126–128
Smith RR (1985) Reaction of sexually derived tetraploid red clover to the northern anthracnose disease. Agron Abstr, p 71
Snaselova V (1982) The regeneration of plants in callus culture derived from the apical meristems of red clover. Sbornik UVTIZ Genetika a Slechteni, Ceskoslovenska Akademie Zemedelska, Ustav vedeckotechnickych informaci pro zemedelstvi 18:175–181 (in Czech; English summary)
Speer GS, Allinson DW (1985) Kura clover (*Trifolium ambiguum*): Legume for forage and soil conservation. Econ Bot 39:165–176
Staba EJ (1969) Plant tissue culture as a technique for the phytochemist. Recent Adv Phytochem 2:75–106
Takebe I, Otsuki Y (1969) Infection of tobacco mesophyll protoplasts by tobacco mosaic virus. Proc Natl Acad Sci, USA, 64:843–848
Tan BH (1983) Plant tissue culture with special reference to forage legumes. Agron Dep, Univ Western Australia, Miscellaneous Publ No 83/1
Tan BH (1985) Genetic control of flowering time in subterranean clover. Aust J Agric Res 36:275–284
Tan BH, Stern WR (1985) Induction of somaclonal variation in subterranean clover via in vitro shoot culture. Personal communication and abstract of results presented at the Pasture Production and Utilization Conference, Leura, NSW, Australia, December 15–19, 1985
Taylor NL (1985) Clover Science and Technology. Agron Monogr No 25, Am Soc Agron, Crop Sci Soc Am, Soil Sci Soc Am, Madison, WI, pp 1–616
Taylor NL, Smith RR (1979) Red clover breeding and genetics. In: Brady NC (ed) Advances in agron, vol 31. Academic Press, NY, pp 125–154
Taylor NL, Quesenberry KH, Anderson MK (1979) Genetic system relationships in *Trifolium*. Econ Bot 33:431–441
Taylor NL, Collins GB, Cornelius PL, Pitcock J (1981) Differential interspecific compatibilities among genotypes of *Trifolium sarosiense* and *T. pratense*. Proc 14th Int Grassland Congr, Lexington, Kentucky, pp 165–168
Teoule E (1983) Somatic hybridization between *Medicago sativa* L. and *Medicago falcata* L. CR Acad Sci (Paris) 297 (III):13–16
Webb KJ (1986) Transformation of forage legumes using *Agrobacterium tumefaciens*. Theor Appl Genet 72:53–58
Webb KJ, Fay M, Pike LS, Woodcock S, Dale PJ (1983) Forage legumes: tissue culture. Welsh Plant Breed Stat Annu Rep, Univ College of Wales, Aberystwyth, pp 40–42
Webb KJ, Woodcock S, Pike LS, Fay MF, Dale PJ (1984) Forage legumes: tissue culture. Welsh Plant Breed Stat Annu Rep, Univ College of Wales, Aberystwyth, pp 48–49
Webb KJ, Fay MF, Dale PJ (1987) An investigation of morphogenesis within the genus *Trifolium*. Plant Cell Tissue Organ Cult 11:37–46
Weber G, Roth EJ, Schweiger HG (1983) Storage of cell suspensions and protoplasts of *Glycine max* (L.) Merr., *Brassica napus* (L.), *Datura innoxia* (Mill.), and *Daucus carota* (L.) by freezing. Z Pflanzenphysiol 109:29–39
White DWR (1984) Plant regeneration from long-term suspension cultures of white clover. Planta 162:1–7
White DWR, Bhojwani SS (1981) Callus formation from *Trifolium arvense* protoplast-derived cells plated at low densities. Z Pflanzenphysiol 102:257–261
White DWR, Greenwood D (1987) Transformation of the forage legume *Trifolium repens* L. using binary *Agrobacterium* vectors. Plant Mol Biol 8:461–469
Williams EG (1978) A hybrid between *Trifolium repens* and *T. ambiguum* obtained with the aid of embryo culture. NZ J Bot 16:499–506
Williams EG (1980) Hybrids between *Trifolium ambiguum* and *T. hybridum* obtained with the aid of embryo culture. NZ J Bot 18:215–220

Williams EG (1987) Interspecific hybridization in pasture legumes. Plant Breed Rev 5:237–305
Williams EG, DeLautour G (1980) The use of embryo culture with transplanted nurse endosperm for the production of interspecific hybrids in pasture legumes. Bot Gaz 141:252–257
Williams EG, Maheswaran G (1986) Somatic embryogenesis: factors influencing coordinated behaviour of cells as an embryogenic group. Ann Bot 57:443–462
Williams EG, Verry IM (1981) A partially fertile hybrid between *Trifolium repens* and *T. ambiguum*. NZ J Bot 19:1–7
Williams EG, White DWR (1976) Early seed development after crossing of *Trifolium ambiguum* and *T. repens*. NZ J Bot 14:307–314
Williams EG, Verry IM, Williams WM (1982) Use of embryo culture in interspecific hybridization. In: Vasil IK, Scowcroft WR, Frey KJ (eds). Plant improvement and somatic cell genetics. Academic Press, NY, pp 119–128
Williams EG, Maheswaran G, Hutchinson JF (1987) Embryo and ovule culture in crop improvement. Plant Breed Rev 5:181–236
Wright RL, Somers DA, McGraw RL (1986) Somatic hybridization between birdsfoot trefoil (*Lotus corniculatus*) and *L. coimbrensis*. In: 6th Int Congr plant tissue and cell cult, Abstract. Univ Minnesota, Minneapolis, p 86
Yamada T, Fukuoka H (1985) Application of ovule culture to interspecific hybridization between *Trifolium repens* and *T. ambiguum*. In: Proc 15th Int Grassland Congr, pp 241–243
Yamada T, Fukuoka H (1986) Production of interspecific hybrids between *Trifolium ambiguum* M. Bieb and *T. repens* L. by ovule culture. Jap J Breed 36:233–239
Zakrzewski J, Zakrzewska E (1976) In vitro breeding of red clover callus as a medium for propagating nematode – *Ditylenchus dipsaci* (Kuhn). Hodowla Roslin, Aklimatyzacja I Nasiennictwo 20:97–104 (in Polish; English summary)

III.3 Lupins (*Lupinus* spp.)

C. SATOR[1]

1 Introduction

The genus *Lupinus* comprises several hundreds of species, but only a few of them have achieved agricultural importance. These are mainly three of the so-called Old World species, i.e., *Lupinus albus*, the white lupin; *L. angustifolius*, the narrow-leafed, formerly called blue lupin, and the yellow lupin, *L. luteus*. All of the them originated in the Mediterranean region (Table 1). One New World species is also of agricultural importance: *L. mutabilis*, the Andean lupin, locally called Tarhui or Tarwi. Some other species are mainly of horticultural value as ornamentals, like *L. polyphyllus* and its hybrids, or *L. hartwegii* (Hackbarth and Troll 1959; Hanelt 1960; Gladstones 1970; Sator 1979; Hondelmann 1984a, b; Williams 1986) (Table 1).

1.1 Importance of Lupins

The agricultural and nutritional value of lupins is mainly due to the high protein content (and oil in *L. mutabilis*) (Table 2) and that they belong to the leguminous family by their ability to fix nitrogen via symbiosis with *Rhizobium*, thus being not only almost self supplying with one of the most expensive fertilizers, but also leaving nitrogen for the following crop as well as a good-textured and phytosanitary improved soil (Hanelt 1960; Rowland et al. 1986; Ewing et al. 1986; Mohr 1986; Armstrong 1986; Mock 1986). Lupins could thus be grown instead of other grain legumes in regions where these cannot be grown successfully. Nevertheless, nowadays lupins make only a minor contribution to world grain legume production (Table 3).

1.2 Historical Overview

Lupins have been cultivated in the Mediterranean region probably for more than 3000 years (Becker-Dillingen 1929; Gladstones 1970; Wells 1984; Pate et al. 1985). In Greek and Roman times *L. albus* was used as forage for stock as well as for

[1] Institut für Produktions- und Ökotoxikologie, Bundesforschungsanstalt für Landwirtschaft, 3300 Braunschweig, FRG

Biotechnology in Agriculture and Forestry, Vol. 10
Legumes and Oilseed Crops I (ed. by Y. P. S. Bajaj)

Table 1. Origin, distribution, and use of the cultivated lupin species. (After Sator 1979; see also Klinkowski 1938; Hackbarth and Troll 1956; Zeven 1982)

Species	Origin	Use	Distribution
Lupinus albus white lupin	Medit. region? not known exactly, since no true wild plant has been found	(Green manure) (fodder), grain crop	Greece, Albania, Yugoslavia, Europe, Medit. region, Australia, Chile, New Zealand
L. angustifolius narrow-leafed lupin	Iberean Penin., South Italy, Greece, Palestine, North Africa	Green manure, fodder, grain crop	Southern France, Medit. region, northern Europe, South Africa, Australia, New Zealand, North America
L. luteus yellow lupin	North Africa, South Italy, Iberean Penin., Greece, Palestine	Green manure, fodder, grain crop	North Med. region, Northern Europe, eastern Europe, North America, South Africa, Australia, USSR
L. hartwegii	Central America	Ornamental	North America, Europe
L. mutabilis	Andean region	Oil, protein crop	South America, Andean region
L. polyphyllus	Central America	Ornamental, fodder for wild animals	North America, northern Europe

Table 2. Gross composition of lupin seed. (After Hill 1977)

Species	Percent of oven-dry matter				
	Crude protein	Ether extract [a]	Crude fiber	Ash	NFE [b]
L. albus	34.3 – 44.9	9.9 – 14.5	3.3 – 10.0	2.9 – 4.7	34.7 – 46.1
L. angustifolius	28.0 – 37.9	5.3 – 6.6	13.0 – 16.8	2.4 – 3.9	37.0 – 46.1
L. luteus	36.0 – 47.6	4.0 – 7.1	14.6 – 17.6	4.0 – 5.2	29.3 – 39.0
L. mutabilis	31.7 – 45.9	13.1 – 23.1	7.4 – 11.3	3.0 – 4.5	24.5 – 26.3

[a] Ether extract equals oil content.
[b] NFE = nitrogen-free extract.

human consumption, and ever since then periodically when alternative foods were scarce (Becker-Dillingen 1929; Hanelt 1960). From Italy it was brought to Central Europe, where some cultivation trials were carried out under the advice of Frederick II of Prussia at the end of the 18th century. Later in the 19th century, *L. luteus* was introduced and grown more successfully as the "yellow gold of sandy soils" (Hanelt 1960). At that time lupins were used not only for the improvement of light sandy soils, but also for feeding sheep. At the beginning of the 20th century, when the woollen industry collapsed in Central Europe, the interest in growing lupins declined drastically. The narrow-leafed lupin was first introduced as an ornamental plant from its original habitat into northern Europe, but nowadays it is still grown as green manure, like *L. luteus*.

Table 3. Relative world status of the grain legumes – 1983 harvest (Williams 1986, based on FAO Production Year Book Vol. 38, 1984 and FAO Trade Year Book Vol. 38, 1984)

		Percent grain production	Percent area
Oilseed	Soybean	58	37
	Groundnut	10	14
	(Total oilseed)	68	51
Pulses	*Phaseolus vulgaris*	9.5	17.7
	Pea	6.9	6.2
	Chickpea	5.0	7.2
	Faba bean	3.1	2.6
	Lentil, Mung bean, Cowpea, Pigeon pea, Lima bean	7.0	14.5
	Lupin	0.5	0.8
	Total pulses	32	49
Total production (1000 mt)		136,082	132,568 (per 1000 ha)

Cultivation of *L. mutabilis* dates back to Inca times and this species is still grown in the Andean region, contributing towards better nutrition (de Mayolo 1980). At the beginning of the 19th century it was brought to England, principally for use as an ornamental (Masefield 1975), and so far it cannot be grown as an agricultural crop in any other part of the world other than in its original habitats.

1.3 Distribution and Area Under Cultivation

In contrast to *L. mutabilis*, the Old World lupins have been introduced to many other parts of the world, such as North and South America (Decker et al. 1953), and South Africa (Preller 1949). Especially the narrow-leafed lupin has adapted well to New Zealand (Hudson 1934; Stocker 1974) and Australia (Gladstones 1959; Gladstones 1980) for use as green manure, fodder, or for seed production.

Table 4. Regional distribution of lupin production 1948 – 1985 (per 1000 ha). (Williams 1986, sources FAO Production Year Books Vol. 20 – 29 and pers. commun.)

	1948/1952	1961/1965	1966/1970	1971/1975	1981/1984	1985
USSR	252^{e}	607	620	522	310	280^{e}
Europe	315	239	185	127	124	148
South Africa	87	185	200	104	20	20
Other Africa	11	9	5	5	6	7
Latin America	2	4	5	4	10	12
Australia	–	2	2	56	242	606
World total	667	1046	1007	818	712	1073

^{e} = estimated.

Table 5. Average yield of grain lupin (per 100 kg) per hectare 1961 – 1985. (Williams 1986, source: FAO Production Year Books Vol. 20 – 29, and pers. commun.)

Country	1961 – 65	1971 – 75	1980 – 84	1985
Argentina [a]	8.6	11.8	14.7	16.0
Australia [b]	3.4	6.2	8.2	9.7
Chile [a]	–	–	24.7	20.2
Ecuador [d]	8.7	5.6	5.6	6.1
Egypt [a]	16.4	13.2	16.4	12.7
Germany DDR	7.7	8.1	8.9	8.9
Greece [a]	6.2	11.1	14.4	9.4
Hungary [a]	4.9	5.8	8.0	10.7
Italy [a]	9.5	12.9	14.1	13.6
Marocco	5.0	9.5	9.8	9.6
Peru [d]	12.9	9.4	12.5	12.5
Poland [c]	11.1	10.9	8.5	9.8
Portugal [a,c]	7.8	4.4	4.6	4.7
South Africa [b]	2.3	1.2	3.6	3.5
Spain [a,c]	5.7	6.4	4.9	5.0
Syria	–	9.9	9.4	10.4
USSR [a,b,c]	5.3	9.1	8.8	14.3
Av. world yields [a,b,c]	5.7	8.3	8.5	10.8

[a,b,c,d] refer to principal species cultivated (Bellido 1984).
[a] *L. albus*, [b] *L. angustifolius*, [c] *L. luteus*, [d] *L. mutabilis*.

Thus it seems to be clear that lupins can be grown between the 65° northern and 42° southern latitude (Sator 1979; Duke 1981; Baylis and Hamblin 1986) on sandy soils, sandy loam, or loamy soils of poor to adequate fertility with pH 4 to 8 (Dieckmann 1968; Gladstones 1970; Duke 1981; Baylis and Hamblin 1986; Pate et al. 1985) in a wide range of climates from 6° to 26 °C. The area under cultivation from the late 1940's to the present has been summarized by Williams (1986) (Table 4). It is clearly shown that Australia now has the leading position in lupin production, even though yields per hectare are much higher in South American countries (Table 5; Williams 1986).

1.4 Common Pests and Diseases

As can be seen from Table 5, yields are rather unstable in many parts of the world. This is partly due to several fungi, viruses, and insects, which can cause drastic losses in yield, up to 90% (Duke 1981). Unlike the Andean lupin, all the Mediterranean lupins are highly susceptible to several *Fusarium* species and *Verticillium alboatrum*, which cause wilting diseases. This can best be controlled by appropriate crop rotations (Duke 1981; Pate et al. 1985). Brown leaf spot caused by *Pleiocheta setosa* is particularly aggressive to winter-grown crops and therefore especially affects *L. albus*, being the main winter-grown lupin. *L. angustifolius*, *L. luteus*, and *L. mutabilis* are attacked to a much lesser extent (Pate et al. 1985). Severe yield losses of *L. angustifolius*, especially in South Africa (Wassermann 1981), are caused by the common fungus *Erysiphe* (mildew disease). In contrast to this, large numbers of *L. angustifolius* plants in Australia may be killed by *Stemphylium vesicarum*, which causes gray-brown necrotic spots on all above-ground plant parts (Wood and McLean 1982). Meanwhile, some resistant cultivars have been developed by means of which this problem might be overcome. Besides this, crop hygienic practices like rotation and restriction of cultivation to well—drained soils are beneficial.

Some fungal diseases cause loss of seedlings. *Pythium, Phytophtora, Botrytis,* and *Rhizoctonia* species are in this regard especially noticeable in cool temperatures, but they can be controlled to some extent by dressing seeds with fungicides.

Two main virus diseases are known. The bean yellow mosaic virus affects the yellow lupin in particular. Wilting of apical leaves and stem tips are the first symptoms. Later, the whole plant turns black and dies. Since insects are vectors for transmission, mainly low-alkaloid cultivars are infected. This is also true for the infection of all Mediterranean lupin species – primarily *L. angustifolius* – with cucumber mosaic virus (Hackbarth and Troll 1959).

Alkaloid-poor cultivars are also preferred by the red-legged earth mite (*Halotydeus destructor*) and the lucerne flea (*Smynthurus viridis*), which cause damage especially to young plants. In contrast, the aphid *Macrosiphon albiform* is a widespread disease today for both sweet and bitter forms of *L. albus* and *L. mutabilis*. The budworm moth (*Heliothis* sp.) is most harmful to winter-grown lupins in Australia, South Africa, Portugal, and southeast parts of North America (Gladstones 1970). Since these moths attack young developing seeds, they can be kept under control by using early maturing cultivars, early sowing dates, and insecticides (Anonymous 1984).

Most troublesome when feeding lupins to animals – both seeds and vegetative plant parts – is infection with *Phomopsis leptostromiformis*. Due to this infection, plants develop toxic substances which cause lupinosis when used as fodder. The symptoms of lupinosis, in milder cases, are decrease of appetite and body weight, in acute cases liver, gall bladder and kidney are damaged, and the animals may die. Australia has the most experience with *Phomopsis* disease at present, and breeding of *Phomopsis*-resistant cultivars is being carried out there, with rotational grazing and feeding rules being recommended (Allen et al. 1980; Allen 1985; Anonymous 1986).

1.5 Conventional Practices for Lupin Improvement

Modern lupin breeding began in the late 1920's with the pioneering work of R. von Sengbusch in Germany. The situation he was faced with was summarized by von Sengbusch and Zimmermann (1947) (Table 6).

Von Sengbusch set up a list of priorities, starting with alkaloid content to be eliminated first, and other wild-type characteristics like shattering and shedding pods as well as nonpermeable seed coat second. This was followed by improving seed yield. He started to develop a method for determination of alkaloid content usable for mass screening. Thus, von Sengbusch was able to find plants having only low alkaloid contents which he called sweet lupins (since alkaloids are bitter in taste). During 1927, 1928, 1931, and 1932 in all important species, *L. luteus*,

Table 6. Plant characteristics before modern lupin breeding started. (After von Sengbusch and Zimmermann 1947)

Characteristics	*Lupinus luteus*	*Lupinus angustifolius*	*Lupinus albus*	*Lupinus mutabilis*
Alkaloid content	+	+	+	+
Shattering pods	+	+	–	–
Shedding pods	+	+	–	–
Non permeable seed coat	+	+	–	–
Slow early growth	+	–	–	–
Dark seeds	+	+	–	–
Hairiness	+	–		–

+ = character present, – = character not present.

L. angustifolius, L. albus, and *L. mutabilis* respectively, a few sweet genotypes were found in vast numbers of screened plants. He and his collaborators also succeeded in finding permeable seed coats in the yellow lupin (von Sengbusch 1928), faster juvenile growth (von Sengbusch 1929), nonshedding pods (von Sengbusch and Zimmermann 1935/36), nonhairiness of plant and pods (Troll 1939), as well as resistance to mildew and fusarium infection (Klinkowski 1941) (see von Sengbusch and Zimmermann 1947 for these citations). A similar development of *L. angustifolius* was achieved in the same time interval. All this was done by means of basic selection methods.

Starting in the late 1950's, lupin breeding work, especially of *L. angustifolius*, concentrated in Australia under the leadership of J. S. Gladstones. Breeding aims were directed towards producing low alkaloid content, disease resistance, high protein, and high seed-yielding cultivars, adapted to the specific environments and soil types in Australia to make this continent self-sufficient in respect to high protein feed (Gladstones 1970, 1982, 1984).

Besides the activities in Australia, breeding work with all agronomically important lupin species is going on in the Soviet Union (Dyubin 1978; Anokhina et al. 1979; Chakalin 1979; Savvichev 1980). A long tradition also exists in Poland on improving the Mediterranean lupin species for better adaptation to Polish environmental conditions, raising yields, and producing resistance to several diseases (Swiecicki 1984, 1986a, b). The adaptation of *L. albus* to the climatic conditions of Central Europe can be regarded as a major break-through (Beltky and Kovacs 1984).

Great emphasis was placed on producing interspecific hybrids. This work started in Germany (Gollmick 1937) and the Netherlands (Lamberts 1958) and concentrated later in Poland (Kazimierski 1961 a, b; Jaranowski 1962; Kazimierski 1962, 1964, 1980). Similar efforts were made in England by Williams (1979) and Williams et al. (1980), to cross the European lupin species with *L. mutabilis* to obtain lupins with both high protein and high oil content able to be grown under European climatic conditions.

Selection for the same purpose is being effected in Germany with *L. mutabilis* (Römer and Jahn-Deesbach 1986) and in France (Vuillaume pers. commun.), while breeding efforts with this species in the Andean countries mainly concentrate on obtaining alkaloid-free mutants (von Baer and von Baer 1986) besides adapting other lupin species to South America (Brücher 1970).

Lupins usually flower profusely, setting seeds and side branches at the same time, and using great amounts of energy and organic compounds to achieve this. Because many flowers and unripe seeds abort as a result of building new branches, research is being done on other plant architecture to reduce branching types. In this respect, Forbes in the USA was the first to select reduced branching types of *L. angustifolius*, with which further breeding and crop management trials are going on in Australia at the moment (Delane et al. 1986).

1.6 Need to Incorporate Unconventional Methods into Lupin Breeding

Formerly, lupin breeding mainly concentrated on the selection of single plants. Having found a plant with the desired characters, multiplication through seed followed. Production of enough seeds for production on a larger scale may take several years, as was shown by von Sengbusch selecting for low alkaloid content. Having found the first promising plants of *L. luteus* and *L. angustifolius*, in 1927 and 1928 it took 4 years to obtain enough seeds for a multiplication on a 2-ha area and another 3 for the seeds to be marketable in 1934 (von Sengbusch 1942). It would thus be helpful to have other methods to speed up the process of mass propagation. On the other hand, lupins have high chromosome numbers of $n = 24$ (*L. polyphyllus* and *L. mutabilis*) or $n = 26$ (*L. luteus*) (Hackbarth and Troll 1959; Zeven 1982). They are therefore considered as polyploids, even though they behave like diploids. Basic chromosome numbers are $x = 6$ for New World species and $x = 5$ for Old World species (Brücher 1970).

The high chromosome numbers complicate breeding work and a lower ploidy level might be of some help, since some characters such as alkaloid content might be governed by several genes.

True interspecific crosses have not been successful in producing viable seeds so far because of incompatibility problems. Since Williams succeeded in obtaining partly developed seeds (Williams et al. 1980) the incompatibility is probably due to physiological reasons. It thus seems possible to cross plants and raise seeds under artificial conditions, thus obtaining viable interspecific seeds.

In some self-pollinating lupin species, the genetic variability is rather low. Radiation experiments to increase variability for crop improvement did not bring the great break-through (Hackbarth 1957). Therefore it seems worthwhile to try other methods to increase genetic variability of lupins, as has been recommended by Bajaj and Gosal (1982), using tissue culture techniques for the induction of genetic variability in grain legumes.

2 In Vitro Approaches

Recent advances in tissue culture techniques have provided the basis for novel breeding techniques as adjuncts to conventional crop improvement (see Bajaj

1986). It should, however, be kept in mind that tissue culture techniques should be used only when there is no conventional method available, or these prove to be unsuccessful or too time- and labor-consuming. These include rapid mass propagation of selected genotypes from plant species which are normally propagated by seeds, production of haploids, which is not possible other than by anther or microspore culture, or culture of unfertilized ovules (except for potatoes and barley), embryo rescue, whenever embryos are not fully developed to maturity on the plants themselves, induction of genetic variability and selection of mutants on a broader scale, or production of hybrids via protoplast fusion in cases of cross-incompatibility. The chances to use and the need to incorporate tissue culture techniques into lupin breeding work have been outlined by Sator (1982, 1984). Nevertheless, practical application of tissue culture techniques always requires collaboration with conventional breeding.

2.1 Summary of in Vitro Work So Far Done on Lupins

Lupin tissue culture started with some morphological studies by Ball (1946) on *L. albus* (see also Ball 1960a, b). Next experiments were reported by Lee (1955a, b) on *L. hartwegii*. More efforts were put into lupin tissue culture work starting from 1980 onwards. A summary of known experiments is given in Table 7.

2.2 Methods of Sterilization of the Explants

Callus cultures were started from a variety of tissues such as stem segments and shoot tips, hypocotyl pieces, leaf tissue, petiole, roots, anthers, and ovules (see Table 7 for details). Several sterilization methods are mentioned. Plant parts which are not directly exposed to the environment but prepared for use as inoculum do not need to be sterilized. However, surface sterilization of the surrounding plant organs is of great help for clean preparation or dissection of the explants.

Ball (1960a), isolating shoot apices, mentioned a surface sterilization of the intact bud prior to dissection by a 5-min immersion in 1.75% sodium hypochlorite solution. Lee (1955a) used 1% bromine water for 5 to 10 min followed by several rinses in sterile water and soaking the seeds for 2 h in sterilized water before removing the embryos. Since commercial bleaches – containing hypochlorite – are available, they can also be used, as has been shown by Sator et al. (1982) and Sator (1985a–d). Thus, sterilization can be started by a short rinse in 70% ethanol followed by 10–30 min sterilization in 5% commercial bleach and several rinses in sterilized water. Sterilization time depends on the plant part being used. For vegetative parts like stem segments or hypocotyl pieces, as well as for leaf tissue from soil-grown plants, 10 min of sterilization is sufficient. To raise plants aseptically, seeds must be sterilized for about half an hour. Because commercial bleaches contain substances which produce foam, the following washing with water is prolonged to remove the adhering bleach, which might be toxic for the plant tissue.

The process can thus be started with sterilization in the bleach followed by a short rinse in 96% ethanol. After this, three rinses in sterile water are sufficient.

Table 7. Summary of in vitro work on lupins (*Lupinus* spp.)

Lupin species	Inoculum/donor material	Medium	Growth response	Reference
L. albus	Shoot tip with leaf primordia, shoot segments	Pfeffer's culture fluid modified	Elongation of the tips and root formation	Ball (1946)
L. hartwegii	Shoot tips, shoot segments	Same as Ball 1946 plus NAA	Elongation and very few roots	Lee (1955a)
	Shoot segments	Gautheret's and White's	Callus formation 'unlimited growth'	Lee (1955b)
L. albus	Shoot tip without leaf primordia	Heller's plus micronutrients of Ball 1946 plus CM and GA	Shoots, no roots	Ball (1960a)
	Shoot apices	Heller's plus micronutrients of Ball 1946	Cell division (purpose of study)	Ball (1960b)
L. spp.	Leaves for protoplasts	MS plus 2,4-D and kin	Not mentioned	Davey et al. (1974)
L. polyphyllus	Stem segments	MS plus 2,4-D and kin kin + IAA + NAA	Callus induction Suspension culture	Wink and Hartmann (1980)
L. spp.	Leaf and apex for protoplasts		Low yields	Binding et al. (1981)
L. polyphyllus	Anthers and microspores	MS plus 2,4-D and kin	(Callus)	Palada and Sator (1981)
L. polyphyllus	Anthers	MS plus 2,4-D and kin	Callus	Sator et al. (1982, 1983)
L. angustifolius	Hypocotyl	LS modified plus 2,4-D + kin	Callus, suspension culture	Sroga (1983)
L. polyphyllus *L. hartwegii* *L. albus* *L. luteus*	Immature embryos and ovules	MS plus 2,4-D kin, NAA, BA	Plantlets, multiple shoots, callus with shoots	Sator (1985a)
L. polyphyllus *L. angustifolius* *L. hartwegii* *L. luteus*	Mature half embryos	MS plus BA and IAA/NAA	Multiple shoots and plantlets	Sator (1985b)
L. polyphyllus *L. hartwegii* *L. albus* *L. angustifolius* *L. luteus*	Anthers	MS plus 2,4-D BA + NAA	Callus, plants of *L. polyphyllus*	Sator (1985c)

L. polyphyllus *L. luteus* *L. hartwegii* *L. angustifolius*	Leaf, hypocotyl, root, petiole	MS plus 2,4-D and kin IAA/NAA + BA	Plantlets, *L. polyphyllus*, shoots *L. hartwegii*, *L. luteus*, and callus	Sator (1985d)
L. albus *L. luteus* *L. angustifolius*	Nodal segments	MS or B5 plus CA B9 or GA	Rooted shoots	Schäfer-Menuhr (1985, 1986a)
L. polyphyllus *L. angustifolius*	Embryo axis, hypocotyl	LS modified plus NAA, BA/kin	Plantlets	Sroga (1987)
L. mutabilis *L. angustifolius* *L. nanus* *L. ornatus*	Shoot tips	Skoog	Rooted shoots *L. mutabilis*	Vuillaume et al. (1986)
L. mutabilis *L. albus*	Immature embryos	Skoog without hormone	Developed further	Vuillaume et al. (1986)
L. luteus *L. angustifolius* *L. hartwegii* *L. polyphyllus*	Immature embryos	Liquid medium without hormones	Developed further	Schäfer-Menuhr (1986b)
L. spp.	Leaves and suspension culture for protoplasts	Not mentioned	Callus and root formation	Schäfer-Menuhr (1986c)

Abbreviations: BA = benzyladenine, CA = casein hydrolysate, GA = gibberellic acid, IAA = indole acetic acid; kin = kinetin, NAA = naphthyl acetic acid, 2,4-D = 2,4 dichlorphenoxy acetic acid, MS = medium of Murashige and Skoog (1962), LS = medium of Linsmaier and Skoog (1965).
Note: All literature cited from 1986 was posters presented at the 4th International Lupin Conference held at Geraldton, Australia (1986).

The water used after sterilization should be of good quality for plant parts which are placed on culture medium directly. For surface sterilization of surrounding plant organs, as done for culture of apices, anthers, ovules, or isolated embryos, sterilized tap water can also be used.

2.3 Media Composition

The media compositions of Ball (1946) and Lee (1955a) are now no longer used. As can be seen from Table 7, well-known media like MS (Murashige and Skoog 1962), LS (Linsmaier and Skoog 1965), or B5 (Gamborg and Wetter 1975) can also be used for lupin tissue culture. The compositions of these media are listed by Gamborg and Wetter (1975) or Thomas and Davey (1975) and in other laboratory manuals besides in the original papers.

2.4 Meristem Culture

Apical meristems are the preferred tissues to start in vitro culture as well as for mass propagation. The main reason for using meristem for direct plant formation is the genetic stability of the cloned plants if plant development occurs without callus formation. Raising one plant from an apex of a selected genotype is not very helpful for mass propagation unless this process is done also with side buds and repeated several times (Schäfer-Menuhr 1985, 1986a). Von Sengbusch (1942) has developed the half-grain method, by means of which one and the same grain serves for testing the quality of the grain as well as for plant regeneration. This process can also be combined with multiple shoot formation to speed up the process of mass propagation (Sator 1985a) (Figs. 1 A, B). The induction of multiple shoot formation can be done with 0.5–2 mg BA/l culture medium.

2.5 Embryo Culture

Williams et al. (1980) succeeded in obtaining ovules of about 1.5 to 2.4 mm after crossing experiments with New World and Old World lupin species. For several reasons hybrids of these lupin species are desirable. Thus embryo rescue or ovule culture might overcome the inviability of these underdeveloped seeds and help to raise them to viable plants. Experiments for embryo rescue are reported by Sator

Fig. 1. A, B Multiple shoot formation from half grains of *L. hartwegii*. **A** 3 weeks after induction on MS medium plus 1.5 mg BA/l. **B** 7-week-old shoots. **C, D** Embryo rescue of *L. polyphyllus*. **C** 4-week-old embryos after rescue on MS medium plus 1.5 mg BA and 2 mg NAA/l. **D** 4-week-old plantlet after embryo culture on MS medium plus 2 mg NAA/l. **E, F** Anther culture of *L. polyphyllus* and *L. albus*. **E** Regenerated plants of *L. polyphyllus* after callus induction on MS medium plus 1 mg 2,4-D, 1 mg BA and 1 mg IAA/l, shoot induction on MS medium plus 0.5 mg NAA and rooting of the shoots on hormone-free medium. **F** Callus formation on an anther of *L. albus* on MS medium plus 1 mg 2,4-D, 1 mg BA and 1 mg IAA/l

Fig. 1A–F

Table 8. Culture and development of several lupin species (absolute figures). (Sator 1985a)

Species Size of embryos in mm	Medium EO1		EO2		EO3		EO4		EO5		EO6	
	1	2	1	2	1	2	1	2	1	2	1	2
L. polyphyllus												
1–3.5	13	0	6	2	14	6	7	0	12	5	–	–
3–5	10	4	9	0	9	0	12	3	11	5	–	–
4–6	13	13	12	12	15	10	15	7	10	7	–	–
L. hartwegii												
1–2	11	1	10	5	12	3	12	3	12	0	–	–
1–4	6	6	8	6	10	7	7	7	8	8	–	–
L. albus												
1–5.5	4	0	3	2	3	3	3	0	2	1	–	–
6.5–9	5	4	5	0	4	4	5	0	4	4	–	–
L. luteus												
3–10	5	0	6	0	5	0	5	0	4	3	–	–
Palfa												
1	–	–	–	–	–	–	–	–	–	–	9	6[a]

EO1 = MS plus 2 mg NAA and 1.5 mg BA/l.
EO2 = MS plus no hormones but 2 g casein hydrolysate/l.
EO3 = MS plus 1.5 mg BA and 2 g casein hydrolysate/l.
EO4 = MS plus 2 mg NAA and 2 g casein hydrolysate/l.
EO5 = MS plus 2 mg NAA, 1.5 mg BA and 2 g casein hydrolysate/l.
EO6 = MS plus 2 mg 2,4-D, 0.25 mg kin and 2 g casein hydrolysate/l.
1 = total number of embryos.
2 = number of further developed embyros.
[a] Abnormal development.

(Table 8, 1985a) and Vuillaume et al. (1986) using solid media, and by Schäfer-Menuhr (1986b) using liquid medium. Schäfer-Menuhr (1986b) and Vuillaume et al. (1986) used no hormones in the culture media, thus obtaining one plant per embryo. Sator (1985a) used BA and NAA in the culture media for better shoot and root development. As can be seen from Fig. 1C, embryos from one pod cultured in one Petri dish showed a different response to the hormones (1.5 mg BA and 2 mg NAA/l). Using NAA as the only additive, very vigorous roots are developed (Fig. 1D). Using 2,4-D as auxin leads to callus formation, as has been shown by Sator (1985a) (Fig. 2A) when culturing small underdeveloped ovules.

2.6 Anther Culture

Palada and Sator (1981) cultivated isolated microspores of *L. polyphyllus* in liquid medium without much success. Callus induction has been achieved in many cases in anthers cultured (Sator et al. 1982, Table 9; Sator 1985b, Table 10) (Fig. 1F), but regeneration occurred only twice with *L. polyphyllus* (Sator 1985b) (Fig. 1E). After callus induction in darkness on MS plus 1 mg 2,4-D, 1 mg BA, and 1 mg IAA at 24 °C, callus was transferred to MS plus 0.5 mg BA and 0.05 mg

Table 9. Callus induction of anthers of ***Lupinus polyphyllus*** on different culture media after incubation in different temperature conditions. (Sator et al. 1982, 1983)

Medium no./temp.	Total no. of anthers	No. of anthers with callus development	Rate of callus development (%)
4/24 °C	3720	635	17
4/26 °C	520	189	36
4/28/22 °C	600	138	23
4 all temps.	4840	962	20
5/24 °C	3160	371	12
5/26 °C	360	70	26
5/28/22 °C	600	110	18
5 all temps.	4120	551	13
7/24 °C	7905	3074	39
7/26 °C	640	223	35
7/28/22 °C	480	120	25
7 all temps.	9025	3417	38
All med. 24 °C	14785	4080	27
All med. 26 °C	1520	482	32
All med. 28/22 °C	1680	368	22

4 = MS plus 0.5 mg 2,4-D and 0.01 mg kin plus 50 g sucrose/l.
5 = MS plus 0.5 mg 2,4-D and 0.01 mg kin plus 100 g sucrose/l.
7 = MS plus 1 mg 2,4-D and 0.5 mg kin plus 50 g sucrose/l.

NAA and kept in a 20 °/15 °C day-night cycle with a daylength of 16 h and about 5000 lx light intensity. Plants were shown to be diploid according to chromosome counts, and might have developed from callus originating from wall tissue instead of from microspore callus, or because the rather instable haploids have doubled the chromosomes during culture.

2.7 Establishment of Callus and Suspension Culture

Callus culture may be started using different plant parts as has been done with anthers, pieces of leaf tissue, roots, small ovules, petiole tissue, stem cuttings, or hypocotyl pieces (see Tables 7 and 11, Sator 1985 d). For induction of callus formation auxins are necessary, preferably 2,4-D, if callus itself is required, or NAA, if regeneration of plants is the final aim (Sator 1985 d). Starting from stem cuttings of *L. polyphyllus*, Wink and Hartmann (1980) induced callus formation and transferred the callus later to liquid medium. Sroga (1983) did the same with *L. angustifolius*, starting from hypocotyl tissue and using different media. Usually callus formation occurs in the dark at about 25 °C. Suspension cultures are kept in the dark to obtain white cultures or kept under continuous light (Wink and Hartmann 1980) with 60 g sucrose/l, but 1.5 mg 2,4-D and 0.2 mg NAA/l suspension cultures have been established for other lupin species also (Sator unpubl., see also Wink 1984). Culture vessels are usually kept on a rotary shaker with 100–120 rpm for better oxygen supply to the cells.

Table 10. Callus induction rates of anthers of several lupin species and cultivars on different culture media. (Sator 1985b)

Species cv Culture medium	*L. hartwegii*		*L. polyphyllus*		*L. albus*		*L. angustifolius*		*L. angustifolius* Kubesa		*L. luteus*		*L. luteus* Barpine		*L. luteus* Palfa		*L.luteus* Topaz	
	1	2	1	2	1	2	1	2	1	2	1	2	1	2	1	2	1	2
A1	1325	61	545	50	295	69	135	100	180	61	–	–	340	46	635	42	405	42
A2	140	74	–	–	–	–	–	–	–	–	–	–	–	–	–	–	–	–
A3	455	70	–	–	–	–	–	–	–	–	–	–	–	–	–	–	–	–
A4	3010	46	1095	79	450	70	130	81	145	64	225	45	985	47	835	47	950	42
A5	745	76	265	32	–	–	135	97	–	–	–	–	480	73	480	42	–	–

A1 = MS plus 2 mg BA and 0.2 mg NAA/l.
A2 = MS plus 2 mg kin and 4 mg IAA/l.
A3 = MS plus 2 mg 2,4-D and 2 mg kin/l.
A4 = MS plus 1 mg 2,4-D, 1 mg BA and 1 mg IAA/l.
A5 = MS plus 1.5 mg BA and 0.2 mg NAA/l.
1 = total number of anthers.
2 = rate of callus development in %.

Table 11. Callus induction of different plant parts from several lupin species on different culture media (absolute figures)[a]. (Sator 1985d)

Medium	Species											
	L. luteus			*L. angustifolius*			*L. polyphyllus*			*L. hartwegii*		
	L	H	R	L	H	R	L	H	R	L	H	R
1	40/37	40/40	40/39	40/40	40/40	40/36	40/40	–	–	40/38	40/40	60/50
2	40/35	40/35	40/40	40/40	40/40	40/36	40/24	40/40	40/37	40/39	40/40	50/49
3	320/313	310/288	160/150	130/130	70/70	90/90	120/90	100/46	110/87	180/175	250/122	260/176
4	40/39	40/40	40/40	40/40	40/40	40/40	40/32	40/40	40/33	40/35	40/32	60/58
5	40/36	40/40	40/38	–	–	40/13	40/36	–	–	40/39	40/34	40/18

[a] The first figure indicates the number of cultivated plant parts, the second figure gives the number of plant parts showing callus.
L = leaf, H = hypocotyl, R = root.
1 = MS plus 1 mg IAA, 1 mg BA and 1 mg 2,4-D/l.
2 = MS plus 2 mg 2 mg 2,4-D and 2 mg kin/l.
3 = MS plus 2 mg BA and 0.2 mg NAA/l.
4 = MS plus 1.5 mg BA and 2 mg NAA/l.
5 = MS plus 2 mg 2,4-D and 1.25 mg kin plus 2 g casein hydrolysate/l.

2.8 Isolation, Fusion, and Culture of Protoplasts

First trials to isolate and cultivate protoplasts of lupins were reported by Davey et al. (1974) using leaf tissue of surface-sterilized leaves incubated in an enzyme mixture of 4% Meicelase, 0.4% Macerozyme plus 13% mannitol and a special salt mixture. Results on cultivation of the protoplasts are not reported. Later, Binding et al. (1981) obtained low protoplast yields from fully expanded young leaves and from apex tissue of aseptically raised plants of an unknown lupin species. Low but adequate protoplast yields were achieved in other experiments (Sator EG-report 1983, project no. 0423) using the method of Davey et al. (1974) with the addition of 2% Rhozyme, and also by using an enzyme mixture of 1% Cellulysin, 1% Macerozyme, and 0.1% Pectolyase in the isolation medium. Thus, protoplasts of *L. luteus, L. hartwegii*, and *L. polyphyllus* have been obtained (other lupin species have not been under trial) from sterile-grown plants using leaf tissue and roots as donor material. Promising results have been reported by Schäfer-Menuhr (1986c) for isolation and culture of lupin protoplasts starting from leaves and suspension culture, but no details are given about species, enzyme mixture used for isolation, or composition of culture medium.

2.9 Regeneration and Transfer of Plants

Recently, plant regeneration has been reported by Sroga (1987) for *L. angustifolius* and *L. polyphyllus* starting from callus derived from hypocotyl and transferring to media containing 0.54 μM NAA and 4.4 μM BA. Shoot buds were rooted on the same media without cytokinin. Rooted shoots were transferred to perlite moistened with two strengths of the microelements. Regeneration conditions were 25 °C and 2000 lx continuous light. Schäfer-Menuhr (1985) also used perlite moistened with nutrient solution of lower concentration than used during culture to root shoot tips as a step between tissue culture and soil culture to strengthen the plants. For better survival, plants should be covered with glass beakers or plastic bags, which can be removed gradually. Reduction of ionic strength can also be carried out during culture on agar medium, as was done by Sator (1985d) to obtain better root formation (Fig. 1E). Vigorous roots are also developed using tap water, as can be seen in Fig. 2F, but requirements for shoot rooting are different (Sator 1985d), obviously depending on the species as well as on the mode of shoot recovery. Rooting was possible after multiple shoot formation for *L. polyphyllus. L. hartwegii*, and *L. luteus* (Sator 1985c, Figs. 3B,C). After shoot formation via callus only shoots of *L. polyphyllus* could be rooted (Table 12, Sator 1985d). Shoots derived from ovule callus (Fig. 2A), from petiole tissue (Fig. 2B) or from leaf material (Fig. 2C,D) of *L. hartwegii* could not be rooted as well as shoots regenerated from leaf tissue of *L. luteus* (Fig. 3). The shoots had all been regenerated after callus induction in the dark at 25 °C on MS medium plus 2 mg BA and 0.2 mg NAA and transfer to medium containing either 2 mg BA plus 2 mg IAA or 0.2 mg IAA in 20°/15 °C day-night cycle with 16 h daylength and about 5000 lx light intensity.

Fig. 2. A Starting shoot formation after callus induction of an ovule of *L. hartwegii* after 4 weeks on MS medium plus 2 mg 2,4-D and 0.25 mg kin/l. **B** Shoot formation from a petiole piece of *L. hartwegii* after callus induction on MS medium plus 2 mg BA and 0.2 mg NAA/l and transfer to MS medium plus 2 mg BA and 0.2 mg IAA/l 3 weeks after transfer. **C** Multiple shoot formation of *L. hartwegii* leaf tissue (conditions see **B**). **D** Direct shoot formation from a leaf of *L. hartwegii* after callus induction and transfer to fresh medium (conditions see **B**). **E** Multiple shoots from *L. hartwegii* leaf (conditions see **B**) 6 weeks after passage. **F** Rooted shoot tips of *L. angustifolius* after 5 weeks on tap water medium

Table 12. Shoot regeneration of *Lupinus* spp. from different plant parts (results after elimination of infected cultures). (Sator 1985d)

Species/cv.	Plant part	Differentiation	Medium
L. polyphyllus	Hypocotyl	1 plant (from 6 pieces)	D1
L. hartwegii	Leaf	2 bulks of shoots (from 19 pieces)	D2
	Leaf petiole	1 bulk of shoots (from 5 pieces)	D2
L. luteus Palfa	Leaf	1 shoot (from 15 pieces)	D1
		2 shoots (from 15 pieces)	D2
L. angustifolius Kubesa	Leaf	1 unknown differentiation (from 3 pieces)	D2

D1 = MS plus 2 mg BA and 2 mg IAA/l.
D2 = MS plus 2 mg BA and 0.2 mg IAA/l.

2.10 Miscellaneous

Wink and Hartmann (1980) established suspension culture of *L. polyphyllus* and later also of other lupin species (Wink 1984) to study lupin alkaloids and their biosynthetic pathways. Harms and Langebartels (1985) used suspension cultures of *L. polyphyllus* as part of a model system in ecotoxicological studies of xenobiotics.

3 Conclusions and Prospects

Lupins are said to belong to the recalcitrant plant species in respect to their behavior in tissue culture, but from recent developments reported here it can be concluded that lupin improvement via tissue culture techniques is possible. Success seems to be correlated with the number of scientists working on the problem. The number of tissue culturists working with lupins is increasing; besides the abovementioned groups there are plans to work on lupin tissue culture in South Africa and Portugal (pers. commun.). Experiments of scientists in Russia to up-grade pathogen resistance of lupins and in genetic engineering work (Gleba et al. 1986a, b) also show promising prospects.

Lupins are attractive crops in many respects, since their potential use is of multiple choice, as was shown, even to excess, by the Vereinigung für Botanik in October 1918 in Hamburg, Germany, when they served a dinner, on a lupin-fiber tablecloth, consisting of lupin soup, lupin steak roasted in lupin oil and seasoned with lupin extract, bread containing 20% lupin flour, lupin margarine, and cheese of lupin albumin, topped with lupin liqueur and coffee. Hands could be washed with lupin soap and dried with lupin-fiber towels. Lupin paper and envelopes with lupin adhesive were also provided (Becker-Dillingen 1929).

The techniques of tissue culture are considered a promising tool for crop improvement, but it should not be forgotten that its right place is in combination with conventional plant breeding methods.

Fig. 3. A Ten-week-old regenerated shoot tip of *L. luteus* leaf tissue after transfer to rooting medium. Callus induction on MS medium plus 2 mg BA and 0.2 mg NAA, shoot induction on MS plus 2 mg BA and 0.2 mg IAA/l (arrow indicates abnormal side branching). **B** Plantlets of *L. hartwegii* transferred to soil after multiple shoot formation of half a seed on MS medium plus 1.5 mg BA and rooting on MS medium plus 0.5 mg/l. (*Steps F and XS5 proved to be inefficient and have been eliminated in later experiments). **C** *L. polyphyllus* plant transferred to the field after multiple shoot formation of half a seed, showing flower and seed production in the first year, which normally shows up only in the second year with this species

Protocol

This protocol is based on the experience of the author and data given by Sroga (1983, 1987).

1. Sterilization of Explants

Rinse in 5% commercial bleach for 10 min (seeds 30 min) followed by a short rinse in 96% ethanol and washing in sterilized water three times.

2. Callus Induction

Use MS or B5 medium supplemented with 1 mg 2,4-D, 1 mg BA and 1 mg IAA/l or 4 mg 2,4-D and 0.1 mg kin/l medium. Place in the dark at 24°/25 °C.

3. Plant Regeneration

Shoot induction on MS medium supplemented with 2 mg BA and 0.2 mg–(2 mg) IAA or 1 mg BA and 0.10 mg NAA.

For rooting shoots, supplement medium with above-mentioned auxin MS-based medium or use basal media without hormones. Physical conditions might be 20/15 °C with 16 h light of about 5000 lx.

4. Embryo Culture

Use basal media with high sugar concentration up to 120 g/l if embryos are not fully developed. For stronger shoot or root development BA and NAA might be added and 2 g casein hydrolysate/l in any case.

5. Multiple Shoot Formation

Use shoot tips of apical or side bud position or start with seeds or half seeds, add 1–2 mg BA/l to the basal medium.

Acknowledgment. The author thanks Prof. Dr. W. Hondelmann for his critical review of the manuscript.

References

Allen JG (1985) Prevent lupinosis in sheep grazing on lupins, Farmnote 90/85
Allen JG, Wood PMcR, Crocker KP and Hamblin J (1980) Lupinosis, Farmnote 80/80
Anakhina VS, Kaptosova AG, Kozlova LS, Fedorva (1979) Induction of changes in the alkaloid complex and recombination of characters in *Lupinus luteus* and *L. angustifolius* by means of intervarietal hybridization. Referantivnyi Zhurnal 12:252–265
Anonymous (1984) Insect pests of lupins. West Austr Dep Agric Farmnote 115/84
Anonymous (1986) Feeding lupins to sheep. West Austr Dep Agric Farmnote 12/86
Armstrong EL (1986) Grain legume effects on wheat yield in Central Western NSW Proc 4th Int Lupin Conf, Geraldton, Western Australia, p 305
Baer E von, Baer D von (1986) *Lupinus mutabilis* sweet and stable. Proc 4th Int Lupin Conf, Geraldton, Western Australia, p 283
Bajaj YPS (ed) (1986) Biotechnology in agriculture and forestry 2. Crops I. Springer, Berlin Heidelberg New York Tokyo
Bajaj YPS, Gosal S (1982) Induction of genetic variability in grain-legumes through tissue culture. In: Rao AN (ed) Tissue culture of economic important plants. Singapore, COSTED, pp 25–41
Ball E (1946) Development in sterile culture of stem tips and subjacent regions of *Tropaeolum majus* L. and *Lupinus albus* L. Am J Bot 33:301–318
Ball E (1960a) Sterile culture of the shoot apex of *Lupinus albus*. Growth 24:91–110

Ball E (1960b) Cell division in living shoot apices. Phytomorphology 10:377–396
Baylis JH, Hamblin J (1986) World lupin rotation survey. Proc 4th Int Lupin Conf Geraldton, Western Australia p 317
Becker-Dillingen J (1929) Handbuch des Hülsenfruchterbaues und Futterbaues. Parey, Berlin
Bellido LL (1984) Situation de la production de lupin dans le monde. Proc 3rd Int Lupin Conf La Rochelle, France, pp 465–487
Beltiky B, Kovacs I (1984) Lupin the new break. Panagri, Bradford on Avon
Binding H, Nehls R, Koch R, Finger J, Mordhorst G (1981) Comparative studies on protoplast regeneration in herbaceous species of the Dicotyledonae class. Z Pflanzenphysiol 101:119–130
Brücher H (1970) Beitrag zur Domestikation proteinreicher und alkaloidarmer Lupinen in Südamerika. Angew Bot XLIV:7–27
Chekalin NM (1979) Main trends in breeding leguminous crops. Sel'scokhozyaistvennaya Biol 14:695–704
Davey MR, Bush E, Power JB (1974) Cultural studies of a dividing legume protoplast system. Plant Sci Lett 3:127–133
Decker P, Webb TE, Edwardson JR (1953) Soil and Crop Science. In: Pate JS, Williams W, Farrington (eds) Lupin (*Lupinus spp.*) Soc Florida Proc 13:75–78
Delane R, Hamblin J, Gladstones JS (1986) Reduced branching lupins. J Agric 2:47–48
Diekmann K (1968) Unsere Nutzpflanzen. Parey, Berlin
Duke JA (1981) Handbook of legumes of world economic importance. Plenum Press, New York
Dyubin VN (1978) Agronomic and climatic basis for the cultivation of different lupin species for seed. Byull Vses Ordena Lenina i Ordena Druzhby Narodov Inst Rastenievodstva Imeni NI Vavilova 76:55–60
Ewing MA, Panell DJ, Morrison DA (1986) The place of lupins in the farm rotation. A whole-farm modelling approach. Proc 4th Int Lupin Conf Geraldton, Western Australia pp 152–160
Gamborg OL, Wetter LR (1975) Plant Tissue Culture Methods. Nat Res Counc Can, Prairie Laboratory Saskatoon, Canada
Gladstones, JS (1959) Grain legumes crops. In: Pate JS, Williams W, Farrington P (eds) Lupin (*Lupinus spp.*). Empire J Exp Agric, pp 27:33–342
Gladstones JS (1970) Lupins as crop plants. IAB Field Crop Abstr 23:123–148
Gladstones JS (1980) Recent development in the understanding, improvement and use of lupins. In: Summerfield RJ, Bunting AH (eds) Advances in legume science. Royal Botanic Gardens, Kew, England p. 603–611
Gladstones JS (1982) Breeding lupins in western Australia. J Agric 23:73–76
Gladstones JS (1984) Present stituation and potential of Mediterranean African lupins for crop production, Proc 3rd Int Lupin Conf, La Rochelle, France, pp 17–37
Gleba YuYu, Cherletti P, Bogdanov GA, Golovchenko VI (1986a) Lupinus breeding for pathogen resistance and increased protein quality: New approaches opened up by biochemistry of storage proteins and genetic engineering. Proc 4th Int Lupin Conf, Geraldton, Australia, p 286
Gleba YuYu, Kaleda V, Bogdanov GA (1986b) Experiments on cellular and genetic engineering in legumes (including *Lupinus*). Proc 4th Int Lupin Conf, Geraldton, Australia, p 286
Gollmick F (1937) Über Artkreuzungen bei Lupinen. Züchter 9:65–68
Hackbarth J (1957a) Die Gene der Lupinenarten I. Z Pflanzenzuecht. 37:1–26
Hackbarth J (1957b) Die Gene der Lupinenarten II. Z Pflanzenzuecht. 37:81–95
Hackbarth J (1957c) Die Gene der Lupinenarten III. Z Pflanzenzuecht. 37:185–191
Hackbarth J, Troll HJ (1959) Lupinen als Körnerleguminosen und Futterpflanzen. In: Kappert H, Rudorf W (eds) Handbuch der Pflanzenzüchtung Bd IV. Parey, Berlin, pp 1–51
Hanelt P (1960) Die Lupinen. Die neue Brehm Bücherei 265, Ziemsen, Wittenberg
Harms H, Langebartels C (1985) Pflanzliche Zellsuspensionkultur als System zum Studium des Abbauverhaltens von Chemikalien in Pflanzen In: Kongreßband VDLUFA 1985, pp 445–455
Hill GD (1977) The composition and nutritive value of lupin seed. Nutri Abstr Rev 47:511–529
Hondelmann W (1984a) Die Lupine, alte und neue Kulturpflanze. Giessener Universitätsblätter 1:57–68
Hondelmann W (1984b) The lupin, ancient and modern crop plant. Theor Appl Genet 68:1–9
Hudson AW (1934) Grain legume crops In: Pate JS, Williams W, Farrington P (eds) Lupin (*Lupinus* spp.). NZ J Agric 129:40–43

Jaranowski J (1962) Fertilization and embryo development with genus *Lupinus* Town. Part II. Fertilization and embryo development following reciprocal species hybridization. Genet Pol 3:19–25
Kazimierski T (1961 a) An interspecific hybrid: *Lupinus mutabilis* sweet×*Lupinus Douglasi* Lindl. Genet Pol 2:106–111
Kazimierski T (1961 b) Certain variations in flower structure of interspecific hybrids of lupins. Genet Pol 2:119–127
Kazimierski T (1962) Certain abnormalities in the hybrid: Russel lupin×*Lupinus hartwegii* Lindl. Genet Pol 3:19–25
Kazimierski T (1964) Interspecific crossing in the Mediterranean and American lupin species. Wiss Z Karl Marx Univ 13:697–698
Kazimierski T (1982) Cytogenesis of species and hybrids in the lutei section of the genus *Lupinus*. In: Gross R, Bunting ES (eds) Agricultural and nutritional aspects of lupins. Proc 1st In Lupin Workshop, Lima, Peru 1980, pp 51–68
Klinkowski M (1938) Das iberisch-nordafrikanische Heimatgebiet von *Lupinus luteus* L., *Lupinus angustifolius* L., *Lupinus hirsutus* L., *Lupinus hispanicus* Boiss. and *Lupinus Rothmaleri* Klink. Züchter 10:113–126
Lamberts H (1958) Ein neuer Typus mit nicht abbrechenden Hülsen und ein neuer Artbastard bei Lupinen. Züchter 28:15–17
Lee AE (1955a) Growth in culture of excised portions of lupin embryos. Bot Gaz 116:359–364
Lee AE (1955b) Potentially unlimited growth of lupin callus. Bot Gaz 116:364–368
Linsmaier EM, Skoog F (1965) Organic growth factors requirements of tobacco tissue cultures. Physiol Plant 18:100–127
Masefield GB (1975) Grain legume crops In: Pate JS, Williams W, Farrington P (eds) 1985. Lupin (*Lupinus* spp.) Exp Agric 11:113–118
Mayolo AS de (1980) Tarwi in ancient Peru. In: Gross R, Bunting ES (eds) Agricultural and nutritional aspects of lupins. Eschborn 1982, Proc 1st Int Lupin Workshop, Lima Peru, pp 1–11
Mock IT (1986) The influence of lupins on the disease and nutritional status of a subsequent wheat crop grown on calcareous sand. Proc 4th Int Lupin Conf, Geraldton, Australia, p 306
Mohr U (1986) Fringe benefits of farming with lupins in the Andes. Proc 4th Int Lupin Conf, Geraldton, Australia, p 317
Murashige T, Skoog F (1962) A revised medium for rapid growth and bioassays with tobacco tissue cultures. Physiol Plant 15:473–497
Palada M, Sator C (1981) Observatii asupra incipients ale evolutiei androgenetice a polenului in vitro. Probl Genet Theor App XM/2:115–119
Pate JS, Williams W, Farrington P (1985) Lupin (*Lupinus* spp.). In: Summerfield RJ, Roberts EH (eds) Grain Legume Crops. Collins, London, pp 699–746
Preller JH (1949) Farming in South Africa 24:25–29 In: Summerfield RJ, Roberts EH (eds) Grain Legume Crops
Robbins WJ (1922) Cultivation of excised root tips and stem tips under sterile conditions. Bot Gaz 73:376–390
Römer P, Jahn-Deesbach W (1986) Developments in breeding of *Lupinus mutabilis*. Proc 4th Int Lupin Conf, Geraldton, Western Australia pp 31–39
Rowland JC, Mason MG, Hamblin J (1986) Effect of lupins on soil fertility. Proc 4th Int Lupin Conf, Geraldton, Australia pp 96–111
Sator C (1979) Lupinenanbau zur Körnerproduktion unter besonderer Berücksichtigung der Gelben Lupine (*Lupinus luteus* L.). Garten Organ 3:74–77
Sator C (1982) In vitro breeding of lupins. In: Thompson R, Casey R (eds) Perspectives for peas and lupins as protein crops. Proc Int Symp Protein Production from Legumes in Europe, Sorrento, Italy, pp 79–87
Sator C (1983) Abschlußbericht zum EG-Forschungsvorhaben Nr. 0423 Protein Production, Braunschweig
Sator C (1984) Chances of crop improvement in lupins using tissue culture techniques. Proc 3rd Int Lupin Conf, La Rochelle, France, pp 180–190
Sator C (1985a) Regeneration von Lupinenpflanzen aus Embryonen. Landbauforsch Völkenrode 35:1–4
Sator C (1985b) Regeneration von Lupinenpflanzen aus Antheren. Landbauforsch Völkenrode 35:5–7

Sator C (1985c) Induktion einer 'in vitro-Vielfachsproßbildung' an Samen von Lupinen. Landbauforsch Völkenrode 35:8–10
Sator C (1985d) Studies on shoot regeneration of lupins (*Lupinus* spp.) Plant Cell Rep 4:126–128
Sator C, Mix G, Menge U (1982) Antherenkulturversuche mit *Lupinus polyphyllus*. Landbauforsch Völkenrode 32:37–42
Sator C, Mix G, Menge U (1983) Investigations on anther culture of *Lupinus polyphyllus*. Plant Res Dev 18:37–46
Savvichev KI (1980) Methods for increasing the effectiveness of breeding *Lupinus luteus*. Referativnyl Z 5:259–265
Schäfer-Menuhr A (1985) Propagation of lupins. In: In Vitro Techniques. Advances in Agricultural Biotechnology. Commission of the European Communities, Nijhoff, Dordrecht, pp 23–28
Schäfer-Menuhr A (1986a) In vitro propagation of lupin species. Proc 4th Int Lupin Conf, Geraldton, Australia, p 288
Schäfer-Menuhr A (1986b) Embryo culture of lupin species. Proc 4th Int Lupin Conf, Geraldton, Australia, p 288
Schäfer-Menuhr A (1986c) Somatic hybridization of lupin species. Proc 4th Int Lupin Conf, Geraldton, Australia, p 288
Sengbusch R von (1942) Züchterisch brauchbare Alkaloidbestimmungsmethode. Landwirtsch Jahrb 91:723–880
Sengbusch R von, Zimmerman K (1947) 20 Jahre Süßlupinenforschung und -züchtung in Deutschland. Forsch Fortschr 21/23:249–255
Songin W, Polanski S (1978) Investigation of the white lupins for cultivation as a stubble after crop in the Sczen region. Zezy Naukowe Akad Ronliczej w Szczecinie 72:289–296
Sroga GE (1983) Callus induction and suspension culture of *Lupinus angustifolius* cv. Turkus. Plant Sci Lett 32:183–192
Sroga GE (1987) Plant regeneration of two *Lupinus* spp. from callus cultures via organogenesis. Plant Science 51:245–249
Stocker R (1974) Grain legumes crops NZ J Agric 129:40–43
Swiecicki W (1984) Breeding methods for forage and grain lupins. Proc 3rd Int Lupin Conf, LaRochelle, France, pp 192–205
Swiecicki W (1986a) Developments in *L. albus* breeding. Proc 4th Int Lupin Conf, Geraldton, Australia, pp 14–19
Swiecicki W (1986b) Development in *L. luteus* breeding. Proc 4th Int Lupin Conf, Geraldton, Australia, pp 20–24
Thomas E, Davey MR (1975) From single cells to plants. Wykeham, London
Vuillaume E, Verchere J, Hoff T (1986) In vitro culture of lupins. Proc 4th Int Lupin Conf, Geraldton, Australia, p 287
Wassermann VD (1981) Grain legume crops. Lupin Newslett 2:18–20
Wells HD (1984) Breeding aims and criteria for lupins. Proc 3rd Int Lupin Conf LaRochelle, France, pp 165–178
Williams W (1979) Studies on the development of lupins for oil and protein. Euphytica 28:481–488
Williams W (1986) The current status of the crop lupins. Proc 4th Int Lupin Conf, Geraldton, Australia, pp 1–13
Williams W, Akhtar MA, Faluyi M (1980) Cross compatibility between European and American *Lupinus species*. Bot J Linn Soc 81:225–232
Wink M (1984) Biochemistry and chemical ecology of lupin alkaloids. Proc 3rd Int Lupin Conf, LaRochelle, France, pp 325–343
Wink M, Hartmann T (1980) Alkaloid pattern of cell suspension cultures and differentiated plants of *Lupinus polyphyllus*. Planta Med 38:238–245
Wood PM, McLean GD (1982) Grain legumes crops
Zeven AC (1982) Dictionary of cultivated plants and their region of diversity. Cent Agric Publ Document, Wageningen

III.4 Cell and Tissue Culture of *Stylosanthes* spp.

E. G. M. MEIJER[1] and L. SZABADOS[2]

1 Introduction

The taxonomically complex genus *Stylosanthes* (family Leguminosae, subfamily Papilinoidae, tribe Stylosanthinae or Aeschynomeneae) consists of approximately 50, mainly herbaceous species and subspecies (Allen and Allen 1981; 't Mannetje 1984). Although the genus is native to Central and South America, *Stylosanthes* species can now be found in the tropical and subtropical regions of the Americas, Africa, Asia, and Australia. Many members of this genus exhibit vigorous growth, a deep-rooting habit, and considerable tolerance to both drought and waterlogging. These species are used as forage crops, as green manure, for soil conservation and as a cover crop in the areas listed above (e.g., Edye et al. 1973; Burt et al. 1980). The most commonly cultivated species is *S. guianenis* (Aubl.) Sw. (sometimes spelled *S. guianensis*, see 't Mannetje (1977) and Burt et al. (1980) for discussions), also known as stylo or Brazilian lucerne. *S. guianensis* is widely grown in Africa, Asia, Australia, South and Central America, and extensive breeding programs are under way in Australia and, more recently, also in South America. Other economically impartant species are *S. capitata, S. erecta, S. gracilis* (all Taub.), *S. hamata* (L.) Taub., *S. humilis* H.B.K. (also known in Australia as Townsville stylo), *S. mucronota* Willd., and *S. sundaica* Taub. The cultivated *Stylosanthes* species generally have coarse foliage, a tendency to root at the nodes, and often possess profusely nodulated root systems. A rather interesting aspect of the biology of *Stylosanthes* species (as well as of other members of the Stylosanthinae or Aeschynomeneae such as *Arachis* and *Aeschynomene* spp.) is the somewhat unusual process of rhizobial infection and root nodule development, which proceeds, unlike in other legumes without the formation of infection threads (Chandler et al. 1982; see also review by Meijer and Broughton 1982).

The various plant breeding programs of the cultivated *Stylosanthes* species are mainly aimed at the development of persistent, vigorous and fertile genotypes, and resistance to diseases such as anthracnose (*Colletotrichum gloeosporoides*), blight (*Sclerotium rofii*), bacterial wilt (*Pseudomonas solanacearum*), and other fungal or bacterial diseases (Lenne and Calderon 1984). Resistance to a number of insects, such as *Caloptillia* spp. and *Platyomopsis pedicornis*, is also an important breeding objective (Lenne and Calderon 1984).

[1] Dept. of Plant Molecular Biology, Leiden University, 2333 AL Leiden, The Netherlands
[2] C.I.A.T., Cali, Colombia

Biotechnology in Agriculture and Forestry, Vol. 10
Legumes and Oilseed Crops I (ed. by Y. P. S. Bajaj)

Cell and tissue culture research could complement certain *Stylosanthes* breeding programs and should be particularly useful for the production of valuable somaclonal variants, for interspecific hybridizations of sexually incompatible species (such as *S. capitata* and *S. guianensis*) through protoplast fusion or through the production of hybrids of partially sexually compatible species through embryo rescue techniques, and perhaps for rapid production of homozygous lines through anther or microspore culture. In the context of interspecific hybridization, one could also mention the possible transfer of the genes coding for (proteins involved in the formation of) glandular epidermal hairs. These glandular hairs can secrete a sticky substance which kills insect larvae (Sutherst et al. 1982). Transfer of the genes for anthracnose resistance between *Stylosanthes* species would also be desirable (Miles pers. commun). The introduction of a callus phase could be used for the induction of exchange of chromosomal fragments ("somatic crossing over") in interspecific hybrids which show incomplete pairing of homeologous chromosomes during meiosis (for a discussion of this in vitro-induced cytogenetic modification see review by Larkin and Scowcroft 1981), such as *S. guianensis* × *S. humilis* hybrids (Cameron 1974). In addition, the rapid in vitro clonal multiplication of important and unique genotypes for use in breeding programmes and the isolation of certain, e.g., symbiotic, mutants could be desirable.

2 Cell and Tissue Culture

To date, nine *Stylosanthes* species have been regenerated from callus cultures derived from various plant parts, and it appears therefore that *Stylosanthes* spp. are amongst the least recalicitrant legumes in terms of in vitro regenerability (Table 1). In most cases callus formation is induced on MS medium (Murashige and Skoog 1962) supplemented with synthetic auxins and cytokinins. Subsequent shoot organogenesis is then obtained with the same growth regulators as used for callus production, by increasing the cytokinin/auxin ratio or by omitting auxin from the culture medium altogether (Table 1). Callus and shoot formation have also been observed in cultures of petiole, leaf, and stem sections of *S. guianensis* cv. Cook grown on MS medium devoid of any plant growth sustances, but tissue proliferation and shoot organogenesis were not as profuse as on medium with auxin and cytokinin (Meijer unpubl. obs.). Regeneration of whole plants from tissue cultures of *Stylosanthes* species is always through shoot organogenesis, and to date no reports on the occurrence of somatic embryogenesis in tissue cultures of members of this genus have been published. Explants from practically all plant parts, particularly of *S. guianensis*, have regenerative ability (Table 1). In *S. guianensis* cultures numerous shoots can be seen only 2 (hypocotyl explants) to 3 (leaf explants) weeks after the explants are put into culture, with benzyladenine (BA) being the most effective cytokinin for the induction of shoot organogenesis (Meijer and Broughton 1981; Meijer, unpubl. obs.). Shoot organogenesis in tissue cultures of *Stylosanthes* spp. is clearly promoted by cytokinin and is inhibited by exogenously supplied auxin, which in turn stimulates root formation (e.g., Meijer and Broughton 1981; Meijer 1982b). Inhibitory effects of auxin on shoot organo-

Table 1. Summary of in vitro studies on *Stylosanthes* spp.

Species	Tissue	Culture medium + plant growth regulators (mg/l)	Used for	Reference
S. capitata	Leaf	MS + NAA(1) + BAP(3) MS + NAA(0.01) + BAP(3)	Callus Shoots	Mroginski and Kartha (1984)
S. guianensis	Hypocotyl, leaf, root	MS + 2,4-D(2) + kin(2) MS + NAA(1) + BAP(2)	Callus Shoots	Meijer and Broughton (1981)
S. guianensis	Leaf	MS + NAA(2) + BAP(0.5) MS + NAA(0.01) + BAP(3)	Callus Shoots	Mroginski and Kartha (1981)
S. guianensis	Hypocotyl, leaf	MS + 2,4-D(2) + kin(2) MS + BAP(1)	Callus Shoots	Meijer (1982a)
S. guianensis	Hypocotyl, leaf, cell suspension-protoplasts	MS + NAA(0.5) + BAP(0.5) D2a, V-KM, + NAA(1) + 2,4-D(1) + BAP(0.1) MS + BAP(1)	Cell suspension Colonies Shoots	Meijer and Steinbiss (1983)
S. guianensis	Hypocotyl, leaf	MS + 2,4-D(2) + kin(2) MS + NAA(1) + BAP(2)	Long-term callus Shoots	Meijer (1984)
S. guianensis	Cotyledon, leaf, hypocotyl	MS + NAA(0,5 – 2) + BAP(0.5 – 3) MS + NAA(0.02) + BAP(0.04 – 2)	Callus, cell suspension Shoots	Szabados and Tabares (unpubl.)
S. guianensis	Leaf mesophyll-, cell suspension-protoplasts	V-KM + NAA(1) + 2,4-D(0.2) Zea(0.5) MS + NAA(0.02) + BAP(0.04)	Protoplasts, cell colonies Shoots	Szabados and Roca (1986)
S. hamata	Cotyledons, radicle	CS5 + 2,4-D(2) + kin(0.05) CS5 + kin(3)	Callus Shoots	Scowcroft and Adamson (1976)
S. hamata	Leaf	MS + NAA(1) + BAP(3) MS + NAA(0.01) + BAP(3)	Callus Shoots	Mroginski and Kartha (1984)
S. hippocampoides	Leaf	MS + NAA(1) + BAP(2)	Callus	Rey et al. (1985)
S. humilis	Hypocotyl, leaf	MS + NAA(1) + BAP(1) MS + BAP(1)	Callus Shoots	Meijer (1982b)
S. leiocarpa	Leaf	MS + NAA(1) + BAP(3) MS + NAA(0.01) + BAP(3)	Callus Shoots	Mroginski and Kartha (1984)
S. macrocephala	Leaf, cotyledon	MS + NAA(1) + BAP(1) MS + BAP(1)	Callus Shoots	Szabados and Roca (unpubl.)
S. macrosoma	Leaf	MS + NAA(1) + BAP(2) MS + NAA(0.01) + BAP(2)	Callus Shoots	Rey et al. (1985)
S. montevidensis	Leaf	MS + NAA(1) + BAP(2) MS + NAA(0.01) + BAP(2)	Callus Shoots	Rey et al. (1985)
S. scabra	Leaf	MS + 2,4-D(1 – 2) + kin or BAP (1 – 2) MS + BAP(2)	Callus Shoots	Godwin et al. (1986)

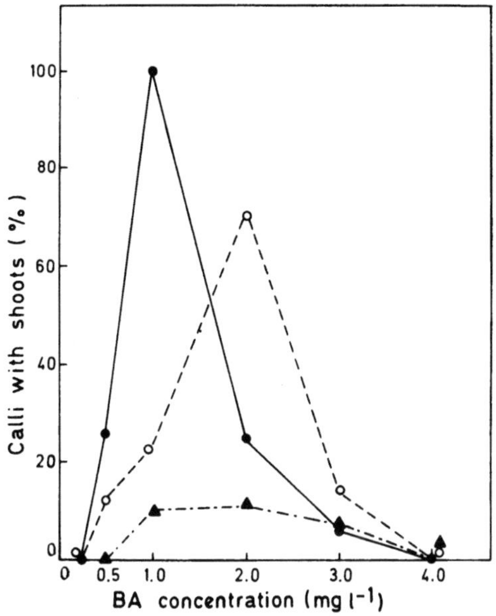

Fig. 1. Effect of BAP and NAA on shoot regeneration in callus cultures of *S. humilis*. (Meijer 1982b)

genesis can, however, be partially offset by increased cytokinin (BA) concentrations in the culture medium. These effects of the auxin/cytokinin balance are clearly illustrated by the examples shown in Table 1 and Fig. 1 for callus cultures of *S. guianensis* and *S. humilis* respectively. Shoots excised from morphogenic *Stylosanthes* readily form roots and eventually develop into plantlets on half- or quarter-strength MS medium without plant growth regulators (e.g., Figs. 2C, D; 3E); regenerated plantlets can be transferred to the soil with survival rates of 50–70% (Meijer 1982b, 1984). In Fig. 2 this callus to plant sequence of *S. humilis* is shown; other *Stylosanthes* species have a similar pattern of in vitro development.

Callus (and cell suspension) cultures of several *Stylosanthes* species are able to maintain their morphogenetic potential over long periods (Scowcroft and Adamson 1976; Meijer 1984; Mroginski and Kartha 1984); this is an advantageous situation for research on mutant isolation. Callus cultures of *S. guianensis* can become auxin autotrophic after a prolonged period of in vitro culture, but such long-term callus lines have a considerably reduced regenerative capacity, possibly due to increased levels of endogenous auxin (Meijer 1984). Curiously, this auxin autotrophy is expressed only when the callus cultures are grown in the light (Meijer 1984). Other long-term *S. guianensis* callus cultures can, however, be highly regenerable.

Experimental protocols for the regeneration of whole plants from cell suspension cultures have been published only for *S. guianensis*. Meijer and Steinbiss (1983) and Szabados and Roca (1986) grew cell suspensions, which were derived from hypocotyl and leaf callus of *S. guianensis* in MS medium supplemented with 0.5 mg/l each of benzyladenine (BA) and α-naphthaleneacetic acid (NAA). These cell suspensions grew rather slowly, with dry weight doubling times of ap-

Fig. 2a–e. Shoot and plantlet formation in tissue cultures of *S. humilis*. (Meijer 1982b) Callus formation was induced on MS containing 0.5 mg/l BAP and 0.5 mg/l NAA. **a, b** After 2 months the callus was transferred to MS containing 1.0 mg/l BAP for shoot formation. **c** After about 4 weeks shoots are transferred to 1/2 MS for further development **d** and after another 2–4 weeks shoots are transferred to 1/4 MS for rooting. **e** Rooted plantlets are then transferred to the soil

proximately 5 days (Meijer and Steinbiss 1983). Plants were regenerated readily, however, following transfers to callus and regeneration media, and plants have been obtained from a 18-month-old cell suspension line of *S. guianensis* (Meijer unpubl. obs.). Mroginski and Kartha (1984) reported on cell suspension cultures

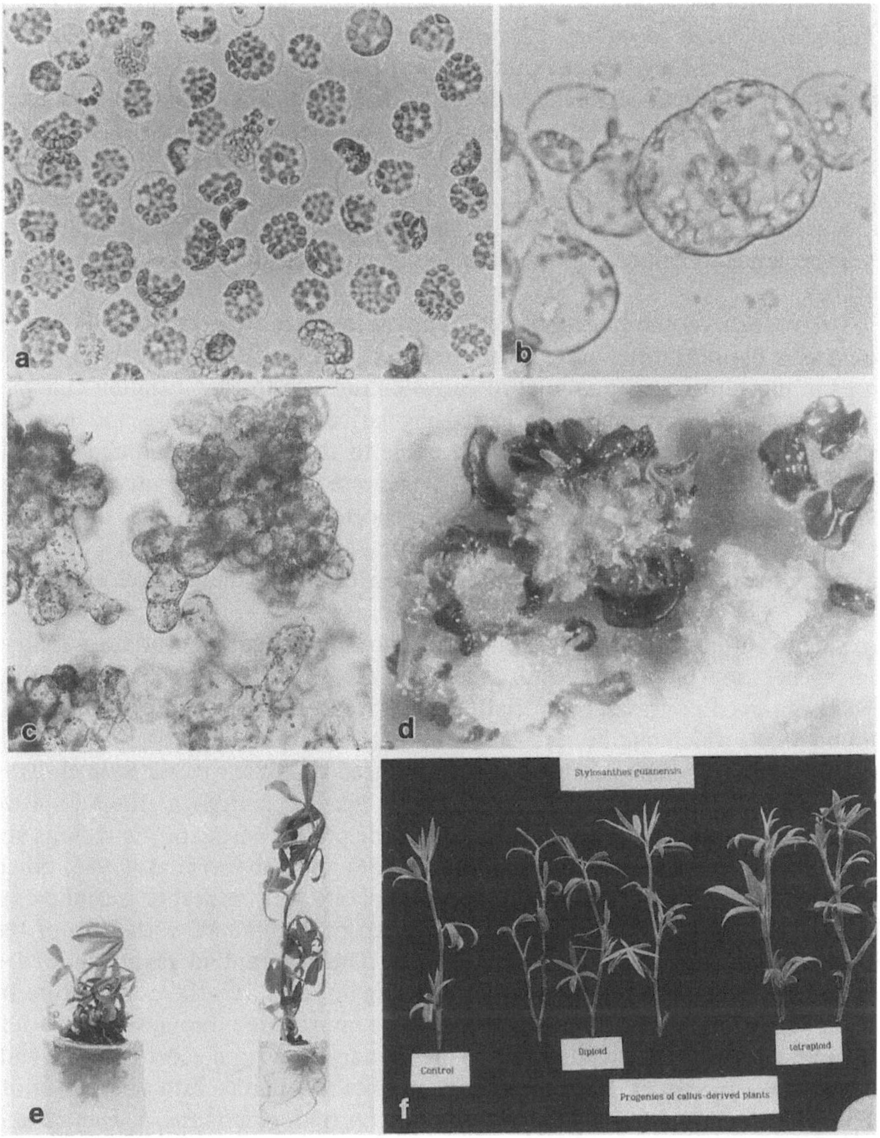

Fig. 3. a–d from mesophyll protoplasts to shoot bud (*S. guianensis*; Szabados and Roca 1986). **e** Shoot and plantlet development (*S. capitata*) on 1/2 MS. **f** tetraploid and diploid progenies of callus-derived *S. guianensis* plants

of several *Stylosanthes* species in a review of legume tissue culture, but they gave no details about explant source and culture media used in those studies.

Regeneration of whole plants from protoplasts has been achieved from cell suspension-derived protoplasts (Meijer and Steinbiss 1983; Szabados and Roca 1986), as well as from mesophyll protoplasts (Szabados and Roca 1986), of *S. guianensis* (Fig. 3A–D). In addition, plants have been regenerated from mesophyll protoplast cultures of *S. macrocephala* (Szabadas and Roca, unpubl). Plating efficiencies of cultured *Stylosanthes* protoplasts, however, are generally low, although most regenerated calli are able to produce shoots (Meijer and Steinbiss 1983; Szabados and Roca 1986). Somatic hybrid cells of *S. guianensis* and *S. capitata* have been obtained through polyethyleneglycol-induced fusion of protoplasts, but no hybrid plants have been regenerated to date (Szabados unpubl.). Studies of other types of in vitro manipulation with *Stylosanthes* species such as anther or microspore culture, in vitro transformation of protoplasts or tissue explants, or embryo culture have not been reported to the knowledge of the authors.

Two reports on the effects of genotype on in vitro regenerative ability of *S. guianensis* tissue cultures have been published. Meijer (1982a) observed differences in morphogenetic potential of callus cultures of three Australian cultivars (cvs. Cook, Endeavour, and Schofield) using various explant sources and medium protocols. Cv. Cook was the most amenable to the in vitro manipulations, although tissue explants from all three cultivars were regenerable to varying degrees (Meijer 1982a). Szabados and Roca (1986) reported genotype-specific differences in the ability of calli which had been regenerated from mesophyll protoplast cultures of two Colombian *S. guianensis* accessions to produce shoots and plantlets.

Several authors have reported the occurrence of somacolonal variation following regeneration of plants from tissue cultures of *Stylosanthes* species. Meijer (1984) observed considerable variation with respect to morphology and vigor among plants which had been regenerated from 2-year-old hypocotyl-derived callus cultures of *S. guianensis*. This in vitro-induced variability persisted in clonally propagated (through cuttings) plants, but no genetic analysis of these putative somaclonal variants was carried out, so that the precise nature of the various abnormalities is not known. Szabados et al. (1986) and Tabares et al. (1986) noted variation among callus-derived *S. guianensis* plants with respect to morphology, fertility, resistance to anthracnose, and chromosome numbers, with 20% of the in vitro regenerated plants being tetraploid. These tetraploid plants showed a somewhat different morphology (Fig. 3F). Again, however, the exact nature of most of the somaclonal variation remains to be investigated through genetic studies. Godwin et al. (1986) reported some preliminary data from genetic studies with somaclonal variants of *Stylosanthes*. They obtained plants with aberrant morphology and reduced fertility from callus cultures of *S. guianensis*, *S. hamata* and *S. scabra*, as well as tetraploid *S. guianensis* plants. In some cases segregation for the morphological characteristics (leaf shape, plant size) and reduced fertility took place in the SC2 generation.

In the course of preliminary experiments aimed at the isolation of disease-resistant mutants, Szabados et al. (1986) found a positive correlation between in vivo resistance of *S. guianensis* genotypes to anthracnose and in vitro resistance of calli of this species to culture filtrate of the fungus and causal agent of this disease *Colletotrichum gloeosporoides*. These experiments suggest that it might be possible to select useful *Stylosanthes* variants in vitro.

3 Selected Methods for *S. guianensis*

3.1 Source and Preparation of Explants for Tissue Culture

Plants are easily grown in a greenhouse or growth chamber, either in soil or in a hydroponic system, but will not always reach the reproductive stage, as many *S. guianensis* varieties require special sequences of different photoperiods before they flower. Seeds or tissue explants may be surface sterilized in 70% ethanol (approximately 1 min) followed by 8–10 min in 5% sodium hypochloride or 0.2% mercuric chloride containing a drop of detergent and three to five rinses in sterile distilled water. It is recommended to scarify seeds lightly with sandpaper prior to surface sterilization. Explants can also be taken from aseptically grown (on half-strength MS) shoot cultures. This is advantageous for mesophyll protoplasts or for other research requiring a carefully controlled environment. Such shoot cultures are easily maintained by periodic subculture of nodal explants.

3.2 Morphogenesis, Plant Regeneration

Callus and shoot formation can be induced on virtually any explant, i.e., hypocotyl, leaf, root, stem (Meijer and Broughton 1981; Meijer 1982b), cultured on MS medium supplemented with 0.5–2 mg/l BAP and 0.1–1 mg/l NAA; explants should be approximately 5 mm long. Shoot buds usually appear after 2–3 weeks, sometimes prior to or simultaneously with callus. Alternatively callus formation may be obtained on MS supplemented with 1.5–2 mg/l 2,4D and 2 mg/l kin. This callus can be subcultured on this medium in the dark at monthly intervals for up to 2 years with retention of the morphogenetic capacity (Meijer 1984). Shoot organogenesis may then be induced efficiently by transferring small pieces of callus to MS medium containing 1 mg/l BAP (no auxin). These regenerating cultures should be grown in continuous light or a photoperiod of 16/8–10/14 light/dark; approximately 1 W/m^2). No shoot formation will occur in the dark but some rhizogenesis may take place on the callus medium in the dark, more so in the light, but root formation will usually cease after two or three subcultures.

Shoots should be excised from the regenerating callus cultures and transferred to half-strength MS medium devoid of plant growth regulators. Plantlets can be allowed to develop further on quarter-strength MS. Well-developed plantlets can then be transferred to the soil or a sand/vermiculite mixture and a nutrient solution. It is advisable to remove the roots which were formed in vitro so that vigorous new roots, which are well adapted to the new conditions, can develop. Treatment of the cut ends with talc-based indolebutyric acid (IBA) helps to speed up the process of root formation. Regenerated plants are easily propagated vegetatively in a similar way by treating cuttings with IBA. Greenhouse or growth chamber temperatures of 28–30 °C are favourable. It is recommended to maintain a high relative humidity during the in vitro to in vivo transition period by using a mist chamber or by growing the plantlets under an inverted and perforated glass tube (Fig. 2e; Meijer 1982b). A very similar protocol can be followed for the regeneration of *S. humilis*, except that callus formation is induced on MS with 0.5 mg/l each of BAP and NAA (Meijer, 1982b).

Table 2. Effect of BAP and NAA on shoot organogenesis in leaf explants of *S. guianensis* cv. Cook after 30 days in culture. (Meijer and Broughton 1981; Meijer, unpubl.)

Growth regulators (mg/l)	No. of days to first shoot	Explants with shoots (%)	Relative shoot density
None	27	22	+
1 NAA	–	0	–
1 BAP 1 NAA	23	64	+ +
2 BAP 1 NAA	23	80	+ + +
2 BAP 2 NAA	23	56	+ +

3.3 Cell Suspensions; Protoplast Isolation and Culture

Cell suspensions can be initiated by transferring small pieces of fast-growing callus to liquid MS medium with 0.5 mg/l each of BAP and NAA (Meijer and Steinbiss 1983; Szabados and Roca 1986). A fine cell suspension can be obtained by subculturing periodically and through sieving or pipetting of the fine fraction. Growth studies (mitotic indices, fresh/dry weight increase, packed cell volume and/or no. of cells/ml) should be carried out to characterize each individual cell line in order to establish the optimal times for subculturing and for protoplast isolation, as the timing of protoplast isolation in relation to the "quality" of the cell suspension can be critical (Meijer and Steinbiss 1983; Szabados and Roca 1986; see below).

Protoplasts can be isolated from cell suspensions, containing mainly relatively small cell clumps with numerous meristematic cytoplasmically dense cells, by using a (phosphate or MES buffered) 1/1 mixture of 2% Cellulase R10, 0.6% hemicellulase, 0.2% pectolyase Y23 and culture medium. Glucose and mannitol (0.3 M each are used as osmoticum). Mesophyll protoplasts may be isolated by incubating leaf strips (from shoot cultures) in the same enzyme mixture but without hemicellulase (Szabados and Roca 1986). Incubation of 5 – 8 h is recommended. Meijer and Steinbiss (1983) and Szabados and Roca (1986) used the Kp8 (Kao 1977) based VK-M medium (Binding and Nehls 1978) with 0.6 M glucose as osmoticum and reduced carbon source. The protoplasts (initial density 10^5 cells/ml) should be diluted at weekly intervals, initially once or twice with protoplast culture medium and later with cell suspension medium, with a gradual reduction of the glucose concentration. Macroscopically visible calli should be transferred to solid MS medium with 0.5 mg/l each of BAP and NAA; shoot formation can then eventually be obtained after transfer to MS containing 1 – 2 mg/l BAP. It has to be emphasized, however, that there is still considerable room for improvement of the protoplast culture procedure and additional experiments with various culture media, plant growth regulators and culture techniques (agarose, feeder cells) are therefore necessary.

4 Conclusions

A priority in cell and tissue culture research of *Stylosanthes* species must be the improvement of the protoplast culture technology. This is an absolute necessity for interspecific hybridization through protoplast fusion and in vitro transformation of protoplasts through co-cultivation with *Agrobacterium tumefaciens* or employing a direct DNA uptake method. The development of low density plating techniques should also be a priority in this context. The rapidity and ease with which shoots can be obtained from cultured explants, particularly of *S. guianensis*, warrant investigations into the potential of this system for in vitro transformation by co-cultivation with *Agrobacterium* (cf. Horsch et al. 1985). The simple regeneration system could also possibly be used in fundamental studies of plant development biology, e.g., of the role of exogenous and endogenous factors controlling shoot organogenesis. Finally, serious efforts should be made to develop anther and microspore culture techniques so that homozygous lines can become available for use in breeding programmes.

References

Allen ON, Allen EK (1981) The Leguminosae. A source book of characteristics, uses and nodulation. Univ Press, Madison, Wisconsin

Binding H, Nehls R (1978) Regeneration of isolated protoplasts of *Vicia faba*. Z Pflanzenphysiol 88:327–332

Burt RL, Williams WT, Grof B (1980) *Stylosanthes*-structure, adaptation and utilisation. In: Summerfield RJ, Bunting AH (eds) Advances in legume science. R Bot Gard, Kew, pp 553–558

Cameron DF (1974) Novel variation from wide crosses in the *Stylosanthes* genus. Proc 11th Int Grassland Congr, Moscow, pp 726–731

Chandler MR, Date RA, Roughley RJ (1982) Infection and root nodule development in *Stylosanthes* species by *Rhizobium*. J Exp Bot 33:47–57

Edye LA, Burt RL, Williams WT, Grof B (1973) A preliminary agronomic evaluation of *Stylosanthes* species. Austr J Agric Res 24:511–525

Godwin ID, Gordon GH, Cameron DF (1986) Callus culture derived somaclonal variation in the tropical pasture legumes *Stylosanthes guianensis*, *S. hamata* and *S. scabra*. In: 6th Int Congr Plant tissue cell culture Abstr, Univ Minnesota, Minneapolis, pp 253

Horsch RB, Fry JE, Hoffmann NL, Eichholtz D, Rogers SG, Fraley RT (1985) A simple and general method for transferring genes into plants. Science 227:1229–1231

Kao KN (1977) Chromosomal behaviour in somatic hybrids of soybean-*Nicotiana glauca*. Mol Gen Genet 150:225–230

Larkin PJ, Scowcroft WR (1981) Somoclonal variation- a novel source of variability from cell culture for plant improvement. Theor Appl Genet 60:197–214

Lenne JM, Calderon MA (1984) Disease and pest problems of *Stylosanthes*. In: Stace HM, Edye LA (eds) The biology and agronomy of *Stylosanthes*. Academic Press, New York, pp 279–295

Li Z (1981) Plantlet regeneration from mesophyll protoplasts of *Digitalis lanata* Ehrh. Theor Appl Genet 60:345–347

't Mannetje L (1977) A revision of the varieties of *Stylosanthes guianensis* (Aubl.) Sw. Austr J Bot 25:347–362

't Mannetje L (1984) Considerations on the taxonomy of the genus *Stylosanthes*. In: Stace HM, Edye LA (eds) The biology and agronomy of *Stylosanthes*. Academic Press, New York, pp 1–12

Meijer EGM (1982a) Shoot formation in tissue cultures of three cultivars of the tropical pasture legume *Stylosanthes guianensis* (Aubl.) Sw. Z Pflanzenzuecht 89:169–172

Meijer EGM (1982 b) High-frequency plant regeneration from hypocotyl- and leaf-derived tissue cultures of the tropical pasture legume *Stylosanthes humilis*. Physiol Plant 56:381–385

Meijer EGM (1984) Some properties of long-term tissue cultures of *Stylosanthes guianensis* (Aubl.) Sw. (Leguminosae). J Plant Physiol 117:131–135

Meijer EGM, Broughton WJ (1981) Regeneration of whole plants from hypocotyl-, root- and leaf-derived tissue cultures of the pasture legume *Stylosanthes guianensis*. Physiol Plant 52:280–284

Meijer EGM, Broughton WJ (1982) Biology of legume-*Rhizobium* interactions in nodule formation. p. 107–129. In: Kahl G, Schell JS (eds) Molecular biology of plant tumors. Academic Press, New York, pp 107–129

Meijer EGM, Steinbiss HH (1983) Plantlet regeneration from suspension and protoplast cultures of the tropical pasture legume *Stylosanthes guianensis* (Aubl.) Sw. Ann Bot 52:305–310

Mroginski LA, Kartha KK (1981) Regeneration of plants from callus tissue of the forage legume *Stylosanthes guianensis*. Plant Sci 23:245–251

Mroginski LA, Kartha KK (1984) Tissue culture of legumes for crop improvement. Plant Breed Rev 2:215–264

Murashige T, Skoog F (1962) A revised medium for rapid growth and bioassays with tobacco tissue cultures. Physiol Plant 15:473–497

Rey HY, Bovo OA, Mroginski LA (1985) Cultivo in vitro de tejidos de tres especies de *Stylosanthes* (Leguminosae) Agronomie 5:819–824

Schenk RU, Hildebrandt AC (1976) Medium and techniques for induction and growth of monocotyledonous and dicotyledenous plant cell cultures. Can J Bot 50:199–204

Scowcroft WR, Adamson JA (1976) Organogenesis from callus cultures of the legume *Stylosanthes hamata*. Plant Sci Lett 7:39–42

Sutherst RW, Jones RJ, Schitzeling HJ (1982) Tropical legumes of the genus *Stylosanthes* immobilize and kill cattle ticks. Nature (Lond) 295:320–321

Szabados L, Roca WM (1986) Regeneration of isolated mesophyll and cell suspension protoplasts to plants in *Stylosanthes guianensis*. Plant Cell Rep 5:174–177

Szabados L, Tabares E, Lopez P, Miles J, Lenne J, Roca W (1986) Variability in *Stylosanthes* cell and tissue culture. In: 6th Int Congr Plant tissue and cell culture. Abstr, Univ Minnesota, Minneapolis, p 271

Tabares E, Pineda O, Lopez P, Szabados L, Roca W (1986) Eusqueda de variacion somaclonal an *Stylosanthes guianensis* usando cultivo de tejidos. IV Congr Latinoam Bot, Medellin

Williams EG, De Lautour G (1980) The use of embryo culture with transplanted nurse endosperm for the production of interspecific hybrids in pasture legumes. Bot Gaz 141:252–314

III.5 Birdsfoot Trefoil (*Lotus corniculatus* L.)

E. B. SWANSON[1], D. A. SOMERS[2], and D. T. TOMES[3]

1 Introduction

1.1 Description and Distribution

The genus *Lotus* (Leguminosae) consists of over 200 species, including xerophytic desert and alpine perennials and salt-tolerant annuals (Larsen 1958; Urbanska 1984). The regional center of origin for *Lotus* is probably the Mediterranean basin, where the greatest diversity of the species occurs. This complex genus has three basic chromosome numbers, 5, 6, and 7 (Angulo and Real 1977). The cytogenetic analysis of 108 species has been reported, including 25 tetraploids, 71 diploids, and 12 with both diploid and tetraploid cytotypes (Grant 1986). Most of the species can be classified as weeds; however, some, (e.g., *Lotus corniculatus* (4×), *Lotus tenuis* (2×) and *Lotus uliginosus* (2×) have attained important agronomic status. Of these *Lotus corniculatus* (birdsfoot trefoil) is the most widely grown and the focus for the majority of the biotechnology research in the genus. Elucidation of the genetic origins of *L. corniculatus* has been the subject of debate for over a quarter of a century (Grant 1986). It is postulated that *L. corniculatus* arose through hybridization of *L. uliginosus* and *L. alpinus* and/or *L. tenuis* followed by chromosome doubling in the hybrid (Ross and Jones 1985), although this explanation may still be incomplete (Grant 1986).

Birdsfoot trefoil is grown in various areas across North America, throughout much of Europe and parts of Asia, in several countries of South America including Brazil, Argentina, Chile, and Uruguay, and in areas of India, Australia, and New Zealand (Grant and Marten 1985). The name was derived from the inflorescence, which is a typical umbel having four to eight florets attached at the end of a relatively long peduncle. During seed pod maturation, the pods branch out in a birdsfoot type pattern. Birdsfoot trefoil was first described in Europe about 1597 and entered North America around the turn of the century (MacDonald 1946). In North America it has grown from a minor nuisance weed to a forage crop of over one million hectares (Grant and Marten 1985).

[1] Department of Plant Science, Allelix Crop Technologies Inc., 6850 Goreway Drive, Mississauga, Ontario L4V1P1, Canada
[2] Department of Agronomy and Plant Genetics, University of Minnesota, St Paul, Minnesota 55108, USA
[3] Department of Biotechnology Research, Pioneer Hi-bred International, Box 38, Johnston, Iowa 50131, USA

Biotechnology in Agriculture and Forestry, Vol. 10
Legumes and Oilseed Crops I (ed. by Y. P. S. Bajaj)

The most serious diseases in birdsfoot trefoil are caused by a group of organisms associated with crown and root rots. The particular complex of organisms making up these diseases may vary with the environment, but have included *Fusarium, Verticillium, Macrophomina, Rhizocttonia,* and *Sclerotina* (Berkenkamp et al. 1972; Murphy et al. 1985; Grant and Marten 1985). The warm humid conditions found in parts of the southern US provide for excellent development of several of these diseases, and this has precluded the use of trefoil in these areas.

1.2 Agronomic Strengths and Weaknesses of *L. Corniculatus* (Birdsfoot Trefoil)

Birdsfoot trefoil grows under a wide range of soil conditions (pH, fertility, and moisture), the feed value is similar to alfalfa, it can persist when heavily grazed, and it does not cause bloat (Seaney and Henson 1970; Smith 1975; Tesar 1977). Some of its major limitations are: it lodges easily (a feature reduced when planted with a companion crop such as timothy) (Anderson and Metcalfe 1957), it is difficult to establish, seed production costs are high because of seed shattering (McGraw and Beuselinck 1983), and it is a poor competitor with weeds. Breeding efforts have been hampered by the fact that birdfoot trefoil is tetraploid, relatively self-incompatible, and has an indeterminate growth habit.

2 In Vitro Approaches

Plant breeding efforts in birdsfoot trefoil are confined to a very small proportion of the land alloted to forage breeding, and tissue culture methods have been exploited to reduce the space, time, and cost for future cultivar improvement (Tomes and Swanson 1983; Swanson and Tomes 1983; Arcioni et al. 1988). Many of the goals of biotechnology research in this species are directed towards helping the breeding programs. These include: haploidy, the transfer of non- or low seed shattering genes, selection for herbicide and disease tolerance, and the maintenance and multiplication of selected genotypes. Biotechnological research in *Lotus* has increased significantly over the past decade, and has resulted in reports on the isolation of callus and suspension cultures, in vitro storage and multiplication of germplasm, plant regeneration, anther culture and embryo rescue, somaclonal variation, in vitro selection, and protoplast isolation and fusion (Table 1.).

2.1 Callus and Suspension Culture

Callus initiation has been produced from a variety of tissue sources including anthers, leaves, roots and stem segments of a number of genotypes of birdsfoot trefoil. Tremendous diversity occurs between and within cultivars for the color, growth rate, and type (nodular, friable etc.) of the callus (Fig. 1) (Swanson and Tomes 1980; Swanson 1983). As birdsfoot trefoil is a relatively self-incompatible

Table 1. In vitro studies in *Lotus*

Explant/study	Species	Response	Reference
1. Embryo	Interspecific	Embryo rescue, interspecific hybrids	Grant et al. (1962) Somaroo and Grant (1971)
2. Endosperm	Interspecific	Endosperm transplant interspecific hybrids	De Lautour et al. (1978)
3. Anthers and Microspores	Interspecific	Aneuploids, euploids, no haploid plants	Niizeki and Grant (1971) Sequin and Grant (1978) Tomes and Peterson (1980)
4. Node sections	*L. corniculatus*	Propagation and cold maintanance	Tomes (1979)
5. Callus	*L. corniculatus*	Callus initiation leaf, root, stem	Swanson and Tomes (1980)
6. Protoplasts	Interspecific	Leaf, root, hypocotyl Fusions	Ahuja et al. (1983b) Webb and Woodcock (1986) Wright (1986) Wright et al. (1987) Webb et al. (1987)
7. Plant Regeneration	*L. corniculatus*	Callus and suspensions Protoplasts	Swanson and Tomes (1980) Orshinsky et al. (1983) Ahuja et al. (1983a) Webb and Woodcock (1986) Wright (1986)
8. Ultra-structure	*L. corniculatus*	Callus	Howarth et al. (1981) Howarth et al. (1983)
9. Stress metabolism	*L. corniculatus*	Callus	Keith and McKersie (1986) Swanson et al. (1983)
10. Selection	*L. corniculatus*	Callus and suspensions	Swanson (1979) Swanson and Tomes (1980) Swanson and Tomes (1983) Tomes and Swanson (1983) Swanson (1983) MacLean and Grant (1986)
11. Somaclonal variation	*L. corniculatus*	Callus	Mariotti et al. (1984) Orshinsky and Tomes (1984) Orshinsky and Tomes (1985) Damiani et al. (1985) Pezzotti et al. (1985)

cross-pollinating tetraploid, every seed is a distinct genotype. A procedure for the maintenance and multiplication of genotypes of birdsfoot trefoil has been reported (Tomes 1979). A number of different media have been used to propagate the species in vitro, including B5 (Gamborg et al. 1968), MS (Murashige and Skoog 1962), and UM (Uchimiya and Murashige 1974). A B5-based medium with 1 mg/l 2,4-D is routinely used for callus initiation and maintenance. The cultures are usually maintained at 25 °C with a 16-h photoperiod of 100-200 $\mu E/m^2/s$. However, the very broad genetic base of the species may require researchers to explore other media to obtain optimum growing conditions for specific genotypes.

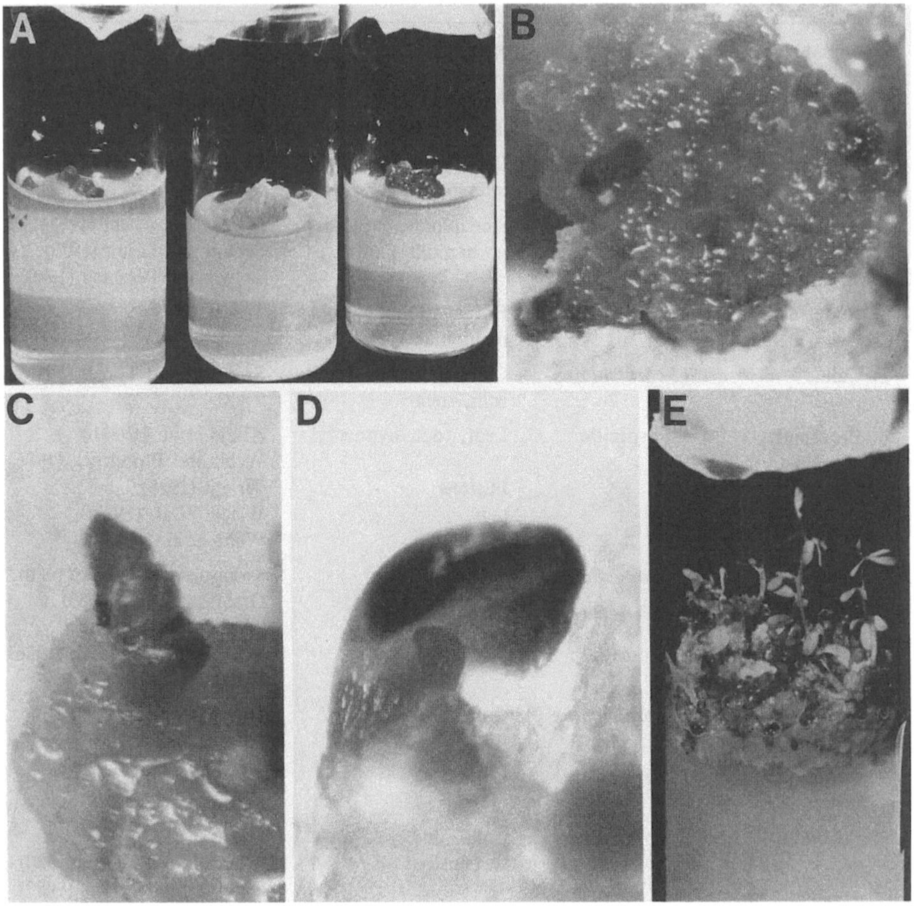

Fig. 1 A–E. Distinct genotype specific callus types are produced from birdsfoot trefoil (**A**). Differentiation to plants through organogenesis is initiated on BA medium from small organized centers (**B**), which give rise to shoots (**C**, **D**), which develop into plants (**E**)

While actively growing suspension have been characterized, these also vary with the genotype (Swanson and Tomes 1980; Ahuja et al. 1983a). Cell suspensions have been characterized for cell morphology, growth rate, and plant regeneration potential. The proportion of particular cell types within a callus may differ substantially; however, actively growing suspensions rapidly develop a higher proportion of small (round cells of up to 20 μM in diameter) and medium (primarilly oval-shaped cells of 20–45 μM in diameter) and reduce the percentage of larger, more vacuolated cells (Table 2). While trefoil calli may retain a good regeneration potential for several years in culture, the number of plants recovered from suspension cultures drops rapidly with culture age (Swanson and Tomes 1980). The establishment of an actively growing single-celled suspension in *Lotus* has not been reported.

Table 2. The growth rate of suspension cultures of two genotypes of birdsfoot trefoil characterized according to cell size. (Swanson and Tomes 1980)

Day of culture	Genotype A		Genotype B	
	Cell size	No. of cells/ml × 1000	Cell size	No. of cells/ml × 1000
0	Small	40d*	Small	9f
	Medium	43d	Medium	60d
	Large	17e	Large	31e
6	Small	600a	Small	100c
	Medium	460b	Medium	150b
	Large	85c	Large	50d
10			Small	260a
			Medium	280a
			Large	50d

* Means within a column followed by the same letter are not significantly different at the 5% level according to Duncan's multiple range test.

2.2 Plant Regeneration

Plant regeneration has been observed in birdsfoot trefoil by both organogenesis and embryogenesis. Regeneration rates are generally excellent within birdsfoot trefoil and have achieved with callus, suspensions, and protoplasts isolated from a variety of different explant sources (Fig. 1) (Swanson and Tomes 1980 Ahuja et al. 1983a). Culture genotype is the most critical variable in determining callus growth rate, type, and number of plants regenerated per gram of callus, while culture origin has little effect (Table 3). Plant regeneration appears to be controlled by only a few genes and improvements to regeneration have proven responsive to phenotypic selection (Swanson and Tomes 1980; Glover and Tomes 1982). A procedure has been described which increases the number of plants regenerated per

Table 3. The affect of genotype and origin on callus growth and number of plants regenerated per gram of callus. (Data from Swanson 1983)

Genotype	Origin	Fresh weight (g) from 2 of callus after 20 days	Plant regeneration (plants/g) B5 medium	
			No hormones	0.05 mg/l BA
A	Leaf	13.0a*	4c	17cd
A	Stem	12.7a	3c	21c
A	Root	12.9a	5c	23c
B	Stem	5.6c	14ab	46a
C	Stem	9.6b	19a	38b
D	Stem	4.0c	10b	14d

* Means within a column that are followed by the same letter are not significantly different at the 5% level according to Duncan's multiple range test.

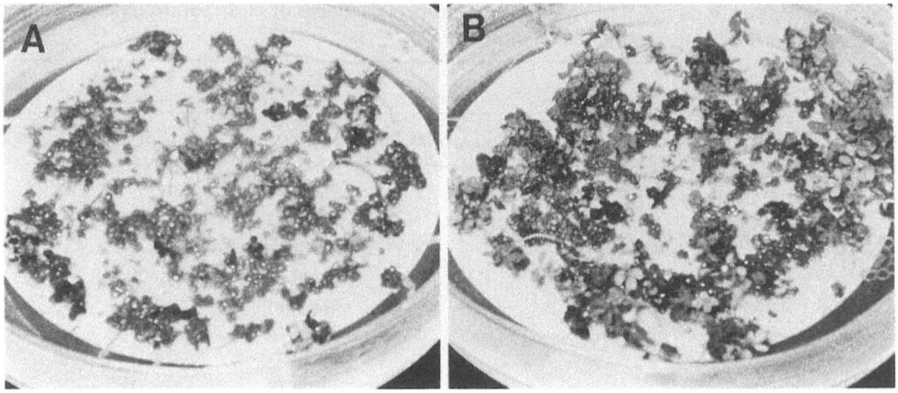

Fig. 2A,B. Plant regeneration from non homogenized (**A**) and homogenized (**B**) callus of birdsfoot trefoil

Table 4. Summary of plant regeneration from callus of six genotypes grown on 2,4-D, glyphosate and metribuzin for six subcultures prior to plant regeneration. (Orshinsky et al. 1983)

Genotype	Plants/g of Callus	
	Nonhomogenized	Homogenized
L-48	2.0	65.3
L5-73	3.8	96.8
L5-98	10.0	89.5
L5-108	8.7	91.6
L5-118	5.3	96.8
L5-129	1.0	34.2
Mean	5.1	79.0

gram of callus by as much as 15-fold (Fig. 2) (Orshinsky et al. 1983). The procedure involves the homogenization of callus tissue in a blender, producing a population of small callus masses and plating these masses onto a B5-based medium usually containing 0.05 mg/l benzyladenine (BA). This procedure increases both the number of shoots produced per gram of callus and the rate of shoot regeneration. When this regeneration procedure was applied to callus isolated after in vitro selection the number of plants regenerated per gram of callus was increased even more dramatically (Table 4). No adverse phenotypic effects caused by the procedure could be detected in the regenerated plants.

2.3 Anther and Embryo Culture

Anther, pollen, and ovule culture has not been successful in *Lotus* (Niizeki and Grant 1971; Sequin and Grant 1978; Tomes and Peterson 1981). These authors did observe various aneuploids, euploids, and morphological variant cells from these cultures; however, no polyhaploids of *Lotus* were obtained.

Embryo culture and endosperm transplants have been used to recover interspecific hybrids in *Lotus* (Grant et al. 1962; Somaroo and Grant 1971; De Lautour et al. 1978). The development of these techniques may lead to the isolation of improved cultivars, help elucidate the evolution of the genus, and facilitate the ability to transfer specific traits to improve birdsfoot trefoil. For example, seed shattering is absent or reduced in the diploid species *L. coimbrensis*, *L. ornithopodioides*, and *L. burttii*. It has been estimated that as much as 50% of the seed harvest of birdsfoot trefoil may be lost to preharvest seed pod dehiscence (McGraw and Beuselinck 1983). Unfortunately, the forementioned species do not cross easily with birdsfoot trefoil (Phillips and Keim 1968; Gershon 1961). However, these diploids will cross with the diploids *L. alpinus* and *L. japonicus*, which are more closely related to birdsfoot trefoil (Somaroo and Grant 1972). Grant et al. (1986) believe that after colchicine doubling and embryo rescue the non-seed shattering character of the diploids may be transferred to birdsfoot trefoil via the more compatible amphidiploids.

2.4 Genetic Stability and Somaclonal Variation

Several general features are evident from a summary of research into somaclonal variation in birdsfoot trefoil. There is significant variation produced in callus cultures; this variation increases with time in culture, gross morphological mutants are rare, most regenerants are euploid, most of the variation is negative in relation to agronomic fitness, the amount of variation may be underestimated as the presence of recessive mutations is not revealed, and the genetic nature of this variation has not been confirmed in many studies (Orshinsky and Tomes 1984, 1985; Mariotti et al. 1984; Damiani et al. 1985; Pezzotti et al. 1985). In one study (Orshinsky and Tomes 1985), high shoot regeneration remained for up to 2.5 years in culture before it rapidly declined. This study reported variation in a number of quantitative traits (Table 5). The authors concluded that the prolonged culture periods may have primarily been responsible for this variation. An earlier report (Orshinsky and Tomes 1984), concluded that the variation in plants produced by in vitro node culture compared to cutting-derived plants was primarily due to

Table 5. Variation in quantitative traits among plants regenerated in five progressive subcultures of 1.5-year-old callus. (Orshinsky and Tomes 1985)

Subculture	Days to first flower	Stems/plant	Florets on first umbel	Yield (g dry wt.)
1	98.5ab[a]	27.0a	4.4a	10.9a
2	92.2b	18.7b	3.4ab	8.7ab
3	106.1a	19.8ab	2.7b	7.2b
4	98.4ab	18.4b	2.5b	8.0ab
5	113.8a	19.7ab	2.6b	7.9b

[a] Means within a column followed by the same letter are not significantly different at the 5% level according to Duncan's multiple range test.

Fig. 3 A–D. Somaclonal variation has resulted in plants of dwarf stature (**B**), different maturity dates (**C** all plants are somaclonal variants from one genotype, *arrow* denotes early flowering of one line), and albinism (**D**). Some somaclonal variants have more than one mutation (**A**)

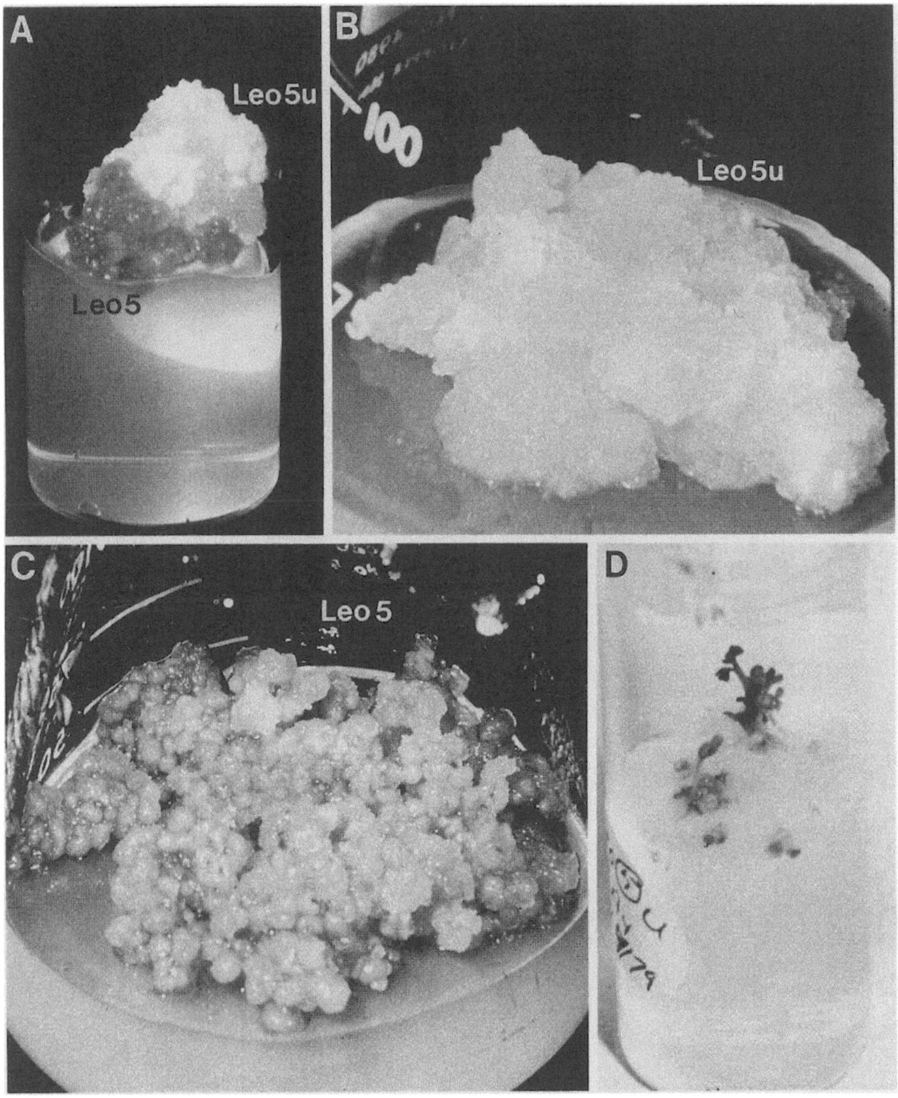

Fig. 4A–D. An achlorophylous mutant (Leo 5U) arose spontaneously (**A, B**) from a dark green parental callus (Leo 5) (**C**). Plant regeneration was obtained from small green islands which denoted areas of reversion (**D**)

the influence of hormones in the medium and was epigenetic in nature. Ethyl methanesulfonate (EMS)-treated node cultures produced a number of chlorophyll-deficient variants. Due to the lack of good genotypic markers in birdsfoot trefoil, many of the reported variants relate principally to stature and albinism (Fig. 3). Genetic analysis of quantitative traits is hampered by the self-incompatible and tetraploid status of the species. Some researchers have reported higher levels of variation after in vitro selection (Swanson 1983; Swanson and Tomes 1983;

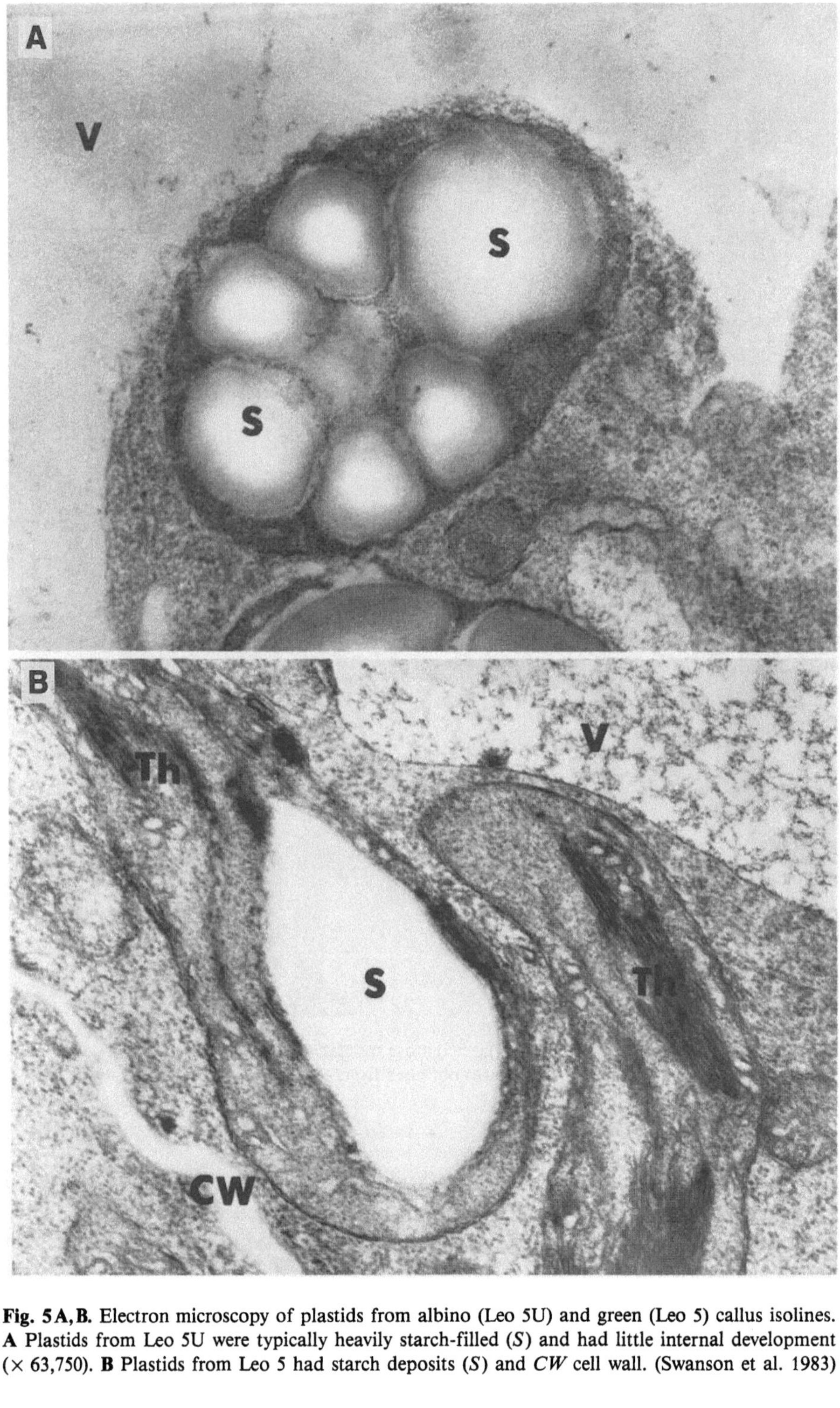

Fig. 5 A, B. Electron microscopy of plastids from albino (Leo 5U) and green (Leo 5) callus isolines. **A** Plastids from Leo 5U were typically heavily starch-filled (*S*) and had little internal development (× 63,750). **B** Plastids from Leo 5 had starch deposits (*S*) and *CW* cell wall. (Swanson et al. 1983)

Table 6. The affect of several factors on the growth (% of controls) of callus isolines containing functional and defective plastids. (Data from Swanson 1983)

Treatment	Leo 5 (green)	Leo 5U (albino)
0.5% Sucrose	65	5
0.5% Sucrose and 0.8% mannitol	66	5
Metribuzin (93 μM)	18	100
Aluminum (2 mM)	35	0
Dark Incubation	25	100
Benzyl Adenine (0.05 mg/ml)	58	0
Glyphosate (2.5 mM)	40	10
Streptomycin (0.05%)	35	15

Controls were grown in the light on B5 medium with 2% sucrose and 1 mg/l 2,4-D.

MacLean and Grant 1986) or from suspension culture (Swanson 1983). However, genetic analysis of the recovered variants is generally not reported.

An achlorophyllous callus isoline was isolated from birdsfoot trefoil cv. Leo (Swanson 1979). Besides being white, this callus line could be distinguished from its green parent by a very rapid growth rate (Fig. 4), absence of any plastid ribosomes or ribosomal RNA, altered chloroplast ultrastructure (Fig. 5), and by different responses to a number of xenobiotic agents and growth conditions (Table 6) (Swanson et al. 1983). This line was 100% inhibited by benzyladenine at 0.05 mg/l, and isolated green islands could be observed in the white callus, indicating sites of reversion. Plants were recovered from many of these islands, and this feature was used to study cellular differentiation from birdsfoot trefoil callus (Howarth et al. 1981, 1983). These studies elucidated that changes in different organelles, the plasmalemma, and the appearance of transfer cells were associated with reversion to green callus and the initiation of plant regeneration. The authors also reported that some form of physiological isolation appears to precede plant differentiation from birdsfoot trefoil cell cultures.

2.5 In Vitro Selection

In vitro selection in birdsfoot trefoil has focused principally on herbicide tolerance (Swanson and Tomes 1980, 1983; Swanson 1983; MacLean and Grant 1986). Thes studies included selection for 2,4-D, chlorsulfuron, glyphosate, and metribuzin tolerance. Genetic characterization of the herbicide tolerance isolated in these studies has been complicated by the self-incompatible tetraploid nature of the species. While several selected regenerants have improved herbicide tolerance (Swanson 1983) (Fig. 6), the only report of genetic transmission of herbicide tolerance after in vitro selection was with 2,4-D (Swanson and Tomes 1983) and this resulted in only a portion of the tolerance obtained through whole plant selection (Devine et al. 1975). We have observed spray chamber tolerance in some of the progeny from selections for metribuzin and glyphosate tolerance (Fig. 6) and some field tolerance in the progeny of one selection for glyphosate tolerance (unpubl.). The results from selection for chlorsulfuron (Glean) tolerance indicated

Fig. 6 A, B. Birdsfoot trefoil plants (cv. Leo) are killed by 1 kg/h glyphosate in a spray chamber; however, some selections survived and produced seed (**A**). All plants are depicted at 10 days post spraying. Tolerant clones in (**B**) are denoted by *arrows*

Fig. 7. a The in vitro toxicity of 5 mM glyphosate to callus growth from six genotypes of birdsfoot trefoil. **b** The response to glyphosate of the same six genotypes if divided into high (above 7% of fr. wt.) and low (below 4.5% of fr. wt.) dry weight categories. (Swanson 1983)

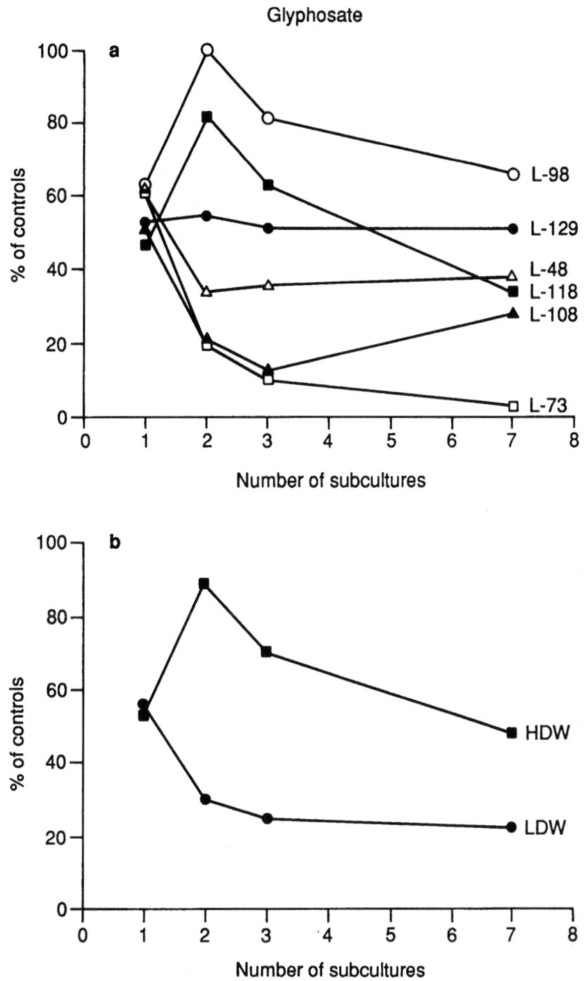

that not only did the cells surviving the herbicide in vitro not regenerate tolerant plants, but these calli frequently regenerated plants more sensitive than the unselected controls (MacLean and Grant 1986).

The in vitro effect of a particular xenobiotic (e.g., herbicides) has varied not only between cultivars but also between individual genotypes within a cultivar (Swanson 1983). Callus morphology had pronounced effects on the in vitro sensitivity of a particular culture to a xenobiotic. The callus response of six genotypes of birdsfoot trefoil to glyphosate are shown in Fig. 7. After seven subcultures the genotypes ranged from a high of 97% inhibition (genotype 73) to a low of 33% (genotype 98). Much of the apparent difference between genotypes could be reduced to differences in callus density. When the the genotypes were divided into either low (LDW) or high dry weight (HDW) callus a clear separation in response was evident, with LDW genotypes more susceptible to glyphosate. Similarly, HDW callus was more sensitive to metribuzin (Swanson 1983). As indicated earlier in this review, the callus photosynthetic potential also had significant effects on the in vitro response of a callus to a number of different growth conditions. These factors and others (callus morphology and growth rate, etc.) will need to be considered in the evaluation of the in vitro selection potential for particular compounds in the future. These factors will be critical in establishing correlations between in vitro and whole responses in birdsfoot trefoil.

The effect of growth temperature, light, and ABA (abscisic acid) levels was assessed for their ability to induce freezing tolerance in callus cultures of birdsfoot trefoil (Keith and McKersie 1986). This study reported that freezing tolerance in the callus approached the plant tolerance when the callus was exposed to B5 medium containing 10 M ABA for 7–14 days. The results of this work should provide more details to researchers interested in selection for cold tolerance in vitro.

2.6 Protoplast Isolation and Culture

Protoplasts of birdsfoot trefoil divide to form callus, which exhibits the prolific plant regeneration characteristics of the species. Ahuja et al. (1983a) reported that protoplasts from roots, hypocotyls, cotyledons, and suspension cultures of regenerated plants when cultured on the enriched media developed by Kao and Michayluk (1975) and Kao (1977). The division frequencies were between 18% and 42% and suspension culture-derived protoplasts divided at lower frequencies than seedling tissues (Ahuja et al. 1983a). Recently, mesophyll protoplasts of mature plants have also been cultured to form plant-regenerating callus (Webb and Woodcock 1986). A number of critical factors have been associated with optimum protoplast isolated and culture conditions in *Lotus*, including: age and condition of the donor material, incubation of tissues or protoplasts at low temperatures prior to plating, careful selection of protoplast plating density and dilutions and use of appropriate nurse cultures (Ahuja et al. 1983a; Davey et al. 1984). Recently, Webb et al. (1987) reported that increased colony formation could be achieved by plating isolated mesophyll protoplasts in a soft agarose medium. Plants were regenerated from the resulting colonies.

2.7 Protoplast Fusion and Somatic Hybridization

Embryo rescue has been used to obtain interspecific hybrids between *Lotus* species with a high degree of sexual incompatibility (Grant 1965; Somaroo and Grant 1972). As pointed out in this chapter, a major problem with seed production in birdsfoot trefoil is preharvest seed shattering. *L. coimbrensis* species has completely indehiscent seed pods. However, hybridization between birdsfoot trefoil and *L. coimbrensis* has been particularly difficult (Gershon 1961; Grant 1965; Phillips and Keim 1968). Recently, Wright et al. (1987) undertook to transfer the indehiscent seed pod character of *L. coimbrensis* to birdsfoot trefoil by protoplast fusion. They isolated protoplasts from etiolated hypocotyls of birdsfoot trefoil and from suspension cultures of *L. coimbrensis* (Wright 1986; Wright et al. 1986). Although a number of different media combinations were tested, no plant regeneration was obtained from the callus colonies recovered from protoplasts derived from the *L. coimbrensis* cell cultures. Protoplasts of these two species can be fused using a polyethylene glycol (PEG), high pH/Ca method described by Ahuja et al. (1983b) and the PEG, dimethyl sulfoxide method described by Menczel and Wolfe (1984). Both methods resulted in about 2% heterokaryon formation on average. The primary limitation to somatic hybridization in *Lotus* is that the genus is almost entirely unexplored for chemical, auxotrophic, or visible phenotypes, which have proven useful for the selection of somatic hybrids in other species (Evans 1983). A potential source of markers may come from some of the selection research described earlier in this chapter; however, further work in that area is required.

In the absence of selectable markers, birdsfoot trefoil and *L. coimbrensis* somatic hybrids were selected using two procedures (Wright 1986). The first method involved the fluorescent labeling of *L. coimbrensis* protoplasts (derived from suspension culture) with fluorescein diacetate and the chlorophyll autofluorescence of green cotyledonary tissue (Wright 1986). Double-stained heterokaryons were visuable selected and these did initiate division. However, they did not develop further, possibly due to the absence of an appropriate nurse culture to facilitate the recovery of colonies from very low plating densities. The use of cell sorting of heterokaryons may alleviate the problem of low protoplast plating density through the isolation of larger numbers of heterokaryons (Galbraith 1984). A second selection method has proven more successful. Etiolated hypocotyl protoplasts of birdsfoot trefoil were inactivated with 10 mM iodoacetate prior to fusion to inhibit division (Medgyesy et al. 1980). These protoplasts were fused to the *L. coimbrensis* suspension protoplasts (which did not regenerate plants on the media tested) and the appearance of regenerated plants used to identify putative hybrids (Wright et al. 1987). From 32 isolated cell lines plants were regenerated, of which three appeared to be hybrids and 29 were like their parents. The hybrids were identified by having light yellow flowers (trefoil flowers are bright yellow and *L. coimbrensis* flowers are white) and purple keel tips (*L. coimbrensis*). The hybrids had complimentary isozymes of esterase and phosphoglucomutase and the hybrid euploid chromosome number of 36. The plants were male sterile and backcrossing has not succeeded, indicating that the plants may also be female sterile. The 29 trefoil regenerants are either selection escapes or are cytoplasmic hybrids ("cybrids"). These regenerants may warrant some further characterization.

Future approaches to somatic hybrid selection in *Lotus* may exploit the observation that birdsfoot trefoil is susceptible to infection by the transforming bacterium *Agrobacterium tumefaciens* (Webb 1986). Antibiotic or herbicide resistances encoded on disarmed Ti plasmids that are expressed in plant cells will provide useful selectable markers that can be engineered into plants or cell lines for somatic hybrid selection.

One obvious candidate for intergeneric fusion would be to transfer the "bloat safeness" of birdsfoot trefoil to bloat-causing forages such as alfalfa. The high tannin content of birdsfoot trefoil is one explanation for the observation that ruminants consuming the forage do not develop bloat.

3 Conclusions and Prospects

Lotus is a genetically complex genus which contains enormous variability. Conventional breeding approaches are limited by economic constraints wich restricts the amount of genetic improvement which is being attempted in the genus. Tissue culture methods have developed rapidly and successfully, and it is likely that biotechnology will be of great value to this species in the future by supplementing the conventional plant breeding approaches.

References

Ahuja PS, Hadiuzzaman S, Davey MR, Cocking EC (1983a) Prolific plant regeneration from protoplast-derived tissues of *Lotus corniculatus* L. (birdsfoot trefoil). Plant Cell Rep 2:101–104

Ahuja PS, Hadiuzzaman S, Rybczynski JJ, Davey MR (1983b) Factors influencing the frequency of heterokaryon formation in forage legume protoplasts. Poster Proc 6th Int Protoplast Symp, Birkhauser, Basel, pp 72–73

Anderson, SR, Metcalfe DS (1957) Seed yields of birdsfoot trefoil as affected by preharvest clipping and by growing in association with three adapted grasses. Agron J 49:52–55

Angulo MD, Real MC (1977) A new basic chromosome number in the genus *Lotus*. Can J Bot 55:1848–1850

Arcioni S, Mariotti D, Damiani F, Pezzotti M (1988) Birdsfoot (*Lotus corniculatus* L.), crownvetch (*Coronilla varia* L.) and Sainfoin (*Onobrychis viciifolia* scop.). In: Bajaj YPS (ed) Biotechnology in agriculture and forestry 6. Crops II. Springer, Berlin Heidelberg New York Tokyo, pp 548–572

Berkenkamp B, Folkins L, Meeres J (1972) Crown and root rot of birdsfoot trefoil in Alberta. Can Plant Dis Surv 52:1–3

Damiani F, Mariotti D, Pezzotti M, Arcioni S (1985) Variation among plants regenerated from tissue culture of *Lotus corniculatus* L.. Z Pflanzenzuecht 94:332–339

Davey MR, Jothan JP, Gilmour DM, Cocking EC, Ahuja PS, Arcioni S, Elkington JSH, Ghose TK, Gosal SS, Hadiuzzaman S, Lu DY, Mariotti D, Rhybczynski JJ, Thomzik J, Xu Z-H (1984) The development of protoplast systems for gene transfer in forage legumes. Proc Int Symp Plant Tissue Culture and its Agricultural Application, Univ Nottingham, England

De Lautour G, Jones WT, Ross MD (1978) Production of interspecific hybrids in *Lotus* aided by endosperm transplants. NZ J Bot 16:61–68

Devine TE, Seaney RR, Linscott DL, Hagin RD, Brace N (1975) Results of breeding for tolerance to 2,4-D in birdsfoot trefoil. Crop Sci 15:721–724

Evans DA (1983) Protoplast fusion. In: Evans DA, Sharp WR, Ammirato P, Yamada Y (eds) Handbook of plant cell culture. Vol 1. Macmillian, New York, pp 291–321

Galbraith DW (1984) Selection of somatic hybrid cells by fluorescence-activated cell sorting. In: Vasil IK (ed) Cell culture and somatic cell genetics in plants. Vol 1. Academic Press, NY, pp 433–477

Gamborg OL, Millar RA, Ojima K (1968) Nutrient requirements of suspension cultures of soybean root cells. Exp Cell Res 50:151–158

Gershon D (1961) Breeding for resistance to pod dehiscence in birdsfoot trefoil, (*Lotus corniculatus* L.) and some studies on the anatomy of pods, cytology and genetics of several *Lotus* species and their interspecific hybrids. Ph D Thesis, Cornell Univ Ithaca, NY

Glover K, Tomes DT (1982) Genetic control of differentiation in birdsfoot trefoil. Can J Plant Sci 62:255

Grant WF (1965) A chromosome atlas and interspecific hybridization index for the genus *Lotus* (Leguminosae). Can J Genet Cytol 7:457–471

Grant WF (1986) The cytogenetics of *Lotus* (Leguminosae) J Natur Hist 20(6):1461–1466

Grant WF, Marten GC (1985) Birdsfoot trefoil. In: Heath ME, Barnes RF, Metcalfe DS (eds) Forages: the science of grassland agriculture. Iowa State Univ Press, Ames, pp 98–108

Grant WF, Bullen MR, Nettancourt D de (1962) The cytogenetics of *Lotus*. 1. Embryo-cultured interspecific diploid hybrids closely related to *L. corniculatus* L. Can J Genet Cytol 4:105–128
Grant WF, O'Donoughue L, St-Marseille P, Realson J (1986) Genetic manipulation takes many forms for crop improvement. Macdonald J. 47(2):2–5
Howarth MJ, Peterson RL, Tomes DT, Swanson EB (1981) Ultrastructural characteristics of an achlorophyllous callus line of *Lotus corniculatus*. In Vitro 17:891–899
Howarth MJ, Peterson RL, Tomes DT (1983) Cellular differentiation in small clumps of *Lotus corniculatus* callus. Can J Bot 61:507–517
Kao KN (1977) Chromosomal behaviour in somatic hybrids of soybean-*Nicotiana glauca*. Mol Gen Genet 150:225–230
Kao KN, Michayluk MR (1975) Nutritional requirements for the growth of *Vicia hajastana* cells and protoplasts at a very low population density in liquid media. Planta 126:105–110
Keith CN, McKersie BD (1986) The effect of abscisic acid on the freezing tolerance of callus of *Lotus corniculatus* L. Plant Physiol 80:766–770
Larsen K (1958) Cytotaxonomical studies in *Lotus* IV. Some cases of polyploidy. Bot Tidskr 54:44–56
MacDonald HA (1946) Birdsfoot trefoil (*Lotus corniculatus* L.). Its characteristics and potentialities as a forage legume. Cornell Univ Agr Exp Sta Mem 261.
MacLean NL, Grant WF (1986) Evaluation of birdsfoot trefoil (*Lotus corniculatus*) regenerated plants following in vitro selection for herbicide tolerance. Can J Bot 65(6):1275–1280
Mariotti D, Pezzotti M, Falistocco E, Arcioni S (1984) Plant regeneration from leaf-derived callus of *Lotus corniculatus* L. cv. Franco. Genet Agrar 38:219–232
McGraw RL, Beuselinck RP (1983) Growth and seed yield characteristics of birdsfoot trefoil. Agron J 75:443–446
Medgyesy P, Menczel L, Maliga P (1980) The use of cytoplasmic streptomycin resistance; chloroplast transfer from *Nicotiana tabacum* into *Nicotiana sylvestris*, and isolation of their somatic hybrids. Molec Gen Genet 179:693–698
Menczel L, Wolfe K (1984) High frequency of fusion induced in freely suspended protoplast mixtures by polyethylene glycol and dimethylsulfoxide at high pH. Plant Cell Rep 3:196–198
Murashige T, Skoog F (1962) A revised medium for rapid growth and bioassays with tobacco tissue cultures. Physiol Plant 15:473–497
Murphy WM, Gotlieb AR, Dugdale DT (1985) The effects of *Fusarium* wilt and weed control on survival of birdsfoot trefoil. Can J Plant Science 65:329–334
Niizeki M, Grant WF (1971) Callus, plantlet formation, and polyploidy from cultured anthers of *Lotus* and *Nicotiana*. Can J Bot 49:2041–2051
Orshinsky BR, Tomes DT (1984) Comparison of plants derived from cuttings, node cultures, and ethyl methanesulfonate-treated node cultures of birdsfoot trefoil (*Lotus corniculatus*). Can J Bot 62:1501–1504
Orshinsky BR, Tomes DT (1985) Effect of long term culture and low temperature incubation on plant regeneration from callus lines of birdsfoot trefoil (*Lotus corniculatus* L.). J Plant Physiol 119:389–397
Orshinsky BR, Swanson EB, Tomes DT (1983) Enhanced shoot regeneration from homogenized callus cultures of birdsfoot trefoil (*Lotus corniculatus* L.). Plant Cell Tissue Organ Cult 2:341–347
Pezzotti M, Arcioni S, Damiani F, Mariotti D (1985) Time-related behaviour of phenotypic variation in *Lotus corniculatus* regenerants under field conditions. Euphytica 34:619–623
Phillips RL, Keim WF (1968) Seed pod dehiscence in *Lotus* and interspecific hybridization involving *L. corniculatus*. Crop Sci 8:18–21
Ross MD, Jones WT (1985) The origin of *Lotus corniculatus*. Theor Appl Genet 71:284–288
Seaney RR, Henson PR (1970) Birdsfoot trefoil. Adv Agron 22:119–157
Sequin G, Grant WF (1978) Anther culture in the genus *Lotus* (*Fabaceae*), In: 4 Int Congr plant tissue and cell culture. Univ Calgary, Alberta, Canada, p 55
Smith D (ed) (1975) Forage management in the north. Kendall/Hunt, Dubuque, Iowa, pp 117–124
Somaroo BH, Grant WF (1971) Interspecific hybridization between diploid species of *Lotus* (Leguminosae). Genetica 42:353–367
Somaroo BH, Grant WF (1972) Crossing relationships between synthetic *Lotus* amphidiploids and *L. corniculatus*. Crop Sci 12:103–105
Swanson EB (1979) A white-colored isoline in *Lotus* which regenerates green plants. Lotus Newslett 10:5

Swanson EB (1983) Isolation and characterization of variant callus cultures of *Lotus corniculatus* L. and the in vitro selection of herbicide tolerant plants. Ph D Thesis, Univ Guelph, Ontario
Swanson EB, Tomes DT (1980) Plant regeneration from cell cultures of *Lotus corniculatus* L. and the selection and characterization of 2,4-D-tolerant cell lines. Can J Bot 58:1205–1209
Swanson EB, Tomes DT (1983) Evaluation of birdsfoot trefoil plants and their progeny after in vitro selection for 2,4-diclorophenoxyacetic acid. Plant Sci Lett 29:19–24
Swanson EB, Tomes DT, Hopkins WG (1983) Modifications to callus culture characteristics and plastid differentiation by the formation of an albino callus of *Lotus corniculatus*. Can J Bot 61:2500–2505
Tesar MB (1977) Productivity of birdsfoot trefoil in Michigan. Pub. CS-LC-7402, Michigan State Univ, USA
Tomes DT (1979) A tissue culture procedure for the propagation and maintenance of *Lotus corniculatus* genotypes. Can J Bot 57:137–140
Tomes DT, Peterson RL (1981) Isolation of a dwarf plant responsive to exogenous GA3 from anther cultures of birdsfoot trefoil. Can J Bot 59:1338–1342
Tomes DT, Swanson EB (1979) "Test tube" breeding in forage legumes. Notes Agric 15:21–22
Tomes DT, Swanson EB (1983) Application of in vitro selection to plant improvement. In: Tomes DT, Ellis BE, Harney PM, Kasha KJ, Peterson RL (eds) Application of plant cell and tissue culture to agriculture and industry. Univ Press, Guelph, pp 25–43
Uchimiya H, Murashige T (1974) Evaluation of parameters in the isolation of viable protoplasts from cultured tobacco cells. Plant Physiol 54:936–944
Urbanska KM (1984) Polymorphism of cyanogenesis in *Lotus alpinus* from Switzerland. II Phenotypic and allelic frequencies upon acid silicate and carbonate. Ber Geobot Inst Eth, Stiftung Rubel 51:132–163
Webb KJ (1986) Transformation of forage legumes using *Agrobacterium tumefaciens*. Theor Appl Genet 72:53–58
Webb KJ, Woodcock S (1986) Plant regeneration from cells and protoplasts in the herbage legumes. In: Somers DA, Gengenbach BG, Biersboer DD, Hackett WP, Green CE (eds) 6th Int Congr Plant tissue and cell cult. Univ Minnesota, Minneapolis, p 229
Webb KJ, Woodcock S, Chamberlain DA (1987) Plant regeneration from protoplasts of *Trifolium repens* and *Lotus corniculatus*. Plant Breed 98(2):111–118
Wright RL (1986) Somatic hybridization between birdsfoot trefoil (*Lotus corniculatus* L.) and *L. coimbrensis* willd. MSc Thesis, Univ Minnesota, St Paul, Minnesota
Wright RL, Somers DA, McGraw RL (1987) Somatic hybridization between birdsfoot trefoil (*Lotus corniculatus* L.) and *L. coimbrensis* willd. Theor Appl Genet 75:151–156

III.6 Crownvetch (*Coronilla varia* L.)

D. L. GUSTINE and B. G. MOYER[1]

1 Introduction

Crownvetch (*Coronilla varia* L.) is a member of the legume family. The plants develop dense stands in the form of a canopy up to 1 meter in height. The leaves are alternate and pinnately compound with 9 to 25 leaflets per leaf (Fig. 1). Stems are angular and hollow, with most leaves at the upper surface of the canopy. Flowers are in stalked clusters (umbels), vary in color from white to dark purple, and are fragrant. The roots form an extensive branching network with deeply penetrating tap roots. The plant spreads by sprouting from adventitious buds on the roots.

1.1 Importance of Crownvetch

Control of soil erosion in roadway cuts and strip-mine spoil banks has been of political and esthetic importance in the past four decades in the United States and other countries. The plant perhaps best adapted for this purpose in the US north of 35° latitude and east of the Mississippi river is crownvetch. This plant has been widely used for erosion control and ground cover (McKee 1964; Ruffner 1964; Ross and Rodgers 1968). Crownvetch stands do not require maintenance and persist for decades, thus providing excellent erosion control and attractive bank stabilization. Crownvetch is also considered valuable as a protective cover for wildlife. It is also used as a forage crop in areas where other forage legumes are difficult to grow because of low soil fertility or wet conditions. It is an acceptable forage when grazed (Burns et al. 1977) or fed as hay (Burns et al. 1972). The seeds contain cardenolide cardiac glycosides, which have a bitter taste (Duke 1981). Crownvetch seeds are sometimes used to flavor flour in the Soviet Union (Duke 1981).

1.2 Distribution and Area Under Cultivation

Crownvetch is native to central and southern Europe and North Africa; it is now common throughout the Mediterranean area (Duke 1981). The plant was probably introduced into North America as a perennial flower and as a contaminant

[1] USDA-ARS, US Regional Pasture Research Laboratory University Park, Pennsylvania 16802, USA

Biotechnology in Agriculture and Forestry, Vol. 10
Legumes and Oilseed Crops I (ed. by Y. P. S. Bajaj)

Fig. 1. Mature crownvetch plants in flower

of alfalfa or clover seed imported from Europe. It was originally cultivated in the US from escapes growing in legume fields in the northeastern and central states. Three commercial cultivars are available in the US; Penngift originated in eastern Pennsylvania, Chemung originated in the Chemung river valley of New York, and Emerald originated in Iowa from plantings of seed from materials originally collected in southern Russia in 1911 (Hawk 1968). No data are available on the total area planted with crownvetch worldwide, but limited information is available for the US. Forage production of the crop in northeastern North America was estimated at 101250ha in 1983 (M.L. Risius pers. commun.). Total area of roadside plantings was estimated at 12000 ha in Pennsylvania (Rodgers et al. 1964; Duke 1981). Crownvetch is widely used for roadside plantings in many states throughout the midwestern and northeastern US.

1.3 Conventional Practices for Propagation and Improvement

1.3.1 Cultivation

Crownvetch may be established in a conventional well-prepared, firmed seedbed by drilling or broadcasting scarified seed (5 to 20 kg/ha) at 6 to 8 mm depth. Good stands are slow to establish, usually requiring 2 to 3 years for complete coverage. The best soils are well-drained medium to heavy silty clay loam or a sandy clay loam with medium fertility levels. Optimum soil pH is 6.5 to 7, but crownvetch will become established on more acid soils. The seed should be mixed with rhizobial inoculant specific to crownvetch. No fungicide is needed at seeding. No-till establishment is possible if the sod is adequately suppressed with a herbicide (Duke 1981).

1.3.2 Genetics and Breeding

Crownvetch is autotetraploid (2n = 24) and cross-pollinated, but is slightly self-compatible (Risius 1968). Present cultivars were developed primarily for erosion control, soil conservation, hay, or forage. Strains vary widely in vigor, plant type, number and fineness of stems, time of maturity, and percentage of tannin (7% to 11% dry wt. basis) (Duke 1981). Concentrations of 3-nitropropanoic acid (NPA), a toxic constituent of crownvetch, were found in the range of 1.5 to 15 mg NO_2/g dry wt. aerial foliage in Penngift. Two generations of selection for high and low NPA-containing plants, followed by progeny testing, indicated genetic control over NPA concentration (Rineer 1977). Breeding for decreased concentrations of NPA may result in greater susceptibility to insect feeding (Byers et al. 1986).

1.4 Susceptibility to Diseases and Insects

Crownvetch is relatively free of insect and disease problems, partly because it is a recently cultivated crop. Wheeler (1974) surveyed insect populations and damage to crownvetch growing in large forage fields, highway plantings, and greenhouses. He reported that 19 of 33 primary pests associated with major forage crops also were observed on crownvetch. However, only grasshoppers (*Melanoplus differentialis*) and a gelechiid moth (*Stomopteryx palpilineela*) were associated with damage to crownvetch, which occurred on south-facing slopes of highway plantings. Other insects found in abundance on crownvetch were the alfalfa plant bug, potato leafhopper, clover stem borer, and a nitidulid beetle. The resistance of crownvetch to insects may be due in part to aliphatic nitro compounds in the plant (Byers et al. 1986).

1.5 Phytochemistry

Crownvetch contains two classes of toxic constituents, aliphatic nitro compounds and cardenolide cardiac glycosides. Investigations of the first group established NPA as the toxic entity. NPA occurs esterified to glucose as mono-, di-, and tri-esters (Gustine 1979). Concentrations of esters in leaves and flowers may be as high as 12 and 20 mg NO_2/g dry wt., respectively. If administered at a sufficiently high dosage, either NPA or the esters are toxic to chickens, pigs, meadow voles, and mice (Gustine 1979; Gould and Gustine 1982). The primary mechanism of toxicity in vivo is the irreversible inhibition of succinate dehydrogenase (EC

1.3.99.1), an enzyme in the tricarboxylic acid cycle (Gould and Gustine 1982; Gustine and Moyer 1983). The NPA compounds are generally not toxic to ruminants because rumen microorganisms hydrolyze the esters and detoxify NPA (Gustine et al. 1977). Little is known about the biosynthesis of NPA and the esters. NPA may be derived from aspartic acid (Gustine 1979), and directly esterified to glucose. Recently, crownvetch tissue cultures have been utilized for NPA metabolism studies (Sect. 2.6).

Cardenolides (Williams and Cassady 1976; Steinmetz 1962) hyrcanoside and coronillin (a cardiac glycoside, not to be confused with the NPA ester) occur in flowers and seeds of crownvetch. These constituents were reported to have potentially beneficial pharmaceutical properties. They have anti-tumor activity and alleviate irregular heart action (Williams and Cassady 1976). No information is available concerning their biosynthesis and pharmacological or toxicological mechanism of action.

2 In Vitro Approaches

2.1 General Account

Mariotti and Arcioni (1983) and Moyer and Gustine (1984) described callus induction, followed by plant regeneration through somatic embryogenesis. The two protocols for regeneration utilize callus cultured on media containing 2,4-D and kin, but differ substantially in the strategy for subsequent induction of embryogenesis. Mariotti and Arcioni (1983) induced somatic embryogenesis by transferring callus tissue to a medium in which 2-iP and IAA replaced the original phytohormones. Alternatively, Moyer and Gustine (1984) induced somatic embryogenesis by sequential transfer of tissue first to a phytohormone-free basal medium, and then to a phytohormone-free medium high in reduced nitrogen. Thus, the first strategy achieved somatic embryogenesis in the presence of phytohormones, while the second induced somatic embryogenesis in their absence.

Moyer and Gustine (1986) established suspension cultures, which they used to demonstrate esterfication of exogenous NPA to glucose (see also Arcioni et al. 1988).

2.2 Methods of Explant Sterilization

Crownvetch callus tissue cultures have been established from both leaves and hypocotyls. Mariotti and Arcioni (1983) used leaflets as the primary explants for establishment of cultures. They surface sterilized leaves with 1.05% sodium hypochlorite for 15 min, rinsed the leaves eight to ten times with sterile distilled water, and then scratched or peeled the epidermis of leaflets before incubating the explants on the callus-inducing medium.

Moyer and Gustine (1984) followed the protocol of Meijer and Broughton (1981) for surface sterilization of leaves from established greenhouse-grown plants, but used discs cut out of leaflets as explants, rather than peeled, scratched, or intact leaflets.

Moyer and Gustine (1984) explanted hypocotyls excised from sterile seedlings. They first surface sterilized seeds in 1.05% sodium hypochlorite under partial vacuum for 15 min. After several rinses with sterile distilled water, seeds were germinated in sterile distilled water in flasks on a rotary shaker. Several times during the germination period, the liquid was poured off and replaced with fresh sterile distilled water.

2.3 Media Composition

Crownvetch callus has been induced from primary explants using modifications of both B5 medium (Gamborg et al. 1976) and MS medium (Murashige and Skoog 1962). Variations of both media have been used by several groups culturing legumes. Mariotti and Arcioni (1983) initiated callus from leaflet explants on UM medium (Uchimiya and Murashige 1974; Table 1). They tested several concentrations of 2,4-D in combination with 0.25 mg kin/l; 2 mg 2,4-D/l was the most effective in inducing callus (Table 2). Calli were maintained on UM medium for 4–5 months prior to transfer to the medium inducing embryogenesis (MS medium modified to contain 1 mg 2-iP/l and 0.1 mg IAA/l; Table 3).

For callus induction, Moyer and Gustine (1984) used G3 medium (B5 modified to contain 1.25 mg 2,4-D/l (5.66 μM), 0.25 mg NAA/l (1.34 μM), and 0.25 mg kin/l (1.16 μM); Table 1). Established cultures were transferred to D20K2

Table 1. Media for callus induction and maintenance

Constituent	Procedure A[a]	Procedure B[b]	
	UM[c], medium for induction and maintenance (mg/l)	G3, medium for induction (mg/l)	D2OK2, medium for maintenance (mg/l)
Mineral salts (reduced N)	Murashige and Skoog (1962) (20.6 mM)	Gamborg et al. (1976) (2.02 mM)	Gamborg et al. (1976) (2.02 mM)
2,4-D	2.0	1.25	4.42
NAA	–	0.25	–
Kin	0.25	0.25	0.43
Thiamine HCl	10	10	10
Nicotinic acid	5	1	1
Pyridoxine HCl	10	1	1
m-Inositol	100	100	100
Glycine	2	–	–
Casein hydrolyzate	2000	–	–
Sucrose	30000	30000	40000
Agar	8000	8500	8500
pH	5.7	5.8	5.5

[a] Mariotti and Arcioni (1983).
[b] Moyer and Gustine (1984).
[c] Uchimiya and Murashige (1974).

Table 2. Growth evaluation of calli from different plant genotypes of *Coronilla varia* on UM media with 0.25 mg/l of kin and various concentrations of 2,4-D. (Data from Mariotti and Arcioni 1983)

Plant identification number	UM medium with 0.25 mg/l of kin and various concentrations of 2,4-D (mg/l)				Plant identification number	UM medium with 0.25 mg/l of kin and various concentrations of 2,4-D (mg/l)			
	1	2	4	8		1	2	4	8
1	–	–	–	–	26	–	+ +	+	–
2	–	+ +	+	–	27	–	–	+	+
3	–	+	–	–	28	–	–	–	+
4	–	+ + +	+	+	29	–	+ + +	+	+
5	–	+	–	–	30	+	+ + +	+	+
6	–	+	+	–	31	–	–	–	–
7	–	+ +	+	–	32	–	–	+	–
8	–	+	+	–	33	–	+ +	+	–
9	–	+	+	–	34	–	+ +	–	–
10	–	–	–	–	35	–	–	+	–
11	+	+ +	+	–	36	–	+	+	–
12	–	+ + +	+	+	37	–	–	–	–
13	–	+	+	–	38	–	–	+	–
14	–	–	–	–	39	–	+ + +	+ +	+
15	–	+	–	–	40	–	+ +	+	+
16	–	–	–	–	41	–	+	+	–
17	–	+ +	+	–	42	–	–	–	–
18	–	–	–	–	43	–	+	+	+
19	–	+ + +	+	+	44	–	+ +	+	+
20	–	–	+	+	45	–	+	+	+
21	–	+ +	+	+	46	–	–	–	–
22	–	–	–	+	47	–	+	+	+
23	–	+ +	+	+	48	–	+ + +	+	+
24	–	+	+	–	49	+	+ +	–	–
25	–	+ +	–	+	50	–	+ +	+	–

$$R = \frac{\text{fresh callus weight after 1 month}}{\text{initial fresh callus weight.}}$$

+ : R ranges from 1 to 2.
+ + : R ranges from 2 to 3.
+ + + : R ranges from 3 to 4.
– : absence of callus.

(B5 modified to contain 4.42 mg 2,4-D/l (20 μM) and 0.43 mg kin/l (2 μM); Table 1). To induce embryogenesis, calli were transferred to basal B5 (Table 3) and then to either basal B5 with increased levels of reduced nitrogen (20 mM ammonium chloride or 10 mM asparagine; Table 3) or to (BOi2Y Blaydes medium: Blaydes 1966) modified to contain 100 mg inositol and 2 g yeast extract/l).

2.4 Establishment of Callus and Suspension Cultures

Mariotti and Arcioni (1983) evaluated 50 5-week-old crownvetch plants for the ability to form callus from leaflets, and found a range of responses; explanted

Table 3. Media for induction of embryogenesis

Constituent	Procedure A[a]	Procedure B[b]	
	Modified MS, induction medium (mg/l)	Basal B5, preinduction medium (mg/l)	Basal B5 + 20 mM reduced N, induction medium (mg/l)
Mineral salts (reduced N)	Murashige and Skoog (1962) (20.6 mM)	Gamborg et al. (1976) (2.02 mM)	Gamborg et al. (1976) (22 mM)[c]
2iP	1.0	–	–
Kin	0.1	–	–
Thiamine HCl	0.1	10	10
Nicotinic acid	0.5	1	1
Pyridoxine HCl	0.5	1	1
m-Inositol	100	100	100
Glycine	2	–	–
Sucrose	30000	30000	30000
Agar	8000	8500	8500
pH	5.7	5.5	5.5

[a] Mariotti and Arcioni (1983).
[b] Moyer and Gustine (1984).
[c] Concentration from mineral salts + 20 mM NH_4Cl or 10 mM asparagine.

leaflets from 7 out of 50 plants (Table 2) yielded callus growing well enough for further examination.

Moyer and Gustine (1984) found that callus lines could be readily established from hypocotyls excised from germinated seeds (see Sect. 2.2); 45% of the hypocotyls yielded callus that could be maintained by serial culture. Hypocotyls excised from 1.5–2-cm-long seedlings served as the primary explants; those from larger seedlings were less likely to form callus. Callus lines were transferred two to four times at 4-week intervals before they were considered established. Callus lines to be examined for embryogenesis or to be put into suspension culture were first transferred to D2OK2 (Table 1). Suspension cultures were initiated with approximately 10 g callus tissue/25 ml liquid D2OK2 medium. An equal volume of fresh medium was added after 7 days. Thereafter, at 7-day intervals, the volume was doubled by addition of fresh medium and the contents of each flask divided between two new flasks to maintain the flask volume at approximately 50 ml. Suspension cultures were maintained in 125-ml Erlenmeyer flasks secured to an orbital shaker operating at 100 rpm.

2.5 Regeneration and Transfer of Plants

Mariotti and Arcioni (1983) induced embryogenesis by replacing 2 mg 2,4-D and 0.25 mg kin/l UM medium with 1 mg 2-iP and 0.1 mg IAA/l, respectively. This constituted both a change from high concentration of a strong synthetic auxin to a low concentration of a mild auxin and a change from a high to a low auxin to cytokinin ratio. Another change in the medium was omission of casein hydrolyzate. Other combinations of phytohormones, as well as their absence were found to be ineffective in promoting embryogenesis.

In contrast, Moyer and Gustine (1984) induced embryogenesis by transfer of calli to a phytohormone-free medium. Calli grown on D2OK2 medium containing 2,4-D and kin were transferred to basal B5 (no phytohormones) and then to a medium high in reduced nitrogen. BOi2Y or basal B5 with either 20 mM ammonium chloride or 10 mM asparagine were equally effective. They noted that in callus lines with the highest embryogenic potential, small embryos formed on basal B5, but did not enlarge or develop further until transfer to a medium high in reduced nitrogen. However, in lines with moderate or limited embryogenic potential, embryo formation was not apparent until transfer to a medium high in reduced nitrogen. This suggests that the concentration of reduced nitrogen in basal B5, which was sufficient for 2,4-D-promoted growth of callus on D2OK2, was insufficient for complete development of embryos. Thus the level of reduced nitrogen was critical. The induction of embryogenesis in crownvetch by removal of phytohormones and elevation of reduced nitrogen levels is similar to the pattern observed by Walker and Sato (1982) in alfalfa (*Medicago sativa*). They suggest that the requirement for reduced nitrogen may be a general phenomenon.

Crownvetch embryogenesis can also be induced without increasing content of reduced nitrogen. Mariotti and Arcioni (1983) diminished the amount of reduced nitrogen by removing casein hydrolyzate when inducing embryogenesis, but the amount of ammonium ion (20.6 mM) in the basal MS medium was nearly as much as the concentration (22 mM) used by Moyer and Gustine (1984) to induce embryogenesis.

In both procedures, only callus tissue showing green areas was transferred to embryogenic media. Moyer and Gustine (1984) maintained callus lines in an embryogenic state through repeated transfers by recurrently selecting green areas, inducing embryogenesis, and then using induced embryos as explants for initiating callus. Both groups reported only the occurrence of embryogenesis; no organogenesis was noted. They also reported variability in frequency of formation of somatic embryos among individual genotypes (Table 4) and among calli from the same genotype (Table 5).

Moyer and Gustine (1984) noted variability in embryo size and morphology. Although some embryos developed two cotyledons and areas analogous to a hypocotyl and radicle, embryos with single or multiple cotyledons were common (Fig. 2A). Highly irregular embryos, which had no axis of symmetry, were also formed.

Table 4. Embryogenesis in callus lines[a] derived from *Coronilla varia* cv. Chemung. (Moyer and Gustine unpubl.).

Expression of embryogenesis	No. of callus lines
No. of embryos	13
Dense green areas	7
Small embryos	6
Large embryos	6

[a] 32 callus lines were evaluated after 4 weeks on BOi2Y.

Table 5. Embryogenic responses of calli derived from different plants of *Coronilla varia* on MS medium supplemented with 1 mg/l of 2-iP and 0.1 mg/l of IAA. (Data from Mariotti and Arcioni 1983)

Calli with embryos/total calli	Plant identification number[a]
3/12	4
10/14	29
5/10	30
4/10	48

[a] These correspond to plant identification numbers in Table 2.

In order to obtain plants from embryos, Mariotti and Arcioni (1983) transferred embryos to filter paper saturated with phytohormone-free basal MS liquid medium. They noted that embryos transferred to the same medium containing agar formed new callus rather than developing into plants. After 1 month, 60% of the embryos on filter paper had developed shoots and roots. Plantlets were transferred to soil-less compost and covered with plastic to prevent desiccation. Their procedure was efficient in that 90% of the plantlets survived the transition to soil.

After formation of embryos, Moyer and Gustine (1984) transferred them to solid basal B5 medium for shoot and root formation. Embryos with two or more cotyledons developed shoots from the cotyledonary node (Fig. 2B). Irregular embryos and those with a single cotyledon first developed roots; adventitious shoots subsequently grew from roots or the areas analogous to the hypocotyl (Fig. 2C). Plantlets were transferred to vermiculite, covered with beakers for 1 week to prevent desiccation, and watered as needed with one-half strength Hoagland's solution. Plants surviving after 4–7 weeks were potted in soil. The survival rate for the plantlets transferred from agar medium to vermiculite and then to soil was low (data not reported). Plants obtained (Fig. 2D) were transplanted to the field, where they subsequently flowered and appeared phenotypically normal. Seed production by regenerated plants in the field was not reported.

2.6 Other Physiological Studies

Among the factors limiting the agronomic use of crownvetch is the presence of NPA esters. These compounds are also present in *Indigofera, Astragalus*, and *Lotus* species, other legumes with potential agronomic value. In addition to Leguminosae, these compounds are found in a limited number of species in other families (Williams 1983).

Little is known concerning the biosynthesis of NPA esters. With the application of in vitro techniques to crownvetch, the use of crownvetch tissue and suspension cultures to study NPA ester biosynthesis seem obvious. The advantages of using plant tissue cultures in biosynthetic studies are largely negated by their frequent inability to produce significant amounts of secondary products (Butcher 1977). Moyer and Gustine (1987) noted the presence of NPA esters in regenerating callus cultures, but at levels far lower than those found in plants. However, they

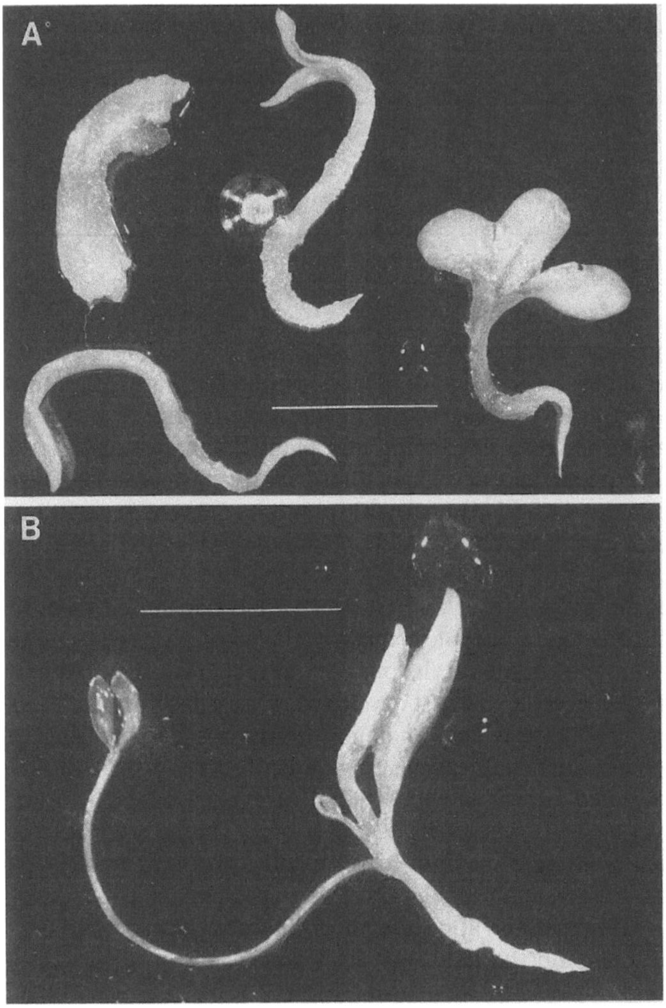

Fig. 2A–D. Stages in production of plants from callus-derived embryos. **A** Embryos formed from tissue grown 4 weeks on BOi2Y. **B** Shoot developing at cotyledonary node of embryo from callus grown 4 weeks on solid basal B5 medium. **C** Shoot development in irregular embryos from callus grown 4 weeks on solid basal B5 medium. **D** Plant obtained through somatic embryogenesis. Bars = 1 cm. (Moyer and Gustine 1984)

found that NPA added to media for either callus or suspension cultures was incorporated into NPA esters, thus increasing NPA ester levels markedly. Suspension cultures were supplemented with 10 mM neutralized, filter sterilized NPA; treated suspensions were incubated 72 h and the NPA and NPA esters extracted. They found that 43% of added NPA had been incorporated into esters, 9% was present as the free acid, and the rest could not be accounted for. The content of NPA esters in suspensions, as indicated by aliphatic NO_2 concentration, was increased 600-fold by the addition of exogenous NPA. Aerial parts of crownvetch contained

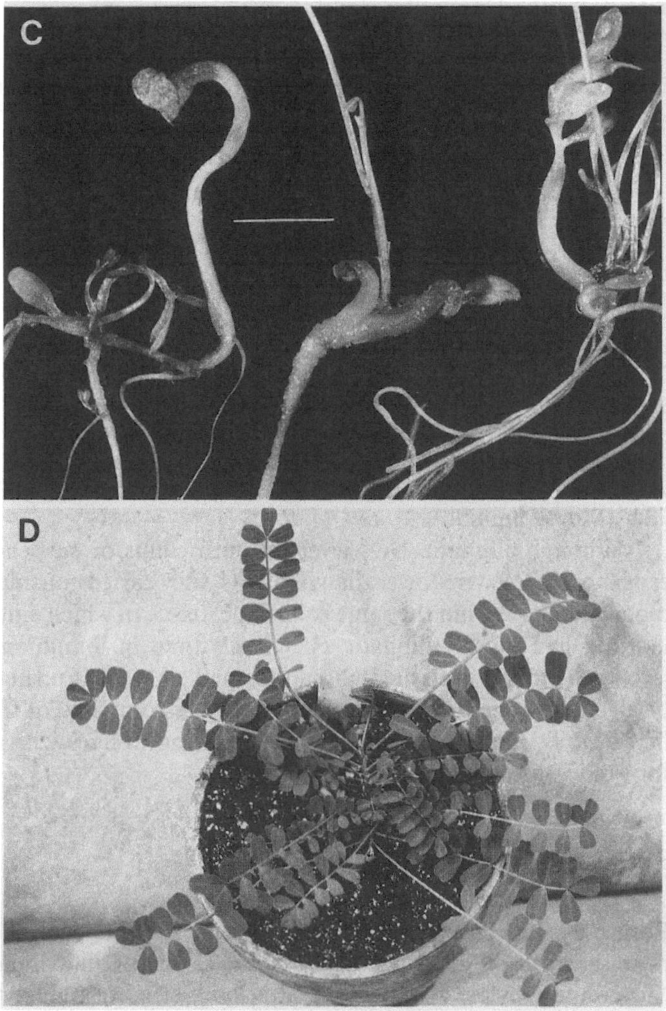

Fig. 2C, D

as high as 63 to 164 μM NO_2/g fresh wt. of tissue (see also Faix et al. 1978). With the addition of exogenous NPA, the NPA ester content in suspension cultures reached 15.6 μM NO_2/g fresh wt., a significant fraction of that in plants. Levels of NPA esters produced in crownvetch suspension cultures seem to be limited by biosynthesis or availability of NPA.

Esters produced in suspension culture upon the addition of exogenous NPA were the tetraester hiptagin and the diesters cibarian and coronarian, the triesters corollin, karakin, and coronillin, and two unidentified NPA esters. The two unidentified compounds were also noted in extracts of the aerial parts of crownvetch, cultivar Chemung, but hiptagin was not found in either Penngift or Chemung (Moyer unpubl.).

Pathways for ester metabolism are operative in cultured crownvetch cells. All the esters produced by suspension cultures upon addition of exogenous NPA are

Table 6. Relationship of embryogenesis to NO_2^- content in callus tissue. (Moyer and Gustine unpubl.).

Embryogenesis[a]	nmol NO_2^-/g tissue grown 4 weeks on		
	D20K2	basal B5	BOi2Y
– – (10 lines)	8[b]	25	60
+ (4 lines)	7	63	150
+ + (5 lines)	10	155	390
+ + + (3 lines)	18	177	418

[a] Expression after 4 weeks on BOi2Y. – –, no embryos; +, dense green areas; + +, small embryos; + + +, large embryos.
[b] Data are the mean of NO_2^- determinations for all callus lines in each group; 3 NO_2^- determinations were done for each line.

also produced if the triester karakin is added to suspension cultures and incubated 72 h (Moyer unpubl.).

Although aliphatic NO_2 levels in both callus or suspensions untreated with exogenous NPA were lower than in plants, they varied considerably with the physiological state. Within the same callus line, tissue in which embryogenesis had been induced had higher aliphatic NO_2 levels than in uninduced tissue; the highest levels were found in tissue that had formed embryos. Further, at all stages of embryogenesis, if calli from different lines were compared at the same stage of embryogenesis, the higher NO_2 levels were found in lines with the greater potential for embryogenesis (Table 6).

3 Conclusions and Prospects

Both published procedures for induction of callus and regeneration of plants appear satisfactory, although direct comparisons of individual steps are difficult. Moyer and Gustine (1984) obtained a more efficient induction of callus from explants, but this may be the result of differences in the explant and crownvetch cultivar used, rather than greater suitability of their medium. The procedures of Mariotti and Arcioni (1983) for obtaining plantlets from embryos and for transferring plantlets to soil are more efficient. It is intriguing that somatic embryogenesis in crownvetch can be induced through two dissimilar procedures, by manipulating either phytohormones or levels of reduced nitrogen. This demonstrates that the same developmental process can be initiated by different physiological mechanisms.

Among the legume species containing aliphatic nitro compounds, there are a number with actual or potential agronomic value; understanding of the biosynthetic pathways for these compounds in one of these species would be useful with regard to the rest.

Published techniques permit further study of NPA ester metabolism in crownvetch suspension culture, as the capability for biosynthesis of NPA esters is retained in culture. As only limited NPA biosynthesis occurs in culture, a means of fully activating NPA biosynthesis is required in order for suspension cultures to be useful for such studies. With a complete understanding of NPA and NPA ester biosynthesis, selection for cells producing low levels of NPA may be possible. As many conventional breeding techniques are difficult to apply to autotetraploid crops (Hanson and Davis 1972), in vitro techniques may be of use with crownvetch.

4 Protocol

The following protocol is suggested for optimal callus formation and subsequent regeneration of plants. The procedure and medium for callus induction are taken from Moyer and Gustine (1984); those for embryogenesis and plant regeneration are taken from Mariotti and Arcioni (1983). Details are set forth both in relevant sections in this chapter (2.2–2.5) and in the two papers cited above. This protocol has not been carried out as a connected whole, but should be satisfactory for studies dealing with large numbers of genotypes, as it is more convenient to obtain sterile seedlings from surface sterilized seeds and excise hypocotyls from them than it is to deal with large numbers of plants and surface sterilize leaves. However, the complete protocol of Mariotti and Arcioni (1983) would be used for studies which have plants as their starting point.

The following steps are suggested. Callus is initiated by explanting excised hypocotyls on G3 medium (Table 1). After two to four transfers at 4-week intervals, enough callus will have formed for induction of embryogenesis. We considered four pieces totalling 5–10 g sufficient, but less may be suitable. Callus is transferred to MS medium with 1 mg 2-iP and 0.1 mg IAA/l (Table 3). After 1 month, embryos formed are transferred to filter paper saturated with phytohormone-free liquid MS medium. Plantlets formed after 1 month are transferred to a commercial potting mixture. The more fragile plantlets, particularly those with few roots, stand a better chance of survival if initially transferred to vermiculite, and watered with half-strength Hoaglands for 1 month to allow root system establishment and then potted in a commercial potting mixture. High humidity at the time of plantlet transfer from filter paper to either vermiculite or commercial potting mix increases the survival rate.

References

Arcioni S, Mariotti D, Damiani M, Pezzotti M (1988) Birdsfoot trefoil (*Lotus corniculatus* L.), Crownvetch (*Coronilla varia* L.) and Sainfoin (*Onobrychis viciifolia* Scop). In: Bajaj YPS (ed) Biotechnology in agriculture and forestry 6. Crops II. Springer, Berlin Heidelberg New York Tokyo, pp 548–572

Blaydes DF (1966) Interaction of kin and various inhibitors in the growth of soybean tissue. Physiol Plant 19:748–753

Burns JC, Mochrie RD, Cope WA (1972) Responses of dairy heifers to crownvetch, *Sericea lespedeza*, and alfalfa forages. Agron J 64:193–195

Burns JC, Cope WA, Barrick ER (1977) Cow and calf performance, per hectare productivity, and persistence of crownvetch under grazing. Agron J 69:77–81

Butcher DN (1977) Secondary products in tissue cultures. In: Reinert J, Bajaj YPS (eds) Applied and fundamental aspects of plant cell, tissue, and organ culture. Springer, Berlin Heidelberg New York, pp 668–669

Byers RA, Gustine DL, Moyer BG, Bierlein DL (1986) 3-Nitropropionate in crownvetch: A natural deterrent to insects? In: Green MB, Hedin PA (eds) Natural resistance of plants to pests. Roles of allelochemicals. Am Chem Soc, Washington, DC, pp 95–105

Duke JA (1981) Handbook of legumes of world economic importance. Plenum Press, New York, pp 59–61

Faix JJ, Gustine DL, Wright MJ (1978) Beta-nitropropionic acid concentration of crownvetch plant parts as affected by growth-temperature-maturation variables. Crop Sci 70:689–691

Gamborg OL, Murashige T, Thorpe TA, Vasil IK (1976) Plant tissue culture media. In Vitro 12:E1–E6

Gould DH, Gustine DL (1982) Basal ganglia degeneration, myelin alterations, and enzyme inhibition induced in mice by the plant toxin 3-nitropropanoic acid. Neuropathol Appl Neurobiol 8:377–393

Gustine DL (1979) Aliphatic nitro compounds in crownvetch: A review. Crop Sci 19:197–203

Gustine DL, Moyer BG (1982) Retention of phytoalexin regulation in legume callus cultures. Plant Cell Tissue Organ Cult 1:255–263

Gustine DL, Moyer BG (1983) Mechanisms of toxicity of 3-nitropropanoic acid in nonruminant animals. In: Smith AJ, Hays VW (eds) Int Grasslands Congr, 14th Proc. Westview, Boulder, pp 736–738

Gustine DL, Moyer BG, Wangsness PJ, Shenk JS (1977) Ruminal metabolism of 3-nitropropanoyl-D-glucopyranoses from crownvetch. J Anim Sci 44:1107–1111

Hanson CH, Davis RL (1972) Highlights in the United States. In: Hanson CH (ed) Alfalfa science and technology. Am Soc Agron, Madison, p 44

Hawk VB (1968) Emerald crownvetch. In: McKee GW, Risius ML (eds) Second Crownvetch Symposium, Agronomy Mimeo. Pennsylvania State Univ, University Park Pennsylvania, p 7

Mariotti D, Arcioni S (1983) Callus culture of *Coronilla varia* L. (crownvetch): Plant regeneration through somatic embryogenesis. Plant Cell Tissue Organ Cult 2:103–110

McKee GW (1964) Cultivated varieties-Origin and description. In: McKee GW (ed) Crownvetch Symposium. Pennsylvania State Univ, University Park, Pennsylvania, pp 3–5

Meijer EGM, Broughton WJ (1981) Regeneration of whole plants from hypocotyl-, root-, and leaf-derived tissue cultures of the pasture legume *Stylosanthes guyanensis*. Physiol Plant 52:280–284

Moyer BG, Gustine DL (1984) Regeneration of *Coronilla varia* L. (crownvetch) plants from callus culture. Plant Cell Tissue Organ Cult 3:143–148

Moyer BG, Gustine DL (1987) Esterification of 3-nitropropanoic acid to glucose by suspension cultures of *Coronilla varia*. Phytochemistry 26:139–140

Murashige T, Skoog F (1962) A revised medium for rapid growth and bioassays with tobacco tissue cultures. Physiol Plant 15:473–497

Rineer K (1977) Variation of 3-nitropropionic acid for crownvetch, *Coronilla varia* L. MS Thesis. Pennsylvania State Univ, University Park Pennsylvania, pp 1–113

Risius ML (1968) Crownvetch breeding. In: McKee GW, Risius ML (eds) Second Crownvetch Symposium, Agronomy Mimeo. Pennsylvania State Univ, University Park, pp 64–66

Rodgers DR, Judd HP, Ross RS (1964) Highway slope stabilization with crownvetch. In: McKee GW (ed) Crownvetch Symposium. Pennsylvania State Univ, University Park, pp 32–35

Ross RS, Rodgers D (1968) Aesthetics and economics resulting from roadside crownvetch treatment. In: McKee GW, Risius ML (eds) Second Crownvetch Symposium, Agronomy Mimeo. Pennsylvania State Univ, University Park, pp 8–14

Ruffner JD (1964) The comparative performance and adaptation of crownvetch on strip mine spoil and other cirtical areas in West Virginia and western Pennsylvania. In: McKee GW (ed) Crownvetch Symposium, Pennsylvania State Univ, University Park, pp 17–31

Steinmetz EF (1962) *Coronillae variae herba*. Q J Crude Drug Res 2:189–192

Uchimiya H, Murashige T (1974) Evaluation of parameters in the isolation of viable protoplasts from cultured tobacco cells. Plant Physiol 54:936–944

Walker KA, Sato SJ (1982) Morphogenesis in callus tissue of *Medicago sativa*: The role of ammonium ion in somatic embryogenesis. Plant Cell Tissue Organ Cult 1:109–121

Wheeler AG (1974) Phytophagous arthropod fauna of crownvetch in Pennsylvania. Can Entomol 106:897–908

Williams MC (1983) Toxic nitro compounds in *Lotus*. Agron J 75:520–522

Williams M, Cassady JM (1976) Potential antitumor agents: a cytotoxic cardenolide from *Coronilla varia* L. J Pharm Sci 65:912–914

III.7 Goat's Rue (*Galega officinalis* L.)

V. NAŠINEC and B. NĚMCOVÁ[1]

1 Introduction

The genus *Galega* (Papilionoideae: Galegeae) comprises six to eight herbaceous and perennial species, which can be found from the warm temperate regions of Mediterranean Europe to Iran, three of them being indigenous to the mountainous areas of tropical East Africa (Allen and Allen 1981). Two agriculturally promising species, *G. officinalis* L. and *G. orientalis* L., have been introduced to many countries throughout the world, including relatively cold regions of Scandinavia and the Soviet Union. The yields reported from Europe are relatively high, ranging from 40 to 80 t/ha and 0.3 to 0.6 t/ha for the green matter and seeds, respectively (e.g., Kondratiuk and Azarach 1984).

Goat's rue, the common name of the type species, indicates its toxicity for goats. The foliage has a bitter taste; as a minor additive it is thought to improve lactation, but in higher quantities causes vomiting in animals and even death under some conditions (Faliu et al. 1985). *G. officinalis*, is a robust and up to 1.5 m tall, plant cultivated to some extent for soil improvement or restoration, for medicinal purposes, and as a bee plant. The seed and foliage contain galegine and various guanidine derivatives, such as smirnovine or spherophysine (see Mears and Mabry 1971), which are known as ganglion-blocking agents. Moreover, plant decoctions show diaphoretic, antispasmodic, anthelmintic, and other pharmacologically important properties.

G. officinalis L. has been grown in Czechoslovakia since the 1950's, mainly for medicinal purposes. The seed and dry matter yields are comparable to, or better than, those in clovers and alfalfa, the most prominent forage legumes. Elimination of alkaloids via plant breeding, or by using a tissue culture methodology, would be of great practical importance.

2 In Vitro Studies

Limited success has so far been achieved in developing efficient techniques for regeneration and in vitro clonal multiplication of forage legumes. Only those

[1] Institute of Experimental Botany, Czechoslovak Academy of Sciences, 370 05 České Budějovice, Czechoslovakia

Biotechnology in Agriculture and Forestry, Vol. 10
Legumes and Oilseed Crops I (ed. by Y.P.S. Bajaj)

methods that are available for several *Medicago*, *Lotus*, and *Trifolium* species (e.g., Orshinski and Tomes 1984; Brown and Atanassov 1985; Pederson 1986) seem to be versatile enough to allow practical implications. As far as we know, no such procedures are available for *Galega* species, and there are also no systematic studies on in vitro metabolism or production of secondary plant products, similar to those published on *Trigonella* sp. (e.g., Radwan and Kokade 1980). The purpose of this investigation is to evaluate morphogenetic potentials of *G. officinalis* L. under various in vitro situations. Preliminary results have already been published elsewhere (Našinec et al. 1984).

2.1 Establishment of Callus Cultures

G. officinalis L. (Czechoslovak cv. Běla) was provided by the Plant Breeding Station Libochovice. The seeds were scarified with concentrated H_2SO_4 (15 min), washed with an excess of sterile distilled water, re-sterilized with 3% H_2O_2 in 70% ethanol (v/v, 15 min) and again rinsed three times in sterile distilled water. Surface sterilized seeds were germinated on the moist filter paper discs, and 2 days later sown into the test tubes with an agar-solidified commercial nutrient solution. Roots, hypocotyls, cotyledons, petioles, first leaves, and shoot tips of 10-day-old seedlings were aseptically cut into segments (5 – 8 mm) and inoculated onto a desired culture medium.

As our preliminary experiments revealed no substantial differences in the explant's responses to MS (Murashige and Skoog 1962) or B5 (Gamborg et al. 1968) media, the latter was chosen the subsequent experiments. The callus-inducing media consisted of inorganic salts and vitamins of B5, 3% sucrose (w/v) and various combinations of 2,4-D, NAA, and kin, all added prior to autoclaving. Explants were cultured in 25-ml glass vials with 10 ml of a medium in darkness (25 °C).

2,4-D alone was found to be able to induce callogenesis (best at concentrations of 1 – 2 mg/l), the presence of kin did not significantly influence yields of callus

Table 1. Relative growth rates of *G. officinalis* L. primary callus cultures during first 4 weeks of cultivation on selected B5 media in darkness

Hormonal levels[a]			Origin of explants[b]						Number of calli[c] Total/rhizogenic
2,4-D	NAA	kin	R	H	C	P	L	S	
1	–	–	+ + +	+ + +	+ +	+ +	+ +	+ + +	656/31
1	0.5	–	+ + +	+ + +	+ +	+ + +	+ +	+ + +	560/37
1	0.5	0.5	+ + +	+ + +	+ + +	+ +	+ +	+ +	579/32
2	0.5	0.5	+ + +	+ + +	+ +	+ +	+ +	+ +	545/26
5	–	–	+ +	+ +	+	+	+	(+)	511/17
10	–	–	+ (N)	+ (N)	N	N	N	–	515/9

[a] mg/l; [b] 6 to 10 segments of root (R), hypocotyl (H), cotyledon (C), petiole (P), first trifoliated leaf (L), and one shoot tip (S) were excised from one seedling; [c] total numbers of calli raised from 15 – 20 individuals per each hormonal treatment and numbers of calli with 3 – 8 spontaneously developed roots (regardless of the origin of explants); + + + = normal growth; + + = moderate growth; + = poor growth, N = symptoms of necrosis on most explants within 4 weeks

or the regeneration responses on the subsequent media. Levels of 2,4-D above 5 mg/l were rather inhibitory. The primary calli were most easily initiated from the cut ends of root and hypocotyl sections or from shoot tips. These explants started callusing within 1–2 weeks, and a conspicuous callus mass could be seen within 3–4 weeks. Under the same culture conditions, callogenesis of leaf, cotyledonary, and petiole sections was usually slower; the former two explant types frequently did not respond at all, and turned necrotic within a few weeks. Spontaneous rhizogenesis occurred on about 5% of explanted tissues without any clearcut correlation with the explant origin (type) or concentrations of 2,4-D. Typical responses of *G. officinalis* L. explants to several selected B5 media are summarized in Table 1.

2.2 Callus Maintenance and Regeneration

After 4–5 weeks in culture, the primary calli were separated from the original plant tissues when possible and inoculated parallel on the maintenance and regeneration media. Long-term maintenance of *G. officinalis* calli was always problematic. Generally, calli could proliferate in the presence of 2,4-D in the dark. Under such conditions, most of them were yellowish or beige in color, and soft. As compared with 2,4-D alone, slightly improved callus yields were obtained on the medium containing 2,4-D, NAA, and kin (2, 2 and 1 mg/l, respectively) and when cultured in the light (see below). Due probably to the presence of NAA, light and/or genotype variability of tissues, some calli became greenish, other remained beige, and up to 15% of cultures spontaneously differentiated roots (e.g., genotype G18, Fig. 1 A). Nevertheless, their growth was always slower than desirable and we were unable to select a rapidly growing, friable, and morphogenetically competent culture.

To obtain plant regeneration, the basal B5 was variously supplemented with kin or BA (up to 2 mg/l), either alone or in a combination with NAA (max. 0.5 mg/l). The cultures were incubated under the photoperiod of 16 h

Table 2. Responses of selected *G. officinalis* L. calli to various phytohormone treatments on the regeneration and maintenance B5 media under 16-h photoperiod

	Hormone concentration[a]				Relative[b] growth	Green colour	Nodular bodies	Rhizogenic cultures	Shoot[c] regeneration	Total no. of genotypes[d]
	2,4-D	NAA	kin	BA						
M	1	–	–	–	+ + +	26%	5%	5%	–	53
M	2	2	1	–	+ + + (+)	40%	13%	15%	–	205
R	–	0.5	1	–	+ +	50%	28%	11%	1 (S)	18
R	–	0.2	–	1	+	81%	48%	7%	3 (S)	82
R	–	0.2	–	2	+	83%	51%	3%	3 (S)	73

M, R – maintenance and regeneration media, respectively; [a] mg/l; [b] see Table 1; [c] only shoot-tip-derived calli differentiated vital shoot primordia; [d] total numbers of genotypes tested under the given conditions, all the relative values were summarized regardless of the genotype and/or explant's origin after 4 weeks in culture

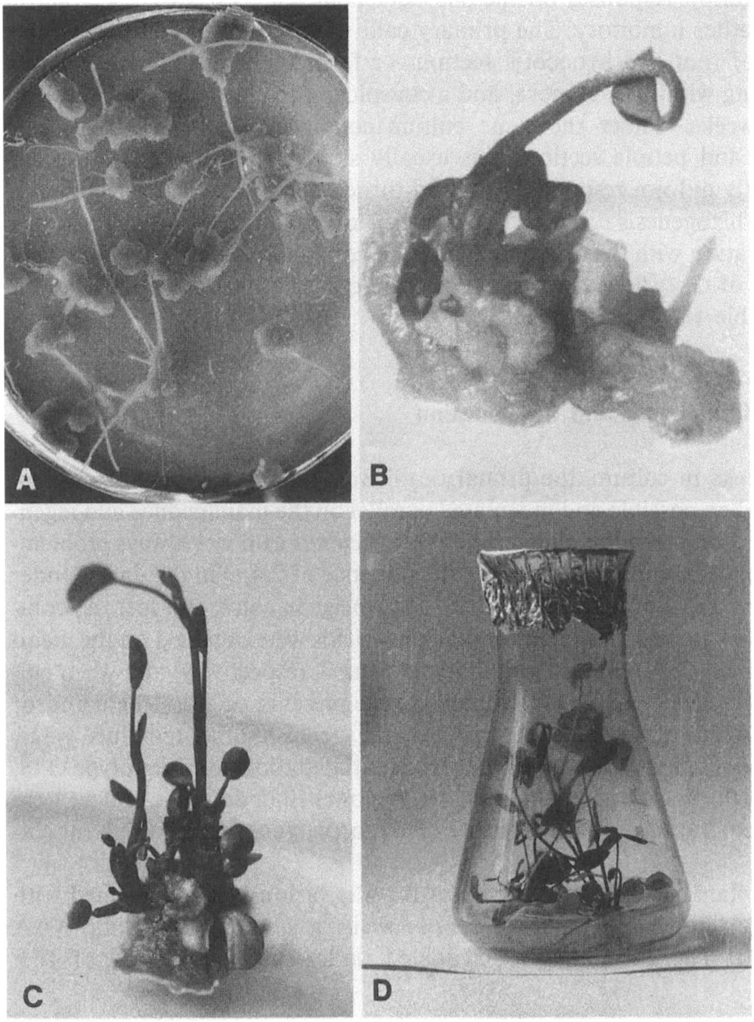

Fig. 1A–D. Differentiation of roots in and shoots callus cultures of *G. officinalis* L. **A** Six-week-old, hypocotyl-derived callus culture (genotype G 18) on a maintenance medium. **B** Early shoot differentiation from a shoot-tip-derived callus (G 89) on a regeneration medium (B 5, BA and NAA, 2 and 0.2 mg/l, respectively, 2nd passage). **C** Further proliferation of the in vitro differentiated shoots (G 75) on the hormone-free B 5 with casamino acids (0.02%), 16 weeks in aseptic culture. **D** Cloning of regenerated *G. officinalis* L. shoots and their rooting on the diluted (half-strength, 1% sucrose) B 5 medium without phytohormones

(180–200 $\mu E/m^2/s$, 27/23 °C) in small vials or Petri dishes containing approximately 7 ml of medium. The most frequent morphogenetic responses were recorded using the medium with BA (1–2 mg/l), NAA (0.2 mg/l) and 3% sucrose (see Table 2). About 50% of the cultures turned light green in 2–5 weeks and within the same time 3–15 dark green nodular bodies or structures resembling bud primordia developed on the surface of approximately 80% of calli. Their further differentiation, however, was noticeably slow, if any. Out of more than 150 genotypes tested in this way, only seven shoot-tip-derived calli regenerated vital buds (Table 2). The conversion of buds to leafy shoots (Fig. 1B, C) was done by transferring them, with a small piece of callus, onto the hormone-free B5 with 0.02% casamino acids and 1% sucrose. The same medium could be used for further propagation and rooting of regenerants (Fig. 1D).

Rooted regenerants were slowly adapted to the nonsterile environment (perlite saturated with a nutrient solution) and 1 month later transferred to the soil. All regenerants were phenotypically indistinguishable from cv. Běla, and fully fertile.

Under the conditions described above, we tested in vitro about 300 plants. Primary calli, derived from more than one half of them, were screened for their ability to regenerate. Genotype effects were always nore visible than the positive responses to hormonal levels or different culture conditions. This is not surprising, because in many plant species the regeneration ability depends upon genotype, and in legumes this may be even more important than the composition of media, regeneration protocol, or the origin of explants (e.g., Brown and Atanassov 1985). All our *Galega* regenerants differentiated from shoot-tip-derived calli. We are not sure whether they arose de novo from fully dedifferentiated cells or from meristematic loci pre-existing in original explants. The latter possibility is more probable, as the calli that had been re-initiated from regenerants, exhibited no improvement in totipotency. On the other hand, the immature leaf tissues within the shoot tip might be the source of morphogenetically competent cells, and their age might be the critical factor, as in pea and peanut (Mroginski and Kartha 1981; Mroginski et al. 1981).

2.3 Clonal Propagation

Plant breeders are usually interested in elite genotypes that had been selected for a desirable trait and then must be maintained and vegetatively cloned for a long time under greenhouse conditions. This involves the hazards of pest and pathogen attacks, through which the germplasm may be lost. Instances are also known where rooting of disease-attenuated clonal propagules is partly or totally inhibited. The following in vitro procedure for cloning and diluting some pathogens (bacterial, viral) or inhibitory factors has been developed by us, and used for propagation of greenhouse-grown goat's rue plants:

Stems and vegetative shoots were defoliated, washed under the running tapwater, cut into segments in internodes, surface sterilized with 3% H_2O_2 in 70% ethanol (v/v, 2–5 min) and again cut on both sides after rinsing three times in sterile water. Segments (1.5–2 cm long) containing axillary buds were vertically stuck into the semi-solid medium (0.5% agar, half-strength B5, 0.5% sucrose, hormone-free, 7 ml per tube) containing 0.5 g/l of the filter-sterilized antibiotic ticarcilin. Stem segments, when cultivated under the light and temperature conditions specified above, easily produced lateral shoots and, in healthy plants, most of the primary explants differentiated roots within 2–3 weeks (Fig. 2). When plants were diseased, or when rooting was retarded, the new lateral shoots were excised after 3–5 weeks and transferred to the same fresh medium to promote

Fig. 2. Cloning of nonsterile *G. officinalis* plants. Early shoot and root differentiation on the surface sterilized stem segments in the presence of ticarcilin (0.5 mg/l, diluted B5, 4-week-old plantlets)

Fig. 3. Well-adapted clonal progenies can be transferred to the soil or can be used as the new source of explants for the next cloning cycle

Fig. 4. A mature *G. officinalis* L. plant just before the oneset of flowering in the greenhouse conditions

root formation. In some cases, omitting sucrose and supplying NAA (0.1 mg/l) further promoted rooting. Rooted shoots were then transplanted to pots with perlite, watered with a commercial nutrient solution, covered in transparent plastic boxes, and gradually (within 3–4 weeks) adapted to the normal air moisture before returning to the soil (Figs. 3, 4). Practically the same procedure has successfully been used by us for the propagation of several important parental clones of alfalfa (Němcová et al. 1987). A schematic representation of the cloning procedure is given in Fig. 5. The whole cycle can be repeated several times in order to dilute pathogenic factors progressively and/or to obtain desirable number of healthy plants. The basic advantages of this simple procedure are its versatility, genetic stability of clonal progenies, and the remarkable independence on the variability of genotypic responses in vitro.

3 Conclusions and Prospects

G. officinalis L. provides high yields of herbage mass and can fix nitrogen when it is properly inoculated. Cultivars are available now that are well adapted to acid soils and cold climate. The agricultural use is restricted mainly because of its rela-

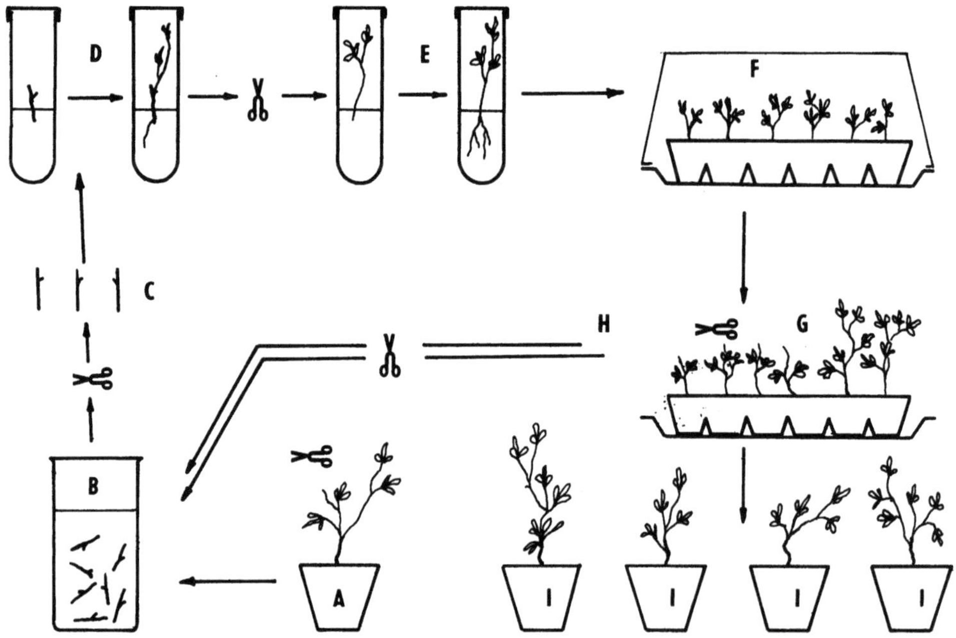

Fig. 5. A schematic representation of *G. officinalis* L. cloning cycle. *A* original plant; *B* surface sterilization of stem segments; *C* sterilized segments with axillary buds; *D* aseptic cultivation in the presence of an antibiotic; *E* aseptic shoot excision and rooting; *F* acclimation and regrowth of in vitro cloned plantlets; *G* transfer to the soil and/or (*H*) repetition of the whole cycle; *I* clonal progenies

tive toxicity. A better understanding of the biochemistry and genetics of alkaloid production in *Galega* is required before developing satisfactory strategies for their elimination (or increase for medical purposes). Although in vitro selections, best at cellular levels, and subsequent regeneration are attractive possibilities, there are many technical problems that have yet to be solved. Our experiments demonstrated high genotypic variation in morphogenetic responses in vitro and relative recalcitrance of *G. officinalis* calli against exogenous hormonal stimuli, but also the ease with which individual plants can be vegetatively cloned. Therefore, any sophisticated tissue culture approach should only be considered by practical breeders when conventional procedures have failed to produce desirable characteristics.

Primary explants of the Czechoslovakian cv. Běla could easily be induced to form calli on B5 medium with 2,4-D (1 mg/l). Regeneration of intact plants was invariably achieved only from shoot-tip-derived callus cultures, again on B5 media supplemented with BA and NAA, best at the concentration of 1 – 2 mg/l and 0.2 mg/l, respectively. All in vitro differentiated or seed-borne plants could be propagated vegetatively from stem cuttings. Their rooting was almost entirely spontaneous on diluted and hormone-free B5.

Acknowledgments. The authors thank Dr. V. Vacek (Res. Breeding Inst. Fodder Plants, Troubsko) for his interest and constructive help.

References

Allen ON, Allen EK (eds) (1981) The Leguminosae. A source book of characteristics, uses, and nodulation. Wisconsin Univ Press, Madison

Brown DCW, Atanassov A (1985) Role of genetic background in somatic embryogenesis in *Medicago*. Plant Cell Tissue Organ Cult 4:111–122

Faliu L, Puyt JD, Jean-Blein C (1985) Intoxication végetable: *Galega officinalis*. Point Vet 17:453–455

Gamborg OL, Miller RA, Ojima K (1968) Nutrient requirements of suspension cultures of soybean root cells. Exp Cell Res 50:151–158

Kondratiuk EN, Azarach LR (1984) *Galega officinalis* – a promising fodder plant for the Donbass region. Introdukciya Aklimatizaciya 2:24–26

Mears JA, Mabry TJ (1971) Alkaloids in the *Leguminosae*. In: Harborne JB, Boulter D, Turner BL (eds) Chemotaxonomy of the Leguminosae. Academic Press, New York, pp 73–178

Mroginski LA, Kartha KK (1981) Regeneration of pea (*Pisum sativum* L. cv. Century) plantlets by in vitro culture of immature leaflets. Plant Cell Rep 1:64–66

Mroginski LA, Kartha KK, Shyluk JP (1981) Regeneration of peanut (*Arachis hypogaea*) plantlets by in vitro culture of immature leaves. Can J Bot 59:826–830

Murashige T, Skoog F (1962) A revised medium for rapid growth and bioassays with tobacco tissue culture. Physiol Plant 15:473–497

Našinec V, Rakouský S, Němcová B (1984) Regeneration of *Galega* plants from primary callus. In: Novák FJ, Havel L, Doležal J (eds) Proc Int Symp Plant tissue and cell culture. Application to crop improvement. Czech Acad Sci Prague, pp 169–170

Němcová B, Našinec V, Chloupek O (1987) In vitro cloning of lucerne. Rostl Vyr 33:1207–1213

Orshinski BR, Tomes DT (1984) Comparison of plants derived from cuttings, node cultures, and ethyl methanesulfonate treated node cultures of bird's-foot trefoil (*Lotus corniculatus*). Can J Bot 62:1501–1504

Pederson GA (1986) In vitro culture and somatic embryogenesis of four *Trifolium* species. Plant Sci 45:101–104

Radwan SS, Kokade CK (1980) Production of higher levels of trigonelline by cell cultures of *Trigonella foenum-graecum* than by the differentiated plant. Planta Medica 147:340–344

Section IV Oilseed Crops

IV.1 Interspecific Hybridization in *Brassica* Through Ovary Culture

N. INOMATA[1]

1 Introduction

The cultivated plants of genus *Brassica* are important as fodder, oil plants, and vegetables throughout the world. They consist of three fundamental species, *B. campestris*, *B. nigra*, and *B. oleracea*. Between these species, there are three natural amphidiploid plants, *B. carinata*, *B. juncea*, and *B. napus* (Table 1) (Morinaga 1934). *B. napus* is the amphidiploid between *B. campestris* and *B. oleracea* (Morinaga 1929).

There are eight subspecies and seven varieties in *B. campestris* and *B. oleracea*, respectively (Hosoda 1961). It is very important for breeding of *B. napus* to be able to produce hybrids easily between *B. campestris* and *B. oleracea*. The production of these hybrids, however, proved very difficult by artificial pollination and embryo culture (Table 2) (Hosoda et al. 1969). If we can produce these hybrids easily, we can produce the new cultivated plants and introduce the genes of *B. campestris* and *B. oleracea* into *B. napus* plants, and also exchange the genes between *B. campestris* and *B. oleracea*.

Håkansson (1956) made embryological studies on seed development in reciprocal crosses between *B. campestris* and *B. oleracea*; usually no viable seeds were obtained in the reciprocal crosses, because of the failure of their normal seed development. These phenomena were similar to the failure of seed development in the reciprocal crosses between diploid and autotetraploid *B. campestris* (Nishiyama and Inomata 1966).

When embryo culture is carried out, healthy embryos must be taken from the degenerating ovules and cultured on the medium. Instead of embryo culture technique, ovule and ovary culture can be done in vitro for producing the new plants. When excised ovules and ovaries are cultured in vitro on the medium, and the normal development of embryo, endospern, and seed coat are observed viable seeds may be obtained. On the basis of this idea, excised ovules and ovaries were cultured and new triploid hybrids were obtained in ovary culture in reciprocal crosses between diploid and autotetraploid *B. campestris* (Inomata 1968).

In recent studies, the production of hybrids between *B. campestris* and *B. oleracea* was carried out, and many interspecific hybrids between them were obtained comparatively easily by ovary culture, and the progenies of hybrids were

[1] Department of Biology, College of Liberal Arts and Sciences, Okayama University, Tsushima Okayama 700, Japan

Biotechnology in Agriculture and Forestry, Vol. 10
Legumes and Oilseed Crops I (ed. by Y.P.S. Bajaj)

Table 1. Relationship between six species, chromosome numbers, and genomic constitution in cultivated *Brassica*. (Morinaga 1934)

Species	Chromosome number (2n)	Genomic constitution
B. campestris	20	aa
B. nigra	16	bb
B. oleracea	18	cc
B. carinata	34	bbcc
B. juncea	36	aabb
B. napus	38	aacc

Table 2. Historical results on production of interspecific hybrids between *Brassica campestris* and *B. oleracea*[a]

No. of flowers pollinated (A)	No. of hybrids obtained (B)	B/A ×100	Reference
Diploid *B. campestris* × diploid *B. oleracea*			
732	4	0.546	U (1935)
1300 ca.	2	0.154	Mizushima (1946)
185	0	–	Feng (1955)
45	1	2.222	Feng (1955)[b]
8330	20	0.240	Hoffmann and Peters (1958)
10395	16	0.154	Olsson (1960)
670	2	0.299	Hosoda (1961)
1269	18	1.418	Shiraishi et al. (?)[c]
4925	2	0.041	Ito et al. (?)[c]
4496	8	0.178	Nishi et al. (1962)
143	0	–	Hosoda et al. (1963)
178	14	7.865	Hosoda et al. (1963)[d]
130	6	4.615	Hosoda et al. (1963)[e]
10341	4	0.039	Sarashima (1964)
358	2	0.559	Namai and Hosoda (1965)
99	7	7.071	Namai and Hosoda (1965)[d]
1441	10	0.694	Sarashima (1966)
936	3	0.321	Nishi et al. (1970)
534	17	3.184	Nishi et al. (1970)[f]
Diploid *B. oleracea* × diploid *B. campestris*			
380	0	–	U (1935)
9703	64	0.660	Hoffmann and Peters (1958)
4074	0	–	Olsson (1960)
518	0	–	Shiraishi et al. (?)[c]
3187	0	–	Nishi et al. (1962)
7644	77	1.007	Nishi et al. (1962)[f]
247	0	–	Hosoda et al. (1963)
137	0	–	Namai and Hosoda (1965)
Diploid *B. campestris* × autotetraploid *B. oleracea*			
1317	0	–	Olsson (1960)
392	11	2.806	Hosoda (1961)
485	0	–	Sarashima (1964)

Table 2 (continued)

No. of flowers pollinated (A)	No. of hybrids obtained (B)	B/A ×100	Reference
Diploid *B. oleracea* × autotetraploid *B. campestris*			
2331	0	–	Olsson (1960)
346	0	–	Nishi et al. (1962)
Autotetraploid *B. campestris* × diploid *B. oleracea*			
1271	3	0.236	Olsson (1960)
15	0	–	Hosoda (1961)
154	1	0.650	Sarashima (1964)
Autotetraploid *B. oleracea* × diploid *B. campestris*			
1382	0	–	Olsson (1960)
Autotetraploid *B. campestris* × autotetraploid *B. oleracea*			
3000 ca.	65	2.166	Frandsen (1947)
8 pods	19	?	Rudorf (1950)[g]
3000	7	0.233	Hagberg (1956)[c]
9711	0	–	Hoffmann and Peters (1958)
22984	133	0.579	Olsson (1960)
276	0	–	Hosoda (1961)
525	0	–	Sarashima (1964)
Autotetraploid *B. oleracea* × autotetraploid *B. campestris*			
7140	1	0.014	Hoffmann and Peters (1958)
18874	130	0.689	Olsson (1960)

[a] After modified Hosoda et al. (1969).
[b] Mixture pollination.
[c] After Nishi et al. (1962).
[d] Grafting.
[e] Stigma mutilation.
[f] Embryo culture.
[g] After Röbbelen (1960).

examined. The details of culture method, culture condition, and hybrid progenies are reported here.

2 The Need for Ovary Culture

Table 2 shows the historical results of the production of interspecific hybrids between *B. campestris* and *B. oleracea*. Many workers have tried to obtain the hybrids, but the process is very difficult. In order to obtain hybrids easily, mixed pollination (Feng 1955), grafting, chemical treatment at the peduncle, stigma mutilation in which the upper part of the stigma was removed and pollination was carried out (Hosoda 1961; Hosoda et al. 1963), and embryo culture (Nishi et al. 1959, 1962, 1970) have been attempted, whereby the production rate of the hybrids was slightly increased.

Ovule culture was made in the reciprocal crosses between diploid and autotetraploid *B. campestris*. These ovules, excised 9 days after pollination, were cultured on basal medium (Nitsch 1951) with various combinations of auxins, kin, and yeast extract. No viable seeds were obtained in the reciprocal crosses between them and their control crosses (Inomata unpubl.).

3 Method of Ovary Culture

3.1 Sterilization

When emasculated flower buds bloom, pollination is carried out with fresh pollen grains. The ovaries with pedicel are removed from the plant and are sterilized with 9% hypochlorite solution for 10 min, during which time, the bottle with the hypochlorite solution is shaken for about 2 min, as the ovaries float on the hypochlorite solution. They are rinsed four times with sterile water, and one ovary is explanted in one test tube with slant medium (Fig. 1).

3.2 Time of Culture

Nishiyama and Inomata (1966) reported that no viable seeds were obtained ordinarily in the crosses between diploid and autotetraploid *B. campestris*. It became clear in embryological studies that the cause of cross-incompatibility was the abnormal development of endosperm in the reciprocal crosses. The developmental rate of endosperm nuclei did not change till 4 days after pollination, and it is considered that ovary culture was possible in vitro in reciprocal crosses between *B. campestris* and *B. oleracea* for 4 days after pollination.

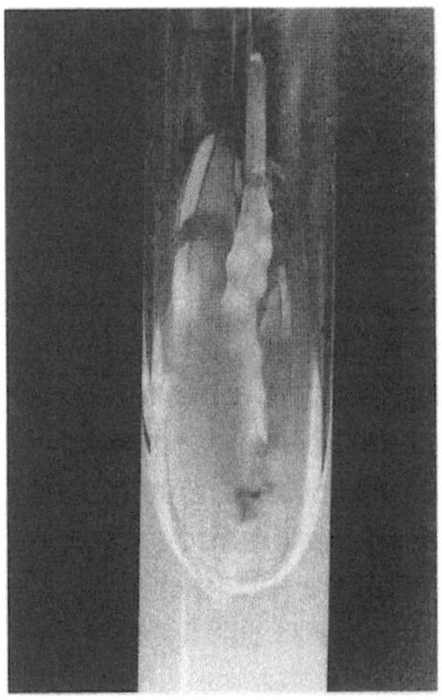

Fig. 1. Ovary culture in the test tube

3.3 Culture Media

Since viable seeds were obtained in culture of excised ovaries of *Cucumis* (Nitsch 1951), successful results of ovule and ovary culture have been reported in *Papaver* (Maheshwari 1958; Maheshwari and Lal 1961 a), *Zephyranthes* (Sachar and Kapoor 1959), *Iberis* (Maheshwari and Lal 1961 b), *Anethum* (Johri and Sehgal 1963), *Abelmoschus* (Bajaj 1965), *Hyoscyamus* (Bajaj 1966), *Allium* (Guha and Johri 1966), *Nicotiana* (Dulieu 1966), *Trifolium* (Nakajima et al. 1969), etc.

The basal medium used above with best results is Nitsch's (1951) or White's (1963) medium. The medium with auxin and kin promoted the growth of cultured ovule and ovary in *Papaver* (Maheshwari 1958) and *Allium* (Guha and Johri 1966), but not in *Zephyranthes* (Sachar and Kapoor 1959). The medium with casein hydrolysate (CH) and yeast extract promoted growth of cultured ovules in *Papaver* (Maheshwari and Lal 1961 a), and *Abelmoschus* (Bajaj 1965) but not in *Nicotiana* (Dulieu 1966).

Inomata (1976) reported studies of macroelements and vitamins on ovary culture made in the sib-crosses of diploid and autotetraploid *B. campestris*. The ovaries, excised 4 days after pollination, were cultured on various culture media. The macroelements used in the medium were Nitsch (1951), Heller (1953), Murashige and Skoog (1962), and White (1963). In each of the macroelements, alternatively no vitamins, Morel's vitamins (Morel 1948), modified White's vitamins (Inomata 1976), and Morel's plus modified White's vitamins were added. To the medium used in the experiment were further added 50 g/l sucrose and 8 g/l agar. The seed setting in the capsules examined in the sib-cross of the diploid *B. campestris* was best in White's minerals and poorest in Murashige and Skoog's minerals. In the sib-cross of autotetraploid *B. campestris*, seed setting was best in Murashige and Skoog's minerals and poorest in Heller's minerals.

To investigate the effect of growth substances, Inomata (1968) undertook ovary culture to obtain the hybrid between diploid and autotetraploid *B. campestris*. The ovaries, excised 4 days after pollination, were cultured on Nitsch's medium with various growth substances; these were casein hydrolysate, coconut milk, gibberellin, IAA, kin, tomato juice, and yeast extract. Hybrid plants in 2x×4x were obtained in the basal medium with coconut milk (CM) and with CM plus gibberellin, and in the reciprocal cross, 4x×2x, hybrid plants were obtained in the basal medium with CH plus CM plus gibberellin and with yeast extract. Inomata (1975) also made embryological studies on the development of embryo and endosperm in cultured ovaries. The development of embryo and endosperm was not promoted in the medium with auxin and kin alone, also not only with combinations of auxin and kin. On the contrary, development of embryo and endosperm was promoted in the medium with CH, CM, and yeast extract.

Temperature and light intensity were investigated in ovary culture of the sib-crosses in diploid and autotetraploid *B. campestris* (Inomata 1976). The seed setting was suitable for 17° and 22 °C, and decreased under both high (32 °C) and low (12 °C) temperature; seed set at light intensity of 300–500 lx continuous illumination by fluorescent lamp, and decreased in the dark and 2000 lx of continuous illumination.

3.4 Production of Interspecific Hybrids by Ovary Culture

Production of interspecific hybrids between *B. campestris* and *B. oleracea* was carried out (Inomata 1977, 1978a). The materials used in the experiment were *B. campestris* ssp. *chinensis*, ssp. *pekinensis*, and *B. oleracea* var. *capitata*. The ovaries were obtained in reciprocal crosses between *B. campestris* and *B. oleracea*, excised 4 days after pollination, and cultured. The basal medium was a solution of White's minerals, modified White's vitamins (Inomata 1976), 50 g/l sucrose and 8 g/l agar. To the culture medium was further added a combination of CH and yeast extract, and of CH and CM. The cultured ovaries were taken from the test tube at 36 days after explantation. When the developed naked embryos protruded from the undeveloped seed coat, they were removed aseptically and cultured on White's medium with 300 mg/l CH. The embryos obtained varied from torpedo-shaped to fully grown. When the embryos developed into plantlets with leaves and roots in the test tube, they were potted.

In *B. campestris*×*B. oleracea*, many seeds and embryos were obtained in the capsules. Many hybrid plants were obtained and showed an intermediate type between *B. campestris* and *B. oleracea* in leaf and inflorescent characteristics (Fig. 2A–K). The chromosome number of the root tip cells was 19 in all hybrids, that is intermediate between *B. campestris* ($2n = 20$) and *B. oleracea* ($2n = 18$) (Fig. 3A). Among the hybrids obtained in one cross combination, there was no difference either in morphological character or in growth rate.

On the contrary, no seeds and no embryos were obtained in the reciprocal cross, *B. oleracea*×*B. campestris*, on all culture media.

The production rate of the hybrids (no. of hybrids obtained/no. of capsules examined×100) was better in the medium with CH or yeast extract than in the basal medium, and best in the medium with CH (Inomata 1977). In the medium with CH and CM, hybrids were also obtained, with no difference in the production rate between the medium with CH and CM. The production rate was best in the medium with CH plus CM (Inomata 1978a). It seemed that CH, CM, and yeast extract in the medium were favorable to, but not indispensable for the production of hybrids in ovary culture.

The production rate of the hybrids in ovary culture was 6.82 in the cross *B. campestris* ssp. *pekinensis*×*B. oleracea* var. *capitata* (Inomata 1977), 3 in the cross *B. campestris* ssp. *pekinensis*×*B. oleracea* var. *capitata*, and 23.21 in the cross *B. campestris* ssp. *chinensis*×*B. oleracea* var. *capitata* (Inomata 1978a).

Production of hybrids between the many subspecies and varieties of *B. campestris* and *B. oleracea* was carried out (Inomata 1978b). The materials used in the experiment were five subspecies in *B. campestris* (ssp. *chinensis*, *japonica*, *oleifera*, *pekinensis*, *rapa*) and seven varieties in *B. oleracea* (var. *acephala*, *alboglabra*, *bullata*, *capitata*, *gemmifera*, *gondylodes*, *italica*). The medium was composed of a solution of minerals and vitamins (White 1963), 300 mg/l CH, 50 g/l sucrose, and 8 g/l agar. The ovaries, excised 4 days after pollination, were cultured in vitro. Seventy hybrids of the 801 capsules explanted were obtained in many cross combinations of *B. campestris*×*B. oleracea*, without a difference in growth rate among the hybrids obtained. The hybrids obtained were of intermediate type between the parents in leaf and inflorescence characteristics (Fig. 2L–R).

Fig. 2A–H

Fig. 2I–P

Fig. 2A–R. Interspecific hybrids between *Brassica campestris* and *B. oleracea*, and their parents. **A** *B. campestris* ssp. *chinensis* used as a female parent; **B** *B. campestris* ssp. *pekinensis* used as a female parent; **C** *B. campestris* ssp. *japonica* used as a female parent; **D** *B. oleracea* var. *capitata* used as a male parent; **E** *B. oleracea* var. *acephala* used as a male parent; **F** *B. oleracea* var. *gemmifera* used as a male parent; **G** *B. oleracea* var. *gongylodes* used as a male parent; **H** *B. oleracea* var. *italica* used as a male parent; **I** F_1 hybrid from ssp. *chinensis* × var. *capitata*; **J** F_1 hybrid from ssp. *chinensis* × var. *acephala*; **K** F_1 hybrid from ssp. *pekinensis* × var. *capitata* (Inomata 1977, 1978a); **L** F_1 hybrid from ssp. *pekinensis* × var. *acephala*; **M** F_1 hybrid from ssp. *pekinensis* × var. *gemmifera*; **N** F_1 hybrid from ssp. *pekinensis* × var. *italica*; **O** F_1 hybrid from ssp. *japonica* × var. *acephala*; **P** F_1 hybrid from ssp. *pekinensis* × var. *gongylodes* (Inomata 1978b); **Q** F_1 hybrid from ssp. *dichotoma* × var. *alboglabra*; **R** F_1 hybrid from ssp. *trilocularis* × var. *alboglabra*. (Inomata 1983b)

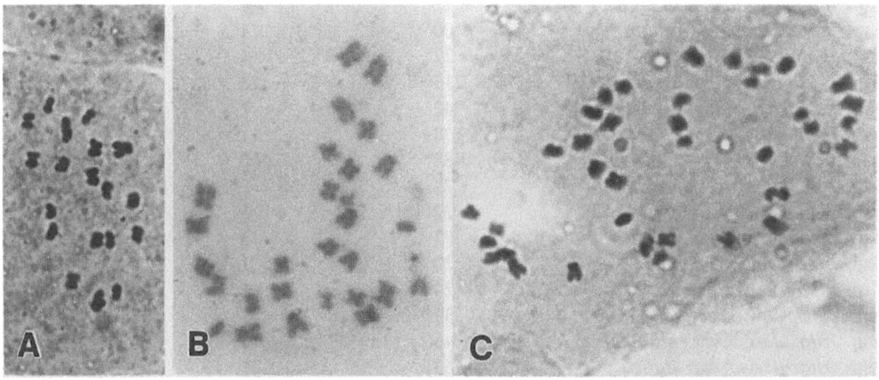

Fig. 3A–C. Chromosome numbers of F_1 hybrids in root tip cells. **A** Amphihaploid chromosome number ($2n = 19$) between *B. campestris* ($2n = 20$) and *B. oleracea* ($2n = 18$); **B** triploid chromosome number ($2n = 28$) between *B. campestris* and autotetraploid *B. oleracea*; **C** amphidiploid chromosome number ($2n = 38$) between *B. campestris* and *B. oleracea*. (Inomata 1978b)

The chromosome numbers of root tip cells was 19 in almost all hybrids. Four hybrids, however, had 38 in root tip cells, presumably produced by spontaneous chromosome doubling (Fig. 3C). When autotetraploid *B. oleracea* var. *acephala* was used as a male parent, the chromosome number in root tip cells was 28 in all hybrids obtained, that is sesquidiploid (Fig. 3B). Many hybrids were obtained in the crosses of various cultivars of *B. campestris* × *B. oleracea* by ovary culture.

As described above, the production rate of the hybrids was increased in the medium with CH, CM, and yeast extract, and the hybrids were also obtained in basal medium only. Further studies on suitable basal medium for ovary culture were carried out (Inomata 1979). The production of hybrids in *B. campestris*×*B. oleracea* was examined in various minerals, Nitsch (1951), Heller (1953), Murashige and Skoog (1962), White (1963), and Nitsch and Nitsch (1969). To these were added 50 g/l sucrose and 8 g/l agar. The donor materials were ovaries of *B. campestris* ssp. *chinensis*×*B. oleracea* var. *acephala*. The ovaries, excised 4 days after pollination, were cultured. Almost all hybrids obtained had 19 chromosomes in root tip cells. Four hybrids of the 120 obtained, however, had 38 chromosomes and one was sesquidiploid, having 28 chromosomes. The production rate

Table 3. Production rate of the hybrids between *Brassica campestris* and *B. oleracea* by ovary culture

Cross-combination	No. of capsules examined (A)	No. of hybrids obtained (B)	B/A×100
Diploid *B. campestris*×diploid *B. oleracea*			
ssp. *chinensis*×var. *acephala*	296	165	55.74
ssp. *chinensis*×var. *bullata*	20	2	10.00
ssp. *chinensis*×var. *capitata*	328[a]	102	31.10
ssp. *dichotoma*×var. *alboglabra*	30	24	80.00
ssp. *dichotoma*×var. *italica*	30	7	23.33
ssp. *japonica*×var. *acephala*	22	1	4.55
ssp. *japonica*×var. *capitata*	100[b,c]	14	14.00
ssp. *japonica*×var. *italica*	22	1	4.55
ssp. *pekinensis*×var. *acephala*	215	28	13.02
ssp. *pekinensis*×var. *alboglabra*	42	2	4.76
ssp. *pekinensis*×var. *bullata*	38	11	28.95
ssp. *pekinensis*×var. *capitata*	1482[a,b]	300	20.24
ssp. *pekinensis*×var. *gemmifera*	22	5	22.73
ssp. *pekinensis*×var. *gongylodes*	22	5	22.73
ssp. *pekinensis*×var. *italica*	97	13	13.40
ssp. *rapa*×var. *acephala*	20	1	5.00
ssp. *rapa*×var. *alboglabra*	21	1	4.76
ssp. *rapa*×var. *capitata*	337[b]	51	15.13
ssp. *trilocularis*×var. *alboglabra*	43	15	34.88
ssp. *trilocularis*×var. *italica*	63	21	33.33
Diploid *B. campestris*×autotetraploid *B. oleracea*			
ssp. *chinensis*×var. *acephala*	43	7	16.28
ssp. *japonica*×var. *acephala*	22	6	27.27
ssp. *oleifera*×var. *acephala*	22	6	27.27
ssp. *pekinensis*×var. *acephala*	67	22	32.84
Autotetraploid *B. campestris*×diploid *B. oleracea*			
ssp. *pekinensis*×var. *acephala*	23	7	30.44

[a] Contained the data of Matsuzawa (1978).
[b] Contained the data of Takeshita et al. (1980).
[c] The hybrids, (*B. campestris* ssp. *japonica*×*B. campestris* ssp. *pekinensis*)×*B. oleracea* var. *capitata*, were included in the cross of ssp. *japonica*×var. *capitata* (Takeshita et al. 1980).

of hybrid plants on each medium was 40.91 (Nitsch 1951), 77.27 (Heller 1953), 127.27 (Murashige and Skoog 1962), 54.55 (White 1963) and 190.91 (Nitsch and Nitsch 1969), with a mean of 90.91.

These results showed that many interspecific hybrids were obtained between many subspecies in *B. campestris* and many varieties in *B. oleracea*. The mean production rate of the hybrids in ovary culture was 6.82 (Inomata 1977), 8.52 (Inomata 1978a), 8.74 (Inomata 1978b), 38.75 (Matsuzawa 1978), 90.91 (Inomata 1979), and 19.48 (Takeshita et al. 1980) in different years and reports.

Many interspecific hybrids between *B. campestris* and *B. oleracea* have been obtained by ovary culture up till now. Table 3 shows the frequency of the production of the hybrids obtained. The results given here contain already published as well as unpublished data. The production rate of the hybrids was better in ovary culture than in artificial pollination and in embryo culture, as shown in Table 2. Hybrids were also obtained in crosses between diploid *B. campestris* and autotetraploid *B. oleracea*, and between autotetraploid *B. campestris* and diploid *B. oleracea*. Two subspecies of *B. campestris* ssp. *dichotoma* and *trilocularis* are self-compatible strains. The former is a Brown Sarson and the latter is a Yellow Sarson. The hybrids between *B. campestris* ssp. *dichotoma* and *B. oleracea* var. *alboglabra*, and between *B. campestris* ssp. *trilocularis* and *B. oleracea* var. *alboglabra* are shown in Fig. 3Q and 3R, respectively (Inomata 1983b).

In recent studies, interspecific hybrids between *B. campestris* and *B. oleracea*, in combinations of diploid and autotetraploid, were made by ovary culture, and many hybrids were obtained (Inomata 1984).

There are many wild forms on the *B. oleracea* group, these are *B. certica*, *B. insularis*, *B. macrocarpa*, etc. (Snogerup 1980). In these wild species, the attempt was made to obtain hybrids between *B. campestris* and *B. cretica* (Inomata 1985a, b), between *B. campestris* and *B. bourgeaui* (Inomata 1986), and between *B. campestris* and *B. montana* (Inomata 1987) by ovary culture, and many F_1 hybrids with 19 chromosomes were obtained and showed the intermediate character of both parents.

Recently, studies on interspecific hybridization in Brassicae using embryo, ovule, and ovary culture have been conducted by Bajaj and co-workers, and the hybrids obtained between *Brassica juncea* and *B. hirta* (Mohapatra and Bajaj 1984, 1987), *B. napus* and *B. juncea* (Bajaj et al. 1986), and *B. juncea* × *B. campestris* (Mohapatra and Bajaj 1988) were transferred to the field.

4 Hybrid Plants and Their Progenies

Three types of F_1 hybrids were obtained in 1978 (Inomata 1978b). The first type had 19 chromosomes in root tip cells, and the second 38 chromosomes, which are presumably produced by spontaneous chromosome doubling during embryo development. The third type of plants had 28 chromosomes, which were produced by the cross diploid *B. campestris* × autotetraploid *B. oleracea*.

Cytological studies of meiotic pairing and pollen fertility on F_1 hybrids were examined (Inomata 1980). Self- and openpollination in the F_1 hybrids and a crossing experiment between F_1 hybrids, and *B. campestris*, *B. oleracea*, and *B. napus* were further investigated (Inomata 1982, 1983a).

4.1 Cytological Studies on F_1 Hybrids

In the first type of plants, with 19 chromosomes, 36 hybrids were examined; pollen fertility ranged from 0% to 26.8%, the mean being 2.6%. The size of normal pollen grains, stained with acetic carmine, was not uniform in some hybrids (Fig. 4A). Monad (Fig. 4B), diad, and unequal division of tetrads (Fig. 4C) were observed during pollen development. The first meiotic division of the hybrids was examined in pollen mother cells. The mode of chromosome pairing was $9_{II}+1_{I}$ in 75% of the hybrids (Fig. 5A). The frequency of PMC's showing $9_{II}+1_{I}$ and $1_{III}+8_{II}$ (Fig. 5B) was 44.7% and 16.9%, respectively. Pentavalent association was the most complex chromosome configuration (3.8%) (Fig. 5C).

In the second type of plants, with 38 chromosomes, two hybrids were examined. Pollen fertility of the hybrids was very good, being the same as a normal *B. napus* plant, and frequency of PMC's showing 19_{II} was 80% in the first meiotic division.

In the third type of plants, with 28 chromosomes, 26 hybrids were examined, and pollen fertility ranged from 0% to 38.0%, with a mean of 14.9%. The mode of chromosome pairing in the first meiotic division was $9_{II}+10_{I}$, observed in 65% of the hybrids. Trivalents (Fig. 5D) and tetravalents were also observed in 38.1% and 1.2% of the PMC's, respectively.

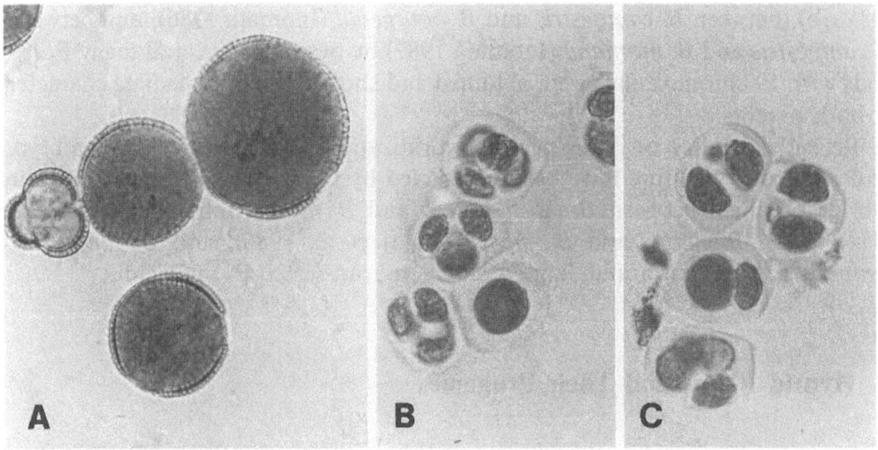

Fig. 4A–C. Pollen development in the F_1 hybrids between diploid *B. campestris* and *B. oleracea*. **A** Large-sized pollen grain; **B** a monad among tetrads; **C** a diad and unevenly divided tetrads. (Inomata 1980)

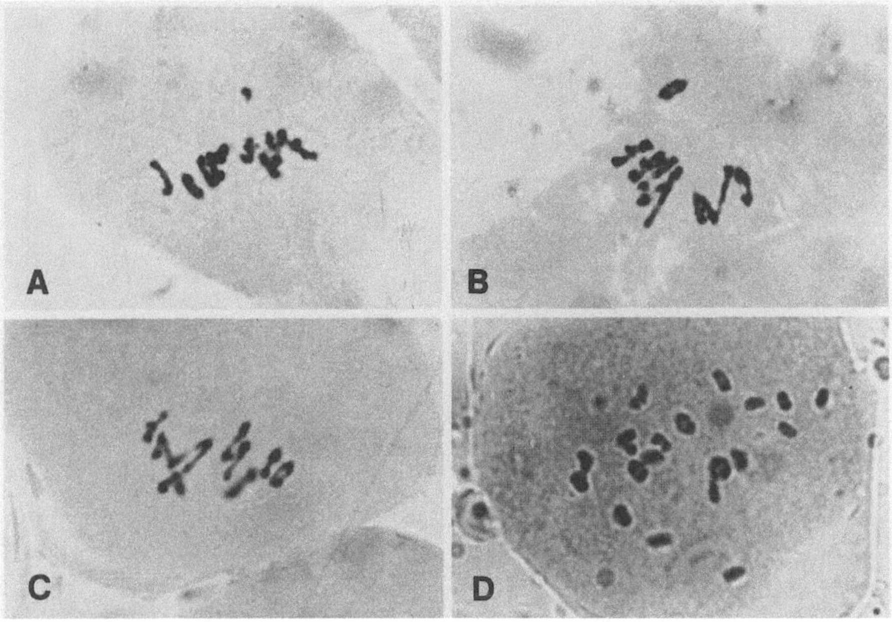

Fig. 5 A–D. First meiotic division of PMC's in the F_1 hybrids between *B. campestris* and *B. oleracea*. **A** $9_{II}+1_{I}$, diploid *B. campestris* × *B. oleracea*; **B** $1_{III}+8_{II}$, diploid *B. campestris* × *B. oleracea*; **C** $1_{V}+7_{II}$, diploid *B. campestris* × *B. oleracea*; **D** $1_{III}+8_{II}+9_{I}$, diploid *B. campestris* × autotetraploid *B. oleracea*. (Inomata 1980)

4.2 Progenies of F_1 Hybrids

When the first type of plants, with 19 chromosomes, were self-pollinated, five of the six F_2 plants obtained had 29 chromosomes. When F_1 hybrids were crossed with *B. campestris*, 131 of the 133 B_1 plants obtained had 29 chromosomes (Fig. 6A, B, C). When F_1 hybrids were crossed with *B. napus*, 88 of the 94 hybrids obtained had 38 chromosomes (Fig. 6D). No hybrid was obtained in the backcross between F_1 hybrids and *B. oleracea*.

In the second type of plants, with 38 chromosomes, two F_1 hybrids were self- and openpollinated, and they were crossed with *B. campestris* and *B. napus*. All progenies obtained had either 29 or 38 chromosomes. Thirty-one of the 39 F_2 plants obtained by self-pollination had 29 chromosomes (Fig. 6E), and in the F_2 plant with 38 chromosomes, the character was somewhat different from the plant with 29 chromosomes (Fig. 6F). When F_1 hybrids were crossed with *B. campestris* and *B. napus*, all progenies had 29 and 38 chromosomes, respectively.

When the third type of plants, with 28 chromosomes, were self-pollinated, three types of hybrids with 20, 29, and 38 chromosomes were obtained. When F_1 hybrids were crossed with *B. campestris*, 14 and 10 of the 28 progenies obtained had 29 and 38 chromosomes, respectively (Fig. 6G, H). When F_1 hybrids were crossed with diploid *B. oleracea*, four of the nine hybrids obtained had 19 chromosomes (Fig. 6I). Other aneuploids were obtained (Fig. 6J). When F_1 hybrids

Fig. 6A–H

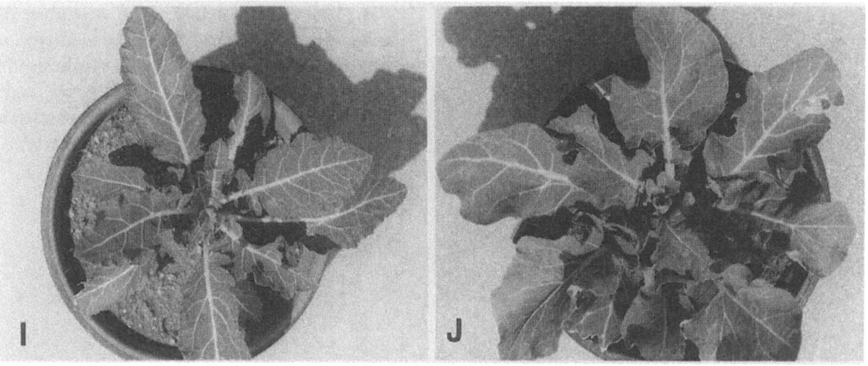

Fig. 6A–J. The F_2, B_1 plants which were obtained from F_1 hybrids (*B. campestris*×*B. oleracea*) and the hybrids which were crossed between F_1 hybrids and diploid *B. oleracea*, and *B. napus*. **A** B_1 plant from F_1 (ssp. *chinensis*×var. *acephala*)×ssp. *chinensis*, which had 29 chromosomes; **B** B_1 plant from F_1 (ssp. *pekinensis*×var. *acephala*)×ssp. *pekinensis*, which had 29 chromosomes; **C** B_1 plant from F_1 (ssp. *japonica*×var. *acephala*)×ssp. *japonica*, which had 29 chromosomes; **D** F_2 plant from F_1 (ssp. *pekinensis*×var. *gemmifera*) self-pollinated plant, which had 29 chromosomes; **E** F_2 plant from F_1 (ssp. *pekinensis*×var. *gemmifera*) self-pollinated plant, which had 38 chromosomes; **F** B_1 plant from F_1 (ssp. *pekinensis*×var. *acephala*, autotetraploid)×ssp. *pekinensis*, which had 29 chromosomes; **G** B_1 plant from F_1 (ssp. *pekinensis*×var. *acephala*, autotetraploid)×ssp. *pekinensis*, which had 38 chromosomes; **H** hybrid plant from F_1 (ssp. *pekinensis*×var. *acephala*)×*B. napus* ssp. *oleifera*, which had 38 chromosomes; **I** hybrid plant from F_1 (ssp. *japonica*×*var. acephala*, autotetraploid)×var. *capitata*, diploid, which had 19 chromosomes; **J** hybrid plant from F_1 ssp. *pekinensis*×var. *acephala*, autotetraploid)×var. *capitata*, diploid, which had 24 chromosomes. (Inomata 1983a)

were crossed with *B. napus*, many seeds were obtained. Half of the 38 hybrids examined had 38 chromosomes, and their characters were similar to those of *B. napus* plants.

It was suggested that the different types of normal gametes might be produced in the three types of F_1 hybrids with 19, 38, and 28 chromosomes. The new types of plants of *B. napus* were obtained in these crosses (Inomata 1983a).

5 Conclusion

Ovary culture was useful for producing interspecific hybrids between *B. campestris* and *B. oleracea*, and new hybrids were obtained in many cultivars between *B. campestris* and *B. oleracea*.

A new headed-type vegetable, made in an artificially synthesized plant between *B. campestris* and *B. oleracea* (Fig. 7), will prove useful for disease resistance breeding in Chinese cabbage (Nishi 1980).

According to the progenies of F_1 hybrids obtained, many new types of *B. napus* plants were obtained in the cross between F_1 hybrids with 19 chromosomes and *B. napus*, and in the self-pollinated F_1 hybrids with 38 chromosomes. The new types of plants of *B. napus* were also obtained in the cross between F_1 hybrid with 28 chromosomes, and *B. campestris* and *B. napus*. The cultivated

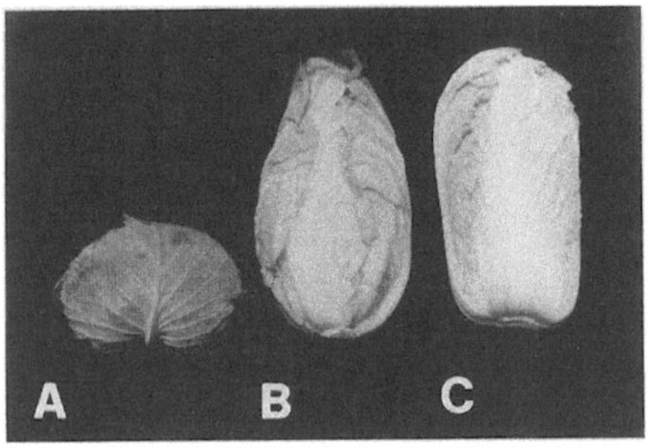

Fig. 7 A–C. New headed-type vegetable. **A** *B. oleracea* var. *capitata*; **B** new headed-type amphidiploid vegetable which was synthesized between *B. campestris* ssp. *pekinensis* and *B. oleracea* var. *capitata*; **C** *B. campestris* ssp. *pekinensis*. (Courtesy of Dr. M. Takada)

B. campestris and *B. oleracea* make use of a new gene source of *B. napus* and are useful for breeding *B. napus*, and also may exchange genes.

Recent study shows that the F_1 hybrids were produced in the cross between *B. campestris* and wild forms of *B. oleracea*, and these progenies resembled the F_1 hybrids between *B. campestris* and *B. oleracea*. Wild forms of *B. oleracea* may be useful for breeding *B. napus*.

References

Bajaj YPS (1965) Development of ovules of *Abelmoschus esculentus* L. var. Pusa Sawani in vitro. Proc Natl Inst Sci (India) 30:175–185

Bajaj YPS (1966) Growth of *Hyoscyamus niger* ovaries in culture. Phyton 23:57—62

Bajaj YPS, Mahajan SK, Labana KS (1986) Interspecific hybridization of *Brassica napus* and *B. juncea* through ovary, ovule and embryo culture. Euphytica 35:103–109

Dulieu HL (1966) Pollination of excised ovaries and culture of ovules of *Nicotiana tabacum* L. Phytomorphology 16:69–75

Feng W (1955) An interspecies cross of *Brassica*. *Brassica pekinensis* Rupr.×*B. oleracea* var. *Fimbriata* Mill. Acta Bot Sin 4:63–70

Frandsen KJ (1947) The experimental formation of *Brassica napus* L. var. *oleifera* DC. and *Brassica carinata* Braun. Dan Bot Ark 12:1–16

Guha S, Johri BM (1966) In vitro development of ovary and ovules of *Allium cepa* L. Phytomorphology 16:353–364

Hagberg A (1956) Synthesis of swede. Ann Rep Agr Coll Norway, 1953–1955, pp 141

Håkansson A (1956) Seed development of *Brassica oleracea* and *B. rapa* after certain reciprocal pollination. Hereditas 42:373–395

Heller R (1953) Recherches sur la nutrition minérales des tissus végétaux cultivés in vitro. Ann Sci Nat Bot Biol Veg 14:1–223

Hoffmann W, Peters R (1958) Versuche zur Herstellung synthetischer und semisynthetischer Rapsformen. Züchter 28:40–51

Hosoda T (1961) Studies on the breeding of new types of *Napus* crops by means of artificial synthesis in genomes of genus *Brassica*. Memo Fac Agr Tokyo Univ Education 7:1–94

Hosoda T, Namai H, Gotoh J (1963) On the breeding of *Brassica napus* obtained from artificially induced amphidiploids. III. On the breeding of synthetic rutabaga (*B. napus* var. *rapifera*). Part 1. Jpn J Breed 13:99–106

Hosoda T, Sarashima M, Namai H (1969) Studies on the breeding of artificially synthesized *Napus* crops by means of interspecific crosses between n = 10 group and n = 9 in genus *Brassica*. Memo Fac Agr Tokyo Univ Education 15:193–209

Inomata N (1968) In vitro culture of ovaries of *Brassica* hybrids between 2× and 4×. I. Culture medium. Jpn J Breed 18:139–148

Inomata N (1975) In vitro culture of ovaries of *Brassica* hybrids between 2× and 4×. II. Histological studies at certain developmental stage. Jpn J Genet 50:1–18

Inomata N (1976) Culture in vitro of excised ovaries in *Brassica campestris* L. I. Development of excised ovaries in culture media, temperature and light. Jpn J Breed 26:229–236

Inomata N (1977) Production of interspecific hybrids between *Brassica campestris* and *Brassica oleracea* by culture in vitro of excised ovaries. I. Effects of yeast extract and casein hydrolysate on the development of excised ovaries. Jpn J Breed 27:295–304

Inomata N (1978a) Production of interspecific hybrids between *Brassica campestris* and *Brassica oleracea* by culture in vitro of excised ovaries. II. Effects of coconut milk and casein hydrolysate on the development of excised ovaries. Jpn J Genet 53:1–11

Inomata N (1978b) Production of interspecific hybrids in *Brassica campestris*×*B. oleracea* by culture in vitro of excised ovaries. I. Development of excised ovaries in the crosses of various cultivars. Jpn J Genet 53:161–173

Inomata N (1979) Production of interspecific hybrids in *Brassica campestris*×*B. oleracea* by culture in vitro of excised ovaries. II. Development of excised ovaries on various culture media. Jpn J Breed 29:115–120

Inomata N (1980) Hybrid progenies of the cross, *Brassica campestris*×*B. oleracea*. I. Cytogenetical studies on F_1 hybrids. Jpn J Genet 55:189–202

Inomata N (1982) Production of interspecific hybrids between *Brassica campestris* and *B. oleracea* by culture in vitro of excised ovaries and their progenies. In: Fujuwara A (ed) Plant tissue culture 1982. Maruzen, Tokyo, pp 773–774

Inomata N (1983a) Hybrid progenies of the cross, *Brassica campestris*×*B. oleracea*. II. Crossing ability of F_1 hybrids and their progenies. Jpn J Genet 58:433–449

Inomata N (1983b) New interspecific hybrids between *B. campestris* and *B. oleracea* by ovary culture in vitro. Cruciferae Newslett 8:18–19

Inomata N (1984) Production of interspecific hybrids between *Brassica campestris* and *B. oleracea* in combinations of diploid and autotetraploid by ovary culture in vitro and their progenies. Cruciferae Newslett 9:30–31

Inomata N (1985a) Interspecific hybrids between *Brassica campestris* and *B. cretica* by ovary culture in vitro. Cruciferae Newslett 10:92–93

Inomata N (1985b) A revised medium for in vitro culture of *Brassica* ovaries. In: Chapman GP, Mantell SH, Daniels RW (eds) The experimental manipulation of ovule tissues. Longman, New York, London, pp 164–176

Inomata N (1986) Interspecific hybrids between *Brassica campestris* and *B. bourgeaui* by ovary culture in vitro. Cruciferae Newslett 11:14–15

Inomata N (1987) Interspecific hybrids between *Brassica campestris* and *B. montana* by ovary culture in vitro. Cruciferae Newslett 12:8–9

Johri BM, Sehgal CB (1963) Growth of ovaries of *Anethum graveolens* L. In: Maheshwari P, Ranga Swamy NS (eds) Plant tissue and organ culture. Symp. Int Soc Plant Morpholo, Delhi, pp 245–256

Maheshwari N (1958) In vitro culture of excised ovules of *Papaver somniferum*. Science 127:342

Maheshwari N, Lal M (1961a) In vitro culture of excised ovules of *Papaver somniferum* L. Phytomorphology 11:307–314

Maheshwari N, Lal M (1961b) In vitro culture of ovaries of *Iberis anara* L. Phytomorphology 11:17–23

Matsuzawa Y (1978) Studies on the interspecific hybridization in genus *Brassica*. I. Effects of temperature on the development of hybrid embryos and the improvement of crossability by ovary culture in interspecific cross, *B. campestris*×*B. oleracea*. Jpn J Breed 28:186–196

Mizushima U (1946) Some amphidiploids in *Cruciferous*. Ikushu to Nogei 1:31–32, 67–68

Mohapatra D, Bajaj YPS (1984) In vitro hybridization in an incompatible cross – *Brassica juncea*×*Brassica hirta*. Curr Sci 53:489–490

Mohapatra D, Bajaj YPS (1987) Interspecific hybridization in *Brassica juncea*×*Brassica hirta* using embryo rescue. Euphytica 36:321–326

Mohapatra D, Bajaj YPS (1988) Hybridization in *Brassica juncea*×*Brassica campestris* through ovary culture. Euphytica 37:83–88

Morel G (1948) Recherches sur la culture associée de parasites obligatoires et de tissus végétaux. Ann Epiphyt N S 14:123–234
Morinaga T (1929) Interspecific hybridization in *Brassica*. I. The cytology of F_1 hybrids of *B. napella* and other species with ten chromosomes. Cytologia 1:16–27
Morinaga T (1934) Interspecific hybridization in *Brassica*. VI. The cytology of F_1 hybrids of *B. juncea* and *B. nigra*. Cytologia 6:62–67
Murashige T, Skoog F (1962) A revised medium for rapid growth and bioassays with tobacco tissue cultures. Physiol Plant 15:473–497
Nakajima T, Doyama Y, Matsumoto H (1969) In vitro culture of excised ovules of white clover, *Trifolium repens* L. Jpn J Breed 19:373–378
Namai H, Hosoda T (1965) On the breeding of *Brassica napus* obtained from artificially induced amphidiploids 6. On the breeding of synthetic oil rape from interspecific cross by means of grafting. Jpn J Breed 15:213
Nishi S (1980) Differentiation of *Brassica* crops in Asia and the breeding of Hakuran a newly synthesized leafy vegetable. In: Tsunoda S, Hinata K, Gómez-Campo C (eds) *Brassica* crops and wild allies. Jpn Sci Tokyo, pp 133–150
Nishi S, Kawata J, Toda M (1959) On the breeding of interspecific hybrids between two genomes "c" and "a" of *Brassica* through the application of embryo culture techniques. Jpn J Breed 18:215–222
Nishi S, Kawata J, Toda M (1962) Studies on the embryo culture in vegetable crops. Part 2. Breeding of interspecific hybrids between cabbage varieties and Chinese cabbage varieties through the application of embryo culture techniques. Bull Hort Res Sta, Jpn A 1:111–156
Nishi S, Toda M, Toyoda T (1970) Studies on the embryo culture in vegetable crops. III. On the conditions affect to embryo culture of interspecific hybrids between cabbage and Chinese cabbage. Bull Hort Res Sta, Jpn A 9:75–100
Nishiyama I, Inomata N (1966) Embryological studies on cross-incompatibility between 2× and 4× in *Brassica*. Jpn J Genet 41:27–42
Nitsch JP (1951) Growth and development in vitro of excised ovaries. Am J Bot 38:566–577
Nitsch JP, Nitsch C (1969) Haploid plants from pollen grains. Science 163:85–87
Olsson G (1960) Species crosses within the genus *Brassica* II. Artificial *napus* L. Hereditas 46:351–386
Röbbelen G (1960) Über die Kreuzungsverträglichkeit verschiedener *Brassica* Arten als Folge eines gehemmten Pollenschlauchwachstums. Züchter 30:300–312
Rudorf W (1950) Über die Erzeugung und Eigenschaften synthetischer Rapsformen. Z Pflanzenzuecht 29:35–54
Sachar RC, Kapoor M (1965) In vitro culture of ovules of *Zephyranthes*. Phytomorphology 9:147–156
Sarashima M (1964) Studies on the breeding of artifically synthesized rape (*Brassica napus*). I. F_1 hybrids between *B. campestris* group and *B. oleracea* group and the derived F_2 plants. Jpn J Breed 14:226–237
Sarashima M (1966) On the artificially induced G_1- plants between *Brassica campestris* subsp. *oleifera*×*B. oleracea* var. *botrytis*. Jpn J Breed 16, separate vol 1:157–158
Snogerup S (1980) The wild forms of the *Brassica oleracea* groups ($2n = 18$) and their possible relations to the cultivated ones. In: Tsunoda S, Hinata K, Gómez-Campo C (eds) *Brassica* crops and wild allies. Jpn Sci Tokyo, pp 121–132
Takeshita M, Kato M, Tokumasu S (1980) Application of ovule culture to the production of intergeneric or interspecific hybrids in *Brassica* and *Raphanus*. Jpn J Genet 55:373–387
U N (1935) Genome analysis in *Brassica* with special reference to the experimental formation of *B. napus* and peculiar mode of fertilization. Jpn J Bot 7:389–452
White PR (1963) The cultivation of animal and plant cells. Ronald, New York

IV.2 Regeneration of Plants from Protoplasts of Oilseed *Brassica* Crops

K. VAMLING[1] and K. GLIMELIUS[2]

1 Introduction

1.1 Importance of Oilseed *Brassica* Crops

Along with soybean, cotton, and sunflower, rapeseed is one of the most important commercial oilseeds in the world. It is grown in temperate zones where most other oil crops do not thrive. There are several species from the family Brassicaceae used as oil crops but the most important are *Brassica napus* (rape), *B. campestris* (turnip rape, yellow sarson) and *B. juncea* (leaf mustard, rai). In Europe most of the oilseed crop consists of winter rape. In India and Pakistan, leaf mustard and sarson dominate, while in China, winter and summer rape are the most important oilseed crops.

Rapeseed oil, is used for human consumption (margarine) and industrial purposes (grease). An intensive breeding program selecting simultaneously for increased levels of the nutritionally desirable linoleic acid and a decrease in the undesirable erucic acid content was first required before the oil was suitable for human consumption. After removing the oil, a meal is left which could provide an excellent source of protein for animals. However, the meal may contain up to 8% glucosinolates depending on cultivar, and if it is used in high amounts in the animal diet, the breakdown products of the glucosinolates can have negative effects on growth and reproduction.

Strong efforts have thus been made to decrease the glucosinolate concentration through plant breeding, i.e., to produce "double-low" rapeseed cultivars containing low levels of both erucic acid in the oil and glucosinolates in the meal. These "double-low" cultivars have made, and will make, rapeseed cultivars still more important in modern agriculture by making them suitable for food and fodder purposes. However, to fulfill industrial demands, plant breeders must also breed oilseed cultivars with high levels of fatty acids with long carbon chains, e.g., erucic acid, since this gives an oil with better greasing and temperature stability properties.

[1] Department of Crop Genetics and Breeding, University of Agricultural Sciences, 26800 Svalöv, Sweden
[2] Department of Crop Genetics and Breeding, University of Agricultural Sciences, Box 7003, 750 07 Uppsala, Sweden

Biotechnology in Agriculture and Forestry, Vol. 10
Legumes and Oilseed Crops I (ed. by Y.P.S. Bajaj)

1.2 Significance and Use of *Brassica* Protoplasts

The species within the *Brassica* group are among the few crops from which it is possible to regenerate plants from single cells or protoplasts. Breeding goals for the *Brassica*, such as broadening the gene pool, introducing disease resistance and stress tolerance, and the production of hybrid seeds, may be obtained through the genetic manipulation of protoplasts. One example is the resynthesis of *Brassica napus* from *B. campestris* and *B. oleracea* through protoplast fusion (for references see Sect. 1.3). Production of atrazine-resistant and cytoplasmic male-sterile *B. napus* (Pelletier et al. 1983; Yarrow et al. 1986; Barsby et al. 1987a; Chuong et al. 1988; Robertson et al. 1987) is another promising example of the potentials of protoplast manipulation. Somatic hybridization may also lead to the production of new crops through intergeneric fusions, e.g., *Raphanobrassica*, which is a potential fodder crop (McNaughton and Ross 1978).

Protoplasts in in vitro selection systems may be useful for identifying single mutant cells from which plants can be regenerated. Swanson et al. (1988) isolated a herbicide (chlorsulfuron)-tolerant plant by using haploid *B. napus* protoplasts in an in vitro selection system.

Culturing, microinjection, or "one-to-one" protoplast fusion experiments have also been applied to *Brassica* protoplasts (Spangenberg et al. 1986b; Schweiger et al. 1987).

Recently developed DNA manipulation techniques can be combined with single cell and protoplast culture techniques. Different methods for transferring DNA into *Brassica* protoplasts have already been developed, e.g., microinjection (Rasmussen et al. 1984; Spangenberg et al. 1986a), using *Agrobacterium* as a vector (Laser and Eriksson 1985; Sacristán et al. 1988), liposome-mediated transfer (Maley and Pauls 1985), direct DNA transfer by polyethylene glycol (Köhler et al. 1988), or electroporation (Guerche et al. 1987). It has also been possible to infect *Brassica* protoplasts with cauliflower mosaic virus (CaMV), one of a few DNA plant viruses that has the potential of being used as a gene vector for plants (Hohn et al. 1985).

1.3 Brief Review of Studies on *Brassica* Protoplasts

The first successful isolation of *Brassica* protoplasts was reported by Wenzel (1973) and shortly after, Kartha et al. (1974) reported that callus formation and plant regeneration from mesophyll protoplasts of *B. napus* was possible. The isolation and culture of protoplasts from other *Brassica* species was demonstrated by Quazi (1975), although no sustained cell growth and callus formation was observed in any of these species.

Mesophyll tissue has been the most common source of isolated protoplasts from *Brassica* (Wenzel 1973; Kartha et al. 1974; Quazi 1975; Thomas et al. 1976; Schenck and Hoffmann 1979; Ulrich et al. 1980; Li and Kohlenbach 1982; Bidney et al. 1983; Guo and Schieder 1983; Vamling and Heneen 1983; Glimelius 1984; Chatterjee et al. 1985; Fu et al. 1985; Jourdan and Earle 1985; Robertson and Earle 1986a; Nishio et al. 1987) and successful plant regeneration from this tissue has also been reported for; *B. napus* (e.g., Kartha et al. 1974; and many of those cited above), *B. oleracea* (e.g., Bidney et al. 1983; Glimelius 1984; Robertson et al. 1984; Jourdan and Earle 1985; Nishio et al. 1987) and *B. juncea* (Chatterjee et al. 1985; Kao and Seguin-Swartz 1987). Plant regeneration has also been reported in *B. nigra* (Klimaszewska and Keller 1986b), *B. carinata* (Chuong et al. 1987b) and at low frequencies in *B. campestris* (Glimelius 1984). Other *Brassica* species seem to be more recalcitrant and only callus growth and root differentiation was obtained in *B. chinensis* (Guo and Schieder 1983) and *B. rapa* (Ulrich et al. 1980).

The increasing number of reports of successful plant regeneration of different *Brassica* species could in some part be due to the use of in vitro-grown material which often gives more stable protoplasts than field- or greenhouse-grown plants (Vatsya and Bhaskaran 1982; Glimelius 1984; Chuong et al. 1985). In vitro-grown hypocotyls are a suitable material for releasing totipotent protoplasts. It has been possible to regenerate plants from a wide range of *Brassica* species from callus obtained from isolated hypocotyl protoplasts (*B. napus, B. oleracea* and *B. campestris* (Glimelius 1984); *B. juncea, B. carinata, B. nigra* (Sjödin 1989) and *Sinapis alba* (Glimelius unpubl.). It is, however, also possible to regenerate plants from callus derived from mesophyll protoplasts of these species.

Other sources of totipotent protoplasts are in vitro-cultured roots from 2-day-old seedlings (Xu et al. 1982; Lillo and Shahin 1986), cotyledons excised from 4-6-day-old seedlings (Lu et al. 1982) or precultured cotyledonary leaves (Vatsya and Bhaskaran 1982), stem cortex (Chuong et al. 1987a; Klimaszewska and Keller 1987), and cell suspensions (Klimaszewska and Keller 1986b). Another in vitro system, probably unique for Brassicaceae, is the stem embryo system, which is useful for protoplast isolation and plant regeneration in *B. napus* (Thomas et al. 1976; Kohlenbach et al. 1982a, b; Swanson et al. 1988). The protoplasts isolated from plants obtained from such stem embryos developed into somatic embryos when cultured (Li and Kohlenbach 1982). Embryogenesis has also been reported from mesophyll protoplasts from ordinary *B. oleracea* plants (Jourdan and Earle 1985), in hypocotyl protoplast culture of *B. napus* (Kirti 1988; Kranz 1988), and from suspension-derived protoplasts in *B. nigra* (Klimaszewska and Keller 1986b).

Somatic hybridization has great potential as a complement to conventional plant breeding in the *Brassica* crops. Members of this group have remarkable tolerance to interspecific crossing, e.g., production of amphidiploids from the diploid *Brassica* species. Moreover, they have a large variation within the species, e.g., the cabbage complex. Since several desirable traits often exist in closely related wild or cultivated species (Namai et al. 1980), there have been attempts to introduce these into the cultivated oilcrops (Olsson and Ellerström 1980). However, some of the attempts have failed, probably due to sexual incompatibility between the species. In these cases, protoplast fusion could be used to circumvent the barrier. Resynthesis of oilseed rape, through fusion of protoplasts from *B. campestris* and *B. oleracea* has been reported by several authors (Schenck and Röbbelen 1982; Robertson et al. 1985; Sundberg and Glimelius 1986; Jourdan et al. 1986; Taguchi and Kameya 1986; Robertson et al. 1987; Terada et al. 1987; Sundberg et al. 1987; Rosén et al. 1988), indicating that cell fusions and subsequent plant regeneration can be made between closely related species. Other examples of interspecific crosses within *Brassica* are the somatic hybridization between *B. napus* and *B. hirta* (Primard et al. 1986), *B. napus* and *B. nigra* (Sjödin and Glimelius 1988); and *B. napus* and any one of *B. juncea, B. carinata,* (Sjödin 1989) and *B. oleracea* (Glimelius et al. 1989). Somatic *Brassica* hybrids have also been produced between more distantly related species. Gleba and Hoffmann (1980) fused protoplasts from *Arabidopsis thaliana* and *B. campestris* and obtained hybrid calli, which after some time in culture, during which a sorting out of chromosomes occurred, finally regenerated into shoots and produced "*Arabidobrassica*" (Hoffmann and Adachi 1981). Other more recent reports of successful plant regeneration of intergeneric somatic hybrids are crosses between *Eruca sativa* and *B. napus* (Fahleson et al. 1988), *Barbarea vulgaris* and *B. napus* (Glimelius et al. 1989), *Sinapis turgida* and *B. oleracea/B. nigra* (Toriyama et al. 1987b), *Moricandia arvensis* and *B. oleracea* (Toriyama et al. 1987a), *Diplotaxis harra/Diplotaxis ibicensis* and *B. napus* (Klimaszewska and Keller 1986a, 1988), *Diplotaxis muralis* and *B. napus* (McLellan et al. 1988), *Diplotaxis muralis* and *B. juncea* (Chatterjee et al. 1988).

Protoplast fusion can also be useful when desired traits are controlled by an organellar genome, e.g., atrazine resistance coded by chloroplast DNA (Souza-Machado et al. 1978) and cytoplasmic male sterility coded by the mitochondrial genome (Vedel et al. 1982). These organellar traits can be transferred using cytoplasts (protoplasts without the nuclei) or irradiated protoplasts (inactivated nuclei) as one of the partners in the fusion experiment. The initial result of such a transfer is a protoplast with one nucleus A and a mixed cytoplasm A+B. Pelletier and his co-workers were the first to show how elegantly these techniques can be used for constructing a functional male sterile *B. napus* (Pelletier et al. 1983; Chetrit et al. 1985). A male-sterile *B. oleracea* was produced by sexual crossing between *B. oleracea* and *Raphanus sativus* (Bannerot et al. 1974), which, after repeated backcrossing to *B. napus*, resulted in a combination of the organelles from *Raphanus sativus* and the nucleus from *B. napus* (Vedel et al. 1982). However, a chlorophyll deficiency is obtained in these plants at temperatures below 15 °C, which probably arises from an incompatibility between the nucleus of *Brassica* and the chloroplasts from *Raphanus*. A way of overcoming this problem, with the "conventionally produced" plants, was to substitute the *Raphanus* chloroplasts with *Brassica* chloroplasts.

Pelletier et al. (1983) achieved this by fusing protoplasts of the male-sterile *B. napus* with protoplasts from *B. campestris* carrying atrazine-resistant chloroplasts. The possibility of manipulating the cytoplasmic traits within *Brassica* species has been described in several publications. The exchange of chloroplasts in Ogura cms lines to correct the chlorophyll deficiency has been performed by several groups (Menczel et al. 1987; Robertson et al. 1987; Jarl and Bornman 1988). Two other cytoplasmic male sterility systems in *B. napus*, napus-cms and Polima-cms (for review, Erickson et al. 1986), have been combined with atrazine resistance, with the help of protoplast fusion. Yarrow et al. (1986) fused a napus-cms *B. napus* line with a *B. napus* line carrying *B. campestris* chloroplasts with atrazine resistance. The fusion products were selected by manual micromanipulation and three plants out of 19 somatic hybrid plants had the desired organelle combination. The Polima-cms system was combined in a *B. napus* line with atrazine resistance, using protoplast fusion (Barsby et al. 1987a). To facilitate recovery of the desired organelle combination, one fusion partner was inactivated with iodoacetate and the other with gamma-irradiation. A total of eight plantlets were regenerated in the two protoplast fusion experiments (one without inactivation of the protoplasts). Out of the four plants which developed to maturity, two were identified as cybrids carrying the desired combination. Another example of manipulation with heritable cytoplasmic traits is the transfer of a cms system from a spring *B. napus* cultivar to a winter *B. napus* cultivar (Barsby et al. 1987b). They introduced the Polima-cms system from a spring Polima-cms cultivar into a winter *B. napus* cultivar in approximately 9 months which should be compared to 3 years by means of conventional backcrossing. A similar transfer of a cytoplasmic trait from a spring type to a winter type was performed by Thomzik and Hain (1988). They transferred atrazine-tolerant chloroplasts from a spring oilseed rape variety to some winter oilseed rape varieties by means of protoplast fusion. Another possibility is that recombination between organellar genomes could occur (preferentially between the two mitochondrial genomes), producing a plant with totally new heritable cytoplasmic traits. This was the case in the material produced by Pelletier and his co-workers (Pelletier et al. 1983; Vedel et al. 1986), where recombination was found between the two types of mitochondria combined. Similar results with extensive mtDNA recombination in their *Brassica* fusion products were reported by Morgan and Maliga (1987) and Robertson et al. (1987). In contrast to these reports, Barsby and her co-workers have not observed mtDNA recombination in their plants (Kemble et al. 1988; Yarrow et al. 1988).

2 Protoplast Isolation

The protoplast techniques described in this chapter (Sects. 2–5) are the techniques used and developed by Dr. Glimelius and co-workers. It should be pointed out, however, that several other protoplast systems are available in *Brassica* (for further details, see Table 1).

2.1 Material

Depending on the purpose, different types of materials can be used for protoplast isolation. If the prerequisite is a high yield of protoplasts from one single individual plant, mesophyll material is the only realistic choice. This can be obtained from: (1) greenhouse- or growth cabinet-grown plants or (2) in vitro-cultured shoots (shoot culture). Our experience is that in vitro-cultured shoots give mesophyll protoplasts with high viability and good experimental reproducibility. The shoots are cultured on MS medium (Murashige and Skoog 1962) without hormones in cabinets with a light intensity of 12 W/m^2 (Osram halogen lamps HQ 1, 400 W) and night/day regime of 16/8 h at 25/20 °C. The shoots are then subcultured from cuttings after 1 month of growth. Plants kept in greenhouses or growth cabinets should be grown under as stable and favorable growing condi-

Table 1. Publications of work with *Brassica* protoplasts – callus initiation – plant regeneration

Species; *Brassica*	Tissue	Transfer[a]	Medium	Hormones (mg/l)											Org. add.[b]	Growth response	Reference
				2,4-D	NAA	IAA	BAP	Z	kin	2ip	AS	GA3	ABA	IBA			
napus	Leaves	1	B5 (m)	0.51	0.19	–	0.23	–	–	–	–	–	–	–	–	Plants	Kartha et al.
	(plants, GH	2	MS (m)	0.51	–	–	–	–	–	–	–	–	–	–	CM(10%)		(1974)
	1 month)	3	MS (m)	–	–	–	1.1	–	–	–	–	0.35	–	–	–		
		4	MS (m)	–	–	–	–	–	–	–	–	–	–	–	–		
napus	Leaves (plants, GH 30–50 days)	1	NT	–	3.0	–	1.0	–	–	–	–	–	–	–	–	Single divisions	Quazi (1975)
napus	Leaves	1	NN (m)	0.25	0.5	–	0.1	–	–	–	–	–	–	–	–	Plants	Thomas et
	(shoot cult.	2	NN (m)	0.1	–	–	0.25	–	–	–	–	–	–	–	–		al. (1976)
	(n) in vitro/ cult. cabinet. From stem embryos)	3	MS	–	–	–	–	–	–	–	–	–	–	–	–		
oleracea	Leaves	1	GK-2	0.25	0.5	–	0.1	–	–	–	–	–	–	–	–	Roots	Schenck and
campestris	(plants in		NT-1	0.25	0.5	–	0.1	–	–	–	–	–	–	–	–	*B. campestris*	Hoffmann
nigra	vitro	2	LS	–	1.0	–	0.5	–	–	–	–	1.0	–	–	–	*B. oleracea*	(1979)
	8 weeks)		LS	–	0.1	–	–	2.0	–	–	–	–	–	–	–	Callus *B. nigra*	
rapa	Leaves	1	KM8p	0.2	1.0	–	–	0.5	–	–	–	–	–	–	–	Roots	Ulrich et al.
	(plants, GH 4–6 weeks)	2	MS (m)	–	–	2.0	1.0	–	–	–	–	–	–	–	–		(1980)
napus	Leaves and	1	N (m)	1.0	–	–	–	0.5[c]	–	–	–	–	–	–	–	Plants	Kolenbach et
	stem em-	2	MS (m)	–	–	–	–	–	–	–	–	–	–	–	–		al. (1982b)
	bryos	3	PCW	1.75	–	–	–	–	0.5	–	–	–	–	–	PE(20%)		
	(plants (n),	4	MS (m)	–	–	0.5	0.05	–	–	–	–	–	–	–	–		
	in vitro-	5	MS (m)	–	–	0.05	0.01	–	–	–	–	–	–	–	–		
	grown)	6	MS (m)	–	–	–	–	–	–	–	–	–	–	–	–		

napus	Leaves	1	N (m)	0.5	0.5	–	0.5	–	–	–	–	–	–	–	–	Plants	Li and
	(shoot cult.		N (m)	1.0	0.5	–	1.0	–	–	–	–	–	–	–	–		Kolenbach
	from stem	2	MS-13	0.2	–	–	–	–	3.0	–	–	–	–	–	–		(1982)
	embryos, in	3	MS-13	0.01	–	–	1.0	–	–	–	–	–	–	–	–		
	vitro-	4	MS	–	–	–	–	–	–	–	–	1.0–3.0	0.05	–	–		
	grown)		MS	–	–	–	–	–	–	–	–	–	–	–	–		
napus	Cotyledons	1	K	0.2	1.0	–	–	0.5	–	–	–	–	–	–	CW(1%)	Roots	Lu et al.
oleracea	(seedlings,	2	MS	–	–	2.0	1.0	–	–	–	–	–	–	–	–	*B. napus*	(1982)
campestris	in vitro-															*B. campestris*	
	grown 4–6															Plants	
	days)															*B. oleracea*	
oleracea	Cotyledons/	1	KM8p	0.2	1.0	–	–	0.5	–	–	–	–	–	–	–	Plants	Vatsya and
	leaves	2	B3	–	–	4.0	–	–	2.56	–	–	–	–	–	–		Bhaskaran
	(10–12		MS-B	–	–	0.88	–	–	2.8	–	–	–	–	–	–		(1982)
	days old/	3	MS-B	–	–	0.88	–	–	2.8	–	–	–	–	–	–		
	shoot cult.,																
	20 days old																
	or GH-																
	grown,																
	30–40																
	days)																
napus	Roots	1	KM8p/	0.1	1.0	–	–	0.2	–	–	–	–	–	–	–	Callus	Xu et al.
oleracea	(seedlings,		KM8	0.2	1.0	–	–	0.5	–	–	–	–	–	–	–	*B. alba*	(1982)
campestris	in vitro-	2	B5	–	0.02	–	2.0	–	–	–	–	–	–	–	–	Roots	
alba	grown	3	MS	–	–	2.0	1.0	–	–	–	–	–	–	–	–	*B. campestris*	
	2 days)		MS	–	0.05	–	0.5	–	–	–	–	–	–	–	–	Plants	
		4	MS	–	–	–	–	–	–	–	–	–	–	–	–	*B. napus*	
																B. oleracea	
oleracea	Leaves	1	CL/R	1.0	–	–	0.4	–	–	–	–	–	–	–	–	Plants	Bidney et al.
	(plants, EC	2	C-1	–	–	–	1.1	–	–	–	–	0.035	–	–	–		(1983)
	30 days	3	T.5	–	–	–	–	–	–	–	–	–	–	–	–		
	old)																

Footnotes to Table 1 see page 397.

Table 1 (continued)

Species; *Brassica*	Tissue	Trans-fer[a]	Medium	Hormones (mg/l)											Org. add.[b]	Growth response	Reference
				2,4-D	NAA	IAA	BAP	Z	kin	2ip	AS	GA3	ABA	IBA			
chinensis	Leaves (plants, in	1	NN (m)	0.25 −0.5	0.1 −0.2	–	0.1 −0.2	–	–	–	–	–	–	–	–	Roots	Guo and Schieder (1983)
	vitro-grown	2	NN (m)	0.1	–	–	0.25	–	–	–	–	–	–	–	–		
	4 – 6 weeks)	3	MS	–	–	1.0	–	–	3.0	–	–	–	–	–	–		
			MS	–	–	2.0	–	–	6.0	–	–	–	–	–	–		
			MS	–	0.2	–	0.5	–	–	0.5	–	–	–	–	–		
napus	Leaves	1	B	0.25	1.0	–	1.0	–	–	–	–	–	–	–	–	Plants	Pelletier et al. (1983)
	(plants, in	2	C	–	0.2	–	1.0	–	–	–	–	–	–	–	–		
	vitro-	3	D	1.0	–	–	–	–	–	–	30.0	–	–	–	–		
	grown)	4	E	–	1.0	–	–	–	–	1.0	–	0.02	–	–	–		
		5	F	–	0.1	–	0.5	–	–	–	–	–	–	–	–		
		6	G	–	0.01	–	–	–	–	–	–	–	–	–	–		
napus oleracea campestris	Leaves/hyp-ocotyls	1	A	1.0	0.1	–	0.5	–	–	–	–	–	–	–	–	Plants	Glimelius (1984)
			KM8p(m)	1.0	0.1	–	0.5	–	–	–	–	–	–	–	–		
	(plants, in	2	K_3 (m)	1.0	–	–	0.1	–	–	–	–	–	–	–	–		
	vitro grown	3	K_3 (m)	–	–	0.1	1.0	1.0	–	–	–	–	–	–	–		
	30 days/ seedlings, in vitro-grown 4 – 5 days)	4	MS	–	–	–	–	–	–	–	–	–	–	–	–		
juncea	Leaves	1	BG	0.3	0.3	–	1.0	–	–	–	–	–	–	–	CM (5%)	Plants	Chatterjee et al. (1985)
	(plants, in	2	BG	0.3	0.3	–	1.0	–	–	–	–	–	–	–	CM (5%)		
	vitro-	3	BG	0.3	0.3	–	1.0	–	–	–	–	–	–	–	CM (5%)		
	grown, 4	4	MS	0.3	0.3	–	1.0	–	–	–	–	–	–	–	–		
	week old)	5	MS	–	1.0	–	2.0	–	–	–	–	–	–	–	–		
		6	CSS	0.5	0.5	–	3.0	–	–	–	–	0.1	–	–	–		
		7	MS	–	–	–	–	–	–	–	–	–	–	–	–		

napus	Hypocotyl	1	NN (m)	0.5	1.0	–	0.5	–	–	–	–	–	–	–	–	Plants	Chuong et
	hooks	2	MS	–	0.5	–	5.0	–	–	–	–	–	–	–	–		al. (1985)
	(seedlings,																
	in vitro-																
	grown, 2																
	days old)																
oleracea	Leaves	1	DPD (m)	0.5	–	–	–	–	1.0	–	–	–	–	–	–	Plants	Fu et al.
	(plants, GH	2	MS (m)	0.1	–	–	–	–	3.0	–	–	–	–	–	–		(1985)
	2–4 weeks)			–0.5					–4.0								
		3	MS (m)	–	–	–	–	–	3.0	–	–	0.1	–	–	–		
		4	MS (m)	–	–	0.1	–	–	0.1	–	–	0.1	–	–	–		
napus	Hypocotyls	1	A/	1.0	0.1	–	0.4	–	–	–	–	–	–	–	–	Plants	Barsby et al.
	(seedlings,		S	1.0	0.1	–	–	–	–	–	–	–	–	–	–		(1986)
	in vitro-	2	S	1.0	–	–	–	–	0.1	–	–	–	–	–	–		
	grown 3–5	3	MS (m)	1.0	–	–	–	–	0.1	–	–	–	–	–	–		
	days)		MS (m)	–	–	0.1	–	2.0[b]	2.0	–	–	–	–	–	–		
		4	B5	–	–	–	–	–	–	–	–	0.03	–	–	–		
nigra	Stem peels	1	NN(m)/	0.5	–	–	0.25	–	–	–	–	–	–	–	–	Plants	Chuong et
			MS (m)														al. (1986)
		2	MS (m)	–	0.5	–	0.625	0.625	0.625	0.625	–	–	–	–	–		
oleracea	Hypocotyls	1	KM8p(m)	1.0	0.1	–	0.5	–	–	–	–	–	–	–	–	Plants	Harney and
	(seedlings,	2	K_3	0.1	–	–	1.0	–	–	–	–	–	–	–	–		Tsai (1986)
	in vitro-	3	MS (m)	–	–	0.05	–	1.0	1.0	–	–	–	–	–	–		
	grown, 5	4	B5	–	–	–	–	–	–	–	–	–	–	–	–		
	days old)																
nigra	Cell suspen-	1	K	0.2	1.0	–	0.5	–	–	–	–	–	–	–	–	Somatic em-	Klimaszewska
	sion (2 days		S_1	0.2	1.0	–	0.5	–	–	–	–	–	–	–	–	bryos	and Keller
	after	2	M III	–	0.01	–	1.0	–	–	–	–	–	–	–	–		(1986b)
	transfer)		MS	–	0.01	–	1.0	–	–	–	–	–	–	–	–		
		3	MS	–	–	–	–	–	–	–	–	–	–	–	–		
			B5	–	–	–	–	–	–	–	–	–	–	–	–		

Table 1 (continued)

Species; *Brassica*	Tissue	Trans-fer[a]	Medium	Hormones (mg/l) 2,4-D	NAA	IAA	BAP	Z	kin	2ip	AS	GA3	ABA	IBA	Org. add.[b]	Growth response	Reference
oleracea	Hypocotyls	1	TM-2 (m)	1.0	0.1	–	0.5	–	–	–	–	–	–	–	–	Plants	Lillo and
	(seedlings,	2	TM-2 (m)	1.0	0.1	–	0.5	–	–	–	–	–	–	–	–		Shahin (1986)
	in vitro-		TM-2 (m)	0.2	–	–	0.5	–	–	–	–	–	–	–	–		
	grown, 6		TM-2 (m)	–	–	0.1	–	1.9	–	–	–	–	–	–	–		
	days old)		TM-2 (m)	–	–	0.1	–	–	–	–	–	–	–	–	–		
		3	TM-2 (m)	–	–	0.1	–	1.9	–	–	–	0.1	–	–	–		
			TM-2 (m)	–	–	–	1.0	–	–	–	–	–	–	–	–		
		4	–[d]	–	–	–	–	–	–	–	–	–	–	–	–		
oleracea	Leaves	1	B	0.25	1.0	–	1.0	–	–	–	–	–	–	–	–	Plants	Robertson
	(plants, in	2	C	–	0.2	–	1.0	–	–	–	–	–	–	–	–		and Earle
	vitro-	3	D	1.0	–	–	–	–	–	–	30.0	–	–	–	–		(1986a)
	grown, 3	4	E	–	1.0	–	–	–	–	1.0	–	0.02	–	–	–		
	weeks old)	5	F	–	0.1	–	0.5	–	–	–	–	–	–	–	–		
		6	G	–	0.01	–	–	–	–	–	–	–	–	–	–		
napus	Hypocotyls	1	K (m)	–	1.5	–	0.1	–	–	–	–	–	–	–	–	Callus	Spangenberg
	(seedlings,	2	K (m)	0.5	–	–	0.05	–	–	–	–	–	–	–	–		et al. (1986)
	in vitro- grown, 4–14 days old)																
carinata	Hypocotyl	1	NN (m)[e]	0.5	0.5	–	0.25	–	–	–	–	–	–	–	–	Plants	Chuong et
	hooks	2[f]	MS (m)	–	–	0.1	–	0.8	–	–	–	–	–	–	–		al. (1987b)
	(seedlings,	3	MS (m)	–	0.5	–	5.0	–	–	–	–	–	–	–	–		
	in vitro-	4[f]	MS (m)	–	–	0.02	1.1	–	–	–	–	–	–	–	–		
	grown, 2 days old)																
napus	Haploid	1	MS (m)	1.0	–	–	0.5	–	–	–	–	–	–	–	–	Plants	Chuong et
	stem peels	2	MS (m)	–	0.5	–	0.625	0.625	0.625	0.625	–	–	–	–	–		al. (1987a)
	(plants, EC)	3	MS (m)	–	0.5	–	0.625	0.625	0.625	0.625	–	–	–	–	–		
		4	MS (m)	–	0.5	–	0.625	0.625	0.625	0.625	–	–	–	–	–		
			MS (m)	–	0.5	–	5.0	–	–	–	–	–	–	–	–		
			MS (m)	–	–	0.1	0.25	0.25	0.25	0.25							

napus	Stem cortex (plants, EC)	1	S_1	0.2	1.0	–	0.5	–	–	–	–	–	–	–	–	Plants	Klimaszewska and Keller (1987)
			B	0.25	1.0	–	1.0	–	–	–	–	–	–	–	–		
			L (m)	–	0.5	–	0.06	–	–	–	–	–	–	–	–		
		2	E	–	1.0	–	–	–	–	1.0	–	0.02	–	–	–		
			B5 (m)	–	–	–	3.0	–	–	–	–	–	–	–	–		
			B5 (m)	–	–	–	–	3.0	–	–	–	–	–	–	–		
		3	B5	–	–	–	–	–	–	–	–	–	–	–	–		
oleracea	Leaves (shoot cult. in vitro-grown)	1	MS (m)	–	5.0	–	–	–	1.0	–	–	–	–	–	–	Plants	Nishio et al. (1987)
			MS (m)	2.0	–	–	–	–	1.0	–	–	–	–	–	–		
			MS (m)	2.0	0.5	–	–	–	1.0	–	–	–	–	–	–		
		2	MS (m)	–	1.0	–	–	–	1.0	–	–	–	–	–	–		
			MS (m)	–	5.0	–	–	–	1.0	–	–	–	–	–	–		
			MS (m)	–	1.0	–	–	–	5.0	–	–	–	–	–	–		
			MS (m)	0.5	–	–	–	–	1.0	–	–	–	–	–	–		
			MS (m)	2.0	–	–	–	–	1.0	–	–	–	–	–	–		
			MS (m)	0.5	–	–	–	–	5.0	–	–	–	–	–	–		
			MS (m)	–	–	2.0	–	–	1.0	–	–	–	–	–	–		
			MS (m)	–	–	10.0	–	–	1.0	–	–	–	–	–	–		
			MS (m)	–	–	2.0	–	–	5.0	–	–	–	–	–	–		
		3	MS	–	–	–	1.0	–	–	–	–	0.035	–	–	–		
		4	MS	–	–	–	–	–	–	–	–	–	–	–	–		
alboglabra	Stem segments, hypocotyls, cotyledons, leaves (in vitro-grown)	1	MS (m)	–	0.2–1.0	–	0–0.4	–	–	–	–	–	–	–	–	Plants	Pua (1987)
			MS (m)	0.2–1.0	–	–	0–0.4	–	–	–	–	–	–	–	–		
		2	MS (m)	–	–	–	0.125–4.0	–	–	–	–	–	–	–	–		
			MS (m)	–	–	–	–	–	–	0.125–4.0	–	–	–	–	–		
			MS (m)	–	–	–	–	0.125–4.0	–	–	–	–	–	–	–		
			MS (m)	–	–	–	–	–	0.125–4.0	–	–	–	–	–	–		
			MS (m)	0/0.25	0/0.25	0/0.25	0.25	–	–	–	–	–	–	0/0.25	–		
		3	MS (m)	–	–	–	–	–	–	–	–	–	–	0.1	–		

Table 1 (continued)

Species; *Brassica*	Tissue	Transfer[a]	Medium	Hormones (mg/l) 2,4-D	NAA	IAA	BAP	Z	kin	2ip	AS	GA3	ABA	IBA	Org. add.[b]	Growth response	Reference
napus	Hypocotyls	1	KM8p(m)	1.0	0.1	–	0.4	–	–	–	–	–	–	–	CW(2%)	Plants	Kirti (1988)
	(seedlings,	2	MS (m)	3.0	–	–	–	–	–	–	–	–	–	–	CW(10%)		
	in vitro-	3	MS (m)	–	–	–	3.0	–	–	–	–	0.1	–	–	CW(10%)		
	grown, 4–5 days old)	4	MS	–	–	–	–	–	–	–	–	–	–	–			
napus	Hypocotyls	1[g]	K (m)	–	1.5	–	0.1	–	–	–	–	–	–	–	–	Plants from	Kranz (1988)
	(seedlings,	2	MS (m)	1.0	–	–	–	–	0.02	–	–	–	–	–	–	somatic em-	
	in vitro-	3[h]	MS (m)	–	0.1	–	2.0	–	–	–	–	–	–	–	–	bryos induc-	
	grown,	4	K_3 (m)	–	–	0.1	0.5	–	–	–	–	–	–	–	–	ed in suspen-	
	4–14 days		MS (m)	–	1.0	–	2.0	–	–	–	–	–	–	–	–	sion culture	
	old)		MS (m)	–	0.1	–	0 –3.0	–	–	–	–	0 –1.0	–	–	–		
napus	Leaves	1[e]	MS	0.25	0.5	–	–	–	–	–	–	–	–	–	–	Organogenic	Millam et al.
	(plants, in	2	MS	0.25	0.25	–	0.5	–	–	–	–	–	–	–	–	callus	(1988)
	vitro-grown)	3	MS	–	0.5	–	5.0	–	–	–	–	–	–	–	–		
napus oleracea	Leaves/ hypocotyls	1	KM8p(m)	1.0	0.1	–	0.1	–	–	–	–	–	–	–	CW (2%)	Plants	Glimelius
campestris	(plants, in	2	K_3 (m)	0.25	0.025	–	0.025	–	–	–	–	–	–	–	–		this chapter
nigra	vitro-	3	K_3 (m)	–	–	0.1	1.0	1.0	–	–	–	–	–	–	–		
juncea	grown, 30 days/seedlings in vitro-grown, 4–5 days old)	4	MS	–	–	–	–	–	–	–	–	–	–	–	–		

[a] The numbers under "Transfer" indicates the order of mediachanges when material (callus and shoots) is transferred to a new combination of medium.
[b] Organic additives; coconut water/milk and potato extract.
[c] Zeatin riboside.
[d] Medium composition not reported.
[e] Ficoll addition.
[f] "Rescue medium".
[g] Medium for established cell suspension.
[h] Induction medium; 2 passages on liquid, 3–4 passages on solidified.

Media; hormones; abbreviations

A	Kao and Michayluk (1981)
B	Pelletier et al. (1983)
B3	Sacristán and Melchers (1969)
BG	Chatterjee et al. (1985)
B5	Gamborg et al. (1968)
C	Pelletier et al. (1983)
C-1	Bidney et al. (1983)
CL	Bidney et al. (1983)
CSS	Chatterjee et al. (1985)
D	Pelletier et al. (1983)
DPD	Durand et al. (1973)
E	Pelletier et al. (1983)
F	Pelletier et al. (1983)
G	Pelletier et al. (1983)
GK2	Schenck and Hoffmann (1979)
K	Kao (1977)
K_3	Nagy and Maliga (1976)
KM8	Kao and Michayluk (1975)
KM8p	Kao and Michayluk (1975)
L	Lichter (1981)
M III	Li and Kohlenbach (1982)
LS	Linsmaier and Skoog (1965)
MS	Murashige and Skoog (1962)
MS-B	Binding (1974)
MS-13	Kohlenbach et al. (1982b)
N	Nitsch (1969)
NN	Nitsch and Nitsch (1969)
NT	Nagata and Takebe (1971)
NT-1	Gleba and Hoffmann (1978)
PCW	Wenzel et al. (1977)
R	Bidney et al. (1983)
S_1	Schenck and Röbbelen (1982)
S	Shepard (1980)
T.5	Bidney et al. (1983)
TM-2	Shahin (1985)
ABA	abscisic acid
BAP	6-benzylaminopurine
2,4-D	2,4-dichlorophenoxyacetic acid
GA3	gibberelic acid
2,i-p	N^6-(Δ^2-isopentenyl) adenin
IAA	3-indoleacetic acid
IBA	3-indolbutyric acid
Kinetin	6-furfuryl amino purine
NAA	1-naphtaleneacetic acid
Zeatin	6-hydroxy-methylbut-ethylamino-purine
AS	Adenine sulfate
CM	coconut milk
CW	coconut water
PE	potato extract
GH	greenhouse
EC	environmental chamber

tions as possible [in our experiments: light 35 W/m^2 (warm-white fluorescent tubes, 40 W and tungsten lamps), 12 h photoperiods, temperature daytime 25 °C and night time 20 °C]. The temperature may not exceed 30 °C, since the yield of viable protoplasts under these conditions usually decreases. When using leaf material, a dark treatment (24–48 h), is recommended for achieving an optimal mesophyll protoplast isolation.

If different individuals can be accepted for protoplast isolation, young, in vitro-grown material is the best source. Cotyledons (Lu et al. 1982), hypocotyls (Glimelius 1984) and roots (Xu et al. 1982) usually give a high reproducible yield of protoplasts with a high regeneration potential. In our experience, young hypocotyls are the best source.

The culture conditions for mesophyll and hypocotyl protoplasts were investigated with *Brassica napus* L. ssp. *oleifera* cv. Olga as the model material, but the cultivars Hanna and Topas and the oilseed crops *B. campestris, B. nigra, B. juncea, B. carinata*, and *Sinapis alba* have also been studied.

To obtain sterile hypocotyls, seeds are surface sterilized with 7.5% calcium hypoclorite for 1 h with continuous shaking, and thereafter treated with 70% EtoH for 1 min, then washed twice in sterile distilled water. The seeds are germinated on MS-medium without hormones in the dark at 25 °C.

2.2 Enzyme Treatment and Purification

Hypocotyl and mesophyll material are treated almost identically for protoplast isolation (mesophyll modifications in brackets). Hypocotyls 4 to 5 (30) days old, are cut transversely into 0.5–1.0 (1–2) mm segments, plasmolyzed for 1 h in 0.3 M sorbitol and 0.05 M $CaCl_2$ $2H_2O$, and then treated with 1% Cellulysin and 0.1% Macerase (Calbiochem-Behring) in the medium K3 according to Nagy and Maliga (1976), but modified as described in Section 7, containing 0.4 M sucrose. The enzyme incubation for hypocotyls is usually performed overnight 16–18 h at 20–22 °C in the dark on a rocking tray with low speed and a low degree of tilting. Mesophyll material is incubated under the same conditions but without shaking. The time for enzyme treatment of mesophyll material can, if necessary, be shortened to 4–5 h without any marked decrease in protoplast yield. The enzyme-treated material is filtered through a nylon mesh of 50 µm to separate undegraded material from the protoplasts. A salt solution, CPW 16 (Banks and Evans 1976), containing 0.47 M sucrose at a volume approximately equal to the enzyme and protoplast volume, is then added, and the resulting mixture is centrifuged in a swing-out rotor at 100 *g* for 7 min. Viable protoplasts with intact cell membranes float on the surface and can be removed with a Pasteur pipette. The protoplast suspension is diluted about ten times, washed once with a salt solution, W5 (Menczel et al. 1981), and finally pelleted by centrifugation in culture medium at 75 *g* for 5 min.

3 Culture of Protoplasts

Protoplasts are cultured at a density of 2.5 – 5.0 10^4/ml. It is often preferable to culture mesophyll protoplasts at the higher cell concentration. When the cell division has started, after 2 – 5 days in culture, a volume of culture medium (without hormones) which is three times the initial medium volume is added. Of the different media tested, a modified KM8p medium (Glimelius et al. 1986) resulted in the best growth. The medium was further modified by the addition of 0.4 M glucose, as a carbon source and as an osmoticum, instead of mannitol and sucrose. When starting the protoplast culture, it is important to use a rather high concentration of 2,4-D, which, however, should be reduced after a few days of culture. The initial hormone concentration is 4.5 µM 2,4-D, 0.5 µM NAA and 0.4 µM BAP. The protoplasts/cells are cultured in dim light with an irradiance of 0.2 W/m^2 (Thorn fluorescent tubes, 40 W) at 25 – 27 °C during a culture period of 2 – 3 weeks.

After 24 h, the first cell division can be observed in hypocotyl material (48 h in mesophyll material). In 2 – 3 weeks the cells have developed into calli with the characteristic star shaped colonies often seen in liquid culture of *Brassica* protoplasts (Fig. 2). The cell aggregates are then transferred with a Pasteur pipette to an agarose-solidified K3 (m) medium, and solidified with 0.25% agarose (Type 1, low EEo, Sigma) and with a sucrose concentration of 3%. Balance and concentration of hormones are the same as in the "diluted" KM8p medium, i.e., 1.120 µM 2,4-D, 0.125 µM NAA and 0.100 µM BAP. These cultures are kept under fluorescent light (35 W/m^2, warm-white fluorescent tubes, 40 W and tungsten lamps) with a daylength of 16 h and a temperature of 25 °C, for about 2 – 5 weeks before transfer to differentiation medium (Figs. 1, 2).

4 Regeneration of Plants

After a growth period of approximately 14 days on the solidified medium (4 – 8 weeks after isolation) the calli have reached a size of 2 – 5 mm in diameter (Fig. 3 A). They are then transferred to a medium with reverse hormone balance, i.e., high cytokinin and low auxin. Plant regeneration has been obtained in all *Brassica* species tested: *B. napus, B. campestris, B. oleracea, B. nigra, B. carinata*, and *B. juncea*, but with considerable differences in the capacity of organogenesis; some are very recalcitrant to develop into shoots, e.g., *B. campestris*. The differences are not limited only to those between species; genotypes within a species also differ. Organogenesis capacity also varies according to the type of tissue used for protoplast isolation. Differences were found in regeneration potential in an experiment (Glimelius 1984) comparing protoplasts isolated from hypocotyl and mesophyll material both from greenhouse-grown plants and in vitro-grown shoots, from *B. napus* cv. Olga. Differentiation from the hypocotyl protoplasts was 40% higher than from the two mesophyll materials, which had similar regeneration capacities.

The highest organogenesis capacities have, up to now, occurred in the cv. Olga of *B. napus* (70%, percentage of callus on regeneration medium which produce

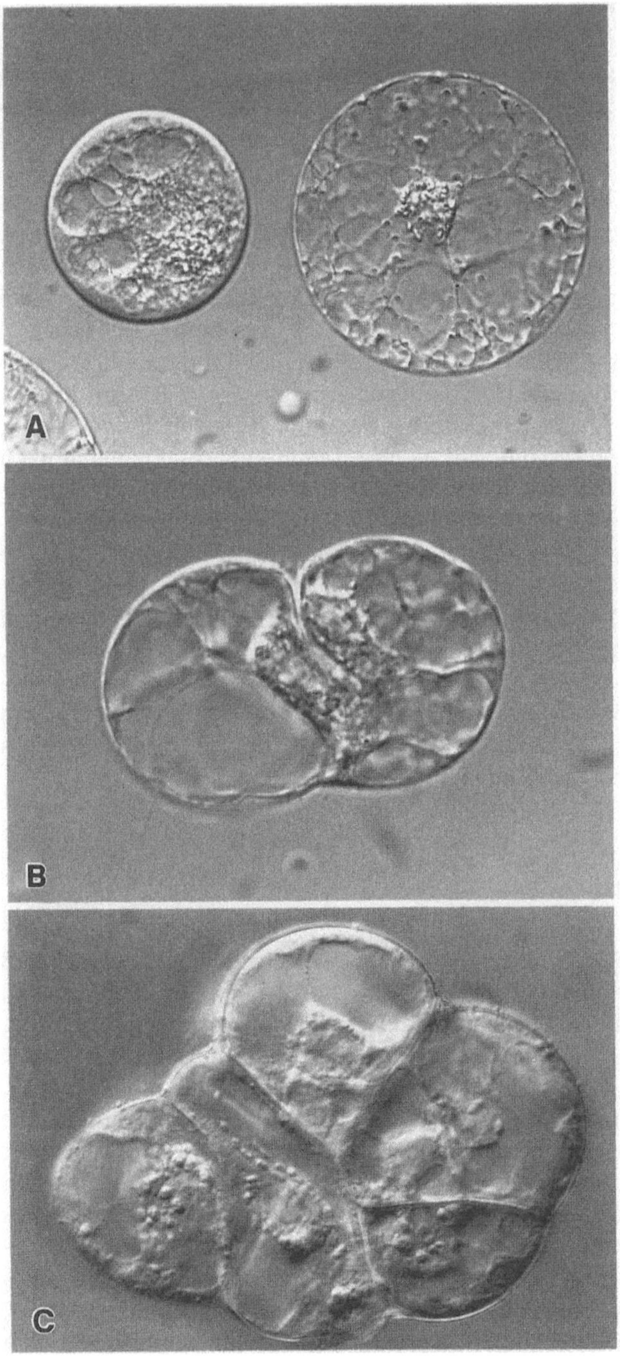

Fig. 1. A Hypocotyl protoplasts from *B. napus.* **B** First cell division (*B. napus*) 24 h after the start of the hypocotyl protoplast culture. **C** "Micro-callus", 6 days old, originating from a mesophyll protoplast culture of *B. napus*

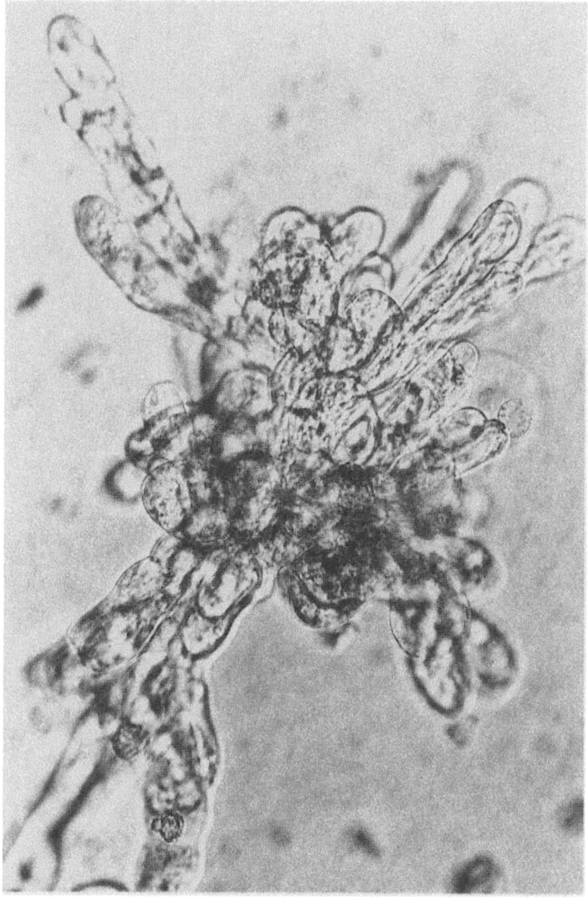

Fig. 2. Two weeks old *B. napus* callus showing the characteristic star shape which often can be seen in liquid cultures originating from *Brassica* protoplasts

shoots) and the cv. Broccoli of *B. oleracea* (50%) using the techniques described above. Shoot differentiation was also observed in the summer oilseed rape cvs. Hanna and Topas, although with slightly lower efficiencies (40–50%). Differentiation experiments using *B. nigra*, *B. carinata*, and *B. juncea* indicate high regeneration capacities, although the conditions have not been optimized. *B. campestris* seems to be the most recalcitrant. In our experiment only about 1% of the calli transferred to the differentiation medium produced shoots which developed into plantlets (Glimelius 1984).

The medium used to differentiate shoots is the K3 (m) medium, modified with 1% sucrose and solidified with 0.35% agarose and culture conditions as described in Section 3. A high regeneration frequency can be obtained with BAP as the only cytokinin, but the highest reproducibility in organogenesis was achieved when combining BAP with zeatin. The best hormone combination for regeneration proved to be 0.6 μM IAA, 2.2 μM BAP and 2.2 μM zeatin. This combination has

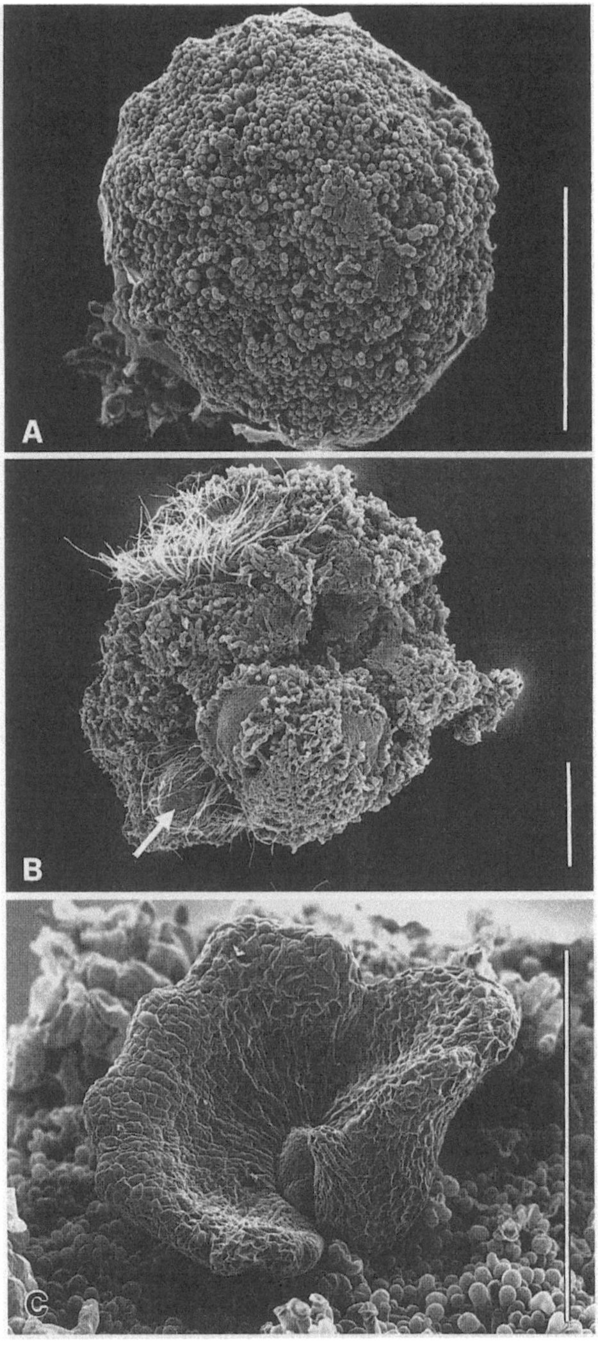

Fig. 3 A–C. Scanning electron microscope pictures of *B. napus* calli. **A** A 40-day-old callus with apparently undifferentiated cells at the surface. **B** A 75-day-old differentiated callus with a variety of cell types at the surface. A spot which was intense green (*arrow*) is an early stage in the development of a shoot primordium. **C** A later stage in the development of a shoot primordium observed in a 90-day-old callus. Bar = 1 mm

induced regeneration in all *Brassica* species tested, although this is not the optimal concentration for all species.

Calli must be at least 0.5 mm (optimal 2–5 mm) in diameter to continue growth on the differentiation medium. The growth rate on this regeneration medium is very slow, but several small green nodules develop in the otherwise light green and friable callus tissue. Root hairs (see Fig. 3B) and tiny roots are often formed on the calli. After about 1 month of culture on the regeneration medium, shoots develop from the green meristematic nodules (Fig. 3B, C). The shoots grow very quickly and some can have a "glassy", light green appearance. These shoots require a rapid transfer to a hormone-free medium (MS-medium with 1% sucrose solidified with 0.35% agarose), when the nodules/shoots start to elongate, to ensure their survival. After a culture period of 1–2 months on hormone-free medium, most shoots can be transferred to soil and gradually adapted to greenhouse environments. This last transfer can be done provided that the shoots have developed leaves with a normal appearance.

5 Protoplast Fusion and Selection of Hybrids

Several different agents and methods (Constable and Cutler 1985) can be used to induce protoplast fusion. The most common agent is polyethylene glycol (PEG) which has also been used for production of somatic hybrids within the Brassicaceae. The most serious problem in producing somatic hybrids does not lie in the efficiency of obtaining fusion products, but rather in selecting them from the protoplast mixture consisting of heterokaryons, homokaryons, and parental protoplasts. PEG-induced protoplast fusion is performed as described by Sundberg and Glimelius (1986). Protoplasts are isolated from hypocotyls and in vitro grown plants as described above (see Sect. 2), with the exception that the last washing in culture medium is excluded. The protoplasts are suspended in 0.15 M sorbitol, 0.03 M $CaCl_2$ $2H_2O$, 0.075 M KCl and 0.05 M Tris-HCl pH 7.2 to a final concentration of 6 10^5/ml hypocotyl protoplasts and 3 10^5/ml mesophyll protoplasts. Droplets, approximately 0.04 ml/droplet, of the mixed protoplast suspension are spaced at regular intervals in each plastic Petri dish. The protoplasts are left to settle for 5 min. To each droplet, two droplets (0.06 ml) of 40% PEG (w/v, mol. wt. 1500, pH 7.0) in 0.3 M glucose and 50 mM $CaCl_2$ $2H_2O$ are added. After 3 to 5 min, the solution is removed with a Pasteur pipette after tilting the Petri dishes. Since the protoplasts are attached to the bottom of the Petri dish, three continued washings (5 min each) can be made in this way. The first washing solution contains 13.3% PEG (w/v) 0.1 M glucose, 0.067 M sorbitol and 0.067 M $CaCl_2$ $2H_2O$ (pH 7.2), the second contains 6.7% PEG (w/v), 0.05 M glucose, 0.082 M sorbitol and 0.083 M $CaCl_2$ $2H_2O$ (pH 7.2), and the culture medium is used for the third washing.

A bottleneck in recovering somatic hybrids from protoplast fusion experiments is the selection of heterokaryons. Most often, nutritional markers are not available for the identification and selection of somatic hybrids. However, it might be possible to introduce different "selection markers", e.g., genes for antibi-

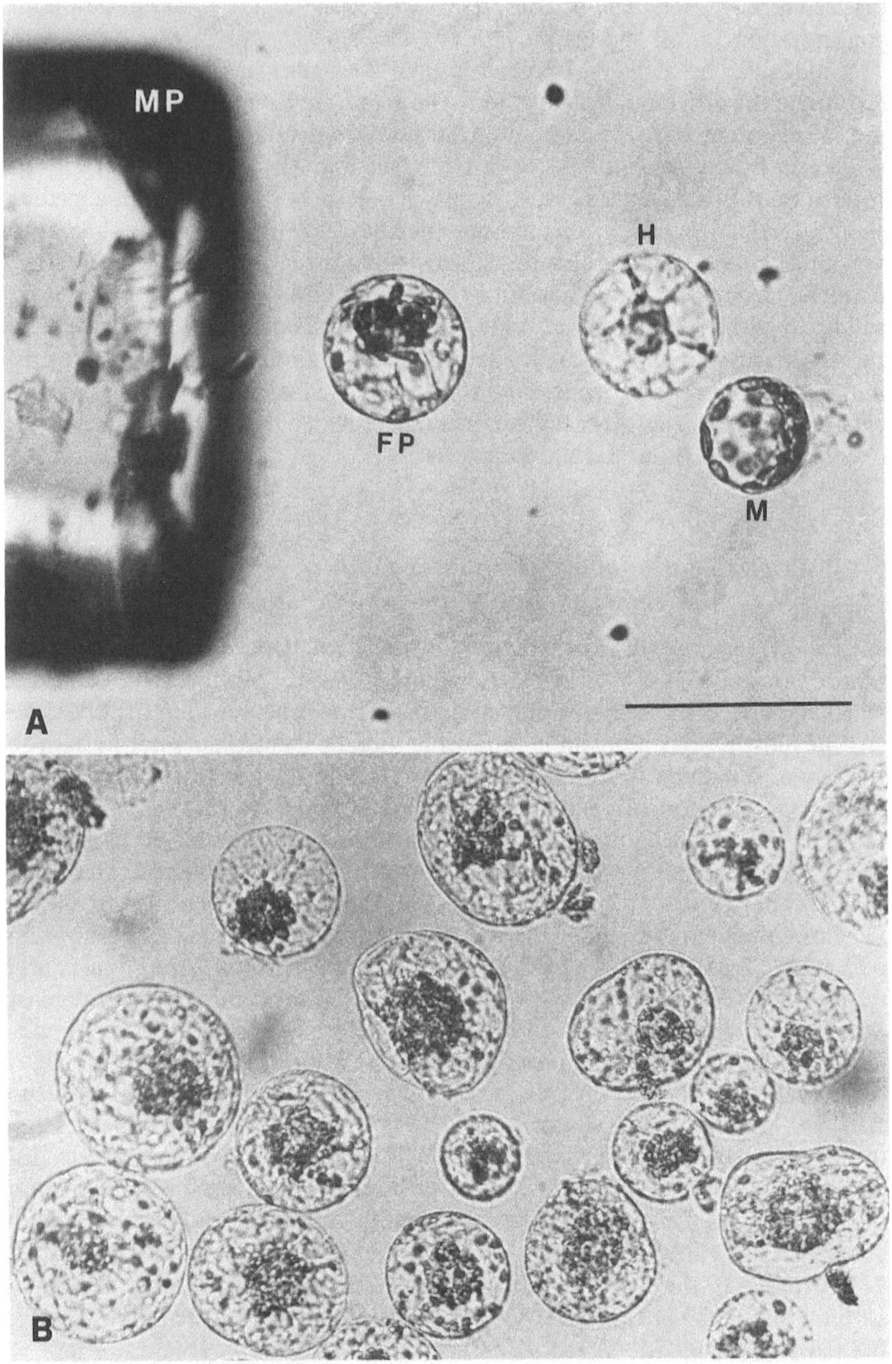

Fig. 4 A, B. Fusion between *B. oleracea* and *b. campestris* in an experiment for resynthesis of *B. napus*. **A** Isolation of a fusion product (*FP*) with the aid of a micropipette (*MP*) handled by a micromanipulator. The two types of parental protoplasts are also shown (*H* hypocotyl; *M* mesophyll). **B** Isolated somatic hybrid, cells. Bar = 100 µm. (Photo: Eva Sundberg)

otic resistance, into the two protoplast "parents" by transformation. This would allow an efficient recovery of somatic hybrids, but requires an efficient transformation system.

Different forms of mechanical selection systems are still most commonly used to isolate the fusion products in protoplast fusion experiments. One technique is to "harvest" the hybrids with micropipettes operated by a micromanipulator. By combining protoplasts isolated from different tissues in the plants, differences in morphology can be used for identification of the hybrids in the microscope. Fusions of green mesophyll protoplasts with hypocotyl protoplasts make it possible to identify the hybrids which have chloroplasts from the green material and visible intervacuolar cytoplasmic strands from the hypocotyl protoplasts. In this way the heterokaryons can be easily identified and collected with a micropipette (Fig. 4). When using this selection method it is best to culture the PEG-treated protoplasts for 24 h in darkness to release the protoplasts from the bottom of the Petri dish. In this way it is possible to isolate the hybrids and transfer them to another culture vessel. The fusion products are cultured for 48 h in small volumes of KM8p medium (10 µl) with hormones, and 100 protoplasts in each well. After 48 h, the cells are diluted with 3 volumes of KM8p medium without hormones. Further culture conditions are in accordance with the methods presented in Section 3 with slight modifications (Sundberg and Glimelius 1986). It has been shown that all obtained shoots were true hybrids (confirmed by using isozyme markers, Sundberg et al. 1987).

Another efficient and rapid method is the selection of hybrids using flow cytometry, as proposed by Galbraith and Galbraith (1979) and Galbraith and Harkins (1982), but modified to suit the material within Brassicaceae by Glimelius et al. (1986). The sorting of hybrids by flow cytometry requires that the parental protoplasts can be distinguished from the hybrids on the basis of fluorescense characteristics. Hypocotyl protoplasts are stained with carboxyfluoresceindiacetate (0.1 mg/l) (peak emission wavelength 520 nm), which differentiates the emission wavelength from mesophyll protoplasts, which have a peak emission wavelength of 690 nm (chlorophyll). By using these fluorophores, a good separation of the parental material and the fusion products (double fluorescence) can be obtained. Protoplasts with the double fluorescence can be enriched and sorted into a multivial dish. On average, about 80% of the protoplast population are fusion products.

Oilseed rape (*B. napus*) has also been produced in protoplast fusion experiments [*B. campestris* (+) *B. oleracea*] without applying selection at the protoplast stage. This system of producing synthetic oilseed rape was first proposed by Schenck and Röbbelen (1982), later used by Rosén et al. (1988), Taguchi and Kameya (1986) and Terada et al. (1987).

6 Miscellaneous Studies

6.1 Plant–Virus Interaction

The protoplasts from different *Brassica* species have been used as experimental material in a wide range of different studies. In many studies of plant virus-host interactions, the use of efficient protoplast systems has led to a better understanding of some aspects of virus replication and its implications for the host. The uptake and replication of DNA viruses, e.g., cauliflower mosaic virus (CaMV) and RNA viruses, e.g., turnip rosette virus (TRosV) and turnip yellow mosaic virus (TYMV), have been studied in *Brassica* protoplast material.

The interaction between TYMV and chinese cabbage has been most thoroughly investigated, and has also been used in several physiological studies of plant-virus interactions (Renaudin et al. 1975; Matthews and Sarkar 1976; Fraser and Matthews 1979; Fernandez-Gonzalez 1980; Cohen et al. 1981 a, b; Sugimura and Matthews 1981; Bove et al. 1982; Renaudin et al. 1982; Sindhu and Cohen 1984; Balint and Cohen 1985 a, b; Cohen et al. 1985; Greenberg and Cohen 1985; Candresse et al. 1987). Infection and replication of TRosV has been studied by Morris-Krisinich and Hull (1983) in a turnip protoplast system.

Methods for increasing the efficiency of protoplast infection by virus or virus DNA have been developed in several laboratories, using CaMV or isolated CaMV DNA. Maule (1983) showed that it was possible to infect protoplasts from a range of *Cruciferous* species with CaMV, using a polyethylene glycol-mediated uptake procedure. However, this procedure works best with intact virions; when isolated DNA from CaMV was used, the efficiency was low (Yamaoka et al. 1982). In order to increase the efficiency of infection with isolated DNA, Hussain et al. (1985) showed that it was possible to use liposome-encapsulated CaMV DNA to infect turnip protoplasts. To achieve replication of the virus DNA, it was necessary to use evacuolated protoplasts.

6.2 Transformation

The increasing interest in gene transfer to protoplasts has led to a frequent use of protoplasts isolated from *Brassica* species, since these protoplasts are easily handled and often possess a high regeneration ability.

Paszkowski et al. (1986) reported a successful genetic transformation of *Brassica campestris* protoplasts with an engineered CaMV genome. Stable, genetically transformed cell lines were obtained in this transformation experiment. Another system used to transfer DNA to protoplasts is described in Tanaka et al. (1984), who used *E. coli* spheroplasts to mediate a transfer of cloned CaMV DNA into *Brassica chinensis* mesophyll protoplasts. Several other methods for transferring DNA into *Brassica* protoplasts have been published recently. Direct DNA transfer methods, electroporation, and polyethylene glycol (PEG), have been used in several examples of transformation in *Brassica* (electroporation; Guerche et al. 1987; Denecke et al. 1988; PEG; Golz et al. 1988; Köhler et al. 1988). The possibility of obtaining stable transformed *Brassica* plants through direct gene transfer to protoplasts was demonstrated by Guerche et al. (1987). The incorporated gene (kanamycin resistance) was also shown to be transmitted to the progeny of the transformed plant as a single Mendelian dominant character. This, to our knowledge, is the only publication reporting a successful transformation of *Brassica* protoplasts which leads to transformed plants. The transformation system with protoplast and *Agrobacterium* co-cultivation has been used by Laser and

Eriksson (1985) to achieve transformation of *B. campestris*. This system was also used by Sacristan et al. (1988) in transformation experiments with *B. nigra*. A liposome-mediated transfer of a *B. campestris* mitochondrial plasmid to *B. napus* was made by Maley and Pauls (1985). *Brassica* protoplasts have also been used in experiments with microinjection of DNA (Rasmussen et al. 1985; Schweiger et al. 1987).

Another system for transferring genetic information is to transplant nuclei into protoplasts. Saxena et al. (1987) used nuclei isolated from cell suspension cultures of *B. napus*, among some other species, and introduced the isolated nuclei into protoplasts of two auxotrophic mutants of *Datura innoxia*.

6.3 Plant Biochemistry and Physiology

Brassica protoplasts have been used in physiological studies concerning hormones and their influence on the protoplast. Schubert and Matzke (1985) have investigated the role of phytohormones and fusicoccin on H^+ extrusion, and compared results from protoplasts and leaf segments of the same plant. In another study (Dekhuijzen et al. 1978), root protoplasts from *B. campestris* were compared to root segments in their ability to take up cytokinins. It was shown that segments took up cytokinins far more rapidly than protoplasts. The reason for this inequality was suggested to be due to differences in the metabolism of cytokinins or that the plasmalemma and cell wall influence the uptake. Root protoplasts of *Brassica* have also been used in a study concerning morphology and myrosinase activity in protoplasts (Iversen et al. 1983; Bones and Iversen 1985).

Protoplasts from *Brassica* species have also been used in other types of physiological studies. The biosynthesis of polyamines and 1-aminocyclopropane-1-carboxylate in chinese cabbage leaf protoplasts was studied by Greenberg (1985). When using nitro-blue tetrazolium as a specific stain for photosynthetic activity in protoplasts Robertson and Earle (1986b) were able to distinguish between leaf protoplasts from atrazine-resistant and -susceptible biotypes of *B. napus*.

The interest in the quality of the plasmalemma has increased since protoplast techniques have become more and more important as a tool in modern biotechnology. Protoplast adhesion, which is a prerequisite for cell fusion, is mainly governed by the surface conditions of the protoplasts. Therefore Nagata and Melchers (1978) have made an effort to characterize the surface charge of protoplasts (*Brassica rapa*, among other species) and their significance in cell-cell interaction. Hahn-Hägerdal et al. (1986) studied surface properties, stability of membrane components, and leakage of intracellular components after PEG and electric field treatments of *B. napus* protoplasts. Their results indicated that the molecular mechanisms of the two protoplast fusion methods are similar, but that the effects of the electric field treatments appeared less harmful. The effects of applying an electric field pulse to protoplasts were also studied by Muller et al. (1986). They concluded that the reduced viability of protoplasts fused with electrical discharges is not due to the electric pulse field strength.

Brassica protoplasts have also been used as bioassay to identify toxic fractions from the pathogen *Phoma lingam* (Sjödin et al. 1988). The main toxic compound was identified as sirosdesmin PL. The toxic effects of this compound were tested on protoplasts, cells, and intact plants of susceptible *B. napus* lines and of nonhost species. The results from the study implies that the nonhost species acquire some degree of resistance during the formation of cell walls, whereas *B. napus* cells remain sensitive.

6.4 Cytology

Cytogenetic analysis in *Brassica* species is difficult due to the small chromosomes. Therefore it is most important to obtain cytological preparations with a good chromosome spread. In *Brassica* this was achieved by Murata (1983), using a technique with air-dried preparations of *Brassica* protoplasts.

Plants regenerated from protoplasts often exhibit alterations in their chromosome number. Newell et al. (1984) showed a higher frequency of chromosome alterations among *B. napus* plants regenerated from protoplasts than among plants regenerated from seedling explants.

Acknowledgments. We wish to express our gratitudes to Dr. Christina Sjödin, Jan Fahleson and Eva Sundberg for providing results and valuable information to this article. We thank Dr. Barbara Giles for her valuable help with the English, and Prof. Waheeb Heneen for stimulating discussions during the preparation. We also thank the Swedish seed companies, W. Weibull AB, Landskrona, and Svalöf AB, Svalöv, for providing us with seed material.

7 Protocol for Isolation, Culture, Regeneration, and Fusion

Mesophyll material. Shoots grown on MS-medium without hormones (see below), culture cabinets; light intensity 12 W/m^2 (Osram halogen lamps HQ1, 400 W, night/day 16/8 h, 25 °C/20 °C

Hypocotyl material. Surface sterilized seeds (7.5% calcium hypochlorite, 1 h continuous shaking) germinated on MS-medium without hormones at 25 °C in dark.

Plasmolyse:
54.65 g sorbitol
7.35 g $CaCl_2$ $2H_2O$
in 1000 ml H_2O

Enzyme treatment:
1% Cellylysin and 0.1% Macerase (Calbiochem-Behring)
in K3 (m) medium with 0.4 M glucose (72.06 g/l) (see below).

Isolation: CPW 16

27.2	mg	KH_2PO_4
101.0	mg	KNO_3
1480.0	mg	$CaCl_2$ $2H_2O$
246.0	mg	$MgSO_4$ $2H_2O$
0.16	mg	KJ
0.025	mg	$CuSO_4$ $5H_2O$
160.0	g	sucrose

in 1000 ml H_2O, pH 5.6–5.8

W 5
18.4 g $CaCl_2$ $2H_2O$
9.0 g NaCl
1.0 g glucose
0.8 g KCl
in 1000 ml H_2O pH 5.6–5.8

Culture medium
8p (m) medium (see below) with hormones

Culture: Culture medium 8p (m)

1900	mg	KNO_3
600	mg	NH_4NO_3
300	mg	$MgSO_4$ $7H_2O$
170	mg	KH_2PO_4
600	mg	CaCl $2H_2O$
300	mg	KCl
10	ml	microelements (stock solution)
		75.0 mg/l KJ
		1000.0 mg/l $MnSO_4$ $4H_2O$
		300.0 mg/l H_3BO_3
		200.0 mg/l $ZnSO_4$ $2H_2O$
		25.0 mg/l Na_2MoO_4 $2H_2O$
		2.5 mg/l $CuSO_4$ $5H_2O$
		2.5 mg/l $CoCl_2$ $6H_2O$
5	ml	Fe-EDTA (fresh) mixed 100 ml;
		557.0 mg $FeSO_4$ $7H_2O$
		in 50 ml H_2O
		745.0 mg Na_2EDTA
		in 50 ml H_2O
1	mg	nicotinic acid
1	mg	pyrodoxine HCl (B_6)
10	mg	thiamine HCl (B_1)
2	mg	ascorbic acid
20	mg	sodium pyruvate
40	mg	citric acid
40	mg	malic acid
40	mg	fumaric acid
250	mg	fructose
250	mg	ribose
250	mg	xylose
250	mg	mannose
250	mg	rhamnose
250	mg	cellobiose
250	mg	sucrose
250	mg	sorbitol
250	mg	mannitol
250	mg	inositol
250	mg	casamino acid (vitamin-free)
20	ml	coconut water
68.4	g	glucose

Hormones:
1.0 mg 2,4-D
0.1 mg NAA
0.1 mg BAP
in 1000 ml H_2O pH 5.6–5.8

Callus: K3 (m)-medium

2500 mg KNO_3
250 mg NH_4NO_3
250 mg $MgSO_4$ $7H_2O$
300 mg $CaCl_2$ $2H_2O$
150 mg NaH_2PO_4 H_2O
134 mg $(NH_4)_2SO_4$
1 mg KJ
10 mg microelements (stock solution);
  1690.0 mg/l $MnSO_4$ H_2O
  25.0 mg/l Na_2MoO_4 $2H_2O$
  2.5 mg/l $CuSO_4$ $5H_2O$
  2.5 mg/l $CoCl_2$ $6H_2O$
  620.0 mg/l H_3BO_3
5 ml Fe-EDTA (see above under culture medium 8p (m))
15 mg Zn-EDTA
10 mg thiamine HCl (B_1)
1 mg pyridoxine HCl (B_6)
1 mg nicotinic acid
250 mg xylose
100 mg inositol
30.0 g sucrose
2.5 g agarose

Hormones:
0.25 mg/l 2,4-D
0.025 mg/l NAA
0.025 mg/l BAP
in 1000 ml H_2O pH 5.6–5.8

Regeneration: K3 (m) medium (see above)
10.0 g sucrose
3.5 g agarose

Hormones:
0.1 mg/l IAA
1.0 mg/l BAP
1.0 mg/l Zeatin

Culture of regenerated shoots: MS medium

1900 mg KNO_3
1650 mg NH_4NO_3
370 mg $MgSO_4$ $7H_2O$
170 mg KH_2PO_4
440 mg $CaCl_2$ $2H_2O$
10 ml microelements (stock solution)
  620.0 mg/l H_3BO_3
  1564.0 mg/l $MnSO_4$ H_2O
  1234.0 mg/l $ZnSO_4$ $7H_2O$
  83.0 mg/l KI
  25.0 mg/l $NaMoO_4$ $2H_2O$
  2.5 mg/l $CuSO_4$ $5H_2O$
  2.5 mg/l $CoCl_2$ $6H_2O$
5 ml Fe-EDTA (see above under culture medium 8p (m))
2 mg glycine
0.5 mg nicotinic acid
0.5 mg pyrodoxine HCl (B_6)
0.1 mg thiamine HCl (B_1)
1 g casein hydrolysate (vitamin-free)

100 mg inositol
10.0 g sucrose
3.5 g agarose
in 1000 ml H_2O pH 5.6–5.8

Protocol for Fusion

Suspension medium
27.32 g/l sorbitol
4.41 g/l $CaCl_2$ $2H_2O$
5.59 g/l KCl
6.06 g/l Tris-HCl
pH 7.2
PEG treatment
400.00 g/l PEG mol. wt. 1500
54.05 g/l glucose
7.35 g/l $CaCl_2$ $2H_2O$
pH 7.2
First washing
133.0 g/l PEG
18.01 g/l glucose
12.20 g/l sorbitol
9.85 g/l $CaCl_2$ $2H_2O$
pH 7.2
Second washing
67.0 g/l PEG
9.0 g/l glucose
15.12 g/l sorbitol
12.20 g/l $CaCl_2$ $2H_2O$
pH 7.2
Third washing
Culture medium 8p (m), with hormones (see above)

References

Balint R, Cohen SS (1985a) The incorporation of radiolabeled polyamines and methionine into turnip yellow mosaic virus in protoplasts from infected plants. Virology 144:181–193

Balint R, Cohen SS (1985b) The effects of dicyclohexylamine on polyamine biosynthesis and incorporation into turnip yellow mosaic virus in chinese cabbage protoplasts infected in vitro. Virology 144:194–203

Banks MS, Evans PK (1976) A comparison of the isolation and culture of mesophyll protoplasts from several *Nicotiana* species and their hybrids. Plant Sci Lett 7:409–416

Bannerot H, Boulidard L, Cauderon Y, Tempe J (1974) Transfer of cytoplasmic male sterility from *Raphanus sativus* to *Brassica oleracea*. Proc Eucarpia Meet Cruciferae, pp 52–54

Barsby TL, Yarrow SA, Shepard JF (1986) A rapid and efficient alternative procedure for the regeneration of plants from hypocotyl protoplasts of *Brassica napus*. Plant Cell Rep 5:101–103

Barsby TL, Chuong PV, Yarrow SA, Wu S-C, Coumans M, Kemble RJ, Powell AD, Beversdorf WD, Pauls KP (1987a) The combination of Polima cms and cytoplasmic triazine resistance in *Brassica napus*. Theor Appl Genet 73:809–814

Barsby TL, Yarrow SA, Kemble RJ, Grant I (1987b) The transfer of cytoplasmic male sterility to winter-type oilseed rape (*Brassica napus* L.) by protoplast fusion. Plant Sci 53:243–248

Bidney DL, Shepard JF, Kaleikau E (1983) Regeneration of plants from mesophyll protoplasts of *Brassica oleracea*. Protoplasma 117:89–92
Binding H (1974) Cell cluster formation by leaf protoplasts from axenic cultures of haploid *Petunia hybrida*. L. Plant Sci Lett 2:185–188
Bones A, Iversen TH (1985) Myrosin cells and myrosinase. Isr J Bot 34:351–376
Bove JM, Renaudin J, Mouches C (1982) Inhibitory effects of virazole and actinomycin D on turnip yellow mosaic virus replication in chinese cabbage protoplasts. Phytoparasitica 10:124–125
Candresse T, Batisti M, Renaudin J, Mouches J, Mouches C, Bove JM (1987) Immunodetection of turnip yellow mosaic virus non-structural proteins in infected Chinese cabbage leaves and protoplasts. Ann Inst Pasteur/Virol 138:217–227
Chatterjee G, Sikdar SR, Das S, Sen SK (1985) Regeneration of plantlets from mesophyll protoplasts of *Brassica juncea* (L.) Czern. Plant Cell Rep 4:245–247
Chatterjee G, Sikdar SR, Das S, Sen SK (1988) Intergeneric somatic hybrid production through protoplast fusion between *Brassica juncea* and *Diplotaxis muralis*. Theor Appl Genet 76: 915–922
Chetrit P, Mathieu C, Vedel F, Pelletier G, Primard C (1985) Mitochondrial DNA polymorphism induced by protoplast fusion in Cruciferae. Theor Appl Genet 69:361–366
Chuong PV, Pauls KP, Beversdorf WD (1985) A simple culture method for *Brassica* hypocotyl protoplasts. Plant Cell Rep 4:4–6
Chuong PV, Pauls KP, Beversdorf WD (1986) Plant regeneration from *Brassica nigra* (L.) Koch protoplasts. Crucifer genetics workshop III. Proceedings, May 29–30, 1986 Univ Press Guelph, 57 pp
Chuong PV, Beversdorf WD, Pauls KP (1987a) Plant regeneration from haploid stem peel protoplasts of *Brassica napus* L. J Plant Physiol 130:57–65
Chuong PV, Pauls KP, Beversdorf WD (1987b) Protoplast culture and regeneration from *Brassica carinata* Braun. Plant Cell Rep 6:67–69
Chuong PV, Beversdorf WD, Powell AD, Pauls KP (1988) Somatic transfer of cytoplasmic traits in *Brassica napus* L. by haploid protoplast fusion. Mol Gen Genet 211:197–201
Cohen SS, Balint R, Sindhu RK (1981a) The synthesis of polyamines from methionine in intact and disrupted leaf protoplasts of virus-infected chinese cabbage. Plant Physiol 68:1150–1155
Cohen SS, Balint R, Sindhu RK, Marcu D (1981b) Polyamine biosynthesis and metabolism in normal and virus infected plant protoplasts. Med Biol 59:394–402
Cohen SS, Sindhu RK, Greenberg M, Yamanoha B, Balint R, McCarthy K (1985) Cellular systems for the study of the biosynthesis of polyamines and ethylene, as well as of virus multiplication. Plant Growth Regul 3:227–238
Constable S, Cutler AJ (1985) Protoplast fusion. In: Fowke LC, Constable S (eds) Plant protoplasts. CRC, Boca Raton, pp 53–65
Dekhuijzen HM, Bossen ME, Vonk CR (1978) The uptake of cytokinins by protoplasts and root sections of *Brassica campestris*. Physiol Plant 43:142–147
Denecke J, Reynaerts A, D'Halluin K, Botterman J (1988) Quantification of transient expression levels of genes transferred to plant protoplasts by electroporation. In: Puite KJ, Dons JJM, Huizing HJ, Kool AJ, Koornneef M, Krens FA (eds) Progress in plant protoplast research. Kluwer, Dordrecht, pp 337–338
Durand J, Potrykus I, Donn G (1973) Plantes issues de protoplastes de pétunia. Z Pflanzenphysiol 69:26–34
Erickson L, Grant I, Beversdorf W (1986) Cytoplasmic male sterility in rapeseed (*Brassica napus* L.), 1. Restriction patterns of chloroplast and mitochondrial DNA. Theor Appl Genet 72:145–150
Fahleson J, Dixelius J, Sundberg E, Glimelius K (1988) Correlation between flow cytometric determination of nuclear DNA and chromosome number in somatic hybrids within Brassicaceae. Plant Cell Rep 7:74–77
Fernandez-Gonzalez O, Renaudin J, Bove JM (1980) Infection of chlorophyll-less protoplasts from etiolated chinese cabbage *Brassica sinensis* cultivar Cantonner hypocotyls by turnip yellow mosaic virus. Virology 104:262–265
Fraser L, Matthews REF (1979) Strain specific pathways of cytological change in individual chinese cabbage protoplasts infected with turnip yellow mosaic virus. J Gen Virol 45:623–630
Fu Y-Y, Jia S-R, Lin Y (1985) Plant regeneration from mesophyll protoplasts culture of cabbage (*Brassica oleracea* var capitata). Theor Appl Genet 71:495–499

Galbraith DW, Galbraith JEC (1979) A method for identification of fusion of plant protoplasts derived from tissue cultures. Z Pflanzenphysiol 93:149–158
Galbraith DW, Harkins K (1982) Cell sorting as a means for isolating somatic hybrids. In: Fujiwara A (ed) Plant tissue culture 1982. Maruzen, Tokyo, pp 617–618
Gamborg OL, Miller RA, Ojima K (1968) Nutrient requirements of suspension cultures of soybean root cells. Exp Cell Res 50:151–158
Gleba YY, Hoffmann F (1978) Hybrid cell lines *Arabidopsis thaliana* + *Brassica campestris*: no evidence for specific chromosome elimination. Mol Gen Genet 165:1257–1264
Gleba YY, Hoffmann F (1980) "*Arabidobrassica*": A novel plant obtained by protoplast fusion. Planta 149:112–117
Glimelius K (1984) High growth rate and regeneration capacity of hypocotyl protoplasts in some Brassicaceae. Physiol Plant 61:38–44
Glimelius K, Djupsjöbacka M, Fellner-Feldegg H (1986) Selection and enrichment of plant protoplast heterokaryons of Brassicaceae by flow sorting. Plant Sci 45:133–144
Glimelius K, Fahleson J, Sjödin C, Sundberg E, Landgren M (1989) Fusion of protoplasts as a method to move nuclei and organelles. Proceedings of the 6th International Congress on Genetic Manipulation in Plant Breeding, organized by EUCARPIA, Sept 11–16, 1988, Elsinore, Denmark. Plenum Press (in press)
Golz C, Köhler F, Sacristan MD, Schieder O (1988) Transformation of *Brassica* species via direct gene transfer. In: Puite KJ, Dons JJM, Huizing HJ, Kool AJ, Koornneef M, Krens FA (eds) Progress in Plant Protoplast Research. Kluwer, Dordrecht, pp 353–354
Greenberg ML (1985) Biosynthesis of polyamines and 1-aminocyclopropane-1-carboxylate in chinese cabbage leaf protoplasts. Diss State Univ New York, Stony Brook, USA
Greenberg ML, Cohen SS (1985) Dicyclohexylamine-induced shift of biosynthesis from spermidine to spermine in plant protoplasts. Plant Physiol 78:568–575
Guerche P, Charbonnier M, Jouanin L, Tourneur C, Paszkowski J, Pelletier G (1987) Direct gene transfer by electroporation in *Brassica napus*. Plant Sci 52:111–116
Guo BJ, Schieder O (1983) Callus and root formation from mesophyll protoplasts of chinese cabbage (*Brassica chinensis* L.). Z Pflanzenphysiol 110:375–377
Harney PM, Tsai AH (1986) Plant regeneration from hypocotyl protoplasts of *Brassica oleracea* var. gemmifera. Crucifer genetics workshop III. Proceedings, May 29–30, 1986 Univ Press, Guelph, 57 pp
Hahn-Hägerdal B, Hosono K, Zachrisson A, Bornman CH (1986) Polyethylene glycol and electric field treatment of plant protoplasts: characterization of some membrane properties. Physiol Plant 67:359–364
Hoffmann F, Adachi T (1981) "*Arabidobrassica*": Chromosomal recombination and morphogenesis in asymmetric intergeneric hybrid cells. Plants 153:586–593
Hohn T, Hohn B, Pfeiffer P (1985) Reverse transcription in CaMV. Biol Sci TIBS 10:205–209
Hussain MM, Melcher U, Essenberg RC (1985) Infection of evacuolated turnip protoplasts with liposome-packaged cauliflower mosaic virus. Plant Cell Rep 4:58–62
Iversen T-H, Myhre S, Evjen K, Baggerud C (1983) Morphology and myrosinase activity in root protoplasts of Brassicaceae. Z Pflanzenphysiol 112:391–401
Jarl CJ, Bornman CH (1988) Correction of chlorophyll-defective, male-sterile winter oilseed rape (*Brassica napus*) through organelle exchange: Phenotypic evaluation of progeny. Hereditas 108:97–102
Jourdan PS, Earle ED (1985) Embryogenesis and plant regeneration from mesophyll protoplasts of male fertile and cytoplasmic male sterile cauliflower (*Brassica oleracea* spp. *botrytis*). Plant Physiol 77:165 Suppl 4
Jourdan PS, Mutschler MA, Earle ED (1986) Production and characterization of somatic hybrids between *Brassica oleracea* and *B. campestris*. EUCARPIA Cruciferae Newslett 11:84–85
Kao KN (1977) Chromosomal behaviour in somatic hybrids of soybean-*Nicotiana glauca*. Molec Gen Genet 150:225–230
Kao KN, Michayluk MR (1975) Nutritional requirements for growth of *Vicia hajastana* cells and protoplasts at very low population density in liquid media. Planta 126:105–110
Kao KN, Michayluk MR (1981) Embryoid formation in alfalfa cell suspension cultures from different plants. In Vitro 17:645–648

Kao H-M, Seguin-Swartz G (1987) Study of factors affecting the culture of *Brassica napus* L. and *B. juncea* Coss. mesophyll protoplasts. Plant Cell Tissue Org Cult 10:79–90

Kao KN, Constable F, Michayluk MR, Gamborg OL (1974) Plant protoplast fusion and growth of intergeneric hybrid cells. Planta 120:215–227

Kartha KK, Michayluk MR, Kao KN, Gamborg OL, Constable F (1974) Callus formation and plant regeneration from mesophyll protoplasts of rape plants (*Brassica napus* L. cv. Zephyr). Plant Sci Lett 3:265–271

Kemble RJ, Yarrow SA, Wu S-C, Barsby TL (1988) Absence of mitochondrial and chloroplast DNA recombinations in *Brassica napus* plants regenerated from protoplasts, protoplast fusion and anther culture. Theor Appl Genet 75:875–881

Kirti PB (1988) Somatic embryogenesis in hypocotyl protoplast culture of rapeseed (*Brassica napus* L.). Plant Breed 100:222–224

Klimaszewska K, Keller WA (1986a) Protoplast fusion of *Brassica napus* with other cruciferous species. Crucifer genetics workshop III. Proceedings, May 29–30, 1986, Univ Press, Guelph, 59 pp

Klimaszewska K, Keller WA (1986b) Somatic embryogenesis in cell suspension and protoplast culture of *Brassica nigra* (L.) Koch. J Plant Physiol 122:251–260

Klimaszewska K, Keller WA (1987) Plant regeneration from stem cortex protoplasts of *Brassica napus*. Plant Cell Tissue Org Cult 8:225–233

Klimaszewska K, Keller WA (1988) Regeneration and characterization of somatic hybrids between *Brassica napus* and *Diplotaxis harra*. Plant Sci 58:211–222

Kohlenbach HW, Körber M, Li L-C (1982a) Cytodifferentiation of protoplasts isolated from a stem embryo system of *Brassica napus* to tracheary elements. Z Pflanzenphysiol 107:367–371

Kohlenbach HW, Wenzel G, Hoffmann F (1982b) Regeneration of *Brassica napus* plantlets in cultures from isolated protoplasts of haploid stem embryos as compared with leaf protoplasts. Z Pflanzenphysiol 105:131–142

Kranz E (1988) Somatic embryogenesis in stationary phase suspension cultures derived from hypocotyl protoplasts of *Brassica napus* L. Plant Cell Tissue Org Cult 12:141–146

Köhler F, Afsar K, Golz C, Eapen S, Schieder O (1988) Transformation studies in tobacco, moth bean and *Brassica* species. In: Puite KJ, Dons JJM, Huizing HJ, Kool AJ, Koornneef M, Krens FA (eds) Progress in Plant Protoplast Research. Kluwer, Dordrecht, pp 359–362

Laser M, Eriksson T (1985) Transformation of protoplasts from *Brassica campestris* with *Agrobacterium tumefaciens*. Physiol Plant 64:13A

Li L-C, Kohlenbach HW (1982) Somatic embryogenesis in quite a direct way in cultures of mesophyll protoplasts of *Brassica napus* L. Plant Cell Rep 1:209–211

Lichter R (1981) Anther culture of *Brassica napus* in a liquid culture medium. Z Pflanzenphysiol 103:229–237

Lillo C, Shahin EA (1986) Rapid regeneration of plants from hypocotyl protoplasts and root segments of cabbage. Hortscience 21:315–317

Linsmaier EM, Skoog F (1965) Organic growth factor requirements of tobacco tissue cultures. Physiol Plant. 18:100–127

Lu DY, Pental D, Cocking EC (1982) Plant regeneration from seedling cotyledon protoplasts. Z Pflanzenphysiol 107:59–63

Maley ME, Pauls KP (1985) Studies on the transfer of DNA into *Brassica* protoplasts by liposome/protoplast fusion and its detection by flow cytometry. Plant Physiol 77:71 Suppl 4

Matthews REF, Sarkar S (1976) A light induced structural change in chloroplasts of chinese cabbage cells infected with turnip yellow mosaic virus. J Gen Virol 33:435–446

Maule AJ (1983) Infection of protoplasts from several *Brassica* species with cauliflower mosaic virus following inoculation using polyethylene glycol. J Gen Virol 64:2655–2660

McLellan MS, Olesen P, Power JB (1988) Towards the introduction of cytoplasmic male sterility (cms) into *Brassica napus* through protoplast fusion. In: Puite KJ, Dons JJM, Huizing HJ, Kool AJ, Koornneef M, Krens FA (eds) Progress in plant protoplast research. Kluwer, Dordrecht, pp 187–188

McNaughton IH, Ross CL (1978) Interspecific and intergeneric hybridization in the *Brassicae* with special emphasis on the improvement of forage crops. Scott Plant Br Sta 57th Annu Rept, Invergowrie, pp 75–100

Menczel L, Nagy I, Kizz ZR, Maliga P (1981) Streptomycin-resistant and sensitive somatic hybrids of *Nicotiana tabacum* + *Nicotiana knightiana*: correlation of resistance of *N. tabacum* plastids. Theor Appl Genet 59:191–195

Menczel L, Morgan A, Brown S, Maliga P (1987) Fusion-mediated combination of Ogura-type cytoplasmic male sterility with *Brassica napus* plastids using X-irradiated CMS protoplasts. Plant Cell Rep 6:98–101

Millam SR, Burns ATH, Hocking TJ, Kattell KJ (1988) Effect of a Ficoll-agarose system on early cell division and development of mesophyll protoplasts of winter oilseed rape (*Brassica napus* L.). Plant Cell Tissue Org Cult 12:285–290

Morgan A, Maliga P (1987) Rapid chloroplast segregation and recombination of mitochondrial DNA in *Brassica* cybrids. Mol Gen Genet 209:240–246

Morris-Krsinich B, Hull R (1983) Replication of turnip rosette virus RNA in inoculated turnip protoplasts. J Gen Virol 64:2661–2668

Muller B, Bauer R, Glaser R (1986) Viability of *Brassica* mesophyll protoplasts, following electric field pulse application. Arch Zuechtungsforsch 16:245–248

Murashige T, Skoog F (1962) A revised medium for rapid growth and bioassays with tobacco tissue cultures. Physiol Plant 15:473–497

Murata M (1983) Staining air dried protoplasts for study of plant chromosomes. Stain Technol 58:101–106

Nagata T, Melchers G (1978) Surface charge of protoplasts and their significance in cell-cell interaction. Planta 142:235–238

Nagata T, Takebe T (1971) Plating of isolated tobacco mesophyll protoplasts on agar medium. Planta 99:12–20

Nagy JJ, Maliga P (1976) Callus induction and plant regeneration from mesophyll protoplasts of *Nicotiana sylvestris*. Z Pflanzenphysiol 78:453–455

Namai H, Sarashima M, Hosoda T (1980) Interspecific and intergeneric hybridization breeding in Japan. In: Tsunoda S, Hinata K, Gomez-Campo C (eds) Brassica crops and wild allies. Jpn Sci Soc, Tokyo, pp 191–201

Newell CA, Rhoads ML, Bidney DL (1984) Cytogenetic analysis of plants regenerated from tissue explants and mesophyll protoplasts of winter rape, *Brassica napus* L. Can J Genet Cytol 26:752–761

Nishio T, Yamagishi H, Takayanagi K (1987) Shoot regeneration capacity destined at an early stage of protoplast culture in cabbage. Jpn J Breed 37:22–28

Nitsch JP (1969) Experimental androgenesis in *Nicotiana*. Phytomorphology 19:389–404

Nitsch JP, Nitsch C (1969) Haploid plants from pollen grains. Science 163:85–87

Olsson G, Ellerström S (1980) Polyploidy breeding in Europe. In: Tsunoda S, Hinata K, Gomez-Campo C (eds) Brassica crops and wild allies. Jpn Sci Soc, Tokyo, pp 167–187

Paszkowski J, Pisan B, Shillito RD, Hohn T, Hohn B, Potrykus I (1986) Genetic transformation of *Brassica campestris* var. *rapa* protoplasts with an engineered cauliflower mosaic virus genome. Plant Molec Biol 6:303–312

Pelletier G, Primard C, Vedel F, Chetrit P, Remy R, Rousselle P, Renard M (1983) Intergeneric cytoplasmic hybridization in Cruciferae by protoplast fusion. Mol Gen Genet 191:244–250

Primard C, Martin D, Vedel F, Pelletier G (1986) Cp and mt genome constitution of different somatic hybrids between *Brassica napus* and *Brassica hirta*. In: Horn W, Jensen CJ, Odenbach W, Schieder OJ (eds) Genetic manipulation in plant breeding. Proc Int Symp EUCARPIA, Sept 8–13, 1985, Berlin (West). de Gruyter, Berlin, pp 719–722

Primard C, Vedel F, Mathieu C, Pelletier G, Chèvre AM (1988) Interspecific somatic hybridization between *Brassica napus* and *Brassica hirta* (*Sinapis alba* L.). Theor Appl Genet 75:546–552

Pua E-C (1987) Plant regeneration from stem-derived protoplasts of *Brassica alboglabra* Bailey. Plant Sci 50:153–160

Quazi HM (1975) Isolation and culture of protoplasts from *Brassica*. N Z J Bot 13:571–576

Rasmussen R, Ward B, Knauf V (1985) Microinjection of immobilized protoplasts. Plant Physiol 75:141 Suppl 1

Renaudin J, Bové JM, Otsuki Y, Takebe I (1975) Infection of *Brassica* leaf protoplasts by turnip yellow mosaic virus. Mol Gen Genet 141:59–68

Renaudin J, Mouches C, Bove JM (1982) Inhibitory effects of virazole and aktinomycin D on turnip yellow mosaic virus replication in chinese cabbage *Brassica sinensis* protoplasts. Ann Virol 133:295–314

Robertson D, Earle ED (1986a) Plant regeneration from leaf protoplasts of *Brassica oleracea* var. *italica* CV Green Comet broccoli. Plant Cell Rep 5:61–64

Robertson D, Earle ED (1986b) Nitro-blue tetrazolium: a specific stain for photosynthetic activity in protoplasts. Plant Cell Rep 6:70–73

Robertson D, Earle ED, Mutschler MA (1984) Enhanced rates of plant regeneration from protoplasts of cultivar Green Comet Broccoli *Brassica oleracea*. Plant Physiol 75:134 Suppl 1

Robertson D, Earle ED, Mutschler MA (1985) Synthesis of atrazine resistant *Brassica napus* through protoplast fusion. Eucarpia Cruciferae Newslett 10:88–89

Robertson D, Palmer JD, Earle ED, Mutschler MA (1987) Analysis of organelle genomes in a somatic hybrid derived from cytoplasmic male-sterile *Brassica oleracea* and atrazine-resistant *B. campestris*. Theor Appl Genet 74:303–309

Rosén B, Halldén C, Heneen WK (1988) Diploid *Brassica napus* somatic hybrids: Characterization of nuclear and organellar DNA. Theor Appl Genet 76:197–203

Sacristán MD, Melchers G (1969) The caryological analysis of plants regenerated from tumorous and other callus cultures of tobacco. Molec Gen Genet 105:317–333

Sacristán MD, Gerdemann-Knörck M, Schieder O (1988) Transformation of *Brassica nigra* through protoplast cocultivation with *Agrobacterium tumefaciens*. In: Puite KJ, Dons JJM, Huizing HJ, Kool AJ, Koornneef M, Krens FA (eds) Progress in Plant Protoplast Research. Kluwer, Dordrecht, pp 351–352

Saxena PK, Liu Y, King J (1987) Nuclear transplantation into plant protoplasts: optimal conditions for induction and determination of nuclear uptake. J Plant Physiol 128:451–460

Schenck HR, Hoffmann F (1979) Callus and root regeneration from mesophyll protoplasts of basic *Brassica* species: *B. campestris, B. oleracea* and *B. nigra* Z Pflanzenzuecht 82:354–360

Schenck HR, Röbbelen (1982) Somatic hybrids by fusion of protoplasts from *Brassica oleracea* and *B. campestris*. Z Pflanzenzuecht 89:278–288

Schubert S, Matzke H (1985) Influence of phytohormones and other effectors on proton extrusion by isolated protoplasts from rape leaves. Physiol Plant 64:285–289

Schweiger H-G, Dirk J, Koop H-U, Kranz E, Neuhaus G, Spangenberg G, Wolff D (1987) Individual selection, culture and manipulation of higher plant cells. Theor Appl Genet 73:769–783

Shahin EA (1985) Totipotency of tomato protoplasts. Theor Appl Genet 69:235–240

Shepard JF (1980) Mutant selection and plant regeneration from potato mesophyll protoplasts. In: Rubenstein I, Gengenbach B, Phillips RL, Green CE (eds) Genetic improvement of crops. Minnesota Univ Press, Minneapolis, pp 185–219

Sindhu RK, Cohen SS (1984) Subcellular localization of spermidine synthase in the protoplasts of chinese cabbage leaves. Plant Physiol 76:219–223

Sjödin C (1989) Transfer of resistance against *Phoma lingam* to *Brassica napus* L. via somatic hybridization in combination with in vitro selection. Thesis. Department of Plant Breeding, Swedish University of Agricultural Sciences

Sjödin C, Glimelius K (1988) *Brassica naponigra* a somatic hybrid resistant to *Phoma lingam*. Theor Appl Genet 77:651–656

Sjödin C, Nyhammar T, Glimelius K (1988) Effects of toxic metabolites produced by *Phoma lingam* (Tode ex Fr.) Desm. on protoplasts, cell and plants of hosts and non-hosts of the pathogen. Physiol Molec Plant Pathol 32:301–312

Souza-Machado VS, Bandeen JD, Stephenson GR, Lavigne P (1978) Uniparental inheritance of chloroplast atrazine tolerance in *Brassica campestris*. Can J Plant Sci 58:977–981

Spangenberg G, Neuhaus G, Schweiger H-G (1986a) Expression of foreign genes in a higher plant cell after electrofusion-mediated cell reconstitution of a microinjected karyoplast and a cytoplast. Europ J of Cell Biol 42:236–238

Spangenberg G, Koop H-U, Lichter R, Schweiger (1986b) Microculture of single protoplasts of *Brassica napus*. Physiol Plant 66:1–8

Sugimura Y, Matthews REF (1981) Timing of the synthesis of empty shells and minor nucleo proteins in relation to turnip yellow mosaic virus synthesis in *Brassica* protoplasts. Virology 112:70–80

Sundberg E, Glimelius K (1986) A method for production of interspecific hybrids within Brassicaceae via somatic hybridization, using resynthesis of *Brassica napus* as a model. Plant Sci 43:155–162

Sundberg E, Landgren M, Glimelius K (1987) Fertility and chromosome stability in *Brassica napus* resynthesised by protoplast fusion. Theor Appl Genet 75:96–104

Swanson EB, Coumans MP, Brown GL, Patel JD, Beversdorf WD (1988) The characterization of herbicide tolerant plants in *Brassica napus* L. after in vitro selection of microspores and protoplasts. Plant Cell Rep 7:83–87

Taguchi T, Kameya T (1986) Production of somatic hybrid plants between cabbage and chinese cabbage through protoplasts fusion. Jpn J Breed 36:185–189

Tanaka N, Ikegami M, Hohn T, Matsui C, Watanabe I (1984) *E. coli* spheroplast-mediated transfer of cloned cauliflower mosaic virus DNA into plant protoplasts. Mol Gen Genet 195:378–380

Terada R, Yamashita Y, Nishibayashi S, Shimamoto K (1987) Somatic hybrids between *Brassica oleracea* and *B. campestris*: selection by the use of iodoacetamide inactivation and regeneration ability. Theor Appl Genet 73:379–384

Thomas E, Hoffmann F, Potrykus I, Wenzel G (1976) Protoplast regeneration and stem embryogenesis of haploid androgenetic rape. Mol Gen Genet 145:245–247

Thomzik JE, Hain R (1988) Transfer and segregation of triazine tolerant chloroplasts in *Brassica napus* L. Theor Appl Genet 76:165–171

Toriyama K, Hinata K, Kameya T (1987a) Production of somatic hybrid plants, "Brassicomoricandia", through protoplast fusion between *Moricandia arvensis* and *Brassica oleracea*. Plant Sci 48:123–128

Toriyama K, Kameya T, Hinata K (1987b) Selection of a universal hybridizer in *Sinapis turgida* Del. and regeneration of plantlets from somatic hybrids with *Brassica* species. Planta 170:308–313

Ulrich TH, Chowdhury JB, Widholm JM (1980) Callus and root formation from mesophyll protoplasts of *Brassica rapa*. Plant Sci Lett 19:347–354

Vamling K, Heneen WK (1983) Mesophyll protoplasts and plant regeneration of *Brassica napus* cytology and morphology of callus growth. Experientia Suppl 45:218–219

Vatsya B, Bhaskaran S (1982) Plant regeneration from cotyledonary protoplasts of cauliflower (*Brassica oleracea* L. var. *botrytis* L.). Protoplasma 113:161–163

Vedel F, Mathieu C, Lebacq P, Ambard-Bretteville F, Remy F, Pelletier G, Renard M, Rouselle P (1982) Comparative macromolecular analysis of the cytoplasms of normal and cytoplasmatic male sterile *Brassica napus*. Theor Appl Genet 62:255–262

Vedel F, Chetrit P, Mathieu C, Pelletier G, Primard C (1986) Analysis of the protoplast fusion-induced molecular events responsible for the mitochondrial DNA polymorphism in the rapeseed *Brassica napus*. In: Mantell SH, Chapman GP, Street PFSJ (eds) The chondriome: chloroplast and mitochondrial genomes. Longman, Harlow, UK, pp 192–210

Wenzel G (1973) Isolation of leaf protoplasts from haploid plants of petunia, rape and rye. Z Pflanzenzuecht, 69:58–61

Wenzel GF, Hoffmann F, Thomas E (1977) Increased induction and chromosome doubling of androgenetic haploid rye. Theor Appl Genet 51:81–86

Xu Z-H, Davey MR, Cocking EC (1982) Plant regeneration from root protoplasts of *Brassica*. Plant Sci Lett 24:117–121

Yamaoka N, Furusawa II, Yamamoto M (1982) Infection of turnip protoplasts with cauliflower mosaic virus DNA. Virology 122:503–505

Yarrow SA, Wu SC, Barsby TL, Kemble RJ, Shepard JF (1986) The introduction of CMS mitochondria to triazine tolerant *Brassica napus* L., var. Regent, by micromanipulation of individual heterokaryons. Plant Cell Rep 5:415–418

Yarrow SA, Kemble RJ, Barsby TL (1988) Protoplast fusion as a method of increasing cytoplasmic heterogeneity in *Brassica napus*. In: Puite KJ, Dons JJM, Huizing HJ, Kool AJ, Koornneef M, Krens FA (eds) Progress in Plant Protoplast Research. Kluwer, Dordrecht, pp 183–184

IV.3 Cybrids in Oilseed *Brassica* Crops Through Protoplast Fusion

G. PELLETIER[1]

1 Introduction

In higher plants, sexual reproduction generally prevents any exchange of cytoplasmic information between the two parents, and chloroplastic as well as mitochondrial genomes are maternally inherited. However, induced fusion of isolated protoplasts followed by regeneration of entire plants allows the mixing in the same cell of organelles and the screening or selection in resulting individuals of recombined forms of cytoplasmic genetic characteristics. These forms are called cybrids, for cytoplasmic hybrids. The originality of cybrids resides in their possibilities to associate in the regenerated plant, after a more or less stochastic sorting-out process, chloroplasts from one parent with mitochondria from the other, or to create new cytoplasmic genomes resulting from interparental recombinations.

Studies with cybrids started with model species belonging to the Solanacea (Gleba et al. 1975; Belliard et al. 1977). Since that time, a number of different methods have been employed to obtain cybrids, and reports concerning several species have shown their importance in cytoplasmic genetics, principally in the study and manipulation of an important trait in plant breeding, cytoplasmic male sterility or cms (Pelletier 1986). The application of these methods of cell genetics to plant breeding is of particular interest when new cytoplasms are needed to improve the production capacity of a crop. This is especially the case of species belonging to the *Brassica* genus. In this genus no agronomically suitable cms system exists, and for this reason the commercial production of F_1 hybrid seeds is only possible through other systems of pollination control, the use of self-incompatibility, or genic male sterility in a limited number of cases. The consequence is that, in these crops, varieties are generally either pure lines (when the inbreeding depression is acceptable) or heterozygous but heterogeneous populations.

The attempts made to apply the methodologies of somatic hybridization to the problems of plant breeding for the production of cybrids are discussed here.

[1] Laboratoire de Biologie Cellulaire, INRA, 78000 Versailles, France

Biotechnology in Agriculture and Forestry, Vol. 10
Legumes and Oilseed Crops I (ed. by Y.P.S. Bajaj)

2 Genetic Constitution of Cybrids Obtained Through Protoplast Fusion

Concerning the fate of the chloroplast genomes (cp) after fusion, the first results (Chen et al. 1977) showed that in *N. glauca*×*N. langsdorfii* somatic hybrids a mixture of both chloroplast genomes seldom occurred at the plant level. Several subsequent reports have confirmed these findings in somatic hybrids as well as in cybrids: (Belliard et al. 1978; Melchers et al. 1978; Aviv et al. 1980; Iwai et al. 1980; Glimelius and Bonnet 1981; Glimelius et al. 1981; Menczel et al. 1981; Bonnet and Glimelius 1983). The prevalence of one parental cpDNA in a plant regenerated from protoplast fusion is not necessarily based upon elimination of the other parental cpDNA, but on the sorting-out process during mitotic division (Akada et al. 1983; Hirai and Akada 1986). Interspecific chloroplast recombination has been described only once in a *Nicotiana* somatic hybrid (Medgyesy et al. 1985).

The fate of mitochondrial (mt) genomes after protoplast fusion has been mainly studied in three genera, *Nicotiana*, *Petunia*, and *Brassica*. The main objective of these experiments was the understanding, through joint genetic and molecular approaches, of cytoplasmic male sterility in higher plants. In contrast with the plastid genomes, mitochondrial genomes from cybrids and somatic hybrids which have been analyzed generally differ from the mitochondrial genomes in the original parental lines. The hybrids' mtDNA contains fragments characteristic of both parents as well as novel fragments not observed in parental mtDNA's (Belliard et al. 1979; Nagy et al. 1981; Galun et al. 1982; Boeshore et al. 1983; Fluhr et al. 1983; Chetrit et al. 1985; Matthews and Widholm 1985; Kemble et al. 1986). In the case of *Nicotiana*, new cytoplasmically inherited flower morphologies which occurred in almost all cases of protoplast fusion between male sterile and male fertile lines, together with new patterns of mtDNA, were considered as convergent evidence for mtDNA recombination (Belliard et al. 1979; Pelletier et al. 1985). The physical proof of intergeneric recombination was given by Rothenberg et al. (1985) in *Petunia*. They showed that a novel fragment in one somatic hybrid mtDNA was constituted by mtDNA restriction fragments derived from both parents.

The apparent rule concerning the chloroplast genome of cybrids can be stated as follows: cybrids possess either one or the other parental type of chloroplast genome, without any changes, the consequence being that protoplast fusion allows the rapid and definitive exchange of chloroplast types between two lines. In contrast, mitochondrial genomes generally recombine, leading to a new variability which can be observed when male sterility is concerned, at the flower morphology level. The ability of the mitochondrial genome to recombine is now well established in several species, including maize (Lonsdale et al. 1984), wheat (Quetier et al. 1985), turnip (Palmer and Shields 1984), and cauliflower (Chetrit et al. 1984), and explains its multipartite structure. Intergenomic recombinations observed after protoplast fusion must be considered at least in part as consequences of this natural process of intramolecular recombination (Vedel et al. 1986).

3 Importance of the Cytoplasm in *Brassica* and the Production of Male-Sterile Lines

Cross-pollination is the most frequent reproductive system in Cruciferae. In *Brassica napus* (rapeseed) self- and cross-pollinations coexist in 2/3 – 1/3 proportions respectively. Even in this species, F_1 hybrids obtained by hand pollination are generally more vigorous than their parents, the heterotic effect reaching 50% (compared to the best parent) for some combinations (Lefort-Buson 1982). For commercial hybrid seed production, a genetic system of control of pollination based on male sterility would be preferred to other systems like self-incompatibility, which is more difficult to handle. Genic (nuclear recessive) male sterility is used in China to produce F_1 hybrids, but this system requires the hand elimination of male-fertile segregants in the female line, which is aided by the fact that rapeseed is planted out in this country (Lee and Zang 1983).

Several systems of cytoplasmic male sterility are known in rapeseed, although none of them is in practical use because of difficulties encountered either with maintainer or restorer genes or morphological and physiological defects. Among them, at least four are known to result from alloplasmic situations (the nucleus of one species into a foreign cytoplasm). They correspond to the cytoplasm of *B. nigra* (Pearson 1972), *B. juncea* (Anand et al. 1985), *Diplotaxis muralis* (Hinata and Konno 1979; Pelban-Delourme 1986) and *Raphanus sativus* Ogura type (Ogura 1968; Bannerot et al. 1974). The latter is of interest because it leads to complete male sterility, and restorer genes have been introduced in rapeseed from radish (Heyn 1978). Nevertheless, *Brassica* plants bearing radish cytoplasm are known to exhibit yellowing, especially at low temperature, and to maintain a low level of chlorophyll at higher temperature (Bannerot et al. 1977; Rousselle 1981; McCollum 1981). In addition, the flowers of these plants have less developed nectaries and a reduced production of nectar. This defect is of importance in hybrid seed production, since the cms plants are less attractive for bees, the most important vectors in cross-pollination (Mesquida and Renard 1978).

Another type of alloplasmic situation is of interest for *Brassica* crops. An atrazine-resistant *B. campestris* biotype has been discovered (Maltais and Bouchard 1978; Souza-Machado et al. 1978). This resistance has been proved to be maternally inherited and associated with the reducing side of photosystem II (suggesting that this trait is encoded in plastid DNA). Indeed, alloplasmic *B. napus* lines with *B. campestris* atrazine-resistant cytoplasm have been shown to be fully resistant to this herbicide (Beversdorf et al. 1980). Atrazine resistance could be useful as a weed control strategy.

Protoplast fusion offers the possibility to improve the phenotypical characteristics of Ogura-type cytoplasmic male sterility and to combine resistance to triazines and male sterility in the same cytoplasm, allowing the commercial production of F_1 hybrid plants resistant to these herbicides.

4 Protoplast Isolation, Fusion, and Culture Until Plant Regeneration and Cybrid Selection

Protoplasts are isolated from leaves of axenic shoot cultures. These shoot cultures are initiated from seedlings and propagated by subculture on medium A (Table 1). Plantlets are grown in a growth room under about 50 μE m^2s light intensity with a 16 h photoperiod and a temperature of 22 °C during the day and 17 °C during the night. Subcultures are repeated at 3-week intervals. Expanded leaves of these 3-week-old plantlets are placed in a maceration medium containing 0.2% cellulase R 10; 0.1% pectolyase Y23; 0.88 g/l $CaCl_2$ $2H_2O$; 0.084 g/l NaH_2PO_4; 100 g/l mannitol; 1 mg/l NAA and 1 mg/l BA. After overnight maceration in this medium, the suspension is filtered through a 40 μ steel mesh and centrifuged at about 100 g. The protoplast pellet is washed twice by a solution containing 2.5% KCl and 0.2% $CaCl_2$ $2H_2O$, and the mixture of parental protoplast is subjected to fusion treatment: 6.5×10^5 protoplasts in 0.3 ml of suspension are put onto a 0.5 ml drop of a solution containing 25% PEG 6000, 3% $CaCl_2$ $2H_2O$ at pH 8. After about 30 min, the mixture is diluted with 10 ml of culture medium B, without moving the protoplasts which generally adhere to the bottom of the plastic Petri dish (Greiner, TC plastic type, 10-cm diameter). Cultures are placed in the dark during the first 3 to 5 days and then in the light (100 μE m^2s, 16 h, 27 °C). After 8 to 12 days (according to the growth rate of protoplasts), when the dividing protoplast population is at the four to eight cell stage, an equal volume of medium C is added (the contents of each Petri dish being divided into two Petri dishes) and at about the 18th day, another dilution by medium D is carried out in the same manner. On the 28th day, colonies are green and of about 1 to 2 mm in size. They are plated on solid medium E (20 colonies on 25 ml solid medium in a 10-cm Petri dish). Buds generally appear after 2 to 6 weeks on this medium. They are transferred to medium F to enhance shoot formation, and then rooted on medium G before transplanting to soil. Routinely, several hundred plants are grown after each fusion experiment.

Cybrid selection in our system is based on the screening at the plant level of "recombined" cytoplasmic characters: male fertility (F) and male sterility (S), which are mitochondrial traits, and yellowing (Y), normal greening (G), and atrazine resistance (A), which are chloroplastic traits born by *R. sativus*, *B. napus*, and *B. campestris* genomes respectively. In fusion experiments, parents differ by both mitochondrial and chloroplastic characters. Several combinations were carried out according to Table 2, allowing the screening of about 20 cybrids. To distinguish between Y and G traits, young plants in pots are grown in a growth chamber at low temperature (10 °C during the night and 18 °C during the day). Under these conditions yellowing is clearly apparent on young developing leaves.

To distinguish between atrazine-resistant (A) and normal chloroplast, a fluorescence test is performed on small leaves of young plantlets according to Fig. 1. This test (Ducruet and Gasquez 1978) needs less than 1 min per plant.

Table 1. Composition (mg/l) of the different media used for plant regeneration from mesophyll protoplasts of in vitro shoot culture of *B. napus* and *B. oleracea*. (Pelletier et al. 1983; Guerche et al. 1987)

	A	B	C	D	E	F	G
NH_4NO_3	–	–	200	200	1650	1650	825
KNO_3	2500	2500	1250	1250	1900	1900	950
$(NH_4)_2SO_4$	134	134	67	67	–	–	–
NaH_2PO_4	150	150	75	75	–	–	–
KH_2PO_4	–	–	35	35	170	170	85
$CaCl_2 \cdot 2H_2O$	750	750	525	525	440	440	220
$MgSO_4 \cdot 7H_2O$	250	250	250	250	370	370	185
H_3BO_3	3	3	3	12.4	12.4	6.2	6.2
$MnSO_4 \cdot 4H_2O$	10	10	10	33.6	33.6	22.3	22.3
$ZnSO_4 \cdot 7H_2O$	2	2	2	21	21	8.6	8.6
KI	0.75	0.75	0.75	1.66	1.66	0.83	0.83
$Na_2MoO_4 \cdot 2H_2O$	0.25	0.25	0.25	0.5	0.5	0.25	0.25
$CuSO_4 \cdot 5H_2O$	0.025	0.025	0.025	0.05	0.05	0.025	0.025
$CoCl_2 \cdot 6H_2O$	0.025	0.025	0.025	0.05	0.05	0.025	0.025
$FeSO_4 \cdot 7H_2O$	27.8	27.8	27.8	27.8	27.8	27.8	27.8
Na_2EDTA	37.3	37.3	37.3	37.3	37.3	37.3	37.3
Inositol	100	100	100	100	100	100	100
Nicotinic acid	1	1	1	1	0.5	1	0.5
Pyridoxine HCl	1	1	1	1	0.5	1	0.5
Thiamine	10	10	10	10	–	10	–
Glycine	–	–	–	–	2	–	2
Glucose	10000	20000	–	–	10000	–	10000
Sucrose	10000	–	20000	20000	10000	10000	–
D Mannitol	–	70000	40000	–	10000	–	–

NAA	–	1	0.2	–	1	0.1	0.01
2,4-D	–	0.25	–	1	–	–	–
BA	–	1	1	–	0.5	0.5	–
Adenine sulfate	–	–	–	30	–	–	–
2iP	–	–	–	–	0.5	–	–
GA3	–	–	–	–	0.02	–	–
Agar	2000	–	–	–	–	4000	8000
Agarose LSM	3000	–	–	–	10000	4000	–

pH adjusted at 5.8 with NaOH

Table 2. Screening of cybrids among regenerated plants at random from fusion experiments. (Pelletier et al. 1983; Pelletier unpubl.)

Experiment	Parent 1		Parent 2		Expected cybrids		No. of cybrid observed	Cybrid genomes	
	cp	mt	cp	mt	cp	mt		cp	mt
1	Y	S(Ogu)	G	F(Nap)	Y	F	0		
					G	S	6	Nap	Ogu/Nap
2	Y	S(Ogu)	A	F(Cam)	Y	F	0		
					A	S	1	Cam	Ogu
3	A	S(Ogu) (Cybrid 77)	G	F(Nap)	A	F	3	Cam	Ogu/Nap
					G	S	4	Nap	Ogu/Nap
4	G	S(Ogu/Nap) (Cybrid 58)	A	F(Cam)	G	F	3	Nap	Ogu/Nap/Cam
					A	S	1	Cam	Ogu/Nap/Cam
5	Y	S(Ogu)	A	F(Ogu/Nap)	Y	F	0		
					A	S	2	Cam	Ogu/Nap

Cybrids are screened at the plant level according to their chloroplastic (cp) and mitochondrial (mt) traits. Y: yellowing, G: normal greening, A: atrazine resistance, F: male fertility, S: male sterility. Ogu, Nap, Cam: Ogura radis, *B. napus, B. campestris* atrazine resistant cytoplasmic genomes respectively. Ogu/Nap: recombined mt genome

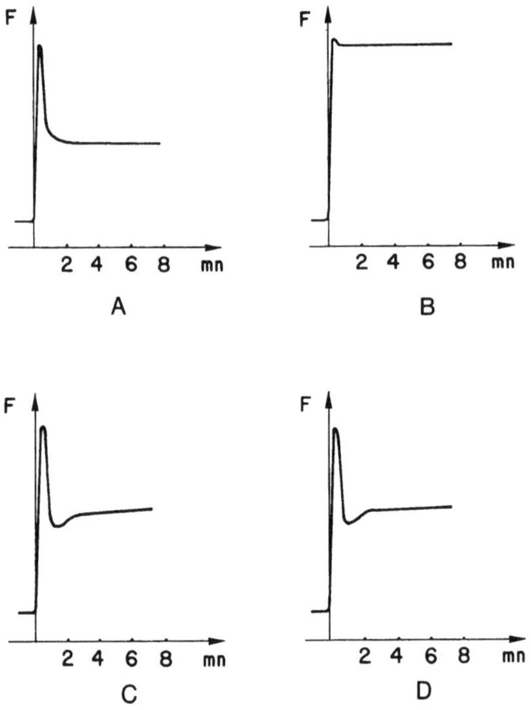

Fig. 1A–D. Fluorescence test of atrazine resistance. **A** Leaf of a susceptible plant immersed in water (control). **B** Leaf of a susceptible plant immersed in atrazine (10^4 g/ml). **C** Leaf of a resistant variety with *B. campestris* cytoplasm, immersed in atrazine (the response is similar to the control). **D** Leaf of a progeny of the cms atrazine-resistant cybrid immersed in atrazine (the response is identical to **C**). *F* red fluorescence (arbitrary units). (Pelletier et al. 1985)

5 Molecular Characterization of Organelle Genomes of Cybrids

A simple and rapid procedure used to isolate cpDNA of cybrids from no more than 1 – 3 g fresh leaves was reported previously (Pelletier et al. 1983). Differences in restriction patterns of cpDNA between *B. napus*, *R. sativus*, and *B. campestris* are easily recognizable (Vedel et al. 1982; Pelletier et al. 1983). Among cybrids, as well as among parental types, a strict correlation is found between cp genomes (*R. sativus*, *B. napus*, *B. campestris*) and the previously described cytoplasmic characters Y, G, and A, respectively.

mtDNA is isolated by using CsCl-ethidium bromide gradients (Vedel et al. 1982). Figure 2 diagrammatically shows parental (Ogu, Nap, Cam) and cybrid (numbers) restriction patterns obtained after agarose gel electrophoresis of mtDNA digests by Sal 1. It is clear on this figure that each cybrid except 77, identical to Ogu, possesses a new pattern different from its parents and from each other (Fig. 2). Novel restriction fragments observed in these cybrids were analyzed (Vedel et al. 1986). One of them was chosen to give molecular evidence that it actually represents intergenomic molecular recombination. This fragment (* on Fig. 2), present in several cybrids, was found to be constituted by a recombination between fragment 1 (33 kb) from *B. napus* and fragment 6 (12.3 kb) from *R. sativus*.

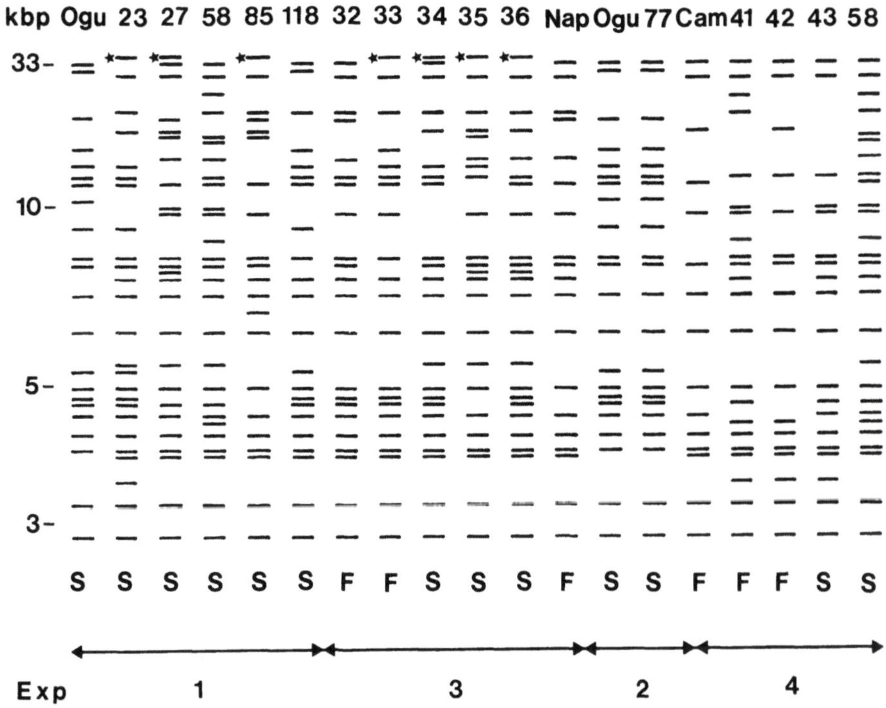

Fig. 2. Diagramatic representation of parental (Ogu, Nap, Cam) and cybrids (numbers) mtDNA digests by Sal1. Cybrids correspond to some of those obtained in experiments, *1, 2, 3, 4* (see Table 2), *F* fertile; *S* sterile. (Chetrit et al. 1985, Vedel unpubl.)

Sites involved in mt recombination in cybrids were compared by molecular hybridization to sites supposedly implicated in intragenomic mt recombination. The data reveal that intergenomic recombination arising through protoplast fusion involves different mtDNA regions, and that some events occur at regions which are homologous to regions supposed to play a role in natural recombination.

Mitochondrial genomes from related species differ mainly by their structure after extensive rearrangements during the course of evolution (Palmer and Herbon 1986). A consequence of these differences is that cybrid mt genome are generally more complex than parental genomes.

6 Agronomical Value of Cybrids

Male sterile (S) cybrids are of particular interest if the agronomical defects previously described (i.e., yellowing and absence of nectar production) are eliminated. Concerning the yellowing, cybrids having *B. napus* or *B. campestris* chloroplast recovered their normal chlorophyll synthesis. Cybrids with *B. campestris* chloroplast are fully resistant to atrazine, allowing the use of this cytoplasm to commercially produce triazine-resistant hybrid seeds. The stability of these traits was observed in the sexual progeny: no changes were observed during the following generations.

Concerning flower morphology, variability was observed. Measurements of nectar production were performed in field conditions on a spring variety (Brutor) with different cytoplasms (*B. napus* and cybrids 27, 58, and 118, Fig. 2). Only mitochondria are different in this comparison, since these cybrids and *B. napus* have the same *B. napus* chloroplasts. Nectar production was 81% of the fertile control with recombinant no. 27, 60% with no. 58, and 38% with no. 118, proving that this important agronomical character is controlled by a mitochondria-nucleus interaction (Mesquida and Renard unpubl.). These observations emphasize the practical importance of mitochondrial rearrangements, since these differences will have an impact on cross-pollination by bees, and finally on the price of hybrid seeds.

At least three kinds of flower morphologies are visible among cms cybrids: the first, similar to the Ogura type, has short stamens and a very low production of nectar (cybrid no. 77). In the second, plants have more normal anthers and a partially restored production of nectar but no pollen (cybrid no. 58). In the last case, plants have short anthers, a partially restored nectar production, and some flowers at the top of stems producing some pollen (cybrid no. 23). Generally speaking, a cms character is qualitatively defined by a specific genetic (nuclear) system of fertility restoration. The original cms Ogura radish cytoplasm used here is restored by genes existing only in radish nucleus (Bonnet 1975; Rousselle 1981). They were introduced by sexual crosses into *B. napus* by Heyn (1978). Using a heterozygous line for these restorer genes, crosses made with the three kinds of cybrids previously described gave different segregation patterns of restoration (Table 2, Fig. 3). In order to interpret these results, taking into account mtDNA recombination as shown by molecular analysis, we must assume that Ogura mito-

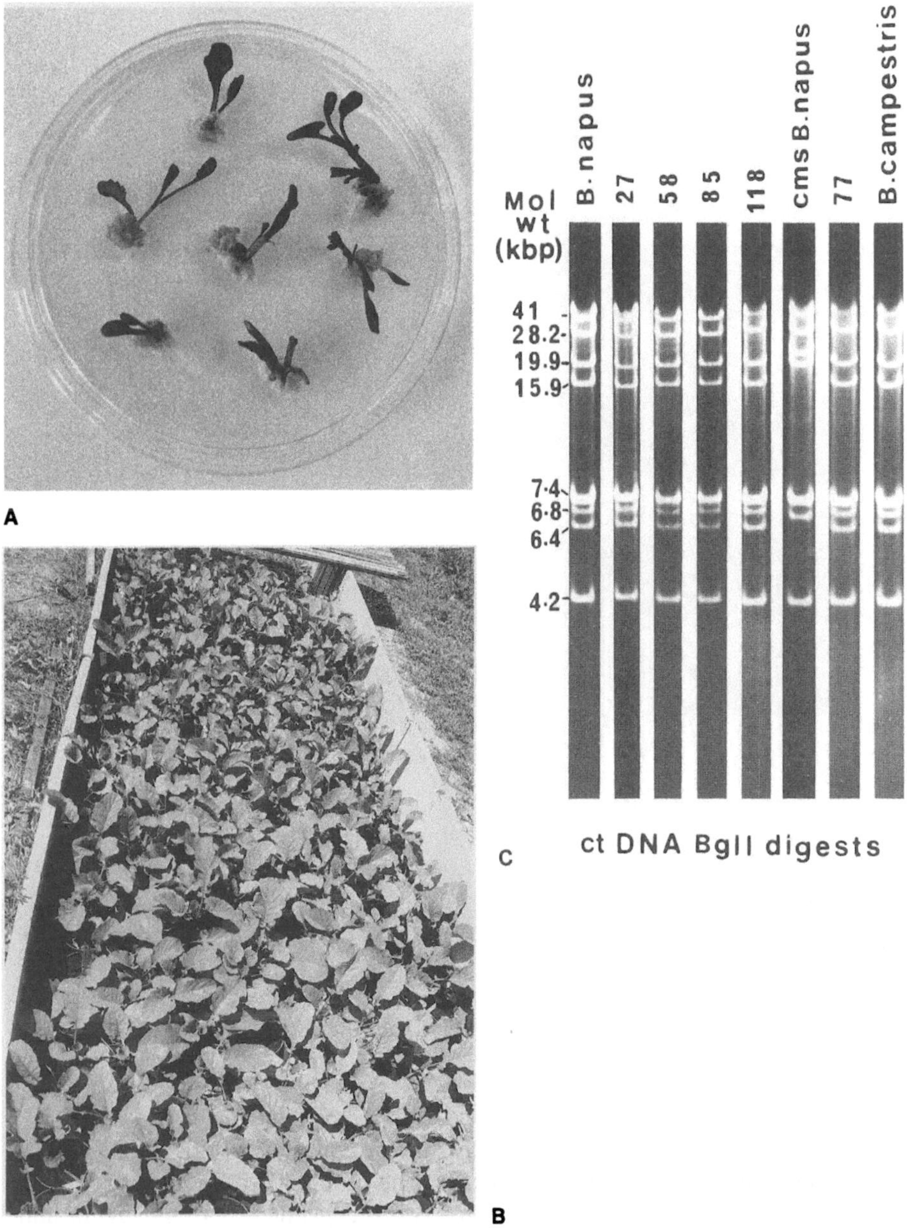

Fig. 3. A Plantlet regeneration on protoplast derived colonies of *B. napus* (medium E, Table 1; 40 days after protoplast isolation). **B** Regenerated plants from protoplasts transplanted in soil (4 months after protoplast isolation). **C** ctDNA Bgl1 digests of parents and some cybrids corrected for chlorophyll deficiency (Pelletier et al. 1983). **D** mtDNA Sal1 digests of parents and some male-sterile cybrids. *B* = *B. napus*; *C* = *R. sativus*; *T* = *B. campestris* mitochondria. (Vedel et al. 1986). **E** Flower morphology *from left to right*, of male-fertile, male-sterile cybrid, and male-sterile Ogura *B. oleracea* plant a more normal flower morphology concerning the pistil is observed in the cybrid plant. **F** Progeny of crosses between *B. napus* cybrids and the restorer material for the genetic analysis of fertility restoration. (Parts **D–F** of this figure see following pages)

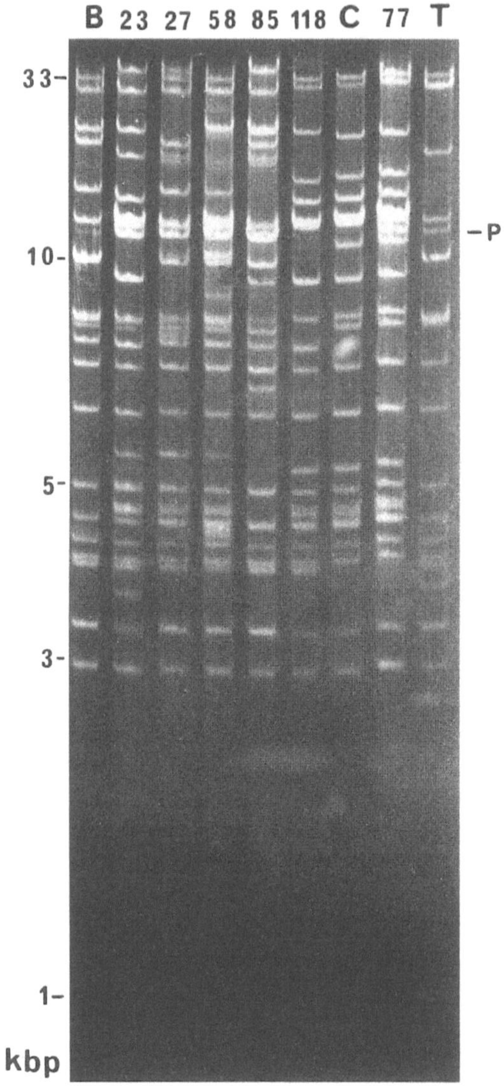

Fig. 3D

chondrial DNA bears more than one cms factor and that cybrids no. 23 and no. 58 recombinants possess fewer cms factors. The simplest hypothesis is to consider that Ogura mitochondria lead to cms for two independent reasons: Ogura male sterility is already expressed in radish, and an alloplasmic male sterility is expressed when radish mitochondria are in the presence of a *Brassica* nucleus (McCollum 1981). mtDNA recombination makes it possible to separate these two factors genetically, and to give a simpler system of restoration as in the case of cybrid no. 58. In fact, this male fertility restoration system raised an unexpected problem, which has made it up to now of no use for hybrid seed production. Among segregating progeny of self.pollinated restored plants, a decrease of female fertility is observed, male-fertile plants always being less female-fertile, as

Fig. 3E

Fig. 3F

Table 3. Segregation of fertility/sterility and flower color in crosses betwen different cybrids and the same heterozygous restorer. (Pellan-Delourme 1986; Pelletier and Primard unpubl.)

Phenotype crosses	W, F	W, S	y, F	y, f	y, S	% F
♀77 × ♂RF	9	9	3	6	30	21
♀23 × ♂RF	11	0	14	4	11	62
♀58 × ♂RF	10	13	11	0	13	44
♀Ogu × ♂RF	14	9	7	15	47	23

A heterozygous restorer (RF) with white flowers crossed with different cytoplasms gives progenies which show segregation for flower color (w: white, y: yellow) and for male fertility (F: fully fertile, f: some pollen produced, S: sterile). Segregations are similar in crosses 77 × RF and Ogu × RF, cybrid 77 having Ogu mitochondria. Frequencies of fully fertile plants in the other crosses are significantly higher (44% and 66%), showing that these cybrids need less restorer genes. Two phenotypes are not represented (W, S) in the case of cybrid 23 and (y, f) in the case of cybrid 58. This indicates different causes of sterility in these two cybrids.

if male fertility restorer genes themselves introduce female sterility (Pellan-Delourme 1986).

On the other hand, the productivity of lines having male-sterile cybrid cytoplasms was compared in plots cross-pollinated by neighboring plots, to the same lines on the normal fertile cytoplasm (Renard unpubl.). It was observed that some cybrid cytoplasms are able to increase the productivity of the variety significantly, a result probably to be compared with physiological observations in other species (Musgrave et al. 1986).

On the basis of these first results on the agronomical value of cybrids, the best recombinants were selected and multiplied. In 1986, about five metric tons of seeds from this material have been produced, and subjected to multilocal and multiannual comparison tests. In addition, male-sterile cybrids cytoplasms obtained in *B. napus* were transferred into *B. oleracea* by sexual crosses (Boulidard and Dore unpubl.) and have been evaluated since 1986 by several seed companies.

7 Conclusion

During the last few years, constant progress in protoplast culture methods (Liu et al. 1982; Xu et al. 1982; Bidney et al. 1983; Glimelius 1984; Terada et al. 1987; Robertson and Earle 1986), selection of hybrids or cybrids (Primard et al. 1986; Sundberg and Glimelius 1986; Glimelius et al. 1986; Yarrow et al. 1986; Jourdan et al. 1986), transformation with *Agrobacteria* (Ooms et al. 1985; Guerche et al. 1987b), or by direct gene transfer (Potrykus et al. 1985; Guerche et al. 1986, 1987a), and finally the selection of rapid cycling populations for the major cultivated species in the genus *Brassica* (Williams and Hill 1986), make this genus a model one in which joint cellular and molecular biology open practical as well as fundamental perspectives. The examples of cybridization described here have already been reproduced with the same (Menczel et al. 1987; Jarl and Bornman 1987), or similar (Jourdan et al. 1986; Yarrow et al. 1986) systems. The creation of novel cytoplasmic combinations is indeed a first concrete example of direct application of protoplast fusion for *Brassica* breeding, but other aspects (disease resistances, chemical composition and, more generally speaking, a better knowledge and handling of genetic variability) would benefit from these advantages.

References

Akada S, Hirai A, Uchimiya H (1983) Studies on mode of separation of chloroplast genomes in parasexual hybrid calli. I. Fraction I protein composition in unseparated hybrid callus. Plant Sci Lett 31:223–230

Anand I, Mishra P, Rawat D (1985) Mechanism of male sterility in *Brassica juncea*. Manifestation of sterility and fertility restoration. Cruciferae, Newsl Eucarpia 10:44–46

Aviv D, Fluhr F, Edelman M, Galun E (1980) Progeny analysis of the inter-specific somatic hybrids: *Nicotiana tabacum* (cms)+*Nicotiana sylvestris* with respect to nuclear and chloroplast markers. Theor Appl Genet 56:145–150

Bannerot H, Boulidard L, Cauderon Y, Tempe J (1974) Transfer of cytoplasmic male sterility from *Raphanus sativus* to *Brassica oleracea*. Proc Eucarpia Meet Cruciferae, p 52

Bannerot H, Boulidard L, Chupeau Y (1977) Unexpected difficulties met with the radish cytoplasm. Eucarpia Cruciferae Newsl 2:16

Belliard G, Pelletier G, Ferault M (1977) Cytoplasmic hybridization in *Nicotiana* by protoplast fusion and transfer of cytoplasmic male fertility and sterility factors. Proc 8th Congress of Eucarpia Madrid pp 237–242

Belliard G, Pelletier G, Vedel F, Quetier F (1978) Morphological characteristics and chloroplast DNA distribution in different cytoplasmic parasexual hybrids of *Nicotiana tabacum*. Mol Gen Genet 165:231–237

Belliard G, Vedel F, Pelletier G (1979) Mitochondrial recombination in cytoplasmic hybrids of *Nicotiana tabacum* by protoplast fusion. Nature (Lond) 281:401–403

Beversdorf WD, Weiss-Lerman J, Erickson LR, Souza-Machado V (1980) Transfer of cytoplasmically inherited triazine resistance from birds-rape to cultivated oilseed rape (*B. campestris* and *B. napus*). Can J Genet Cytol 22:167–172

Bidney DL, Shepard JF, Kaleikau E (1983) Regeneration of plants from mesophyll protoplasts of *Brassica oleracea*. Protoplasma 117:89–92

Boeshore ML, Lifshitz I, Hanson MR, Izhar S (1983) Novel composition of mitochondrial genomes in *Petunia* somatic hybrids derived from cytoplasmic male sterile and fertile plants. Mol Gen Genet 190:459–467

Bonnet A (1975) Introduction et utilisation d'une stérilité mâle cytoplasmique dans des variétés précoces européennes de radis (*Raphanus sativus*). Ann Amelior Plant 25 (4):381–397

Bonnet HT, Glimelius K (1983) Somatic hybridization in *Nicotiana*: behaviour of organelles after fusion of protoplasts from male fertile and male sterile cultivars. Theor Appl Genet 65: 213–217

Chen K, Wildmann SG, Smith HH (1977) Chloroplast DNA distribution in parasexual hybrids as shown by polypeptide composition of fraction I protein. Proc Natl Acad Sci USA 74:5109–5112

Chetrit P, Mathieu C, Muller JP, Vedel F (1984) Physical and gene mapping of cauliflower (*Brassica oleracea*) mitochondrial DNA. Curr Genet 8:413–421

Chetrit P, Mathieu C, Vedel F, Pelletier G, Primard C (1985) Mitochondrial DNA polymorphism induced by protoplast fusion in Cruciferae. Theor Appl Genet 69:361–366

Ducruet JM, Gasquez J (1978) Observation de la fluorescence sur feuille entière et mise en évidence de la résistance chloroplastique à l'atrazine chez *Chenopodium album* et *Pao annua*. Chenosphere 8:691–696

Fluhr R, Aviv D, Edelman M, Galun E (1983) Cybrids containing mixed and sorted-out chloroplasts following interspecific somatic fusions in *Nicotiana*. Theor Appl Genet 65:289–294

Galun E, Arzee-Gonen P, Fluhr R, Edelman M, Aviv D (1982) Cytoplasmic hybridization in *Nicotiana*: Mitochondrial DNA analysis of progenies resulting from fusion between protoplasts having different organelle constitutions. Mol Gen Genet 186:50–56

Gleba YY, Butenko RG, Sytnik KM (1975) Protoplast fusion and parasexual hybridization in *Nicotiana tabacum* L. Dokl Acad Nauk USSR 221:1196–1198

Glimelius K (1984) High growth rate and regeneration capacity of hypocotyl protoplasts in some Brassicaceae. Physiol Plant 61:38–44

Glimelius K, Bonnet HT (1981) Somatic hybridization in *Nicotiana*. Restoration of photoautotrophy to an albino mutant with defective plastids. Planta 153:497–503

Glimelius K, Chen K, Bonnet HT (1981) Somatic hybridization in *Nicotiana*: segregation or organellar traits among hybrid and cybrid plants. Planta 153:504–510

Glimelius K, Djupsjobacka M, Fellner-Feldegg H (1986) Selection and enrichement of plant protoplast heterokaryons of Brassicaceae by flow sorting. Plant Sci 45:133–141

Guerche P, Charbonnier M, Jouanin L, Pelletier G (1986) Direct gene transfer in *Brassica napus*. Abstr Int Symp Plant Mol Biol Strasbourg, France, July 1986, p 19

Guerche P, Charbonnier M, Jouanin L, Tourneur C, Paszowski J, Pelletier G (1987a) Direct gene transfer by electroporation in *Brassica napus*. Plant Sci 52:111–116

Guerche P, Jouanin L, Tepfer D, Pelletier G (1987b) Genetic transformation of oilseed rape (*Brassica napus*) by the Ri T-DNA of *Agrobacterium rhizogenes* and analysis of inheritance of the transformed phenotype. Mol Gen Genet 206:382–386

Heyn FW (1978) Introgression of restorer genes from *Raphanus sativus* into cytoplasmic male sterile *Brassica napus* and the genetics of fertility restoration. Proc 5th Int Rapeseed Conf, Malmoe, pp 82–83

Hinata K, Konno N (1979) Studies on a male sterile strain having the *Brassica campestris* nucleus and the *Diplotaxis muralis* cytoplasm. Jpn J Breed 29 (4):305–311

Hirai A, Akada S (1986) Chloroplast genomes in hybrid calli derived from cell fusion. In: Mantell SM, Chapman GP, Street PFS (eds) The chondriome. Pitman, London, pp 290–298

Iwai S, Nagao T, Nakata K, Kawastrima N, Matsuyama S (1980) Expression of nuclear and chloroplastic genes coding for fraction I protein in somatic hybrids of *Nicotiana tabacum*+*N. rustica*. Planta 147:414–417

Jarl CI, Borman CM (1987) Correction of chlorophyll-defective male-sterile winter oilseed rape (*Brassica napus*) through organelle exchange. 7th Int rapeseed Congr, Poznan Poland 11–14 May 1987, abstr, p 344

Jourdan PS, Mutschler MA, Earle ED (1986) Production and characterization of somatic hybrids between *Brassica oleracea* and *B. campestris*. Cruciferae Newsl 11:84–85

Kemble RJ, Barsby TL, Wong RSC, Shepard JF (1986) Mitochondrial DNA rearrangements in somatic hybrids of *Solanum tuberosum* and *Solanum brevidens*. Theor Appl Genet 72:787–793

Lee L, Zhang Y (1983) The utilization of genetic male sterility (*Brassica napus*) in Shangai. Proc 6th Int Rapeseed Conf, Cetiom, Paris, pp 360–364

Lefort-Buson M (1982) Heterosis with summer rapeseed (*Brassica napus*). Cruciferae Newsl Eucarpia 7:16–17

Liu DY, Pental D, Cocking EC (1982) Plant regeneration from seedling cotyledon protoplasts. Z Pflanzenphysiol 107:59–63

Lonsdale DH, Hodge TP, Fauron CMR (1984) The physical map and organisation of the mitochondrial genome from the fertile cytoplasm of maize. Nucl Acid Res 12:9249–9261

Maltais B, Bouchard C (1978) Une moutarde des oiseaux (*B. rapa* L.) résistante à l'atrazine. Phytoprotection 59:117–119

Matthews BJ, Widholm JM (1985) Organelle DNA composition and isoenzyme expression in an interspecific somatic hybrid of Daucus. Mol Gen Genet 198:371–376

McCollum G (1981) Induction of an alloplasmic male sterile *Brassica oleracea* by substituting cytoplasm from early scarlet globe radish (*Raphanus sativus*). Euphytica 30:855–859

Medgyesy P, Fejes E, Maliga P (1985) Interspecific chloroplast recombination in a *Nicotiana* somatic hybrid. Proc Natl Acad Sci USA 82:6960–6964

Melchers G, Sacristan MD, Holder AA (1978) Somatic hybrid plants of potato and tomato regenerated from fused protoplasts. Carlsberg Res Commun 43:203–218

Menczel L, Nagy F, Kiss ZR, Maliga P (1981) Streptomycin resistant and sensitive somatic hybrids of *Nicotiana tabacum*+*Nicotiana knightiana* correlation of resistance to *N. tabacum* plastids. Theor Appl Genet 59:191–196

Menczel L, Morgan A, Brown S, Maliga P (1987) Fusion-mediated combination of Ogura-type cytoplasmic male sterility with *Brassica napus* plastids using X-irradiated CMS protoplasts. Plant Cell Rep 6:98–101

Mesquida I, Renard M (1978) The entomophilous pollination of male sterile strains of winter rapeseed (*Brassica napus* L.) and a preliminary study of alternating devices. In: Caron TA (ed) 4th Int Symp on Pollination, Washington, Maryland, pp 49–57

Musgrave M, Antonovics J, Siedow J (1986) Is male sterility in plants related to lack of cyanide-resistant respiration in tissue? Plant Sci 44:7–11

Nagy F, Torok I, Maliga P (1981) Extensive rearrangements in the mitochondrial DNA in somatic hybrids of *Nicotiana tabacum* and *N. knightiana*. Mol Gen Genet 183:437–439

Ogura H (1968) Studies of the new male-sterility in Japanese radish with special reference to the utilization of this sterility towards the practical raising of hybrid seeds. Mem Fac Agric Kagoshima Univ 6:39–78

Ooms G, Bains A, Burrell M, Karp A, Twell D, Wilcox E (1985) Genetic manipulation of cultivars of oilseed rape (*Brassica napus*) using Agrobacterium. Theor Appl Genet 7:325–329

Palmer JD, Herbon LA (1986) Tricircular mitochondrial genomes of *Brassica* and *Raphanus*: reversal of repeat configurations by inversion. Nucl Acid Res 14:9755–9764

Palmer JD, Shields CR (1984) Tripartite structure of the *Brassica campestris* mitochondrial genome. Nature 307:437–440

Pearson O (1972) Cytoplasmically inherited male sterility characters and flavor components from the species cross *B. nigra* × *B. oleracea*. J Am Soc Hort Sci 97 (3):397–402

Pellan-Delourme (1986) Etude de deux système de stérilité mâle génocytoplasmique introduits chez le colza (*B. napus*) per croisement intergénériques avec *Raphanus* et *Diplotaxis*. Thesis, Université de Rennes

Pelletier G (1986) Plant organelle genetics through somatic hybridization. In: Miflin B (ed) Oxford. Surveys of plant molecular and cell biology 3:97–121

Pelletier G, Primard C, Vedel F, Chetrit P, Remy R, Rousselle P, Renard M (1983) Intergeneric cytoplasmic hybridization in Cruciferae by protoplast fusion. Mol Gen Genet 191:244–250

Pelletier G, Vedel F, Belliard G (1985) Cybrids in genetics and breeding. Hereditas Suppl 3:49–56

Primard C, Martin D, Vedel F, Pelletier G (1986) Cp and Mt genome constitution of different somatic hybrids between *Brassica napus* and *B. hirta*. In: Horn W, Jensen CJ, Odenbach W, Schieder O (eds) Genetic Manipulation in Plant Breeding. Walter de Gruyter, Berlin

Quetier F, Lejeune B, Delorme S, Falconet D (1985) Molecular organization and expression of the mitochondrial genome of higher plants. In: Douce R, Day DA (eds) Encyclopedia of plant physiology. News series. Springer, Berlin Heidelberg New York Tokyo 18:25–36

Robertson D, Earle E (1986) Plant regeneration from leaf protoplasts of *Brassica oleracea* var. *italica* cv. Green Comet Broccoli. Plant Cell Rep 5:61–64

Rothenberg M, Boeshore ML, Hanson MR, Izhar S (1985) Intergenomic recombination of mitochondrial genomes in a somatic hybrid plant. Curr Genet 9:615–618

Rousselle P (1981) Etudes de systèmes d'androstérilité chez le colza (*Brassica napus*). Thesis, ENSA, Rennes

Souza Machado V, Bandeen J, Stephenson G, Lavigne P (1978) Uniparental inheritance of chloroplast atrazine tolerance in *Brassica campestris*. Can J Plant Sci 58:977–981

Sundberg E, Glimelius K (1986) A method for production of interspecific hybrids within Brassicaceae via somatic hybridization, using resynthesis of *Brassica napus* as a model. Plant Sci 43:155–162

Terada R, Yamashita Y, Nishibayashi S, Shimamoto K (1987) Somatic hybrids between *Brassica oleracea* and *B. campestris* selection by the use of iodoacetamide inactivation and regeneration ability. Theor Appl Genet 73:379–384

Vedel F, Mathieu C, Lebacq P, Ambard-Bretteville F, Remy R, Pelletier G, Renard M, Rousselle P (1982) Comparative macromolecular analysis of the cytoplasms of normal and cytoplasmic male sterile *Brassica napus*. Theor Appl Genet 62:255–262

Vedel F, Chetrit P, Mathieu C, Pelletier G, Primard C (1986) Several different mitochondrial DNA regions are involved in intergenomic recombination in *Brassica napus* cybrid plants. Curr Genet 11:17–24

Williams PH, Hill CB (1986) Rapid cycling populations of *Brassica*. Science 232:1385–1389

Xu Z, Davey MR, Cocking E (1982) Plant regeneration from root protoplasts of *Brassica*. Plant Sci Lett 24:117–121

Yarrow SA, Wu SC, Barsby TL, Kemble RJ, Shepard JF (1986) The introduction of cms mitochondria to triazine tolerant *Brassica napus* L., var. Regent by micromanipulation of individual heterokaryons. Plant Cell Rep 5:415–418

IV.4 Sunflower (*Helianthus annuus* L.): Establishment of Cultures, Transformation, and the Regeneration of Plants

K. E. P. ROBINSON[1] and N. P. EVERETT[2]

1 Introduction

Sunflower (*Helianthus annuus* L.) is one of the four major annual crops in the world grown for edible oil. It is also grown for the confectionery, birdfood, and petfood markets, but these uses account for only about 10% of total sunflower production. Current oilseed cultivars and hybrids consistently produce seed that contains in excess of 40% oil by weight. The oil is rich in the unsaturated fatty acids linoleic acid and oleic acid, contains large amounts of α-tocopherol (vitamin E) and is easy to refine. The light color, bland flavor, and high polyunsaturated fatty acid content of sunflower oil make it a premium salad oil, while its high smoke point and lack of flavour retention make it an excellent oil for home and commercial frying operations (US Sunflower Crop Quality Report 1984).

In addition to the high quality of the oil, sunflower is an important crop because of its adaptation to marginal land in many areas of the temperate zones of the world. Particularly important in this respect is sunflower's extensive and heavily branched tap root system. Sunflower is not especially resistant to water stress, but its ability to extract moisture from depths exceeding 2 m permits satisfactory yields when other crops are seriously damaged by periods of drought. Other important agronomic characters include short growing season, tolerance of both low and high temperatures, and an ability to grow on low-fertility soils (Robinson 1978).

Sunflower is grown commercially throughout the world (Table 1). The profitability of sunflower production is highly dependent on the conditions in the soybean industry, the dominant source of vegetable oil (Table 2) and protein meal. Sunflower oil usually commands a higher price than soybean oil. However, oil represents approximately 80% of sunflower's overall value, whereas soybean is also a valuable source of protein meal. Therefore sunflower is usually grown on land that is unsuitable for soybean.

The major diseases of sunflower in North America and many other sunflower producing areas are: downy mildew caused by *Plasmopara halstedi* (Farl.) Berl and de Toni, rust caused by *Puccinia helianthi* Schw., Sclerotinia white mold caused by *Sclerotinia sclerotiorum* (Lib.) deBary, and verticillium wilt caused by

[1] DNA Plant Technology Corporation, 6701 San Pablo Avenue, Oakland, CA 94608-1239, USA
[2] EniChem Americas, Inc., 2000 Princeton Park Corporate Center, Monmouth Junction, N.J. 08852, USA

Biotechnology in Agriculture and Forestry, Vol. 10
Legumes and Oilseed Crops I (ed. by Y.P.S. Bajaj)

Table 1. World sunflower seed production (FAO 1984)

	Area harvested 1000 ha	Yield kg/ha	Production 1000 mt	%
World	13431	1187	15941	100
USSR	3907	1152	4500	28.2
Europe	2915	1504	4385	27.5
Asia	2142	1152	2468	15.5
S. America	2092	1078	2255	14.1
North Central America	1584	1130	1790	11.2

Table 2. World vegetable oil production 1984/85 (Bye 1986)

Crop	1000 mt
Soybean	13836
Sunflower	6518
Palm	6433
Rapeseed	5914

Verticillium dahliae klebahn. Other diseases, such as charcoal rot, caused by *Macrophomina phaseoli* (Maubl.) Ashby, are more regional, with sensitivity being determined by climatic factors (Zimmer and Hoes 1978). Substantial control over rust, downy mildew and verticillium wilt has been achieved by the development of resistant hybrids through conventional breeding.

Archeological evidence indicates that sunflower was domesticated as a food crop by the American Indians 5000 years ago. Sunflower arrived in Spain in the early part of the 16th century and gradually moved into Central Europe, arriving in Russia in the 18th century. By 1840 the Russians were using sunflower as a commercial source of edible oil from cultivars developed by Russian peasants. Breeding programs were undertaken in Russia during the early part of the 20th century. The resulting high oil cultivars were then introduced into North America in the 1960's. The discovery of cytoplasmic male sterility and nuclear restorer genes in the later 1960's made possible the production of high-quality sunflower hybrids. The first hybrids were made available for commercial production in the USA in 1972 (Putt 1978; Fick 1978). Today, virtually all of the US sunflower crop is planted to hybrid seed. Benefits that have resulted from hybrid breeding programs include increased yield, disease resistance, high self-compatibility (reduced dependence on insect pollination), and uniformity of plant height and maturity (Lilleboe 1985a).

The source of traits such as disease resistance, cytoplasmic male sterility, and fertility restoration has been the diverse germplasm available among the species of wild sunflowers in the *Helianthus* genus. Problems encountered in making use of this diverse wild germplasm include embryo abortion and sterility in crosses with domesticated sunflower, and the inheritance of undesirable traits along with the desirable ones (Lilleboe 1985b). These problems present a number of opportunities to exploit biotechnology for sunflower improvement.

2 In Vitro Approaches

The first step towards the use of biotechnology for sunflower improvement is the development of technology for in vitro culture of the pertinent *Helianthus* species or inbred lines. If advances are to be made in transferring genetic information via interspecific hybrids between *H. annuus* and exotic species, then the in vitro methodology for embryo rescue needs to be developed. Anther or ovule culture methods might also aid in the development of new inbreds. For most other biotechnology applications (see Bajaj 1988), the development of methods for regenerating plants from callus, suspensions, and protoplasts is required. Callus and suspension cultures can be used for the selection of mutants. Once plant regeneration is routine, the production of new variants through somaclonal variation may be possible. Since *Agrobacterium tumefaciens* naturally infects sunflower, it provides a convenient method for gene transfer. Alternatively, protoplasts could be used for direct gene transfer. Finally, shoot tip or meristem culture would be useful for the clonal propagation of new inbreds, hybrids or varieties.

Crown Gall. Most reports of *H. annuus* tissue culture prior to 1970 concerned crown gall cells (see Table 3). Most of the crown gall work attempted to answer questions regarding the physiology of the tumor cells, in particular factors responsible for the initiation and maintenance of the growth of tumor cells.

White and Braun (1941) confirmed that sunflower tumor cells would continue to proliferate in the absence of the bacteria. A series of papers followed (Hildebrandt) et al. 1945, 1946; de Ropp 1947; Riker and Gutsche 1948; Hildebrandt and Riker 1949; Struckmeyer et al. 1949), which defined the optimal media conditions for sunflower tumor growth in tissue culture. Braun (1958) then suggested that the autonomous growth of sunflower tumor cells was a result of continual synthesis of growth substances. The current knowledge of the effects of the T-DNA genes of *A. tumefaciens*, especially on tobacco, (Leemans et al. 1982; Buchmann et al. 1985; Kemper et al. 1985) can explain the phenomena observed by the early sunflower researchers. Although *A. tumefaciens* will easily induce tumor formation on sunflower, there is no report of shoot regeneration from those tumors. The only organ differentiation as yet observed from sunflower tumors is root formation (Broekaert et al. 1976).

Normal (Nontumor) Callus. By the early 1950's nontumorous callus from *Helianthus annuus* had been induced by growing stem or hypocotyl segments on simple media with auxin (Kandler 1952; Henderson et al. 1952; Henderson 1954). The first attempts to establish sunflower callus that would differentiate into whole plants were in 1974. Rogers et al. (1974) successfully established callus from a cms sunflower line, but only roots were differentiated from this callus. The first report of successful plant regeneration from sunflower callus was by Sadhu (1974). By growing stem pith on a modified White's medium with 1 mg/l IAA, callus was induced. After 10 weeks, one piece of callus differentiated into several plantlets.

Regeneration from Callus. Following Sadhu's report (1974), Georgieva-Todorova et al. (1980) reported regeneration from callus of the sterile hybrid *H. annuus* ×

Table 3. In vitro culture of *Helianthus*. (see also Chap. IV.5, 6, 7, this Vol.)

Growth response	Seed source	Explant	Medium	Hormone	In vitro response	Reference
Crown gall (selected references)	*Helianthus annuus* L.	Secondary tumors	White's (1943)	None	Tumor	White and Braun (1941)
	H. annuus L. var. Giant Russian	Secodary petiolar crown galls	White's (1943) modified (hydrogen-ion concentration, sugars, vitamins)	None	Tumor	Hildebrandt et al. (1945, 1946)
	H. annuus L.	Secondary tumor	White's (1943)	IBA, NAA, IAA	Tumor	de Ropp (1947)
	H. annuus L. var. Giant Russian	Secondary petiolar crown gall	White's (1943) modified (nitrogen)	None	Tumor	Riker and Gutsche (1948)
	H. annuus L. var. Giant Russian	Secondary crown gall	White's (1943) modified (carbon compounds)	None	Tumor	Hildebrandt and Riker (1949)
	H. annuus L. var. Giant Russian	Secondary petiolar crown-gall	White (1943)	IAA, IBA, PCPA	Tumor	Struckmeyer et al. (1949)
	Sunflower	Secondary tumors	White's (1943) modified (inositol, amino acids)	NAA, kin	Tumor	Braun (1958)
	H. annuus cv. nanus fl. pl. lage Sungold cv. giant Russian cv. uniflorus giganteus cv. Dubbele gele van California	Hypocotyl segments	MS (1962)	NAA (5×10^{-6} M) kin (10^{-6} M)	Tumor, roots	Broekaert et al. (1976)
Normal callus-nonregenerating (selected references)	*H. annuus*	Stem	Gautheret (1942)	IAA (1 mg/l)	Callus	Kandler (1952)
	H. annuus var. Giant Russian	Stem	Gautheret (1942), White's (1943), (adenine, vitamins, casein hydrolysate, coconut milk)	IAA (0.01 – 0.1 mg/l)	Callus	Henderson et al. (1952)
	H. annuus	Stem, hypocotyl	Gautheret (1942) (coconut milk, adenine sulfate, casein hydrolysate, vitamins, sugars)	IAA (10^{-5} 10-8M)	Callus	Henderson (1954)
	H. annuus, line 232	Stem sections 2 mo. old plants	MS, 3% sucrose, 10% coconut milk, 5% Eagle TC vitamins (1955), 5 mg/l yeast extract	NAA (1 mg/l) kin (1 mg/l) 2,4-D (5 mg/l)	Callus, roots	Rogers et al. (1974)

Table 3 (continued)

Growth response	Seed source	Explant	Medium	Hormone	In vitro response	Reference
Normal callus-regenerating and direct regeneration	Sunflower	Stem pith 2 mo. old plants	White's (1963) modified	IAA (1 mg/l)	Callus, plants	Sadhu (1974)
	H. annuus × *H. decapetalus*	Stem parenchyma	MS (White's vitamins, adenine sulfate, casein hydrolysate, inositol)	NAA (0.1 mg/l) BA (0.1 mg/l) GA (0.01 mg/l)	Callus, plants	Georgieva-Todorova et al. (1980)
	H. annuus-synthethic, variety, Sannace	Cotyledons, leaves, shoot apices, hypocotyl segments	MS	BA (0.1 – 5.0 mg/l) 2,4-D (1 – 5 mg/l)	Callus, plants	Greco et al. (1984)
	H. arizonensis × *annuus* P21 and *H. annuus* 89	Immature embryos	B5 Gamborg (1968) or MS	NAA (0.05 – 0.3 mg/l)	Callus, plants	J.M. Chandler (pers. commun.)
	H. annuus SS415	Hypocotyls	MS (casamino acids)	NAA (1 mg/l) BA (1 mg/l) GA (0.1 mg/l)	Callus, plants	Paterson and Everett (1985)
	H. annuus × *H. decapetalus* *H. annuus* × *H. hirsutus*	Stem parenchyma shoot apices	MS	NAA (0.1 mg/l) BA (0.1 mg/l)	Callus	Bohorova et al. (1985)
	H. annuus × *H. tomentosus*		MS, glutamine (800 mg/l), asparagine (800 mg/l)	NAA (0.1 mg/l) BA (0.1 mg/l) GA (0.01 mg/l)	Plants from callus	
	H. annuus HA40, HA699, hybrid C204	Immature embryos	MS, B5 vitamins, 12% sucrose	2,4-D (1 mg/l)	Somatic embryos, plants	Finer (1987)
	H. annuus Argentario	Hypocotyls	MS	NAA (0 – 0.01 mg/l) BA (0.1 – 2 mg/l)	Adventitious buds or callus, buds, plants	Lupi et al. (1987)
	H. annuus Mammoth Russian, SIGCO-HO849, 455, 468, 461D	Immature zygotic embryo cotyledons	MS (adenine sulfate, glucose or sucrose)	NAA (0.5 mg/l) BA (0.1 mg/l)	Adventitious shoots, plants	Power (1987)
	H. annuus SS415, SS410	Hypocotyls	MS	2,4-D (0.1 mg/l) GA (0.1 mg/l)	Transformed callus	Everett et al. (1987)
				BA (1 mg/l) GA (0.1 mg/l)	Transformed plants	

	H. annuus T76B, HA89B, HA291B, HA300B, HA303B, RT26	Immature embryo hypocotyls	MS (inositol, alanine, glutamine, serine, tryptophane, cysteine) 9% sucrose	BA (0.5 – 1 mg/l)	Somatic embryos, plants	Freyssinet and Freyssinet (1988)
	H. annuus inbreds	Immature embryos	Chandler and Beard, B5 salts 3 – 12% sucrose	2,4-D (10 μM) ABA (5 μM) ABA (3 μM) BA (1 μM) IAA (0.3 μM) kin (5 μM)	Irregular lobes Meristems Shoots	McCann (1988)
	H. annuus BHA 89, RHA 274, R299, RHA880, CSIRO breeding lines, hybrids – A124, R299	Immature embryos	MS, 3% sucrose	BAP (1 mg/l) 3 weeks BAP (1 mg/l) IAA (0.5 mg/l)	Plants	Witrzens et al. (1988)
	H. tuberosus × *H. annuus*	Tuber	MS (except CS major salts)	BAP (1 mg/l) IAA (0.5 mg/l)	Plants	
Shoot tip culture	*H. annuus* L. var. Mammoth Russian	Shoot tip ± cotyledons	White's (1943) modified	None	Premature flowering	Henrickson (1954)
	Sunflower	1/2 shoot tips ± cotyledons	Hoagland's (1950)	None	Buds, plants	Pawlowski (1963)
	H. annuus 816/b	Shoot tip + hypocotyl	MS	NAA (0.1 – 3.0 mg/l) BA (0.1 – 1.0 mg/l)	Buds, plants premature flowering	Paal et al. (1981)
	H. annuus var. Issanka var. Record Blueberry = F10 × V135	Stem tips	MS	NAA (0.5 μg/l) BA (0.5 μg/l)	Buds, plants	Trifi et al. (1981)
	H. annuus 89, SS405, 109 experimental inbreds	Shoot tip half or whole	MS	kin (1 – 6 mg/l) BA (1 – 6 mg/l)	Multiple shoots, adventitious shoots, plants	Paterson (1984)
	H. annuus Argentario	Shoot tips	MS	kin (2 mg/l) GA (5 mg/l) or BA (2 mg/l) GA (10 mg/l)	Multiple shoots, plants	Lupi et al. (1987)

Table 3 (continued)

Growth response	Seed source	Explant	Medium	Hormone	In vitro response	Reference
Embryo culture	*H. annuus* × *H. hirsutus* *H. scaberimus* × *H. annuus*	Hybrid seed (immature embryos)	White's (1963) (sucrose, glucose, yeast extract, casein hydrolysate)	GA (0.01 mg/l) kin (0.05 mg/l) NAA (0.05 mg/l)	Interspecific hybrid plants	Bohorova (1982) Bohorova et al. (1985)
	Many *Helianthus* interspecific hybrids	Immature embryos	Modified B5	NAA (0.05 mg/l)	Interspecific hybrid plants	Chandler (1979) Chandler and Beard (1983)
	H. annuus 6 inbred lines	Immature embryos (3 – 14 days postpollination)	Modified MS (HD) or B5 (2 mm – 9% sucrose, low nitrogen; rooting – no inositol, 1% sucrose		Plants	Espinasse et al. (1985)
	H. annuus *H. anomalus* *H. resinousus* *H. annuus* × *H. anomalus*	Immature embryos (3 – 7 days old)	1/2 × MS, 3% sucrose (thiamine, pyridoxine, biotin, nicotinic acid, glycine)	NAA (0.01 mg/l) BAP (0.01 mg/l)	Hybrid plantlets	Assad et al. (1986)
Anther or ovule culture	*H. annuus* var. Armavirskii 3497 var. Magak 2 local populations 2 mutant lines	Anthers with mature pollen, not opened	Modified medium H White's MS Nitsch	IAA (0.8 mg/l) kin (2 mg/l) IBA (1.5 mg/l) or NAA (1.5 mg/l)	Plantlets	Plotnikov (1975)
	H. annuus cv. Chernianka cv. Admaniric cv. Jupiter	Anthers	modified Blaydes (1966)	IAA (1 – 2 mg/l) 2,4-D (1 – 2 mg/l) kin (2 – 4 mg/l)	Callus	Tzen and Lin (1975)
	H. annuus × *H. decapetalus*	Anthers (uninucleate microspores)	MS (glutamine, asparagine)	Z (5 mg/l)	Plants	Bohorova et al. (1980)

Interspecific hybrids with *H. annuus*	Anthers	MS	NAA (0.1 – 0.5 mg/l) Z (4 mg/l) BAP (0.1 – 0.5 mg/l) kin (0.5 mg/l) IAA (0.5 mg/l)	Callus, plants	Assad et al. (1985)
Many *H. annuus* species and hybrids	Uninucleate microspores	MS	kin (0.2 mg/l) 2,4-D (1 mg/l)	Callus	Bohorova et al. (1985)
H. annuus × *H. decapetalus* *H. divariratus*	Uninucleate microspores	MS	Z (5 mg/l)	Direct shoot formation	Bohorova et al. (1985)
H. annuus Inra, Lucide 3 French varieties	Anthers, ovaries	MS, thiamine, 6% sucrose	IAA (1 mg/l) NAA (2 mg/l) 2,4-D (0.1 mg/l) BAP (2 mg/l)	Haploid plants	Mix (1985)
H. annuus – 8 cultivars	Young flowers, ovules (pre-fertilized)	N6, 6% sucrose	MCPA (0.5 – 2 mg/l)	Haploid and diploid embryoids	Yang et al. (1986)

Table 4. Nitrogen effect on regeneration from SS415 hypocotyl callus (Paterson and Everett 1985)

NH_4 mM	KNO_3 mM	Regeneration (embryos/segment) (%) ± S.E.
20	40	19 ± 5
20	70	74 ± 7
20	90	100 ± 10[a]
20	110	68 ± 11
0	90	0
10	90	65 ± 7
20	90	100 ± 10
30	90	67 ± 10

[a] One hundred percent corresponds to 9.8 ± 1.0 embryos per hypocotyl segment callus.

H. decapetalus. Their work showed that (1) a more complex regeneration medium was required for regeneration from the sterile hybrids compared to Sadhu's undefined sunflower material, and (2) there were differences between the responses of the hybrids in culture. Bohorova et al. (1985) demonstrated regeneration from callus of *H. annuus* × *H. hirsutus* and *H. annuus* × *H. tomentosus*. The importance of the "genotype" of the sunflower used has also been confirmed using inbred material (Paterson and Everett 1985). In this study a large number of oil-producing inbred lines were screened for regeneration potential and a regeneration medium was optimized for an oil-producing inbred line (SS415) with very high regeneration potential. Table 4 shows the importance of KNO_3 in the regeneration from SS415 callus. Increasing the KNO_3 concentration to 90 mM stimulated regeneration of somatic embryos. Also reported here was evidence for somatic embryogenesis as the means of regeneration in oil-producing inbreds (Fig. 1 B).

Greco et al. (1984) have shown that callus capable of regeneration can be obtained from every part of seedlings (except roots) of the synthetic variety Sannance. Using immature embryos as an explant, plant regeneration has been obtained from the important oil-producing inbred HA89 and an interspecific hybrid, *H. arizonensis* × *H. annuus* P21 (J. M. Chandler 1985, pers. commun; see Fig. 1 A, C).

Plant regeneration in sunflower has now been reported from oil-producing inbreds (Paterson and Everett 1985; Power 1987; Freyssinet and Freyssinet 1988; McCann 1988; Witrzens et al. 1988), synthetic varieties (Greco et al. 1984; Lupi 1987), hybrids (Finer 1987; Witrzens et al. 1988) and interspecific hybrids (Georgieva-Todorova et al. 1980; Bohorova et al. 1985; Witrzens 1988). Most of the procedures used a simple medium (generally MS salts and vitamins) with either only an auxin (NAA), a cytokinin with or without an auxin (6-BAP ± 2.4-D), or an auxin, a cytokinin and a gibberellin (NAA, BA and GA_3) used together. The addition of casamino acids (100–500 mg/l) and adenine sulfate (40 mg/l) or glutamine (800 mg/l) and asparagine (800 mg/l) may also be useful (Georgieva-Todorova et al. 1980; Paterson and Everett 1985; Bohorova et al. 1985).

The explant source for callus capable of plant regeneration has tended to be young tissue. Success has been obtained from immature embryos (McCann

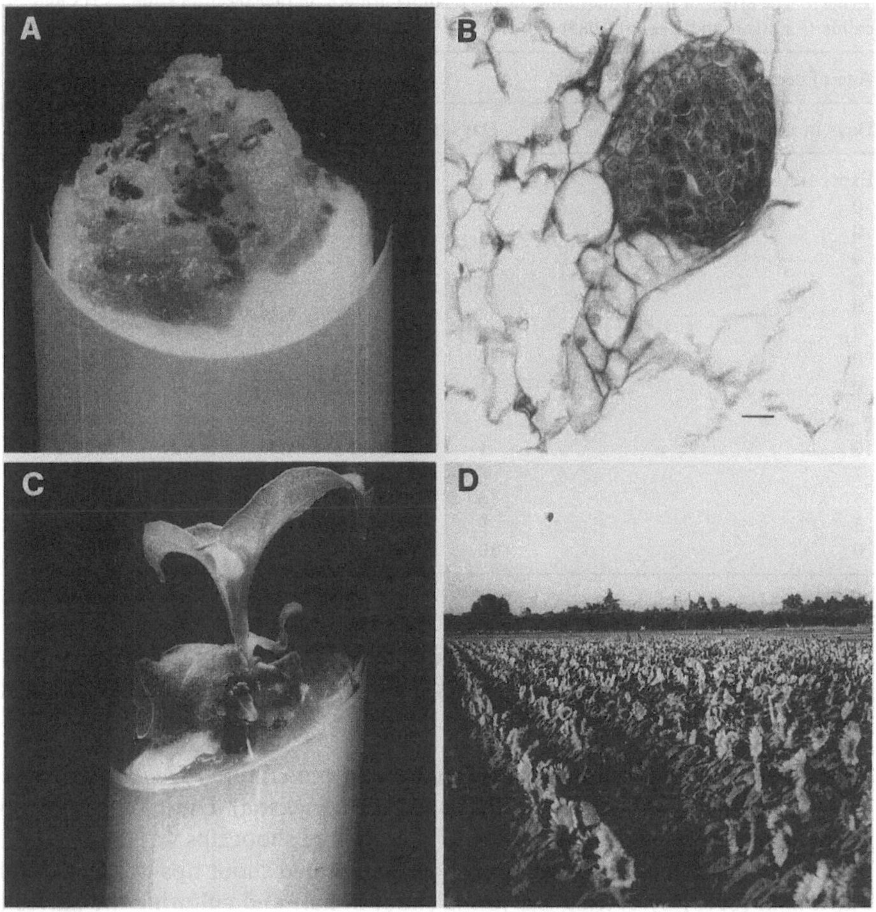

Fig. 1 A–D. Regeneraton of *Helianthus* plants from callus. **(A)** Dark green somatic embryos regeneration from callus (*H. cusickii* × *H. annuus* P21) growing on MS medium containing 2 mg/l IAA, 2 mg/l K and 3% sucrose (×2.5). **(B)** 10 µm section through *H. annuus* SS415 callus including a longitudinal section of a globular stage somatic embryo. (Bar = 100 µm). **(C)** Rooting regenerated plant from *H. mollis*×*H. annuus* p21 (×1.8). **(D)** Field of progeny obtained after two generations of self-pollinating plants regenerated from *H. annuus* HA 89. **(A, C, D)** courtesy of Dr. John Chandler, University of California, Davis. **B** Paterson and Everett (1985)

1988; Witrzens et al. 1988), stem parenchyma (Georgieva-Todorova et al. 1980; Bohorova et al. 1985), seedling hypocotyls (Greco et al. 1984; Paterson and Everett 1985; Lupi et al. 1987; Everett et al. 1987) seedling cotyledons, leaves and shoot apices (Greco et al. 1984; Bohorova et al. 1985) and tubers (Witrzens et al. 1988). In one study (Paterson and Everett 1985), the effect of the physiological state of the explant source on the subsequent regeneration from callus has been demonstrated. If hypocotyl segments were used as explants, then regenerating callus was obtained only if the seedlings were grown in the light for more than 5 days (Table 5). This physiological state appears to be related to ethylene sensitivity in the seedling tissue (Robinson and Adams 1987).

Table 5. The effect of light conditions during seedling growth on regeneration from SS415 hypocotyl callus (Paterson and Everett 1985)

Age of seedling		Regeneration (embryos/piece) ± S.E.
Days in dark	Days in light	
Expt. 1		
0	5	0.1 ± 0.1
0	7	6.9 ± 0.8
0	9	6.3 ± 0.8
0	12	4.6 ± 0.5
0	14	6.0 ± 0.8
0	16	8.0 ± 0.9
0	19	5.8 ± 0.9
0	21	8.5 ± 0.7
Expt. 2		
10	1	0
7	4	0.7
5	6	5.2
3	8	4.7
0	10	11.2

Direct plant regeneration can also be obtained from immature embryos (Finer 1987; Power 1987; Freyssinet and Freyssinet 1988) and cultured leaves (Paterson 1984).

Shoot Tip Culture. Plants were first recovered from cultured *Helianthus annuus* L. var. Mammoth Russian shoot tips in 1954 (Henrickson). During this study, the phenomenon of premature flowering of the cultured shoot tips was observed. The first multiplication of sunflower plants using cultured shoot tips was achieved by Pawlowski (1963) by cutting the young shoot in half and culturing the halves on Hoagland's solution containing agar. The effect of an auxin (NAA) and a cytokinin (BA) on cultured shoot tips was described by Paal et al. (1981). Their study, using *H. annuus* var. 816/b, showed a stimulation of shoot growth by NAA and induction of callus with BA. They also noted the rapid flowering of cultured shoots. Taking advantage of the axillary buds by subculturing nodes from cultured shoot tips resulted in multiplication of two *H. annuus* varieties (Issanka and Record) and one genetic hybrid (Blueberry = $F_{10} \times V_{135}$) (Trifi et al. 1981). This method was successful until the shoots began to flower. Also noted in this study, was the formation of some abnormal rosette plants.

A large number of oil-producing inbred lines can be clonally propagated by exposing half-shoot apices to BA or kin (Paterson 1984). One synthetic variety will produce multiple shoots when exposed to BA or kin plus GA (Lupi et al. 1987). Rooting of the multiple shoots is difficult and premature flowering also occurs. These problems are not due to ethylene (Robinson et al. 1987), but can be overcome by treatment with phlorizin (Witrzens et al. 1988).

Embryo Culture. There are several reported examples of embryo rescue techniques for inbreds and interspecific hybrids of *Helianthus*. Plants were obtained from the two interspecific hybrids, *H. annuus*×*H. hirsutus* and *H. scaberimus*×*H. annuus*, by culturing the rescued embryos on a modified White's medium (1.5% sucrose, 1.5% glucose, 200 mg/l yeast extract, 200 mg/l casein hydrolysate, 0.01 mg/l GA, 0.05 mg/l kin and 0.05 mg/l NAA; Bohorova 1982). Fifty-three different interspecific hybrids of *Helianthus* species were obtained by a different embryo rescue technique (Chandler 1979; Chandler and Beard 1983). This method has two steps: (1) an embryo growth medium and (2) an embryo germination medium. The embryo growth medium, used for the first 1 to 2 weeks of culture, consisted of a modified B5 medium (1 mg/l nicotinic acid, 10 mg/l thiamine·HCl, 1 mg/l pyridoxin·HCl, 4000 mg/l myoinsitol, 1000 mg/l L-alanine, 800 mg/l L-glutamine, 160 mg/l L-serine, 50 mg/l L-tryptophan, 10 mg/l cysteine, 0.05 mg/l NAA, 120 mg/l sucrose and 7 mg/l agar). The embryo germination medium consisted of B5 salts supplemented with 10 mg/l sucrose. Assad et al. (1986) have recovered hybrid plantlets from *H. annuus*×*H. anomalus* using a modified MS medium. Using six inbred lines of *H. annuus*, Espinasse et al. (1985) showed that embryos less than 2 mm in length require a low nitrogen medium with 9% sucrose.

Anther Culture. Callus was obtained from three cultivars of *H. annuus* (Chernianka, Admaniric, and Jupiter) using a modified Blaydes' medium (1−2 mg/l IAA, 1−2 mg/l 2,4-D, 2−4 mg/l kin), but no regeneration was seen (Tzen and Lin 1975). Regeneration of plantlets from cultured anthers of a *H. annuus* variety used a modified H medium, White's, Nitsch or MS medium (Plotnikov 1975). Shoots were regenerated but the plantlets did not survive to maturity. Haploid plantlets have also been regenerated from *H. annuus* Inra and Luciole (Mix 1985).

Plants from *H. annuus*×*H. decapetalus* were obtained through direct androgenesis on MS medium with 5 mg/l zeatin (Bohorova et al. 1980, 1985). Regeneration of plants either directly or through callus from anthers can be accomplished for a number of *Helianthus* species and interspecific hybrids using a modified MS medium (3% sucrose, 0.8% agar, 1 mg/l thiamine, 1 mg/l pyridoxine, 0.01 mg/l biotin, 1 mg/l nicotinic acid, 100 mg/l inositol, 1 mg/l glycine and either 0.1−0.5 mg/l NAA+0.1−0.5 mg/l BAP, 0.5 mg/l IAA+0.5 mg/l kin +0.5 mg/l BAP or 0.5 mg/l NAA+4 mg/l z; Assad et al. (1985); for details see Chap. IV.7, this Vol.).

Ovule Culture. Ovules or ovaries of *H. annuus* can also be a potential source of haploid plants (Mix 1985; Yang et al. 1986; see Chap. IV.6, this Vol.).

Protoplasts. There are several reports of protoplast isolation (Table 6) from sunflower leaves (Wegmann and Muhlbach 1973) or suspension cultures (Bottger et al. 1978) for use in physiological studies.

Some protoplasts are capable of plant regeneration. Binding et al. (1981) have reported successful regeneration of plants from *H. annuus* using cultured shoot tips as an explant source. The shoots were maintained on B5 media with 2.5 μM 6-BA. Protoplasts were isolated using the following solution: 0.3% Mazerozyme, 1.5% Meicellase, 0.5 M mannitol, and 5 mM Ca $(NO_3)_2$. The protoplast-derived plants were regenerated on V-KM or B5 media with 15 μM 6-BA. Flowering in

Table 6. Protoplast isolation and culture

Seed and explant source	Enyzme solution	Culture medium	Hormones	Reference
Sunflower, leaves	Pectinase (0.5%) Cellulase (5%) Sucrose (20%)	None	None	Wegmann and Muhlbach (1973)
H. annuus cv. Bismarckianum, hypocotyl	Mannitol (0.35 M) Gentamycin (6 mg/l) $Ca(NO_3)_2$ (100 mg/l) KCl (50 mg/l) Sucrose (1 mg/l) Cellulase (20 mg/l) Pectinase acid transeleinase (5 mg/l) Pectinol (500 mg/l) Rohamint (2.5 mg/)l	None	None	Bottger et al. (1978)
H. annuus, cultured shoot tips	Mazerozyme (0.3%) Meicellase (1.5%) Mannitol (0.5 M) Ca $(NO_3)_2$ (5 mM)	V-KM or B5	BA (15 μM)	Binding et al. (1981)
H. annuus, TMA, RHA 2 hybrids – Rodeo, Bolero *H. argophyllus*, hypocotyl	KCl (330 mM) $CaCl_2$ (14 mM) Macerozyme R-10 (0.02%) Onozuka R-10 (0.1%) Driselase (0.05%)	T_0	NAA (3 mg/l) BA (1 mg/l)	Lenee and Chupeau (1985)
H. annuus (4 cultivars) *H. praecox* *H. scaberimus* *H. rigidus*	CPW salts Mannitol (13%) Rhozyme HP150 (2%) Meicelase (2%) Macerozyme R10 (0.03%)	K8P	NAA (0.5 mg/l) 6-BAP (0.5 mg/l)	Bohorova et al. (1986)
H. annuus, TMA1 + 51, RHA 274, 2 hybrids – Rodeo, Bolero	KCl (25 mg/l) $CaCl_2$ (2 mg/l) Bromocresol purple (8.1 mg/l) MES (700 mg/l) Cellulase-Onozuka R-10 (0.1%) Driselase (0.05%) Macerozyme R-10 (0.02%)	LB	NAA (0.1 mg/l) BAP (1 mg/l)	Lenee and Chupeau (1986)

vitro was reported. The quality of the protoplast-derived callus is highly dependent on genotype, but sunflower protoplasts with good plating efficiencies can be obtained with responsive genotypes on 8P media (J.–F. Jia and I. Potrykus 1981 pers. commun.).

Hypocotyls from seedlings germinated in vitro were the only source of dividing protoplasts from two lines of *H. annuus* (TMA and RHA), two hybrids (Rodeo and Bolereo) and one wild species (*H. argophyllus*) (Lenee and Chupeau 1985). The protoplasts were isolated in Macerozyme (0.02%), Cellulase Onozuka R-10 (0.1%), Driselase (0.05%), KCl (330 mM) and $CaCl_2$ (14 mM). Using this procedure, a 60% plating efficiency was obtained after 10 days on medium with only reduced nitrogen. The final plating efficiency was 5%. No plants were regenerated. Lenee and Chupeau (1986) modified their procedure using different enzymes with similar results. Using the agarose bead culture method with K8P medium, callus colonies have been recovered from cotyledon and hypocotyl protoplasts (Bohorova et al. 1986). The timing of lowering the osmotic pressure of the media was critical. Callus transferred to medium with 0.5 mg/l NAA and 0.5 mg/l 6-BAP organized into meristem-like structures.

Genetic Stability in Regenerated Plants. No studies have been reported which indicate what the rate of somaclonal variation in *Helianthus* might be. There is one study in progress designed to answer this question using the inbred line HA 89 (J.M. Chandler 1985, pers. commun.). In this study, 41 plants have been regenerated from immature embryos, selfed twice, then field-grown (Fig. 4); nine heritable mutants were identified.

Selection for Herbicide-Resistant Plants. Sunflower plants that are potentially resistant to the herbicide Glean (chlorsulphuron) have been selected using callus from the interspecific hybrid *H. arizonensis* × *H. annuus* P21 (J.M. Chandler 1985, pers. commun.). The callus used was initiated from immature embryos and maintained on MS media with 3% sucrose and NAA (0.3 mg/l). To initiate selections for herbicide resistance, the callus was agitated in liquid medium for 3–5 days and the smaller cell aggregates plated on solid medium with 10 mg/l chlorsulphuron. After 1 month, resistant colonies were removed. These colonies had more than a ten fold increase in tolerance to the herbicide. Preliminary data on plants regenerated from the resistant colonies suggest that the plants also have a ten fold increase in herbicide tolerance.

Developmental Studies. In vitro culture of leaves and leaf primordia of *Helianthus annuus* demonstrated that leaf determination occurs very early in this species (Steeves et al. 1957). Callus cultures from *H. annuus* seedling internodes were used to show that phloem developed prior to xylem (Aloni 1980). The pattern of re-establishing the quiescent cells of the shoot apex was examined by culturing surgically bisected *H. annuus* shoot apices (Davis and Steeves 1979).

The oil content and quality of maturing sunflower seed are sensitive to environmental factors, especially temperature. The effect of temperature on linoleic acid synthesis has been examined by culturing immature seeds of *H. annuus* Sunfola 68-3 (Silver et al. 1984). The results suggest that immature seed culture could be used to select for genotypes with temperature-stable linoleic acid synthesis. Also, using cultured embryos, the temperature effect on fatty acid synthesis was shown to be due to desaturation of newly synthesized oleate (Silver et al. 1984). Variation in fatty acid composition between individual embryos was also observed, suggesting the potential for selection for fatty acid composition.

3 Transformation and Genetic Engineering

Sunflower is naturally susceptible to transformation by *Agrobacterium tumefaciens* and much of the early tissue culture research was performed on sunflower crown gall tissue (see Table 3). There is some evidence for differences between inbred lines in their susceptibility to transformation by strains of *A. tumefaciens* harboring particular Ti plasmids (Buchholz and Thomashow 1984). These differences may be related to the production of factors that can act as enhancers of crown-gall tumor initiation (Bouckaert-Urban and Vendrig 1982). It has been shown that *A. tumefaciens* T-DNA integrates into multiple sites of the sunflower crown gall genome (Ursic et al. 1983), and that the T-DNA is transcribed into at least seven polyadenylated RNAs (Muray and Kemp 1982). The mutation in *A. tumefaciens* strain A66 which leads to slow tumor growth and shoot production in tobacco has no such effect in sunflower (Binns et al. 1982). Because of the ease with which tumors can be induced on sunflower seedlings, sunflower tumors have been used to test the expression of a number of foreign genes in plant cells. As might be expected, neomycin phosphotransferase (*neo*) gene cassettes using promotor sequences from Ti plasmid T-DNA genes are expressed efficiently in sunflower tumors (Gelvin et al. 1985; Nutter et al. 1987). The expression of neo confers antibiotic resistance on the sunflower cells, thus providing a selectable marker of transformation. The expression of a phaseolin seed storage protein gene from bean (*Phaseolus vulgaris* L.) in sunflower tumors provides an example of introduction and expression of a gene from a taxonomically distinct botanical family (Murai et al. 1983). Similarly, the transcription of a zein gene (from maize) in sunflower tumors (Matzke et al. 1984) provides evidence that monocot genes, under the influence of their own transcriptional control sequences, can be transcribed in a dicot host plant. The fact that neither of the proteins encoded by these two genes accumulated to significant levels in sunflower tumors may be more a reflection of post-translational degradation of specialized proteins synthesized in inappropriate tissues than inefficient gene transcription or translation (Sengupta-Gopalan et al. 1985). Thus far there is only one report of regeneration of plants from transformed sunflower cells (Everett et al. 1987). In order to achieve transformation, aseptically grown sunflower seedlings were inoculated with *A. tumefaciens* (LBA 288) containing a chimeric *neo* gene and co-cultured for 7 days. Hypocotyl segments were than excised and cultured on MS medium (0.1 mg/l 2,4-D and 0.1 mg/l GA) supplemented with 200 μg/ml carbenicillin and 200 μg/ml kanamycin. Once sufficient transformed callus had grown, plant regeneration was initiated by transfer to media with 1 mg/l BA and 0.01 mg/l GA. Kanamycin was found to be inhibitory to plant regeneration from kanamycin-resistant callus.

In the analysis of the first backcross generation, both kanamycin resistance and NPT activity showed a 1:1 segregation, as expected for the Mendelian inheritance of a single dominant marker gene. The variation in the level of expression of *neo* between transformants was similar to that observed in other plant species.

The major current limitations to sunflower improvement via genetic engineering are the lack of availability of cloned genes that are likely to confer agronomically useful traits, and the adaptation of tissue culture techniques to a wider range of germplasm in current breeding programs. Another factor that will influence when the first genetically engineered sunflower hybrids are available commercially is the need to carefully assess potential ecological effects on the environment prior to release. This would particularly be the case with an example such as herbicide-resistant sunflower, because cultivated sunflower can interbreed with some of its wild relatives.

4 Conclusions and Prospects

From the information reviewed here, it can be concluded that most of the techniques required for biotechnology to impact on sunflower breeding are now available for at least some inbred lines. Embryo rescue techniques are almost certainly going to assist the use of wild sunflower germplasm as a source of traits such as disease resistance, drought tolerance, and oil quality. Whether the screening of regenerated plants for somaclonal variation can provide an additional source of novel and useful variation is yet to be established. However, it is clear that if the appropriate selection pressure can be applied to tissue cultures (e.g., herbicide resistance), mutants can be obtained.

Because of its natural susceptibility to transformation by *A. tumefaciens*, sunflower could be one of the first crops to benefit from genetic engineering. A number of foreign genes have been expressed in sunflower cells using tumor-forming strains of *A. tumefaciens*. More recently, the production of fertile transgenic plants of an oil-producing inbred line has been demonstrated. The results obtained with these first genetically engineered sunflower plants suggest that, once introduced, the foreign genes will be inherited and expressed according to the same principles established for other dicotyledonous plant species.

The important future areas of research include (1) the introduction of new agronomically useful traits, and (2) the development of a genotype-independent method of transformation.

References

Aloni R (1980) Role of auxin and sucrose in the differentiation of sieve and tracheary elements in plant tissue cultures. Planta 150:255–263

Assad A, Serieys H, Jonard R (1985) Sur les possibilités de régénération d'espèces sauvages et d'hybrides interspecifiques du genre *Helianthus* par androgénèse in vitro. CR Acad Sci Paris t. 300, Ser III:25–30

Assad A, Jonard R, Serieys H, Vincourt P (1986) La culture d'embryons isolés in vitro dans un programme d'amélioration du Tournesol. CR Acad Sci Paris t. 302, Ser III 5:161–164
Bajaj YPS (ed) (1988) Biotechnology in agriculture and forestry, vol 6. Crops II. Springer, Berlin Heidelberg New York Tokyo
Binding H, Nehls R, Kock R, Finger J, Mordhorst G (1981) Comparative studies on protoplast regeneration in herbaceous species of the dicotyledonae class. Z Pflanzenphysiol 101:119–130
Binns AN, Sciaky D, Wood HN (1982) Variation in hormone autonomy and regenerative potential of cells transformed by strain A66 of *Agrobacterium tumefaciens*. Cell 31:605–612
Blaydes DF (1966) Interaction of kinetin and various inhibitors in the growth of soybean tissue. Physiol Plant 19:748–753
Bohorova NE (1982) In vitro plant development of seeds of *Helianthus* interspecific hybrids. Biologie, genetique. C.R. Acad Sci Bulgar 35:105–107
Bohorova NE, Atanassov AI, Antonova GP (1980) In vitro isolation of anthers from interspecific hybrids in the *Helianthus* genus. Biologie, genetique. CR Acad Sci Bulgar 33:1545–1548
Bohorova N, Atanassov A, Georgieva-Todorova J (1985) In vitro organogenesis and embryo culture, in the genus *Helianthus* L. Z Pflanzenzuecht 95:35–44
Bohorova NE, Cocking EC, Power JB (1986) Isolation, culture and callus regeneration of protoplasts of wild and cultivated *Helianthus* species. Plant Cell Rep 5:256–258
Bottger M, Engvild KC, Soll H (1978) Growth of *Avena* coleoptiles and pH drop of protoplast suspensions induced by chlorinated indoleacetic acids. Planta 140:89–92
Bouckaert-Urban A-M, Vendrig JC (1982) Enhancement of the crown-gall tumor initiation by extract fractions from tumor forming cotyledons of *Helianthus giganteus* L. Z Pflanzenphysiol 105:395–405
Braun AC (1958) A physiological basis for autonomous growth of the crown-gall tumor cell. Proc Natl Acad Sci USA 44:344–349
Broekaert D, Cazier C, van Parijs R (1976) Histophotometric DNA measurements in sunflower seedlings, crown gall tumours and habituated callus tissues. J Exp Bot 27:532–540
Buchholz WG, Thomashow MF (1984) Host range encoded by the *Agrobacterium tumefaciens* tumor-inducing plasmid pTiAg63 can be expanded by modification of its T-DNA oncogene complement. J Bacteriol 160:327–332
Buchmann I, Marner F-J, Schroder G, Waffenschmidt S, Schroder J (1985) Tumor genes in plants: T-DNA encoded cytokinin biosynthesis. EMBO J 4:853–859
Bye S (1986) In: The Sunflower, D Lilleboe (ed), National Sunflower Association, Vol. 12, No. 1, p 28
Chandler JM (1979) Sunflower interspecific hybridization using embryo culture. MS Thesis, Univ California, Davis
Chandler JM, Beard BH (1983) Embryo culture of *Helianthus* hybrids. Crop Sci 23:1004–1007
Davis EL, Steeves TA (1979) Experimental studies on the shoot apex of *Helianthus annuus*: the effect of surgical bisection on quiescent cells in the apex. Can J Bot 55:606–614
de Ropp RS (1947) The response of normal plant tissues and of crown-gall tumor tissues to synthetic growth hormones. Am J Bot 34:53–62
Eagle H (1955) Nutrition needs of mammalian cells in tissue culture. Science 122:501–504
Espinasse A, Lay C, Dybing CD (1985) Factors controlling in vitro development of sunflower embryos. Agronomie 5:825–832
Everett NP, Robinson KEP, Mascarenhas D (1987) Genetic engineering of sunflower (*Helianthus annuus* L.) Bio/Technology 5:1201–1204
Fick GN (1978) Breeding and genetics. In: Carter JF (ed) Sunflower Science and Technology. Am Soc Agron, Crop Sci Soc Am Soil Sci Soc Am, Madison, Wisconsin, USA, pp 279–338
Finer JJ (1987) Direct somatic embryogenesis and plant regeneration from immature embryos of hybrid sunflower (*Helianthus annuus* L.) on a high sucrose-containing medium. Plant Cell Rep 6:372–374
FAO (1984) Production Yearbook vol 38, Food and Agriculture Organization of the United Nations, Rome
Freyssinet M, Freyssinet G (1988) Fertile plant regeneration from sunflower (*Helianthus annuus* L.) immature embryos. Plant Sci 56:177–181
Gamborg OL, Miller RA, Ojima K (1968) Nutrient requirements of suspension cultures of soybean root cells. Exp Cell Res 50:151–158
Gautheret RJ (1942) Hetero-auxines et cultures de tissus végétaux. Bull Soc Chim Biol 24:13–41

Gelvin SB, Karcher SJ, Goldsbrough PB (1985) Use of a T_R T-DNA promoter to express genes in plants and bacteria. Mol Gen Genet 199:240–248
Georgieva-Todorova J, Bohorova N, Atanassov A (1980) Utilization of the method of tissue culture in the interspecific hybrids of *Helianthus* L. Proc 9th Int Sunflower Conf, Torremolinos, Spain, pp 122–128
Greco B, Tanzarella OA, Carrozzo G, Blanco A (1984) Callus induction and shoot regeneration in sunflower. (*Helianthus annuus* L.) Plant Sci Lett 36:73–77
Henderson JHM (1954) The changing nutritional pattern from normal to habituated sunflower callus tissue in vitro. Ann Biol 30:229–348
Henderson JHM, Durrell ME, Bonner J (1952) The culture of normal sunflower stem callus. Am J Bot 39:467–473
Henrickson CE (1954) The flowering of sunflower explants in aseptic culture. Plant Physiol 29:536–538
Hildebrandt AC, Riker AJ (1949) The influence of various carbon compounds on the growth of marigold, paris-daisy, periwinkle, sunflower and tobacco tissue in vitro. Am J Bot 36:74–85
Hildebrandt AC, Riker AJ, Duggar BM (1945) Growth in vitro of excised tobacco and sunflower tissue with different temperatures, hydrogen-ion concentrations and amounts of sugar. Am J Bot 32:357–361
Hildebrandt AC, Riker AJ, Duggar BM (1946) The influence of the composition of the medium on growth in vitro of excised tobacco and sunflower tissue cultures. Am J Bot 33:591–597
Hoagland DR, Arnon DI (1950) Calif Univ Agric Exp Sth Cir No 347
Kandler O (1952) Über eine physiologische Umstimmung von Sonnenblumenstengelgewebe durch Dauereinwirkung von –Indolylessigsäure. Planta 40:346–349
Kemper E, Waffenschmidt S, Weiler EW, Rausch T, Schroder J (1985) T-DNA-encoded auxin formation in crown-gall cells. Planta 163:257–262
Leemans J, Deblaere R, Willmitzer L, Degreve H, Hernalsteens JP, Van Montagu M, Schell J (1982) Genetic identification of functions of TL-DNA transcripts in octopine crown galls. EMBO J 1:147–152
Lenee P, Chupeau Y (1985) Callus formation from sunflower (*Helianthus annuus* L.) protolasts. In: Magnien E, de Nettancourt D (eds) Genetic engineering of plants and microorganisms important for agriculture. Junk, Dordrecht, pp 168–169
Lenee P, Chupeau Y (1986) Isolation and culture of sunflower protoplasts (*Helianthus annuus* L.): Factors influencing the viability of cell colonies derived from protoplasts. Plant Sci 43:69–75
Lilleboe D (1985a) Tracing the hybrid saga. Sunflower, March 1985:10–11
Lilleboe D (1985b) Wild species offer germplasm diversity. Sunflower, March 1985:18–20
Lupi MC, Bennici A, Locci F, Gennai D (1987) Plantlet formation from callus and shoot-tip culture of *Helianthus annuus* (L.). Plant Cell Tissue Org Cult 11:47–55
Matzke MA, Susani M, Binns AN, Lewis ED, Rubenstein, I., Matzke AJM (1984) Transcription of a zein gene introduced into sunflower using a Ti plasmid vector. EMBO J 3:1525–1531
McCann Wilcox A, Cooley G, van Dreser J (1988) A system for routine plantlet regeneration of sunflower (*Helianthus annuus* L.) from immature embryo-derived callus. Plant Cell Tissue Org Cult 14:103–110
Mix G (1985) Antheren- und Ovarienkultur von Sonnenblumen (*Helianthus annuus* L.) Landbauforsch Voelkenrode 35:153–156
Murai N, Kemp JD (1982) T-DNA of pTi-15955 from *Agrobacterium tumefaciens* is transcribed into a minimum of seven polyadenylated RNAs in a sunflower crown gall tumor. Nucl Acids Res 10:1679–1689
Murai N, Sutton DW, Murray MG, Slightom JL, Merlo DJ, Reichert NA, Sengupta-Gopalan C, Stock CA, Barker RF, Kemp JD, Hall TC (1983) Phaseolin gene from bean is expressed after transfer via tumor-inducing plasmid vectors. Science 222:476–482
Murashige T, Skoog F (1962) A revised medium for rapid growth and bioassays with tobacco tissue cultures. Physiol Plant 15:473–497
Nutter R, Everett N, Pierce D, Panganiban L, Okubara P, Lachmansingh R, Mascarenhas D, Welch H, Mettler I, Pomery L, Johnson J, Howard J (1987) Factors affecting the level of kanamycin resistance in sunflower cells. Plant Physiol 84:1185–1192
Paal HA, Kurnik E, Szabo L (1981) Napraforgo noveny-regeneracio in vitro hajtascscus-tenyeszetbol. Novenytermeles 30:201–208

Paterson KE (1984) Shoot tip culture of *Helianthus annuus*-flowering and development of adventitious and multiple shoots. Am J Bot 71:925–931

Paterson KE, Everett NP (1985) Regeneration of *Helianthus annuus* inbred plants from callus. Plant Sci 42:125–132

Pawlowski SH (1963) A method of obtaining genetically indentical sunflower plants. Can J Bot 41:743–746

Plotnikov VA (1975) Production of sunflower sprouts in a generative tissue culture. Severo-kavkazskii Nauchnyii Tsentr Vysshei Shkoly. Ives Estestvennye Nauki. Rostov-on-don 3:79–81

Power CJ (1987) Organogenesis from *Helianthus annuus* inbreds and hybrids from the cotyledons of zygotic embryos. Am J Bot 74:497–503

Putt ED (1978) History and present world status. In: Carter JF (ed) Sunflower science and technology. Am Soc Agron, Crop Sci Soc Am Soil Sci Soc Am, Madison, Wisconsin, USA, pp 1–25

Riker AJ, Gutsche AE (1948) The growth of sunflower tissue in vitro on synthetic media with various organic and inorganic sources of nitrogen. Am J Bot 35:227–238

Robinson RG (1978) Production and Culture. In: Carter JF (ed) Sunflower science and technology. Am Soc Agron, Crop Sci Soc Am Soil Sci Soc Am, Madison, Wisconsin, USA, pp 89–143

Robinson KEP, Adams DO (1987) The role of ethylene in the regeneration of *Helianthus annuus* (sunflower) plants from callus. Physiol Plant 71:151–156

Robinson KEP, Adams DO, Lee RY (1987) Differential physiological and morphological responses of inbred lines to the ethylene precursor 1-aminocyclopropane-1-carboxylic acid by cultured *Helianthus annuus* (sunflower) shoot tips. Plant Cell Rep 6:405–409

Rogers MA, Gal HL, Horner Jr HT (1974) Callus formation and differentiation in tissue cultures of normal and cytoplasmic male sterile sorghum, pepper, sunflower, and tobacco. In Vitro 9:463–467

Sadhu MK (1974) Effect of different auxins on growth and differentiation in callus tissue from sunflower stem pith. Indian J Exp Biol 12:110–111

Sengupta-Gopalan C, Reichert NA, Barker RF, Hall TC (1985) Developmentally regulated expression of the bean phaseolin gene in tobacco seed. Proc Natl Acad Sci USA 82:3320–3324

Silver JG, Rochester CP, Bishop DG, Harris HC (1984) Unsaturated fatty acid synthesis during the development of isolated sunflower (*Helianthus annuus* L.) seeds. Exp Bot 35:1507–1514

Steeves TA, Gabriel HP, Steeves MW (1957) Growth in sterile culture of excised leaves of flowering plants. Science 126:350–351

Struckmeyer BE, Hildebrandt AC, Riker AJ (1949) Histological effects of growth regulating substances on sunflower tissue of crown-gall origin grown in vitro. Am J Bot 36:491–495

Trifi M, Mezghani S, Marrakchi M (1981) Multiplication végétative du tournesol *Helianthus annuus* L., par culture in vitro. Physiol Veg 19:99–102

Tzen M., Lin C (1975) Influence of concentrations of IAA, 2,4-D and kinetin on the in vitro anther cultures of radish, broccoli and sunflower. Chung-hua Nung Hsuch Hui Pao Taipei 91:23–31

Ursic D, Slightom JL, Kemp JD (1983) *Agrobacterium tumefaciens* T-DNA integrates into multiple sites of the sunflower crown gall genome. Mol Gen Genet 190:494–503

US Sunflower Crop Quality Report (1984) Edited and published by the National Sunflower Association, Bismarck, North Dakota

Wegmann K, Mulbach H-P (1973) Photosynthetic CO_2 incorporation by isolated leaf cell protoplasts. Biochim Biophys Acta 314:79–82

White PR (1943) Handbook of plant tissue culture. Cattel, Lancaster, PA

White PR (1963) The cultivation of animal and plant cells. 2nd edn. Ronald, New York

White PR, Braun AC (1941) Crown gall production by bacteria-free tumor tissue. Science 94:239–241

Witrzens B, Scowcroft WR, Downes RW, Larkin PJ (1988) Tissue culture and plant regeneration from sunflower (*Helianthus annuus*) and interspecific hybrids (*H. tuberosus*×*H. annuus*). Plant Cell Tissue Org Cult 13:61–76

Yang HY, Chang Z, Detian C, Hua Y, Yan W, Xiaoming C (1986) In vitro culture of unfertilized ovules in *Helianthus annuus* L. In: Hu H, Yang HY (ed) Haploids of higher plants in vitro. Springer, Berlin Heidelberg New York Tokyo, pp 182–191

Zimmer DE, Hoes JA (1978) Diseases. In: Carter JF (ed) Sunflower science and technology. Am Soc Agron, Crop Sci Soc Am Soil Sci Soc Am, Madison, Wisconsin, USA, pp 225–255

IV.5 Sunflower (*Helianthus annuus* L.): In Vitro Interspecific Hybridization

N.E. BOHOROVA[1] and A.I. ATANASSOV[2]

1 Introduction

1.1 Importance and Distribution of Sunflower

The cultivated sunflower (*Helianthus annuus* L.) is one of four most important annual crops in the world grown for the edible oil, confectionery, birdfood, and petfood markets. Dedio and Putt (1980) described how extensive use of sunflower as a source of edible oil began in 1830 after the crop reached Russia. The sunflower has become the main source of edible oil in the USSR, other eastern European countries, and Argentina, and in the last 10 to 15 years, production has increased greatly in some other countries. The introduction of hybrid seed has improved the yield and uniformity of the crop as well as its disease resistance. Sunflower genetics and breeding for high yielding capacity and resistance to the main diseases and pests continues to be the main joint research activity. The applied genetic program focusses its joint scientific efforts on the determination of the main genetic components of sunflower ideotypes for different ecological regions, aiming primarily at achieving a new morphophysiological structure of the plant which would radically improve its yielding ability (Vranceanu 1987).

1.2 Utilization of the Wild Species in the Improvement of the Crop

Wild *Helianthus* germplasm, besides constituting the basic stock from which cultivated sunflower originated, continues to contribute specific characteristics for sunflower improvement (Thompson et al. 1981). Schilling and Heiser (1981) presented intrageneric classification for the genus *Helianthus*. The 49 species of *Helianthus* are placed into four sections and seven series. The classification was derived using phenotypic, cladistic, and biosystematic procedures. Anashchenko and Popova (1985) examined USSR collection of 450 forms of annual and perennial wild species in resistance breeding in the USSR. Some of the promising material obtained from over 2000 crosses made during 1981–1983 is mentioned.

Many species are directly cross-compatible with cultivated *H. annuus*, and many other crosses could be obtained with greater effort (Tables 1, 2, 3, Fig. 1).

[1] Institute of Genetics, Bulgarian Academy of Science, Sofia 1113, Bulgaria
[2] General Laboratory of Genetic Engineering, Agric. Acad. 2232, Kostinbrod-2, Bulgaria

Biotechnology in Agriculture and Forestry, Vol. 10
Legumes and Oilseed Crops I (ed. by Y.P.S. Bajaj)

Table 1. Morphological characteristics, cross-compatibility, and survival status for *Helianthus* species. (Thompson et al. 1981)

Helianthus sp.	Subspecies	Annual (a) perrenial (p)	General plant height (m)	Branching habit	Blooming date[a]	Cross-compatibility with *annuus*	Survival status
niveus	*niveus*	p	<1.0	Mostly below	All year	No	Excellent
	tephrodes	p	<1.0	Mostly below	Sept. – May	No	Excellent
	canescens	a	<1.5	Entire stem	All Summer	Yes	Excellent
debilis	*debilis*	a	<0.5	Entire stem	Jan. – Oct.	Yes	Rare
	vestitus	a	<2.0	Entire stem	March – Sept.	Yes	Good
	tardiflorus	a	<2.0	Mostly below	May – Oct.	Yes	Excellent
	silvestris	a	<2.0	Mostly above	April – Oct.	Yes	Excellent
	cucumeriflorus	a	<2.0	Entire stem	June – Nov.	Yes	Excellent
praecox	*praecox*	a	<0.5	Entire stem	April – Nov.	Yes	Excellent
	runyonii	a	<0.5	Entire stem	June – Nov.	Yes	Excellent
	hirtus	a	<0.5	Entire stem	May – Oct.	Yes	Highly endemic
petiolaris	*petiolaris*	a	<1.5	Entire stem	June – Sept.	Yes	Excellent
	fallax	a	<2.0	Entire stem	June – Sept.	Yes	Excellent
neglectus		a	<2.0	Entire stem	July – Sept.	Yes	Excellent
annuus		a	<3.0	Variable	July – Sept.	Yes	Excellent
argophyllus		a	<3.0	Variable	Aug. – Oct.	Yes	Excellent
bolanderi		a	<1.5	Entire stem	July – Sept.	Yes	Excellent
exilis		a	<1.5	Entire stem	July – Sept.	Yes	Endangered
deserticola		a	<0.5	Reduced banching	May – Oct.	Yes	Rare
anomalus		a	<1.0	Reduced branching	May – Oct.	Yes	Rare
paradoxus		a	<2.0	Mostly above	Sept. – Nov.	Yes	Endangered
agrestis		a	<2.0	Mostly above	July – Dec.	No	Good
gracilentus		p	<2.0	Variable	May – Oct.	No	Excellent
pumilus		p	<1.0	Mostly above	July – Sept.	No	Excellent
cusickii		p	<1.3	Mostly above	April – Aug.	No	Excellent
arizonensis		p	<0.3	Mostly below	June – Aug.	No	Good
laciniatus		p	<1.2	Mostly below	June – Sept.	No	Excellent
ciliaris		p	<1.2	Mostly below	July – Sept.	No	Excellent
mollis		p	<1.2	Reduced branching	July – Sept.	No	Excellent
occidentalis	*occidentalis*	p	<1.5	Reduced branching	July – Sept.	No	Excellent
	plantagineus	p	<1.5	Reduced branching	July – Sept.	No	Excellent

divaricatus		p	<1.5	Mostly above	July – Sept.	Yes	Excellent
hirsutus		p	<2.0	Mostly above	July – Sept.	Yes	Excellent
decapetalus		p	<2.0	Entire stem	July – Aug.	Yes	Excellent
× *multiflorus*[b]		p	<2.0	Estire stem	July – Aug.	?	Endangered
eggertii		p	<2.0	Mostly above	Aug. – Sept.	No	Rare
strumosus		p	<2.0	Mostly above	Aug. – Sept.	Yes	Excellent
tuberosus		p	<2.5	Mostly above	Aug. – Oct.	Yes	Excellent
rigidus	*rigidus*	p	<1.5	Mostly unbranched	Aug. – Sept.	No	Excellent
	subrhomboideus	p	<1.2	Mostly unbranched	July – Aug.	No	Excellent
× *laetiflorus*[b]		p	<1.2	Mostly unbranched	July – Aug.	No	Excellent
giganteus		p	<4.0	Mostly above	Aug. – Oct.	Yes	Excellent
grosseserratus		p	<5.0	Mostly above	Aug. – Oct.	No	Excellent
nuttallii	*nuttallii*	p	<3.0	Mostly above	Aug. – Sept.	No	Excellent
	rydbergii	p	<2.0	Mostly above	Aug. – Sept.	No	Excellent
	parishii	p	<4.0	Mostly above	Aug. – Oct.	No	Probably extinct
maximiliani		p	<3.0	Mostly above	Sept. – Oct.	Yes	Excellent
salicifolius		p	<2.6	Mostly above	Aug. – Sept.	No	Good
californicus		p	<3.0	Mostly above	July – Oct.	No	Excellent
resinosus		p	<3.0	Mostly above	July – Sept.	No	Excellent
schweinitzii		p	<1.5	Mostly above	Aug. – Sept.	No	Rare
microcephalis		p	<2.5	Mostly above	Aug. – Sept.	No	Excellent
glaucophyllus		p	<2.0	Mostly above	Aug. – Sept.	No	Excellent
laevigatus		p	<2.2	Mostly above	Aug. – Sept.	No	Endangered
smithii		p	<1.5	Mostly above	Aug. – Sept.	No	Rare
longifolius		p	<1.0	Mostly above	Sept. – Oct.	No	Good
augustifolius		p	<2.0	Mostly above	Sept. – Oct.	Yes	Excellent
simulans		p	<2.5	Mostly above	Oct. – Nov.	No	Excellent
floridanus		p	<2.0	Mostly above	Sept. – Oct.	No	Excellent
silphioides		p	<2.5	Mostly above	Aug. – Sept.	No	Excellent
atrorubens		p	<2.0	Mostly above	Aug. – Oct.	No	Excellent
heterophyllus		p	<1.0	Mostly above	Aug. – Sept.	No	Excellent
radula		p	<1.0	Unbranched	Sept. – Nov.	No	Excellent
carnosus		p	<0.5	Unbranched	June – Sept.	No	Excellent

[a] Mainly from reference.
[b] Hybrids common enough to be recognized.

Table 2. Oil fatty acid composition of *Helianthus* species. (Thompson et al. 1981)

Species and subspecies	Number of collections	Fatty acid composition									
		Palmitie	Palmitoleic	Stearic	Oleic	Linoleic	Linolenic	Arachidic	Eicosenoic	Behenic	Lignoceric
niveus											
niveus	1	7.7		5.1	25.4	59.9	0.3	0.4		0.7	0.2
tephrodes	1	7.7		6.9	28.9	54.5	0.3	0.6		0.4	0.2
canescens	6	7.2 – 9.4[a]		3.5 – 8.0	17.1 – 26.0	54.7 – 68.4			2.7 – 3.8	0.3 – 0.5	
		8.4[b]		5.1	21.2	63.3			3.3	0.4	
debilis											
debilis	1	8.8		5.7	34.9	48.9	0.2	0.7		0.4	0.2
tardiflorus	1	8.5		4.1	13.3	72.6	0.3	0.5		0.4	0.2
silvestris	3	7.2 – 7.7	0.1 – 0.3	4.6 – 5.5	19.0 – 22.6	64.4 – 67.3				0.1 – 0.2	
		7.5	0.2	5.2	21.3	65.5				0.2	
cucumerifolus	1	7.1	0.1	4.9	40.1	47.2				0.4	
praecox											
praecox	1	8.8	0.1	5.0	28.7	57.2				0.2	
runyonii	4	7.7 – 9.4	4.9 – 7.2	4.9 – 7.2	31.4 – 41.0	44.3 – 52.8				0.1 – 0.3	
		8.6	6.3	6.3	34.6	50.3				0.2	
hirtus	4	8.5 – 9.7		6.2 – 7.0	26.4 – 33.0	49.6 – 57.3			3.4 – 4.6	0.1 – 0.2	
		9.2		6.7	29.0	53.0			4.0	0.2	
petiolaris											
petiolaris	20	4.9 – 8.6	0.1 – 0.2	2.6 – 5.8	11.7 – 35.7	52.9 – 70.5	0.3 – 0.4	0.4 – 0.5	2.2 – 5.7	0.3 – 0.6	0.2 – 0.4
		6.6	0.1	4.3	21.8	64.8	0.4	0.3	4.2	0.4	0.3
fallax	13	5.2 – 7.3		3.0 – 5.4	12.2 – 36.1	52.0 – 74.0	0.3 – 0.4	0.4 – 0.5	2.2 – 5.7	0.3 – 0.5	0.2 – 0.3
		6.0		3.8	25.7	62.0	0.3	0.4	3.6	0.5	0.2
negiectus	8	7.2 – 8.2		4.0 – 6.8	20.2 – 32.3	52.1 – 62.2			2.8 – 4.0	0.3 – 0.7	
		7.6		5.4	27.1	58.0			3.5	0.5	
annuus	188	4.4 – 9.8	0.1 – 0.2	2.4 – 7.1	13.1 – 46.3	40.0 – 72.9	0.2 – 0.6	0.3 – 0.7	2.2 – 7.1	0.1 – 3.6	0.2 – 0.5
		6.7	0.1	4.5	23.7	62.1	0.3	0.4	3.5	0.6	0.3
argophyllus	12	6.5 – 10.0		5.3 – 11.1	16.2 – 47.5	36.9 – 68.4				0.1 – 0.3	
		7.8		7.3	40.1	44.3				0.2	

exilis	2	6.1–6.6		5.3–5.5	12.8–13.7	72.0–73.5	0.3–0.3	0.4–0.5		0.4–0.5	0.2–0.3
		6.4		5.4	13.2	72.8	0.3	0.4		0.5	0.3
paradoxus	1	10.8	0.2	0.7	22.3	58.6				0.5	
agrestis	1	10.8		7.2	17.4	62.7	0.2	0.6		0.6	0.2
pumilus	1	6.1		3.9	18.7	69.3	0.3	0.3		0.4	0.2
ciliaris	2	6.9–8.7		5.3–5.8	17.5–20.3	63.4–63.9			3.0–4.9	0.1–0.2	
		7.8		5.6	18.9	63.6			3.9	0.1	
mollis	5	6.2–9.1	0.0–0.1	2.8–5.1	12.0–35.4	45.3–76.2				0.3–0.4	
		7.7	0.1	4.0	22.9	63.8				0.3	
occidentalis											
plantagineus	1	7.3		4.4	29.3	56.1				0.2	
eggertii	1	5.1		3.4	21.0	68.7	0.4	0.4		0.5	0.2
strumosus	1	6.8		5.0	13.8	73.7				0.5	
tuberosus	3	5.1–7.8		2.9–5.6	15.8–30.5	59.4–74.3				0.2–0.7	
		6.0		3.9	20.9	68.1				0.5	
rigidus											
rigidus	4	5.9–7.3	0.1–0.2	3.9–6.3	13.0–30.7	56.2–75.0				0.3–0.7	
		6.5	0.1	5.2	20.0	67.2				0.5	
nuttallii											
nuttallii	2	4.7–5.5		3.5–4.1	23.3–30.6	58.0–63.6	0.4–0.6	0.7–0.7		0.7–0.7	0.4–0.4
		5.1		3.8	26.9	60.8	0.5	0.7		0.7	0.4
maximiliani	1	9.3		5.6	16.7	67.2			6.1	0.4	
salicifolius	2	5.5–5.7	0.2–0.3	4.3–4.4	17.4–29.4	58.4–72.4				0.1–0.4	
		5.6	0.2	4.3	23.4	65.4				0.2	
californicus	1	5.1		2.5	18.0	71.4	0.5	0.6		0.6	0.4
resinosus	2	4.6–6.6		2.3–5.0	34.0–44.8	44.8–53.4	0.5–0.6	0.3–0.4		0.7–0.8	0.5–0.6
		5.6		3.7	39.4	49.1	0.6	0.4		0.7	0.6
microcephalus	1	6.4		4.7	26.0	60.9	0.3	0.6		0.6	
glaucophyllus	1	5.4		4.5	26.6	61.2	0.7	0.5		0.5	0.2
longifolius	1	6.3		3.5	19.1	68.8	0.6	0.6		0.4	0.1
angustifolius	1	6.7		2.9	18.9	70.0	0.3	0.3		0.4	0.5
radula	1	7.2		4.6	9.3	76.6	0.3	0.5		0.3	1.0
carnosus	1	8.6		3.2	32.0	54.8	0.3	0.3		0.4	

[a] Range.
[b] Mean.

Table 3. Results of hybridization between *H. annuus* and tetra- and hexaploid *Helianthus* species. (Georgieva-Todorova 1984)

Parents	Hybrid seed obtained	Hybrid plants obtained	Plant fertility in F_1	
			Open pollination	Backcrossing with *H. annuus*
H. annuus × *H. decapetalus* (4×)	×	×	–	×
H. annuus × *H. hirsutus* (4×)	×	×	–	×
H. annuus × *H. scaberimus* (4×)	×	–	–	–
H. scaberimus (4×) × *H. annuus*	×	×	0	×
H. annuus × *H. rigidus* (6×)	×	–	–	–
H. annuus × *H. tuberosus* (6×)	×	×	–	×
H. annuus × *H. laetiflorus* (6×)	0	0	0	0
H. annuus × *H. resinosus* (6×)	×	×	×	×
H. annuus × *H. strumosus* (6×)	0	–	–	–

The complexity of the genus is increased by natural interspecific hybridization and polyploidy (Heiser 1978; Whelan 1978). The importance of the wild species in sunflower breeding (Heiser 1978; Whelan 1978; Fick 1978; Georgieva-Todorova 1976, 1984; Vannozzi and Paolini 1984; Skorič and Vannozzi 1984) is well known, and interspecific hybridization is used as a potential source of cytoplasmic male sterility, fertility restoration, insect and disease resistance, early ripening, and high content of linoleic acid and protein.

Genes from some wild relatives can undergo meiotic recombination, producing segregants from which the plant breeder can select improved genotypes; these genes can usually function effectively in cultivars. The limit to utilization of wild relatives depends on the breeder's ability to produce hybrids, but hybrids may be sterile. Ploidy differences between crops and wild relatives are frequent, but manipulating ploidy level can improve gene transfer. When fertile hybrids have been produced, lack of recombination may mean that undesirable characters are always expressed in the desired selection. When the gene has been transferred, gene interaction or pleiotropy may prevent selection of the desired new phenotypes. Alternatively, transferred genes may produce effects not present in the wild species. One of these barriers may be effective, but in some cases many may occur, and the more distantly related the species, the more problems are likely. When a gene has been introduced and has undergone recombination within the cultivated genome, it can then recombine and be used by the breeders (Moss 1985).

1.3 Gene Introduction by in Vitro Methods

In spite of the major economic importance of sunflower, in vitro studies are scarce, and very little attention has been given to the culture of sunflower protoplasts and gene transfer. Paterson and Everett (1985) concluded that most of the techniques required for biotechnology for sunflower breeding are now available, for at least some inbred lines.

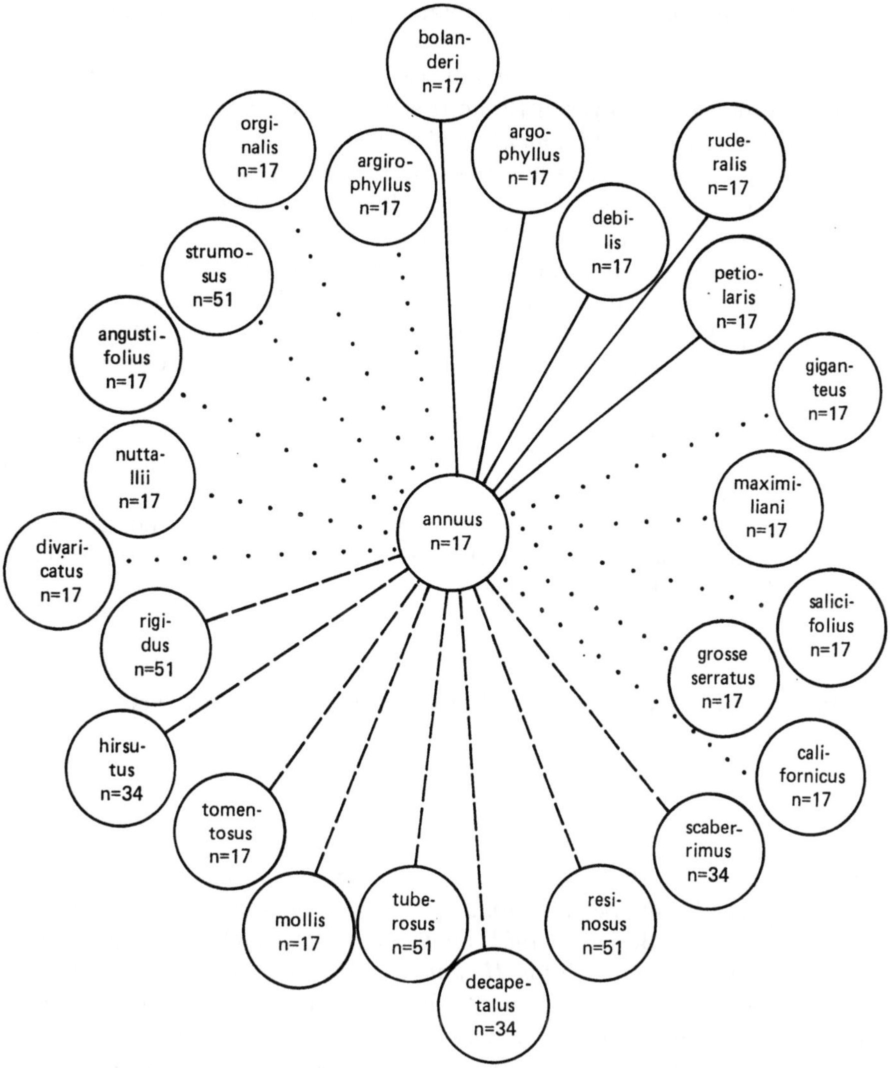

Fig. 1. Interspecific crossability in *Helianthus*. *Solid line* good compatibility; *broken line* crosses are difficult but possible; *dotted line* complete incompatibility (Georgieva-Todorova 1984)

Embryo Culture and Anther Culture. Embryo rescue techniques will almost certainly assist the use of wild sunflower germplasm as a source of traits such as disease resistance, cms genes, drought tolerance, and oil quality (Chandler 1979; Bohorova 1982; Chandler and Beard 1983; Chandler and Jan 1984; Assad et al. 1986). Chandler and Beard (1983) described an embryo culture system using two modifications of Gamborg's B5 medium for producing 53 interspecific *Helianthus* hybrids. Embryos (3 to 7 days after pollination) initially developed on a solid medium containing inorganic components, vitamins, amino acids, and 120 g/l sucrose. For embryo germination and seedling growth, the cultured embryos were transferred to a liquid medium containing inorganic components and 10 g/l sucrose.

There are few reports on successful regeneration of plants from anthers of *Helianthus* interspecific hybrids (Bohorova et al. 1980, 1985; Assad et al. 1985).

Protoplast Cultures and Somatic Hybridization. The first, very brief, report of sunflower protoplast isolation and culture was described by Binding et al. (1981). Recently Lenée and Chupeau (1986) have tested the division capability of sunflower protoplast from mesophyll tissue stems, cotyledons, and hypocotyls, but only the hypocotyl underwent any division. Bohorova et al. (1986) described the isolation, culture, and callus regeneration of protoplasts of wild and cultivated *Helianthus* species. The first report described the condition for cell fusion from green mesophyll protoplasts of *H. praecox* with etiolated hypocotyl protoplasts of cultivated sunflower *H. annuus* was from Bohorova (1987 a).

Gene Transformation. It has been shown that *Agrobacterium tumefaciens* T-DNA integrates into multiple sites of the sunflower crown gall genome (Ursic et al. 1983; Matzke et al. 1984; Goldsbrough et al. 1986; Everett et al. 1987; Tabata et al. 1987) and that the T-DNA is transcribed into at least seven polyadenylated RNA's (Murai and Kemp 1982). In the very near future it should be possible to produce transformed sunflower plants.

2 Interspecific Hybridization Between Cultivated Sunflower and Some Wild Species

2.1 In Vitro Methods for Obtaining Viable Plants

Interspecific hybrids obtained by crosses of *H. annuus* with the tetraploid species *H. hirsutus* and *H. decapetalus* (Bohorova 1977; Georgieva-Todorova et al. 1979), and *H. scaberimus* (Bohorova 1988) are vigorous and combine the characters of both parent species. F_1 hybrids *H. annuus*×*H. hirsutus* and *H. annuus*×*H. decapetalus* had propagation similar to the perennial parent, and pollen and seed sterility. Hybrids of the combinations *H. annuus*×*H. hirsutus*, *H. annuus*×*H. decapetalus*, and *H. scaberimus*×*H. annuus* have been obtained for the first time, and represent valuable initial material for cytogenetic investigation, and genetic and breeding work with the cultivated sunflower. Despite their valuable resistance to various sunflower diseases, the use of tetraploid species actually proved to be very difficult in interspecific hybridization with *H. annuus*. Low crossability and almost complete sterility of the hybrids, resulting from considerable differences in their genomes, appear as a serious barrier for the transfer of valuable characters from the wild species to *H. annuus*.

The use of embryo cultures makes it possible to obtain interspecific hybrids of seeds with not fully developed embryos, in the case of pronounced embryonal incompatibility. Well-shaped seeds, obtained by interspecific hybridization usually do not germinate under ordinary conditions, or do so only with great difficulty, due to the genetic incompatibility of the species participating in the hybridization. Tests revealed that the nutrient media of White (1963) and Murashige and

Fig. 2A, B. In vitro plant development from seeds of *Helianthus* interspecies hybrids. **A** F_1 *Helianthus annuus* (2n = 2× = 34)×*H. hirsutus* (2n = 4× = 68). **B** F_1 *H. scaberimus* (2n = 4× = 68)×*H. annuus* (2n = 2× = 34). (Bohorova 1982)

Skoog (1962) with supplements were the most suitable (Bohorova 1982). A week after being placed in a nutrient medium, the hybrid seeds sprouted and soon grew further. It may be assumed that seeds with a relatively well-developed embryo grew faster than less well-developed ones. Another part of the seeds had a relatively long period of germination in vitro (2–10 weeks). These seeds were twice transferred to a fresh nutrient medium during that period. The germination frequency of the hybrid seeds was 13.04% of F_1 *H. annuus*×*H. hirsutus* and 36.36% of F_2 of the same combination. It is worth noting that under normal field or greenhouse conditions hybrid seeds do not develop fully, and arrest development at various stages of embryo differentiation, so that no hybrid plants are obtained. By applying the in vitro method of this hybrid combination, three hybrid plants were obtained for the first time, two of them remaining in a rosette form and only one continuing to grow. The plant was 120 cm high, with the stem branched from the base, with large leaves, broad at the base and sharp towards the tip (Fig. 2A). Normal anthers with nonviable pollen were formed, and the seed fertility was 7.40%. By in vitro micropropagation and after well-developed shoots, rooted in White's medium, regenerated plantlets were transplanted in the field (Bohorova 1982).

In *H. annuus*×*H. hirsutus* F_1 hybrids, the study of diakinesis showed various disturbances (Georgieva-Todorova et al. 1979). In 12.9% of cells, accumulation of chromatin was recorded, as a result of which a varying number of chromatin

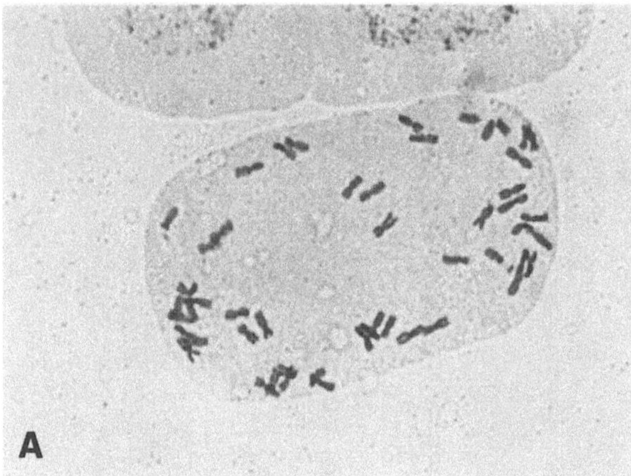

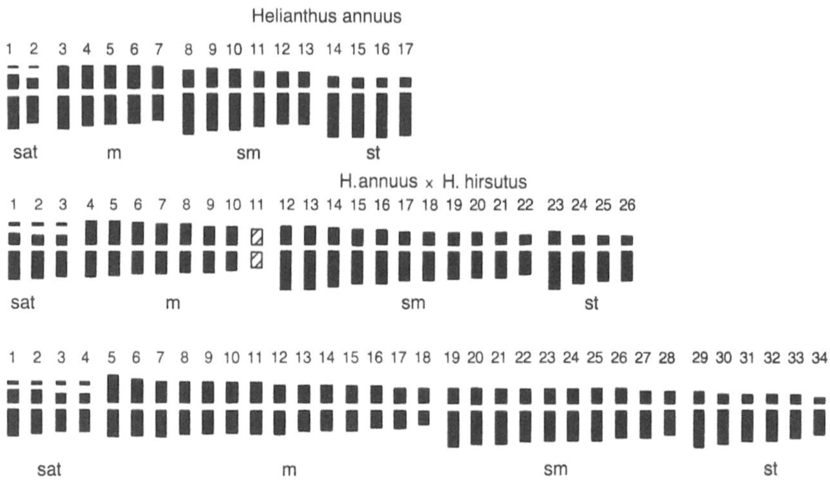

Fig. 3. A Metaphase with 51 chromosomea in F_1 *Helianthus annuus* ($2n = 2\times = 34$)$\times$*H. hirsutus* ($2n = 4\times = 68$), **B** Ideogram of F_1 *H. annuus*$\times$*H. hirsutus*. (Georgieva-Todorova and Bohorova 1980a)

groups were formed. Cells in which part of the chromatin flowed into the cytoplasm were also observed. In this case, meiosis in PMC's was often interrupted at an early stage and no microsporocytes were formed. Various numbers of univalents (7–34) were observed in all the cells at metaphase 1. In anaphase I of some hybrids, lagging chromosomes and chromosome bridges were found in 95% of the cells studied. As a result of these disturbances, inferior microsporocytes were formed with highly disturbed structure. They were mostly shapeless and without clearly defined cell membrane. At telophase II, in 75.3% of the PMC's, triad, pentad, or tetrad with micronuclei were formed instead of regular tetrades. The analysis of the karyotype of the interspecific hybrid *H. annuus*$\times$*H. hirsutus*, shows that it is with a stable karyotype in whose structure only individual chromosomes may have been obtained from the exchange of chromatin material between chromosomes from the different genomes of the parent species, while the remaining ones are similar to the corresponding chromosomes of *H. hirsutus* (Georgieva-Todorova and Bohorova 1980a) (Fig. 3A). This confirms the assumption made on the basis of the meiotic behavior of the chromosomes in the hybrid, that the conjugation has taken place primarily between chromosomes of the tetraploid species. This serves to explain the phenotype of the hybrid.

Backcrossing with cultivated sunflower restored the chromosome number of *H. annuus* (after the fifth backcross) (Bohorova and Georgieva-Todorova 1987). It is possible that some structural chromosomal differences still existed. Therefore, considerable phenotypical diversity was observed among the progeny. Male-sterile plants with high fertility were obtained as a result of this hybridization and backcrossing (Fig. 4A). They are being tested to determine the type of sterility.

The compatibility between *H. annuus* and the tetraploid species *H. decapetalus* was also low. The F_1 hybrids exhibited dominating phenotypic features of the wild-growing parent. In the first and second backcrosses (BC_1, BC_2) (Georgieva-Todorova 1984), the introduction of genetic material from the cultivated sunflower resulted in definite changes in plant type. Summarizing, it may be stated that in BC_4 and BC_5 of this material, the wild-type characters could be attenuated to such an extent that, by careful selection, generations close to the cultivated type were obtained.

Direct bud formation was established in MS nutrient medium with zeatin 5 mg/l for anthers from *H. annuus* × *H. decapetalus* (Bohorova et al. 1985). On the 20th day after isolation, two of the anthers had compact leaf-like green formation, which showed the beginning of three vegetative buds. Secondary callogenesis and organogenesis from vegetative buds obtained from direct androgenesis of anthers was established by placing the stems of the three buds into a nutrient medium MS (1962) with 1320 mg/l L-asparagine or combination of 800 mg/l L-asparagine with 800 mg/l L-glutamin. The average number of newly formed buds in the individual tube in each subculture was 10–15. A total of 300 regenerants was obtained (after three to seven subcultures), which differed in their biologo-morphological characteristics. The cytological characteristics of the regenerants obtained from anthers showed that 33.4% of them have chromosome count 34, at $2n = 51$ for the hybrid, while 66.7% have chromosome count 42–51, 51, 68, and 102. From the cytological investigations carried out, it is not possible to draw a concrete conclusion whether the additives to the nutrient medium induce development of the pollen only, or whether they act on the somatic tissues. The fact that the higher number of regenerants are $2n = 34$ suggests that the pollen cells in the anthers of the tetraploid parent have been the initial principle for obtaining both callus and vegetative buds. Further research will show whether these are "pure" microspores or whether there is a genetic constitution of *H. annuus* present in them. It is more difficult to explain the origin of the regenerants with chromosome count 51–102. The broad variability of the chromosome count in the root cells of the regenerants of these hybrids is probably due to the influence of the nutrient medium with polyploidizing and aneuploidizing role.

Hybridization of the cultivated sunflower with tetraploid species *H. scaberimus* was successful only if the female parent had the higher ploidy level. F_1 hybrids were very vigorous, intermediate in morphological characters, but

Table 4. Cytological characterization of *Helianthus scaberimus* (2n = 4× = 68) × *Helianthus annuus* (2n = 2× = 34) hybrids. (Bohorova 1988)

Generation	No. plants investigated	Type of plant	Diakinesis				Disturbances observed (range %)		
			Bivalents	Polyvalents	Univalents	Disturb. in diak. %	Metaphase II	Anaphase II	Telophase
F_1	5	d	22–32	1–4	2–4	80	80	80	80
	30	b	34	–	1–4				
BC_1	25	c	17–34	1–4	2–10	95	90	90	80
BC_2	5	d	22–34	–	1–4	97	97	67	77
	8	c	17–35	1–4	10	20–97	25–86	25–83	0–67
	3	a	17–22	1–2	1–4	35	9–16	10–20	16–20
BC_3	16	b	32–34	1	1	80	20–40	20–30	20–30
	4	c	14–33	1–5	2–14	50–100	40–80	40–80	20–60
	40	a	14–17	1	1	80	10–20	5–20	2–20
BC_4	20	d	22–30	1	1	8–40	11–52	6–65	4–53
	11	b	30–34	1	2–4	50	51	43	33
	30	a	16–18	1	2–4	100	1–32	1–37	27
BC_5	50	a	16–18	1 fr.	8–40	1–30	1–35	1–30	1–20

a – cultivar; b – wild; c – interm.-cultivar; d – interm.-wild

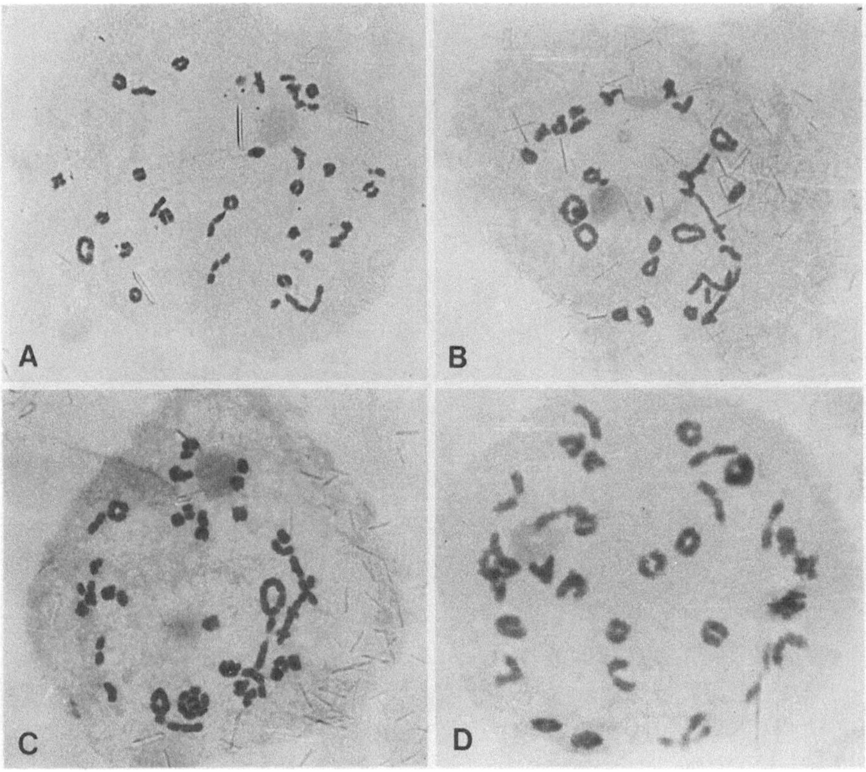

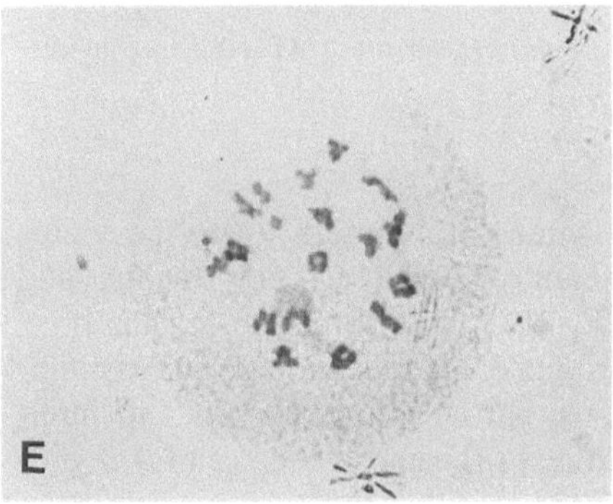

Fig. 4A–E. Diakinesis in F_1–BC_5 (P_2) *H. scaberimus* (2n = 4× = 68)×*H. annuus* (2n = 34). **A** F_1–25II+1III+3I. **B** BC_1–28II+2III+2I. **C** BC_2–31II+1III+3IV+5I. **D** BC_3–28II+1IV. **E** BC_5–17II+1 fr. (Bohorova 1988)

stood closer to the wild parental species. The application of the in vitro method to hybrid seeds of this combination showed that 41.2% of the seeds developed and 25 viable plants were grown on the experimental field. The plants were of an intermediate type of inheritance, the property of one of the initial parent forms predominating (Fig. 2B). They differed in height, branching, size, and shape of leaves and racemes (Bohorova 1982). The pollen viability varied, being quite sterile in 24% of the investigated plants and in these remaining forming microsporacytes varying in size and with a viability from 1.4% to 78.8%. Seed production was relatively low, from 10% to 33%. From the hybrid seeds raised

Fig. 5 A, B. Plants with cytoplasmic male sterility (*H. annuus*×*H. hirsutus*) (**A**) and interesting economic characters (*H. scaberimus*×*H. annuus*) (**B**)

in vitro, 13.9% of the F_2 progeny in the same combination grew, the viable plants being raised in the field experimentally. Bivalents, polyvalents, and univalents were observed in PMC's of the $F_1 - BC_3$ hybrids (Table 4). The presence of configurations of more than two chromosomes substantiates the assumption that the genomes of the two species are partially homologous and interchromosome pairing occurs (Fig. 4A–E). Some of the chromosomes of the wild-growing species do not pair and are represented as univalents which, up to BC_5, are separated in the form of a typical disturbance of the PMC phases (Bohorova 1982, 1988). Populations providing opportunities for the selection of plants possessing valuable characters are obtained as a result of hybridization. Male-sterile plants have been produced and are used for testing, as well as plants with interesting economic characters, such as with short stems of up to 1 m, with inflorescence of diameter 25–28 cm, early blooming (up to July 15), and early ripening with high protein content of the seeds (Fig. 5B).

The ability to regenerate plants from single cells and protoplasts represents a basic step towards the genetic improvement of crop plants in vitro. The effects of different hormonal balances for growth and differentiation from sunflower tissue

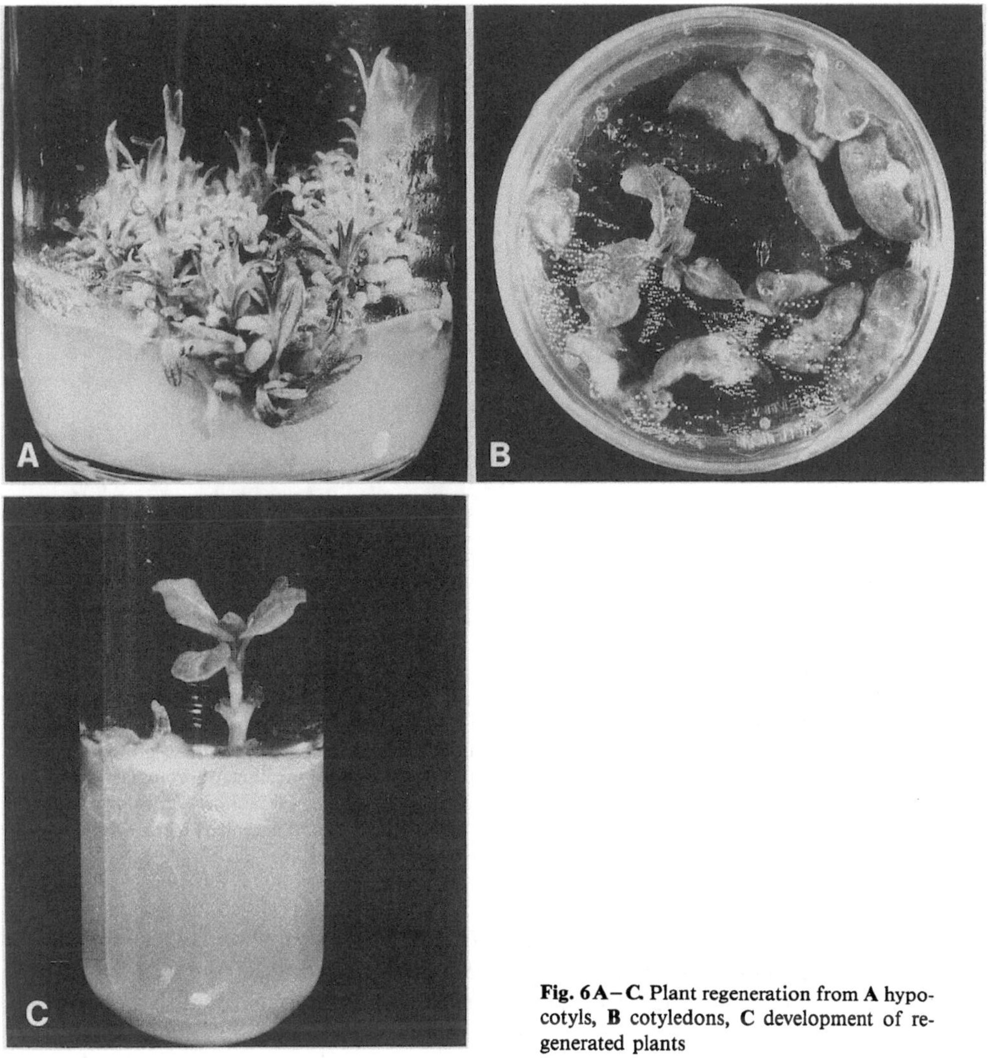

Fig. 6 A – C. Plant regeneration from **A** hypocotyls, **B** cotyledons, **C** development of regenerated plants

are rather limited (Pall et al. 1981; Greco et al. 1984; Paterson and Everett 1985; Bohorova et al. 1985; Piubello and Caso 1986). Very little attention has been paid to the culture of sunflower protoplasts (Binding et al. 1981; Lenee and Chupeau 1986; Bohorova et al. 1986). Different explants of 16 *Helianthus* cultivars, wild species, and interspecific hybrids were tested for direct shoot formation, callus induction, and callus regeneration from tissue and protoplast culture (Bohorova 1987b). Callus induction was obtained on the media supplemented with 2 mg/l NAA and 0.05 mg/l BAP related to all genotypes. The best response with callus induction was the young leaves of *H. resinosus*, *H. scaberimus* × *H. annuus*, and stem species of interspecific hybrids. This explant was suitable for producing friable, light green granular callus. Using the medium with NAA : BAP in ratio 1 : 10, friable callus with numerous green meristematic sections was induced from leaf,

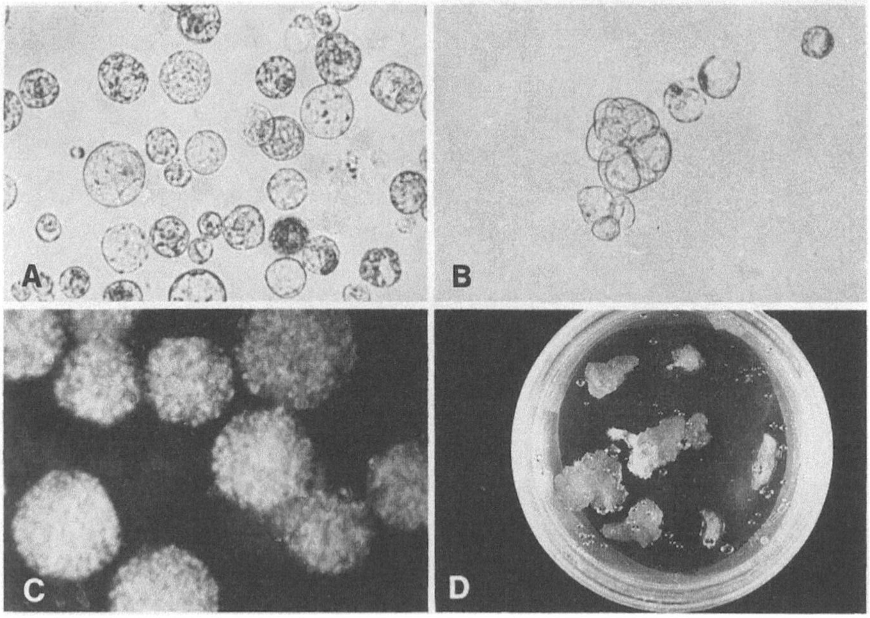

Fig. 7. A Protoplasts isolation from hypocotyls of sunflower (*H. annuus* L.). (Bohorova et al. 1986). Protoplasts were isolated from 7-day-old etiolated seedlings in enzyme mixture –2% w/v Rhozyme HP 150, 2% w/v Meicelase, 0.03% w/v Macerozyme R10 in CPW 13M medium, pH 5.8, incubated at 25 °C for 16 h on a rotary shaker (30 cycles/min). **B** Division of protoplasts. The protoplasts/agarose mixture (5 ml) was dispensed as small beads (0.25 ml) with K8P medium, of protoplast densities 2×10^5 into a 9-cm Petri dish with 6 ml of liquid culture medium. **C** Calli of hypocotyl protoplasts using the agarose bead method of culture. **D** Callus with roots and organized meristematic regions on MS medium with 2 mg/l NAA, 0.5 mg/l 6-BAP

hypocotyl, stem segments of *H. praecox*, *H. scaberimus*, *H. resinosus*, as well as the young leaf and stem explants of *H. scaberimus*×*H. annuus*. As far as the embryogenic tissues are concerned, more than 20 new embryoids per dish were obtained and within 7 days most of these structures developed into small shoots, plantlets, and roots (Fig. 6A–C). Higher regeneration was observed when hypocotyl explants were taken from 7-day-old seedlings (Fig. 6A).

Most plant species need auxin and cytokinin for dedifferentiation, and this effect was obtained by NAA and BAP in different sunflower explants. The different hormonal combinations tested appeared to modify the morphogenetic responses. Responses consisted of changes in the color and consistency of the callus tissues, formation of the nodules and bud-like structures, frequency of the roots and the embryoids. The presence of NAA and BAP (with fixed concentrations) were an essential components of the culture medium for morphogenetic responses, and this regenerational potential was under genetic control (Bohorova 1987a, b).

Bohorova et al. (1986) described protoplast isolation from seedling roots, hypocotyls, and cotyledons of four cultivars of *Helianthus annuus* and from leaves of axenic shoot cultures of the wild species *H. praecox*, *H. scaberumus*, and *H. rigidus*. Optimal culture conditions were established for the respective protoplast systems, using the agarose bead method of culture. Protoplast division was induced in all the species examined. In the cultivars of *H. annuus* hypocotyl (Fig. 7A–D) and cotyledon protoplast division was sustained leading to callus formation, which in turn could induce roots and organized meristematic regions in the presence of NAA and BAP. These studies form the basis of the development of procedures following induced protoplast fusion and utilizing the in-built visual identification of interspecific fusion products between leaf protoplasts of *H. praecox* and colorless etiolated hypocotyl protoplasts of cultivated sunflower. Heterokaryons were recognized by the presence of chloroplasts from the mesophyll protoplast and cytoplasmic strands from the etiolated hypocotyl protoplast (Bohorova 1987a). The beads were microscopically examined and those containing a single heterokaryon were recorded. Such experiments are currently in progress.

3 Conclusions

During the past decade a number of in vitro techniques – tissue culture, cell culture, protoplast fusion, and recombinant DNA technology – have shown promise as plant breeding tools for creating genetic variability or increasing selection efficiency. Embryo, ovule and ovary cultures have yielded hybrids in more sunflower cross-combinations with early abortion of embryo or endosperm, or both. Clonal propagation of hybrids has been achieved. The screening of regenerated plants for somaclonal variation can provide an additional source of novel and useful variation. Somatic hybrid and cybrid plants produced by unconventional genetic manipulation procedures will be of importance in sunflower improvement programs. Experiments on fusion of protoplasts with suitable selection procedure and the development of the procedures for the inactivation of the nuclear genome of one of the species to obtain cybrids are currently in progress. Using Ti-plasmids which have had tumor-inducing gene deleted, it should be possible to produce transformed sunflower plants in the very near future.

4 Protocol for Embryo Culture

Seeds are cultured 50 days after pollination. They were sterilized in a 5% solution of potassium hypochloride for 20 min or in 15% domestic bleach for 30 min, followed by six changes of sterile distilled water. All materials are germinated on the surface of Murashige and Skoog (1962) (MS) medium lacking phytohormones or on the same medium with 0.01 mg/l IAA, 0.03 mg/l kin and 0.001 mg/l folic acid, or on White's (1963) media with 1.5% sucrose, 1.5 glucose, 200 mg/l yeast extract, 200 mg/l casein hydrolysate, 0.01 mg/l gibberelic acid, 0.05 mg/l kin, 0.05 mg/l NAA with agar 0.7%. The cultures are maintained at 26°, relative humidity 70%, 2000 lx fluorescent tubes with 16 h light and 8 h dark.

References

Assad A, Serieys H, Jonard R (1985) Sur less possibilités de régéneration d'espèces sauvages et d'hybride interspecifiques du genre *Helianthus* par androgenèse in vitro. CR Acad Sci Paris t. 300, III 1:25–30

Assad A, Jonard R, Serieys H, Vincourt P (1986) La culture d'embryons isolé in vitro dans un programme d'amelioration du tournesol. CR Acad Sci Paris T. 302, III 5:161–164

Anashchenko N, Popova A (1985) Wild sunflower collection and ways of utilizing it in breeding. Sel'skohozyaistvennaya Biol 10:9–11

Binding H, Nehls R, Kock R, Finger J, Mordhorst G (1981) Comparative studies on protoplast regeneration in herbaceous species of the Dicotyledonae class. Z Pflanzenphysiol 101:119–130

Bohorova N (1977) Study on *Helianthus* interspecific relations and cytogenetic characteristic of F_1 hybrids. Thesis Ph D Inst Genet, Bulg Acad Sci, Sofia (in Bulg)

Bohorova N (1982) In vitro plant development of seeds of *Helianthus* interspecific hybrids. CR Acad Sci Bulg t. 35, 1:105–107

Bohorova N (1983) Hybridization between *H. scaberimus* and *H. annuus*. In: Symposium of distant hybridization. Bulg Acad Sci, Sofia, pp 219–228 (in Bulg)

Bohorova N (1987a) Aspects of somatic hybridization between *Helianthus annuus* L. and *H. praecox* E. and Gray. In: Proc 2nd Int Symp on Experimental mutagens in plants. Plovdiv, pp 247–251

Bohorova N (1987b) Effects of NAA and BAP on protoplasts, callus and direct regeneration from different explants in sunflower. Proc 2nd Int Symp Experimental mutagens in plants. Plovdiv, pp 244–247

Bohorova N (1988) Cytogenetical study of F_1-BC_5 (P_2) hybrids between *H. scaberimus* ($2n = 4\times = 68$) and *H. annuus* ($2n = 2\times = 34$). Genet Breed 1:36–43 (in Bulg)

Bohorova N, Georgieva-Todorova J (1987) Hybridization between *H. annuus* L. ($2n = 34$)$\times$*H. hirsutus* Raf. ($2n = 68$). Genet Breed 3: 204–210 (in Bulg)

Bohorova N, Atanassov A, Antonova G (1980) In vitro isolation of anthers from interspecific hybrids in the *Helianthus* genus. CR Acad Sci Bulg t 33, 11:1545–1548

Bohorova N, Atanassov A, Georgieva-Todorova J (1985) In vitro organogenesis, androgenesis and embryo culture in the genus *Helianthus* L., Z Pflanzenzuecht 95:35–44

Bohorova N, Cocking EC, Power B (1986) Isolation, culture and callus regeneration of protoplasts of wild and cultivated *Helianthus* species. Plant Cell Rep 5:256–258

Chandler J (1979) Sunflower interspecific hybridization using embryo culture. MS Thesis, Univ California

Chandler J, Beard B (1983) Embryo culture of *Helianthus* hybrids. Crop Sci 23:1004–1007

Chandler J, Jan C (1984) Interspecific hybridization via embryo culture. Col, Eval and conserv of wild species and their use in sunfl breed progr July, Novi Sad, pp 24–27

Dedio W, Putt E (1980) Sunflower. In: Hybridization of crop Plants. Amer. Soc. of Agronomy Crop Science Soc. of America. Madison, pp 631–660

Everett N, Robinson K, Mascarenhas D (1987) Genetic engineering of sunflower (*Helianthus annuus* L.), Bio/technology, November 5:1201–1204

Fick G (1978) Breeding and Genetics. In: Carter J (ed) Sunflower science and technology. ASA, CSSA, SSSA, Madison, Wisconsin, pp 279–456

Georgieva-Todorova J (1976) Interspecific relationship within the genus *Helianthus*. Bulg CR Acad Sci Sofia Bulg (in Bulg)

Georgieva-Todorova J (1984) Interspecific hybridization in the genus *Helianthus*. Z Pflanzenzuecht 93:256–279

Georgieva-Todorova J, Bohorova N (1979) Studies on the chromosomes of the two *Helianthus* L. tetraploid species. Caryologia 32 (3):335–347

Georgieva-Todorova J, Bohorova N (1980a) Karyological investigation of the hybrid *Helianthus annuus* L. ($2n = 34$)$\times$*H. hirsutus* Ralf. ($2n = 68$). Acad Sci Bulg 33 (7):961–964

Georgieva-Todorova J, Bohorova N (1980b). Interspecific hybridization in sunflower breeding. 9th Conf sunflower. Malaga, 8–13 Junio, pp 101–106

Georgieva-Torodova J, Lakova M, Bohorova N (1979) Hybridization of diploid sunflower *Helianthus annuus* ($2n = 34$) with some tetraploid *Helianthus* species. Z Pflanzenzuecht 83:340–349

Goldsbrough PB, Gelvin SB, Larkins BA (1986) Expression of maize genes in transformed sunflower cells. Mol Gen Genet 202:374–381

Greco B, Tanzarella G, Carrozzo G, Blanco A (1984) Callus induction and shoot regeneration in sunflower (*Helianthus annuus* L.). Plant Sci Lett 36:73–77
Heiser C (1978) Taxonomy of *Helianthus* and origin of domesticated sunflower. In: Carter J (ed) Sunflower science and technology. ASA, CSSA, Madison, Wisconsin, pp 31–54
Lenée P, Chupeau J (1986) Callus formation from sunflower (*Helianthus annuus* L.) protoplast. Plant Sci 43:69–75
Matzke M, Susani M, Binns A, Lewis E, Rubenstein I, Matzke A (1984) Transcription of a zein gene introduced into sunflower using a Ti plasmid vector. EMBO J 3:1525–1531
Moss J (1985) Wild species in crop improvement. Biotechnology Int Agric Res IRRI, Manila, pp 199–208
Murai N, Kemp J (1982) T-DNA of pTi – 15955 from *Agrobacterium tumefaciens* in transcribed into a minimum of seven polyadenylated RNAs in a sunflower crown gall tumor. Nucl Acid Res 10: 1679–1689
Murashige T, Skoog F (1962) A revised medium for rapid growth and bioassays with tobacco tissue cultures. Physiol Plant 15:473–497
Pall H, Kurnik E, Szobo L (1981) Napraforgo noveny-regeneracio in vitro hajtascscustenyeszetbol. Novenytermeles 30:201–209
Paterson K, Everett N (1985) Regeneration of *Helianthus annuus* inbred plant from callus. Plant Sci 42:125–132
Piubello S, Caso O (1986) In vitro culture of sunflower (*Helianthus annuus*) tissues. Phyton 46 (2):131–137
Schilling E, Heiser Ch (1981) Intrageneric classification of *Helianthus* (Compositae). Taxon 30 (2):393–403
Skorič D, Vannozzi P (1984) Genetic resources in *Helianthus* genus. Int Symp Sci Biotech for an integral sunflower utilization. Oct 25, Bari, Italy
Tabata T, Terayama C, Mikami K, Uchimiya H, Iwabuchi M (1987) An accurate transcription of wheat histone genes in sunflower cells. Plant Cell Physiol 28 (1):73–82
Thompson T, Zimmerman D, Roters C (1981) Wild *Helianthus* as a genetic resource. Field Crops Res 4:333–343
Ursic D, Slighton J, Kemp J (1983) *Agrobacterium tumefaciens* t-DNA integrates into multiple sites of the sunflower crown gall genome. Mol Gen Genet 190:494–504
Vannozzi G, Paoline R (1984) Morphological and biological characters of a wild species and interspecific hybridization. Collec. Eval and Conserv of wild species and their use in sunflower breed progr. July, 24–27, Novi Sad, pp 12–34
Vranceanu A (1987) Research network on sunflower. Progress Report in FAO, July, 1987
Walter D, Putt E (1980) Hybridization of crop plant. Am Soc Agr Crop Sci, Madison, p 631
Whelan E (1978) Cytology and interspecific hybridization. In: Carter J (ed) Sunflower science and technology. Amer Soc Agr, Madison, USA
White P (1963) The culture of animal and plant cells. 2nd edn, Rold, New York

IV.6 In Vitro Production of Haploids in *Helianthus*

H.Y. YANG, H. YAN, and C. ZHOU[1]

1 Introduction

Sunflower (*Helianthus annuus* L.) is one of the most important oilseed crops in the world. As a highly cross-pollinated species, homozygotes are especially needed both for genetic studies and breeding programs. However, so far we lack a method of haploid production for shortening the duration of pure line establishment or the breeding cycle of hybrid offspring. In vitro techniques may offer a useful means for this purpose. In this chapter, the in vitro production of haploids through anther and ovary/ovule culture in sunflower is reviewed. The emphasis is given on the recent work done in our laboratory on the fundamental problems of ovule culture of sunflower which are indispensible for making of a new biotechnology in this crop.

2 Anther Culture

In an extensive study on in vitro organogenesis, androgenesis, and embryo culture in the genus *Helianthus*, Bohorova et al. (1985) tested anther culture of seven diploid, three tetraploid, and two hexaploid species as well as the F, hybrids of *H. annuus* × *H. hirsutus* and *H. annuus* × *H. decapetalus*. Twenty-four growth regulator combinations were tested in modified MS medium. Callus formation was observed in all species and hybrids. After 2 weeks, 70% – 100% of the anthers were covered with friable, light green, granular calli. This was most evident in a medium with 1 mg/l 2,4-D and 0.2 mg/l kin. All attempts to induce shoot formation with this type of callus failed. On a medium containing 5 mg/l zeatin, direct shoot formation occurred in anthers of *H. divaricatus* (on plant) and *H. annuus* × *H. decapetalus* (three plants from two anthers). Secondary callus induction and shoot organogenesis were achieved from regenerants obtained directly from *H. annuus* × *H. decapetalus* (three plants from two anthers). Secondary callus induction and shoot organogenesis were achieved from regenerants obtained directly from *H. annuus* (2 n = 34) × *H. decapetalus* (2n = 68) hybrid anthers by plating stem explants onto a medium enriched with 2 mg/l BA, 0.2 mg/l IAA and

[1] Department of Biology, Wuhan University, Wuhan 430072, China

Biotechnology in Agriculture and Forestry, Vol. 10
Legumes and Oilseed Crops I (ed. by Y.P.S. Bajaj)

20 mg/l adenine. This medium was suitable for producing compact callus with numerous green meristematic sections. After several subcultures, continuous shoot formation was observed. Shoots were successfully rooted after 25 days of culture on white's medium. A total of 276 regenerants differing in morphological characteristics was obtained. Cytological analysis of root tips of 125 regenerated plants showed that one-third of them had 34 chromosomes, while two-thirds had 2n = 45–51, 68 and 102. All plants with chromosome number 34 were produced during the first three subcultures. After eight transfers, all the regenerants had a chromosome count of 2n = 51 identical to the donor hybrid plant. The *H. divaricatus* (2n = 34) anther regenerant was propagated by the same procedure. Only one secondary regenerant survived and developed to maturity. Pollen mother cell analysis revealed 17 bivalents. These authors concluded that more experiments will be required before the technique becomes a feasible method for producing genetically pure lines.

Mix (1985) carried out anther culture experiments with eight sunflower cultivars. 10600 anthers were cultured on MS medium supplemented with 0.4 mg/l thiamine, 1 mg/l IAA, 2 mg/l NAA, 0.1 mg/l 2,4-D, 2 mg/l BAP, 6% sucrose and 0.8% agar and maintained at 24 °C in darkness. There was a high division rate of the microspores. Callus formation was observed in all the tested cultivars with the percentage ranging from 15% to 80%. Most of the calli turned brown and died off, but a few embryoids survived. After 4–5 weeks of culture, the embryoids were transferred onto MS medium with 0.1 mg/l thiamine, 0.5 mg/l BAP and 3% sucrose, and maintained at 24 °C with 12 h illumination every day. Six plantlets were regenerated from the cultivar Inra; two of them were haploids and the other four were diploids. From the cultivar Luciole, one haploid plantlet was obtained. The regenerated plants developed rapidly and flowered at the three- to four-leaf stage. One haploid plant of Inra was transplanted to soil and grew normally.

Recently, Jonard and his coworkers (Assad et al. 1985) have done extensive work on the induction of androgenesis and the utilization of the anther-derived plants in sunflower breeding (see Chap. IV.7, this Vol.).

3 Ovary and Ovule Culture for the Induction of Haploids

The female gamete (egg cell) of angiosperms is harbored in the female gametophyte (embryo sac), which is in turn surrounded by sporophytic tissues of the ovule and ovary. After fertilization, the ovule and ovary serve as a natural "womb" of the developing zygotic embryo. In occasional cases, parthenogenetic or apogamous embryos occur spontaneously in this womb as well. Therefore it is logically possible to induce haploid embryoids by submitting the unpollinated and unfertilized ovaries or ovules to suitable in vitro conditions. However, in contrast to anther culture, little attention has been paid to ovary/ovule culture until recent years. We made a preliminary review of this topic several years ago (Yang and Zhou 1982). Up to now, records scored in this field have shown at least 14 species belonging to seven families in which haploids have been induced by this

Table 1. Records of induction of haploids by ovary or ovule culture

Family	Species	Explant	Reference
Gramineae	*Hordeum vulgare*	Ovary	San Noeum (1976) Wang and Kuang (1981) Huang et al. (1982) Gu and Zheng (1984)
	Triticum aestivum	Ovary	Zhu and Wu (1979)
	Oryza sativa	Ovary	Asselin de Beauville (1980) Zhou and Yang (1980) Kuo (1982)
	Zea mays	Ovary	Ao et al. (1982) Truong-Andre and Demarly (1984)
	Coix lacryma-jobi	Ovary	Li and Zhang (1984)
Solanaceae	*Nicotiana tabacum* and *N. rustica*	Ovary	Zhu and Wu (1979) Wu and Cheng (1982)
	Petunia axillaris	Ovule	De Verna and Collins (1984)
Compositae	*Gerbera jamesonii*	Ovule	Sitbon (1981) Meynet and Sibi (1984)
	Helianthus annuus	Ovary	Cai and Zhou (1984) Gelebart and San (1987)
Liliaceae	*Lilium davidii*	Ovary	Gu and Cheng (1983)
Salicaceae	*Populus × simonigra*	Ovary	Wu and Xu (1984)
Chenopodiaceae	*Beta vulgaris*	Ovule	Hosemans and Bossoutrot (1983) Borman (1985)
Euphorbiaceae	*Hevea brasiliensis*	Ovule	Chen et al. (unpubl.)

Note: Only the first report is listed for works from one laboratory

method (Table 1). It is believable that production of haploids via in vitro gynogenesis will sooner or later become a practical way following that via in vitro androgenesis.

3.1 Induction of Haploids in Sunflower by Ovary/Ovule Culture

Studies on sunflower ovary and ovule culture were begun in our laboratory in 1982. Cai and Zhou (1984) first reported the results obtained by young floret culture and ovule culture with eight cultivars. In these experiments, N6 medium supplemented with 0.125 – 2 mg/l 2-methyl-4-chlorphenoxyacetic acid (MCPA), 100 mg/l inositol and 2% – 6% sucrose was used for float culture of the florets and ovules. After 45 – 55 days of culture, small globular or rod-shaped embryoids could be dissected from some enlarged ovaries and ovules. When transferred onto solid media, four embryoids from the floret culture grew into small plantlets, eight gave rise only to roots, and 96 embryoids from the ovule culture produced roots without shooting. Chromosome counts on root-tip squashes showed both haploids and diploids. In both culture methods, somatic calli often occurred on the surface of ovaries or ovules, especially when the medium was supplemented with kinetin; a few somatic calluses could regenerate into plantlets after transferring on differentiation medium. This original work led to more detailed studies on the origin and development of the gynogenetic embryoids and the role of various factors in sunflower ovule culture. The advances since then are described below.

Mix (1985) also cultured 3590 unpollinated ovaries of eight cultivars of sunflower on MS medium enriched with auxins and cytokinins. Calli formed in 2 weeks with induction frequency ranging from 10% to 85%. When transferred to differentiation media, 19 plantlets were regenerated from the calli of three cultivars. Some of them grew to flowering in the culture flask. In all of these plantlets diploid metaphases were observed.

Gelebart and San (1987) carried out successful work on sunflower ovary culture. Eight genotypes, including male-sterile and restorer lines and their F_1 hybrids, were used. Ovaries 12–24 h prior to anthesis were placed on a medium with half-strength MS macroelements, full MS microelements, Morel vitamins, 10^{-4} M FeEDTA, 6%–10% sucrose, 9% agar and various combinations of exogenous hormones. During the course of culture, ovules were dissected out of the ovaries and placed on the same medium. This intermediate dissection was emphasized as necessary to avoid nutritional shortage of the ovules enclosed inside the ovaries. After 4–6 weeks from the beginning of ovary culture, the ovules were dissected and the calli or embryoids inside were transferred for regeneration to a medium of the same basic components but with 2% sucrose and free of growth substances. Gynogenesis was observed when either 2,4-D or NAA was included in the induction medium. The best results were obtained with 2 mg/l NAA and 10% sucrose. Among 6562 ovaries cultured, 78 embryoids (1.19%) were formed, from which 37 plants (47%) were regenerated. Observations on squashes or sections revealed that the embryoids or calli were located at the micropylar end of the ovules. Gynogenesis was said to originate from the egg cell or the fused polar nuclei. Root-tip examination at different stages of regenerated plants showed that among 37 plants, 20 had haploid metaphases, 8 had diploid metaphases, and 9 had both haploid and diploid cells, i.e., mixoploids. (In a recent personal communication, the authors told us that no more mixoploids were obtained; they were all haploids.) However, during the course of growth, the frequency of haploid metaphases decreased rapidly and all the cells observed became diploid when flowering occurred.

3.2 Embryological Studies

From the above description one can conclude that both haploid and diploid derivatives could be induced in sunflower ovary or ovule culture, so that it is very important to know the precise origin of the products and to distinguish them in order to get precise information on haploid induction. This can only be clarified by embryological observations. In our laboratory embryological observation combined with culture experiments is a routine task of particular significance in a system as complicated as sunflower ovule culture. Yan et al. (1985) made a microscopic observation on the embryoid formation in cultured sunflower ovules. The main results of this study have been described in detail by Yang et al. (1986). Subsequently, Wei and Yang (1986) reported histochemical studies on the in vitro-

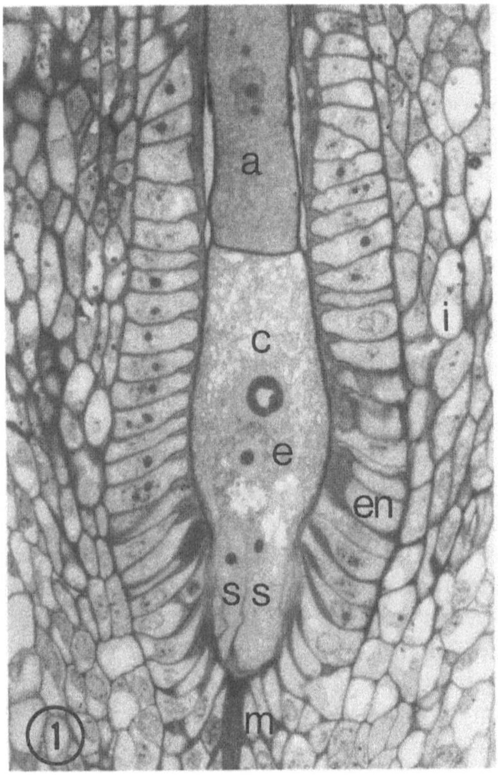

Fig. 1. Semi-thin section of a sunflower ovule 2 days before anthesis, fixed with glutaraldehyde-osmium, embedded in Poly/Bed 812 and stained with toluidine blue O. Note the well-organized embryo sac. *a* antipodal cell; *c* central cell; *e* egg cell; *en* endothelium; *i* integument tissue; *m* micropyle; *s* synergid

induced gynogenetic embryoids as compared with the in vivo zygotic embryos. Here only the main points are summarized.

The sunflower ovule is anatropous, unitegmic, and tenuinucellate. At the time of inoculation, the nearly matured embryo sac is composed of an egg cell, two synergids, a central cell with the secondary nucleus (fused polar nuclei), and two antipodal cells. The embryo sac is surrounded by the endothelium, or integumentary tapetum, a specialized inner cell layer of the integument (Fig. 1).

In cultured ovules, the gynogenetic embryoids were all located at the micropylar end of the embryo sac and undoubtedly derived from the unfertilized egg cells. Process of gynogenesis from the initiation of proembryo till advanced stages has been observed (Fig. 2, 3A, B, C; also see Figs. 2 and 3 in Yang et al. 1986).

Sporophytic derivatives from sunflower ovule culture showed two types of structures: one was embryoid-like structures originating from the endothelium; the other was callus-like structure produced from the remaining part of the integument. These somatic derivatives usually occurred behind the gynogenetic embryoids, but often grew more vigorously and surpassed the latter in competition for nutrition. At advanced stages, the endothelial embryoids often bore a shape, similar to the gynogenetic ones. They could hardly be distinguished by the naked eye at the time of ovule dissection. Even by microscopic judgement, we could

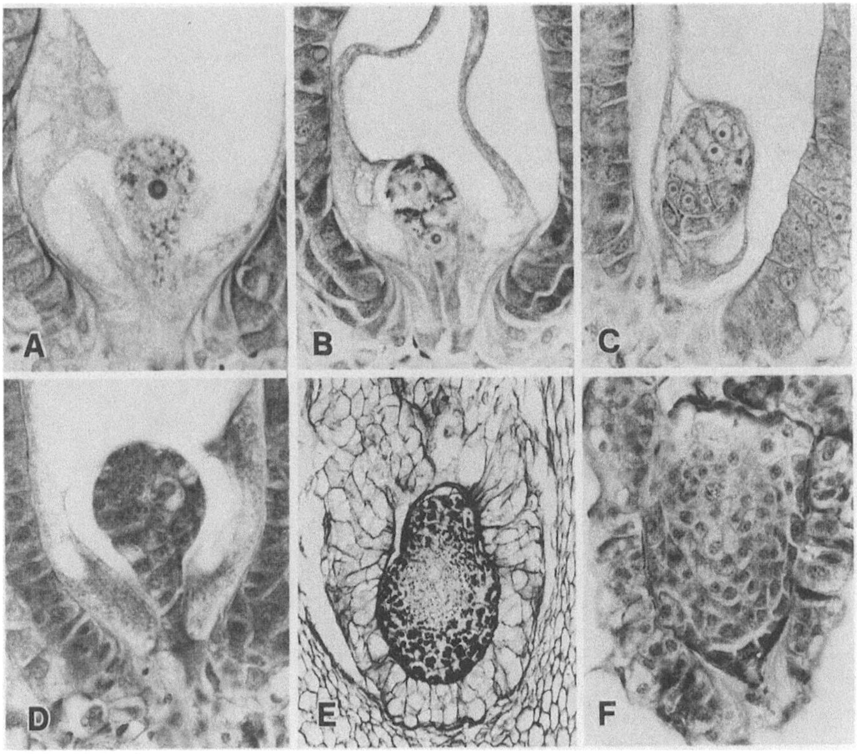

Fig. 2A–F. Histochemical observations on early stages of gynogenetic embryoids in sunflower ovules. Paraffin sections were stained by methyl green-pyronin (**A, B, C**), mercuric bromphenol blue (**D, F**) and periodic acid-Schiff reaction (**E**). **A** An egg cell after 2 days of ovule culture with prominent nucleolus and cytoplasmic pyroninophilic substances. **B** A four-celled proembryo after 4 days of ovule culture, showing cytoplasmic pyroninophilic substances. Only two nuclei were seen in this section. **C** A "partitioned" proembryo after 4 days of ovule culture. Its basal part remained meristimatic and apical part became vacuolated. **D** A proembryo after 4 days of ovule culture. The cells were rich in protoplasmic protein. **E** An ovoid embryoid after 8 days of ovule culture, showing centripetal distribution of starch grains. **F** A gynogenetic embryoid already filling up the embryo sac after 8 days of ovule culture

identify them only with great care, mainly according to their different position in relation to the embryo sac (Fig. 3).

3.3 Regulation of Gynogenesis and Somatic Proliferation

In sunflower ovule culture, how to regulate the balance between gynogenetic and somatic products in favor of the former is an important task. In order to know the effects of various factors concerned; we carried out a large-scale experiment. The factors involved were genotype, kind of explants, embryo sac stage, cold-pretreatment, culture on solid or liquid medium, kind of basal media, concentration of exogenous hormone, concentration of sucrose, culture in light or darkness, etc. More than 20000 ovules were cultured and compared. To identify the induced

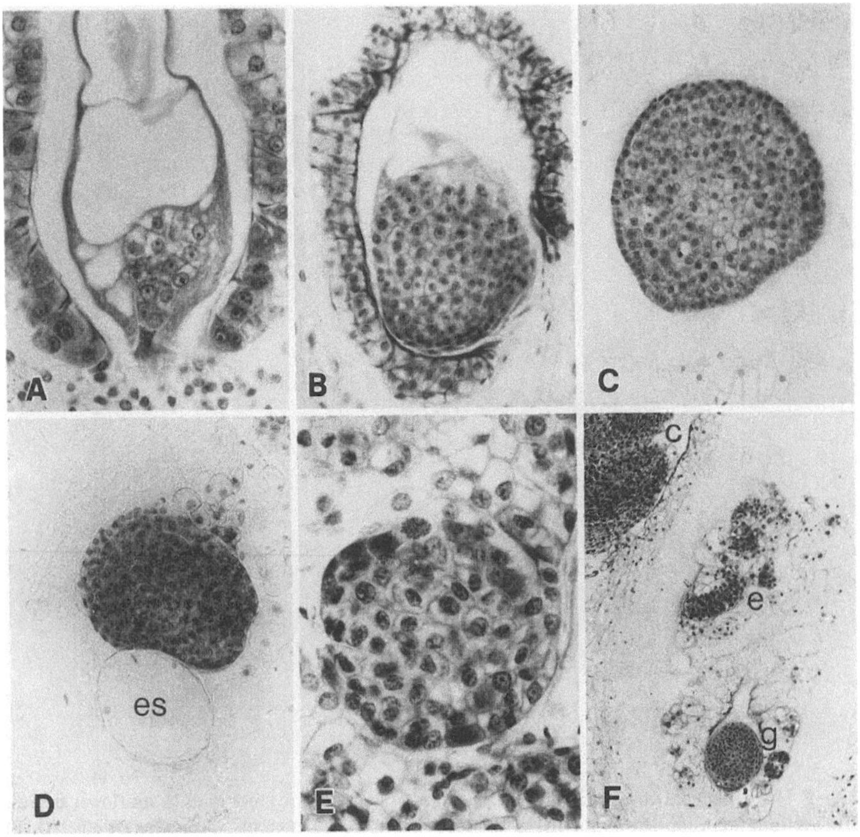

Fig. 3 A–F. A comparison between various derivatives in sunflower ovule culture. The ovules were en bloc stained with Ehrlich's hematoxylin and sectioned by paraffin method. Notice that the gynogenetic embryoids were located inside the embryo sac (**A, B**); the endothelial embryoids were located outside the embryo sac (*es*, **D**). At advanced stage, the endothelial embryoids (**E**) were morphologically similar to the gynogenetic embryoids (**C**). In **F**, a gynogenetic embryoid (*g*), some endothelial derivatives (*e*) and a large integument callus (*c*) were seen in one section at low magnification

products, ovules were sampled after 10 and 30 days of culture, stained en bloc with Ehrlich's hematoxylin, sectioned by the paraffin method and examined microscopically (Fig. 3). Alternatively, an Ehrlich's hematoxylin staining-methyl salicylate clearing technique developed recently (Yang 1986) was adopted, which greatly facilitated the work. For an example of this method see Fig. 4.

Based on these experiments, three factors have been affirmed to play important roles in regulating the balance between gynogenetic and somatic products, i.e., exogenous hormone, cold-pretreatment, and sucrose concentration (Yan et al. 1987). Other factors were, of course, by no means negligible and are summarized elsewhere (Yan et al. 1988).

The presence or absence of exogenous hormone is a key factor. In most of the previous works on unpollinated ovary/ovule culture in various species including sunflower, exogenous hormone has been an indispensable supplement to the in-

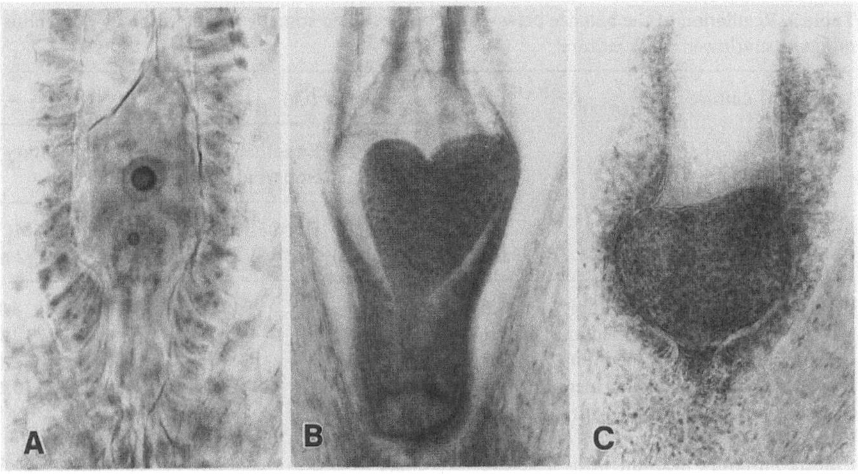

Fig. 4A–C. Whole mounts of sunflower ovules stained with Ehrlich's hematoxylin and cleared in methyl salicylate en bloc, showing a mature embryo sac (**A**), a zygotic heart-shaped embryo surrounded by endosperm (**B**), and an early heart-shaped gynogenetic emboryoid in contact with the degenerating endothelium (**C**)

Table 2. Induction of gynogenetic sunflower embryoids in hormone-free condition as compared with hormone-supplemented condition

Method of culture	Hormone-free			MCPA (2 mg/l)		
	Cultured ovules	Gynogenic embryoids	Induction frequency (%)	Cultured ovules	Gynogenic embryoids	Induction frequency (%)
Float culture on liquid medium[a]	121	9	7.5	120	8	6.8
Vertical culture on solid medium[a]	124	23	18.5	126	8	6.3
Float culture on liquid medium	113	7	6.2	118	12	10.2
Horizontal culture on solid medium	120	9	7.5	109	6	5.5
Culture on shallow liquid medium with 6% sucrose	123	8	6.5	127	2	1.6
Culture on shallow liquid medium with 12% sucrose	134	12	9.0	119	4	3.4
Culture on liquid medium with Polybuffer 74	113	18	15.9	113	4	3.5
Culture on liquid medium with achene extract	111	9	8.1	114	4	3.5

Note: Data were obtained by microscopical observations at 10th day of ovule culture. All treatments were exerted on N6 medium with 100 mg/l inositol and 12% sucrose unless otherwise pointed out. The ovules were inoculated 3 days before anthesis except the treatments marked[a], in which the ovules were inoculated 1 day before anthesis.

Table 3. Regulation of the balance between gynogenesis and somatic proliferation by exogenous hormone in sunflower ovule culture

Method of culture	Induction-frequency (%)					
	Gynogenic embryoid		Endothelial embryoid		Integumentary callus	
	Hormone-free	MCPA (2 mg/l)	Hormone-free	MCPA (2 mg/l)	Hormone-free	MCPA (2 mg/l)
Float culture on liquid medium[a]	6.6	2.4	0	2.4	0	8.1
Vertical culture on solid medium[b]	9.1	3.4	0	5.9	0	12.6
Float culture on liquid medium	3.5	0.8	0	0	0	2.6
Horizontal culture on solid medium	8.7	0.8	0	0.8	0	13.6
Culture on liquid medium with Polybuffer 74	3.5	1.7	0	1.7	0	2.5
Culture on liquid medium with achene extract	3.4	0.9	0	6.0	0	1.7

Notes as for Table 2.

Table 4. Role of sucrose concentration in sunflower ovule culture

Sucrose concentration (%)	Cultured ovules	Gynogenic embryoids	Induction-frequency (%)
1	119	0	0
6	120	2	1.7
12	116	7	6.0
18	118	5	4.2
21	118	3	2.5

Note: Data were obtained by microscopic observations at 10th day of ovule culture on N6 medium with 100 mg/l inositol and 2 mg/l MCPA.

duction medium. It appeared to be different with sunflower. A series of comparative experiments have shown that a medium deprived of exogenous hormone gave even better results than its counterpart – hormone-supplemented – regardless of the culture methods used (Table 2). Furthermore, perhaps more importantly and unexpectedly, the hormone-free condition almost fully inhibited the occurrence of somatic products. This situation was in sharp contrast with the hormone-supplemented condition, in which the somatic derivatives almost invariably took place (Table 3).

Sucrose concentration is also important. Experiments with three cultivars revealed that each kind of the derivatives responded differently to sucrose concentration of a medium: The optimum concentration was the lowest (1%) for integumentary callus, higher (3%–9%) for endothelial embryoid and the highest (above 12%) for gynogenetic embryoid. In the cultivar Dang Yang, the optimum sucrose concentration for gynogenetic embryoid was certified as 12% (Table 4, also see Table 2).

Table 5. Effect of cold-pretreatment on early development of gynogenetic embryoids in sunflower ovule culture

Cold-pretreatment (h)	Cultured ovules	Gynogenic products		Proportion					
				Meristematic cell type		Vacuolated cell type		Free-nucleate type	
		No.	%	No.	%	No.	%	No.	%
0 (control)	113	7	6.2	3	42.9	0	0	4	57.1
24	117	28	23.9	18	64.3	3	10.7	7	25.0

Note: The ovules were cultured in hormone-free and 12% sucrose condition and observed microscopically at 10th day of culture.

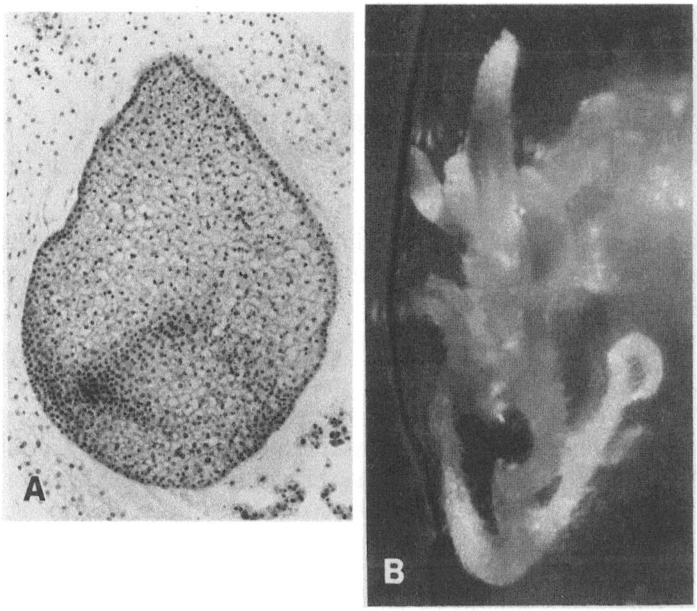

Fig. 5. A Gynogenetic embryoid after 30 days of culture showing root differentiation. **B** Shooting of a gynogenetic embryoid

Cold-pretreatment also plays an important role. When the inflorescence was treated 4 °C for 24–48 h prior to inoculation, the induction frequency of gynogenetic embryoids increased and that of integument calluses decreased. The effect was more evident when cold-pretreatment was coupled with a condition of 12% sucrose and exclusion of exogenous hormone. In this case, not only the yield of dividing egg cells increased, but also cell division proceeded in a more normal, embryogenetic way: The meristimatic cell type predominated over the vacuolated type and the free-nucleate type (Table 5). At an advanced stage, the growth and differentiation of the embryoids was also improved (Fig. 5).

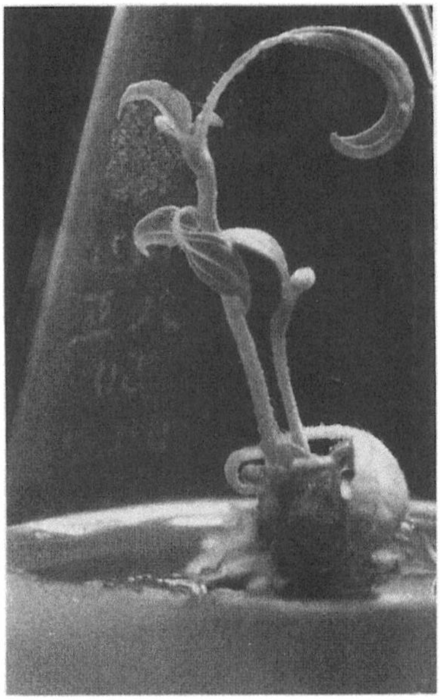

Fig. 6. A regenerated plantlet of gynogenetic origin

On the basis of these experiments, we now approach to final establishment of a procedure for inducing haploid embryoids and simultaneously excluding somatic derivatives:

1. Put the capitula in polyethylene bags and at low temperature (~4 °C) for 24 h.
2. Cut the capitula into several parts, sterilize them in 0.1% $HgCl_2$ solution, and rinse with sterile distilled water several times. Choose suitable tubular florets (1–3 days prior to anthesis) and pick out the ovules aseptically.
3. Inoculate the ovules on the surface of liquid N6 medium (or alternatively solid medium) supplemented with 100 mg/l inositol and 12% sucrose, 10 ovules per tube containing 20 ml medium. Keep the culture at approximately 27 °C in the dark.
4. After 1–2 months of culture, dissect the enlarged ovules and take out the gynogenetic embryoids, if any induced, for further experiments.

Experiments are now being done for refinement of the regeneration procedure. Preliminary results by Hua and Yang (unpubl.) showed that it seemed better to transfer gynogenetic embroids to a medium originally developed by Paterson and Everett (1985) for regeneration of plants from hypocotyl-derived callus. This was composed of MS salts and vitamins supplemented with 5 g/l KNO_3, 100 mg/l inositol, 1 mg/l NAA, 1–2 mg/l BA, 0.1 mg/l GA_3 and 3%–6% sucrose. Under illumination conditions the underdeveloped embryoids cultured on this medium grew vigorousely into calli with green spots. After further transferring and deletion of NAA, green plantlets were regenerated (Fig. 5B, 6).

4 Conclusions

Sunflower has been one of a few crops lacking methods for artificial induction of haploid plants. This situation is now changing, since haploid plants or plantlets have been reared recently via either anther or ovary/ovule culture in several laboratories. Nevertheless, the techniques are still immature due to low induction-frequency as well as usual occurrence of undesirable somatic products. This is perhaps a serious problem in ovary/ovule culture works also in many other plant species, which accounts for the fact that so far this method is still far from practical application. Our experience indicates the necessity of combining culture experimentation with embryological studies in solving these difficulties. To do so, we should first distinguish products of different origin with great care. Next we should know the different responses of these products to various factors involved in the culture. By regulating some key factors, we might find out the best possible means to promote gynogenesis and inhibit or even delete somatic products. It is desirable that the potentiality of unpollinated ovary/ovule culture as a new biotechnology be exploited better in the near future.

References

Ao GM, Zhao SX, Li GH (1982) In vitro induction of haploid plantlets from unpollinated ovaries of corn (*Zea mays.* L.) Acta Genet Sin 9:281–283

Assad A, Serieys H, Jonard R (1985) Sur les possibilités de régénération d'espèces sauvages et d'hybrides interspécifiques du genre *Helianthus* per androgenese in vitro. CR Acad Sci Paris 300 III:25–30

Asselin de Beauville M (1980) Obtention d'haploides in vitro à partir d'ovaires non fécondés de Riz, *Oryza sativa* L. CR Acad Sci Paris 296D:489–492

Bohorova N, Atanassov A, Georgieva-Todorova J (1985) In vitro organogenesis, androgenesis and embryo culture in the genus *Helianthus* L. J Plant Breed 95:35–44

Borman CH (1985) Haploidization of sugar beet (*Beta vulgaris*) via gynogenesis. In Vitro 21(3) part II:103

Cai DT, Zhou C (1984) In vitro induction of haploid embryoids and plantlets from unpollinated young florets and ovules of *Helianthus annuus* L. Kexue Tongbao 29:680–682

De Verna JW, Collins GB (1984) Maternal haploids of *Petunia axillaris* (Lam.) B.S.P. via culture of placenta-attached ovules. Theor Appl Genet 69:187–192

Gelebart P, San LH (1987) Obtention de plantes haploides par culture in vitro d'ovaires et d'ovules non fécondés de Tournesol (*Helianthus annuus* L.). Agronomie 7:81–86

Gu ZP, Cheng KC (1983) In vitro induction of haploid plantlets from unpollinated young ovaries of lily and its embrological observation. Acta Bot Sin 25:24–28

Gu ZP, Zheng GC (1984) In vitro induction of haploid plantlets from unpollinated ovaries of highland barley. Acta Bot Sin 26:549–551

Hosemans D, Bossoutrot D (1983) Induction of haploid plants from in vitro culture of unpollinated beet ovules (*Beta vulgaris* L.). Z Pflanzenzuecht 91:74–77

Huang QF, Yang HY, Zhou C (1982) Embryological observations on ovary culture of unpollinated young flowers in *Hordeum vulgare* L. Acta Bot Sin 24:295–300

Kuo CS (1982) The preliminary studies on culture of unfertilized ovary of rice in vitro. Acta Bot Sin 24:33–38

Li MW, Zhang B (1984) In vitro induction of haploid plantlets from unpollinated ovaries of *Coix lacryma-jobi* L. Hereditas (Beijing) 6:5–6

Meynet J, Sibi M (1984) Haploid plants from in vitro culture of unfertilized ovules in *Gerbera jamesonii*. Z Pflanzenzuecht 93:78–85

Mix G (1985) Antheren- und Ovarienkultur von Sonnenblumen (*Helianthus annuus* L.). Landbauforsch Völkenrode 35:153–156

Paterson KE, Everett NP (1985) Regeneration of *Helianthus annuus* inbred plants from callus. Plant Science 42:125–132

San Noeum LH (1976) Haploides d'*Hordeum vulgare* L. par culture in vitro non fécondés. Ann Amelior Plant 26:751–754

Sitbon M (1981) Production of haploid *Gerbera jamesonii* plants by in vitro culture of unfertilized ovules. Agronomie 1:807–812

Truong-Andre I, Demarly Y (1984) Obtaining plants by in vitro culture of unfertilized maize ovaries (*Zea mays* L.) and preliminary studies on the progeny of a gynogenetic plant. Z Pflanzenzuecht 92:309–320

Wang CC, Kuang BJ (1981) Induction of haploid plants from the female gametophyte of *Hordeum vulgare* L. Acta Bot Sin 23:329–330

Wei CY, Yang HY (1986) Histochemical studies on in vitro parthenogenesis in *Helianthus annuus* L. Acta Bot Sin 28:117–122

Wu BJ, Cheng KC (1982) Cytological and embryological studies on haploid plant production from cultured unpollinated ovaries of *Nicotiana tabacum* L. Acta Bot Sin 24:125–129

Wu KX, Xu MZ (1984) Induction of maternal haploid plants from unpollinated ovaries of poplar in vitro. Acta Genet Sin 11:47–51

Yan H, Wu Y, Chen XM, Wei ZY, Zhou C, Yang HY (1985) Microscopical observations on the embryoid formation in cultured unfertilized ovules of *Helianthus annuus* L. Acta Bot Sin 27:13–18

Yan H, Dong J, Zhou C, Yang HY (1987) Regulation of in vitro parthenogenesis and somatic proliferation in sunflower by several factors. Acta Bot Sin 29:580–587

Yan H, Zhou C, Yang HY (1988) Experimental studies on several factors affecting unfertilized ovule culture in sunflower. J Wuhan Bot Res 6:319–326

Yang HY (1986) The use of a whole stain-clearing technique for observations on embryo sac, embryo, endosperm and embryoid. Acta Bot Sin 28:575–581

Yang HY, Zhou C (1982) In vitro induction of haploid plants from unpollinated ovaries and ovules. Theor Appl Genet 63:97–104

Yang HY, Zhou C, Cai DT, Yan H, Wu Y, Chen XM (1986) In vitro culture of unfertilized ovules in *Helianthus annuus* L. In: Hu H, Yang HY (eds) Haploids of higher plants in vitro. China Acad, Beijing, Springer, Berlin Heidelberg New York Tokyo, pp 182–191

Zhou C, Yang HY (1980) In vitro induction of haploid plantlets from unpollinated young ovaries of *Oryza sativa* L. Acta Genet Sin 7:287–288

Zhu ZC, Wu HS (1979) In vitro induction of haploid plantlets from the unpollinated ovaries of *Triticum aestivum* and *Nicotiana tabacum*. Acta Genet Sin 6:181–183

IV.7 Sunflower (*Helianthus* spp.): Anther Culture and Field Studies on Haploids

R. JONARD[1] and A. MEZZAROBBA[2]

1 Introduction

The genus *Helianthus* (Compositeae family) contains 60 annual and perennial species originating from America. Two among them have been improved for nutritional use, *H. tuberosus* L., for its succulent tubers, and *H. annuus* L., the cultivated sunflower, the oil from whose seed, and protein-rich oil-cakes are utilized (Table 1).

The basic number of chromosomes is $\times = 17$. *Helianthus* contains: diploid species, the cultivated sunflower (*H. annuus*) with $2n = 2\times = 34$, tetraploid species (*H. hirsutus*, *H. laevigatus*) with $2n = 4\times = 68$, hexaploid species (*H. resinosus*, *H. tuberosus*) with $2n = 6\times = 102$ (Heiser 1969; Bonjean 1986).

In 1969, Leclercq, after crossing *H. annuus* with *H. petiolaris* Nutt, obtained a form of cytoplasmic male sterility (cms) more easily utilizable than the genetic one. Two years later, the discovery of fertility gene restoration started the cultivation of sunflower hybrids T1 with improved production.

Sunflower is one of the main sources of oil seeds after soybean, cotton, and rapeseed. It contains 45–55% seed oil and is the world's third vegetable oil after soya and palm oil (Oil World, 1987).

The main producers of sunflower are the USSR with 5.33 mt, followed by Argentina (2.6), China (1.85), France (1.7), USA (1.21), Turkey (0.94), Romania (0.84) and Hungary (0.81).

The species covered 585,904 ha in France in 1985 and with an average production of 23 quintal/ha, produced 1.349,879 t. (Bonjean 1986).

Production of sunflower in 1986–1987.

Production in mt		Mean produce of a plantation
in seeds	in oils	in t/ha
19.04	7.1	1.6

[1] Laboratoire de Cultures de Tissus, Groupe de Physiologie Végétale Appliquée, Université des Sciences et Techniques du Languedoc, 34060 Montpellier Cédex, France
[2] UCST Verneuil, BP 83, 11400 Castelnaudary, France

Biotechnology in Agriculture and Forestry, Vol. 10
Legumes and Oilseed Crops I (ed. by Y.P.S. Bajaj)

Table 1. The species of *Helianthus genus*

Section I: ANNUI	Section II: CILIARES ($2\times$, $4\times$)	
($2n = 2\times = 34$)	Ciliares	Pumili
H. agrestis	*H. arizonensis*	*H. cusickii*
H. annuus	*H. ciliaris*	*H. gracilentus*
H. anomalus	*H. laciniatus*	*H. pumilus*
H. argophyllus		
H. bolanderi		
H. debilis		
H. deserticola		
H. exilis		
H. neglectus		
H. niveus		
H. paradoxus		
H. petiolaris		
H. praecox		
Section III: DIVARICATI ($2\times$, $4\times$, $6\times$)		Section IV: FRUCTICOSI
Angustifolii	Divaricati	(perennial species)
H. angustifolius	*H. decapitalus*	*H. argenteus*
H. floridanus	*H. divaricatus*	*H. acuminatus*
H. simulans	*H. eggertii*	*H. discolor*
Atrorubentes	*H. hirsutus*	*H. grandiceps*
H. atrorubentes	*H. mollis*	*H. hypargyreus*
H. carnosus	*H. occidentalis*	*H. inbaburensis*
H. heterophyllus	*H. rigidus*	*H. jelskii*
H. radula	*H. strumosus*	*H. lanatus*
H. silphioides	*H. yuberosus*	*H. microphyllus*
Microcephali	Gigantei	*H. matthewsii*
H. microcephalus	*H. californicus*	*H. nigrescens*
H. glaucophyllus	*H. giganteus*	*H. lehmanii*
H. laevigatus	*H. grosseseratus*	*H. stenbelii*
H. smithii	*H. maximiliani*	*H. subniveus*
H. longifolius	*H. nuttalii*	*H. serax*
	H. resinosus	*H. verbesinoides*
	H. salicifolius	*H. viridion*
	H. schweinitzii	

Genetic improvement of the oilseed sunflower is relatively new: but *Helianthus*, due to its diversity, has a great potential for variability, especially adaptation to dryness or tolerance to parasite attacks.

Some ecotypes of *H. resinosus*, *H. rigidus*, and *H. tuberosus* have a high level of tolerance to *Sclerotinia* (Serieys 1984). *H. tuberosus*, *H. salicifolius*, and *H. ciliaris* are known also because of their tolerance to different parasites such as

Zygogramma exclamationis, *Bothyrus gibbosus*, and *Masonphis masoni* (Rogers and Thompson 1978). The creation of new male cytoplasmic sterilities utilizing the crossing between *H. annuus* and *H. petiolaris* (Leclercq 1969) was an important factor in the development of cultivated sunflower. But the interspecific crossings between the wild species and the cultivated sunflower (*H. annuus* L.) lead to F_1 hybrids with very reduced fertility, often provoked by a lack of co-adaptation of genomes.

There are two possibilities for androgenesis with *Helianthus*:

- for the wild species and some hybrids, there is genome reduction, which later facilitates hybridization, but there is also selection, from different gametes suitable for synthetic medium, of viable individuals at different degrees of ploidy leading to improved fertility.
- for the cultivated sunflower, there is isolation of haploid plants which, after doubling the chromosome stock, gives homozygotes, and pure stocks, with accelerated fixation of F_1 hybrids.

2 Androgenesis in Wild Species and Interspecific Hybrids of *Helianthus*

2.1 Obtaining Viable Plants by Androgenesis

In Bulgaria, Bohorova et al. (1980, 1985) worked on F_1 hybrids between: *H. annuus* ($2n = 34$)×*H. decapetalus* ($2n = 68$) and *H. annuus* ($2n = 34$)×*H. hirsutus* ($2n = 68$).

The anthers cultured at the uninucleate microspores stage on Murashige and Skoog medium (MS) with added sucrose (3%) and 2 phytohormones, 2,4-D (1 mg/l) and kin (0.2 mg/l) will produce haploid callus – sprouts with leaves are obtained only for the hybrid *H. annuus*×*H. decapetalus*, with zeatin (5 mg/l) and asparagine (1320 mg/l).

In 1985, at Montpellier, France, Assad, Serieys and Jonard used the *Helianthus* collection of "La Station d'Amélionsration des Plantes, INRA", to select wild species and hybrids known for their interesting agronomic characteristics, tolerance to disease, or adaptation to low temperatures or dryness.

1. Perennial wild *Helianthus*: three hexaploid species with $2n = 6\times = 102$ chromosomes:

243 *H. resinosus* (small),
236 *H. rigidus rigidus*,
325 *H. tuberosus*,

and a diploid species with $2n = 2\times = 34$:

231 *H. occidentalis plantagineus*.

The first three species are tolerant to sclerotinia, the last to low temperature.

2. Interspecific F_1 hybrids:

H.21 *H. annuus* L. (CMS-ISS 56)×*H. occid. plantagineus*
$2n = 2\times = 34$

H.27 *H. rigidus rigidus*×*H. annuus* L. (CE 278)
$2n = 4\times = 68$

H.37 *H. annuus* L. (CMS-ISS 56)×*H. resinosus*
$2n = 4\times = 68$

H.38 *H. annuus* L. (CMS-FS 20)×*H. resinosus*
$2n = 4\times = 68$

H.41 *H. annuus*×*H. tuberosus*
$2n = 4\times = 68$

H.61 *H. rigidus rigidus* (236)×*H. annuus* L. (ISS 56)
$2n = 4\times = 68$

H.83 *H. tuberosus* (325)×*H. annuus* L. (BM)
$2n = 4\times = 68$

H.1210 *H. annuus* (CMS-ISS-56)×*H. occid. plantagineus* (231), in free fertilization
$2n = 2\times = 34$.

Anthers were taken from plants in open fields in 1983 at the stage near the first pollen mitosis, when they are pale green in color.

Table 2. Percentages of regenerated plants with the wild species *Helianthus* and their interspecific hybrids according to genotype. (Assad 1986)

Group and genotypes of utilized plants	No. of isolated anthers	No. of formed callus	No. of buds with formed embryos	No. of regenerated plants	% of obtained plants
Cultivated sunflower					
H. annuus B21	80	80	0	0	0
B32	90	90	15	0	0
HA89	100	80	0	0	0
MH1,2	100	71	0	0	0
Primasol	90	82	0	0	0
Wild species					
H. tuberosus 325	126	122	170	44	36.5
H. resinosus 243	115	99	58	29	25
H. occidentalis 231	83	83	47	14	17
H. rigidus rigidus 127	80	64	4	2	2.5
Interspecific hybrids					
H. annuus×*H. tuberosus*: H41	120	110	0	0	0
H. tuberosus×*H. annuus*: H83	100	100	0	0	0
H. annuus×*H. resinosus*: H37	160	155	365	114	71
H. rigidus rigidus×*H. annuus*: H61	87	72	10	7	8
H. rigidus rigidus×*H. annuus*: H27	200	168	10	10	5
H. annuus×*H. occidentalis*: H21	119	119	23	10	8
H. annuus×*H. occidentalis*: R1210	126	126	390	12	9.5

The medium is MS with sucrose (3%) and different associations of phytohormones (auxin, cytokinins).

The most favorable medium for different genotypes is NAA (0.5 mg/l with BAP (0.5 mg/l). This permits the formation of callus (96%) and the regeneration of plants (44%, all genotypes mixed). When the concentration of BAP is reduced to 0.1 mg/l, the percentage of callogenesis falls to 57%. The in vitro result depends on the genotype, as the percentages of regenerated plants shows (Table 2).

The percentage of regenerated plants varies from 0 for the hybrids H.41 and H.83 to 71% for the hybrid H.37, which is of great agronomic interest. The most favorable sampling stage for eight wild species and their interspecific hybrids is that of uninucleate microspores before the first pollen mitosis (pale green anthers).

2.2 Analysis of Plants Regenerated by Androgenesis

2.2.1 Degrees of Ploidy

At Montpellier, viable plants developed on anthers isolated from different wild species, and hybrids of the collection were analyzed. The examination of mitotic

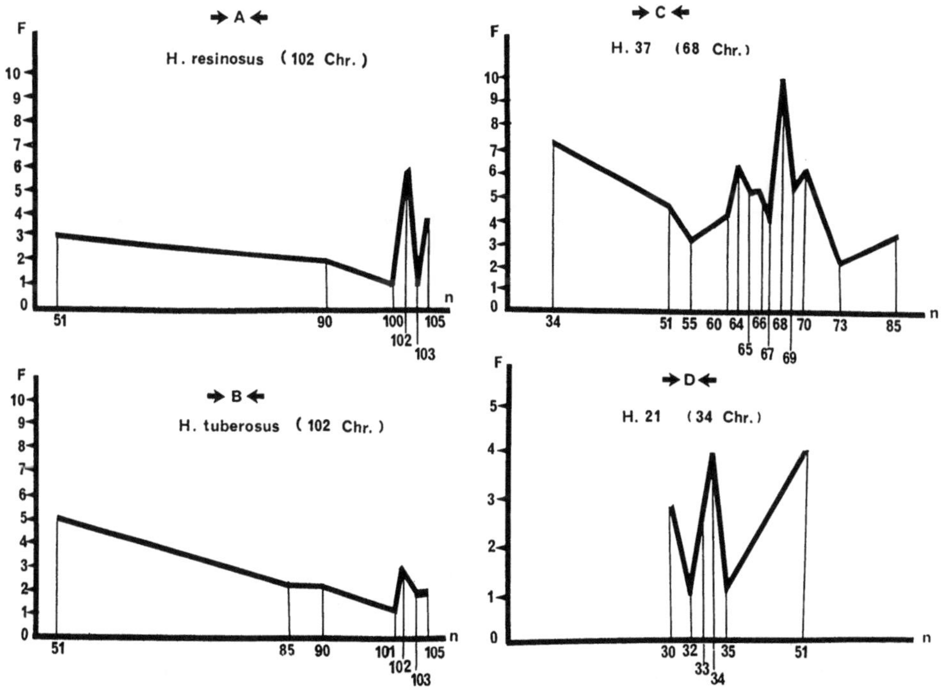

Fig. 1. The androgenetic lineage of two interspecific F_1 hybrids H.37 (68 chromosomes) H21 (34 chromosomes) and two species *H. resinosus*, *H. tuberosus* with 102 chromosomes (Assad 1986)

Table 3. Observations of meioses with the in vitro regenerated plants coming from anthers of the interspecific hybrid H37 (with $2n = 4\times = 68$), 15 of 30 plants gave metaphasic plates which can be counted (Assad 1986)

Numerals of plants	Average of univalent on 5 cells for each plant	Average of bivalent on 5 cells for each plant	Entire number of chromosomes in the analyzed plant	
1	2.4	15.8	34	3 supposed haploid
2	6.0	14.0	34	
3	18.0	8.0	34	
4	21.6	23.2	68	4 supposed diploid
5	14.8	26.2	68	
6	11.6	28.2	68	
7	9.6	29.2	68	
8	27.8	11.6	51	8 supposed aneuploid
9	31.8	9.6	51	
10	19.8	15.6	51	
11	14.8	22.6	60	
12	12.6	24.2	61	
13	16.0	25.0	66	
14	15.0	26.0	67	
15	12.4	28.8	70	

plates of root cells allowed the definition of three genetic states, with the hybrid H.37 with chromosomes:

7 haploid plants with 34 chromosomes,
10 diploid plants with 68 chromosomes,
47 aneuploid plants: 4 plants with 51 chromosomes

may be derived by elimination of a genome with 17 chromosomes with a plant which initially has 68 chromosomes. Three plants with 85 chromosomes may have come by the addition of a genome with 17 chromosomes to the 68 of the original plant (Table 4).

The same variability was observed in the lineage of the hybrid H.21 with 34 chromosomes where there are no haploid plants, and the two wild species *H. resinosus* and *H. tuberosus* with 102 chromosomes. Figure 1, A, B, C, and D shows the frequency (F) of the different degrees of ploidy (n) with the androgenetic plants coming from the two perennial and wild species: *H. resinosus* and *H. tuberosus* with 102 chromosomes and two interspecific F_1 hybrids = H.37 with 68 and H.21 with 34 chromosomes. The observation of meiotic structures in anthers at the beginning of flowering confirmed this genetic variability observed in the androgenetic lineage (Table 3).

Besides the young haploid plants obtained, the presence of aneuploid plants is useful because it may allow the restoration of fertility with a less fertile interspecific hybrid (Georgieva-Tordorava and Bohorova 1980; Georgieva-Todorova 1984) (Fig. 1 – Table 3).

Table 4. Observation on pollen grains of the androgenetic plants of the hybrid H37, showing different degrees of ploidy. 64 plants were analyzed. The hybrid H37 is initially very sterile (10% fertility of the mother plant) (Assad 1968)

No. of chromosomes in each in vitro regenerated plant	Number of plants analyzed	Total number of observed pollen grains	Mean % of colorless pollen grains	Abnormally colored pollen grains[a]	Normally colored pollen grains[b]
34	7	862	18	15.1	66.9
51	4	418	81.3	16.3	2.4
55	3	306	87.9	10.5	1.6
60	4	477	12.0	9	79
64	6	714	14.3	72.8	12.9
65	5	Less quantity	0	0	0
66	5	675	20.8	15.3	63.9
67	4	475	15.2	6.3	78.5
68	10	1209	39.6	20.1	40.3
69	5	564	11.9	8.3	79.8
70	6	686	10.2	7.6	82.2
73	2	227	10.0	11.5	78.4
85	3	320	88.1	11.6	0.3

[a] Abnormally colored or colorless grains of pollen characterize the sterile pollen material.
[b] The normal and colored grains are a fertility sign of the pollen material which is proved in the field.

2.2.2 Restoration of Fertility in the Lineage of the Interspecific F_1 Hybrid H.37

Net restoration of pollen fertility has been shown in the androgenetic line of the hybrid H.37. As the results in Table 4 show, the initial fertility of this hybrid with 68 chromosomes is only from 10% to 20%, but with the androgenetic plants with 34 chromosomes, 67% of the pollen is normal, and with the plants with 68 chromosomes, 40% of the pollen is normal, but with great heterogeneity.

In aneuploid plants with 60, 66, 67, 69, 70, and 74 chromosomes, the percentages of normal pollen vary from 78% to 82%. This restoration of fertility was substantiated in the field. These examples of fertility restoration open the possibility of using genetic material which was not workable before. Plants with restored fertility can be crossed either with the cultivated sunflower or among themselves. Their hybrids can be improved again in a second cycle of in vitro androgenesis fixing the new characteristics introduced during hybridization.

3 Androgenesis in the Cultivated Sunflower, *Helianthus annuus* L.

3.1 Material and Methods

For the cultivated sunflower, except for some data of Mix (1985), no result was published before the publication of Mezzarobba and Jonard (1986) who used the INRA collection at Montpellier-Maugio:

Table 5. Effects of genotype on the in vitro development of the cultivated sunflower anthers at different stages of development (Mezzarobba and Jonard 1986)

Genotypes	No. of anthers sown	Organogenic callus		Embryogenic anthers	
		No.	%	No.	%
Stage before meiosis					
MH 1.2	158	1		0	
M 6.4	202	0		0	
82 R 159	170	0		0	
M 6.4×82 R159	225	0		1	
Do 131	202	0		0	
Luciole	185	0		0	
Remil	192	0		0	
Rodéo	210	0		0	
All genotypes are combined	1544	1	(0.06)	1	(0.06)
Stage between meiotic diad and tetrad					
MH 1.2	192	71	37	0	–
M 6.4	186	15	8	4	2.2
82 R159	203	0	–	1	–
M 6.4×82 R159	215	11	5.1	21	9.8
Do 131	198	6	3	9	4.5
Luciole	231	1	–	6	2.6
Remil	222	8	3.6	2	0.9
Rodéo	197	5	2.5	0	–
All genotypes are combined	1644	117	7.1	43	2.6
Stage of uninucleated microspores with vacuoles					
MH 1.2	162	16	9.9	0	–
M 6.4	203	5	2.5	2	1
82 R159	181	0	–	0	–
M 6.4×82 R159	195	4	2.1	5	2.6
Do 131	211	3	1.5	2	0.9
Luciole	208	0	–	1	–
Remil	153	2	1.3	0	–
Rodéo	162	0	–	0	–
All genotypes are combined	1475	30	2	10	0.7

Three stocks MH 1–2 = INRA stock
M 6–4 = Spanish stock
82 R 159 = INRA stock
and five hybrids
DO 131 = American material
Luciole = INRA material
Remil = INRA material
M6–4×82–R159 = INRA-CST material
Rodeo = INRA-CST material

The anthers were taken from the mother plants, greenhouse- or field-grown, at three stages:

Stage A = premeiotic stage = mother cells

Stage B = méiosis II = cells between the diad and the tetrad stage
Stage C = uninucleate microspores with vesicles.

The cultivation medium contains MS solution with agarose (6 g/l), sucrose (60 g/l) plus a mixture of vitamins of Morel and Wetmore as follows:

meso Inositol	= 1 mg/l
Calcium pantothenate	= 1 mg/l
Nicotinic acid	= 1 mg/l
Pyridoxin	= 1 mg/l
Thiamine	= 1 mg/l
Biotin	= 0.01 mg/l
Vit B12	= 0.1 mg/l

and an amino acid, glycine (1 mg/l), with two phytohormones, NAA and BAP (0.5 mg/l).

3.2 Influence of the Stage of Anther Sampling for Different Genotypes

The examination of the results summarized in Table 5 shows that the percentages of embryogenous anthers vary according to their sampling stage. Stage B, between the two periods of meiosis, is the most favorable for the cultivated sunflower for in vitro development of microspores.

In general, in the in vitro reactivity of cultivated anthers, the genotype M6−4×82−R159 gave the best results. The embryos at the torpedo stage are transferred onto a new medium which is different from the isolation one. The macro-elements are half-strength (except for the Fe EDTA), sucrose reduced (10 g/l); auxin is omitted, and the concentration of BAP is reduced to $5\ 10^{-4}$ mg/l.

When the first leaves appear, the young plants are transferred again onto the rooting medium as given but with NAA at 0.02 mg/l and BAP at 0.01 mg/l; ploidy is shown in root tips after staining with acetocarmin. Several young haploid plants were obtained (Figs. 2, 3).

3.3 Effect of a Thermic Shock on the Development of Anthers

Table 6 indicates the results obtained after different pretreatments applied to in vitro isolated anthers, for all genotypes. The percentage of anthers producing embryos varies according to the thermic pretreatment. The best percentage is obtained after a treatment at 35 °C during the first 12 days of cultivation.

The values in Table 7 show the effects of treatment at 35 °C (during the first 12 days of cultivation in darkness) on the improvement of quantities of embryogenic anthers and regenerated plants. Independently of the genotype, most of the regenerated plants showed precocious flowering. Some of the plants obtained were identified as haploids.

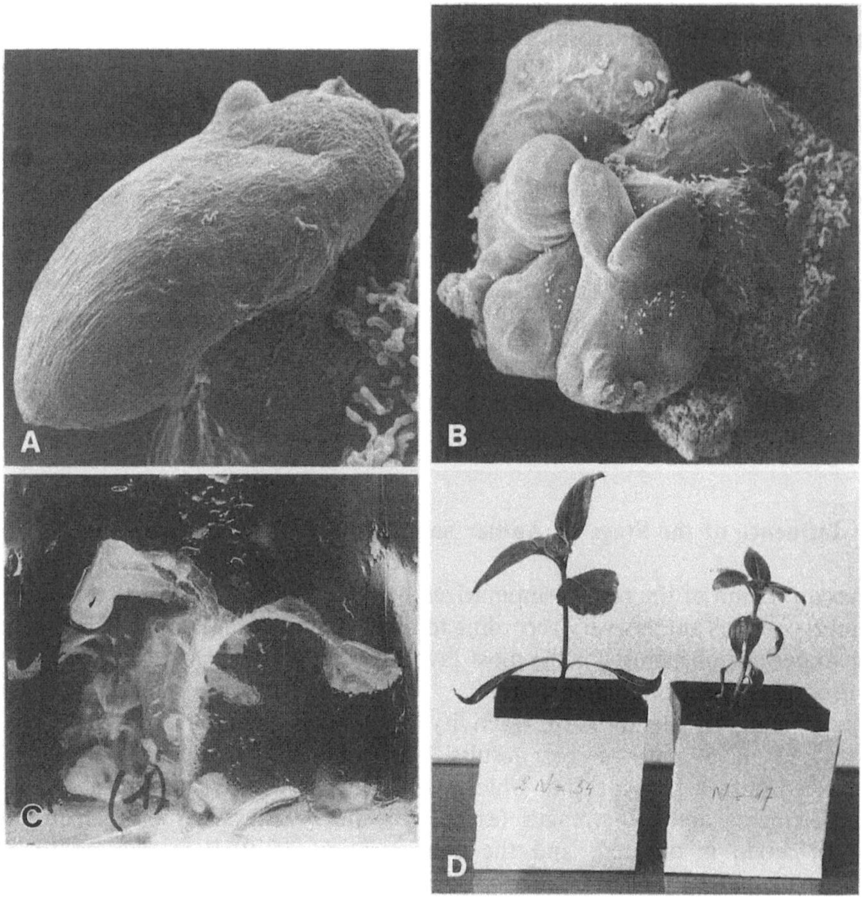

Fig. 2. A Sunflower androgenetic embryo (F_1 hybrid Luciole) at the torpedo stage (EM×40) (Mezzarobba and Jonard 1986). **B** Sunflower embryo F_1 hybrid: M64×82 R159) at the 2 cotyledons stage (EM×40) (Mezzarobba and Jonard 1986). **C** Androgenetic sunflower plantlet from the development of an embryo (stage of transfer in the soil) (Mezzarobba and Jonard 1986). **D** *Left* Mother-plant at $2n = 34$ chromosomes; *right* androgenetic plant at $n = 17$ chromosomes (Mezzarobba 1988)

3.4 Increase in Viable Plants Obtained by Improvements in Cultivation Medium

During recent experiments we utilized four new types of cultivated sunflower:

$B_{11}A_3$ = INRA stock with "anthocyanin" hypocotyl (T/t) as dominant factor. INRA 4701 = hybrid T1 INRA with "anthocyanin" hypocotyl (T/t) H_{11} and H_{12} from the hybridization M64×82 R159. We also utilized two modified cultivation media: one for the initiation of androgenesis, the other for the development of formed embryos.

The solid initiation medium contained MS half-strength, vitamins of Morel and Wetmore with vitamin B12 (0.2 mg/l) and a mixture of amino acids

Fig. 3. A *Left* mother sunflower; *right* androgenetic sunflower plant (Mezzarobba 1988). **B** Meiosis from doubled-haploid plant (2n = 34) (Mezzarobba 1988). **C** Mitosis from haploid plant (n = 17) (Mezzarobba 1988)

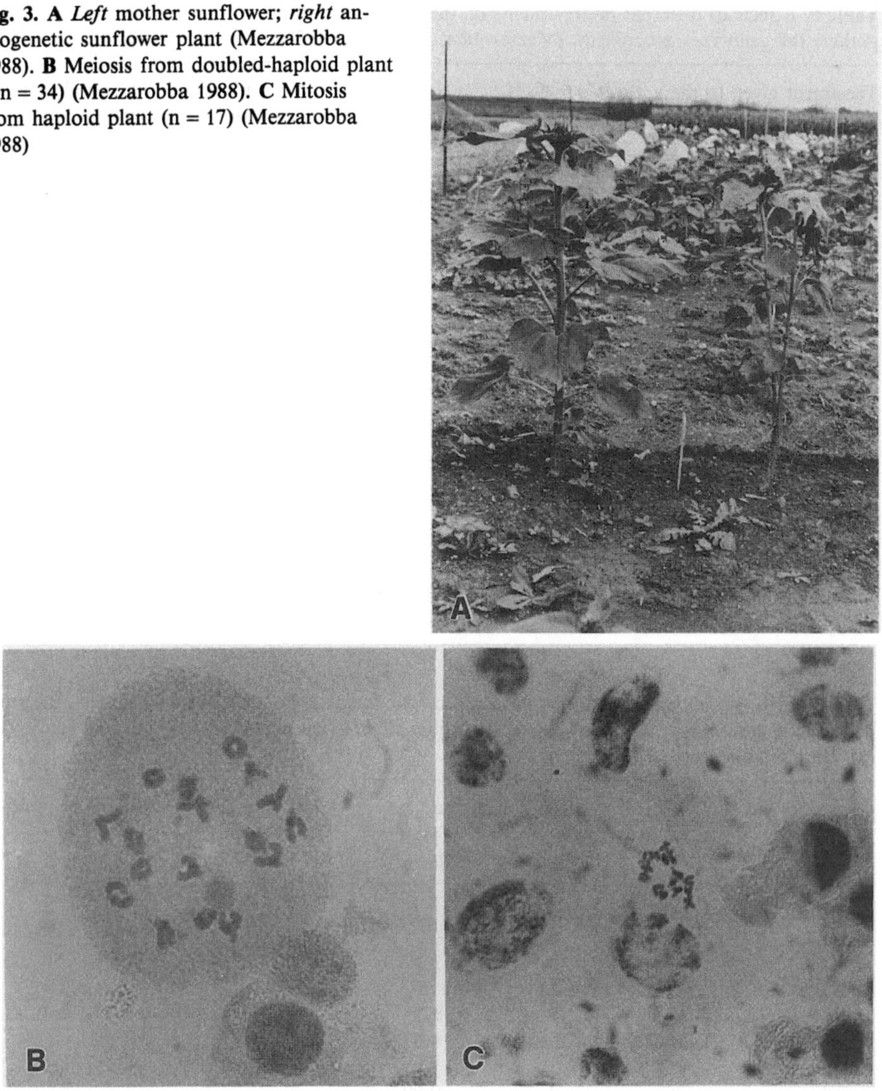

with (mg/l): L-glycine 2, L-glutamine 40, L-serine 25, L-tryptophane 1, L-cysteine 2.5.

The sucrose concentration was 120 g/l supported by Gelrite (3.3 g/l). pH was 5.9; two phytohormones: NAA and BAP 0.5 mg/l were added.

The embryo transfer medium is liquid. Embryos are put on a filter paper bridge. This medium contains:

- macro and microelement solution of Monnier
- vitamins of Morel and Wetmore, associated with the given mixture of amino acids. Sucrose is utilized at weak concentration (15 g/l); pH is 5.9.

The results are collected in Tables 8 and 9. The utilization of two different media leads to a general improvement of the results found before, with 762 embryos

Table 6. Effects of different pretreatments on the in vitro development of the cultivated sunflower anthers (all genotypes are mixed). (Mezzarobba and Jonard 1986)

Treatment given to the anthers	No. of anthers sown	No. of embryogenic anthers	% of embryogenic anthers
1st reference: 25 °C-2000 lx 16 h in light	212	6	2.8
2nd reference: 25 °C-darkness	219	8	3.7
30 °C – Darkness for			
2 days	230	9	3.9
4 days	242	8	3.3
8 days	215	12	6
12 days	202	12	5.9
16 days	188	10	5.3
35 °C – Darkness during			
2 days	190	5	2.6
4 days	205	4	2
8 days	209	21	10
12 days	218	38	**17.4**
16 days	198	34	17.1

The results are taken after 3 weeks of cultivation with storage at 25 °C in the darkness after the pretreatments.

Table 7. Effects of a thermic treatment at 35 °C on the formation of embryogenic anthers, the number and the nature of regenerated plants, according to different genotypes of cultivated sunflower. (Mezzarobba and Jonard 1986)

Genotypes	No. of anthers sown	Embryogenic anthers		No. of regenerated plants	Degree of ploidy
		No.	%		
Treatment of anthers: 25 °C – darkness					
MH 1.2	182	0	–	–	–
M 6.4	178	2	1.1	–	–
82 R 159	192	2	1.0	–	–
M 6.4 × 82 R 159	198	18	**9.1**	1	**n = 17**
Do 131	210	10	**4.8**	1	2n = 34
Luciole	162	3	1.9	–	–
Remil	201	2	0.99	–	–
Rodéo	188	0	–	–	–
Treatment of anthers: 35 °C 12 days in darkness					
MH 1.2	192	8	4.2	1	2n
M 6.4	186	22	11.8	–	–
82 R 159	203	6	2.9	–	–
M 6.4 × R 159	215	118	**54.9**	4	3 at 2n and 1 at **n**
Do 131	198	61	**30.8**	2	2n
Luciole	231	32	13.9	1	**n**
Remil	222	29	13.1	–	–
Rodéo	197	4	2.1	1	2n

Table 8. In vitro development of anthers sown on modified MS medium after isolation from different genotypes of cultivated sunflower. Anthers subjected to a thermic pretreatment at 35 °C for 12 days. (Mezzarobba 1988)

Genotypes	No. of anthers sown	Embryogenic anthers				Organogenic anthers	
		Direct embryogenesis		Indirect embryogenesis		No.	%
		No.	%	No.	%		
MH 1.2	238	0	0	10	4.2	107	45
M 6.4×82 R159	435	252	58	17	3.9	0	–
Do 131	310	107	34.5	19	6.1	4	1.3
B11 A3	258	39	15.1	16	6.2	3	1.2
INRA 4701	322	71	22	7	2.2	7	2.2
H 11	198	53	26.8	22	11.1	0	–
H 12	266	56	21.1	40	15	0	–
Rodéo	251	8	3.2	45	17.9	0	–

Table 9. Growth and development of embryos formed on initial medium after transfer to a new liquid medium. (Mezzarobba 1988)

Genotypes	Direct and indirect sown embryos		Viable plants obtained			
				Degree of ploidy		
	No.	%	No.	n	2n	Aneuploid plants
MH 1.2	10	4.2	2	0	2	0
M 64×82 R153	269	61.8	32	4	25	3
Do 131	126	40.6	13	2	10	1
B11 A3	55	21.3	11	1	10	0
INRA 4701	78	24.2	14	0	8	6
H 11	75	37.9	8	0	5	3
H 12	96	36.1	10	1	8	1
Rodéo	53	21.1	1	0	0	1
All genotypes are mixed	762	30.9	91	8	68	15

(31%) and 91 viable plants with 15 aneuploid, 8 haploid, and 68 diploid plants: among these latter, some haplodiploid must be verified, as all the genotypes are mixed.

3.5 Verification of Haplodiploid Plants with Viable Ones with 2n Chromosomes

By using the "anthocyanin" marker from the two genotypes, $B_{11}A_3$ and INRA 4704, we could verify the spontaneous doubling of haploid plants, as the observations indicated in Table 10.

Table 10. Verification of haploid plants in 2 genotypes with "anthocyanin" genetic marker. (Mezzarobba 1988)

Genotypes	Number of viable plants with 2n chr.	$\frac{T}{T}$ or $\frac{T}{t}$ "Anthocyanin" containing plants among the 2n chr.	$\frac{t}{t}$ No "anthocyanin" containing plants among the 2n chromosomes
B11 A3	10	4	6
INRA 4701	8	5	3

Table 11. Verification in fields of plants coming from the self-fertilization of plants with 2n chr. from in vitro androgenesis. (Mezzarobba 1988)

Genotypes	No. of diploid regenerated plants	No. of plants giving seeds after self-fertilization	Fixed lineage	Lineage in segregation
MH 1.2	2	0	0	0
M 64×82 R159	25	8	6	2
Do 131	10	4	3	1
B11 A3	10	3	3	1
INRA 4701	8	6	4	2
H 11	5	2	2	0
H 12	8	4	3	1
Rodéo	0	–	–	–
All genotypes combined	68	27	21	6

This table shows the presence of a plant without "anthocyanin", thus with recessive genomes t/t which are certainly doubled haploid plants.

3.6 Verification in the Field of the Lineage Coming from the Self-Pollinated Plants with 2n Chromosomes

For all the genotypes tested, we obtained 21 haplodiploid plants fixed to the stable genome (Table 11). However, other plants are now being segregated which indicate either the creation of variants induced by in vitro culture with haplodiploid plants, or a somatic origin of the embryo coming from diploid cells of the anther.

4 Conclusions

With wild species with high chromosome number, the reduction by half of genomes will facilitate the crossings with the cultivated species. So the androgenetic descendants of hybrid *H. annuus*×*H. resinosus* ($2n = 4\times = 68$) can

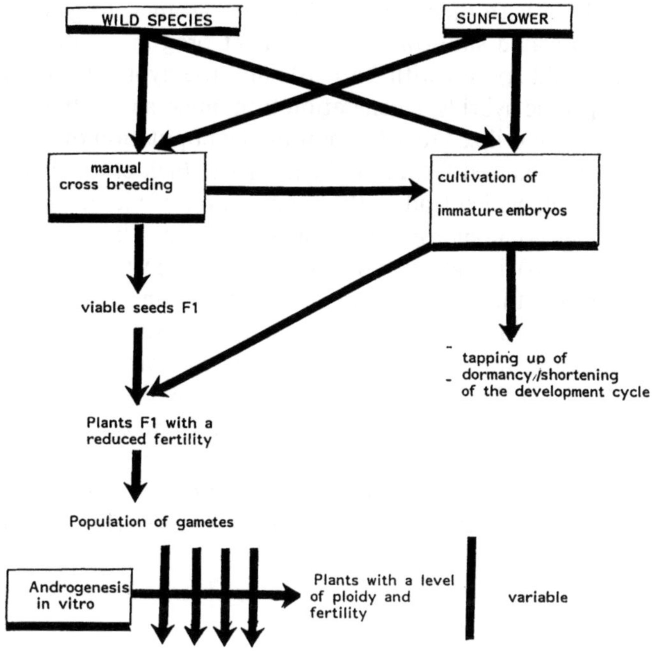

Fig. 4. The possibilities of incorporation of in vitro culture techniques: androgenesis and isolation of isomature embryos in a diagram of improvement of the sunflower (Assad 1986)

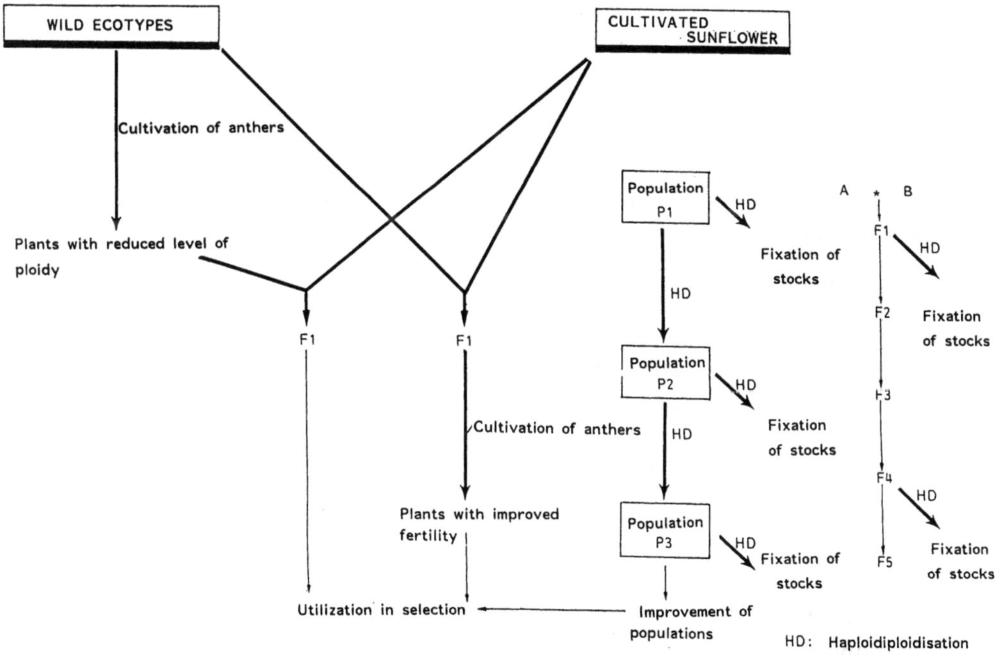

Fig. 5. The haplodiploidization for improvement of cultivated sunflower (Mezzarobba 1988)

be utilized with their form at 34 chromosomes, to give tolerance to *Sclerotinia*, in a cultivated sunflower improvement program, with more efficiency than the parent wild species with $2n = 102$ and the hybrid from which they derive. With interspecific hybrids, regeneration from microspores but also from the anther wall cells can be utilized to select among the population of gametes produced by a hybrid with reduced fertility (Whelan 1976; Fick et al. 1976), the chromosomic arrangements which lead to the regeneration of viable plants with normal fertility.

For us, some androgenetic individuals coming from the hybrid H.37 (*H. annuus* × *H. resinosus*), initially very sterile, show an important restoration (80%) of their pollen fertility; this improvement can be utilized in sunflower improvement programs.

In the cultivated sunflower, by androgenesis (creation of haplodiploid plants), it is possible to rapidly obtain stocks fixed from heterozygous individuals, allowing better exploitation of the genetic variability to create varieties.

Associated with the in vitro isolation technique of immature embryos, which permits shortening the development cycles of cultivated and wild species, and restoration of viability to some hybrids, androgenesis can affect important improvement in the selection scheme of the sunflower (Figs. 4 and 5).

5 Protocol

Best Experimental Conditions. The most favorable **stage of sampling** for the cultivated sunflower is the period B between the two phases of meiosis; for the wild species this is stage C with pale green anthers, which are rich in uninucleated microspores with vesicles.

A net effect of the genotype is noted in the in vitro reactivity of anthers: the hybrid genotype of cultivated sunflower M6−4×82−R159 is the most reactive.

Anther pretreatments at 35 °C in darkness during the first 12 days of culture increase the percentage of embryogenic anthers, whatever may be the genotype.

6 Conditions for Culture

The first medium for initiation of androgenesis contains half-strength MS solution with agar (Gelrite: 3.3 g/l), sucrose (120 g/l) and a mixture of vitamins:

meso inositol	= 100 mg/l
calcium pantothenate	= 1 mg/l
nicotinic acid	= 1 mg/l
pyridoxine	= 1 mg/l
thiamine	= 1 mg/l
biotin	= 0.01 mg/l
vitamin B_{12}	= 0.2 mg/l

and with several amino acids, in mg/l:

L-glycine = 1
L-glutamine = 40
L-serine = 25
L-tryptophan = 1
L-cysteine = 2.5

Two phytohormones are added with this basal medium: naphthalene acetic acid (NAA) and benzylamino purine (BAP) at 0.5 mg/l.

A second liquid medium for transfer of embryos contains Monnier solution, with the given mixtures of vitamins and amino acids, but with only 15 g/l sucrose.

The development of embryos in vitro on a filter paper bridge till the young plant period occurs at 25 °C, 1000 lx, till their transfer to the greenhouse.

References

Assad A (1986) Application de la culture in vitro à l'amélioration du Tournesol, *Helianthus annuus* L. Thèse USTL, Montpellier, p 158

Assad A, Serieys H, Jonard R (1985) Sur les possibilités de régénérations d'espèces sauvages et d'hybrides interspécifiques du genre *Helianthus* par androgénèse in vitro C.R. Acad Sci, Paris 300 III (1):25–30

Bohorova N, Atanossov AI, Antonov BP (1980) In vitro isolation of anthers from interspecific hybrids in the *Helianthus* genus. C.R. Acad Bulg. Biol genet 33 (11):1545–1548

Bohorova N, Atanossov AI, Georgieva-Todorova J (1985) In vitro organogenesis, androgenesis and embryo culture, in the genus *Helianthus* L. Z Pflanzenzuecht 95:35–44

Bonjean A (1986) Tournesols de France. CST-Union des coopératives agricoles de semences de Tournesol Paris, 190 pp

Chandler JM, Beard BH (1983) Embryo culture of *Helianthus* hybrids. Crop Sci, 23 (5):1004–1007

Chandler JM, Jan CC (1983) Production of interspécific sunflower hybrids through tissue culture. Agron Abstr 58

Fick GN, Zimmer DE, Dominquez-Gimenez J, Rehder DA (1976) Fertility restoration and variability for plants and seed characteristics in wild sunflowers. Proc 6th Int Sunflowers Conf, Alex Viorel Vranceau, Romania, pp 334–338

Georgieva-Tordorova J (1984) Interspécific hybridization in the genus *Helianthus*. Z Pflanzenzuecht 93:265–279

Georgieva-Tordorova J, Bohorova NE (1980) Interspécific hybridization in sunflower breeding. Proc Int Sunflower Conf, Torremolinos, Spain 1:101–106

Heiser CB (1969) The North American sunflower (*Helianthus*). Mem-Torrey Bot Club 22 (3):218

Leclercq (1969) Une stérilité mâle cytoplasmique chez le Tournesol. Ann Amelior Plant 19 (2):99–106

Mezzarobba A (1988) L'androgenèse chez le Tournesol. Thèse USTL. Montpellier, 150 pp

Mezzarobba A, Jonard R (1986) Effet du stade de prélèvement et des prétraitements sur le développement in vitro d'anthéres prélevées sur le Tournesol cultivé (*Helianthus annuus* L.). CR Acad Sci Paris, 303 III (5):181–186

Mix G (1985) Antheren und Ovarienkultur von Sonnenblumen (*Helianthus annuus* L.). Landbauforsch Volkenrode 35:153–156

Monnier M (1976) Culture in vitro de l'embryon immature de *Capsella bursa-pastoris* Moench. Rev Cytol Biol Veg 39:1–20

Morel G, Wetmore RH (1951) Tissue culture of monocotyledons. Am J Bot 38:138–140

Murashige T, Skoog F (1962) A revised medium for rapid growth and bioessays with tobacco tissue cultures. Physiol Plantarum 15:473–497

Oil World (1987) (Oleagineux) Mars 1987, 8 pp

Rogers CE, Thompson TE (1978) Evaluation of *Helianthus* species for resistance to insects pests. Proc 8th Int Sunflower Conf, Minneapolis, pp 320–327

Serieys (1984) Collection, evolution and conservation of wild species and their use in sunflower breeding programs FAO. Progr Rep Subn Work, Novissad, pp 109–113

Thompson TE, Rogers CE, Zimmerman DC, Huang HC, Whelan EDP, Miller JF (1978) Evaluation of *Helianthus* species for disease resistance and oil content. Proc 8th Int Sunflower Conf, Minneapolis, pp 501–509

Whelan EDF (1976) Sterility problems in interspecific hybridization of *Helianthus* species. Proc IIth Sunflower Forum, pp 5–7

IV.8 Flax (*Linum usitatissimum* L.): In Vitro Studies

A. McHUGHEN[1]

1 Introduction

1.1 Importance

The genus *Linum* contains one species of economic importance as a crop (*usitatissimum*), although some other species have limited value as ornamentals. *Linum usitatissimum* has been cultivated by man for thousands of years; its importance even in ancient times is evidenced by its importation by the Egyptians over 6000 years ago (Trump 1980) and it has been found in remains of prehistoric settlements. The plants were useful as a source of fibre for weaving, oil from the seeds and feed from the meal, hence the species name, *usitatissimum* – meaning "most useful". In addition to being versatile, the plant has been adapted to grow in many parts of the world, and is cropped on all arable continents.

World production of oilseed flax ("linseed") in 1982 was 2.9 million tons, grown on 5.4 million hectares. Fibre flax production for 1982 totalled 0.6 million tons, grown on 1.3 million hectares worldwide. Total *Linum* area was 6.7 million hectares in 1982 (FAO 1982).

Decreased demand for fibre flax is due to the introduction of synthetic fibres easier to care for and cheaper than natural linen in many parts of the world. Flax oil is a high-quality drying oil used industrially in the manufacture of paint, varnish, linoleum, oil cloths, and other such products because it forms a tough, durable film on drying. This property also makes linseed oil useful as a sealant for concrete, to keep moisture and salts out, and aids the curing process. Unfortunately, the same component of the oil that makes it such a good drying oil, linolenic acid, also makes it unsuitable as an edible oil. Linolenic acid (C18:3) comprises about 50% of the oil, but this varies according to genotype, growth conditions, etc. The linolenic acid is not inedible in itself, as indeed, there is evidence that the ancient civilizations consumed linseed oil (Dillman 1936) as do modern societies, albeit in small quantities. The acid does, however, metabolize rather rapidly into toxic and semi-toxic products. It is this rapid rancidity that makes flax oil unsuitable as a commercially viable edible oil.

The market for flax, at present, is dependent upon the industrial application of its oil as a drying product. While such products as linoleum and nonacrylic

[1] Crop Development Centre, University of Saskatchewan, Saskatoon, Saskatchewan S7N 0W0, Canada

Biotechnology in Agriculture and Forestry, Vol. 10
Legumes and Oilseed Crops I (ed. by Y.P.S. Bajaj)

conventional paints are diminishing, other uses, such as concrete sealant and anti-spalling agent, are being promoted. Overall, the market for flax is relatively small but stable. The introduction of a low-linolenic acid linseed oil variety would greatly expand the use of flax, as it would then be a high quality edible oil, approximately equal in quality to sunflower oil. This is one primary breeding objective, and has recently been achieved by Green (1986).

1.2 Distribution

Linum is grown on every continent, but does best in areas with cool evenings; high altitudes or high latitudes are preferred for highest quality of oil. Argentina, Canada and India are the major producers of linseed oil, together accounting for 70% of world production. About 40 other countries around the world produce the remainder, including such places as Australia, USSR, Ethiopia and France (FAO 1982). In spite of its obviously wide distribution, it does not appear to be a primary crop anywhere.

1.3 Diseases

Linum is susceptible to many diseases, including rust (*Melampsora lini* (Ehrenb.) Lev.), wilt (*Fusarium oxysporum* f. sp. *Lini* (Bolley) Snyd. and Hans) and pasmo (*Septoria linicola* (Speg.) Garass.); those of less importance include several *Rhizoctonia* species, Polyspora lini Laff., aster yellows, and crinkle virus (Hoes 1975; Dybing and Lay 1981). Wilt was primarily responsible for the pioneering aspect of flax production in North America; as wilt attacked the flax crops on the east coast, flax was forced further and further west, until it reached its present growing area, west of the Great Lakes, east of the Rocky Mountains. Wilt was overcome by selective breeding by H.L. Bolley (Dillman 1936).

Rust can be a major problem for flax growers. Although cultural practice (rotation) and breeding for resistant varieties helps to control losses from rust, the fungus evolves new, virulent races quickly, so flax breeders must expend effort on maintaining resistant germplasm to keep ahead of the disease (Hoes 1975).

The third disease of importance to flax growers worldwide is pasmo. This fungal disease is rarely a major problem because fungicidal seed treatments and crop rotations give adequate control.

The other, more minor diseases are either fungal (*Rhizoctonia* ssp.; *Polyspora lini*, etc.), mycoplasma (aster yellows) or viral (crinkle), the latter two being insect transmitted by the same insect, the six-spotted leafhopper (Hoes 1975). Overall, seed treatment with fungicides, good cultural practices and breeding for disease resistance have been successful in protecting the *Linum* crops. Although occasional disease outbreaks occur in different areas, major pandemic losses are unlikely.

1.4 Conventional Breeding and Need for Novel Techniques

Linum usitatissimum is self-pollinating and very little outcrossing occurs in the field (~2% – Martin and Leonard 1949). Conventional flax breeders have limited interspecific germplasm available for crossing. While successful crosses with *L. pallescens* and *L. africanum* have been reported (Bari and Godward 1970), and several crosses with related species were reported by Gill and Yermanos (1967) and by Seetharam (1972), *L. usitatissimum* is easily crossed only with *L. angustifolium*, the probable ancestor of our flax (Tammes 1928).

In spite of these limited germplasm resources, breeders have a fairly wide selection of traits such as rust and wilt resistance, seed size, oil content, oil quality and other important traits (Kenaschuk 1975). Breeders expend effort in maintaining resistance to rust and other diseases, to developing differences in fatty acid composition to "fine tune" the oil for specific purposes and to develop earlier maturity, higher yields, and other "motherhood" issues.

However, some traits are difficult or beyond the technical capability of the flax breeder due to the limited crossable germplasm; for example good genetic resistance to pasmo has not been attained (Kenaschuk 1975). Resistance to the other major diseases is acceptable at present, but the rapid evolution of rust races means that eventually a rust race will rise for which there is no resistance in *Linum*. Unless other sources of resistance are developed and incorporated into flax, this eventuality could devastate the world flax market.

Resistance to herbicides is a desirable breeding objective because flax does not compete well with weeds and major losses can occur without weed control. Fortunately, *Linum* does have natural tolerance to several types of herbicides, so weed control is usually acceptable where herbicide use is practicable. But as new herbicides are developed, if flax is not already resistant, there will be little chance of incorporating it through breeding.

Salt tolerance is a major concern in some flax growing areas, especially in North America and Egypt. Conventional means are stymied here because there are simply no suitable sources of salt tolerance available to the flax breeder.

Novel breeding methods will probably make a major impact on these three areas:

- disease resistance; especially to new, virulent races of rust.
- herbicides; especially modern, highly specific compositions. The flax market is probably too small for herbicide manufacturers to design such a herbicide for the flax crop exclusively.
- stress tolerance; there is no natural source of salt tolerance in *Linum*, so without novel methods, flax could never be productively grown on saline soil. Similarly to other stress problems in many flax-growing regions of the world – cold tolerance, early maturity, etc. may be incorporated into flax using novel breeding methods.

2 In Vitro Approaches

2.1 General

Linum is relatively simple to culture in vitro: to develop callus or suspension cultures, and to regenerate shoots from callus. The ability of flax to regenerate shoots from decapitated hypocotyls was reported years ago (Link and Eggers 1946). This capability of the flax hypocotyl is maintained in vitro (Gamborg and Shyluk 1976). Complete plant regeneration has been reported by several workers using various media, tissues, etc. (Table 1).

The versatility of *Linum* cells in vitro is also indicated by the range of objectives of different groups; Table 1 provides several examples of using *Linum* cul-

Table 1. Utility of in vitro cultures of *Linum usitatissimum* (and related species, where noted)

Inoculum	Objective	Response	Reference
Anther	Haploid plants	Haploid plants	Sun and Fu (1981)
Shoot meristem	Regeneration	Shoots, roots	Lane (1979)
Shoot (protoplasts)[a]	Regeneration	Shoots, roots	Barakat and Cocking (1984)
Stem pieces	Regeneration	Shoots, roots	Chlyah et al. (1980)
Stem pieces	Regeneration	Shoots, roots	Murray et al. (1977)
Stem pieces	Nucleic acid analysis	Callus	Cullis (1981)
Cotyledon	Regeneration	Shoots, roots	Rybczyński (1975)
Cotyledon	Regeneration	Roots	Ibrahim (1971)
Cotyledon	Regeneration	Roots	Mathews and Narayanaswamy (1976)
Cotyledon (suspension)	Phenolic analysis	Callus and suspension	Ibrahim and Phan (1978)
Cotyledon (protoplast)	Regeneration	Shoots, roots	Barakat and Cocking (1983)
Seedling	Regeneration	Shoots, roots	McHughen and Swartz (1984)
Hypocotyl	Regeneration	Shoots, roots	Lane (1979)
Hypocotyl	Regeneration	Shoots, roots	Mathews and Narayanaswamy (1976)
Hypocotyl	Regeneration	Shoots, roots	Gamborg and Shyluk (1976)
Hypocotyl	Regeneration	Shoots, roots	Ibrahim (1971)
Hypocotyl (suspension)	Freeze preservation	Suspension	Quatrano (1968)
Hypocotyl (protoplast)	Regeneration	Roots	Barakat and Cocking (1983)
Root (protoplast)	Regeneration	Shoots, roots	Barakat and Cocking (1983)
Epicotyl	Genetic transformation	Tumors	Hepburn et al. (1983)
Embryos	Ultrastructure	Embryos	Ciamporova and Pretǒvá (1981)
Immature embryos	Regeneration	Somatic embryos	Pretǒvá and Williams (1986)
Seedlings, shoot tips[b] (protoplasts)	Regeneration	Shoots, roots	Ling and Binding (1987)
Leaf	Regeneration	Shoots, roots	McHughen (in prep.)

[a] In *L. lewisii* only.
[b] In several *Linum* species.

tures to study biochemical pathways (Ibrahim and Phan 1978), cryopreservation (Quatrano 1968), and genetic transformation using a Ti plasmid system (Hepburn et al. 1983).

Infecting flax cells with *Agrobacterium* is relatively simple. The bacteria containing the T-DNA of choice are grown to log phase in an appropriate liquid or solid medium. Flax inocula freshly excised or pre-cultured are immersed briefly (less than 5s) into the liquid bacterial culture, or colonies from the solid medium are smeared onto the injured flax tissue, then the inoculum plated onto the flax culture medium. The bacteria must be in contact with freshly injured tissue in order to activate the *vir* genes and effect transformation. We use fresh or pre-cultured hypocotyl segments as inocula, after removing the epidermis. After 2–3 days on MS medium with BA or BA and NAA, the tissues are transferred to similar media containing, in addition, 400 μg/ml kanamycin and 500 mg/l of Cefotaxime. The kanamycin retards and eventually kills most flax cells not transformed, and so serves as a selection mechanism. the Cefotaxime retards growth of remaining bacteria. If not included, the bacteria will outgrow the flax tissue and eventually smother and kill the flax. Transformation is indicated by continued growth and shoot formation along the length of the peeled hypocotyl (Jordan and McHughen 1988a). Presumptive transformants must be ascertained by progeny testing or Southern blot analysis; growth of callus tissue on kanamycin medium is not definitive of transformation because of the possibility of SCV conferring resistance, or normal escapes, or of cross-protection of nearby transformed cells (Jordan and McHughen 1988b).

While complete plant regeneration in vitro is a requisite for crop improvement protocols involving cell and tissue culture, these other in vitro studies provide a greater utility and make more of the novel techniques practicable for *Linum*. While most crop species have protocols for plant regeneration from tissues in vitro, few have also been shown capable of regeneration from protoplasts (Barakat and Cocking 1983), infection and genetic transformation in vitro by wild-type *Agrobacterium tumefaciens* (Hepburn et al. 1983), as well as by disarmed strains (McHughen et al. 1986; Basiran et al. 1987) and direct regeneration from organized tissues (hypocotyls). *Linum* cultures are also amenable to direct selection in vitro for variant phenotypes, followed by plant regeneration from the selected cells. Flax plants have been produced from chlorsulfuron-resistant cell lines (Jordan and McHughen 1987) and from salt-tolerant cell lines (McHughen and Swartz 1984). In both cases, the variant types were generated and selected in vitro.

In general, the culturability of *Linum* is ranked as somewhat less than that of Solanaceous species, but much better than legumes or Gramineae. The range of culture techniques to which *Linum* is amenable makes this genus exploitable by the novel methods being developed for crop improvement.

2.2 Protocol

Sterilization of explants/organs: the majority of *Linum* in vitro studies use juvenile tissues-hypocotyls, cotyledons, etc., and are provided by germinating surface sterilized seeds on aseptic media. Seeds are easily surface sterilized. One way is to rinse seeds in 70% ethanol for 1 min, then transfer

Table 2. Summary of media used for various inocula

Callus formation	
Norstog's B_{11} + 2,4-D	McHughen and Swartz (1984)
MS + 2,4-D	McHughen and Swartz (1984)
MS + BA + NAA	Cullis (1981)
White's + 2,4-D	Mathews and Narayanaswamy (1976)
MS + 2,4-D	Mathews and Narayanaswamy (1976)
White + kin + NAA	Ibrahim (1971)
MS (Major salts) + B5	Gamborg and Shyluk (1976)
MS + kin + IAA	Ibrahim and Phan (1978)
MS + kin + 2,4-D	Quatrano (1968)
Shoot or bud formation	
MS + BA + NAA	Murray et al. (1977)
MS + BA + NAA	Barakat and Cocking (1983)
MS + kin + IBA	McHughen and Swartz (1984)
MS	Mathews and Narayanaswamy (1976)
MS + IBA	Mathews and Narayanaswamy (1976)
MS + BA	Lane (1979)
MS[a] + BA	Gamborg and Shyluk (1976)
LS + kin + NAA	Rybczynski (1975)
Shoot elongation	
MS	Murray et al. (1977)
Rooting	
MS	Lane (1979)
MS + NAA	Mathews and Narayanaswamy (1976)
B5	Murray et al. (1977)
1/2 B5 + IAA	Barakat and Cocking (1983)
	Gamborg and Shyluk (1976)
B5	McHughen and Swartz (1984)

Abbreviations: 2,4-D (2,4-dichlorophenoxyacetic acid), BA (benzyladenine), NAA (naphthalene acetic acid), kin (kinetin (6-furfurylamino purine)), IAA (Indole acetic acid), IBA (Indole butyric acid), Norstog's B_{II} medium (Norstog 1973), MS medium (Murashige and Skoog 1962), White's medium (White 1963), B5 medium (Gamborg et al. 1968) and LS medium (Linsmaier and Skoog 1965).

[a] Modified; see reference.

to 1% sodium hypochlorite for 20 min, then rinse in sterile water (Gamborg and Shyluk 1976). Seeds can then be germinated on agar (with or without nutrients) in sealed Petri dishes. The juvenile tissues need no direct sterilization.

Other tissues, such as immature embryos and anthers, are naturally pathogen free and so do not require a sterilization procedure. Flax tissue will form callus and regenerate shoots on any of several media, as indicated in Table 2. Callus will form from many tissues if plated on any standard basal medium (MS, B5, White's, etc.) supplemented with an auxin. Best response will be affected by auxin concentration, genotype, physical culture conditions (T°, light), etc. Shoot, or shoot bud, formation directly from hypocotyls can occur on a simple hormone-free medium, but these buds remain quiescent (Mathews and Narayanaswamy 1976). Typically, MS with BA is most effective for evoking shoot formation (Table 2), rooting of such shoots occurs on transfer to as basal medium (MS or B5) with low or no auxin (Table 2).

Transferring plantlets to soil is more difficult with flax than with many species. Plantlets are very tender and require careful attention or they will fail to establish in oil. One protocol with reasonable success involves transferring washed plantlets to sterile vermiculite in pots in a high humidity chamber (i.e., partially covered with plastic film) (McHughen and Swartz 1984). Now, we transfer washed plantlets directly to sterile sand in a mist chamber for ca. 2 weeks, then into peat pots with sterile soil in a high humidity chamber, gradually lifting off the plastic cover over a period of several days. Even with this elaborate regime, many plantlets wilt or otherwise fail to establish.

2.3 In Vitro Improvement Programs for *Linum*

The culturability of *Linum* is being exploited to produce genetically superior crop varieties. Cells growing in culture, especially in a callus or suspension culture, undergo spontaneous genetic changes ranging from single-gene mutations to gross chromosomal aberrations. By analysing progeny from regenerants from such cells, different genotypes can be identified. These novel genotypes are usually detrimental, but sometimes are potentially valuable to breeders. While many of the genetic variants are similar to those produced through the use of physical or chemical mutagens, others are not. While mutagens are typically destructive, it has been suggested that somaclonal variation affects gene regulation, perhaps through the activity of transposable elements, and this could explain the differences observed between SCV (somaclonal variation) and mutation breeding. For a review of somaclonal variation in general, see Larkin and Scowcroft (1981) and Chaleff (1983), and for an example in flax, see Cullis and Cleary (1986).

We have identified several variant plants derived from flax cells in vitro, including an earlier maturing type that appears to segregate as a single dominant gene and a large-seeded polyploid type (data not shown). We are also interested in producing a variant incapable of metabolizing linolenic acid to linolenic acid in the seed. Such a line has been produced via mutation breeding in Australia (Green 1986). This mutation makes flax oil a high quality edible product and promises to revitalize the flax industry wherever it can be adapted.

A more powerful method using cell cultures for crop development also makes use of the spontaneous genetic changes, but selection occurs in vitro, instead of in progeny of regenerated plants. The disadvantage is that the desired phenotype must be pre-decided and have a selectable marker system. This eliminates selection in vitro for such traits as oil quality, quantity, maturing time, or morphological characteristics. The in vitro system is useful for selecting cell lines that are resistant to or tolerant of some kinds of herbicides, environmental stresses or diseases. Work in my laboratory has focused on selecting flax (and wheat) cell lines tolerant to an excess concentration of salts as found in highly saline local soils (McHughen and Swartz 1984). This is accomplished by including the mixture of salts as found in the local saline soil (rarely NaCl alone) at a normally cyto-lethal concentration in the typical callus medium omitting the hormonal component. By transferring small ($\sim 1\ mm^3$) pieces of green callus to this medium, presumptive salt-tolerant cell lines can be isolated by their survival amid necrotic cells. Green flax callus will be maintained on hormone-free medium for about 5 weeks in the light before necrosing, but will begin to die within 2 weeks on the high salt medium. Any green colonies remaining after 4–5 weeks are presumed to be (either physiologically or genetically) salt-tolerant. These surviving colonies can be transferred to regeneration medium to produce plants (see Fig. 1). One line, tested in normal and saline soil in the greenhouse (Tables 3 and 4) as well as in the field,

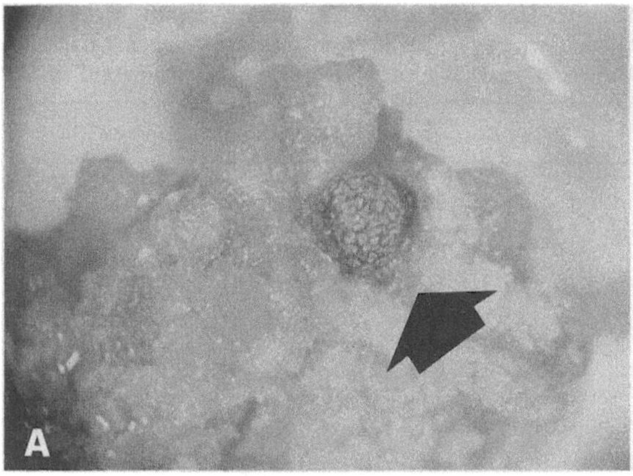

Fig. 1. A Discrete spot of green, healthy cells (*arrow*), indicating a colony derived from a cell (or cells) resistant to or tolerant of a lethal level of the selection agent (in this case, salts), growing amid brown, necrotic cells. Approx.×30. **B** Two progeny of a plant regenerated from a similar green spot, growing in saline soil; the middle plant is a control grown from seed of the parent cultivar. The plants derived from the salt-tolerant cell line grow more vigorously, flower earlier and more prolifically and yield more than the control under similar, salt-stressed conditions. Pot is 15 cm in diameter

appears to be salt-tolerant by nature of a more vigorous habit (McHughen 1987). That is, the plants, by whatever direct mechanism, are more vigorous as seedlings and earlier to mature than the parent controls. Such plants have undergone a genetic change, but the change is not necessarily in a gene directly related to osmotic or ionic metabolism. So, while the cells giving rise to this line were selected directly for saline survival, and while the plants are apparently more salt-tolerant in saline soil conditions, they might not be, strictly speaking, salt-tolerant. Another line tested in saline soil did appear to be salt-tolerant in that it was dramatically superior to the control under saline soil conditions; however, under normal soil conditions the selected line yielded much worse than standard checks (data not shown). Over 90 presumptive salt-tolerant lines have now been produced and are

Table 3. Test of germination of normal control McGregor flax seeds against two siblings of salt-tolerant selected plants in normal and saline soils. Each pot (24 in each soil type) contained two McGregor seeds and one seed from each salt-tolerant sibling. (McHughen 1987)

		No. seeds sown	No. plants	% germination
	McGregor	48	34	70.8
E.C. 1	STS IIA	24	20	83.3
	STS IIB	24	20	83.3
	McGregor	48	26	54.2
E.C. 12	STS IIA	24	13	54.2
	STS IIB	24	14	58.3
	McGregor	48	16	33.3
E.C. 16	STS IIA	24	12	50
	STS IIB	24	10	41.6

Table 4. Mean heights of plants of McGregor control and STS IIA and B just prior to floral evokation, indicating seedling-juvenile vigor. Note that both control and experimental lines lose vigor in increasing salinity, and that the experimental line is more vigorous even under normal, nonstressed soil conditions. Height of mature plants is similar in normal, nonsaline soil. (McHughen 1987)

	McGregor (n) $\bar{x}$ht. (cm)	STS IIA (n) $\bar{x}$ht. (cm)	STS IIB (n) $\bar{x}$ht. (cm)
E.C. 1	34 24.4±3.6	20 35.2±6.6	20 34.2±6.6
E.C. 12	26 16.6±3.5	13 26.3±5.6	14 20.8±5.6
E.C. 16	16 11.4±2.7	12 18.7±4.3	10 16.1±4.7

awaiting further tests to determine the genetic nature, if any, of their apparent tolerance. One line has been accepted for registration as a commercial cultivar, not on the basis of salt-tolerance but because of its earlier maturity (McHughen and Rowland 1988).

This protocol has also been used to isolate flax cell lines resistant to the herbicide chlorsulfuron and several plants have been regenerated from these cell lines (Jordan and McHughen 1987). Progeny from some of these appear to be segregating for the trait.

In addition to selecting spontaneous somaclonal variant plants with potentially improved traits, and to selecting directly in cell culture for desired traits, *Linum* is amenable to direct genetic transformation using *Agrobacterium tumefaciens.* Using a bacterium with a 'disarmed' Ti plasmid carrying chimeric genes for kanamycin resistance (NPTII from Tn5), glyphosate resistance (from a modified EPSP synthase) and nopaline synthesis (NOS) in the T-DNA, flax hypocotyls were peeled, inoculated and plated onto a regeneration medium containing the selection agent kanamycin at 400 mg/l. Noninoculated hypocotyls fail to grow, but hypocotyls inoculated with the bacterium produce callus and shoots (Fig. 2) (Jordan and McHughen 1988a). This transformation system can be used to insert other genes of agronomic value directly into elite germplasm to produce high quality locally adapted varieties with one or two novel, valuable genes. We have

Fig. 2. A Shoots proliferating along lengths of hypocotyls inoculated with *Agrobacterium tumefaciens* containing a disarmed Ti plasmid. Almost all such shoots also contain nopaline, thus suggesting transformation, but few actually are transgenic, unless the epidermis is removed from the hypocotyl at the time of inoculation, or the hypocotyl was pre-cultured for several days prior to inoculation, or both (Jordan and McHughen 1988a, b). Putative transgenic shoots should be subjected to further testing, such as leaf growth or shoot rooting on selection medium, but ultimately transformation must be confirmed by Southern analysis or, preferably, progeny analysis (Jordan and McHughen 1988b). No shoots were produced on noninoculated hypocotyls on this medium containing kanamycin. (×2.5). **B** Shoots transferred to vials with B5 or MS basal medium will elongate and also form roots. (×1.5)

used this method to produce transgenic chlorsulfuron-resistant plants as well, transferring a mutant ALS gene from Arabidopsis (Haughn et al. 1988).

2.4 Genetic Stability

Linum is a classic example of a genetically labile organism; indeed, the apparent Lamarckian variability of flax gave rise to the term "plastic genotroph" (Durrant 1958). Durrant's work suggested that stable genetic changes in flax arose through differential fertilizer treatments. As this finding appears contrary to modern genetic theory, many people rejected Durrant's work, usually on the basis that the initial seed lots were not genetically uniform to begin with (Flor 1982). It is certainly true that plant breeders' "genetically uniform" is different from that of molecular geneticists in that no population of seed is completely genetically uniform, except perhaps for a few identifiable genes or traits.

The enigma of the "plastic genotroph" was more recently analyzed from a molecular perspective by Cullis (1977, 1981, 1984) who implicated transposable elements as the unstable genetic agent in the plasticity (Cullis 1977), and suggested that flax genome can be divided into two parts: a stable portion and a labile portion where, we presume, the "inducible" changes occur (Cullis 1984).

Whether or not Durrant's work was valid, the concept of the plastic genotroph is not inconsistent with modern genetic theory. *Linum* does appear genetically unstable, at least in vitro, as a result of somaclonal variation, and this is certainly not unique to *Linum*. Other genera are susceptible, although some are quite stable. For example, Singh et al. (1975) grew suspension cultures of *Haplopappus* for more than 4 years to obtain poly- and aneuploid cell lines. Environmental stimuli do, then, induce genetic changes, probably in all organisms, but *Linum* seems more susceptible to these environmental stimuli, perhaps mediated by transposable elements, than most others. It is this genetic lability that makes somaclonal variation feasible for producing variant strains. For a genus as labile as *Linum*, a relatively short culture period is indicated to produce a desirable variant (few genes affected); for a relatively stable genus – *Haplopappus* – a longer culture period will be needed to achieve a reasonable level of variation to select for improved types.

There are few examples of genetic stability over sexual generations in flax derived from tissue culture. The STS II line of flax (McHughen 1984) has maintained its variant character over 11 generations to date, so there is no reason to suspect other such variants to be necessarily unstable, but this will depend on the transposable element activity hypothesis.

3 Conclusions and Prospects

Linum is ripe for exploitation using novel techniques, for the following reasons.

- It is an ancient crop capable of growth in most parts of the world; the basic agronomy is well worked out for each region.
- There is plenty of room for improvement. Flax it not a major crop anywhere, so conventional breeding of flax is nowhere a major effort. Flax improvement has not developed at the same rate as for other crops because of this, thus many valuable traits remain to be incorporated into the cultivars.

With the advent of low linolenic acid linseed oil, the world market will increase dramatically for the edible product. While the demand for conventional (high linolenic acid) linseed oil is relatively stable, the introduction of low linolenic acid linseed oil would enable many world regions to tap the increasing world market for high-quality edible vegetable oils. The "new" technologies will aid in producing the edible-oil cultivars adapted to particular regions.

- *Linum* is readily manipulated in vitro, including capability of regeneration to whole plants, even from protoplasts (see Table 1).

- *Linum* is amenable to somaclonal variation in vitro; desired phenotypes have been selected in vitro for stress tolerance (McHughen and Swartz 1984) and herbicide resistance into the flax genome (Jordan and McHughen 1987), with fertile plants having been produced from the selected cells. Other traits are likely to be selectable at the cell level.
- *Linum* is amenable to infection by wild-type *Agrobacterium* (Hepburn et al. 1983) and by disarmed plasmid vector transformation, incorporating glyphosate resistance, among other things, into the flax genome (Jordan and McHughen 1988a).

Linum is one of the few crop genera to combine the requirements for crop improvement using novel techniques of regenerability in vitro, genetic variation in vitro, and transformability with genetic engineering vectors. With the prospect of dramatically increased production, and the need for rapid cultivar development, *Linum* should be a prime target for crop improvement using biotechnological methods.

References

Barakat MN, Cocking EC (1983) Plant regeneration from protoplast-derived tissue of *Linum usitatissimum* L. (flax). Plant Cell Rep 2:314–317

Barakat MN, Cocking EC (1984) An assessment of the cultural capabilities of protoplasts of some wild species of *Linum*. Plant Cell Rep 4:164–167

Bari G, Godward MBE (1970) Interspecific crosses in *Linum*. Euphytica 19:443–446

Basiran N, Armitage P, Scott R, Draper J (1987) Genetic transformation of flax (*Linum usitatissimum*) by *Agrobacterium tumefaciens*: Regeneration of transformed shoots via a callus phase. Plant Cell Rep 6:396–399

Chaleff RS (1983) Isolation of agronomically useful mutants from plant cell cultures. Science 219:676–682

Chlyah H, Squalli-Khalil M, Chlyah A (1980) Dimorphism in bud regeneration in flax (*Linum usitatissimum*). Can J Bot 58:637–641

Ciamporova M, Pretová A (1981) Ultrastructural changes of plastids in flax embryos cultivated in vitro. New Phytol 87:473–479

Cullis CA (1977) Molecular aspects of the environmental induction of heritable changes in flax. Heredity 38:129–154

Cullis CA (1981) DNA sequence organisation in the flax genome. Biochim Biophys Acta 652:1–15

Cullis CA (1984) The flax genome – its plasticity and response to stress. (Abstr) Heredity 53:576

Cullis CA, Cleary W (1986) DNA variation in flax tissue culture. Can J Plant Sci 28:247–251

Dillman AC (1936) Improvement in flax. Yearbook of Agriculture. USDA, Washington, pp 745–784

Durrant A (1958) Environmental conditioning in flax. Nature (Lond) 181:928–929

Dybing CD, Lay C (1981) Flax. In: McClure TA, Lipinsky ES (eds) Handbook of Biosolar Resources, vol II. CRC, Boca Raton, pp 71–85

FAO (1982) Production Yearbook, FAO, Rome

Flor HH (1982) Plastic vs. stable flax lines. In: Proc 49th Ann Flax Institute of the US, Fargo, pp 74–93

Gamborg OL, Shyluk JP (1976) Tissue culture, protoplasts and morphogenesis in flax. Bot Gaz 137:301–306

Gamborg OL, Miller RA, Ojima K (1968) Nutrient requirements of suspension cultures of soybean root cells. Exp Cell Res 21:359–368

Gill KS, Yermanos DM (1967) Cytogenetic studies on the genus *Linum*. 1. Hybrids among taxa with 15 as the haploid number. Crop Sci 7:623–627

Green AG (1986) A mutant genotype of flax (*Linum usitatissimum* L.) containing very low levels of linolenic acid in its seed oil. Can J Plant Sci 66:499–503

Haughn G, Smith J, Mazur B, Somerville C (1988) Transformation with a mutant *Arabidopsis* acetolactate synthase gene renders tobacco resistant to sulfonylurea herbicides. Mol Gen Genet 211:266–271
Hepburn AG, Clarke LE, Blundy KS, White J (1983) Nopaline Ti-plasmid, pTiT37, T-DNA insertions into a flax genome. Molec Appl Genet 2:211–224
Hoes JA (1975) Diseases of flax in Western Canada. In: Harapiuk JT (ed) Oilseed and pulse crops in Western Canada. Western Co-operative Fertilizers, Calgary, pp 415–424
Ibrahim RK (1971) Media for growth of flax tissue culture. Can J Bot 49:295–298
Ibrahim RK, Phan CT (1978) Phenolic synthesis in relation to chloroplast ultrastructure in flax callus and cell suspension cultures. Biochem Physiol Pflanz 172:199–212
Jordan MC, McHughen A (1987) Selection for chlorsulfuron resistance in flax (*Linum usitatissimum*) cell cultures. J Plant Physiol 131:333–338
Jordan MC, McHughen A (1988a). Glyphosate tolerant flax plants from *Agrobacterium*-mediated gene transfer. Plant Cell Rep 7:281–284
Jordan MC, McHughen A (1988b). Transformed callus does not necessarily regenerate transformed shoots. Plant Cell Rep 7:285–287
Kenaschuk EO (1975). Flax breeding and genetics. In: Harapiuk JT (ed) Oilseed and pulse crops in Western Canada. Western-Co-operative Fertilizers, Calgary, pp 203–221
Lane WD (1979) Influence of growth regulators on root and shoot initiation from flax meristem-tips and hypocotyls in vitro. Physiol Plant 45:260–264
Larkin PJ, Scowcroft WR (1981) Somaclonal variation – a novel source of variability from cell cultures for plant improvement. Theor Appl Genet 60:197–214
Ling HQ, Binding H (1987) Plant regeneration from protoplasts in *Linum*. Plant Breed 98:312–317
Link GKK, Eggers V (1946) Mode, site and time of initiation of hypocotyledonary bud primordia in *Linum usitatissimum* L. Bot Gaz 107:441–454
Linsmaier EM, Skoog F (1965) Organic growth factor requirements of tobacco tissue cultures. Physiol Plant 18:100–127
Martin JH, Leonard WH (1949) Principles of field crop production. MacMillan, New York
Mathews VH, Narayanaswamy S (1976) Phytohormone control of regeneration in cultured tissues of flax. Z Pflanzenphysiol 80:436–442
McHughen A (1984) Salt-tolerant flax plants regenerated from cells surviving a highly saline culture medium. In: Symp Genetic manipulation in Crops. 22–26 October, 1984, Beijing, China
McHughen A (1987) Salt tolerance through increased vigor in a flax line (STS-II) selected for salt tolerance in vitro. Theor Appl Gen 74:727–732
McHughen A, Rowland G (1988) Canada's first somaclonal variant crop cultivar. 2nd Canadian Plant Tissue Culture Genetic Engineering Workshop. May-June, 1988, Ottawa, Canada, p 43
McHughen A, Swartz M (1984) A tissue-culture derived salt-tolerant line of flax (*Linum usitatissimum*). J Plant Physiol 117:109–117
McHughen A, Browne R, Kneeshaw D, Jordan M (1986) Ti-mediated transformation and regeneration of flax plants in vitro. 6th Int Congr IAPTC; Minneapolis, USA, p 130
Murashige T, Skoog F (1962) A revised medium for rapid growth and bioassays with tobacco tissue cultures. Physiol Plant 15:473–497
Murray BE, Handyside RJ, Keller WA (1977) In vitro regeneration of shoots on stem explants of haploid and diploid flax (*Linum usitatissimum*). Can J Genet Cytol 19:177–186
Norstog K (1973) New synthetic medium for the culture of premature barley embryos. In Vitro 8:307–308
Preťová A, Williams EG (1986) Direct somatic embryogenesis from immature zygotic embryos of flax (*Linum usitatissimum* L.). J Plant Physiol 126:155–161
Quatrano R (1968) Freeze-preservation of cultured flax cells utilizing dimethyl sulfoxide. Plant Physiol 43:2057–2061
Rybczyński JJ (1975) Callus formation and organogenesis of mature cotyledons of *Linum usitatissimum* L. var. szokijskij in vitro culture. Genet Pol 16:1–12
Seetharam A (1972) Interspecific hybridization in *Linum*. Euphytica 212:489–495
Singh BD, Harvey BL, Kao KN, Miller RA (1975) Karyotypic changes and selection pressure in *Haplopappus gracilis* suspension cultures. Can J Genet Cytol 17:109–116
Sun H, Fu W (1981) Induction of pollen plants in flax (*Linum usitatissimum*) and preliminary observations on performance of their progenies. Acta Genet Sin 8:369–374
Tammes T (1928) The genetics of the genus *Linum*. Bibliogr Genet 4:1–36
Trump DH (1980) The prehistory of the Mediterranean. Yale Univ Press, New Haven, London
White PR (1963) The cultivation of animal and plant cells. 2nd ed. Ronald, New York

IV.9 Flax *(Linum usitatissimum* L.): Embryo Culture Studies

A. PREŤOVÁ[1]

1 Introduction

Embryo culture in general serves to study factors influencing embryo nutrition under both in vitro and in situ conditions. It is useful in studying embryo-endosperm relationships (Erdelská 1985 a, b, c, 1988) and the process of embryo growth and differentiation, and offers new and unconventional methods of characterizing the process of embryogenesis and the dynamism of embryo development (Monnier 1978). It can also be utilized in studying first germination processes and various questions – including biochemical ones connected with this process – and in studying problems of dormancy. It is also an excellent method for rescuing embryos originating as a result of a distant hybridization (Pickering and Morgan 1983). Culture of embryos can also be useful in the modern practices of genetical engineering and in producing progeny with high fertility and resistant to diseases and pests.

Some of the above questions have been studied in our research on embryos of *Linum usitatissimum* L. cv. Viera and cv. Glenelgand are detailed in this chapter.

1.1 In Situ Development of *Linum* Embryo

The embryogenesis of the genus *Linum* (flax) was lucidly elaborated by Souegés (1924) for the species *Linum cartharticum* L. *Linum* embryo development was also described by Poddubnaja-Arnoldi (1952), Iljina (1954), Kantor (1957), and others.

Linum usitatissimum L. is a self-pollinated flax species. Its pollen begin to germinate on the stigma in the course of ca. half an hour (Kantor 1957). Following some 3.5 to 4 hours, the pollen tube enters the embryo sac. Double fertilization takes place in ca. 4.5 to 5 h after pollination. First division of the zygote occurs on the second day following pollination, at the time when several endosperm nuclei have already been formed. Endosperm of the species *Linum usitatissimum* L. is of nuclear type.

The embryo of the species *Linum usitatissimum* L. develops according to the type Solanad, var. *Linum* (Johansen 1950). The apical cell of the bicellular em-

[1] Institute of Experimental Genetics of Slovak Academy of Sciences, Hlohovská 2, 94992 Nitra, Czechoslovakia

Biotechnology in Agriculture and Forestry, Vol. 10
Legumes and Oilseed Crops I (ed. by Y.P.S. Bajaj)

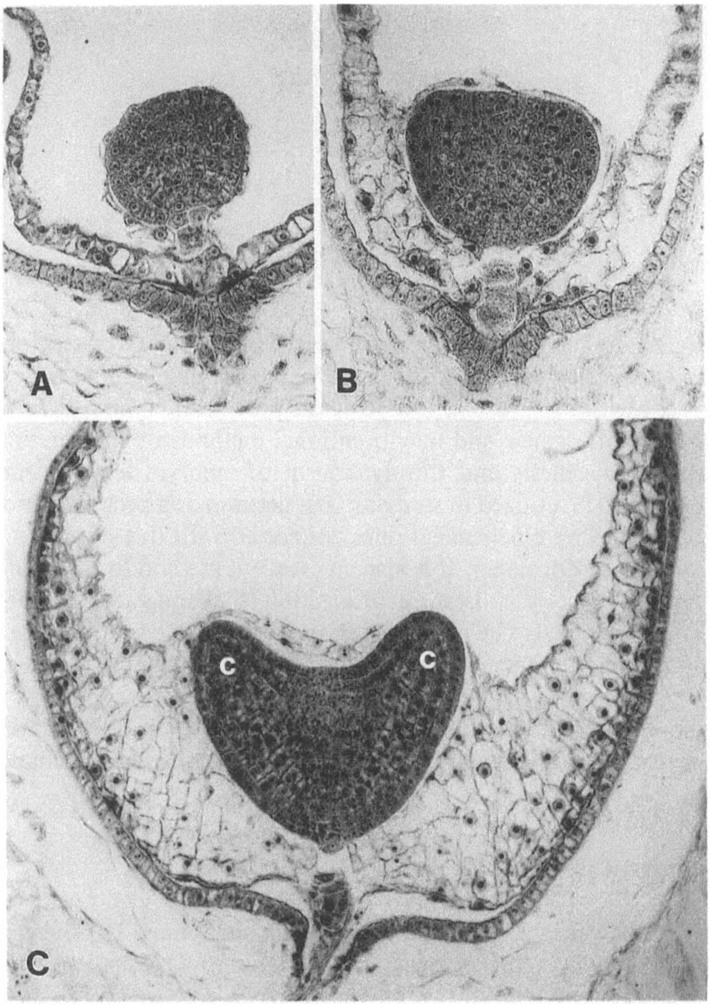

Fig. 1. A Globular stage of flax embryo development 5-day-old embryo (×200). **B** Widened globular stage of flax embryo. The cotyledons start to form. 7-day-old embryo (×200). **C** Young heart-shaped flax embryo *c* forming cotyledons. (Erdelská et al. 1973)

bryo produces cotyledons, plumule, hypocotyl, and the basis of the central root cylinder. The basal cell of the bicellular embryo gives rise to the root apex and the suspensor. Both apical and basal cells of the bicellular embryo differ also in structure (Deschamps 1968). The apical cell and cells developing from it have numerous ribosomes. The basal cell and the cells derived from it have fewer ribosomes and are vacuolar.

Flax embryo develops first as a linear formation. Later it assumes a globular shape (Fig. 1 A). Embryo growth at this stage is rather slow, the embryo harboring intensive cell division without obvious width or length growth. The mean duration of the mitotic cycle in the globular phase of embryo development is about

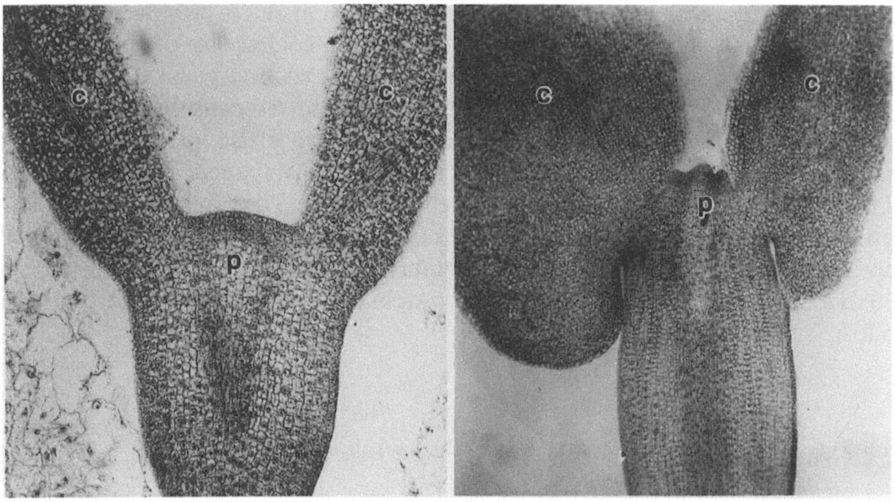

Fig. 2. A Torpedo stage of flax embryo. Plumule is without leaf primordia. 10-day-old embryo (×100). **B** Plumule of 14-day-old flax embryo with two leaf primordia (×110) *p* plumule; *c* cotyledon

13 h (Erdelská 1986). The coefficient of embryo differentiation at this stage equals 1. The embryo is at the first crucial point of its development. As early as in the globular phase, the flax embryo is green. Later, cotyledons begin to develop in the apical part of the embryo (Fig. 1 B), expanding it in the process. The coefficient of differentiation decreases temporarily to a value lower than 1, later rising again. The embryo enters the heart-shaped phase of development (Fig. 1 C), and cotyledons are established at 45° angle to the longitudinal embryo axis. In one moment in this phase of development the coefficient of embryo differentiation once again equals 1. The embryo has reached its second crucial point. The late heart-shaped phase of development shows fast exponentially elongating embryo growth, its width, however, extending only in linear proportion. At the end of this phase, the coefficient of embryo differentiation achieves the value 8 to 10. When the flax embryo enters the exponential phase of growth, its suspensor (a polycellular structure composed of one layer of cells) degenerates. On ca. the 10th day of development, the embryo is pear-shaped, with two differentiated cotyledons and a radicle. The plumule is so far without leaf primordia (Kantor 1957) as can be seen in Fig. 2 A. In about 16 days from the beginning of embryogenesis, the embryo attains its final size, entering the stage of intensive accumulation of reserve materials. This process, however, started much sooner, as early as in the exponential phase of embryo development, but now it becomes more intensive. The embryo shows full grenn coloration. (Dillman 1928) noticed that the embryo is completely developed at this time, having two leaf primordia (Fig. 2 B).

The embryos contain a large quantity of starch and lipid globules, starch being only a transitory reserve material in flax embryos. Between 8 to 22 days following fertilization, fat is accumulated in flax embryos, and between 10 to 14 days they begin to store proteins in the form of aleurone grains. The formation of aleurone grains in flax embryo cells grown in situ shows the following phases observable under the light microscope: (1) accumulation of precursors in cytoplasm, (2) their accumulation in vacuoles in the form of grains, (3) coalescence of the grain-like shapes in crystalloids, (4) differentiation of the matrix and globoid (Ostapenko 1968, Erdelská 1978). At the time of accumulation of reserve substances, the intensive green color of the embryo vanishes, the embryo is also losing water, which can lead to reduced size. A ripe flax embryo is evenly shaped. It has two large spatulate cotyledons with stomata in their epidermis. The plumule has two leaf primordia, and the radicle is thick and straight. The cells of the ripe flax embryo are filled with lipids and aleurons. The embryo ripens, together with the seed, in 30 to 35 days. A ripe seed is elongated, brown in color, glossy, with a mucous layer on the surface.

2 Growth of Flax Embryos Under in Vitro Conditions

The flax embryo was one of the first embryos cultivated in vitro (Dietrich 1924; Laibach 1925). Dietrich succeeded in growing flax embryos under artificial conditions on a very simple medium (agar + sucrose). Laibach utilized the in vitro cultivation method of flax embryos to obtain fertile progeny from otherwise sterile hybrids. However, he did not describe in detail the character of embryo growth under artificial conditions. Later, flax embryos received little attention from this aspect, until our early experiments from the years 1973 and 1974, which considered the possibility of flax embryo cultivation and tried to describe the character of their growth in vitro (Erdelská et al. 1973; Pretová 1974).

We proceeded from the assumption that embryo transfer from the maternal environment into the artificial medium, differing so much from ambient tissues of the maternal plant both in chemical and physical properties, will certainly cause a number of differences in the development of the excised embryo. The character and degree of these differences will depend partly on the age of the excised embryo and on the degree of its differentiation at the time of excision, and partly on the physiological condition of the embryo. Younger embryos will find it more difficult and will show greater differences in completing their embryogenesis under in vitro conditions. However, older embryos too, as shown later, will undergo certain changes in this process.

2.1 Cultivation of Very Young Globular and Heart-Shaped Flax Embryos

At first, when only very simple medium was used for the cultivation of isolated flax embryos (White 1942), it was very difficult to achieve successful cultivation of globular and heart-shaped embryos. We avoided, for obvious reasons, the use of nondifferentiated medium compounds such as coconut milk or various extracts.

In 1978, Monnier published a report on successful cultivation of very young embryos of the species *Capsella bursa-pastoris* L. We began to use the cultivation

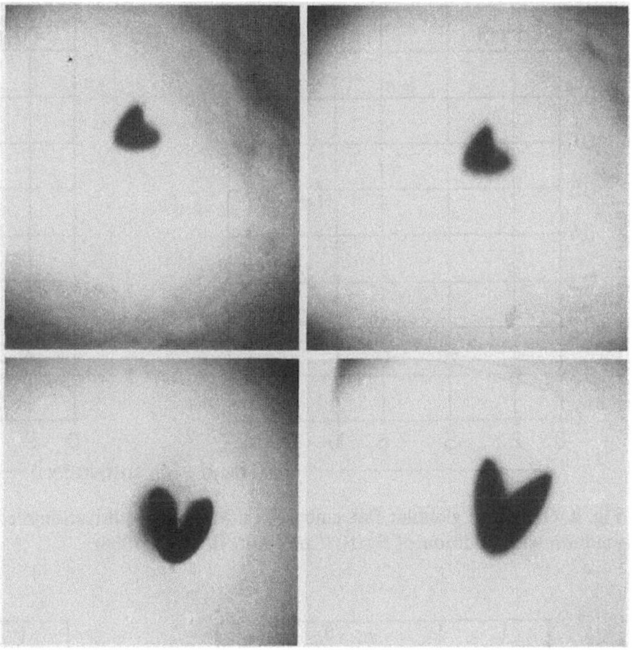

Fig. 3. Growth of very young flax embryos in vitro. (Prečová 1983)

medium advocated by this author, designated as a "morphogenetic medium" with balanced content of mineral substances and added glutamine, also for very young flax embryos. The medium appeared to be very effective. The growth of young embryos in vitro can be seen in Fig. 3.

If flax embryos are excised at the globular stage (i.e., in the first crucial stage when the differentiation coefficient equals 1 for the first time), they first enlarge their volume by 25 to 30% as a result of belancing their water deficit and osmotic conditions. This phase lasts only 2 to 4 h. Then the second growth phase sets in, lasting also several hours only. A limited proportional growth takes place as influenced by reduced cell division. After several days lag phase, intensive growth occurs of a character similar to that of in situ embryo growth. The result of this growth (3rd phase of in situ embryo development) is the establishment and growth of cotyledons (Prečová 1985 a, c; Erdelská 1986). Thus the embryo gradually proceeds from radial symmetry into bilateral symmetry. This developmental stage under in vitro conditions is crucial for the embryos. If no cotyledons set in, the embryos do not develop further. After successful establishment of cotyledons, the embryos passed to the heart-shaped phase of their development. The embryo's width at first grew faster than its length. For the second time the embryos reached the moment when their differentiation coefficient equals 1, which happened within about 10 days of embryo cultivation. Then the exponential phase of embryo development set in, with embryo differentiation coefficient higher than 1.

If a flax embryo was excised in the young heart-shaped phase, the development of already established cotyledons continued, achieving the second crucial point

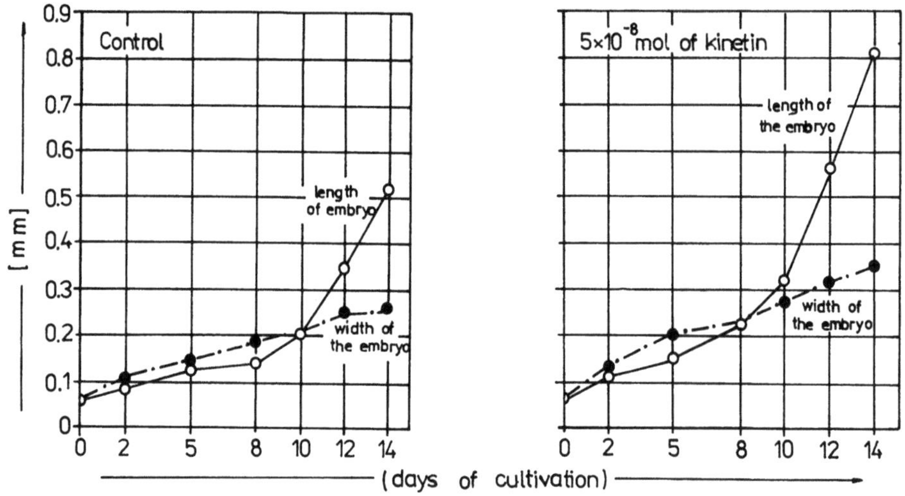

Fig. 4. Growth of globular flax embryos on Monnier's cultivation medium (control) and on the same medium with addition of 5×10^{-8} mol kin. (Pretová 1986)

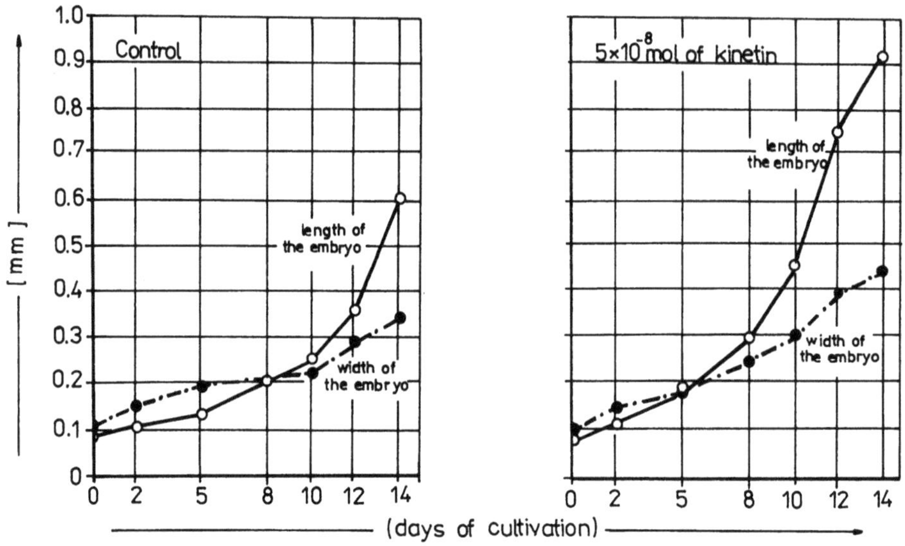

Fig. 5. Growth of widened globular flax embryos on Monnier's cultivation medium (control) and on the same medium with addition of 5×10^{-8} mol kin. (Pretová 1986)

of development in vitro with in about 8 days, then resuming its development by the exponential growth phase.

Heart-shaped embryos isolated close to the second crucial point reached it very soon (within 2 to 4 days of cultivation). Embryos isolated just at the moment of the second crucial stage continued their longitudinal growth very rapidly. It is almost a rule that the cotyledons established under in vitro conditions do not attain the length of those grown in situ (Veen 1963).

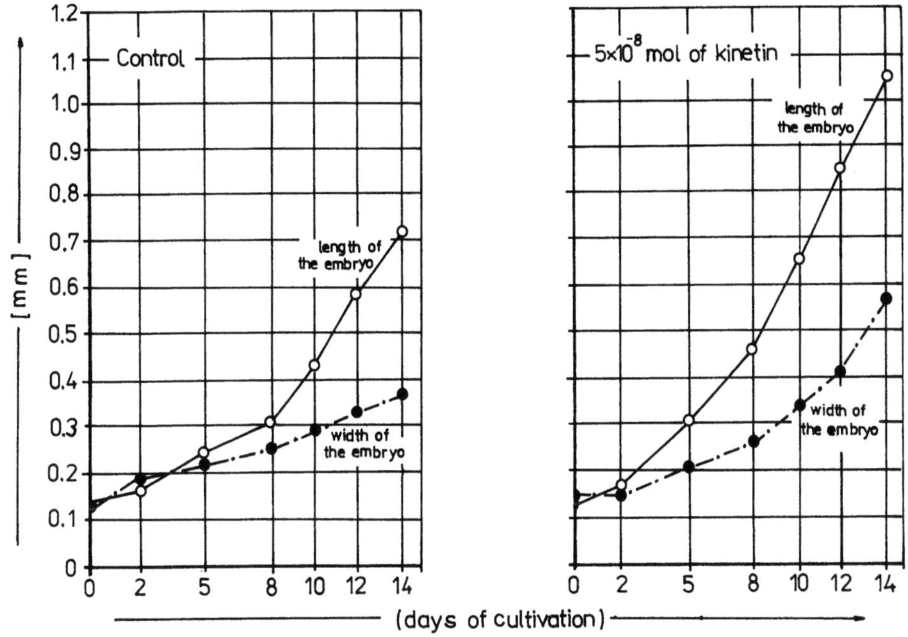

Fig. 6. Growth of heart-stage flax embryos on Monnier's cultivation medium (control) and on Monnier's medium with 5×10^{-8} mol kin. (Prečová 1986)

Postembryonal development of in vitro embryos was characterized by the regreening of embryo cotyledons and by the elongation of both radicle and hypocotyl, including the appearance of the plumule with first leaves, together with the emergence of lateral roots (Prečová 1974, 1978, 1983; Prečová and Erdelská 1977, Prečová et al. 1981).

The embryonal and postembryonal development of embryos under in vitro conditions requires different cultivation media and other physical conditions of cultivation. The embryonal development requires higher osmotic values of the medium and cultivation in darkness, although the embryo concerned is green. The postembryonal development of flax embryos requires lower osmotic value of medium and light.

Under our cultivation conditions, about 20%to 25% globular embryos and about 60% to 65% heart-shaped embryos were capable of completing their embryogenesis in vitro. It seems, however, that the emergencc of cotyledons is a crucial moment in the growth and development of globular embryos in vitro (Fig. 4, 5, 6).

2.2 Influence of Some Growth Substances on in Vitro Cultivation of Embryos

The growth substances added into media can considerable influence the growth and development of embryos under in vitro conditions. For unripe embryos, the addition of cytokinins proved particularly useful. Cytokinins stimulate cell division and total proteosynthesis, influencing at the same time the accumulation of reserve materials (Letham 1963). It has been proved that the synthetizing sugars and amino acids are transported into the sites with high cytokinin activity (Khan 1982).

Growth of embryos as early as in the globular and very young heart-shaped stages was favourably influenced by adding 5×10^{-6} mol, but especially 5×10^{-8} mol kin/l of medium. This led to increased frequency in the completion of embryo's development as early as in 26 days. Also the percentage of embryos successfully completing their embryogenesis increased from 20%–25% to 35% in glubular embryos, and from 60% to 90%–95% in older embryos (Figs. 4, 5, 6). Similarly, kin helped to preserve the green pigment in embryos during the whole cultivation period. Higher kin concentration stimulated nondifferentiated growth and callus production (Pretová 1986).

However, different developmental stages of embryos can differ in their sensitivity to added exogenous growth substances (Millborow 1974). Realizing this, we tried to determine both the level and quality of growth substances in flax embryos at selected stages (Vizárová et al. 1987). It has been proved that, depending on the stage of embryo development, it accumulates also various forms of cytokinins. The addition of cytokinins at later stages of embryo development speeds up the maturing of the embryos under in vitro condition and improves both embryo germination and the growth of the germinating plantlet. Other growth substances (IAA and GA) did not show such an unambiguous influence on the development of in vitro-cultivated embryos.

2.3 Influence of Some Physical Environmental Factors on Embryo Growth in Vitro

The embryo developing under in vitro conditions has certain requirements also concerning physical environmental conditions, depending on its age and developmental stage. In our laboratory the influence of the osmotic value of the cultivation medium was studied. Many authors (Maheshwari 1950; Rijven 1952; Rietsema et al. 1953; Mauney 1961; Pecket and Selim 1965; Raghavan 1966) consider the osmotic value of the cultivation medium to be a factor decisively influencing embryo development.

Embryos 7, 14, and 28 days old were cultivated on the medium according to White (1942), where the sucrose content was modified according to the needs of the experiments. Sucrose and mannitol (separately or together) were used to vary osmotic value of the nutrient medium. Sucrose was applied in the concentrations 2%, 5%, 10%, and 15%. Mannitol concentrations were adapted to correspond to the osmotic potentials of sucrose-containing media.

Established osmotic potentials of the media:

2% sucrose and other media	1.7 bar
5% sucrose and other media	4.0 bar
10% sucrose and other media	8.2 bar
15% sucrose and other media	12.7 bar

The embryos were cultivated on media with 0.8% agar, or on liquid media (on a bridge of Whatman No.1 chromatography paper).

Sucrose media with the osmotic value 4 bar (5% sucrose) were the most suitable for the cultivation of excised 7- and 14-days-old flax embryos. Nutrient media with the osmotic potential 1.7 bar (2% sucrose) were optimal for 28-day-old embryos. The most convenient osmotic potential for optimal root growth was 4 bar and for hypocotyls and stems 1.7 bar (Fig. 7). A high osmotic value of the cultivation medium prevented early germination of the embryos, and a lower osmotic value promoted germination. Mannitol, compared with sucrose, inhibited development and germination. Various abnormalities often occurred on the media using mannitol. Agar media were superior to liquid media for the cultivation of the embryos.

Our results have shown that 2% to 5% sucrose in the media are sufficient for the cultivation of flax embryos (starting on the 7th day of development). Sucrose in this range fulfills both nutritional and osmoregulative functions. These sucrose concentrations were also successfully used in experiments with flax embryos by Dietrich (1924), and it seems that they are suitable for the cultivation of the embryos of many other plant species (Andronescu 1919 – maize; Haagen-Smit et al. 1945 – maize; Raghavan and Torrey 1963 – *Capsella bursa-pastoris* etc.).

The osmotic value of the medium also influences the relations between the growth rate of the stem and that of the root in young seedlings – higher osmotic values promote root development and lower osmotic values promote hypocotyl and stem development. Honma (1955) and Kobetičová (1971) found similar relations.

The use of mannitol as osmoregulator is based on the assumption that mannitol does not enter living cells or is not metabolized (Ferguson and Street 1958; Thiman et al. 1960). Slatyer (1961) considers the question of the assimilation of mannitol and its penetration to plant tissues as still unanswered. The use of mannitol as osmotic regulator was recommended by Szweykowska et al. (1959). Rietsema et al. (1953) found that mannitol inhibited premature germination of *Datura* embryos and favored their natural pregerminative development. Negative growth effects occurring in our experiments in the media containing mannitol could be caused by plasmolytic and toxic effects of mannitol. The plasmolytic effect of mannitol is mentioned by Linser and Herwig (1963). Kozinka and Klenovská (1965) state that mannitol can cause dehydratation of plant tissues, while Stroganov and Lapina (1964) consider mannitol as a toxic substance.

The physical properties of the agar cultivation media in our experiments were more convenient for flax embryo development than those of liquid media. There are only two more papers dealing with the cultivation of explanted flax embryos. Dietrich (1924) mentions only briefly that flax embryos can be cultured in vitro, and Laibach (1925) reports the cultivation of excised hybrid embryos of flax on filter paper soaked with sucrose solution with concentration up to 10%.

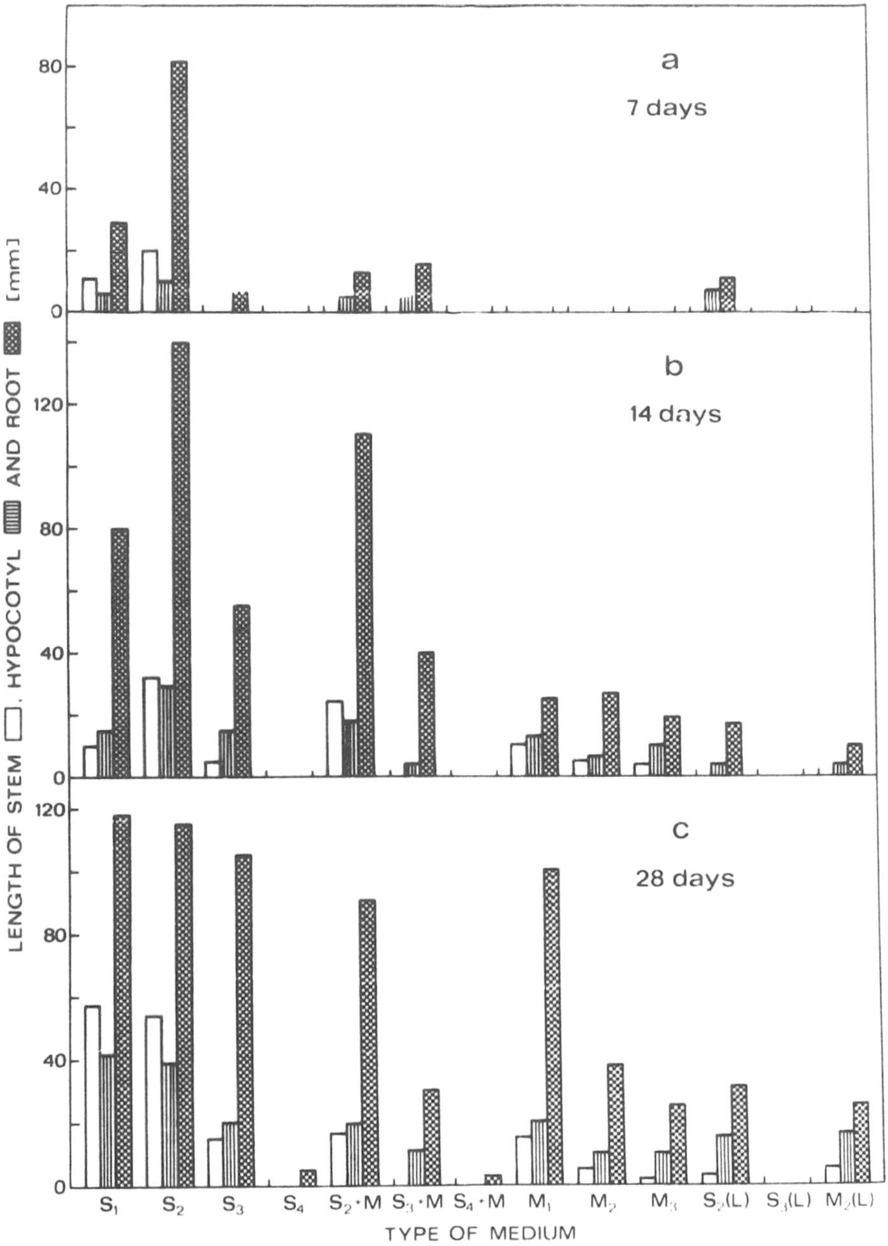

Fig. 7a–c. Effect of osmotic potential of cultivation medium on the growth of flax embryo isolated 7, 14, and 28 days after pollination. Type of media: S_1, S_2, S_3, S_4: 2%, 5%, 10%, 15% sucrose in medium respectively; S_2+M, S_3+M, S_4+M: 2% sucrose in medium and the osmotic potential adjusted with mannitol up to 4.0, 8.2, 12.7 bar, respectively; M_1, M_2, M_3: 2%, 5%, 10% mannitol in medium, respectively; $S_2(L)$, $S_3(L)$: 5%, 10% sucrose in medium, respectively – liquid medium; $M_2(L)$: 5% mannitol in medium – liquid medium. (Pretová 1974)

2.4 Some Differences in Flax Embryogenesis Comparing in Situ and in Vitro Conditions

Cytological Changes. Embryos of *Linum usitatissimum* L. cv. Viera were excised 14 days after pollination. Such embryos were intensively green. They entered the phase of accumulation of reserve substances, and the mitotic activity in the radicle was already in its final phase.

Differentiation processes in the embryos cultivated in vitro were compared with those in embryos of equal age developing on their maternal plants. There are only a few studies comparing in detail the development of embryos of dicotyledonous plants under natural and artificial conditions. Moreover, they have appeared only sporadically and do not concern flax (Rijven) 1952, 1955; Rietsema et al. 1953; Mauney 1961; Raghavan and Torrey 1963; Monnier 1978).

The development of 14-day-old flax embryos under in vitro conditions was divided in three stages in our experiment:

1. Completion of embryogenesis following excision shock under changed conditions.
2. Dormancy.
3. Germination.

Stage 1 lasts about 7 to 10 days. In vitro-grown embryos as compared with in situ-grown embryos (those developing on maternal plant) showed a decrease of mitotic activity in radicular cells and a slight longitudinal growth, especially in the region of the hypocotyl and the base of the radicle. Immediately after transfer of flax embryos onto cultivation medium, they at first enlarge their volume to balance their water deficit and their osmotic conditions. The embryos then went through a lag phase (3 – 4 h) and continued with radicle lengthening. At the end of the first day of cultivation, the radicle had lengthened by about 15%, as shown in Fig. 8. Also planar growth of cells in cotyledons was observed, starting first in the cotyledon in contact with the cultivation medium. This cotyledon also attained a larger final size; weak growth of the plumule was later observed. Elongation in plumule and increase in size of the cotyledons results especially from cell growth parallel with the longer axis of the embryo (Fig. 9). The cells of embryos developing in situ elongate more in perpendicular direction to the longer axis

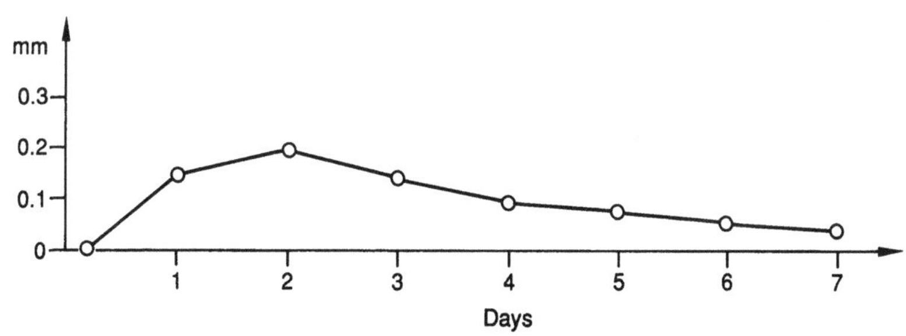

Fig. 8. Growth increase of radicle in a single day of cultivation of flax embryos on the artificial medium. (Erdelská 1978)

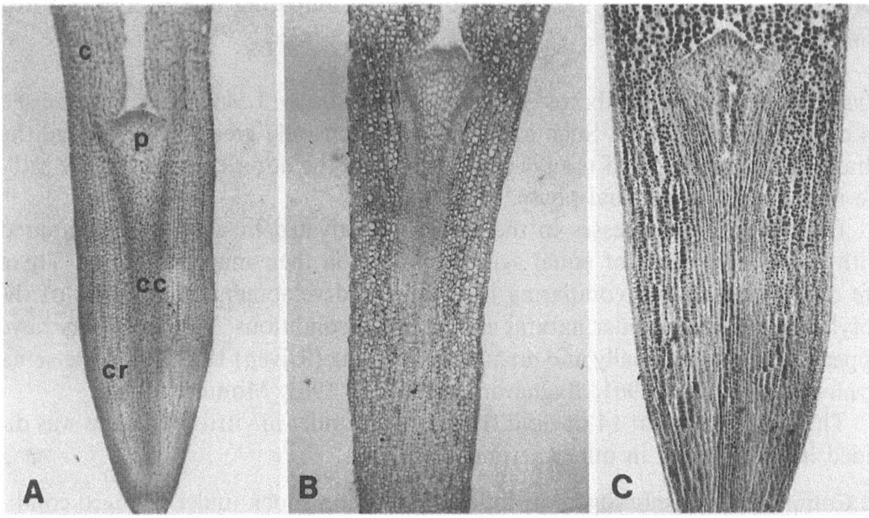

Fig. 9. A Embryonal axis and part of cotyledons after excision *c* cotyledon; *p* plumule; *cc* central cylinder; *cr* cortex of radicle. (×65). **B** Embryonal axis after 5 days of cultivation in vitro (×65). **C** Embryonal axis after 5 days of development in situ (×65). (Erdelská 1978)

of the embryo. Figure 9 shows the changes in embryo axis after 5 days of cultivation.

Under in situ conditions, provascular bands are being formed at that time, remaining in this stage until seed maturity. On the other hand, embryos developing under in vitro conditions showed the maturation of provascular elements usually within 24 h following the transfer of embryos on the cultivation medium, in any case at the latest within 2 to 3 days. The first to mature are the vessels in the distal part of the cotyledons, later also in the proximal part and in the hypocotyl. This is a specific response of the embryo to the changed conditions of the environment and to the changed transport of substances in this environment.

Accumulation of Reserve Substances. Accumulation of reserve substances also varies. Starch as transient reserve substance in flax embryos retreats as early as in 14-day-old embryos. Instead of starch, flax embryos accumulate fats as the main reserve substance. Fat accumulation begins (Sveshnikova 1955; Ostapenko 1968) on the 8th to 10th day of flax embryo development, so that a 14-day-old embryo (i.e., at the time of excision) contains up to 40% of total fats. The accumulation of fat reserves continues also under in vitro conditions, but here the quantity of produced fats represents only about one-fifth of that accumulated during the same time under in situ conditions. Substances from the cultivation medium either could not be utilized in the process of production of reserve substances, or the composition of the cultivation medium was insufficient for full-value production of fats or else the produced fats were consumed immediately after production, for embryo growth process and for their subsequent germination.

Reserve proteins are accumulated in the form of aleurone grains. This leads in situ to the production of granular formations in the vacuoles of cotyledon cells

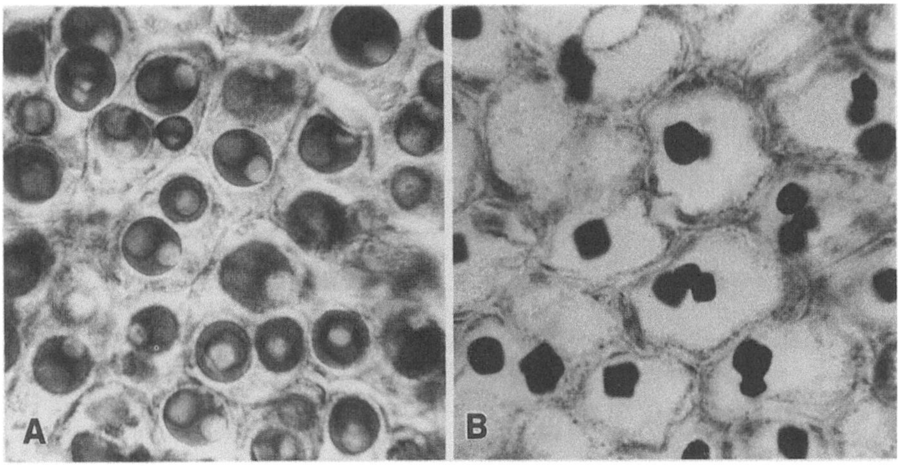

Fig. 10. A Cotyledon cells (development in situ) with complete aleurone grains (with crystaloid and globoid). (×1070). **B** Cotyledon cells after 5 days of cultivation in vitro with crystaloids of incomplete aleurone grains. (Erdelská 1978)

and subsequent production of aleurone grains containing crystalloid and globoid in the matrix. Under in vitro conditions the production of aleurone grains begins on ca. the 2nd to 3rd day following excision (it appears to begin first in the cotyledon which is in contact with cultivation medium), proceeding later differently in the two cotyledons. In the cotyledon lying on the medium, one or more either round or edged formations were found on the cell layer, with pigmentation similar to the crystalloids in the embryos of the maternal plant. In the cotyledon diverted from the cultivation medium, the clustering of proteinaceous parts resulted in formations similar to aleurone grains, yet considerably slower. The embryos completing their embryogenesis in vitro showed incomplete production of aleurone grains (Fig. 10). These consisted of crystalloids only with no visible globoid production, and the matrix was only rarely visible. This may be due to the simplicity of the cultivation medium (Preťová and Erdelská 1977; Erdelská 1978).

Changes in Pigment Content. Many changes during flax embryo cultivation appear in the quality and quantity of pigments and in the ultrastructure of plastids. Young unripe flax embryos isolated from the seed contain green pigment in all tissues. The presence of green pigment in flax embryos at a certain stage of embryogenesis classifies flax as belonging to the group of plants designated as Chloroembryophyta (Yakovlev and Zhukova 1973). Flax embryos contain green pigment from the 5th day of embryogenesis till nearly to maturity (25th day of embryogenesis).

Our analyses of pigment content in flax embryos show that the youngest isolated embryos (12 days old) contain 37.04 μg of chlorophyll a and 28.09 μg of chlorophyll b per 1 g fresh weight. Maximum content of the two pigments was measured on the 15th to 16th day following pollination (69.3 μg of chlorophyll a and 63.02 μg of chlorophyll b per 1 g fresh weight). After this maximum the

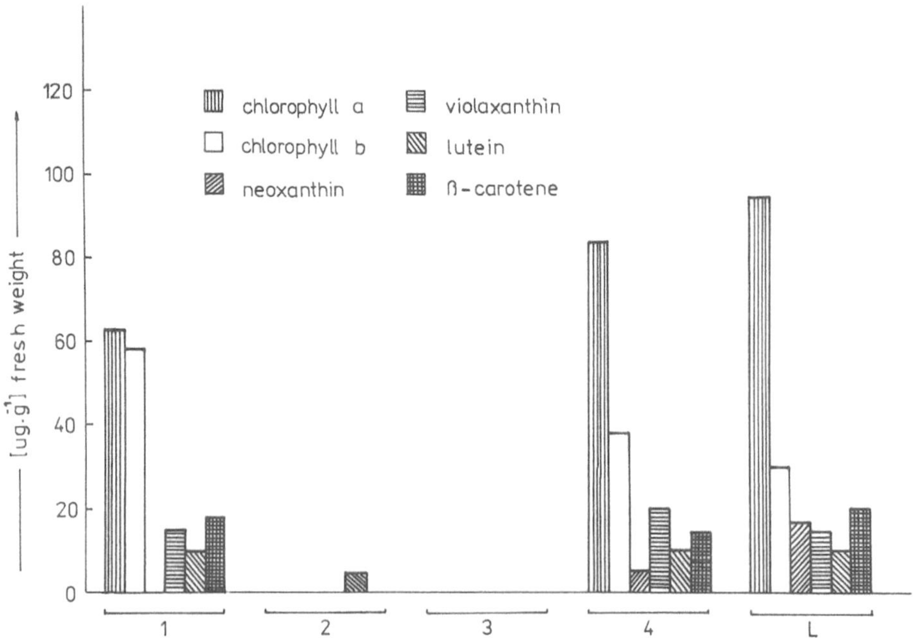

Fig. 11. Changes in the pigment content during in vitro cultivation of flax embryos. *1* stage before the planting of embryos (14 days old) on the cultivation medium; *2* stage in the yellowed embryos after 7–10 days of cultivation in vitro (age of the embryos 24 days); *3* stage in the yellowed embryos on maternal plants (age of embryos 24 days); *4* stage at the germination of embryos after 14–20 day cultivation in vitro (age of embryos 34 days) *L* pigments in the mature healthy leaves. *Abscissa* amount of pigments in μg/g of fresh weight. (Preťová 1985b)

chlorophyll content decreases gradually. Towards maturity the embryo assumes a pale yellow pigmentation. The loss of pigment proceeds from the radicle towards the top of the cotyledons. Total loss of pigments occurs on the 24th to 26th day of embryogenesis. The pigment of the embryos have characteristics slightly different from these of the leaves. Maximum of absorbance is shifted towards longer wave lengths (668 to 670 nm). Chlorophyll ratio a:b is variable during embryogenesis ranging from 1.1:1 to 2:1.

In addition to chlorophylls, flax embryos also contain carotenoids. The quantity of carotenoids during embryogenesis varies considerably. In spite of this, a certain regularity can be observed: with increasing content of lutein the violaxanthin content decreases, especially in connection with light conditions in the seed. Flax embryos contain no neoxanthin during embryogenesis. The course of changes in the content of *β*-carotene is similar to that in the chlorophylls. The content of both chlorophylls and carotenoids is the highest in the period of embryogenesis when the flax embryo reaches the size close to the maximum (0.4 to 0.5 mm) and has the highest content of water. These are the conditions in the embryo developing in situ.

If a 14-day-old flax embryo is placed on a simple medium (White 1942), the number of pigments considerably decreases in the course of its first developmen-

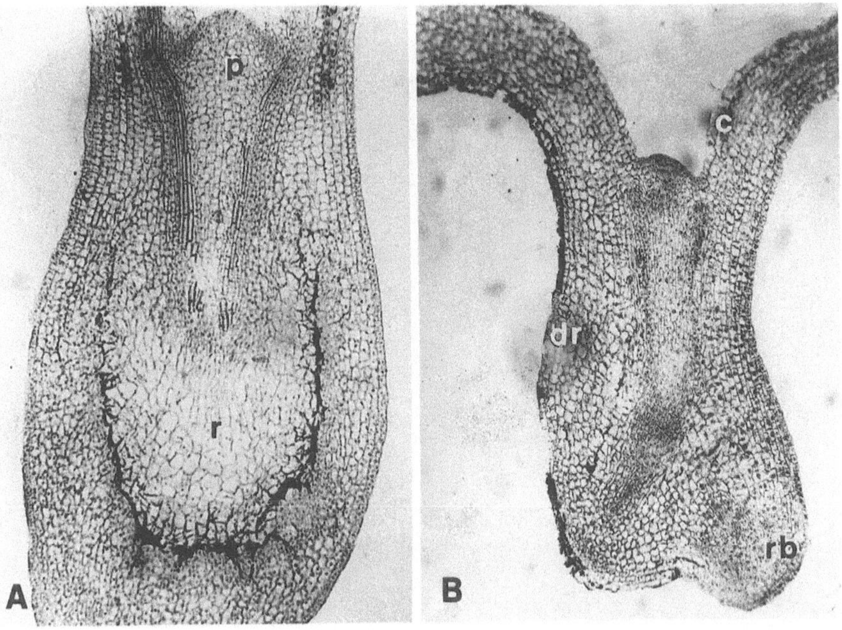

Fig. 12. A Thickened radicle of flax seedling (×70). **B** New root basis formation on degenerated radicle. (×70). *p* plumule; *r* radicle; *c* cotyledons; *dr* degenerated radicle; *rb* new root basis. (Erdelská et al. 1973)

tal phase in vitro taking 7 to 10 days (i.e., on completing its embryogenesis under changed conditions following excision shock). At the end of this phase only residues of lutein were noted. Embryos of equal age in situ present more complete loss of pigments with no lutein residues either (Fig. 11). The loss of pigments in vitro proceeds in the radicle from the top towards the base and from the center of the radicle towards its perimeter, continues by losing pigments in the base of the cotyledons, ending eventually in total yellowing of the whole embryo (Přeťová 1977a, 1978, 1980, 1985; Přeťová and Vojteková 1985; Přeťová et al. 1981).

After yellowing and the succeeding period of ca. 7 to 10 days, the flax embryos under in vitro conditions proceed to the second developmental stage – dormancy in vitro, which also takes 7–10 days, no visible morphological changes occurring in the embryos. The period of dormancy can be identified in time with the final phase of embryo maturing in the seed under in situ conditions.

The period of dormancy in vitro is followed by germination. Flax embryos begin to germinate ca. 15 to 20 days following the beginning of cultivation. The radicle of the embryo elongates intensively and simultaneously, some 2 or 3 days later, the cotyledons of the embryo turning green again. The hypocotyls of the embryos begin to elongate on the 18th to 19th day, with stem growth following on ca. the 23rd day of cultivation. Pigments in regreened cotyledons have the same characteristics as those from healthy mature flax leaves (Fig. 11). During germination some abnormalities were observed, especially in the root (Fig. 12). In some cases the tip of the radicle first lost meristematic character, its cells became enlarged

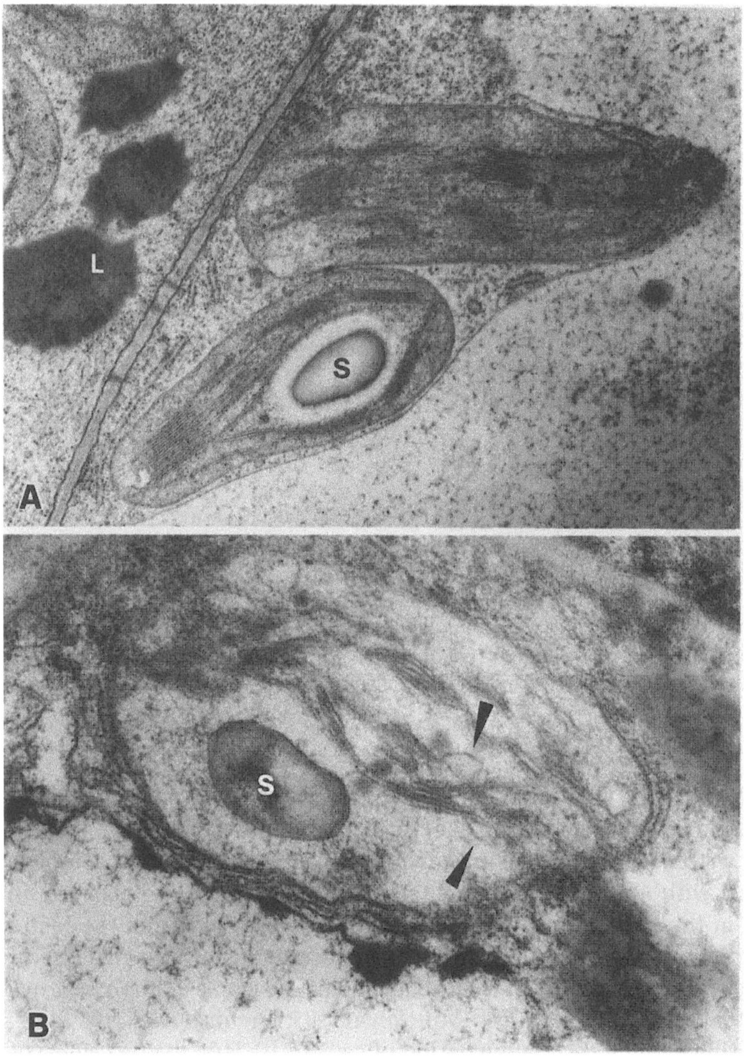

Fig. 13. A Plastids – chloroplasts in 14-day-old embryos. (×20, 560) *S* starch; *L* lipids. (Čiamporová and Preťová 1981, 1983). **B** Plastids in flax embryos cultivated for 4 days in vitro (age of embryos 18 days). Dilatation of thylakoids can be seen (*arrows*) (×26, 660) *S* starch grain (**A**, **B** Čiamporová and Preťová 1981, 1983)

and later died. At other times, the radicle transformed into extended tumor-like formation. Very frequently the original tissues of the radicle degenerated, with new root base being established laterally (Fig. 12). Also abnormalities of the hypocotyl were observed (callus), sometimes even extended number of cotyledons (Erdelská et al. 1973).

Changes in Plastid Structure. Plastids in 14-day-old flax embryos belong to chloroplasts due to their inner structure (Fig. 13A). In some plastids of 14-day-old embryos of flax phytopherritin can also be observed.

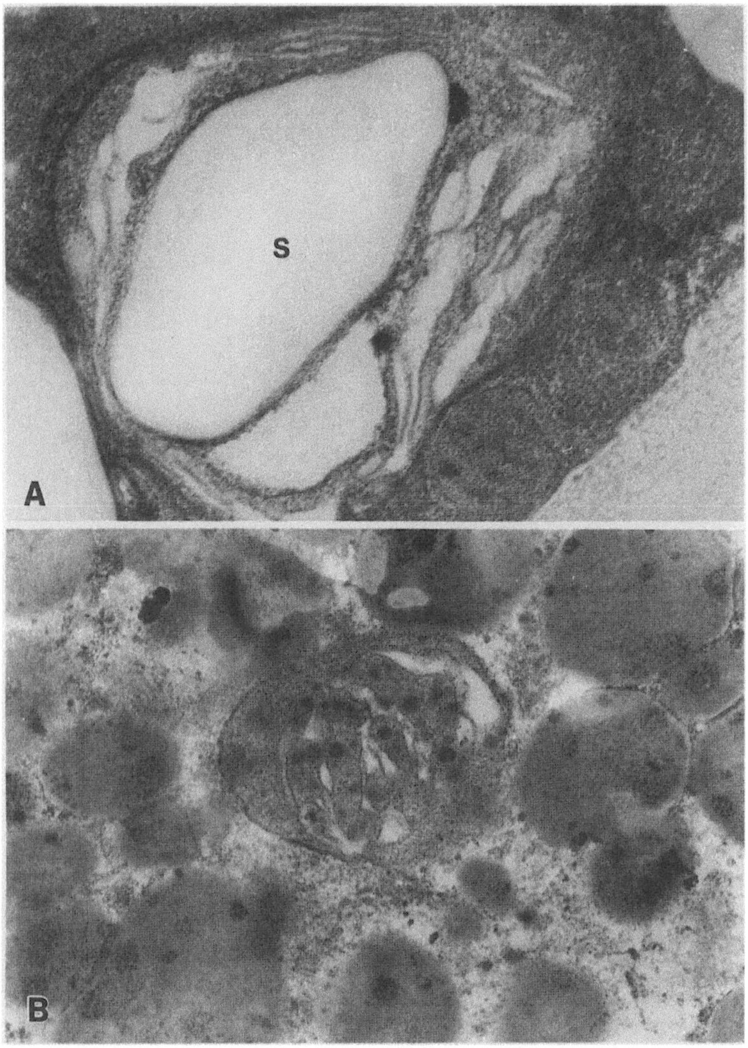

Fig. 14. A Plastids in flax embryos after 7–10 days in vitro (age of embryos 21–24 days). Plastids have a large starch grain and dilatated and irregularly oriented lamellas (×42, 420) *S* starch grain. **B** Plastids in flax embryos cultivated for 7–10 days in vitro (age of embryos 21–24 days). In plastids a number of plastoglobuli arise. Plastids have a spherical shape and are smaller than in 14-day-old embryos (×25, 600). (**A, B** Čiamporová and Pret̆ová 1981, 1983)

Chloroplast structure is preserved unchanged for some 2 to 3 days following the planting of the embryo on cultivation medium. The first changes can be observed on ca. the 4th day of cultivation. They concern the dilation of thylakoids (Fig. 13B). Later appears a gradual degradation of the thylakoids of grana and those of the stroma, so that on the 7th to 10th day of cultivation no chloroplasts are present in the embryos, only plastids. These contain a large starch grain, dilated and irregularly oriented lamellae, along with a number of plastoglobuli

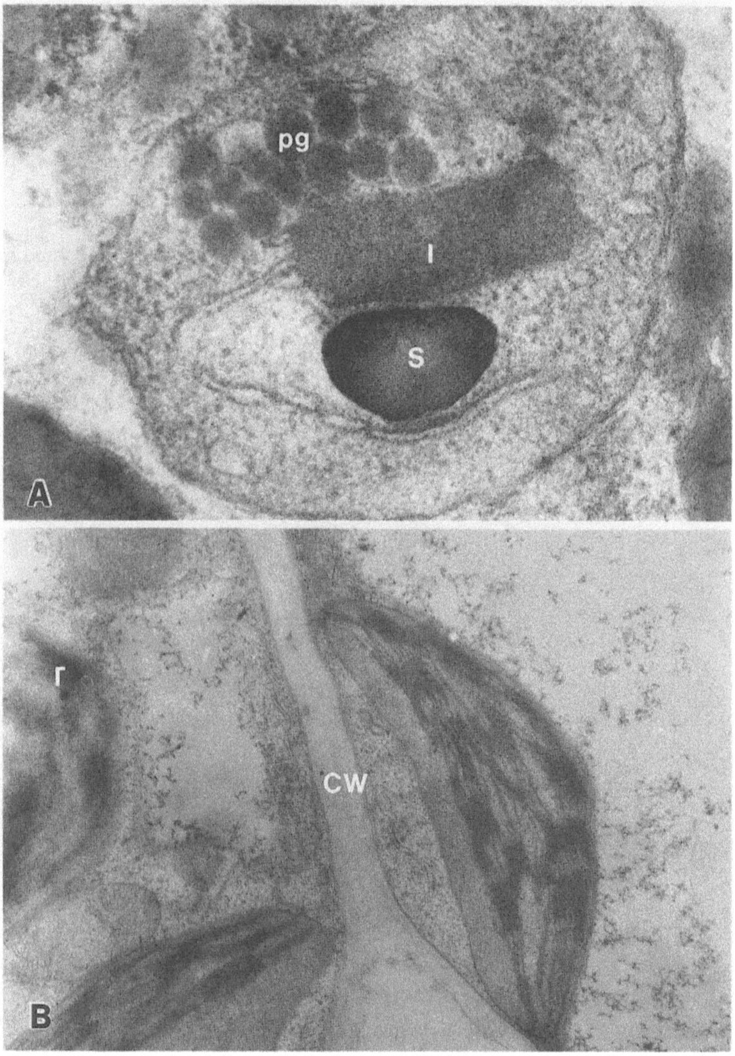

Fig. 15. A Plastids in flax embryos cultivated for 13–15 days in vitro (age of embryos 27–29 days). Such plastids appear sometimes also in 10-day cultivation. Plastids contain protein inclusion, isolated short lamellas, starch grain and numerous plastoglobuli (×79, 100). *I* inclusion; *S* starch grain; *pg* plastogloguli. **B** The shape and structure of plastids from cotyledons of in vitro germinated flax embryos (×26, 700). *cw* cell wall. (**A, B** Čiamporová and Pretová 1981, 1983)

(Fig. 14A). The shape of the plastid changes from elongated to spherical and then diminishes (Fig. 14B). This stage of plastids corresponds with the period of embryos becoming totally yellow on the medium.

Some 24 h later, gradual redifferentiation of plastids can already be observed. On the 13th to 15th day of cultivation of embryos plastids with electrondense protein inclusion are observable having single short lamellae (Fig. 15A). The plastids later increase, gradually forming new lamellae which elongate further. The process of plastid redifferentiation continues until the establishment of a system of granal and intergranal thylakoids that is characteristic for the chloroplasts of assimilating flax leaves. The shape of plastids changes to lenticular (Fig. 15B).

Embryo culture also helped us, to contribute to the understanding of possible ways of dedifferentiation and redifferentiation of plastids, mainly of chloroplasts in plant organs other than leaves, and to assess how this process depends on environmental conditions. Our research also supports the view that the plastids in the embryos of Chloroembryphytes do not take part in photosynthesis (one of the reasons being their envolopment by seed and fruit coats, which reduces the penetration of light to the embryos), nevertheless, they are metabolically active, playing an important role in the synthesis of reserve substances (Preťová 1985a, b; Čiamporová and Preťová 1981, 1983). This can be inferred especially from the fact that flax embryos contained green pigment in the period of their most intensive growth and accumulation of reserve substances.

3 Flax Embryo Cultures Serve to Proliferate Plant Material

Zygotic flax embryos are capable of producing somatic embryoids under certain conditions of cultivation. Morphogenetic response was obtained from flax embryos isolated at the late heart-shaped to torpedo stage and cultivated on the cultivation medium according to Monnier (1978) with added 5% sucrose, 0.05 mg/l BAP and 1 g/l yeast extract at 22 °C and 16 h photoperiod with intensity of illumination ca. 60 μmol/m^2/s, all between the 10th to 21st day of embryo cultivation.

The embryoids and accessory cotyledons were formed as ring around the top of the hypocotyl (Fig. 16). The process of embryoid formation began by the thickening of the hypocotyl and its conspicuous striation. Later embryoids were observed on the hypocotyl tissue of zygotic embryo without callus formation. Therefore the process is called direct somatic embryogenesis (Preťová and Williams 1986a). Somatic embryoids underwent all stages typical for the development of zygotic embryos. From embryoids we obtained normal plants with viable pollen.

The addition of BAP in the process of direct somatic embryogenesis seems to be responsible for the preservation of mitotic stimulus in the hypocotyl cells of immature zygotic embryo, while the yeast extract seems to be responsible for the growth inhibition of the shoot-root axis and for the nondisorganization of the internal pre-embryonic determination state of the initiating cells. The sole addition of BAP to the cultivation medium (without yeast extract) resulted only in abnormal growth of shoots, as well as in the formation and proliferation of calli from the cotyledons and other tissues of the zygotic embryo, probably in consequence of stimulation of cell division. The presence of yeast extract alone (without BAP) strongly inhibited the growth of embryo axis (shoot-root axis). Thus the combination of BAP and yeast extract is responsible for three factors included in the process of direct somatic embryogenesis. Similar results were obtained also in embryos of *Brassica napus* L. (Preťová and Williams 1986b).

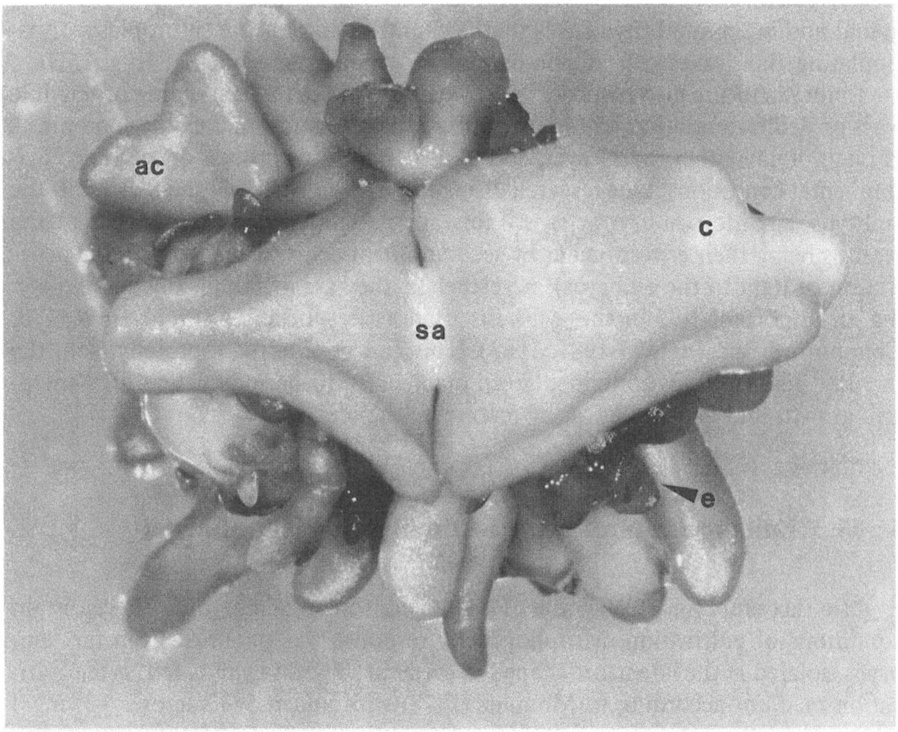

Fig. 16. Flax embryos cultured for 3 weeks on Monnier's basal medium (+glutamine) plus 1 g/l yeast extract and 0.05 mg/l BAP, showing direct somatic embryogenesis from the hypocotyl. *ac* accessory cotyledons; *c* cotyledons; *e* embryoids; *sa* suppressed shoot apex. (Pretová and Williams 1986)

We believe the formation of accessory cotyledons and root poles, a process which might be described as embryo proliferation or "cleavage", to be homologous with the production of discrete, complete embryoids. Embryo cleavage and somatic embryoid formation produce a continuous spectrum of morphogenetic structures, and appear to be a manifestation of the same phenomenon. The production of accessory cotyledons and embryoids at the top of the hypocotyl and accessory roots near the base presumably reflects a gradient along the hypocotyl, either with respect to the early stage of differentiation of the hypocotyl cells initiating morphogenetic structures, or to some other determinative factors affecting cell to cell coordination.

We also suppose that the hypocotyl is the first to give embryogenetic response as a result of its observed priority in activation in the further embryo development under in vitro conditions (Fig. 9B). The hypocotyl also showed the first response (elongated cell growth) after the transfer of an immature (14-day-old) embryo on the cultivation medium.

Flax is an important oilseed and fiber crop with limited means for vegetative propagation. Shoots have been regenerated in vitro from cotyledonary callus (Rybczyński 1975), seedling hypocotyl segments or callus (Gamborg and Shyluk 1976), and stem explants (Murray et al. 1977). Protoplasts have been regenerated to produce roots (Gamborg and Shyluk 1976) and complete plants (Barakat and Cocking 1983; Ling and Binding 1987).

The method of direct somatic embryogenesis will be of great importance in micropropagation of valuable flax hybrids, especially hybrids with low linolenic acid content in seeds, used in effort to make flax oil edible.

4 Conclusions

In conclusion it can be said that the embryos of flax excised in a more mature stage need for their cultivation a simple medium (White 1942) with lower osmotic potential (2% – 5% sucrose). The younger embryos (early heart-shaped and globular stages) need a more complete medium with addition of glutamine (Monnier 1978) and high osmotic potential (12% sucrose) complete embryogenesis in vitro. This process is promoted by addition of 5×10^{-8} mol kin. The temperature is maintained at 25 °C. Embryos can be cultivated in darkness or very low illumination intensity.

The in vitro-ripened embryos need lower osmotic potential for their germination (2% – 5% sucrose), 16 h photoperiod, illumination 3000 lx.

To produce embryoids from zygotic flax embryos it is necessary to use the same medium as for cultivation of younger embryos (Monnier 1978 + 5% sucrose) with addition of glutamine, which promotes the embryoid formation, and BAP and yeast extract. Temperature of cultivation 22 – 25 °C, continuous light or 16-h photoperiod, illumination intensity ca. 60 $\mu mol/m^2/s^1$ are the other conditions.

References

Andronescu D (1919) Germination and further development of the embryo of *Zea mays* separated from the endosperm. Am J Bot 6:443 – 452

Barakat MN, Cocking EC (1983) Plant regeneration from protoplast-derived tissues of *Linum usitatissimum* L. (flax). Plant Cell Rep 2:314 – 317

Čiamporová M, Pret̆ová A (1981) Ultrastructural changes of plastids in flax embryos cultivated in vitro. New Phytol 87:473 – 479

Čiamporová M, Pret̆ová A (1983) Ultrastructural studies of plastids in the flax embryos cultivated in vitro. Fertilization and embryogenesis in ovulated plants. Veda (Bratislava) p 384

Deschamps R (1968) Ĺ embryon du Lin, *Linum usitatissimum* L., étude en microscope électronique. Rev Cytol Biol Veg 31:357 – 360

Dietrich K (1924) Über die Kulturen von Embryonen außerhalb des Samens. Flora 17:379 – 417

Dillman AC (1928) Daily growth and oil content of flax seeds. J Agr Res 37:357 – 377

Erdelská O (1978) Reaction of immature flax embryos to excision. Biologia (Bratislava) 33:17 – 24

Erdelská O (1985a) Some structural features of the embryo / endosperm interaction in angiosperms. In: Willemse MTM, Went JL (eds) Proc 8th Int Symp. Sexual Reproduction in Seed plants, Ferns and Mosses. PUDOC, Wageningen, pp 168 – 170

Erdelská O (1985b) Dynamics of the development of embryo and endosperm I. (*Papaver somniferum, Nicotiana tobacum, Jasione montana*). Biologia (Bratislava) 40:17 – 30

Erdelská O (1985c) Dynamics of the development of embryo and endosperm II. (*Linum usitatissimum, Capsella bursa-pastoris*). Biologia (Bratislava) 40:849 – 857

Erdelská O (1986) Dynamics of flax embryo growth in vitro and in situ. Proc 4th Plant Physiol Day Kosice, August 1986, pp 179–181 (in slovak)
Erdelská O (1988) Dynamics of the development of embryo and endosperm III. (*Pisum sativum*). Biologia (Bratislava) 43:3–9
Erdelská O, Kobetičová D, Freťová A (1973) The in vitro development of excised flax embryos. Biologia (Bratislava) 28:235–239
Ferguson JD, Street HE (1958) The promotion and inhibition of excised root growth by various sugars and sugar-alcohol. Am J Bot 22:513–525
Gamborg OL, Shyluk JP (1976) Tissue culture, protoplasts and morphogenesis in flax. Bot Gaz 137:301–306
Haagen-Smit AJ, Siu R, Wilson G (1945) A method for culturing excised immature corn embryos in vitro. Science 101:234–238
Honma S (1955) A technique for artificial culturing of bean embryos. Proc Am Soc Hort Sci 65:405–409
Iljina AI (1954) Ossobennosti processa formirovanija semeni lna (Differences in seed development in oil seed flax). Sbornik rabot po biologii rozvitija i fiziologii lna. Selkhozgiz, Moscow
Johansen DA (1950) Plant embryology. Waltham, Mass, USA
Kantor TS (1957) K embriologii kulturnogo lna. Byull Gl Bot Sada ZSSR, 29
Khan AA (1982) The physiologiy and biochemistry of seed development, dormancy and germination. Elsevier, Amsterdam
Kobetičová D (1971) Growth of excised *Zea mays* embryos on media with increasing sucrose concentration. Biologia (Bratislava) 26:341–348 (in slovak)
Kozinka V, Klenovská S (1965) The uptake of mannitol by higher plants. Biol Plant 7:285–297
Laibach F (1925) Das Taubwerden von Bastardsamen und die künstliche Aufzucht von früh absterbenden Bastardembryonen. Z Bot 17:417–459
Letham DS (1963) Zeatin, a factor inducing cell division isolated from *Zea mays*. Life Sci 2:569–573
Ling HQ, Binding H (1987) Plant regeneration from protoplasts in *Linum*. Plant Breed 98:312–317
Linser H, Herwig K (1963) Untersuchungen zur Abhängigkeit der Nährungsstoffaufnahme vom osmotischen Druck der Außenlösung. Protoplasma 57:588–600
Maheswari P (1950) An introduction to the embryology of angiosperms 1. ed., McGraw-Hill, New York, 453 pp
Mauney JR (1961) The culture in vitro of immature cotton embryos. Bot Gaz 122:205–209
Millborrow BW (1974) The chemistry and physiology of abscisic acid. Annu Rev. Plant Physiol 25:259–307
Monnier M (1978) Culture of zygotic embryos. In: Thorpe TA (ed) Frontiers of plant tissue culture 1978. Proc 4th Int Congr Plant Tissue Cell Culture, Calgary, Canada, pp 277–286
Murray BE, Handyside RJ, Keller WA (1977) In vitro regeneration of shoots and stem explants of haploid flax (*Linum usitatissimum* L.). Can J Genet Cytol 19:177–186
Ostapenko O (1968) Dostizhenija dinamiki nagromazhdenija zapasnykh pozhivnykh rechovin v nosinni l'nu olijnogo (L.v.L.) ta l'nu bogatorichnogo (*L. perenne* L.) Ukr Bot Zh 25:60–66
Pecket RC, Selim ARAA (1965) Embryo culture in *Lathyrus*. J Exp Bot 16:325–328
Pickering RA, Morgan PW (1983) Plant regeneration from cultured embryos derived from *Hordeum vulgare* L. pollinated with *H. bulbosum* L. Euphytica 32:585–591
Poddubnaya-Arnoldi VA (1952) Issledovanije zarodyshei u pokrytosemyannykh rastenij v zhivom sostoyaniji. Byull Gl Bot Sada 14:3–11
Preťová A (1974) The influence of the osmotic potential of the cultivation medium on the development of excised flax embryos. Biologia Plant (Prague) 16:14–20
Preťová A (1977a) Pigments in young embryos of *Linum usitatissimum* L. Photosynthetica 11:217–219
Preťová A (1977b) Effect of some growth regulators on embryogenesis of flax in vitro condition. Proc Plant Physiol Day, Bratislava, pp 260–263
Preťová A (1978) Flax embryogenesis. Changes in the contents of pigments and fats in vitro and in situ. Biologia (Bratislava) 35:413–416
Preťová A (1980) Effect of streptomycin on growth of immature flax embryos. Biologia (Bratislava) 35:413–416
Preťová A (1983) Cultivation of globular and heart stage flax embryos. Fertilization and embryogenesis in ovulated plants. Proc 7th Int Emb Symp, High Tatras, 1982, pp 307–309

Pretová A (1985a) Chlorophyll and carotenoids in flax embryos during embryogenesis. Photosynthetica 19:194–197
Pretová A (1985b) The assimilation pigments, fats and ultrastructure of plastids during the embryogenesis of flax in situ and in vitro. Veda (Bratislava). p 73
Pretová A (1985c) In vitro embryogenesis of flax (*Linum usitatissimum* L. Proc 8th Int Embryol Symp 24–28 August 1984. Sexual Reproduction in Seed Plants, Ferns and Wageningen, The Netherlands, Mosses, pp 182–183
Pretová A (1986) Influence of kinetin on the growth of zygotic flax embryos in vitro. Plant Cell Rep 3:210–211
Pretová A, Erdelská O (1977) Some differences in the embryogenesis of excised flax embryos. Use of tissue cultures in plant breeding, Olomouc, CSSR, pp 299–307
Pretová A, Vojteková M (1985) Chlorophylls and carotenoids in flax embryos during embryogenesis. Photosynthetica 19:194–197
Pretová A, Williams EG (1986a) Direct somatic embryogenesis from immature zygotic embryos of flax (*Linum usitatissimum* L.). J Plant Physiol 126:155–161
Pretová A, Williams EG (1986b) Zygotic embryo clonning in oil-seed rape (*Brassica napus* L.). Plant Sci 47:195–198
Pretová A, Čiamporová M, Erdelská O (1981) Chlorophyll content and culture of flax embryos in vitro. Acta Soc Bot Pol 50:291–293
Raghavan V (1966) Nutrition, growth and morphogenesis of plant embryos. Biol Rev 41:1–58
Raghavan V, Torrey JC (1963) Growth and morphogenesis of globular and older embryos of *Capsella bursa-pastoris* in culture. Am J Bot 50:22–39
Rietsema J, Satina S, Blakeslee AF (1953) The effect of sucrose on the growth of *Datura stramonium* embryos in vitro. Am J Bot 40:538–545
Rijven AHGC (1952) In vitro studies on the embryo of *Capsella bursa-pastoris*. Acta Bot Neerl 1–2:157–200
Rijven AHGC (1955) Effects of glutamine, asparagine and other related compounds in the in vitro growth of embryos of *Capsella bursa-pastoris*. Proc Kon Ned Akad Wetensch 658:368–376
Rybczyński JJ (1975) Callus formation and organogenesis of mature cotyledons of *Linum usitatissimum* L. cv. Szokijskij in vitro culture. Genet Pol 16:161–172
Slatyer RO (1961) Effects of several osmotic substrates on the water relationships of tomato. Aust J Biol Sci 14:519–540
Souegés R (1924) Développement de l' embryon chez le *Linum catharticum* L. Bull Soc Bot France 71:925–938
Stroganov BP, Lapina LP (1964) A possible method for separately studying the toxic and osmotic action of salts on plants. Fiziol Rast 11:674–680 (in Russ)
Sveshnikova IN (1955) The fat arise in plastides. DAN SSR 102:5–9
Szweykowska A, Gierczak M, Luszczak R (1959) Anthocyanin synthesis in isolated embryos of red cabbage and radish. Acta Soc Bot Pol 28:28–35
Thiman KV, Loos GM, Samuele V (1960) Penetration of mannitol into potato discs. Plant Physiol 35:848–853
Veen H (1963) The effect of various growth regulators on embryos of *Capsella bursa-pastoris* growing in vitro. Acta Bot Neerl 12:129–171
Vizárová G, Pretová A, Vozár I (1987) Free endogenous cytokinins in development of flax embryos. Biologia (Bratislava) 42:39–44
White PR (1942) Plant tissue culture. Annu Rev Biochem 11:615–628
Yakovlev MS, Zhukova GYA (1973) Angiospermous plants with green and colourless embryos. Nauka, Leningrad, p 114 (in Russ)

IV.10 Coconut (*Cocos nucifera* L.): Micropropagation

J. BLAKE[1]

1 Coconut as an Oilseed Crop

1.1 Importance of Coconut Oil

Despite the increasing production of soybean and palm oils over the last two decades, coconut oil remains an important product on the world market as it is one of only two commercial sources of short-chain fatty acids with 8–14 carbon atoms. Coconut oil is thus in direct competition with only one other such source, that being palm kernel oil. Coconut oil has a particularly high percentage (48%) of the C12 compound, lauric acid, which is in demand for the production of soaps and oleochemicals (Schwitzer 1985; Berger and Ong 1985). High quality soaps are particularly dependent on the inclusion of a percentage of coconut oil. A significant proportion of all coconut oil goes into the oleochemicals industry where the hydrophilic and hydrophobic qualities of the fatty acid and its derivatives are of particular importance for use in the production of surfactants and foam stabilizers for detergents, in cosmetics, shampoos and pharmaceuticals, as corrosion inhibitors, emulsifiers and wetting agents, and as plasticizers for PVC products.

In addition to the large domestic demand for drinking nuts, for fresh copra and for oil, exported coconut oil is in demand for edible use because of its useful characteristics (Ohler 1984): relatively high melting point, narrow melting range, good creaming quality for shortening, bland flavour and absence of unpleasant odour, stability and resistance to oxidation and rancidity, and high luster in spray oil use.

Thus it can be used for frying and popping corn, for producing simulated dairy products (filled milk, coffee whitener, ice cream etc.), in confectionery, as a cocoa butter replacer for coating biscuits etc., as a spraying oil to act as a moisture barrier on e.g., biscuits and as a low-percentage addition to margarines (Maarsen 1985). Desiccated coconut is an important product, especially in Sri Lanka.

[1] Unit for Advanced Propagation Systems, Dept. of Horticulture, London University, Wye College, Wye, Ashford, Kent TN25 5AH, United Kingdom

Biotechnology in Agriculture and Forestry, Vol. 10
Legumes and Oilseed Crops I (ed. by Y.P.S. Bajaj)

1.2 World Production of Coconut

The proportion of coconut oil used for food or non-food purposes varies from one country to another. Thus in the USA a little over half (52%) is used for non-edible purposes (Table 1) and non-food uses predominate in W. Germany, the USSR and Italy, whereas food uses have the edge in the UK, France and the Netherlands (Ignacio 1985). Price may be the determinant in regard to usage of a particular oil for a particular purpose and price stability is important to commercial users. Coconut oil is thus at a disadvantage since natural misfortunes such as typhoons and droughts can have a major impact on the supply of copra onto the world market. Although prices of all vegetable oils rose spectacularly in 1984, the rise of the laurics (coconut and palm kernel oil) was most outstanding as the direct result of weather conditions (Daly 1985).

A recent development is in the use of coconut oil as a fuel, and it has been used successfully as a 5–10% addition in diesel fuel (Banzon 1984). An expansion of this type of use will depend upon increased production at lower costs, and use as a fuel additive could be particularly important in producing countries. There are indications that improved management providing adequate manuring and irrigation combined with the use of hybrid palms could result in much in-

Table 1. US average utilization of major vegetable oils (yearly average for 1978–82) in 1000 mt. (After Ignacio 1985)

	Total	Food uses	Non-food uses
Coconut	399	190	209
Palm	107	100	7
Soya	4033	3754	279
Cottonseed	305	275	29
Corn	284	260	24
Total	5128	4579	548

Table 2. Yield of oil from coconut under different management levels. (Bavappa 1982)

Management	Mean yield nuts/palm	Oil yield	
		Tonnes/ha	% Increase over national average
National average	30	0.56	–
West Coast Tall, rainfed			
a) Cultivation alone	38	0.71	26.8
b) Cultivation and manuring	62	1.16	107.1
c) Irrigated and manured (12 years old)	75	1.76	214.3
d) Mixed cropping with cacao, irrigated and manured	111	2.60	364.3
Dwarf × Tall, rainfed, 9 years old	112	2.93	423.2
Tall × Dwarf, irrigated, 20 years old	130	2.89	416.1

Table 3. Coconut area, production and productivity in the main producing countries. (After Das 1985)

	Coconut area (million ha)		Production of coconut (million t/annum)		Productivity[a]		Productivity[b] 1976–80
	1971–75	1976–80	1971–75	1976–80	1971–75	1976–80	
World	6.81	8.26	30.60	34.99	–	–	–
Of which:							
India	1.09	1.08	4.43	4.28	4.05	4.01	3.97
Indonesia	1.42	1.69	8.16	10.67	5.76	6.30	8.21
Malaysia	0.31	0.33	0.97	1.25	3.09	3.46	3.79
Papua New Guinea	0.19	0.22	0.73	0.77	3.74	3.49	4.18
Philippines	1.78	2.16	8.46	10.44	3.92	3.67	5.28
Sri Lanka	0.47	0.47	1.85	1.63	3.96	3.60	3.51
Thailand	0.28	0.31	0.69	0.75	2.19	1.94	2.47
Mexico	0.09	0.13	0.81	0.88	6.11	5.84	7.37
World average					4.51	4.47	5.51

[a] Production of a given year divided by the area of that year.
[b] Production of a given year divided by the area of 8 years earlier.

Table 4. World production of coconuts and copra (FAO vol. 39, 1986) in 1000 mt

	Coconuts			Copra		
	1983	1984	1985	1983	1984	1985
World	34193	33115	34661	4536	4084	4507
Africa	1577	1587	1567	193	194	198
Of which:						
Ivory Coast	260	290	300	38	43	44
Mozambique	410	400	400	67	65	65
Tanzania	320	320	320	29	29	29
America	1965	1900	1890	218	195	194
Of which:						
Mexico	815	700	655	143	120	120
Oceania	2149	2317	2452	306	334	361
Of which:						
Fiji	234	234	240	24	24	25
Asia	28503	27310	28752	3818	3361	3754
Of which:						
India	4315	5117	4550	345	370	380
Indonesia	10097	10355	10754	1070	1101	1160
Malaysia	1708	1716	1721	218	214	216
Philippines	8881	6959	7793	1935	1500	1700
Sri Lanka	1758	1350	2100	138	62	180
Thailand	1090	1128	1121	35	35	35
Vietnam	360	380	400	59	60	64

creased yields (Bavappa 1982). Costs would also be increased, but Table 2, for conditions in India, suggests that the benefit should far outweigh the cost.

Productivity does not appear to be increasing, except in Indonesia and Malaysia (Table 3 a), but this may not indicate the true situation because the calculations do not allow for the increased area which has been recently planted to coconuts (Table 3) and which is not yet yielding. This factor has been taken into account by Das (1985) in calculating yield data based on the planted area 8 years earlier (Table 3 b), which then suggests there has been increased productivity in all countries except India and Sri Lanka.

Recent data of annual yields given by FAO (1986) (Table 4) shows the variation which can occur from year to year. There was an overall drop in 1984 but this was mostly attributable to a drop in Philippines production in 1984 which began to show some recovery in 1985.

Although coconut represents only about 4–4.5% of the world production of fats and oils (FAO 1986), Das (1985) concludes that on a global perspective coconut is making a come-back as a commercially strategic crop in spite of rapid growth in alternative oil crops such as soybean and oil palm. High-yielding hybrids have increased yields and reliability of coconut palms, and Das suggests that the crop holds great promise both as a food crop as a powerful source of energy for the tropics.

2 Conventional Propagation

Conventional propagation of coconut is carried out entirely by the use of seed nuts. With open pollination, the seed nuts are so variable that some plantations may contain palms producing as few as 10 nuts a year whilst others produce over 100. The average is often not more than 30–40 nuts per palm per year, whereas individual palms have been recorded which produce up to 400 nuts per year (Iyer 1982). Hybrids between two varieties usually have a higher yield than that of either parent, and these are particularly required in order to obtain improved stock for field planting. In this case, it is essential to control pollination, which can be done by one of two methods.

The simplest method is by using directed natural pollination, which allows open pollination to occur in a mixed planting of the two chosen parents. The success of this method relies entirely on the complete emasculation of the female parent, which is often difficult, or even impossible, to achieve without a very reliable labour force. Since the stocks of both parents are raised from seed, and are therefore heterozygous, the hybrids produced will still be very variable but are generally an improvement on the parents. An improved method is assisted pollination which involves the collecting, storing and transporting of pollen from an isolated plantation of the male parent to that of the female parent. This method is highly labour-intensive as all the female parents must be efficiently emasculated and then pollinated with the required pollen. An advantage of this method is that with certain combinations of parents, illegitimate seedlings can be recognized in the nursery by their off-colour. Although expensive, this method produces the most reliable true-to-type seed nuts.

There has been considerable genetic improvement of coconut yield by use of techniques developed by Meunier et al. (1986). Several of their hybrid varieties produced at Port Bouet in the Ivory Coast (such as PB 121) are now being planted on a worldwide basis. These hybrids may have a considerable impact on indigenous varieties when planted in countries such as the Philippines. However, Santos (1986) gives the case for the exploitation of indigenous genetic resources with strict selection of best individual performers for use as parents. Meunier et al. (1986) discuss the possible use of in vitro multiplication and suggest that cloning of the best 13% of trees in a PB 121 plantation would provide a theoretical copra yield increase of 24% from a variety which already has a higher than average yield.

3 Micropropagation

3.1 Introduction

Various reviews of palm regeneration and micropropagation have discussed the three most important commercial palms, date, oil and coconut palms (Blake 1983; Brackpool et al. 1986; Tisserat 1984), but the only review relating specifically to coconut is that of Pannetier and Buffard-Morel (1986) who included detailed accounts of their own work up to 1983 in France and of the early work at Wye College in the UK by Blake, Branton and Eeuwens. The present chapter will not be concerned with the early literature so comprehensively reviewed by Pannetier and Buffard-Morel, but will deal with that from 1983 onwards. The paucity of publications between 1983 and 1987 seems to indicate the difficulties of the research in this area, rather than the fact that commercial development has inhibited publication.

In the early days of coconut tissue culture, there was an implicit assumption that the coconut would respond in a similar manner to that of the oil palm, i.e., that a relatively slow-growing callus would be produced which could be multiplied, as could the succeeding embryogenic callus, with the regeneration of embryoids to produce plantlets. In the event, it has proved difficult to obtain plantlets in coconut and the production and organization of callus is sufficiently different from oil palm for Branton and Blake (1986) to use the term calloid to describe the type of proliferation which occurs. Nyman et al. (1983) suggested the term calloid for the callus they observed on Taro which is more organized than a true callus and appears to resemble orchid protocorm initiation. In coconut the calloid arises from meristematic tissue which becomes partly but not totally dedifferentiated before regeneration occurs.

3.2 Calloid Development and Embryogenesis

The establishment of explants on a medium with a high level of 2,4-D (10^{-4} M) and activated charcoal has proved most successful. Branton and Blake (1983b) used primarily inflorescence tissue (not roots as often quoted) whereas Pannetier and Buffard-Morel (1982) used leaf tissue. Under the influence of a high level of 2,4-D Branton and Blake (1983b, 1986) observed files of cells produced in the outer layers of the flower meristem (Fig. 1A) followed by the development of nodular meristematic structures in the superficial layers (Fig. 1B, C). It is not clear whether the callus obtained by Pannetier and Buffard-Morel (1986) which emerged at the level of the vascular bundle had a similar origin to the calloid obtained by Branton and Blake (1986) which arose from the flower meristems. However, after further development and subculture for 3–4 months, the nodular callus obtained in both laboratories appears similar, as do cultures at the early embryoid stage (cf. Branton and Blake 1986, Fig. 5; Pannetier and Buffard-Morel 1986, Fig. 3E).

The nodules which various workers have observed, and which have been optimistically described as embryoids, may easily become roots (Pannetier and Buffard-Morel 1986, Fig. 2; Gupta et al. 1984) but in our own laboratory, under certain conditions which have not yet been defined, the calloid nodules will form embryoidal structures (Fig. 1D) which may develop comparably with a zygotic embryo (Fig. 1E, cf. Fig. 3B). However, a variety of embryo-like and abnormal structures are commonly formed (Pannetier and Buffard-Morel 1986). These include leaf-like structures, often with leaves fused (Fig. 1F), as well as the development of haustorial-type tissue, with or without roots (Fig. 1G). The development of a haustorium gives a good indication that embryogenesis, rather than organogenesis of shoots or roots, has been observed, although root organogenesis can also occur. At the present time, it is not clear whether shoot organogenesis can occur, or whether shoots are produced from embryoids in which the root has not developed. There is, in any case, a certain amount of difficulty in determining the exact pathways since the normal root system in coconut is almost entirely adventitious.

An interesting report has appeared (Raju et al. 1984) suggesting that a callus stage may not be necessary for regeneration in coconut, but that direct embryogenesis in cultured leaf tissue could lead to the formation of clonal plants. Over 2000 combinations of media ingredients were used but there is no indication which was successful except that the Mg : K ratio and the hormone levels were optimized. After embryoids were formed, the authors state that "withdrawal of NH_4^+, HCHO and casein hydrolysate and simultaneous increase in auxin cytokinin ratio through several transfers (each time auxin was reduced to 75% and cytokinin increased to 125%) has resulted in the development of shoot (having green scale leaves with sheathing leaf base), and primary root with root-cap along with a primordial haustorium". Embryoids are stated to arise from the vascular tissue but this would be unusual as it is the area where root primordia normally originate. One of the auxins used (NAA) has been shown to induce root initiation in coconut (Blake and Eeuwens 1980) so that the level needed for inducing embryogenesis could be very critical. Details of the protocol used by Raju et al. (1984) are not given and no further information has been published.

Embryogenesis has been observed at a very low rate in the culture of coconut pollen (Monfort 1985) with the aid of activated charcoal, TIBA and glutamine added to the basic medium. Development of the embryoids has not yet been achieved.

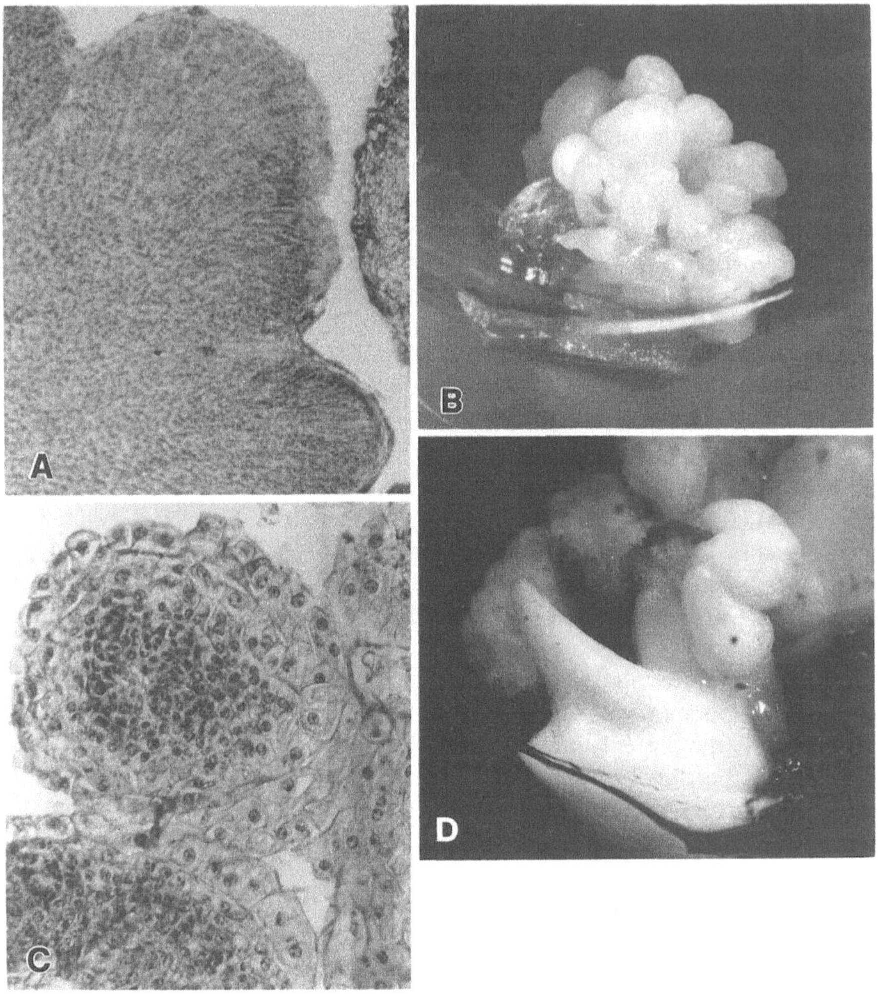

Fig. 1 A–H. Culture of coconut inflorescence tissue var. Malayan Dwarf. **A** Section of early stage of calloid formation from young flower meristem. **B** Calloid nodules (after 5 months in culture) from flower meristem cultured with 10^{-4} M 2,4-D. **C** Section of nodule. **D** Embryoids developing from calloid nodules (after 7 months in culture). **E** Shoot development from embryoid (after 9 months). **F** Abnormal leafy developments from fused embryoids. **G** Haustorial tissue and roots developed from embryoid. **H** Plantlet developed from embryoid (after 11 months)

3.3 Embryoid and Plantlet Development

The protocol for embryoid and plantlet development has not yet been defined but the gradual lowering of the 2,4-D level is associated with embryoid production. Complete elimination of auxin results in a haustorial type of development (Fig. 1 G) with or without roots, and low levels of other auxins may be preferable in the later stages of culture. The role of cytokinin has not been determined but it

Fig. 1E–H

appears that a moderate level (10^{-5}M to 10^{-6}M with activated charcoal) is usually incorporated throughout the culture period.

Although the early stages of culture are best carried out in the dark, as soon as the embryoids become leafy they should be transferred to the light. If only shoots are observed, a rooting treatment will be required.

The first clonal plants were produced at Wye (Branton and Blake 1983a) and since then small numbers have been produced regularly from at least 20 mother

Fig. 2. Plant regenerated from culture of inflorescence tissue. Photograph taken 3 1/2 years after establishment in the Solomon Is. (Courtesy Unilever Ltd)

palms (Fig. 1 H), but none have been successfully established, probably due to inadequately functional root systems. The only success appears to be that of one plant produced in Unilever's UK laboratory (Smith 1986) and now established in the Solomon Islands (Fig. 2). The culture of zygotic embryos and their establishment ex vitro (see Sect. 3.4) should provide information which can be related to establishment of clonal plants.

The cause of the development of abnormal embryoidal structures is not known and may be attributable to sub- or supra-optimal levels of auxin and/or cytokinin. Fasciations can be induced in other species by variations in levels of growth substances and abscisic acid has been shown to influence normal embryoid germination (Ackerson 1984; Vasil and Vasil 1981). However, there is little useful information on the causes of such abnormal developments and none which relates to palms. Fundamental investigations of the fate of growth substances in cultured tissues, and of their morphological effects on the organs produced, is badly needed. It is particularly important since Armstrong and Francis (1987) have reported that long-term culture of coconut callus on 2,4-D may produce some chromosomal instability. They found that one particular callus line which had been in culture for several years consisted of tetraploid cells. In our laboratory we are aware of the danger of continued culture on 2,4-D and have found that regeneration occurs only after rapid calloid production and that embryoids should appear in 6–9 months from the culture of the original explants. Long-term subculture of callus produces material which will not regenerate (unpubl.) and in view of Armstrong and Francis' results, this may be because doubling of chromosomes has occurred. It will be necessary to determine the ploidy of plants which have been regenerated through culture to carry out genetic fingerprinting analyses.

3.4 Zygotic Embryo Culture

Most of the early work on zygotic embryo culture was carried out in de Guzman's laboratory in the Philippines with the makapuno variety (see Sect. 3.5) but more recently, several laboratories have published work on the culture of other varieties. Normal development of the zygotic embryo can be achieved, but no laboratory has reported consistent success through all the stages from excision of the embryo to establishment in the nursery. Several laboratories have obtained proliferation of callus or calloid on zygotic embryos, but consistent success in induction or proliferation of the callus or in regeneration has not been reported.

Culture of zygotic embryos is an important area for research since exchange of coconut germplasm by embryos rather than by nuts would be easier, cheaper and safer. In our laboratory Watayanothin (unpubl.) has shown that embryos can be harvested from nuts in Thailand and sent in a small package through the normal postal channels to the UK. One hundred percent germination has been obtained after embryos have spent 5 days in transit; longer periods are not completely inhibitory but may cause delayed germination. This is not surprising as the coconut has a recalcitrant seed which loses viability soon after harvest. Therefore, cryopreservation of coconut embryos could have an important future role in the preservation and exchange of germplasm (Bajaj 1983, 1984).

Considerable variation in success with the culturing of zygotic embryos has been reported and a wide variety of media and additives have been used. Thus Fisher and Tsai (1978) had problems with contamination and inhibition of growth with sodium hypochlorite, whereas Watanayothin had few contamination problems with embryos, and these were easily eliminated without inhibitory effects by the use of sodium hypochlorite. Gupta et al. (1984) harvested mature and immature embryos of West Coast Tall which were cultured on Eeuwens (1976) Y 3 liquid medium with variations in auxins, cytokinins and coconut water. They obtained normal germination, but without root elongation. When moved to Y 3 medium without additives the roots elongated and some produced plantlets which could be transplanted to a soil : sand : compost (3 : 3 : 1) mixture. Immature embryos could not be cultured through to the plantlet stage but 60% of the embryos of 7×4 mm size yielded plantlets. However, none of the plantlets survived for over 4 weeks. Iyer (1982) reported the use of MS medium for successful embryo culture but also failed to obtain successful establishment. Pannetier and Buffard-Morel

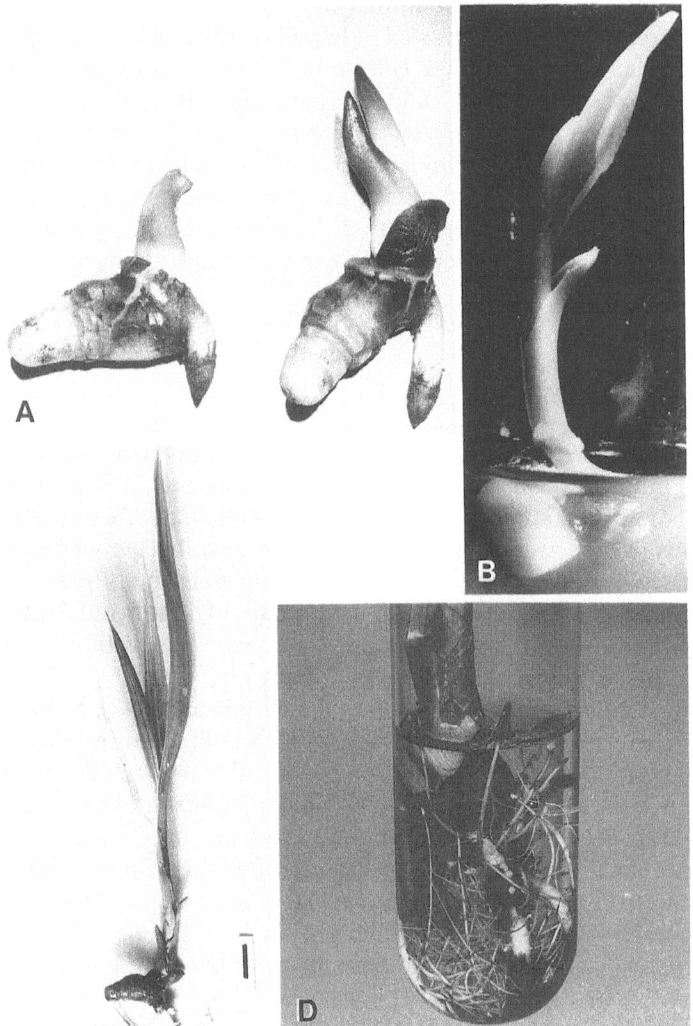

Fig. 3 A–H. Culture of zygotic embryos. **A** Early shoot and root development of embryos on basal medium with 15% coconut water. **B** Shoot development from embryo (*bottom left*). **C** Development of shoot with primary and secondary roots. **D** Good root system before establishment. **E** Established plants from zygotic embryos, 7 months after transfer to compost. **F** Abnormal shoot and root development after germination on 10^{-4} M 2,4-D with activated charcoal. **G** Calloid initiation (*right*) on zygotic embryo (undeveloped haustorium on *left*). **H** Multiplying calloid from embryo

(1986) quote unpublished results of some success with embryos of the hybrid PB 121 (Malayan Yellow Dwarf×West African Tall). Dwarf coconut embryos have been cultured by Karunaratne (1985).

The most comprehensive published work on embryo culture has been carried out by Assy Bah (1986) who used mature (11–12 months after fertilization) em-

bryos of 20 varieties. The basic medium was MS and germination was obtained only in the presence of agar and activated charcoal. Germination was often slow especially on liquid medium or agar medium without charcoal. Shoot development occurred first and some development of the haustorium was observed with both ages of embryo. Root development was obtained by adding 20 mg/l of NAA. After 6–7 months in culture, the plants were successfully established in soil. All varieties which were tested showed germination but Tall varieties were slower than Dwarf varieties. Thus at 4 months, over 90% of the embryos of most of the Dwarf varieties showed shoot development whereas most of the Tall varieties had only reached 45–86% germination.

In our laboratory, embryos of the varieties Malayan Dwarf, Thai Dwarf and Malayan Dwarf×West African Tall have been cultured in a range of treatments (Watanayothin unpubl.). Initial growth during the first few weeks was often good

(Fig. 3A, C), but it was difficult to maintain growth, especially that of the roots (Fig. 3B). A few embryos in each batch would form good plantlets in less than 6 months, but these were exceptional. The basic medium was MS/Y3 (Branton and Blake 1983b) either liquid or with agar at 0.55% and sucrose levels between 4 and 12%. Additives to the media included auxins, cytokinins, gibberellic acid, coconut water, hydrolyzed casein, some amino acids (glutamine, arginine and asparagine), potassium chloride and activated charcoal. Of the treatments used, none produced significant or consistent effects, although the use of growth regulators, especially IBA, NAA, 2iP, BAP and zeatin, had some promotive effects. Provided that plantlets had two good leaves and more than 15 lateral roots (Fig. 3D), they could be established in the glasshouse in a mixture of coir, sand and loam (Fig. 3E). Feeding of nutrients could take place once new leaf growth had started, which presumably indicated that new root growth had occurred. Malijan and del Rosario (1986) have shown that embryo-cultured seedlings take 8 weeks to achieve photosynthetic competence.

3.5 Makapuno Embryo Culture

The culture of makapuno embryos is an example of the technique of embryo rescue. The makapuno variety of the Philippines has a jelly endosperm instead of the normal endosperm of solid copra and water. Although a normal-looking embryo is present, this never germinates. De Guzman and her associates (reviewed by del Rosario and de Guzman 1982), isolated embryos from these nuts and successfully produced plants which are now planted in the field and are yielding nuts. The occurrence of makapuno nuts is a chance event in a particular variety (Laguna Tall) and on any one tree only a small proportion of the total nuts are makapuno type. In vitro culture of embryos has ensured transmission of the makapuno characteristic to the extent that the resulting plants have produced a much higher proportion of makapuno-type nuts, although it is never (or rarely) as high as 100%.

A comparison of the metabolism of developing makapuno nuts with normal nuts has shown the involvement of galactomannan metabolism in the formation of the makapuno phenotype. As normal nuts mature, α-D-galactosidase has a major role in converting galactomannans to water-insoluble mannans. This enzyme is deficient in makapuno nuts, and viscous galactomannans accumulate so that the endosperm remains as a jelly instead of being deposited as copra within the nut (Mujer et al. 1984a, b). β-mannosidase may have some involvement in normal embryo germination, and is also present at a very low level in makapuno nuts (Mendoza et al. 1985).

The successful culture of makapuno embryos has encouraged the Philippine Coconut Research and Development Foundation to establish a laboratory for mass propagation of this type of coconut (del Rosario and de Guzman 1982). However, there are still problems at the establishment stage which need to be resolved before the technique can be fully exploited.

3.6 Embryo Callus

The multiplication of zygotic embryos resulting from the crossing of elite palms could be of interest on a commercial basis if desirable plants could be produced more easily or more economically than by the method discussed in Sections 3.2 and 3.3 above for mature palm tissues. One possible method would be to induce direct secondary embryogenesis on the zygotic embryo itself without an intervening callus stage, but this has not yet been successfully achieved.

The alternative method would be by production of callus on embryo tissue which would yield plants through embryogenesis. One of the earliest records of the successful production of callus was that of de Guzman et al. (1978). The use of 2,4-D in the culture medium gave a variety of developments from the cotyledonary sheath tissue which included nodular callus, root-like structures and a spongy tissue which was suggestive of haustorial tissue. The authors suggested that the callus nodules resembled the protocorms of orchids.

Twenty-day-old germinated embryos were sliced into sections and were cultured more successfully on Y3 medium than on MS or SH media (Jagadeesan and Padmanabhan 1982). The portion containing the cotyledonary sheath was the only one to survive and in the presence of NAA and BAP produced callus which formed roots and pneumatophores. Tahardi (1985) also used Y3 medium with activated charcoal and 2,4-D or 2,4,5-T and sliced mature embryos before culturing. Nodular proliferations were obtained which were thought to be similar to orchid protocorms as had been observed by D'Souza (1982). When transferred to a reduced auxin level, callus or further nodules suggesting shoots were produced. Nodular callus transferred to low auxin was observed to produce somatic embryos of heart and cylindrical shape resembling zygotic embryos (Thanh-Tuyen and Dionzon 1986). However, it is doubtful if heart-shaped embryos are as characteristic of monocotyledons as they are of dicotyledons, and these may not have been true embryos.

A more novel approach has been taken by Bhalla-Sarin et al. (1986) who cultured immature embryos, 8 – 10 months from pollination of West Coast Tall, on Gamborg's B5 medium with a variety of additions. They obtained a white, compact, nodular callus, but only with the addition of IAA conjugates (IAA-asp and IAA-ala). After transfer to NAA, profuse rooting was observed. If callus was transferred onto a medium with IAA-asp and kinetin, roots were induced in 60% of the cultures, but shoot induction in less than 5%. Three plantlets were obtained but they had poor roots and could not be established. The authors attribute the effect of IAA conjugates to their persistence in the medium being greater than that of IAA itself.

In our laboratory, Petchpirun (unpubl.) has cultured embryos in media with and without 2,4-D and activated charcoal. Normal shoot development of the embryo occurred with low levels of 2,4-D although root development was inhibited up to 10^{-5} M, but at higher levels the shoot was deformed and a nodular callus was formed at the base which eventually produced a mass of roots, which were possibly non-functional. With charcoal present in the medium, a level of 2,4-D as high as 10^{-4} M sometimes gave nearly normal shoot development, though root development was disrupted (Fig. 3F). When embryos were cultured on 2,4-D medium without charcoal, normal development occurred at very low levels, whilst at higher levels death usually resulted. Slicing of the embryos did not always prevent normal shoot development, but encouraged calloid development, especially on 11-month-old embryos as compared with 12-month-old embryos. The calloid came from the end of the embryo opposite the haustorium (Fig. 3G) and thus may have arisen from the cotyledonary sheath tissue. Culturing with 2×10^{-6} M 2,4-D without charcoal gave good calloid development and at the time of writing, the calloid is multiplying (Fig. 3H) but has not yet shown signs of regeneration.

The production of callus from the cellular endosperm has been reported by Kumar et al. (1985). The endosperm was taken from 6-7-month-old germinating nuts and placed on Eeuwens (1976) Y3 medium with a high level of 2,4-D (50 mg/l). Thirty percent of the explants initiated callus which continued to proliferate on transfer to a lower level (1 – 5 mg/l) of 2,4-D. The embryo could be separated from the endosperm callus and did not itself give rise to callus.

4 Conclusions

Although many laboratories around the world have been culturing coconut with the aim of obtaining clonal plants, progress has been slower than anticipated. This may be because of expectations that cultured coconut tissues would behave like those of oil and date palm. However, if coconut produces a completely dedifferentiated callus, regeneration fails to occur. Even when a good calloid is obtained it is difficult to obtain successful regeneration although progress is being made. It is hoped the work on production of somatic embryos from zygotic embryos may contribute information on the general requirements for embryogenesis in coconut.

Culture of zygotic embryos, especially of makapuno type, has been fairly successful, but there is still need for considerable improvement of the various protocols, particularly at the establishment stage.

Acknowledgments. I am grateful to the Deutsche Gesellschaft für Technische Zusammenarbeit (GTZ) and to Booker Agriculture International for financial support for myself and for the coconut project at Wye College. I am glad to acknowledge the help of those who contributed to the work described. Dr. R.L. Branton, Dr. H.F. Taylor, Mrs. N. Wilson, Mrs. M. Wrigley, Mrs. J. Spicer, Mrs. E.C. Campling, MR. S. Watanayothin and Mr. C. Petchpirun. The Coconut Industry Board of Jamaica kindly supplied coconut material.

References

Ackerson RC (1984) Regulation of soybean embryogenesis by abscisic acid. J Exp Bot 35:403 – 413

Armstrong SW, Francis D (1987) Cell and nuclear area RNA and protein content and chromosome counts in meristematic cells of secondary roots and callus of *Cocos nucifera*. Can J Bot 65:547 – 552

Assy Bah B (1986) In vitro culture of coconut zygotic embryos. Oleagineux 41:321 – 328

Bajaj YPS (1983) Cryopreservation and international exchange of germplasm. In: Sen SK, Giles KL (eds) Plant cell culture in crop improvement. Plenum Press, NY, pp 19 – 41

Bajaj YPS (1984) Induction of growth in frozen embryos of coconut and ovules of citrus. Curr Sci 53:1215–1216

Banzon JA (1984) Harvestable energy from the coconut palm. Energy Agric 3:337–344

Bavappa KVA (1982) Perennial oilseed plants, coconut and oil palms. Indian Farming November:100–103

Berger KG, Ong SH (1985) The industrial uses of palm and coconut oils. Oleagineux 40:613–624

Bhalla-Sarin N, Bagga SS, Sopory K, Guha-Mukherjee S (1986) Induction and differentiation of callus from embryos of *Cocos nucifera* L by IAA conjugates. Plant Cell Rep 5:322–324

Blake J (1983) Tissue culture propagation of coconut, date and oil palm. In: Dodds JH (ed) Tissue culture of trees. Croom Helm, London, pp 29–50

Blake J, Eeuwens CJ (1980) Inflorescence tissue as source material for vegetative propagation of the coconut palm. Proc Int Conf Cocoa and Coconuts, Kuala Lumpur 1978, pp 549–556

Brackpool AL, Branton RL, Blake J (1986) Regeneration in palms. In: Vasil IK (ed) Cell culture and somatic cell genetics of plants, Vol 3. Academic Press, New York, pp 207–222

Branton RL, Blake J (1983a) A lovely clone of coconuts. New Sci 98:554–557

Branton RL, Blake J (1983b) Development of organized structures in callus derived from explants of *Cocus nucifera* L. Ann Bot 52:673–678

Branton RL, Blake J (1986) Clonal propagation of coconut. In: Pushparajah E, Chew Poh Soon (eds) Cocoa and coconuts: progress and outlook. Kuala Lumpur: Inc Soc Plant, pp 771–780

Daly C (1985) Commercial aspects of the laurics market. Proc Symp Palm kernel and coconut oils. Leatherhead June 1985, Food Res Assoc FOSFA Int, pp 85–91

Das PK (1985) Trends in coconut oil production and trade in the world. Oleagineux 40:85–93

D'Souza L (1982) Organogenesis in coconut embryo callus. In: Fujiwara A (ed) Plant tissue culture 1982. Marzuen, Tokyo, pp 179–180

Eeuwens CJ (1976) Mineral requirements for growth and callus initiation of tissue explants excised from mature coconut palms (*Cocos nucifera*) and cultured in vitro. Physiol Plant 36:23–28

FAO (1986) Production Yearbook Vol 39:1985. FAO, Rome

Fisher JB, Tsai JH (1978) In vitro growth of embryos and callus of coconut palm. In Vitro 14:307–311

Gupta PK, Kendurka SV, Kulkarni VD, Shirgurkar MV, Mascarenhas AF (1984) Somatic embryogenesis and plants from zygotic embryos of coconut (*Cocos nucifera* L.) in vitro. Plant Cell Rep 3:222–225

de Guzman EV, del Rosario AG, Ubalde EM (1978) Proliferative growths and organogenesis in coconut embryo and tissue cultures. Philipp J Coconut Stud 3:1–10

Ignacio LF Jr (1985) Present and future position of coconut oil in world supply and trade. J Am Oil Chem Soc 62:197–203

Iyer RD (1982) Embryo and tissue culture for crop improvement, especially of perennials, germplasm conservation and exchange-relevance to developing countries. In: Rao AN (ed) Proc COSTED Symp on Tissue culture of economically important plants, Singapore 1981, pp 219–230

Jagadeesan M, Padmanabhan D (1982) Induction of rooting in cotyledon callus of coconut. Curr Sci 51:567

Karunaratne S, Kurkulaarachoh C, Gamage C (1985) A report on the culture of embryos of dwarf coconut. *Cocos nucifera* L. var. *nana in vitro*. Cocos 3:1–8

Kumar P Prakash, Raju CR, Chadra Mohan M, Iyer RD (1985) Induction and maintenance of friable callus from the cellular endosperm of *Cocos nucifera* L. Plant Sci 40:203–207

Maarsen JW (1985) Edible and uses of coconut and palm kernel oils. In: Proc Symp Palm kernel and coconut oils. Leatherhead, June 1985, Food Res Assoc FOSFA Int, pp 53–60

Malijan LC, del Rosario AG (1986) Photosynthetic capacity in embryo-cultured coconut seedlings during acclimation to greenhouse conditions. In: Int Congr plant tissue and cell culture. Abstr Univ Minnesota, Minneapolis, p 280

Mendoza EMT, Samonte JL, Mujer CV (1985) Galactomannan metabolism in developing normal and makapuno coconut endosperms. Trans Nat Acad Sci Tech (Phils) 7:237–247

Meunier J, Le Saint JP, Gascon JP, de Nuce de Lamothe M (1986) Recent advances in genetic improvement of coconut yield. In: Pushparajah E, Chew Poh Soon (eds) Cocoa and coconuts: progress and outlook. Kuala Lumpur: Inc Soc Plant, pp 719–732

Monfort S (1985) Androgenesis of coconut, embryos from anther culture. Pflanzenzuecht 94:251–254

Mujer CV, Ramirez DA, Mendoza EMT (1984a) α-D-galactosidase deficiency in coconut endosperm: its possible pleiotropic effects in makapuno. Phytochemistry 23:893–894

Mujer CV, Ramirez DA, Mendoza EMT (1984b) Coconut α-D-galactosidase isoenzymes: isolation, purification and characterization. Phytochemistry 23:1251–1254
Nyman LP, Gonzales CJ, Arditti J (1983) Reversible structural changes associated with callus formation and plantlet development from aseptically cultured shoots of Taro (*Colocasia esculenta* var *antiquorum*). Ann Bot 51:279–286
Ohler JG (1984) Coconut, tree of life. FAO Plant production and protection Paper 57: FAO Rome
Pannetier C, Buffard-Morel J (1982) First results of somatic embryo production from leaf tissue of coconut, *Cocos nucifera* L. Oleagineux 37:349–354
Pannetier C, Buffard-Morel J (1986) Coconut Palm (*Cocos nucifera* L.) In: Bajaj YPS (ed) Biotechnology in agriculture and forestry Vol 1. Trees. Springer, Berlin Heidelberg New York Tokyo, pp 430–450
Raju CR, Kumar PP, Chandramohan M, Iyer RD (1984) Coconut plantlets from leaf tissue cultures. J Plant Crops 12:75–78
del Rosario AG, de Guzman EV (1982) The status of plant tissue culture in the Philippines. In: Rao AN (ed) Proc COSTED Symp on Tiss Cult of Economically Important Plants. Singapore 1981, pp 293–294
Santos GA (1986) Genetic improvement of coconut genotypes: a review of past and present methods. Philipp J Coco Stud XI:24–30
Schwitzer MK (1985) Technical end uses of palm kernel and coconut oils. Proc Symp Palm kernel and coconut oils. Leatherhead June 1985, Food Res Assoc FOSFA Int, pp 42–51
Smith BG (1986) Tissue culture of *Cocos nucifera* – biochemical changes preceding embryogenesis. In: Pushparajah E, Chew Poh Soon (eds) Cocoa and coconuts: progress and outlook. Kuala Lumpur: Inc Soc Plant, pp 781–786
Tahardi S (1985) Organogenesis in hybrid coconut embryo and tissue cultures. Poster Abstract Symp In vitro problems related to mass propagation of horticultural plants. Gembloux, Belgium 16–20 Sept, 1985
Thanh-Tuyen NJ, Dionzon MN (1986) Callus initiation and development of somatic embryos from cultured embryos of coconut (*Cocos nucifera* L.) In: Int Congr plant tiss and cell culture. Abstr, Univ Minnesota, p 193
Tisserat B (1984) Clonal propagation: Palms. Vasil IK (ed) Cell culture and somatic genetics of plants. Academic Press 1:74–81
Vasil V, Vasil IK (1981) Somatic embryogenesis and plant regeneration from suspension cultures of pearl millet (*Pennisetum americanum*.) Ann Bot 47:669–678

IV.11 Coconut (*Cocos nucifera* L.): Anther Culture

NGUYEN T. THANH-TUYEN[1]

1 Introduction

Coconuts are grown extensively in the tropics, and millions of people depend on the crop for their livelihood. However, the productivity of coconut trees is generally low, despite the fact that many efforts have been made to improve their yield potential. The long life span and heterozygosity seem to be the limiting factors in coconut improvement through breeding. Thus the production of inbred lines for this purpose is difficult and time-consuming. Unconventional methods must therefore be evolved as a means of shortening the inbreeding cycle (Thanh-Tuyen and de Guzman 1983a).

The success in haploid plant production by anther and pollen culture has aroused the desire to obtain homozygous lines within one or two generations in perennial crops through the use of haploids. A naturally occuring haploid embryo in coconut palm has been reported with authentic evidence of its haploid chromosome number (Ninan and Raveendranath 1965). It was suggested, therefore, that haploid formation in coconut could be induced by in vitro culture techniques. Recently, anther culture of coconut has been conducted in an attempt to induce haploid formation in large quantities. Reports by Thanh-Tuyen and de Guzman (1983a) and Monfort (1985) on the formation of pollen embryos at advanced stages in culture anthers of *Cocos nucifera* L. are a strong indication of the eventual possibility of the regeneration of coconut haploid plants via in vitro culture.

2 Ontogeny of Coconut Pollen

As early as 1926, Santos reported a cytological study on the reduction division of coconut microspore mother cells (MMC) and subsequently, Juliano and Quisumbing (1931) briefly described the various developmental stages in the formation of coconut pollen grains, starting from MMC to mature pollen, observed from microtome sections. Almost 30 years later with the evolution of the squash techniques, a detailed study of meiosis in pollen mother cells and of chromosome morphology in some coconut cultivars was undertaken by Nambiar and Swami-

[1] Tissue Culture Laboratory, Department of Horticulture, Visayas State College of Agriculture, Leyte, Philippines 6521-A

Biotechnology in Agriculture and Forestry, Vol. 10
Legumes and Oilseed Crops I (ed. by Y.P.S. Bajaj)

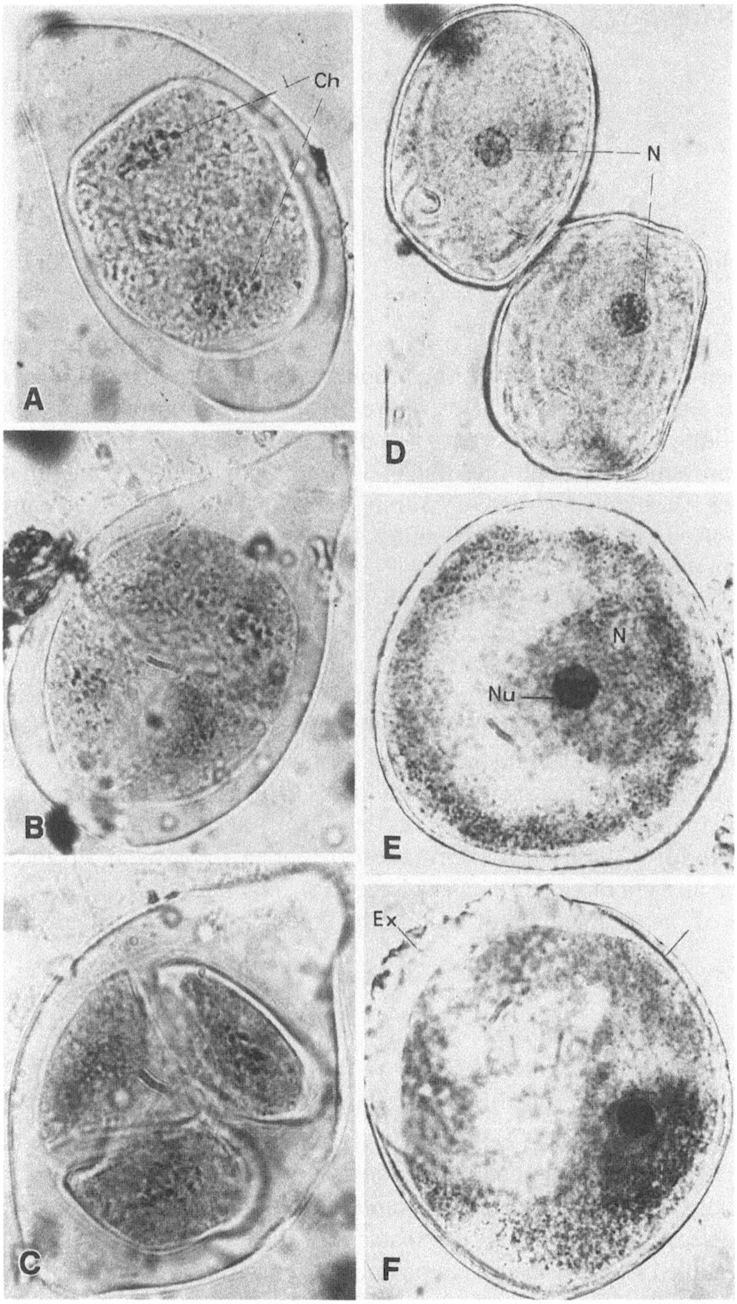

Fig. 1A–F. Various stages in the microsporogenesis of coconut. **A** Microspore mother cell (MMC) at meiotic anaphase I. **B** Tetrad. **C** Triad. **D** Early uninucleate microspore with oblong shape, thin wall, and small nucleus at the center. **E** Mid-uninucleate microspore with spherical shape, thin wall, and large nucleus. **F** Late uninucleate microspore with thick exine and darkly stained nucleus located nearest the spore wall. *Ch* chromosome; *N* nucleus; *Nu* nucleolus; *Ex* exine; *In* intine

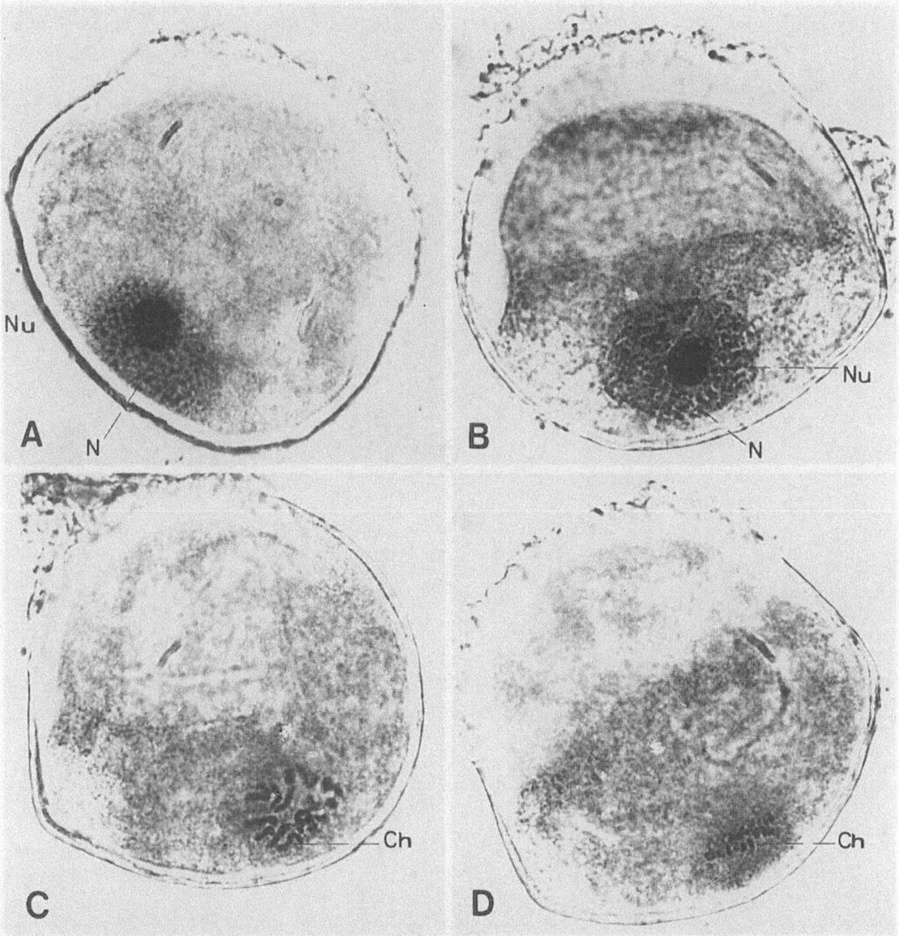

Fig. 2A–D. Microspores at first mitosis. **A** and **B** Prophase showing the condensed chromatin network. **C** Metaphase showing 16 chromosomes arranged on an equatorial plane. **D** Metaphase showing chromosomes arranged on the side, view of an equatorial plane. (Abbreviations as in Fig. 1)

nathan (1960a, b). The microsporogenesis of coconut starting from the formation of pretetrads and ending with the shedding of pollen grains, has been reinvestigated by Thanh-Tuyen and de Guzman (1983b) in an attempt to correlate the stage of pollen development with the age of the bearing spadix in order to facilitate the determination of the optimum responsive stage of pollen for coconut anther culture.

The process of pollen development in coconut starts with the simultaneous division of microspore mother cells which appear as thin-walled, and lenticular to spherical in shape (Fig. 1A), to give rise to microspore tetrads (Fig. 1B), and occasionally to triads (Fig. 1C). After being released from MMC the microspores undergo a series of morphological and cytological changes which involve an increase in size and volume, thickening of the exine, expansion of the vacuoles, and en-

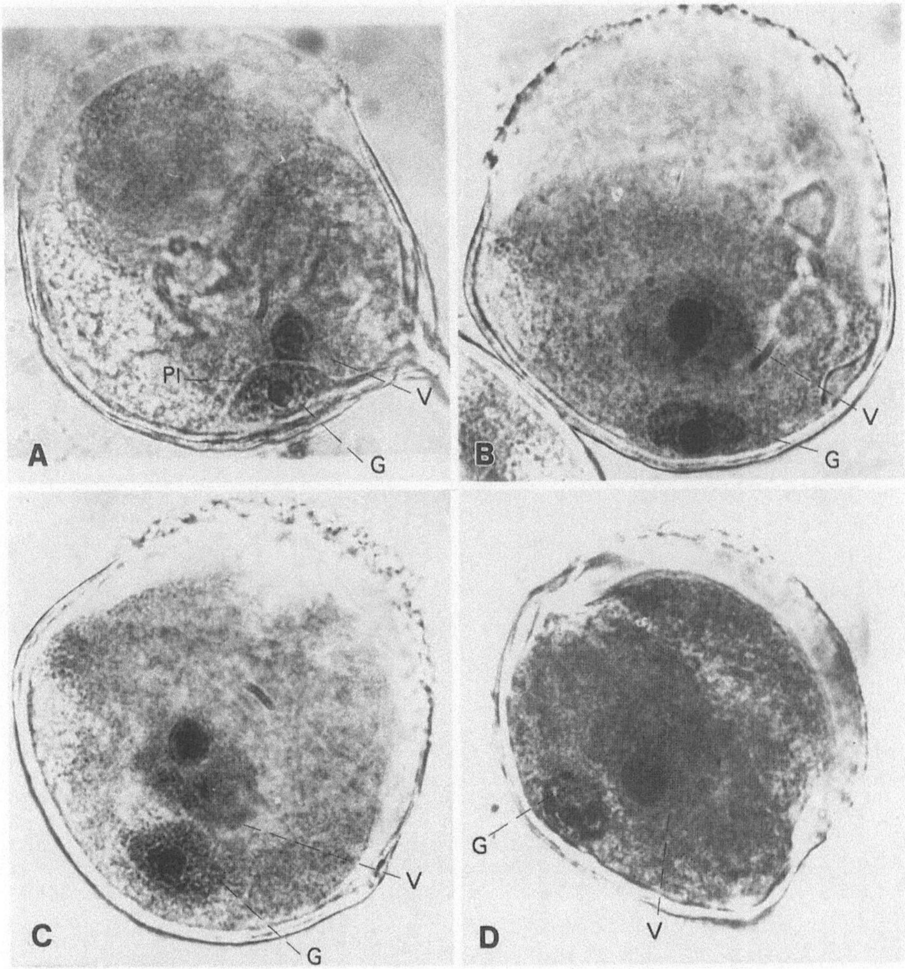

Fig. 3A–D. Binucleate and mature pollen grains. **A** and **B** Early binucleate grain with a small lenticular generative nucleus attached to the intine and separated from a large vegetative one by a curved plate. **C** Late binucleate grain with generative nucleus detached from the intine. **D** Mature pollen grain with deposits and obscure nuclei. *Pl* curved plate; *G* generative nucleus; *V* vegetative nucleus

largement of the nucleus with a distinct nucleolus (Fig. 1 D, E, F). At the position nearest the spore wall, the nucleus appears compact, stains darker (Fig. 2 A, B), and divides mitotically (Fig. 2 C, D) to give rise to two unequal nuclei (Fig. 3).

It is interesting to note that the mitotic division of the pollen nucleus takes place within a short period of 1 to 2 days, compared with other stages in coconut pollen development. By contrast, the maturation of coconut pollen grains is a long-lasting process, which starts with the formation of two gametophytes and ends with the shedding of bicellular pollen. The latter was also observed by Juliano and Quisumbing (1931) and Brewbaker (1967).

Adopting the staging nomenclature by Sunderland (1974) with some modifications, the correlation between the stage of pollen development and the age of

Table 1. The correlation between the stage of pollen development in coconut and the age of the bearing spadix which is reckoned as number of weeks before splitting (WBS)

Description of developmental stages	Stage[a]	Age (WBS)
a) MMC at meiosis and tetrad	0	9–8
b) Early uninucleate microspores with thin wall, oblong shape and small nucleus at the center	1	7
c) Mid-uninucleate microspores with thin walls, spherical shape and large nucleus moving to the spore side	2	6
d) Late uninucleate microspores with thick exine, vacuolated, and nucleus located nearest to the spore wall	3	5–4
e) Mitotic microspores and early binucleate grains with the generative nucleus still attached to the intine	4	4
f) Binucleate grains with the generative nucleus detached from the intine	5	3
g) Mature pollen grains	6	2–1 and newly opend

[a] Staging nomenclature adopted from Sunderland (1974) with some modifications.

the bearing spadix, which is reckoned as weeks before splitting (WBS), is presented in Table 1. This correlation can be used as a marker for the collection of coconut inflorescences at a suitable stage for anther culture.

3 Androgenesis in Coconut

3.1 The Current Status

Anther culture has been successful in a large number of species belonging to dicots and monocots (Maheswari et al. 1983; Bajaj 1983); however, only a few species of angiosperm trees have been attempted successfully with the induction of androgenic embryos/plants. Apparently, woody species are exceptionally difficult material for in vitro manipulation. The coconut is an example of such recalcitrant material.

An attempt to induce embryo or callus formation from pollen in cultured anthers of coconut by Iyer and Raina (1972) was reported to be unsuccessful; however, almost a decade later, Iyer (1982) was able to obtain multicelled proembryoids within the exine of coconut microspores in cultured anthers, but no further developments have been reported. Radojevic, as cited in Kovoor (1981) obtained callus from coconut anther in culture. Unfortunately, this was the only one of the thousand cultured anthers which produced callus. Another "symptom" of androgenesis was also observed (Haibou, cited in Kovoor 1981) with the extrusion of globular pollen from cultured coconut anthers. In the Philippines, Thanh-Tuyen and de Guzman (1983a) attempted to induce haploid formation in coconut via anther culture. Various stages in the development of pollen embryos were observed in cultured anthers. More interestingly, the androgenic origin of the embryos was demonstrated. Emergence of an embryo from cultured anther has been described briefly in a publication by Monfort (1985). Despite the low frequency (less than 1%) of pollen embryo formation, the reported evidence indicates strongly the possibility of inducing androgenic haploids in coconut through direct embryogenesis.

3.2 Experimental Androgenesis

3.2.1 The Responsive Pollen Stage for Coconut Anther Culture

The stages of tetrad, uninucleate, and approaching the first pollen mitosis have been observed as optimum for coconut anther culture by Haibou (cited in Kovoor 1981), Iyer (1982), and Monfort (1985), respectively. On the other hand, Thanh-Tuyen and de Guzman (1983a) reported that the stage just before, during or immediately after the first pollen mitosis, is most responsive for the formation of pollen embryos.

The literature has indeed shown that in most species, the suitable stage for culture is when the pollen is at the verge of mitosis. The peak of androgenic response seems to coincide with the intense metabolic activities in the microspores at this stage (Vasil and Nitsch 1975). In-depth studies on the control point of differentiation of a pollen grain into an embryogenic cell have revealed pollen dimorphism, that is, according to Heberle-Bors and Reinert (1980), the formation of pollen in situ which are competent for androgenesis besides the normal gametophytic pollen in nearly mature anthers. The former are generally smaller and lightly staining while the latter are larger and densely staining. The two types of pollen can be separated by density gradient centrifugation (Wernicke et al. 1978). It has been confirmed from isolated pollen culture experiments in tobacco that it is the former type which forms embryos (Rashid and Reinert 1980, 1983).

Further experiments on tobacco have shown a correlation of pollen dimorphism with pollen sterility and an in vivo shift in sex balance towards femaleness (Heberle-Bors 1982a). The phenomena can be regulated by temperature changes at early stage in flower development, that is, before meiosis, and thereafter the pollen become irreversibly determined (Heberle-Bors 1982b).

Pollen dimorphism has been shown to occur in rye (Wenzel et al. 1975; Sunderland 1978), barley (Dale 1975), tobacco (Horner and Street 1978), *Anemone coronaria*, *Tradescantia bracteata* (Sunderland 1977), oats, *Nicotiana knightiana* (Sunderland 1978), and wheat (Tan and Halloran 1982).

In a study at the biochemical level in various classes of pollen grains in henbane, Raghavan (1981) reported evidence for the transcription of new informational type RNA during the induction of pollen embryogenesis which is lacking in other grains ready to germinate.

3.2.2 Nutritional Factors

The nutrient requirements for inducing a switch from the gametophytic to androgenetic development in coconut are relatively simple. Modified Blaydes' (1966) and B5 (Gamborg et al. 1968) media supplemented with 6%–9% sucrose, 15% coconut water, 2 mg/l naphthaleneacetic acid and 0.5% activated charcoal were found suitable for coconut pollen embryo formation (Thanh-Tuyen and de Guzman 1983a). A close look into the mineral composition of these media indicates that a relatively high level of NO_3^- and low concentration of NH_4^+ in their formulations favor the formation of pollen embryos. The beneficial effect of high nitrate and low ammonium concentrations for haploid production of cereals has been reported by Chu et al. (1975). B5 medium also includes high content of chlorine, which is known to promote the growth of cultured embryos and in vitro seedlings of coconut (Miniano and de Guzman 1978). In addition, the high level of potassium ions in Blaydes' medium seems, to some extent, to favor the induction of pollen embryo formation. The Y3 medium formulated by Eeuwens (1976) using a very high concentration of iodine was also tested. Anthers cultured on Y3 medium responded very well with regard to the percentage of embryonic pollen grains formed, but the latter did not develop further into embryos. Surprisingly, there was no significant differences in terms of anther response to the variations in the availability/concentration of various organic substances, e.g., coconut water, potato extract, auxins, etc. added to any of one of the mineral formulations mentioned above. Further studies on the role of growth hormones and other organic constituents in coconut haploid plant regeneration are being undertaken.

3.2.3 *Anther Pre-Treatments*

Cold treatment of excised anthers or flower buds is believed to improve anther response in culture (Nitsch and Norreel 1972; Bajaj et al. 1977) and has been successfully employed in a good number of species (Maheshwari et al. 1980). In coconut, cold treatment given to flower buds before the excision of anthers was more effective with regard to the retention or preservation of pollen viability in culture, than when given to cultured anthers. In anthers exposed to 7 °C, the peak response was observed after 7 days of cold treatment, while at 14 °C, anthers cold-treated for 4 days gave the maximum response (Thanh-Tuyen 1981). A similar trend was also observed with tobacco anthers by Sunderland and Roberts (1979). The latter thus concluded that the enhancing effect of cold treatment is restricted not only at temperature as low as 3–5 °C but still higher temperature, e.g., 10–15 °C, could also be effective as long as the duration of pre-treatment is reduced appropriately.

The stimulatory effect of cold pre-treatment to androgenesis has been attributed to a number of factors, such as dissolution of microtubules (Jensen 1974), alteration in the first pollen mitosis (Nitsch 1974), or preservation of higher ratio of viable pollen capable of pollen embryogenesis (Sunderland 1978). Cold treatment of flower buds may also act to help the differentiation of the embryogenic class of pollen grains of tobacco by repressing their gametophytic differentiation (Rashid 1983) or by lowering the abscisic acid content of anthers (Johansson et al. 1982), which is considered to be inhibitory for androgenesis. Bernard (1980) and Huang and Sunderland (1982) reported an increase in the relative yield of green versus albino plantlets after cold treatment.

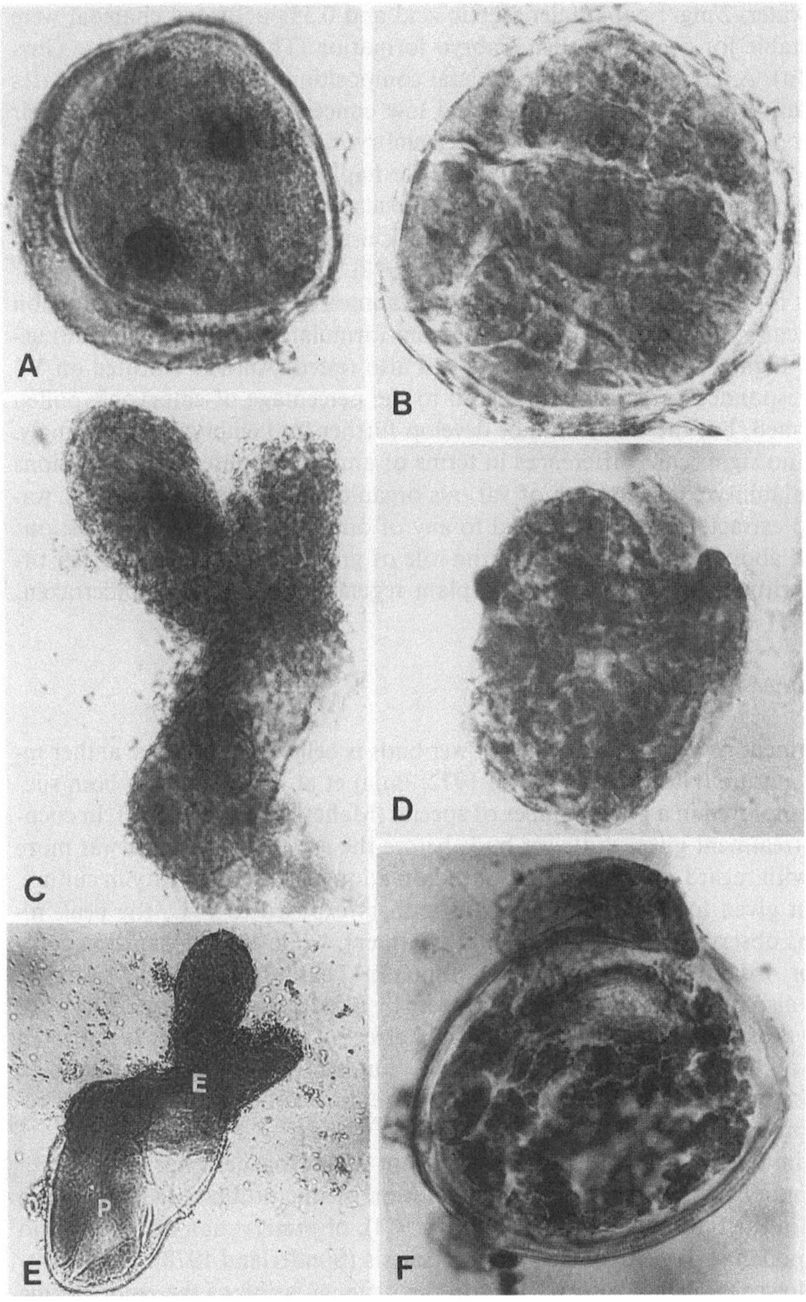

Fig. 4A–F. Pollen embryogenesis in cultured anthers of coconut. **A** A two-celled microspore derived from a symmetrical mitotic division. **B** A pro-embryo with more than ten cells still enclosed in the exine. **C** A heart-shaped embryo. **D** A torpedo embryo showing distinct shoot and radicular poles. **E** A torpedo embryo rupturing the exine. *E* embryo; *P* pollen. **F** An unorganized multicellular pollen grain

3.3 Development of Pollen Embryos

As reported by Thanh-Tuyen and de Guzman (1983 a), coconut pollen embryogenesis starts with symmetrical pollen mitosis to give rise to two equal and identical cells (Fig. 4A). This is considered to be one of the more common pathways of androgenesis. Through this pathway, no vegetative and generative cell is differentiated; however, the staining characteristics of the two identical cells are similar to those of a typical vegetative cell. The continued division of the two identical cells led to the formation of multicellular proembryos. Figure 4B shows a proembryo with more than ten cells still enclosed within the exine. Still at the microscopic level, the exine of embryonic grains burst out to expel embryos at the heart-shaped (Fig. 4C) and torpedo (Fig. 4D) stages. The rupture of the exine to liberate the embryo, as shown in Fig. 4E, is strong evidence to support the androgenic origin of the embryos. Morphologically, coconut pollen embryos at the torpedo stage possess a distinct bipolarity with shoot and radicular poles. The development of one cotyledon is consistently suppressed in the heart- and torpedo-shaped embryos. This obviously is one of the characteristics of monocot embryos.

Direct embryogenesis appears to be the principal pathway of haploid formation in coconut. Although unorganized multicellular grains were occasionally observed, as shown in Fig. 4F, probably indicating the early stage of pollen callus formation, no further developments were noted. In terms of genetic stability, plants derived from pollen via direct embryogenesis are preferable to those regenerated from callus. The former is mostly limited to the Solanaceae, while the indirect mode of development is quite common in other plant taxa (Reinert and Bajaj 1977). In coconut, the successful development of pollen embryos via direct embryogenesis is a very encouraging sign in attempts to produce haploid plants by anther culture. It is believed that once a cell has been turned into an embryo, plant regeneration mostly follows (Jensen 1980), yet the regeneration of pollen-derived coconuts from cultured anthers still remains a challenge.

4 Problems and Prospects

One of the most serious problems encountered in coconut tissue culture is the variation in culture response from one tree, or batch of collelction, to another. Commonly, in coconut tissue and embryo culture, the desired growth proliferation is not always obtained in different batches of cultures which have been subjected to exactly the same culture conditions (de Guzman 1980). In coconut anther culture, Radojevic (cited in Kovoor 1981) reported callus formation in only one out of a few thousand cultured anthers. A similar phenomenon was also observed by Thanh-Tuyen (1981) in experiments subsequent to the initial success obtained in the induction of microspore embryo formation.

The formation of embryos in cultured anthers of coconut as reported by Thanh-Tuyen and de Guzman (1983 a) is considered to be a response of the microspores in vitro culture conditions. This statement is supported by the observation that among anthers extracted from one inflorescence, pollen embryogenesis was observed in those cultured on B5 medium, while those on Bourgin and Nitsch (1967) medium degenerated (Thanh-Tuyen 1981). Moreover, the formation of coconut pollen embryos was induced in anthers of two inflorescences of the same age, collected in the same batch, and inoculated onto the suitable media of Blaydes' and Gamborg et al. These facts rule out the possibility of the existence of spontaneous or in vivo pollen embryogenesis which naturally occurs in anthers, as observed in *Narcissus biflorus* (Koul and Karihaloo 1977).

Thanh-Tuyen and de Guzman (1983a) speculated regarding the factors governing the switch from gametophytic to sporophytic pathway of pollen development in cultured anthers of coconut. From the results, the authors attributed the triggering of androgenesis to internal factors within the anther-supplying palm or within the anthers themselves. The effects of genotype and physiological conditions of the donor plants on androgenesis are well known in other plant species (Maheshwari et al. 1980, 1982). Such effects are manifested in the differential behavior of species, cultivars, and even individuals. Attempts to induce haploid formation have thus been recently geared towards manipulations of growth conditions of the donor plants.

Increase in androgenic response has been reported after various treatments to the donor plants. Among these, light intensity (Foroughi-Wehr and Mix 1979), light duration (Heberle-Bors and Reinert 1979) and low temperature (Foroughi-Wehr and Mix 1979) are well-known regulatory factors. Further studies with *Nicotiana tabacum* var. Badischer Burley have confirmed the environmental control on pollen embryogenesis, and that the response is predetermined before excision and culture of the pollen or anthers (Heberle-Bors and Reinert 1981). Short-day and/or low temperature (15 – 18 °C) treatments of the donor plants result in differentiation of embryonic pollen in vivo at higher frequency (Heberle-Bors 1982a; Rashid 1983). Likewise, formation in situ embryonic pollen grains and subsequent in vitro pollen embryogenesis have been induced by treatments with feminizing agents (Heberle-Bors 1983).

For coconut, unlike herbaceous species, the bulk of the tree and its long lifespan render the exposure of the palm to the controlled environment in the greenhouse or the growth chamber impractical, if not impossible. To a great extent, this fact accounts for variation in anther response among individual trees, not only in coconut but perhaps also in other perennial woody species where successful manipulation in vitro cultures is still very limited.

Nevertheless, as more and more attention is given to in-depth studies on regulation of haploid induction, physiology, and biochemistry of haploid production, it should be possible to predict conditions favorable for androgenic haploid formation from even recalcitrant species, including coconut.

The significance and value of haploids for study of fundamental and applied genetics and for crop improvement have long been recognized (Kimber and Riley 1963). In general, haploids can be useful in three ways: (1) shortening the time requirement for the production of homozygous lines, (2) speeding up the fixing of desirable genotypes, and (3) facilitating the induction and selection of mutants.

Despite the fact that several problems remain to be tackled before the techniques of anther and pollen cultures become a common practice, coconut haploid plants would be desirable in hastening production of homozygous lines of selected characters. Successful crossing of anther-derived pure lines will guarantee hybrid vigor.

Hybridization of heterozygous parents, which is quite common in coconut, usually results in a variety of gene combination. New combinations of characters which otherwise can not be detected until the F_2 and in homozygous form, until the F_3 or even later generations, may already be selected among plants which develop from haploid pollen grains of an F_1 hybrid. When these are diploidized, the new combination is available as a fertile pure line. Once desirable genotypes are fixed, they can be propagated rapidly by asexual methods. Success in clonal propagation of coconut is not far-reaching. Progress has been reported recently with regard to development of organized structures/embryos in callus derived from coconut explants (Branton and Blake 1983; Pannetier and Buffard-Morel 1986) and regeneration of coconut plantlets (Raju et al. 1984). Reproducible initiation of callus formation and somatic embryogenesis on cultured zygotic embryos of coconut has been reported (Thanh-Tuyen et al. 1988). Development of globular, heart-shaped, torpedo and cotyledonary somatic embryos, which remarkably resemble the zygotic ones, was followed by regeneration of shoots, polarized structures with shoot and root ends and a small plantlet.

Another advantage of haploids is the expression of genes within a single genome which eliminates the complexity of diplont. Indeed, pollen embryogenesis is a better system to induce, detect, and analyze mutants, as large numbers of haploid cells are available which have the potential of developing into plantlets. Notable success has been reported in obtaining several nitrate reductase deficient and auxotrophic mutants utilizing protoplasts and cultured cells from pollen-derived haploids of *Datura innoxia*, *Hyoscyamus muticus*, *Nicotiana plumbaginifolia*, and *N. tabacum* (King 1983). Selection has also been made for several useful resistant variants/mutants employing haploid cells, including those for better nutritional quality in rice (Schaeffer and Sharpe 1981). Such an advantage could be exploited in induction and selection of desirable mutants of coconut which are resistant to important pests and diseases, with high oil and protein content of the endosperm, or with fatty acid composition particularly high in a certain fatty acid, e.g., lauric acid.

References

Bajaj YPS (1983) In vitro induction of haploids. In: Evans DA, Sharp WR, Ammirato PV, Yamada Y (eds) Handbook of plant cell culture vol 1. Macmillan, New York, pp 228–287

Bajaj YPS, Reinert J, Heberle E (1977) Factors enhancing in vitro production of haploid plants in anthers and isolated microspores. In: Gautheret RJ (ed) A la mémoire de Georges Morel. Masson, Paris, pp 47–58

Bernard S (1980) In vitro androgenesis in hexaploid triticale: determination of physiological conditions increasing embryoid and green plant production. Z Pflanzenzuecht 85:308–321

Blaydes DF (1966) Interaction of kinetin and various inhibitors in the growth of soybean tissue. Physiol Plant 19:747–753

Bourgin JP, Nitsch JP (1967) Obtention de *Nicotiana* haploids à partir d'étamine cultivée in vitro. Ann Physiol Veg 9:377–382

Branton RL, Blake J (1983) Development of organized structures in callus derived from explants of *Cocos nucifera* L. Ann Bot 52:673–678

Brewbaker JL (1967) The distribution and phylogenetic significance of binucleate and trinucleate pollen grains in the angiosperms. Am J Bot 54:1069–1083

Chu CC, Wang CC, Sun CS, Chen H, Yin KC, Chu CY, Bi FY (1975) Establishment of an efficient medium for anther culture of rice through comparative experiments on the nitrogen sources. Sci Sin 18:569–666

Dale PJ (1975) Pollen dimorphism and anther culture in barley. Planta 127:213–220

Eeuwens CJ (1976) Mineral requirements for growth and callus initiation of tissue explants excised from mature coconut palms (*Cocos nucifera* L.) and cultured in vitro. Physiol Plant 42:173–178

Foroughi-Wehr B, Mix G (1979) In vitro response of *Hordeum vulgare* L. anthers cultured from plants grown under different environments. Environ Exp Bot 19:303–309

Gamborg OL, Miller RA, Ojima K (1968) Nutrient requirements of suspension cultures of soybean root cells. Exp Cell Res 50:151–158

de Guzman EV (1980) In vitro culture studies on coconut: progress and problems. Professorial Lecture. Univ Philippines, Los Baños

Heberle-Bors E (1982a) In vitro pollen embryogenesis in *Nicotiana tabacum* L. and its relation to pollen sterility, sex balance, and floral induction of the pollen donor plants. Planta 156:396–401

Heberle-Bors E (1982b) On the time of embryogenic pollen grain induction during sexual development on *Nicotiana tabacum* L. plants. Planta 156:402–406

Heberle-Bors E (1983) Induction of embryogenic pollen grains *in situ* and subsequent in vitro pollen embryogenesis in *Nicotiana tabacum* by treatments of the pollen donor plants with feminizing agents. Physiol Plant 59:67–72

Heberle-Bors E, Reinert J (1979) Androgenesis in isolated pollen cultures of *Nicotiana tabacum*: Dependence upon pollen development. Protoplasma 99:237–245

Heberle-Bors E, Reinert J (1980) Isolated pollen cultures and pollen dimorphism. Naturwissenschaften 67:311

Heberle-Bors E, Reinert J (1981) Environmental control and evidence for predetermination of pollen embryogenesis in *Nicotiana tabacum* pollen. Protoplasma 109:249–255

Horner M, Street HE (1978) Pollen dimorphism. Origin and significance in pollen plant formation by anther culture. Ann Bot 42:763–777

Huang B, Sunderland N (1982) Temperature-stress pretreatment in barley anther culture. Ann Bot 49:77–88

Iyer RD (1982) Embryo and tissue culture for crop improvement, especially of perennials, germplasm conservation and exchange - Relevance to developing countries. In: Rao AN (ed) Proc COSTED Symp on Tissue Culture of Economically Important Plants. National Univ, Singapore, pp 219–230

Iyer RD, Raina SK (1972) The early ontogeny of embryoids and callus from pollen and subsequent organogenesis in anther cultures of *Datura metel* and rice. Planta 104:146–156

Jensen CJ (1974) Chromosome doubling techniques in haploids. In: Kasha KJ (ed) Haploids in higher plants.: advances and potential. Univ Press, Guelph, pp 153–190

Jensen CJ (1980) Uses of cell and tissue culture techniques in plant breeding and genetics. In: Rachie KO, Lyman JM (ed) Genetic engineering for crop improvement. Rockefeller, New York, pp 87–104

Johansson L, Anderson B, Eriksson T (1982) Improvement of anther culture technique: Activated charcoal bound in agar medium in combination with liquid medium and elevated CO_2 concentration. Physiol Plant 54:24–30

Juliano JB, Quisumbing E (1931) Morphology of the male flower of *Cocos nucifera* Linnaeus. Philipp J Sci 45:449–458

Kimber G, Riley R (1963) Haploid angiosperms. Bot Rev 29:480–531

King PJ (1983) Some reflections on the isolation of plant auxotrophs *in vitro*. IAPTC Newslett 39:2–7

Koul AK, Karihaloo JK (1977) In vivo embryoids from anthers of *Narcissus biflorus* Curt. Euphytica 26:97–102

Kovoor A (1981) Palm tissue culture: State of the art and its application to the coconut. FAO Plant Production and Protection Paper 30. FAO, Rome

Maheshwari SC, Tyagi AK, Malhotra K, Sopory SK (1980) Induction of haploids from pollen grains in angiosperm – the current status. Theor Appl Genet 58:193–206

Maheshwari SC, Rashid A, Tyagi AK (1982) Haploids from pollen grains – retrospect and prospect. Am J Bot 69:865–879

Maheshwari SC, Rashid A, Tyagi AK (1983) Anther/pollen culture for production of haploids and their utility. IAPTC Newslett 41:2–9
Miniano AP, de Guzman EV (1978) Responses of non-makapuno embryos *in vitro* to chloride supplementation. Philipp J Coconut Stud 3:27–44
Monfort S (1985) Androgenesis of coconut: embryos from anther culture. Z Pflanzenzuecht 94:251–254
Nambiar MC, Swaminathan MS (1960a) Meiosis in pollen sterile and pollen fertile varieties of coconut. Curr Sci 29:234–236
Nambiar MC, Swaminathan MS (1960b) Chromosome morphology, microsporogenesis and pollen fertility in some varieties of coconut. Indian J Genet Plant Breed 20:200–211
Ninan CA, Raveendranath TG (1965) A naturally occurring haploid embryo in the coconut palm (*Cocos nucifera* L.). Caryologia 18:619–623
Nitsch C (1974) Pollen culture – A new technique for mass production of haploid and homozygous plants. In: Kasha KJ (ed) Haploids in higher plants. Advances and potential. Univ Press, Guelph, pp 123–135
Nitsch C, Norreel C (1972) Factors favoring the formation of androgenetic embryos in anther culture. In: Srb AR (ed) Genes, enzymes and populations. Plenum, New York, pp 129–144
Pannetier C, Buffard-Morel J (1986) Coconut palm (*Cocos nucifera* L.) In: Bajaj YPS (ed) Biotechnology in agriculture and forestry 1. Trees I. Springer, Berlin Heidelberg New York Tokyo, pp 430–450
Raghavan V (1981) Distribution of poly (A)-containing RNA during normal pollen development and during induced pollen embryogenesis in *Hyoscyamus niger.* J Cell Biol 89:593–606
Raju CR, Prakash Kumar P, Chandramohan M, Iyer RD (1984) Coconut plantlets from leaf tissue cultures. J Plantation Crops 12:75–78
Rashid A (1983) Pollen dimorphism in relation to pollen plant formation. Physiol Plant 58:544–548
Rashid A, Reinert J (1980) Selection of embryonic pollen from cold-treated buds of *Nicotiana tabacum* var. Badischer Burley and their development into embryos in cultures. Protoplasma 105:161–167
Rashid A, Reinert J (1983) Factors affecting high frequency embryo formation in ab initio pollen cultures of *Nicotiana.* Protoplasma 116:155–160
Reinert J, Bajaj YPS (1977) Anther culture: Haploid production and its significance. In: Reinert J, Bajaj YPS (eds) Applied and fudamental aspects of plant cell, tissue, and organ culture. Springer, Berlin Heidelberg New York Tokyo, pp 251–267
Santos JK (1926) A cytological study of *Cocos nucifera* Linnaeus. Philipp J Sci 37:417–437
Schaeffer GW, Sharpe FT (1981) Lysine in seed protein from S-amino-ethyl-L-cysteine resistant anther-derived tissue cultures of rice. In Vitro 17: 345–352
Sunderland J (1974) Anther culture as a means of haploid induction. In: Kasha KJ (ed) Haploids in higher plants: advances and potential. Univ Press, Guelph, pp 91–122
Sunderland N (1977) Comparative studies of anther and pollen culture. In: Sharp WR, Larsen PO, Paddock EF, Raghavan V (eds) Plant cell and tissue culture – Principles and Applications. Ohio State Univ Press, Columbus, pp 203–219
Sunderland N (1978) Strategies in the improvement of yields in anther culture. Proc Symp Plant tissue culture, Science Press, Peking, pp 65–68
Sunderland N, Dunwell JM (1977) Anther and pollen culture. In: Street HE (ed) Plant tissue and cell culture. Blackwell, Oxford, pp 223–265
Sunderland N, Roberts M (1979) Cold pretreatment of excised flower buds in float culture of tobacco anthers. Ann Bot 43:404–405
Tan HH, Halloran GM (1982) Pollen dimorphism and the frequency of inductive anthers in anther cultures of *Triticum monococcum.* Biochem Physiol Pflanz 177:197–202
Thank-Tuyen NT (1981) Anther culture of coconut. Ph D Dissertation, Univ Philippines, Los Baños, 177p
Thanh-Tuyen NT, Apurillo DI, Dionzon MN (1988) Somatic embryogenesis in coconut. Proc First Nat Symp Plant Tissue Culture in Philippine Agriculture and Forestry, May 25–28, 1988, Univ Philippines, Los Baños (in press)

Thanh-Tuyen NT, de Guzman EV (1983a) Formation of pollen embryos in cultured anthers of coconut (*Cocos nucifera* L.). Plant Sci Lett 29:81–88

Thanh-Tuyen NT, de Guzman EV (1983b) Pollen developmental stages for coconut anther culture. Kalikasan, Philipp J Biol 13:135–144

Vasil IK, Nitsch C (1975) Experimental production of pollen haploids and their uses. Z Pflanzenphysiol 76:191–212

Wenzel G, Hoffmann F, Potrykus I, Thomas E (1975) The separation of viable rye microspores from mixed populations and their development in culture. Molec Gen Genet 138:293–297

Wernicke W, Harms CT, Lorz H, Thomas E (1978) Selective enrichment of embryogenic microspore populations. Naturwissenschaften 65:540–541

IV.12 Oil Palm (*Elaeis guineensis* Jacq.): Tissue Culture and Micropropagation

Wooi Kheng Choo[1]

1 Introduction

The oil palm belongs to the family Palmaceae and the genus *Elaeis*. There are two important species in the genus *Elaeis*, *E. guineensis*, which is African in origin and *E. oleifera*, which is South American in origin. This chapter reviews the in vitro studies on the *E. guineensis*. (2 n = 32).

The palm has only one terminal growing point. The crown of the palm may carry as many as 50 unopened leaves. In a mature palm, a new leaf primordium is produced every 2 weeks and becomes fully developed after 2 years (Corley and Gray 1976). A mature leaf is pinnate, i.e., leaflets each consisting of a lamina and midrib are attached to a central rachis. The rachis extends into the part of the leaf stalk called the petiole, which attaches the pinnate leaf to the trunk. Frond production rate normally averages 18 to 24 per annum.

The average trunk diameter is 40 cm and height increment is in the order of 35–75 cm per year (Corley and Gray 1976). The trunk, usually covered by old leaf bases, is erect and carries the vascular bundles. In 10- to 15-year-old palms, the old leaf bases may begin to abscise, leaving the palm with a smooth appearance. The oil palm has an adventitious root system. The primary roots, ca. 8 mm in diameter, arise at the base of the trunk and descend into the soil to anchor the palm into the ground. The primary roots branch to give rise to the secondaries, which in turn branch into the tertiaries. Lastly, quarternary roots branch from the tertiaries.

Each leaf axil carries an inflorescence, which may be male or female. Male and female inflorescences are produced in cycles. Pollination of female flowers will result in fruit set. Fruits are formed on spikelets held together to form a bunch. It takes approximately 5 months for a bunch to ripen from the time of pollination. The fruits are orange when ripe. The mean bunch weight ranges from 10 to 30 kg, with an average of 1500 fruits. The fruit is a drupe, and consists of the mesocarp from which palm oil is extracted, and the endosperm from which palm kernel is extracted (Fig. 1).

[1] Sime Darby Plantations, PO Box 147, 41910 Klang, Selangor, Malaysia

Biotechnology in Agriculture and Forestry, Vol. 10
Legumes and Oilseed Crops I (ed. by Y.P.S. Bajaj)

Fig. 1. A Mature oil palm in field, 12 years old. **B** Oil palm bunches in leaf axils. **C** Oil palm fruit and longitudinal section through fruit showing mesocarp (*M*) from which palm oil is extracted, endocarp or shell (*En*), and endosperm (*E*) from which palm kernel oil is extracted. Bar = 1 cm

1.1 Distribution and Area under Cultivation

The oil palm originated in West Africa in wet areas along river banks and in swamps. Today, it is extensively cultivated in large plantations as well as smallholdings in South East Asia, West Africa, and Central and South America (Moll 1987). Oil palm was first grown as a smallholders' crop back in its homeland. Its development as a plantation crop began in Sumatra, where it was first introduced as an ornamental palm (Hardon 1984). The palm's high productivity (Table 1), together with improved cultivation techniques and availability of good planting material, have all boosted its development into a profitable plantation crop. It now ranks as a very important export commodity for countries in South East Asia, especially Malaysia and Indonesia.

The largest areas under cultivation are now found in Malaysia and Indonesia. As at the end of 1986, Malaysia had about 1.54 million ha under oil palm cultivation (Palmoil Update 1986). Indonesia currently has about 0.47 million ha, and

Table 1. Best average yields of oilseeds (Jones 1983 a)

	Tonnes/h seed	% Oil	Tonnes/ha oil
Soya	2.5	20	0.5
Groundnut	4.5 (in shell)	20	0.9
Cotton seed	2.5	20	0.5
Sunflower	2.0	40	0.8
Rapeseed	2.5	40	1.0
Sesame	1.0	50	0.5
Coconut	4.0 (copra)	65	2.6 New hybrids
Oil palm	20.0 (bunch)	28	5.6 Best Malaysia

aims to increase the area under oil palm by 1 million ha by 1989. A new large acreage in Kalimantan, Indonesia, has been identified as suitable for growing oil palm.

The world total area under oil palm cultivation is also expected to rise in countries in South America, especially Brazil. Even Costa Rica has also increased the area cultivated from 13,400 ha in 1980 to 20,100 ha in 1986 (Palmoil Update 1987).

1.2 Diseases

With improved nursery growing techniques and good estate practice, disease incidence in the oil palm is usually kept within acceptable limits (Turner 1968). Varying climatic and environmental differences for the different areas under oil palm cultivation would naturally result in the incidence of different disease disorders. Most diseases are caused by fungal pathogens. In seed production, the embryos can be affected by the fungus *Schizophyllum commune* causing brown germ if the seed moisture level is high (Turner and Bull 1967). Affected embryos will result in abnormal seedling development or even not germinate. Nursery leaf and root diseases can be easily controlled by the application of relevant fungicides. During wet seasons, spear rot may occur in nursery seedlings. Young palms in the field, 1 to 3 years old, may be affected by crown disease, which has a genetic basis (Blaak 1970).

The most important disease in South Asia and the South Pacific is basal stem rot of mature palms caused by *Ganoderma*, which can result in high losses if left unchecked (Turner 1976). In Africa, *Fusarium* or vascular wilt can cause heavy losses of palms in oil palm stands (Renard 1976). The other economically important disease in Africa is blast disease, which affects nursery palms. Spear rot, which affects young palms, and sudden wilt (Marchitez), which affects mature palms, are economically important diseases in South America (Renard 1976).

Table 2. Fatty acid composition (%wt. as methyl esters) of palm, palm kernel and coconut oils. (Timms 1986)

Fatty acid	Palm oil	Palm kernel oil	Coconut oil
6:0 Caproic	–	0.3	0.3
8:0 Caprylic	–	3.3	7.8
10:0 Capric	–	3.5	6.7
12:0 Lauric	0.2	47.5	47.5
14:0 Myristic	1.1	16.4	18.1
16:0 Palmitic	44.0	8.5	8.8
19:0 Stearic	4.5	2.4	2.6
18:1 Oleic	39.2	15.3	6.2
18:2 Linoleic	10.1	2.4	1.6

1.3 Oil Yield

Most major oilseed crops are annuals; the oil palm and coconut being perennials are therefore comparatively more productive, as can be seen from Table 1 (Jones 1983a). The oil palm is, however, by far the most efficient producer of oil in terms of yield per hectare.

1.4 Oil Composition and Uses

Palm oil has about 44% saturated palmitic acid. However, the presence of about 39% oleic and 10% linoleic acids makes the oil as a whole less unsaturated than that of palm kernel and coconut oils (Table 2). Other important minor constituents of palm oil are (1) the carotenoids, precursors of vitamin A, which give the oil its characteristic red color; and (2) the tocopherols (which funtion as vitamin E) which are anti-oxidants.

Palm oil is used mainly in the manufacture of cooking oil, vanaspati (vegetable ghee), margarine, soap and detergents, shortenings, confectionery fats, and ice cream (Berger 1986).

Palm kernel oil resembles coconut oil because of its high lauric acid content (48%), and the two oils are therefore interchangeable. The other major constituent is myristic acid (16%). Palm kernel oil is used in the making of soap, methyl esters, fatty alcohols, confectionery fats, imitation dairy products, margarines and cooking oils (Knaut and Richtler 1985; Timms 1986).

1.5 Production and Trade

During the period 1958 to 1968, palm oil products formed only a minor part of the world production of oils and fats. This profile has changed significantly after 1968 (Mielke 1985). World production of palm oil and palm kernel oil have been increasing steadily since then. Palm and palm kernel oils averaged about 10% of world production of oils and fats from both vegetable and animal sources for the period 1980 to 1984. By the year 2000, palm and palm kernel oils are expected

Table 3. World production of 17 selected oils and fats. (Mielke 1985)

	1958	1968	1984 (E)	2000 (F)
Palm oil	1325	1406	5839	17950
Palm kernel oil	446	365	812	2117
Soybean oil	2985	4641	13442	22253
Other	22608	30741	41627	52838
17 oils and fats	27364	37153	61720	95158

E = estimated, F = forecasts

Table 4. World production of palm and palm kernel oil (×1000). (FAO 1986)

	1979–81	1984	1985	1986
Palm oil				
Malaysia	2528.8	3714.8	4131.8	4542.2
Indonesia	720.8	1132.4	1230.0	1298.0
Ivory Coast	152.5	150.6	164.0	180.0
Nigeria	666.7	700.0	730.0	760.0
Cameroun	76.9	77.0	80.0	85.0
Brazil	14.7	23.0	29.0	37.0
Costa Rica	31.3	40.0	40.0	40.0
Papua New Guinea	47.4	108.0	120.0	125.0
Other countries	785.3	986.4	1104.2	1159.5
World total	5024.4	6932.2	7629.0	8226.9
Palm kernel				
Malaysia	540.6	1045.6	1212.3	1336.0
Indonesia	129.7	280.6	252.2	243.7
Ivory Cost	36.0	32.8	34.6	43.0
Nigeria	343.3	340.0	360.0	355.0
Cameroun	42.8	44.0	46.0	52.3
Brazil	265.6	223.1	250.0	250.0
Costa Rica	9.2	11.8	11.8	11.8
Papua New Guinea	19.2	39.2	48.8	51.2
Other countries	365.3	392.7	403.7	409.6
World total	1751.7	2409.8	2619.4	2752.6

to rise to 20 million tons and contribute to 21% of world production of oil and fats (Table 3).

Table 4 summarizes world production of palm and palm kernel oil for the years 1979 to 1986. During 1986 Malaysia and Indonesia produced 71 and 53% of the world production of palm and palm kernel oil respectively. The increase in production of palm kernel oil can be attributed mainly to these two countries, whereas production by the other countries remains rather consistent. With the expansion of oil palm cultivation in various producing countries, it is not surprising that palm and palm kernel oils will capture 44% of world export by the year 2000, thus making them major commodities in world oils and fats (Mielke 1985).

Table 5. World exports of palm oil annual five year averages (100 t). (Oil World 1985)

	1988–92	1983–87	1978–82	1973–77	1968–72
Nigeria	–	–	1	7	11
Indonesia[a]	1270	371	357	348	192
Malaysia	5570	3389	2230	1124	462
Other countries[b]	1170	1010	798	505	379
Total	8010	4770	3386	1984	1044

[a] Crude, refined and stearin.
[b] Includes reexports (especially largely from Singapore, but also from the EEC).

The bulk of exported palm oil comes from countries in South East Asia (Table 5). Domestic consumption in Malaysia of palm and palm kernel oil is projected bo be about 13% of the country's production in the year 2000 (Mielke 1985). Most of the production is therefore exported, making Malaysia the leading exporter in the world. The domestic consumption of Indonesia is estimated to be 46% because of its large population. There is very little or no export from Nigeria, once a major producing country.

Countries which import a substantial portion of palm oil are India, European Community Countries, Pakistan, and the Soviet Union, ranging from 9 to 17% of the world total (Anonymous 1986a). As much as 50% of palm kernel oil is imported by the European Community Countries.

1.6 Conventional Practices for its Propagation and Improvement

The main aim in oil palm breeding is to maximize oil yield per hectare. Thus factors which affect the production of bunches and the amount of oil in the oil-bearing mesocarp and kernels will affect oil yield. A comprehensive review which traces the history of the development of oil palm productiom and early selections based on field observations of a limited population to current day selections in breeding programs can be found in Hardon (1984).

Until the early 1960's, the dura palms, with characteristic thick shells, were the main source of planting material. Discovery that the inheritance of shell thickness is controlled by a single gene has had a significant effect in increasing oil yield. Crosses made between the thick-shelled dura as mother parent (homozygous Sh^+Sh^+ for shell thickness) and the shell-less pisifera as pollen parent (homozygous Sh^-Sh^- for absence of shell) give rise to the hybrid, tenera, with thin shell (Sh^+Sh^-). Today, commercial planting materials are mainly teneras.

The oil palm is an outcrossing species, hence individual palms are highly heterozygous and vary widely in yield, oil quality, and disease resistance. Controlled pollination between selected dura mother palms and the pisiferas is the practice in oil palm breeding to obtain seedling progenies. Generally, the male parents are usually female-sterile pisiferas and progeny testing is used to evaluate the potential of the pisiferas for use in breeding programs. Duras are selected for their fruit and bunch characters. Currently, the genetic base of most oil palm breeding and planting materials is limited. The introduction of new genetic material into the gene pool will greatly enhance future breeding programs. Therefore, the search for new germplasm from palms growing in their natural habitat has become a part of oil palm breeding.

1.7 Need to Incorporate Unconventional Methods

In spite of selection of superior parents for breeding, good seedling progenies from these parents still show a wide range of variation in oil yield per palm (Corley 1982). The results from a selected D×P progeny trial showed that the average mean yield is about 6 t/ha, with outstanding individual palms yielding more than 10 t/ha. The variation in vigor and productivity are often expressed in the palms only many years after field planting. The generation time in a breeding program of the oil palm is about 8 years, hence breeding for new lines is indeed a very long-term project.

Also, the oil palm is a nonsuckering palm and has only one growing point. Conventional horticultural practices for its vegetative propagation are therefore not possible. Tissue culture has thus provided a means whereby selected high-yielding palms for the field can be vegetatively propagated nondestructively. This therefore enables circumvention of long-term breeding programs, as uniform plantlets from any selected elite palm can be produced in vitro without having to breed for uniformity of desirable traits.

2 In Vitro Approaches

The advantages of producing clones of an individual selected palm using tissue culture have been described by several authors (Jones 1974; Corley et al. 1977; Noiret 1981). These include the supply of uniform planting materials from out-

Table 6. In vitro studies on oil palm (*Elaeis guineensis* Jacq.)

Explant	Medium	Morphogenetic-response	Reference
Meristem tissue (2–6-year-old palms)	Modified Miller's medium	Leaves; roots and callus nodules	Staritsky (1970)
	+0.1 mg/l kin +5 mg/l NAA or IAA		
Seedling embryos	Liquid medium Heller's salt +1 μM 2,4-D and 1 μM kin	Nodular cell colonies	Rabechault et al. (1970)
	+IAA	Roots produced	
Stem apex and leaf bases (15-month-old palms)	Liquid medium Knop, Heller or MS salts		Rabechault et al. (1972)
	+2 μM 2,4-D+1 μM kin or BA	Callus	
	+NAA or IAA	Roots, shoots but no plantlets	

Table 6 (continued)

Explant	Medium	Morphogenetic-response	Reference
Seedling roots Shoot apex from aseptically germinated embryos	MS salts + growth regulators (details not available)	Callus and embryoids Callus and embryoids	Jones (1974)
Leaf tissues	MS + 10 – 50 μM 2,4-D or 2,4,5-T MS + 2.5 μM NAA + 10 μM kin BA or 2ip MS + 0.05 to 0.1 mg/l NAA	Callus Emryoids and shoots Rooted shoots	Rabechault and Martin (1976)
Young leaf sections (2-year-oil palm)	Not available	Callus, embryoids and plantlets (from fast growing callus)	Loiret (1981)
Roots from mature palms	MS salts + growth regulators	Callus, embryoids and plantlets	Wooi et al. (1981)
Young leaf sections from nursery and mature palms	Modified MS + growth regulators (details not available)	Callus, embryoids and plantlets (from slow growing callus)	Hanower and Pannetier (1982)
Roots and young leaf from aseptically grown seedlings. Also tertiary roots, inflorescence and young leaves from mature palms	MS + 2,4-D or NAA + lowered levels auxin + cytokinins + NAA or IBA + BAP or kin Liquid MS + NAA	Callus Embryoids Shoots from embryoids Rooted shoots	Paranjothy and Othman (1982)
Leaf and roots from aseptically grown seedlings (pisifera). Also leaf and root from nursery seedlings	MS + 5 – 10 mg/l NAA or 2,4-D + lowered auxin + 20 mg/l NAA + activated charcoal + 20 mg/l NAA; 1 mg/l GA3	Callus Embryoids Shoots Root or both shoots and root formation	Nwanko and Krikorian (1982)
Young leaves and meristemmatic leaf bases (6-month-old palm)	MS + 50 – 70 mg/l 2,4-D + 10 mg/l 2,4-D + 0.01 mg/l 2ip + 0.5 g/l activated charcoal + 0.01 – 0.5 mg/l 2ip 1/2 strength MS + 1 mg/l NAA + 1 mg/l GA3	Callus Callus proliferation Embryoid and shoots Rooted shoots	Thomas and Rao (1985)

standing individuals and cloning of parents for use in breeding programs for improvement of planting materials.

Reviews on palm tissue culture are numerous (Corley 1982; Reynolds 1982; Jones 1983b; Blake 1983; Paranjothy 1984, 1986a; Wooi 1984; Jones and Hughes 1989); the present chapter is an attempt to update the current status of oil palm tissue culture.

In the early 1970's, oil palm tissue culture was dominated by two groups working independently – the British (Smith and Thomas 1973; Jones 1974; Corley et al. 1977) and the French (Rabechault et al. 1970, 1972; Rabechault and Martin 1976). The envisaged commercial potential of the oil palm tissue culture has led several other groups to set up tissue culture laboratories to specifically clone oil palms. Currently, there are at least a dozen such laboratories in existence and about ten of these can be found in Malaysia alone. Most of the Malaysian laboratories, except for the Unilever laboratory, have been set up over the last 4 to 5 years. However, publications from these commercially oriented groups are not easily available. A summary of micropropagation activities in oil palm is presented in Table 6.

2.1 Sterilization of the Explants

Many sources of explants, such as zygotic embryos, apical meristem, roots, young leaves, rachis, and immature inflorescences, have been used for in vitro studies to initiate cultures.

Root explants are heavily contaminated and difficult to decontaminate (Smith and Thomas 1973). Normally, treatment of mercuric chloride at 0.1% (w/v) for 15 min followed by washing with a 10% (v/v) sodium hypochlorite solution for 10 min and sterile distilled water rinse is essential (Wooi et al. 1981). About 50% of explants still become infected even after surface sterilization, and the callusing rate from remaining explants is about 5–10%. Using 2–6-year-old greenhouse palms, Staritsky (1970) immersed the whole oil palm "cabbage", consisting of apical meristem and very young leaf primordia, in a saturated solution of calcium hypochlorite for 1–2 h. Explants were taken after removal of the outermost layers and planted. Contamination was reported to be high except for vigorously growing palms.

A 0.1% (w/v) mercuric chloride solution has also been tried for sterilization of the "cabbage" (Rabechault et al. 1972). Callus initiation was obtained for young leaf explants disinfected with 1% mercuric chloride solution for 20 min (Rabechault and Martin 1976).

However, when satisfactory precautions to prevent entry of microorganisms are taken during field sampling, young leaves still enclosed by leaf bases and inflorescence still enclosed within the spathes need no sterilization. These can be sampled without damaging the apex. Swabbing the external enclosing leaf bases or enclosing spathes with a 70% (v/v) ethanol and careful removal of inner tissues under aseptic conditions is adequate to establish sterile explants.

2.2 Media Composition

Exact media formulations are usually never published because of the intense commercial interests in oil palm tissue culture. Basal media, however, are generally modified Murashige and Skoog medium. It is generally known that different protocols are used by the various laboratories to produce oil palm clones. Plant growth hormones, usually auxins and cytokinins, may be used. Generally it would appear that auxins, at least, are necessary for the initiation of callus from all explant sources and also for the rooting of shoots produced in vitro. The medium used for proliferating embryoids (the multiplication phase in oil palm tissue culture) may or may not contain plant growth hormones.

2.3 Callus Culture

Callus cultures can be initiated from roots of aseptically grown seedlings within 3 weeks of culture in the presence of 2.5×10^{-5} M 2,4-dichlorophenoxyacetic acid (2,4-D) or α-naphthaleneacetic acid (NAA) (Smith and Thomas 1973). Root explants from mature palms normally produce callus within 2 months after explanting (Wooi et al. 1981). It is generally agreed that growth of callus from root explants is slow (Smith and Thomas 1973; Paranjothy 1982), whereas callus from leaf explants has a faster growth rate (Loiret 1981).

Ease of callusing is also generally dependent on a number of factors. Factors which will affect the ease and rate of callusing from explants of mature palms are the genotype of the palm, the distance tissues are located from the meristem in the case of young leaf tissues, and also the auxin used (Rabechault et al. 1972; Loiret 1981; Duval et al. 1987).

Apart from the explant source and culture conditions, the physiological stage of the palm from which leaf explants are extracted also showed that in the younger and more vigorously growing palms, the callus production was easier (Paranjothy 1986b). This phenomenon has also been reported for palms grown under greenhouse conditions (Nwanko and Krikorian 1983).

Growth rates of oil palm callus cultures from root explants have also been reported to be significantly variable for the various genotypes studied (Wooi et al. 1981).

Callus tissues are induced on the veins of young leaf tissues (Pannetier et al. 1981). Root callus appears to originate from the pericycle and central stele region of the root (Paranjothy and Othman 1982; Jones 1983b). Initiation of callus on the explants is normally done in the dark at 28 ± 1 °C. Callus cultures are also normally incubated under the same growing conditions.

The type of auxin also influenced the appearence of the callus. The presence of NAA tends to give a rooty callus, and 2,4-D a more nodular appearance (Smith and Thomas 1973; Jones 1974). Pretreatment of explants in liquid basal medium enhanced rate of callusing on explants (Rabechault and Martin 1976), although other workers have found this unnecessary.

Browning of explant and medium is often a common complaint among oil palm tissue culturists (Reynolds 1982; Paranjothy 1984), but this need not necessarily be harmful, as successful cultures have been grown from explants which had turned almost black (Jones 1974). That production of phenolics is not necessarily harmful is confirmed by work involving other plant species (Krikorian 1987).

A reduction in the level of auxin is normally followed for maintenance of callus cultures (Smith and Thomas 1973; Paranjothy and Othman 1982; Rabechault and Martin 1976; Thomas and Rao 1985).

Young male inflorescences harvested at the right stage provide excellent explants for callus initiation. Often, calli initiated do not have the capacity to maintain rapid growth or continue growth as callus tissues (unpubl.).

2.4 Embryogenesis and Plant Regeneration

Direct embryogenesis from explant cultures has not been reported, although personal communications with people involved in oil palm tissue have revealed that it is sometimes possible to obtain embryoids formed directly on leaf explants. Frequency, however, is rare and sporadic, and confined to rather young palms. Embryoids have been obtained from parts of aseptically grown seedlings (Jones 1974); roots (Corley et al. 1977), and young leaves of selected field palms (Rabechault and Martin 1976).

Inflorescence explants can also serve as a source of explants from which embryoids can be obtained. Successful multiplication of these embryoids with subsequent plant regeneration has also been achieved (unpubl.).

Generally, a callus phase seems unavoidable in the tissue culture of oil palm. Given the right culture conditions and growth regulator concentrations, embryoids are formed on the callus. Observations of embryoids formed have led to the conclusion that these are somatic embryoids (Pannetier et al. 1981).

Normally, embryoids are observed when callus cultures on nutrient medium rich in auxin are transferred onto medium with decreasing amounts of phytohormones or onto a medium devoid of phytohormones. Paranjothy and Othman (1982) have obtained embryoids by lowering the auxin levels plus the incorporation of cytokinins in the culture medium. Lowering the NAA level (40 – 0 mg/l) in the presence of activated charcoal gives rise to embryoids (Nwanko and Krikorian 1983). These embryoids produced shoots and roots in the presence of 20 mg/l NAA plus 1 mg/l GA_3. Nodular callus formed in the presence of 2,4-D became embryogenic when transferred onto MS medium supplemented with 0.01 – 0.5 mg/l 2-ip (Thomas and Rao 1985).

Embryogenesis has been reported from both primary or slow-growing callus (Jones 1974; Corley et al. 1977; Wooi et al. 1981) and also secondary or fast-growing callus (Rabechault and Martin 1976; Loiret 1981). White friable or fast-growing callus was reported to be more highly embryogenic (Loiret 1981).

The occurrence of unbalanced embryogensis has also been reported. Production of roots on the callus cultures inhibits further caulogenesis (Rabechault et al. 1970; Corley 1980). Prolonged incubation of callus cultures was reported to give rise to embryonal structures called neomorphs, which do not develop further (Nwanko and Krikorian 1983).

Development of fast-growing callus, however, appeared to be random and may take at least six months after the formation of primary callus (Hanower and Pennetier 1982). It therefore became obvious to induce embryogenesis on primary callus in order to shorten the time taken for embryogenesis (Pannetier et al. 1981; Hanower and Pannetier 1982).

Results of the French workers showed that the age of the ortets from which callus was derived affects the success of embryogenesis in culture (Table 7).

Table 7. Age effect on the success of embryogenesis in vitro

Age of ortet (year)	Sample size	Percent embryogenesis time in culture (months)				Reference
		4	12	24	40	
20–25	Ω 200		44	75	Ω 100	Duval et al. (1987)
9–10	Ω 30	70	Ω 100			Duval et al. (1987)
2	4	100				Pannetier et al. (1981)

Once embryoids are formed, they are transferred to light at an incubating temperature of 28 ± 1 °C. Light intensity is in the region of 1500 lx and the lighting duration may vary from 16 to 24 h.

Maintenance of embryoids was reported to be on medium devoid of growth hormones (Hanower and Pannetier 1982). Multiplication of embryoids appeared to be through adventive embryogenesis (Hanower and Pannetier 1982; Paranjothy and Othman 1982).

Plant regeneration often accompanies embryoid formation, though the shoot production ability is variable for different genotypes. For shoot production, embryoids are maintained on medium with low concentrations of α-NAA or IBA, together with low concentrations of BAP or kinetin (Paranjothy and Othman 1982). Growth of shoots is enhanced in agitated liquid cultures (Thomas and Rao 1985). The occurrence of terminal inflorescence in culture has also been observed by various laboratories. Sometimes expression of such terminal inflorescences may be delayed until after transplanting plantlets in the prenursery.

Rooting of shoots does not seem to pose any difficulty, although again efficacy varies with genotypes. Rooting is often accompanied by subculturing shoots regenerated onto a rooting medium. A short exposure of shoots to a relatively high concentration of NAA and then to basal medium, or a continuous exposure of shoots to low levels of an auxin is usually adequate for successful rooting.

2.5 Establishment of Regenerated Plantlets

Establishment of in vitro plantlets does not appear to be problematic (Rabechault and Martin 1976; Corley et al. 1977; Loiret 1981; Paranjothy and Othman 1982). Tissue culture plantlets do not develop a normal cuticle (Wardle et al. 1979) and therefore require gradual weaning from the in vitro conditions which are nearly 100% humid. Various ways of hardening plantlets have already been reported (Wooi et al. 1981). Acclimatization of the plantlets by spraying leaf surface with antitranspirants to reduce transpiration is often harmful and has also been reported for ornamentals (Sutter and Hutzell 1984). Maintenance of high humidity around newly transplanted plantlets is by far the most commonly practiced method for hardening plantlets, and success rates above 90% have been reported (Wooi et al. 1981). No doubt, survival of plantlets is directly correlated with the vigor and size of the plantlets transplanted. Similar results have been reported for date plantlets (Tisserat 1982).

Table 8. Relative leaf area growth rates of seedlings with and without endosperm and clonal plants. (Wooi et al. 1981)

Weeks after planting/ germination	R_A (percent per day)		
	5–10	10–14	16–24
Intact seedlings	3.20±0.27	2.41±0.16	1.93±0.14
Seedlings without endosperm	2.09±0.42	2.13±0.17	2.09±0.23
Clone JB 20.1C	–	2.26±0.15	2.26±0.13
Clone 171.97	–	2.05±0.35	2.22±0.28

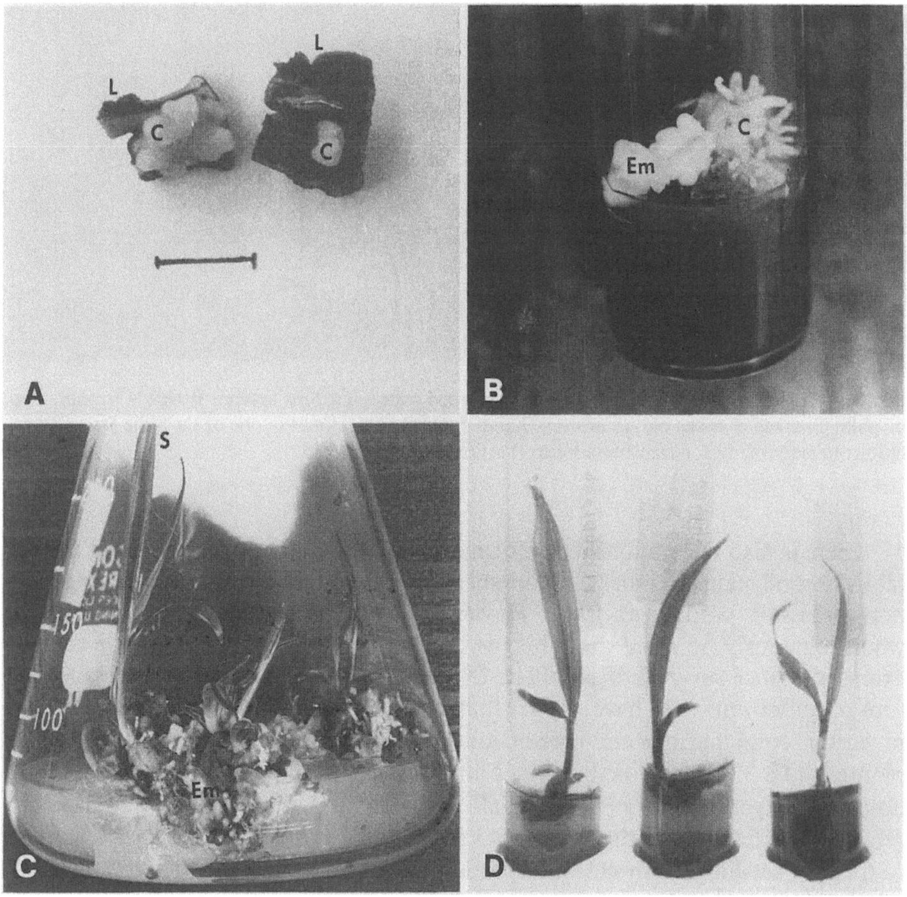

Fig. 2. A Callus (*C*) developing on leaf explants (*L*), 5 months old. Bar = 1 cm. **B** Embryoids (*Em*) developing on callus tissues (*C*). **C** Proliferating embryoids (*Em*) with shoots (*S*). **D** Isolated single shoots on rooting medium

Fig. 3. **A** Transplantable plantlet with well-developed shoot and root system. **B** Newly transplanted oil palm plantlets, 3 weeks old, in misting chamber. **C** Oil palm clones in one of the Sime Darby Plantations nursery ready for transplanting to the field, 16 months old

There is also little published information on the potting medium used for establishing oil palm plantlets. Good survival was obtained by transplanting into nonsterile soil (Wooi et al. 1981). However, a more homogenous potting mixture would obviously be more desirable, as different sources of soil will to a certain extent influence survival of plantlets. Observations from visiting other tissue culture oil palm nurseries have shown that peat, vermiculite, sand, and soil mixed in various combinations and proportions are also utilized. Growth rates of clonal plantlets (JB 32) in a growth cabinet are generally comparable to those of seedlings in a polybag nursery (Corley et al. 1976, 1977). A time-scale trial to monitor growth rates has shown that the initial slow growth rates of plantlets are similar to those of seedlings with the endosperm removed after emergence of the first leaf (Table 8). Comparable growth rates of clonal plantlets with intact seedlings are, however, observed after 3 months. This is probably because the endosperm in the intact seedlings would have been utilized by then.

Oil palm plantlets are normally grown for about 1 year in the nursery before field planting. Figures 2 and 3 show the various stages in the process of in vitro propagation of oil palm.

2.6 Embryo Culture

Early studies in oil palm tissue culture have utilized embryos in fundamental studies to formulate defined media and culture conditions. Callus cultures are easily established from zygotic embryos, and aseptically germinated seedlings provide a ready source of explants for in vitro studies. A review on embryo culture of oil palm can be found in Paranjothy (1984).

Other recent applications of oil palm embryo culture lie in rescuing embryos from poorly germinating seeds, and for germplasm conservation (Paranjothy et al. 1985).

Seeds of pisiferas (shell-less) and very thin-shelled teneras germinate poorly, embryos die probably due to infection in the soil; hence such embryos can be rescued by culturing them in vitro. Embryo culture will also enable sampling from immature as well as mature fruits during prospection for new germplasm, and movement of sterile cultures internationally should minimize the risk of introduction of diseases (Paranjothy et al. 1985).

2.7 Genetic Stability

For successful implementation of oil palm tissue culture on a practical scale it is important to establish that the plantlets regenerated are stable. To date, all work published on palm tissue culture has involved an intervening callus stage. It is widely kown that callus is generally unstable (D'Amato 1978).

In the oil palm, genetic stability of cultures was illustrated by the fact that in the field two batches of plantlets regenerated from callus cultures of a clone 2 years apart showed similar fruit characters (Table 9).

The distinctive features of clone 932; high fruit set and fruit to bunch, and high shell and kernel content of the fruit, are shown by both sets of data. Chromosome counts made from root tip squashes of regenerated plants showed the normal diploid number $2n = 32$ (Corley et al. 1977; Jones et al. 1982). Quantitative measurements of the amount of DNA present in callus and embryoid tissues were made (Jones et al. 1982). Their results showed that, although callus cultures exhibited a widespread of ploidy levels, embryoids that were produced were shown to be predominantly diploid. The presence of a few tetraploid cells in em-

Table 9. Fruit characters of clone 932. (Wooi et al. 1982)

Date plantlets regenerated	1977	1979
Fruit set (%)	80.0 ± 22.2	81.8 ± 10.5
Fruit/bunch (%)	72.3 ± 5.9	67.8 ± 6.4
Mesocarp/fruit (%)	73.3 ± 4.3	66.4 ± 8.2
Kernel/fruit (%)	10.0 ± 1.7	11.4 ± 1.4
Shell/fruit (%)	14.7 ± 4.2	15.1 ± 2.6

Data from palms at 30 months after field planting.

bryoid tissues is to be expected during multiplication in a growing culture. Thus, there appears to be a natural selection for the normal karyotype.

Most other evidence for genetic uniformity of oil palm cultures comes from early clone trial results (Corley et al. 1977, 1981; Wooi et al. 1981, 1982), which will be discussed in the following section.

2.8 Field Testing

Field testing of clonal material is important both to test that selected palms are genetically superior and that clones produced are genetically stable. Oil yield of an individual palm is genetically determined and also influenced by environmental conditions. The efficient selection of a palm for cloning would ensure the expression of yield characters that are under genetic control, i.e., the components in the oil extraction ratio (Hardon 1976) in the clone. Field testing of clones would therefore eliminate ortets that are not genetically superior.

The first clone trials comprising seedling-derived materials were first planted in Malaysia during 1977 to 1980 by the Unilever group. Clones produced from selected ortets were first planted in 1981. To test the performance of clones in varying environments, clone trials were also later established in Cameroun and Colombia. Altogether some 29,500 plantlets, representing at least 50 clones, had been planted on 200 ha for field evaluation (Anonymous 1982, 1985).

Early field trial results of oil palm clones indicated that clones are more uniform than seedling materials with respect to leaf production (Corley et al. 1979). Clonal material also showed less variation than seedling material in petiole width×depth, fruit composition such as mesocarp/fruit and kernel/fruit, oil/bunch, and kernel/bunch (Wooi et al. 1981). Detailed statistical analyses of early field trial data showing uniformity within clones are given by Corley et al. (1981). The early results showed that for highly hetitable characters such as fruit and bunch characters (Hardon 1976), the variation between palms within a clone is not significantly higher than the variation between repeated measurements on the same palm (Table 10). The variation between succesive measurements on the same palm could be attributed to environmental influences. These early results also showed that an expectation of 30% increase in oil yield using clonal materials is possible.

Uniformity in flowering patterns was also reported in small plots of individual clones (Corley 1980; Corley et al. 1981). Clones exhibited abrupt changes in sex ratio from 100% of one sex to 100% of another, rather than the gradual change seen in seedling materials.

Fruit cross-sections of various clones showed the distinctness in fruit characters among the various clones and uniformity of fruit characters within the same clones (Wooi et al. 1982). One of the clones also consistently exhibits polyembryony.

An incident citing four abnormal palms derived from a single culture of a clone representing 1.8% of the plants produced was reported (Corley et al. 1981). The leaf morphology was unusual, with few, broad, thick leaflets, and occasionally branched rachises. No other gross abnormalities were reported among 13000 plantlets produced.

Table 10. Analysis of variance for fruit characters from three clones (936, 932 and 997) in observation plots[a]. (Corley et al. 1981)

Fruit characters	Mean squares		
	Between clones	Between palms within clones	Between bunches within palm
Fruit to bunch (%)	2005.7[c]	170.46	107.31
Mesocarp to fruit (%)	1244.8[d]	10.62	5.29
Dry matter in mesocarp (%)	499.0[b]	88.27	45.09
Oil to dry mesocarp (%)	32.0	22.17	22.86
Kernel to fruit (%)	108.6[d]	2.74	1.57
Shell to fruit (%)	802.1[d]	2.49	1.05
Degrees of freedom	2	8	22

[a] Three bunches per palm sampled between 30 and 36 months after field planting.
[b] Significantly greater than between palms mean square ($p<0.05$).
[c] Significantly greater than between palms mean square ($p<0.01$).
[d] Significantly greater than between palms mean square (<0.001).

Later, field results reported abnormal flower production on clonal materials (Corley et al. 1986). Androgynous male inflorescence, mantled and parthenocarpic fruit development were described for three clones produced from selected palms. Planting of these clones was done in 1981, 1982, and 1983. It would appear that the abnormalities reported were absent from the 1981 planting, but with increasing incidence in 1982 and 1983 plantings. Possible causes of such abnormal development were also discussed (Corley et al. 1986; Soh 1987). The presence of plant growth regulators such as 2,4-D, which is a potent mutagen, was ruled out, as two of the abnormal clones had never been cultured on 2,4-D. Time in culture may be an important factor, as prolonged culture on media supplemented with phytohormones may induce epigenetic changes, thus disrupting normal hormone metabolism of the cultures.

Field observations have shown that some abnormal palms, not more than 5%, have apparently recovered, and it is too early to tell whether recovery will be permanent (Corley pers. commun.). That recovery occurs at all certainly supports the idea that the abnormality is not a mutation problem.

Abnormality of flower development has also been reported in two clones produced by the French workers (Duval et al. 1987). These clones were produced from fast-growing callus cultures prior to the development of embryoids from primary callus cultures (Pannetier et al. 1981). Smith and Thomas (1973) had found numerous abnormal mitoses occurring within their fast-growing callus cultures. The use of fast-growing callus was implicated to be the probable cause of the abnormality (Duval et al. 1987). However, the three abnormal clones reported by Corley et al. (1986) had been derived from slow-growing callus.

Plantlets are now routinely regenerated from embryoids derived from slow growing primary callus (Duval et al. 1987). Early trial results from such plantlets are reported to be normal. Results from later plantings will be needed to establish that the production protocol is safe for use in large-scale production of clonal materials. Such trials are believed to be underway.

Latest results from oil palm clone trials show that at least 5 years of yield recording would be required in order to estimate a clone's potential (Corley et al. 1987). Variation in yield from year to year and significant clone×location, and climatic ×clone interactions are reported for various oil palm clones. It would be highly speculative at this stage to draw any firm conclusions on the performance of clonal materials. Further field testing are needed to ascertain the stability of clonal materials when these are pushed through the system to produce large numbers.

2.9 Other Investigations

Most published work is rather empirical in nature. Biochemical studies on oil palm tissue culture are rarely described. In one such report, Turnham and Northcote (1982) observed, from electron micrographs, the accumulation of fat droplets within callus maintained on 2.5 mg/l NAA just prior to embryogenesis. This accumulation correlated with an increase in acetyl-CoA carboxylase activity, which is probably a rate-limiting step in fatty acid synthesis in these cells. Acetyl-CoA carboxylase activity could be used as a quantitative marker of somatic embryogenesis.

Isolation of oil palm protoplasts from leaves and rachis of greenhouse palms was attempted by Vouyouklis (1981). Subsequent survival of viable protoplasts was not successful. Using a "nurse" medium containing oil palm cells, protoplasts regenerated cell walls, and divided to form cell colonies (Bass and Hughes 1984). Successful development of protoplasts cultures may make the oil palm more amenable for somatic cell hybridization and genetic engineering studies.

Micropropagation of the oil palm has been well developed, thereby providing a system for in vitro screening for disease resistance. Vascular wilt, a disease caused by *Fusarium oxysporum*, is a serious problem in West Africa. To screen clones for wilt resistance, oil palm callus cultures are grown in the presence of the crude fungal filtrates or purified toxin (Flood 1987). Known resistant as well as oil palm callus lines susceptible to the fungal pathogen have now been identified. The efficacy of the screening method will be tested in vivo for the outcome of practical applications.

2.10 Potential of Oil Palm Micropropagation

The main objective of oil palm micropropagation is to develop a vegetative propagation technique which can be used to produce large numbers of commercial planting material. Exact production figures are not readily available from the various laboratories involved in oil palm tissue culture. The potential number that can be produced in vitro by the system by the more established laboratories is illustrated in Table 11.

Shoot production potential by individual clones varies considerably. Of the 59 clones represented in the La Mé Laboratory, only three produced in excess of 10,000 plantlets, 19 produced more than 1000 plantlets and the rest in the region of 100 plantlets.

Table 11. Production of oil palm plantlet

Laboratory	Year	No. of plantlets	No. of clones represented	Reference
La Mé	1983–1986	137000	59	Duval et al. (1981)
Unifield Tissue Culture Limited, United Kingdom	1984–1985	568000	NA	Anonymous (1985)
Bakasawit Sendirian Berhad, Malaysia	1984–1985	102000	NA	Anonymous (1985)

NA, not available.

The techniques now being used for large-scale propagation are still basically those of the laboratory. However, the scale-up of this laboratory technique has not been without its problems (Jones 1982). The reliable supply of genetically uniform planting material and efficient throughout of quality cultures are important requirements for a commercial operation.

The cost of planting material produced must be economically attractive to the buyer. The cost of producing a three-leaf stage of an oil palm plantlet, which took into account the research and development costs, was estimated at M$5, compared to M$1 for a comparable seedling (Wooi et al. 1981). From data on plantlet production presented by Duval et al. (1987) and from our own experience, it can be envisaged that a considerable reduction in production costs can be made by making most, if not all, clones equally efficient shoot producers.

A simple financial analysis made by Corley et al. (1987) showed that it will be profitable to plant clonal material if there is at least an 80% chance that the required yield increase is realized.

Most laboratories are set up to produce high-yielding clonal oil palm planting materials for their own plantations. Therefore, in this case, profit expected from sale of planting material becomes secondary to ensuring the production of genetically superior planting material.

Currently, faced by the abnormal flower development problem (Corley et al. 1986; Duval et al. 1987), priorities now lie in producing small numbers of a large number of clones for field evaluation.

Close to 20,000 plantlets representing at least 16 clones had been distributed to various commercial and government agencies by the Palm Oil Research Institute of Malaysia for field evaluation (Anonymous 1986b).

3 Conclusions and Prospects

The laboratory technique for in vitro propagation of oil palm has now been mastered with varying degrees of success (Jones 1982; Noiret et al. 1985). Results from early clone trials were indeed encouraging (Corley et al. 1981), however, later trial results reported that abnormality of clonal materials is expressed by abnormal flower development (Corley et al. 1986). Visual detection of these abnormalities is only possible when the clonal palms start to flower and this may be ca. 18 months from time of field planting. Early detection of abnormalitiy is therefore desirable, more so at the test tube level or early ramet stage.

Improving efficiency of the micropropagation process by shortening time in culture for production of embryoids and plantlet regeneration and minimizing use of plant growth regulators in culture appear to be the current thrust of empirical investigations. Any changes in protocols of present techniques and management of the cultures will need to be field-tested to ascertain that plantlets regenerated are normal and stable. Parallel with empirical investigations, biochemical, anatomical, and cytological studies are also in progress in attempts to develop a diagnostic tool for detection of abnormalities with respect to flower development of clonal materials. Until such a tool becomes available, precautions will need to be taken in in vitro propagation of oil palm in order to avoid the abnormal flower development currently observed.

Since field testing of clones is a very long-term process, cryopreservation and successful revival of oil palm embryoids for plant regeneration is an attractive possibility to reduce risk of accumulating aberrant cells in culture through regular subculture for maintenance of clones. This would mean that clones which are no longer required temporarily as field trial data are monitored, can be preserved with minimal maintenance. Successful revival of oil palm enbryos had been reported by Grout et al. (1983) and growth rates of plants regenerated from the retrieved embryos were shown to be comparable to those of plantlets regenerated from enbryoids that had not been cryopreserved (Wooi 1984). Somatic embryos have also been successfully cryopreserved and revived (Engelmann and Duval 1986). Nursery plantlets from revived somatic embryos are reported to develop normally so far (Engelmann et al. 1987). In view of the abnormality problems associated with the flowering of oil palm clones, plantlets regenerated from revived embryos will also need to be field-tested to prove the usefulness of the technology.

It would appear that at this stage a regular reinitiation program for all cultured palms to replace lines that have been in culture for some time would be a good strategy to avoid abnormalities.

Commercialization of oil palm micropropagation will take longer than initially envisaged. Choice of correct micropropagation protocols and careful monitoring of clonal materials before field planting will hopefully weed out abnormal palms. Much of the research and development work on the micropropagation of oil palm is currently being carried out by commercial organizations. There is, therefore, an element of secrecy which makes hard facts difficult to come by. Generally, it would appear that most laboratories share the same basic approach, which has changed little since the early work on oil palm tissue culture in the early 1970's.

Most workers use leaf, root, and inflorescence explants from selected palms as starting material, and opinion differs as to which explant source gives the best results. There are also no published data to date showing any one tissue to be superior. Control of embryogenesis still appears to be the rate-limiting step, and there is some evidence that there may be a genotype effect (Duval et al. 1987). Once embryoids are obtained, they can be easily multiplied, and produce shoots, most of which will root and eventually be transplanted.

There also does not appear to be any specific optimum medium or media for the various developmental stages (Table 6). The presence of an auxin, however, is essential for callus initiation and rooting of shoots. The inclusion of a cytokinin may be beneficial for shoot production. Currently, many workers are inclined to maintain embryoids on a hormone-free medium because of the abnormal flowering problem reported, although there is no concrete evidence to suggest that growth regulators are the cause of the flowering abnormality.

Another application for oil palm tissue culture will be for the production of clonal seeds where outstanding duras and pisiferas can be cloned. Propagation of parents for seed production will certainly not be of the magnitude of elite individuals for commercial production. Therefore there will be little or no risk of producing abnormal palms.

The production of speciality oils would be another possible application whereby selected individuals for specific oil quality, e.g., less saturated oils, can be cloned (Jones 1984).

Acknowledgment. The author is grateful to Sime Darby Plantations, Malaysia, for permission to publish, and to Dr. RHV Corley, Unilever Research, for helpful comments.

References

Anonymous (1982) In: Progress Report 1980–1982. Bakasawit Clonal Oil Palm Research Unit, Malaysia

Anonymous (1985) In: Annual Report for 1984/1985. Unilever PLC Plantations Group, Research and Advisory Programmes

Anonymous (1986a) Oil palm market review. Trop Agric (Trinidad) 63:71–74

Anonymous (1986b) In: Annu Res Rep 1986. Palm Oil Research Institute of Malaysia

Bass A, Hughes W (1984) Conditions for isolation and regeneration of viable protoplasts of oil palm (*Elaeis guineensis*). Plant Cell Rep 3:169–171

Blaak G (1970) Epistasis for crown disease in the oil palm (*Elaeis guineensis* Jacq.). Euphytica 19:22–24

Blake J (1983) Tissue culture propagation of coconut, date and oil palm. In: Dodds JH (ed) Tissue culture of Trees. AVI Westerport, pp 29–50

Berger KG (1986) Food uses of palm oil. Occ Paper No 2. Palm Oil Res Inst Malaysia (Revised from 1981)

Corley, RHV (1980) Vegetative propagation of palms. In: Stanton WR, Flack M (eds) SAGO. The Equatorial Swamp as a Natural Resource. Nijhoff, The Hague, pp 98–109

Corley RHV (1982) Clonal planting material for the oil palm industry. Planter 58:515–528

Corley RHV, Gray BS (1976) Growth and morphology. In: Corley RHV, Hardon JJ, Wood BJ (eds) Oil Palm Research. Elsevier, Amsterdam, pp 7–21

Corley RHV, Barrett JN, Jones LH (1977) Vegetative propagation of oil palm via tissue culture. Oil Palm News 22:2–7

Corley RHV, Barrett JN, Jones LH (1977) Vegetative propagation of oil palm via tissue culture. In: Earp DA, Newall V (eds) International Developments in Oil Palm. Inc Soc Plant, Kuala Lumpur, pp 1–8

Corley RHV, Wooi KC, Wong CY (1979) Progress with vegetative propagation of oil palm. Planter 55:377–380

Corley RHV, Wong CY, Wooi KC, Jones LH (1981) Early results from the first oil palm clone trials. In: Pushparajah E, Chew PS (eds) Oil palm in agriculture in the eighties, Inc Soc Plant, Kuala Lumpur, pp 173–176

Corley RHV, Lee CH, Law IH, Wong CY (1986) Abnormal flower development in oil palm clones. Planter 62:233–240
Corley RHV, Lee CH, Law IA, Cundall E (1987) Field testing of oil palm clones. In: Hassan AH, Chew PS, Wood BJ, Pushparajann E (eds) International Oil Palm Conference: Progress and prospects – agriculture, PORIM, Kuala Lumpur, pp 173–185
D'Amato F (1978) Chromosome number variation in cultured cells and regenerated plants. In: Thorpe TA (ed) Frontiers of plant tissue culture. Int Assoc Plant tissue culture, Calgary, pp 287–295
Duval Y, Gasselin TD, Konan K, Pannetier C (1987) In vitro vegetative micropropagation of oil palm (*Elaeis guineensis* Jacq.): Choice of strategy and results. In: Hassan AH, Chew PS, Wood BJ, Pushparajah E (eds) Int Oil Palm Conf: Progress and prospects – agriculture, PORIM, Kuala Laumpur, pp 191–196
Engelmann F, Duval Y (1986) Cryopreservation of oil palm somatic embryos (*Elaeis guineensis* Jacq.): Results and applications prospects. Oleagineux 41:169–174
Engelmann F, Duval Y, Pannetier C (1987) Use of cryopreservation for setting up a bank of oil palm (*Elaeis guineensis* Jacq.) Somatic embryos. In: Hassan AH, Chew PS, Wood BJ, Pushparajann E (eds) Int Oil Palm Conf: Progress and prospects – agriculture, PORIM, Kuala Lumpur, pp 648–652
FAO (1986) Production Year book, FAO, Rome
Flood J (1987) Screening for disease resistance in oil palm. In: Recent advances in plant tissue culture for crop production and improvement. The British Council Course 758, Bath UK
Grout, BWW, Shelton K, Pritchard HW (1983) Orthodox behavior of oil palm seed and cryopreservation of the excised embryo for genetic conversation Ann Bot 52:381–384
Hanower J, Pannetier C (1982) In vitro vegetative propagation of the oil palm (*Elaeis guineensis* Jacq.) In: Fujiwara A (ed). Plant tissue and cell culture 1982. Maruzen, Tokyo, pp 745–746
Hardon JJ (1976) Oil Palm breeding - introduction. In. Corley RHV, Hardon JJ, Wood BJ (eds) Oil palm research. Elsevier, Amsterdam, pp 89–108
Hardon JJ (1984) The oil palm: Progress through plant breeding in Malaysia. Span 27:59–61
Jones LH (1974) Propagation of clonal oil palms by tissue culture. Oil Palm News 17:1–8, Planter 50:374–381
Jones LH (1982) The use of clonal oil palms in developing countries. In: Van Hemert PA, Lelieveld HLM, la Riviera JWM (eds) Biotechnology in developing countries. Univ Press, Delft, pp 141–158
Jones LH (1983a) The oil palm and its clonal propagation by tissue culture. In: Symp No 29. Proc Leatherhead Food. RA FOSFA Int, p 11–26
Jones LH (1983b) The oil palm and its clonal propagation by tissue culture. Biologist 30:181–188
Jones LH (1984) Novel palm oils from cloned palms. J Am Oil Chem Soc 61:1717–1719
Jones LH, Hughes WA (1989) Oil Palm (*Elaeis guineesis*) Jacq.). In: Bajaj YPS (ed) Biotechnology in agriculture and forestry 5. Trees II. Springer, Berlin Heidelberg New York Tokyo, pp 176–202
Jones LH, Barfield D, Barrett J, Flook A, Pollock K, Robinson P (1982) Cytology of oil palm cultures and regenerant plants. In: Fujiwara A (ed) Plant tissue culture 1982. Maruzen, Tokyo, pp 727–728
Knaut J, Richtler HJ (1985) Trends in industrial uses of palm and lauric oils. J Am Oil Chem Soc 62:317–327
Krikorian AD (1989) Multiplication strategies for tropical woody plants - problems and prospects. In: Rao AN, Yusoff AM (eds) Seminar on Tissue Culture of Forest Species, FRIM and IDRC, Kuala Lumpur, pp 3–44
Loiret C (1981) Vegetative propagation of oil palm by somatic embryogenesis. In: Pushparajah E, Chew PS (eds) Oil palm in the eighties. Inc Soc Plant, Kuala Lumpur, pp 163–172
Mielke S (1985) Present and future position of palm and palm kernel oils in world supply and trade. J Am Oil Palm Chem Soc 62:193–197
Moll HAJ (ed) (1987) The economics of oil palm. Pudoc, Wageningen
Noiret JM (1981) Application of in vitro culture to improvement and production of clonal material in the oil palm. Oleagineux 36:123–126
Noiret JM, Grascon JP, Pannetier C (1985) Oil palm production through in vitro culture. Oleagineux 40:356–372
Nwanko BA, Krikorian AD (1983) Morphogenetic potential of embryo and seedling-derived callus of *Elaeis guineensis* Jacq. var. Pisifera Becc. Ann Bot 51:65–76
Oil World (1985) Statistics update. Oil World, April 1985
Palmoil Update (1986) December 1986, p 1

Palmoil Update (1987) December 1987, p 19

Pannetier C, Arhuis P, Lievoux D (1981) Neoformation of young *Elaeis guineensis* plants from primary callus obtained from leaf fragments cultured in vitro. Oleagineux 36:119–122

Paranjothy K (1982) A review of tissue culture of oil palm and other palms. Occ Paper 5, Palm Oil Research Inst, Malaysia

Paranjothy K (1984) Oil Palm. In: Ammirato PV, Evans DA, Sharp WR, Yamada Y (eds) Handbook of plant cell culture. Macmillan, New York, pp 591–605

Paranjothy K (1986a) Recent developments in cell and tissue culture of oil bearing palms. Occ Paper 19, Paper Oil Research Inst, Malaysia

Paranjothy K (1986b) Some aspects of oil palm tissue culture in relation to ortet selection and clonal evaluation. In: Colloq Breeding and Selection for Clonal Oil Palms. ISOPB, Kuala Lumpur, pp 41–43

Paranjothy K, Othman R (1982) In vitro propagation of oil palm. In: Fujiwara A (ed) Plant tissue culture 1982. Maruzen, Tokyo, pp 747–748

Paranjothy K, Othman R, Hashim AT (1985) Oil palm germplasm conservation: The possible use of in vitro methods. In: Int Workshop on Oil Palm Germplasm and Utilization. ISOPB, PORIM, Bangi, pp 202–205

Rabechault H, Martin J-P (1976) Multiplication vegetative du palmier à huile (*Elaeis guineensis* Jacq.) à l'aide de cultures de tissue foliares. CR Acad Sci Paris Ser D 283:1735–1737

Rabechault H, Ahee J, Guenin G (1970) Colonies cellulaires et formes embryos in vitro à partir de cultures d'embryons de palmier à huile (*Elaeis guineensis* Jacq. Var. *dura* Becc.) CR Acad Sci Paris 270:3067–3070

Rabechault H, Martin J-P, Cass (1972) Recherches sur la culture des tissus de palmier à huile (*Elaeis guineensis* Jacq.). Oleagineux 27:531–534

Renard JL (1976) Diseases in Africa and South America. In: Corley RHV, Hardon JJ, Wood BJ (eds) Oil palm research. Elsevier, Amsterdam, pp 447–466

Reynolds JF (1982) Vegetative propagation of palm trees. In: Bonga JM, Durzan DJ (eds) Tissue culture in forestry. Nijoff, The Hague, pp 182–207

Smith WK, Thomas JA (1973) The isolation and in vitro cultivation of cells of *Elaeis guineensis*. Oleagineux 28:123–127

Soh AC (1987) Abnormal oil palm clones. Possible causes and implications: further discussions. Planter 63:59–65

Staritsky G (1970) Tissue culture of the oil palm (*Elaeis guineensis* Jacq.) as a tool for its vegetative propagation. Euphytica 19:288–292

Sutter EG, Hutzell M (1984) Use of humidity tents and antitranspirants in the acclimatization of tissue cultured plants to the greenhouse. Sci Hortic 23:303–312

Thomas V, Rao PS (1985) In vitro propagation of oil palm (*Elaeis guineensis* Jacq. var. Tenera) through somatic embryogenesis in leaf-derived callus. Curr Sci 54:184–185

Timms RE (1986) Processing of palm oil. Fette. Seifen. Anstrichmittel. 8:294–300

Tisserat B (1982) Factors involved in the production of plantlets from date palm callus cultures. Euphytica 31:201–214

Turner PD (1968) Agronomic practices and disease control in oil palms. In: Turner PD (ed) Oil palm developments in Malaysia. Inc Soc Plant, Kuala Lumpur, pp 83–91

Turner PD (1976) Oil palm diseases in South-East Asia and the South Pacific. In: Corley RHV, Hardon JJ, Wood BJ (eds) Oil palm research. Elsevier, Amsterdam, pp 427–445

Turner PD, Bull RA (eds) (1967) Diseases and disorders of the oil palm in Malaysia. Inc Soc Plant, Kuala Lumpur

Turnham E, Northcote DH (1982) The use of acetyl-CoA carboxylase activity and changes in wall composition as measures of embryogenesis in tissue cultures of oil palm (*Elaeis guineensis*). Biochem J 208:323–332

Vouyouklis GV (1981) Use methode d'isolement de protoplastes du palmier à huile *Elaeis guineensis* Jacq. Phyton 40:169–178

Wardle K, Quinlan A, Simpkins I (1979) Abscisic acid and the regulation of water loss in plantlets of *Brassica oleracea* L. var. *botrytis* regenerated through apical meristem culture. Ann Bot 43:745–752

Wooi KC (1984) Palm tissue culture. In: George EF (ed) Micropropagation of selected rootcrops, palms, citrus and ornamental species. FAO plant production and protection paper 59, FAO, Rome, pp 88–112

Wooi KC, Wong CY, Corley RHV (1981) Tissue culture of palms – a review. In: Rao AN (ed) Tissue culture of economically important plants, COSTED and ANBS, Singapore, pp 138–144

Wooi KC, Wong CY, Corley RHV (1982) Genetic stability of oil palm callus cultures. In: Fujiwara A (ed) Plant tissue culture 1982. Maruzen, Tokyo, pp 749–750

IV.13 Olive (*Olea europaea* L.) as an Oilseed Crop

E. RUGINI[1] and E. FEDELI[2]

1 Introduction

1.1 Classification and Distribution

Olive (*Olea europaea* L.) belongs to the family Oleaceae, comprising about 30 genera with 600 species (Cronquist 1981) which are well distributed in every continent, in both cool and warm climates, and include a large number of garden and economic plants, trees, and shrubs.

The first review of the Oleaceae family was published in the 19th century by Knoblauch (1895) who, on the basis of cytogenetic studies, divided the family as follows: Oleoideae, tribes: Fraxineae (*Fraxinus* L.), Oleeae (*Syringa* L., *Ligustrum*, and *Olea* L.). Subfamily: Jasminoideae, tribes: Jasmineae, Fontanesieae, Forsythieae, Schrebereae, Myxopyreae.

Additional studies carried out by other authors at the beginning of the 20th century (Lingelsheim 1920; Ciferri 1942; Taylor 1945; Johnson 1957), and more recently by Mazzolani and Altamura Betti (1979, 1980) and Altamura Betti et al. (1985), aimed to identify form, cycles and heterotypic synonyms by means of comparative morphobiometric analysis of the genus *Olea* found in Africa, Minor and Western Asia, and Oceania.

Sax and Abbe (1932) found that six genera of two tribes belonging to the subfamily Oleoideae had the basic haploid chromosome number 23 ($2n = 2\times = 46$). Successively, Taylor (1945), after a survey of 14 genera, concluded that the basic chromosome number 23 occurs in all the tribes of Oleoideae, but is universal only in *Olea*.

The genus *Olea*, with its 30–35 species, is well distributed from Africa to Oceania, and belongs to the Oleeae tribe.

According to Ciferri (1942), the genus *Olea* is distributed mainly in three areas:

1. Afro-Mediterranean area (western Africa, Tanganyika, Canary and Azores Islands);
2. Indo-China Malaysia area;
3. Natalense-Malgica area (South Africa, Mauritius, and Mascarene Comore).

[1] Istituto di Ortofloroarboricoltura, Facoltá di Agraria, Universitá Degli Studi della Tuscia, Via Riello, 01100 Viterbo, Italy
[2] (SSOG) Stazione Sperimentale Oli e Grassi, Via G. Colombo 79, 20133 Milano, Italy

Biotechnology in Agriculture and Forestry, Vol. 10
Legumes and Oilseed Crops I (ed. by Y.P.S. Bajaj)

The species *europaea* L. comprising many varieties, belongs to this genus. In Italy, for example, about 500–600 varieties are well known, although many of them are probably only ecotypes. Furthermore, *Olea europaea* L. does not seem to be a species, but rather a cycle of forms having originated by mutation and hybridization (Chevalier 1948). Tropical and subtropical afro-asiatic species, such as *Olea chrysophylla* Lam. and *Olea excelsa* Ait. (Mazzolani and Altamura Betti 1976–1977), probably contributed to the evolution of the crop, as shown by fossil and paleontological studies.

According to De Candolle (1883), the cultivated olive is a native of Asia Minor and the Phoenicians spread it to the Mediterranean Basin. On the basis of archeological finds and historical sources, the olive has been found from the Orient to Spain, and etymologically in almost all languages, olive has a common root.

Wild *oleaster* and cultivated *sativa* are usually recognized among the Euro-Mediterranean olive varieties. *Oleaster* has shoots with thorns, often of quadrangular section; small, oval, almost round leaves on young plants and small, elliptical, black-reddish fruits with low oil content. *Sativa* (common olive) has shoots with cylindrical section, lanceolate leaves; big, oval, or elliptical fruits, purplish-blue, black-reddish or rarely white in color, with high oil content.

According to Chevalier (1948) and Ciferri (1950), wild *oleaster* was not necessarily wild in origin; it probably originated as a product of genetic segregation of the cultivated olive (*sativa*). These plants are called olivastre in French, liastro in Greek, and oleastro in Italian. This assumption, however, seems to be incorrect because in many areas along the shores of the Mediterranean sea, *oleaster* occupies niches not disturbed by the cultivated olive.

Knowledge of the olive tree goes back to 2000 B.C. in Egypt, 1400–1200 B.C. in Libya and Tunisia, 1000 B.C. in Syria and Palestine, 900–800 B.C. in Greece and 600–500 B.C. in Italy. Carbonized olive stones from 3700–3500 B.C. were found north of the Dead Sea, but obviously we do not know whether they belonged to the wild or cultivated olive.

According to Ciferri (1950), from the geographical point of view, the genus *Olea* might be divided into two sectors: an oriental one, having one ancestral form with small drupes, and a more developed occidental one with bigger drupes. Through the centuries, the stages of its path from the Orient to the Occident can be considered as secondary centers of evolution as man has cultivated and spread the best ecotypes.

1.2 Economic Importance

The olive is one of the oldest and most important crops in many Mediterranean countries. Its cultivation is mainly concentrated in this area, but the olive is present in many countries of the world. In China, for example, olive was imported in 1940 from Europe and at present more than 12 million olive plants are cultivated.

It is estimated that there are about 805 million olive trees in the world, distributed on 9.7 million ha (Guerbaa 1987), of which 9 million are located in the Mediterranean Basin with about 750 million plants. World distribution of olive production, both for oil and table olives is as follows: Europe 69.13%; Africa 15.60%; Asia 12.68%; South America 1.7%; North and Central America 0.86%; Oceania 0.03%. Ninety-seven percent of olive production is accounted for by countries in the Mediterranean Basin (FAO 1985), with a production of 8.7 million tons of olives, of which 0.7 million tons are table olives. The most important country for olive oil is Italy, with 606,000 tons, followed by Spain with 457,000 tons, and Greece with 285,500 tons, Tunisia with 119,500 tons, Turkey with 87,250 tons, Syria with 45,000 tons, Portugal with 34,750 tons, Morocco with 28,500 tons, and Libya with 23,000 tons (average from 1979 to 1985).

The annual world production of olive oil was 1,367,000 tons in the period from 1961 to 1965 and it reached an average of 1,749,000 tons in the period from 1979 to 1985 (FAO 1985). The most productive continent is Europe which, with 1,392,000 tons, supplies almost 80% of this product, followed by Africa with 191,000 tons, Asia with 150,000 tons, and South America with 13,500 tons.

Olive oil ranks sixth in the world production of fluid vegetable fats preceded by: soya 89,7 mt, cotton seed 29,7 mt, peanut 19,6 mt, sunflower 16,0 mt, rapeseed 15,0 mt, olive 1,7 mt.

Olive oil is, however, still first for its taste, which makes it very appetizing, and for both its nutritional quality and use in diet therapy. It is the only fluid oil which may be used in the crude state. Unfortunately, production costs are not yet competitive with seed oils. Each year, 608,325,000 dollars worth of olive oil is imported and 572,660,000 dollars is exported.

Earnings from olive cultivation are fairly low, mainly due to the use of old-fashioned cultivation techniques, to unsuitable areas of cultivation, and above all to the lack of renewal of old orchards with new ones, using highly productive cultivars, and finally to the higher costs of labor. However, many countries have begun renewal of olive cultivation. From 1972 to date Spain has established 100,000 ha of new orchards and restored 200,000 ha, making them more suitable for mechanical harvesting. Even Greece, Tunisia, Portugal, Algeria, Morocco, Syria, and finally Italy are following the Spanish example and are growing olive trees in more suitable areas with the aim to obtain: high and stable yield, at least 40–50 quintals of olive per hectare; total mechanization of harvesting methods; high qualitative standard of oil.

High quality oil can be achieved by a variety of technologies whose common objective is to obtain from *Olea europea* L. what is known as "virgin oil" for human consumption without refining being required. Whether this objective may be achieved or not is dependent on many contingencies, the most important of which is the quality of the fruit, followed by the technology used.

The quality of the fruit, in turn, is closely related to agricultural events such as the suitability of the soil, the age of the trees and the care devoted to them, the amount of precipitation during the proper season, aggression by pests, and pest control, to mention but a few. The majority of technological problems relate to the freshness of the fruit from which the oil is extracted and this, in turn, is dependent on the organization and quality of the industrial apparatus employed. In practice, "quality" oils may be obtained only when the fruit is processed within a few days of collection, and extraction has to be effected on undamaged fruit.

1.3 General Plant Biology and Morphology

Olive is a long-lived evergreen plant and adapts quite easily to many and varied environments. Its longevity is due to the presence of big hyperplasies or excrescences (sferoblasts or "ovuli") at the base of the trunk, which are rich in adventitious buds capable of giving rise to new branches (suckers) and new roots with a high degree of juvenility. The suckers are then able to replace the dead trunk if necessary. Furthermore, the trunk can be strengthened (usually not before 6–7 years) by means of the so-called ribbing (vein-like prominences) which goes from the base to the upper part of the trunk. Other vigorous shoots, arising from older branches, together with the basitonic characteristics of the species (prevailing development of basal shoots compared with the apical ones) (Scaramuzzi 1962) give the plants a shrub-like appearance if it is allowed to grow without any pruning or training (Morettini 1972). The branches are upright or pendulous according to the cultivar. The wood is hard and heavy. Leaves are opposite, with a phyllotaxis index of 1/2, and coriaceous; in vigorous shoots it is possible to observe nodes with three leaves due to the noncontemporary formation of the opposite leaves. Leaf shape is usually lanceolate, although notable polymorphism can be observed due to age, position, and vigor of shoots. The abaxial surface of the leaves is covered with well-known, characteristically large (130 μm) pluricellular peltate trichomes ($143/mm^2$). There are only $18/mm^2$ on the adaxial surface. Contrary to previous belief, they do not play any protective role in water vapor loss, as is shown by removing them (Bongi et al. 1987a). Stomata are present only on the abaxial surface ($470/mm^2$). Vegetative, flower, and mixed buds are located at the axil of the leaves. Flower buds are normally located on 1-year-old shoots. Flower induction occurs in the year of flowering, usually from January to March (Wigodsky and Philippis 1937). Morphological differentiation takes place from late February to March.

Flowers are gathered in inflorescences, with an average of 15 to 20 each. Flowers are usually hermaphroditic with 2 stamens and a unique pistil containing a super-bilocular ovary, with each locule containing two ovules. It is uncommon to find flowers having only stamens, while more frequently flowers are found with aborted ovaries. The frequence of ovary abortion is dependent on the cultivar and the season. Although olive trees have hermaphrodite flowers, they are partially or totally self-sterile, and therefore, in both cases, pollinators are required. Sterility is usually of a physiological nature, in that the pollen tube does not penetrate into the style.

Anthesis takes place at the end of May till early June. It is wind-pollinated. Fruit set is very low compared to other crop species, only 5 to 8% of flowers set. Of the four ovules, only one usually gives rise to a seed. It is often possible to find fruit without seed, but integuments are present, so it is possible to speak of apparent parthenocarpy (Morettini 1972), which in some cultivars may reach 50%. However, in some years, characterized by a particularly humid spring, vigorous cultivars can develop parthenocarpic fruit. In this case, fruits are round and small (3–4 mm in diameter), often gathered in bunches that frequently ripen.

The fruit is a drupe with a woody endocarp. At maturity, the embryo fills about 2/3 of the entire seed. The mesocarp occupies 70–80% of the entire fruit. Ripening takes place over a very limited period of time; for industrial apparatus the olive needs to be oversized with respect to the real average requirement, precautions have to be taken in gathering the fruit and sending it to the processing unit situated in the neighborhood of production.

Extraction technology is based mainly upon three possible techniques, namely pressure, centrifugation, and adhesion, which have virtually replaced other methods of extraction. The three types of procedure yield very similar results in terms of the quantities of oil obtained, but there are several differences:

- Centrifugal and adhesion systems are continuous and are comparatively versatile, whereas presses are discontinuous and require much more human action.
- More water is utilized in centrifugal systems and therefore a greater quantity of waste water (vegetation water) is discharged; the equilibria relating to the components of the least concentration and which are responsible for the organoleptic properties may be subject to change.
- In centrifugal systems, the solid residue has a higher moisture content and therefore a lower overall oil concentration. In practice, we observe that all methods leave a certain percentage of oil in the residue (roughly about 8%, which is normally recovered by solvent extraction (Fig. 1). A higher water content leads to higher drying costs.

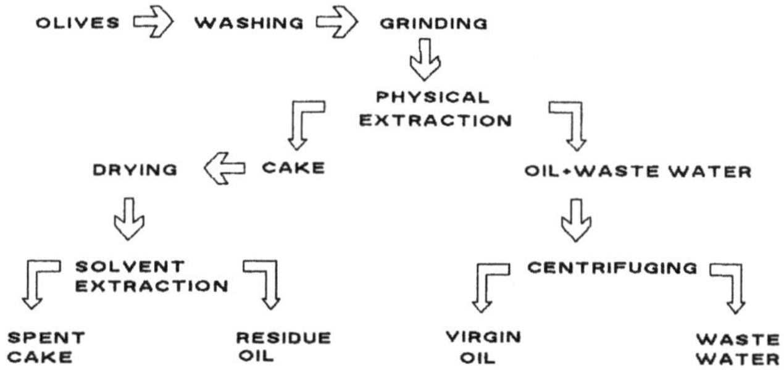

Fig. 1. Extraction of olives

As the result of the whole extraction activity, two oils, a dry, de-oiled residue and a waste liquid, are obtained. These are used primarily as fuel of approximate calorific value 4300 kcal/kg. The waste water poses a number of problems connected with pollution, for the solution of which specific technologies are being developed.

Residue oils are utilized as foods only after they have been refined; the features distinguishing them from virgin oil are described later. Oils extracted using any of the physical techniques mentioned are utilized directly as food when their chemical properties are suitable for human consumption (virgin oil), otherwise they are sent for refining. The properties considered are the free fatty acid content, UV absorption and the organoleptic properties. After refining and mixing with a quantity of virgin oil to restore partially the organoleptic properties, the refined oil is termed olive oil or simply Riviere-type oil or pure olive oil, according to the regulations in force in the country where it is to be sold.

1.4 General Plant Environment and Physiology

Olive is a typical Mediterranean crop and its specific characteristics are summarized below:

1. Winter chilling is not necessary for vegetative bud development, but is essential for flower bud induction (Hartmann 1953); Greenhouse experiments have shown that flowering can be induced by exposing the plants for at least 10 weeks to a constant temperature of 13 °C (about 800–1000 h under 8 °C) with little variation, according to cultivar. In addition, flowering can be inhibited by a constant higher or lower temperature. A temperature above 20 °C, for 2 to 3 weeks, may reverse floral induction if it occurs early (in the period before February), but not during or after February (Lavee 1985). However, chilling requirements are not enough to avoid alternate bearing, which is an inherent characteristic of the species; its causes and remedies have been extensively reviewed by Poli (1979), Milella (1983), and Lavee (1985). In addition, Stutte and Martin (1986) showed that low levels of carbohydrates do not appear to limit flower induction, as was believed earlier. A substance which probably seems to be involved in flower bud induction could be oleuropein, which inhibits flower bud induction in *Kalanchoe blossfeldiana* bioassay (Bongi 1986).

2. Olive cannot tolerate winter temperatures from −6 °C to −10 °C. In northern Mediterranean areas, metereological history shows that every 25–30 years, in some areas very low temperatures from −15 °C to −20 °C cause irremediable damage to the plants. In southern regions (southern Italy, Greece, Spain and the countries of North Africa) the olive is cultivated up to 800 m above sea level, but in northern regions not higher than 500–550 m.

3. In areas of olive cultivation, light intensity does not represent a serious limiting factor even though the branches located in the shade or in the inner parts of the canopy flower vary rarely, and those exposed to light are subjected to photoinhibition at some periods (Bongi and Long 1987). However, light reduction by shading with black net, reduces flower differentiation (Tombesi and Standardi, 1977).

4. Olive can tolerate high summer temperatures, very low humidity, and highly negative water potentials in the soil better than any other fruit tree, although its root system is quite superficial. Irrigation is generally not done, although, under semi-arid conditions with an annual rainfall of less than 450–500 mm, it is necessary to obtain a satisfactory yield.

Olive cultivars possess different drought tolerances, i.e., Nocellara is more tolerant than Coratina. This difference manifests itself in a 2–3 °C lower leaf temperature and LSC (leaf specific conductivity) in cv. Nocellara with respect to cv. Coratina, while leaf stomatal resistance was not found to be different. Hydraulic conductivity of stems was found to be ten times smaller than in poplar, and 600 times less than in grapes (Thompson et al. 1983; Salleo and Lo Gullo 1983).

Water consumption, the effect of water stress on leaf water potential, and diffusive resistance have been studied Natali et al. (1985). According to these authors, olive shows an intensive transpiration rate in low soil water level, considered critical for other species, resisting water deficit conditions not by reducing consumption through closed stomata, but by increasing absorption.

Due to the range of promptness and duration of osmotic adjustment in active elevation of solute concentrations, olive occupies an intermediate position with 0.6 MPa from 0.7 – 0.8 MPa for tropical grasses to 0.3 – 0.4 MPa for apple (Larcher et al. 1981).

Olive has a medium tolerance to salinity, contrasting with secretive halophytes; however, olive tolerance is paid for in terms of drop of leaves (Loreto and Bongi 1987); leaf Na of 0.38% and 0.85% Cl may be excessive (Wallace et al. 1979). Irrigation water with 8 g NaCl/l has been found to be the limit of tolerance for olive trees; salt from 2 to 8 g/l decreases the total lipid content of leaves and roots, while the percentage of oleic acid increases (Zarrouk and Cherif 1981). Increase of SAR causes a decrease in plant growth, but shows a positive interaction between salinity and phosphorous on cytokinin treatment (El-Deen et al. 1979). Salt stress induces more closed and less reactive stomata.

5. Olive prefers light soil (60 – 70% sand, 15 – 20% lime, and 15 – 20% clay) rich in calcium with a subalkaline pH. Olive fertilization has been extensively investigated under field and hydroponic culture conditions both in the USA (Hartmann et al. 1966) and the Mediterranean Basin (Klein and Lavee 1977; Bouat 1960; Gonzales et al. 1973; Crescimanno and Sottile 1976; Ferreira Llamas 1984, Therios and Sakellariadis 1982; Therios 1984).

1.5 Photosynthesis

According to Larcher et al. (1981), olive has a photosynthetic apparatus of medium to low capacity for CO_2 uptake. Regarding the degree of instantaneous and post-stress depression of CO_2 uptake at low water potentials and after recovery, it occupies an intermediate position between mesophytic conifers and broad-leaved trees and semi-desert woody plants.

Photosynthetic activity of olive at low temperature is lower than in many other Mediterranean plants, e.g., at 0 °C, net photosynthesis of olive is restricted to 10% of the optimum rates, in contrast to 25% for *Quercus ilex* (Larcher 1981), and the optimum temperature for olive varies among cultivars (Bongi et al. 1987a).

According to Bongi et al. (1987b), this low photosynthetic activity seems also to be due to the great resistance to CO_2 diffusion caused by the thick mesophyll walls and the very low surface area of mesophyll cells exposed to air, because the palisade cells are held tight by a special structure, "ladders" made of K-pectate (Carr et al. 1980). This structure produces resistance to CO_2 assimilation and an extreme tolerance to low humidity, but as water loss proceeds from stomatal chambers, it decreases water use efficiency (Bongi et al. 1987b). Stomatal reactivity under particular kind of stress (e.g. low humidity) can also affect photosynthetic behaviour, by decreasing intercellular CO_2 partial pressure (Loreto, personal communication).

Olive is a characteristic C_3 plant; it incorporates one quarter of the CO_2 into the glycolate pathway and only a little into C_4 acids (Bongi et al. 1987b).

1.6 Growth Regulator Treatments

Several growth regulators, found to be active in other species, have little or no effect on olive. Foliar spray BAP was found to modify the adverse effect of salinity by decreasing the hazard coefficient (Fadl and El-Deen 1980). To control plant growth in nurseries (Antognozzi and Preziosi 1986) and to control flowering (Porlingis and Voyiatzis 1986), PP333 (paclobutrazol) is promising, but not sufficiently researched.

Cycocel (Chlormequat) at 200 or 500 mg/l, Ethephon at 200 mg/l, and SADH at 2000 mg/l stimulated bud burst and flower bud formation in cv. Blanquetta. A remarkable increase in the percentage of perfect flowers with 100 mg/l kinetin or 200 mg/l Cycocel or Ethephon, according to cultivar, has also been reported (Hegazi and Stino 1982).

High fruit set has been obtained by means of putrescine and its precursor, L-arginine (Rugini et al. 1986), although they require high concentrations due to difficulty in uptake.

Ethylene promoters favored early fruit ripening only if applied on the plant before climacteric pick (Rugini et al. 1982), while on picked fruit, they inhibited anthocyanin accumulation and delayed softening of the pericarp (Shulman et al. 1979). Cytokinin at 50 ppm applied on the picked green fruit enhanced anthocyanin accumulation without affecting fruit maturation; NAA and 2,4,5 TP (200–300 mg/l) apparently induced ripening (Shulman and Lavee 1973). Morphactin (CME–72210) solution paint-banded with a brush on the olive scaffold could replace knife girdling in reducing biennial yield fluctuations (Ben-Tal and Lavee 1985).

1.7 Breeding and Crop Improvement

A well defined standard of cultivars does not exist, but there are "groups" that are a collection of similar clones (Roselli and Scaramuzzi 1974; Scaramuzzi and Roselli 1986).

At present, there are no satisfactory olive cultivars, and clonal selection has not been successful, except in a few cases (Morettini 1961; Berenguer 1978; Fontanazza 1987). However, clonal selection and germplasm preservation are essential before starting serious genetic improvement programs.

Attempts to classify olive clones have been made by using chemical markers (Cantarelli 1962); Pontikis et al. (1980) used a pollen isozyme system to distinguish Greek cultivars. Loukas and Krimbas (1983) calculated the genetic distance between all pairs of cultivars on the basis of protein identity. Olive cultivars have great variability in fruit size and oil content. At present it is still impossible to identify the hereditability of the most important traits in olive, because breeding programs started only a few years ago; the long juvenility phase, which usually exceeds 10 years, is discouraging for the initiation of research projects. Furthermore, olive is a highly heterozygous species, primarily due to its self-incompatibility (Morettini 1972) and to a lesser extent to inter-incompatibility (Griggs 1975) and male sterility (Baldini and Guccione 1952).

Chinese (FAO 1980; AAVV 1984) and Isreali genetists (Lavee et al. 1986) are particularly active in this field. The latter have already selected, by crosses, two cvs., one for table olives (Kadesh) and one for oil (Barnea).

1.8 Productivity

Many factors may determine low productivity, such as low fruit-set, self-incompatibility, high chilling requirement, fruit drop, ovary abortion, low shoot elongation, and also the tendency to differentiate flowers on apical buds of branches, which can predispose to the alternate bearing.

1.8.1. Self-Compatibility

Cultivars with this characteristic, which could be employed in breeding programs, are Frantoio, Biancolilla, Casaliva, Gentile di Chieti, Ogliarola Messinese, Taggiasca, Pizze Carroga, Rossellino Cerretano, Gremignolo, Procanica and Corsicano (Morettini 1972; Jacoboni and Fontanazza 1981), the Chinese selection, YT-14, and a few others.

1.8.2 Low Winter Chilling Requirement

Cultivars Rubra and Arpa (Hartmann and Porlingis 1957), which have this feature, could be used in breeding programs with the aim of extending the species to warmer areas.

1.8.3 Contemporaneous Fruit Ripening

It is advisable to select cultivars having contemporaneous fruit ripening, but during different periods, in order to delay the time of machine utilization (Fiorino and Natali 1985). Cultivars with early fruit ripening should be selected in order to avoid damage of early winter frost and to permit easier flower induction for the subsequent year.

1.8.4 Other Traits

Ovary abortion, the tendency of the apical bud of 1-year-old branches to differentiate inflorescence and the tendency of inflorescence apical flowers (King flower) to set, are also a characteristic of the cultivars. Zui-jun et al. (1984) observed that the number of flowers per inflorescence showed a wide range of variation in offspring.

1.9 Frost Resistance

An important program has been started by a group of Chinese researchers (Zui-jun et al. 1984), who have carried out 36 intervarietal crosses and six interspecific and intergeneric crosses by using excellent geographically distant cultivars (grown in far distant countries). Ying et al. (1984) found that freezing resistance of intervarietal hybrids was intermediate, higher or lower than the parentals one. Of seedlings from seeds collected in some cold Russian regions, 13.7% were resistant to frost, compared to 1.9% of seedlings from seeds collected in Albania. Hybrids from their selections (ZS−558×Berat and ZS−80×Frantoio) are of value for their frost resistance and fruit quality. Zhigarevitch (1958) ascertained that the least susceptible cultivars are Busowna zeituni Nikitsky and Tiflis. In France, Charlet (1965) reported that cvs. Tanche, Sigoise, and Aglandau are less susceptible to frost, as in Italy the cv. Leccio del Corno (Morettini 1972). Cultivars Laurina, Moresca, Ogliarola, Bosana, Leccino, and Dolce Agogia were found to be less susceptible to frost damage, as shown in Italy after the frost of 1985 (Antognozzi and Catalano 1985). Frost-resistant cultivars have higher sucrose and maltose content in their leaves during the winter time (Shamtsyan 1985).

1.10 Precocity

Early flowering of seedlings should considerably facilitate selection programs. In other fruit trees, this trait is positively correlated with a short juvenile period and negatively with leaf area, leaf index, stem diameter, and color of the lower leaf surface. In olive seedlings with brighter color in the abaxial part have a shorter juvenility period (Ying et al. 1984). Seedlings of the cv. Picholine are more precocious than many other cultivars (Scaramuzzi and Roselli 1986). Hereditability of precocity seems to be higher in offspring from maternal parents of precocious cultivars (Zui-jun et al. 1984).

1.11 Tree Habit and Vigor

At present, there are no reports on hereditability of these traits, but compact shape with upright shoots should be selected because this trait permits high density planting, better light interception, and easier mechanical harvesting. Determination of early dwarfishness is correlated positively with the number of branches and negatively with internode length (Ying et al. 1984). Furthermore, bearing habit, fruit shape, color, and drop appear to be linked (Zui-jun et al. 1984).

1.12 Oil Content and Quality

Apart from environment and culture techniques, these characteristics may be genetically controlled. A significant correlation was found between the fruit quality index (fruit weight×oil content) and the leaf index of 1-year-old seedlings. Furthermore, maternal inheritance in fruit quality was found to be more dominant (Ying et al. 1984).

1.13 Pest, Disease, and Water-Logging Resistance

The hereditability of pest and disease resistance has not yet been studied, but susceptibility or resistance to these traits is well known in some cvs. (see Sect. 1.18).

Seedlings resistant to water logging are characterized by a dark color of the outward-turning leaf margin, a thicker and stronger stem, and shorter internodes (Ying et al. 1984).

1.14 Root Ability

Many cultivars root quite well using semi-hardwood cuttings; however, several economically very important cultivars root at a very low percentage. The hereditability of this trait has not been studied yet. A list of cultivars, with their aptitude for rooting, very useful for all geneticists, has been made by Cimato and Fiorino (1981) and Caballero (1981).

1.15 Mutagenesis

Mutagenesis is useful to obtain compact, self-compatible cultivars (Donini and Roselli 1972). By irradiating self-rooted plantlets of a table olive cultivar, Ascolana Tenera, a new compact cultivar called Briscola has been obtained (Roselli and Donini 1983). Recently, by using the same technique has been obtained some mutants from the cvs. Frantoio and Leccino with large leaves or compact plants with smaller leaves have been obtained (Mannino personal communication).

1.16 Conventional Method of Propagation

Propagation by seeds is not advisable, owing to the extreme heterogeneity of the progeny and the extremely long juvenile phase of the offspring. At present, seedlings are used as rootstocks to be grafted with difficult-to-root cultivar scions, and as material for genetic improvement programs.

Seeds used for olive rootstocks usually come from known *Olea europaea* cultivars, and stratified at 2–4 °C before sowing. However, good, immediate seed germination can be obtained without stratification if the fruits are collected at the green stage, before the beginning of seed dormancy, and sown after collecting (Yuce 1979). According to Lagarda et al. (1983), seed maturity influences the storage life of olive seed. Immature seeds harvested in early September germinate well after harvest, but germination decreases sharply after 10–12 months of storage. The best germination temperature is 15 °C, while 35 °C decreases it. Recently, rapid seed germination was obtained by freeing the seeds from their endocarp and stratifying them for 30 days in moist vermiculite at 15 °C in darkness. One hundred percent germination was obtained when embryos reached their maximum dry weight, which was variable from late August to mid-October according to the cultivars (Coballero personal communication).

The easy-to-root cultivars are propagated by leafy subapical semi-hardwood cuttings (Hartmann and Loreti 1965; Cimato and Fiorino 1981). They are prepared with four to six leaves, and are usually immersed for a few hours in tap water and then treated at the base with an auxin (IBA at 2000 to 3000 mg/l) and then usually placed under mist conditions. It is advisable to add 160 mg/Kg putrescine to talc in order to promote an earlier and higher rooting percentage (Rugini et al. 1990).

It would be advisable to use a heating bench covered with transparent polyethylene. This method is simpler and more economic than mist (Fontanazza and Rugini 1978), and is now used by several Italian growers. A heating bench is also very useful to propagate cutting grafts made with easy-to-root cuttings grafted with difficult-to-root cultivars, joined together with parafilm tape. Both rooting and wound healing take place at the same time (Jacoboni and Fontanazza 1976).

Troncoso et al. (1984) studied a new method to root semi-hardwood cuttings. The system consists of vertically hanging black polyethylene tubes containing perlite and heating rods, irrigated by mist.

Olive cuttings have two periods of maximum rooting ability: one before restart of growth (March-April) and the second before the winter rest (September-October).

To increase rooting of cuttings, treatments to the mother plants can be undertaken, such as severe pruning, treatments with GA_3 (Fontanazza and Rugini 1977), boric acid (Briccoli-Bati 1986), or by shading mother plants. However, the last treatment only increases natural rooting ability (Pannelli and Filippucci 1982). Other substances, such as ACC (1-aminocyclopropane-1-carboxylic acid) (Bartolini et al. 1986), *Agrobacterium rhizogenes* (Diana 1986), or dipping in 5% sucrose or sorbitol for 24 h (Caballero and Nahlawi 1979), have also been found to increase rooting.

1.17 Rootstocks

At present, particular rootstocks are almost nonexistent for olive cultivars, so difficult-to-root cultivars are grafted onto 1-year-old seedlings from cultivar seeds. Disformity of plant growth, induced by seedling population, observed by many authors (Hartmann 1958; Crescimanno and Sottile 1971; Guerriero et al. 1974; Guerriero and Scaramuzzi 1979), could be reduced by grafting onto apparent uniform seedlings, as near as possible to the roots, and planting plantlets in the field below the soil surface in order to stimulate root emission from the scion. *Olea cuspidata* seedlings are used in China as rootstocks because they seem to control vigor and yield (Scaramuzzi and Roselli 1986). The cultivars Verdial and Lechin are used in Spain as clonal rootstocks with the cv. Gordal (Jacoboni and Fontanazza 1981). *Olea oblonga* is used as clonal rootstocks for olive cultivars susceptible to *Verticillium dahliae* (Hartmann and Whisler 1975). Further attempts to graft olive onto other species or genera of the family Oleaceae have been unsuccessful.

1.18 Major Disease and Pests

Olive is attacked by numerous parasites: fungi, bacteria, viruses, nematodes, and insects (Circulli et al. 1981; Inserra et al. 1976). The most harmful are reported below.

1.18.1 Fungi

Cycloconium oleaginum Cast. (*Spilocaea oleaginea* Hugh.) causes leaf spots and is particularly harmful in areas having a highly humid and hot climate (Ciccarone 1964). The most susceptible cultivars are: Arbequina, Frantoio, Manzanilla, Picholine Marrocaine, while the least susceptible are Ascolana, Lechin, Nevadillo Blanco, Sevillano and Zorzaleno, Leccino, Cellina di Nardo, Ottobratica, Zaituna, Nocellara etnea, Nocellara del Belice, Mignolo, Leccio de Corno, Piangente, Koronaiki and Farga (Morettini 1972). Resistance seems to be correlated to a thin leaf palisade layer (Morettini 1954).

Verticillium dahliae Kleb. (*Verticillium* wilt) represents a serious problem, particularly in olive orchards, intercropped with susceptible herbaceous or woody crops. The cultivar Oblonga is resistant to this fungus (Hartmann et al. 1971), as are Arbequina and Frantoio (Wilhelm and Taylor 1965), followed by the cvs. Coratina and Frangivento, whereas the cvs. Santagostino, Ascolana, Leccino, and Cellina are very susceptible (Circulli and Montemurro 1975). *Gloesporium olivarum* Alm. (leprosy) is spread mainly in humid and hod areas and affects fruits, shoots, and leaves.

Other fungi causing less damage are: *Sphaeropsis dalmatica* (Thum.) Gigante; *Phoma incompta* Sacc. et Mart., which causes leaf drop; *Cercospora cladosporoides* Sacc., which affects both leaves and drupes; *Oidiopsis taurica* (Lev.) Arn. and *Stictis panizzei* De Not.

Trunks of old plants are susceptible to heart rot, known as decay. Many fungi are associated with this wood deterioration (Circulli et al. 1981).

1.18.2 Bacteria

Pseudomonas syringe pv. savastanoi (Smith) Young et al., also known as *Bacterium savastanoi* and *Phytomonas savastanoi* (Savastano 1886), (olive-knot) is the most harmful bacterium, which induces hyperplasia. The outgrowths are caused by indolacetic acid and cytokinins, produced by the bacterium itself (Hutzinger and Kosuge 1968). Susceptibility of various cultivars expressed according to size rather than number of knots, at the point of artifical inoculation, declined in the order Nocellara Belice, S.Agostino, Leccino, Frantoio, Coratina (Varvaro and Surico 1978). However, olive growers know very well the high susceptibility of cv. Frantoio, particularly in cold and humid areas, and the resistance of cv. Ascolana tenera, probably due to its higher frost resistance. It has been isolated from symptomless olive leaves (Ercolani 1971) on which it is able to survive and to establish resident populations (Varvaro and Ferrulli 1983).

Several strains with different molecular weight of IAA plasmid have been isolated. The capacity of producing high IAA is directly linked to its pathogen virulence and survival (Surico 1986).

Research on the possibility of finding a bacterium able to contrast virulent strains are currently underway (Varvaro pers. commun.).

1.18.3 Virus

Many olive diseases are associated with viruses whose etiology is not yet clear (Martelli 1981). Nevertheless, in many apparently healthy cultivars, viral particles were found in pollen (Pacini and Cresti 1977). Also the following viruses were found in apparently healthy plants: strawberry latent ring spot virus (SLRV) and Arabis mosaic virus (AMV) (Savino et al. 1979), cherry leafroll virus (CLRV) (Savino and Gallitelli 1981), cucumber mosaic virus (CMV) (Savino and Gallitelli 1983), olive latent ring spot virus (OLRV) (Savino et al. 1983), olive latent virus-1 (OLV-1) (Gallitelli and Savino 1985) and olive latent virus-2 (OLV-2) (Savino et al. 1984).

At present, OLRV, OLV-1, and OLV-2 have not been found in hosts other than olive. Furthermore, Marte et al. (1985), in their survey on the table olive cv. Ascolana tenera, observed an evident narrowing of the leafblade, shortening of the internodes, and fruit and endocarp malformation in the presence of a nepovirus (SRLV). These symptoms could be reproduced, on healthy Ascolana scions, grafted on the branches of diseased plants. The scions of several other oil cultivars did not show these symptoms even after two to three growing seasons. Lavee and Tanne (1984) observed an anomaly in cv. Manzanillo that they called it Spherosis and considered a virus infection.

1.18.4 Nematodes

Several species of nematodes, endo-, semi-endo and ecto-parasites alter the root tissue, cause loss of sap, and in some cases are virus-carriers. The most common are the endo-parasites: *Meloidogyne incognita* (Kofoid et White) Chitwood; *Meloidogyne javanica* (Treub.) Chitwood, *Rotylenchulus macrodoratus* Dasgupta. Raski et Sher; *Pratylenchus vandenbrandei* De Grisse; *Paratylencus peraticus* (Raski) Siddiqi et Goodey; *Paratylenchus vulnus* Allen et Jensen; semi-endo-parasites: *Tylenchulus semipenetrans* Cobb; *Helicotylenchus oleae* Inserra, Vovlas et Golden; *Helicotylenchus erythrinae* (Zimmermann) Golden; ectoparasites: *Gracilacus peraticus* Raski.

Nematode-resistant plants could be the best alternative, but at present there is little information regarding resistance and susceptibility of the cultivars (Lamberti and Boines 1969). Inserra et al. (1981) tested seven olive cvs. under glasshouse conditions with *Pratylenchus vulnus*, *Rotylenchulus macrodoratus*, and *Tylenchulus semipenetrans*, but all proved to be susceptible. In contrast, populations of *Meloidogyne incognita* on the roots remained low in all cases, particularly on cv. Moraiolo.

1.18.5 Honey Dew

An alteration in plant metabolism resulting in an increase in sugar substances, like mannose, in the cell juice can occur outside, and cover the leaves, especially in the lower part of the canopy. These substances favor the development of sooty mold, that is black colonies of fungi such as *Capnodium eleophilum* (Mont.) Prill., *Limacinula oleae* (Arn.) Sacc. et Trotter. Sooty mold, however, seems to be of physiological nature (Petri 1927). It is also frequently associated with scale attacks.

1.18.6 Insects

In order of decreasing importance the most harmful olive insects are: *Dacus oleae*, *Saissetia oleae*, and *Prays oleae*. *Dacus oleae* (Gmel.) (Olive fly) is the most harmful olive insect. It overwinters as pupa and lays eggs on each drupe; after 2–4 days, the larvae feed on olive pulp by digging a tunnel. It produces up to seven generations even during the autumn-winter in warm countries such as Greece and North Africa, but not more than two generations in colder areas. Biological control has not been successful.

Saissetia oleae (Oliv.) (olive scale) is a very harmful olive insect, especially due to the disappearance of its parasites after heavy or excessive use of pesticides. It overwinters as a larvea at 2nd and 3rd stage and has one or two generations. This insect sucks the sap of fruits and defecates sugar substances on which fungi such as sooty mold develop.

Prays oleae Bern. (olive moth). This small butterfly overwinters as larva; in spring the adult lays eggs on the flowers and the larva ties the flowers with a silky string. The second generations lays eggs on the small fruits and the larvae feed on the internal part of the fruit endocarp, causing the fruit to drop. Adults of the third generation lay eggs on leaves, on which the larvae make the characteristic galleries. It may be also controlled by many predators and parasites.

Other insects which occasionally damage the olive are: among diphtera (Circulli et al. 1981); *Prolasioptera berlesiana* Paoli, *Thomasiniana oleisuga* Targ.; among lepidoptera: *Zelleria oleastrella* Mill., *Oecophyllembius latifoliellus* Mill., *Zeuzera pyrina* L., *Palpita unionalis* Hb., *Erannis bojaria* Schiff; among coleoptera: *Phloeotribus scarabaeoides* Bern., *Leperisinus taranio* Bern., *Leperisinus fraxini* Panz., *Otiorrhynchus cribricollis* Gyll., *Coenorrhinus cribripennis* Desbr., and among scale: *Pollinia pollini* Costa, *Philippia oleae* Targ., *Parlatoria oleae* Colv..

Other harmful insects are psyllide *Euphyllura olivina* Costa, rhynchota *Calocoris trivialis* Costa and thysanoptera *Liothrips oleae* Costa.

1.18.7 Mites

Among mites, which up to now have not required any artificial control, are: *Aceria olea* Nal, *Oxycenus maxwelli* Keif., and *Ditrymacus athiasellus* Keif.

2 Chemical Composition of Olive Oil

2.1 Virgin Olive Oil (Voo)

2.1.1 Glycerides and Fatty Acids Composition

Glycerides and fatty acids account for over 98% of the components of virgin oils. Virgin olive oil is classified by law into several categories, which may differ from country to country according to the amount of free fatty acids (FFA, expressed as oleic acids) it contains. In Italy, oils having a FFA content less than 1% are termed "extravirgin", whereas those of FFA content between 1.01 and 1.5% are termed "soprafino vergine"; two further categories of oil, "fino vergine" and "vergine" may be marketed without refining, but they are present primarily as bulk products for industry rather than for consumption on the retail market (AAVV 1982).

Fatty acids classified as olive oil glycerides are described in Table 1; they form the triglycerides in accordance with the random-2-random principle. Table 1 reflects the average FA composition of oils produced in Italy. Mono- and diglycerides are also present (Mariani and Fedeli 1985), in addition to triglycerides, and originate partially from lypase action before extraction and partially from the incomplete biosynthesis of triglycerides.

2.1.2 Minor Components

When a VOO is saponified, a neutral part of maximum value 1.5% is obtained. The mixture contains several components which may be grouped into classes in accordance with their chemical properties.

The description of the classes, according to polarity, commencing with the least polar, is as follows: hydrocarbons, tocopherols, aliphatic alcohols, cyclic alcohols.

Table 1. Olive oil fatty acids (Italian production)

Acids	Min %	Max %
C12+C14	0.0	0.1
C16:0	7.0	17.0
C16:1	0.3	3.0
C17:0	0.0	0.5
C17:1	0.0	0.5
C18:0	1.5	4.0
C18:1	63.0	83.0
C18:2	–	13.5
C18:3	0.0	1.5
C20:0	0.0	0.7
C20:1	0.0	0.5
C22:0	0.0	0.3

The latter class may be subdivided into subclasses as follows: pentacyclic triterpenic alcohols, tetracyclic triterpenic alcohols, 4-α-methylsterols, sterols, dihydroxy pentacyclic triterpenic alcohols.

Some of the unsaponifiable components occur in VOO as esters of the fatty acids present in triglycerides but taken in different proportions (Fedeli 1968).

Hydrocarbons. The fraction contains even and odd n-paraffins from C 11 to C 30; small quantities of branched chain iso and anti-iso hydrocarbons are also present. Some aromatic hydrocarbons have also been found among the volatile substances. Squalene is the principal component of the hydrocarbon fraction, accounting for not less than 60% of the total hydrocarbon fraction, which is 2500 ppm. β-carotene (4 – 0.3 mg/kg is also present.

Aliphatic Alcohols. VOO does not contain appreciable amounts of free long-chain aliphatic alcohols (LCAA) which, instead, are esterified with fatty acids (ELCAA); esters of methanol and ethanol are also present (Jacini et al. 1967). Phytol, probably resulting from the degradation of chlorophyll, has also been encountered in VOO. Maximum LCAA concentration in VOO is approximately 200 mg/l.

The oils extracted by solvent from the residue contain a higher concentration of ELCAA (500 – 1500 mg/kg compared to 100 – 500 mg/kg) (Fedeli and Jacini 1967).

Pentacyclic Triterpenic Alcohols. Part of the substances are present as esters of fatty acids; two different sets of structures have been found, namely α-and β-amyrin on one side and (five six-membered rings), cycloartenol, 24-methylen-cycloartanol on the other side (one cyclopropane, one five-membered and three six-membered rings). Additionally, trace quantities of lanosterol have been found (Fedeli et al. 1966; Fedeli 1968; Itoh et al. 1973 a; Fedeli and Mariani 1973, 1974).

4-α Methysterols. Traces of demethylated tetracyclic triterpenic alcohols, namely obtusifoliol, have been identified in unsaponifiable VOO with: 4-α-methyl-24-methylene-δ-cholestene-3β-ol, 4-α-methyl-24-methyl-δ-cholestene-3β-ol, 4-α-methyl-24-ethylidene-δ-cholestene-3β-ol, 4α-methyl-24-ethyl-δ-cholestene-3β-ol.

At the time of writing, no one has shown whether they occur as free or as esterified compounds (Mariani and Fedeli 1989), and probably they are intermediates in the biosynthesis of sterols from squalene as well as lanosterol (Fedeli et al. 1974; Fedeli and Mariani 1974).

Sterols. Free sterols and esterified sterols are among the most important constituents of the cyclic alcohol fraction; their importance has been recognized and the determination of these substances froms the basis for a method of checking the authenticity of the VOO. The sterols mixture, as shown in the unsaponifiable oil, is dominated by sitosterol with campesterol and stigmasterol.

Several other components are present in addition to the components mentioned, but they are not resolved from them by employment of the conventional GLC method.

The capillary column shows a more complex situation which is as follows: 24-ethyl-δ-cholestene-3β-ol; 24-methyl-δ-cholestene-3β-ol; 24-ethyl-δ-cholestene-3β-ol.

As the sequence of the compound shown infers, it is probably that they are bound by a biogenetic relationship, as we have stated previously for squalene. It is likely that polyfunctional cyclic pentacyclic triterpenes described previously are connected with the same sequence (Fedeli et al. 1966; Itoh et al. 1973b; Cortesi et al. 1973, 1977, 1978; Morchio et al. 1987).

Dihydroxy Triterpenes. This class is represented in VOO by two substances, namely: 3β, 17β-dihydroxy-δ-oleanane (erythrodiol), and 3β, 17β-dihydroxy-δ-ursene (uvaol); the concentrations of these substances, calculated relatively to the sterol mixture, are lower than 5% in VOO, but may be as high as 30% in solvent-extracted oils. Higher concentrations of the sum erythrodiol+uvaol indicate mixture with solvent-extracted oils (Fedeli and Mariani 1973; Fedeli 1983).

Hydroxy Triterpenic Acids. It has been demonstrated that the following pentaterpenic acids are present in VOO: 3β-hydroxy-17carboxy, δ-oleanolene (oleanolic); 3β-2α-dihydroxy-17carboxy, δ-oleanolene (maslinic); 3β-hydroxy-dihydroxy-17carboxy-ursene (ursolic); and 3β-2α-dihydroxy-17carboxy-ursene (deoxyursolic).

It is likely that they are constituents of a water-repellant mixture present on the surface of the olives (Jacini and Fedeli 1972).

Tocopherols. The tocopherol content of olive oil is not very high compared to that of other vegetable oils (150–300 mg/kg mostly as α-tocopherol), although it is very well stabilized against auto-oxidation by other phenolic compounds, as described later, and by its fatty acid composition, in which oleic acid is predominant (Taufel and Sevzisko 1961).

2.1.3 Other Minor Components

Other minor components are present besides those described and obtained through research performed on the unsaponifiable mixture from VOO, but they cannot be encountered in the unsaponifiables, on account of their chemical properties. Some of these substances impart the specific characteristics of olive oil. For instance, volatile compounds determine the organoleptic properties because they impart characteristic and specific flavors which may vary from crop to crop; phenolic compounds have specific action on the taste and the keeping properties of olive oil.

Flavoring Components. Hydrocarbons, alcohols, aldehydes, ketones, ethers, furanes, tiophenes, and esters constitute the majority of the volatile components of olive oil; the chemical structure of some has been demonstrated by several authors (Table 2), mostly after isolation and GLC-MS analysis. As we may observe from the description, the mixture is very complex. The complexity is such as to

Table 2. Olive oil flavoring components (selected)

Hydrocarbons		
Hydroxy compounds	1-Pentenol	Nonane-1-ol
	3-Metylbutane-1-ol	2-Phenylethanol
	cis-3-hexene-1-ol	1,8-Cineole
	Hexane-1-ol	Linalol
	Heptane-1-ol	Lavandulol
	Octane-1-ol	
Aldehydes	3-Methylbutanal	trans-Octenal
	2-Methylbutanal	n-Nonanal
	n-Heptanal	trans-2-Nonenal
	Benzaldehyde	2,4-Nonadienal
	n-Octanal	trans-2-Decenal
	2,4-Heptadienal	trans-2-Undecenal
Ketones	3-Methyl-butan-2-one	2-Methylheptenone
	Pentan-3-one	Octan-2-one
	Hexan-2-one	Nonan-2-one
Esters	Ethylproprionate	Methyloctanoate
	Methylbutyrrate	Ethylbenzoate
	Methyl-3-methylbutyrrate	Methylsalycilate
	Ethylbutyrrate	Ethylphenylacetate
	Propylproprionate	Ethylnonanoate
	Methylhexanoate	Ethyldecanoate
	Hexenylacetate	Hexenylacetate
	Methylheptanoate	

Table 3. Some phenolyc compounds present in olive oil

Hydroxy compounds	Phenylethanol
	p-Hydroxyphenylethanol
	3,4-Dihydroxyphenylethanol
	3-Hydroxy-4-methoxyphenylethanol
	3,4-Dihydroxy-5-methoxyphenylethanol
Carboxylic compounds	3,5-Dihydroxycynnamic
	3-Methoxy-4-hydroxycynnamic
	p-Hydroxycynnamic
	3,5-Dihydroxybenzoic
	3,4,5-Trihydroxybenzoic
	p-Hydroxybenzoic
	3,4,5-Trimethoxybenzoic

forbid the assessment of quality by the evaluation of specific components likely to impart specific properties (Fedeli and Jacini 1970; Fedeli et al. 1973a, b; Flat et al. 1973).

Phenolic Components. In the past, several authors have postulated the existence of phenols in VOO and finally Ragazzi and Veronesi isolated β (4-hydroxy phenyl) ethanol and 3,4 dihydroxy ethanol. Fedeli and Cortesi showed the presence of the compounds listed in Table 3 and were able to demonstrate the structures of some complex products. The products of this complex are probably the precursors, by spontaneous hydrolysis, of several of the simple components of Table 3.

The structures discovered thus far constitute only a part of the whole mixture, which contains many individual structures as shown by HPLC analysis.

The complex phenolic components have an influence on some specific taste properties of VOO, such as bitterness and pepper-like action, shown by some VOO (Fedeli 1983; Cortesi and Fedeli 1983).

Chlorophylls. Both chlorophyll a and b, which are readily degraded to pheophytins, are among VOO components; the changes in product color are mostly due to this transformation. Chlorophyll (a+b) concentration ranges from 1 to 10 mg/kg and phenophytin concentration from 0.2 to 24 mg/kg (Fedeli and Brillo 1975a, b; Vitagliano 1960).

Phospholipids. VOO is very poor in phosphorilated glycerol derivatives and most of the phosphorons found are derived from the seed contained in the pith (Vitagliano and Turri 1958; Vitagliano 1960; Vioque and Maza 1961).

2.2 Solvent-Extracted Oil

Mechanical extraction from olives leaves a residue, the oil content of which is still high (6–8% on the dry residue); it is therefore submitted to solvent extraction.

Hexane is almost exclusively used as the extraction solvent, although trichlorethylene is still utilized in some extraction facilities; carbon disulfide, which was used universally at one time, is no longer used.

The oil obtained is not to be confused with VOO or even with the refined oil derived from VOO. Effectively, it has some specific properties because the solvent extracts other components which are not present, or are present in lesser amounts in VOO. For instance, the amount of unsaponifiable matter is more than double (max. 3%) the amount encountered in VOO, the minimum triterpenic alcohol content is at least twice that of VOO.

It follows that residue oil, even when it has been refined, has analytical properties which are different from those of VOO or OO (olive oil = refined olive oil).

The oil extracted from residue with the aid of solvents is termed sansa residue oil (50) and constitutes a product which is commercially different from any other produced from olives.

This assumption is mandatory in the majority of countries, where olive oil is a traditional product and solvent-extracted olive oil is termed sansa oil, even when it is obtained from fresh olives.

2.3 Products and By-Products of Olive Extraction

The main product from olive oil is VOO, the chemical and organoleptic properties of wich are such as to render it apt for direct consumption; the main characteristics of edible VOO are listed in Table 4 and refer to Italian and EEC Regulations.

Table 4. Chemical and physical characteristics of olive oil

	VOO	OO	SOO
Fundamental			
Refractive index	1.4665 – 1.4679	1.4665 – 1.4679	1.4672 – 1.4686
Iodine value	80 – 88	80 – 88	79 – 88
Unsaponifiable	1.5	1.5	3.0
FA composition	see Table 1	see Table 1	see Table 1
C16:0 (position 2)	Max = 1.8	Max = 1.8	Max = 2.2
Bellier index	Max = 16	Max = 16	Max = 16
Sterol composition %			
Cholesterol	Max = 0.5	Max = 0.5	Max = 0.5
Campesterol	Max = 4	Max = 4	Max = 4
Stigmasterol	Less than campesterol	Less than campesterol	Less than campesterol
Sitosterol[a]	Min = 93	Min = 93	Min = 93
d-7-Stigmasterol	Max = 0.5	Max = 0.5	Max = 0.5
Erithrodiol	Max = 5	Max = 5	–
BCB test	Negative	Negative	Negative
Vizern test	Negative	Negative	Negative
I-P test	Negative	Negative	Negative
Quality			
Organoletic	Good	Good	Good
Acid value	Max = 1	Max = 2	Max = 3
Peroxide value	Max = 20	Max = 20	Max = 20
UV absorption			
K.268 nm	Max = 0.20	–	–
DK	Max = 0.010	–	–
Impurities	Max = 0.5	Max = 0.5	Max = 0.5

OO = olive oil, Riviera Type.
SOO = residue oil + virgin.
[a]Sum of all the sterols having the same retention time as sitosterol.

The requirements of other countries may differ from those specified in Table 4, although the values contained therein also constitute the basis for COI (Conseil Oleicole Internationale) specifications which have been adopted by the majority of nations producing olive oil.

Two further categories of VOO are contemplated by Italian regulations, in addition to the two types discussed thus far; they have not been discussed here, since they are used primarily for trading in bulk, as already mentioned. Any VOO having characteristics different from those contemplated is therefore not utilized for direct consumption and is conveyed to a refining plant. It forms the base for a class of olive oil, the label of which may not contain the term "virgin" (see Table 4). SO also has to be sent for refining so that its origin (solvent-extraction) is indicated on the label. The solid residue (exhausted residue or exhausted sansa = ES) consists of olive pulp, skin, and husk, it has no value as cattle feed unless it is fractionated to separate the husk. In this case, the protein content rises to 20–25% ES applications include use as a filler, as fuel, and as a starting compound for the preparation of furfurol (the latter on account of its pantosane content).

The VW's are composed mainly of sugars, proteins, organic acids, and phenols; as pollutants, they set a difficult problem on account of the widespread occurrence of extraction units.

2.4 Refining

The VOO that has chemical or organoleptic properties rendering it unsuitable for consumption in the natural state, and the 50 are submitting to refining. Two main processes should be considered namely the classical alkali refining methods, and neutralization by distillation. The second of these procedures is more suitable for VOO than for S0 refining.

2.4.1 Alkaline Refining (Fig. 2)

Before the oil is submitted to the alkaline treatment, it is usually passed through a degumming process; a Lewis acid (phosphoric, sulfuric or oxalic acid) is added to warm oil at 60–80 °C, followed by the addition of warm water which is poured in after a few minutes, while stirring. The phases are separated by centrifuging the water and gum mixture. An almost stoichiometrical amount (5% excess) of sodium hydroxide at a concentration variable from 5–20% is then added to the oil, and the two phases are subsequently separated centrifugally.

The degumming and neutralization operations may be executed in batches but, more usually, the two steps are accomplished in a continuous unit, while separations are effected by centrifuging.

Bleaching is effected by drying the oil in vacuo, adding the proper amount of earth (sometimes with carbon black) and filtration; whilst bleaching is continuous process, filtration is a semi-continuous operation, accomplished with the aid of a battery of closed filter presses.

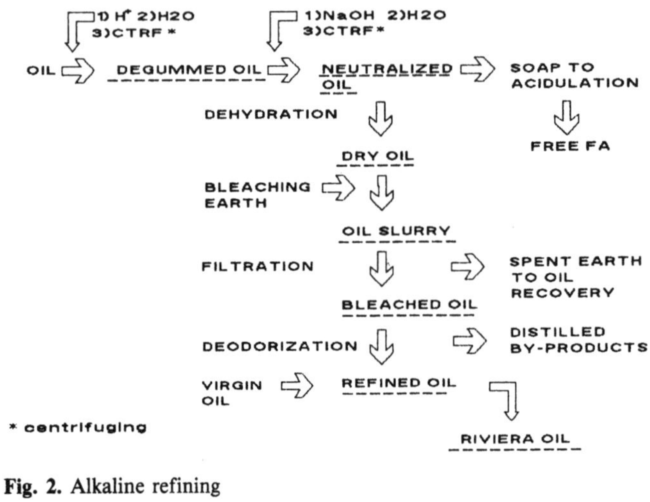

Fig. 2. Alkaline refining

Such a configuration enables the operator to recover, with the aid of solvents, the oil entrained in the bleaching earth; after evaporation, the mixture is returned to the raw oil tank. The bleached oil is deodorized in a variety of ways, employing batch and semicontinuous apparatus in vacuo (residual pressure 2–5 mm Hg); the deodorized oil is cooled and is ready for canning after mixing with 5–20% good quality VO.

When the refined oil is obtained from S0, a dewaxing process is usually required, achieved by crystallization in hexane, in continuous systems.

Yields are highly dependent upon the initial free fatty acid content (FFA) and refining ratios ranging from 1.25 to 1.60 are claimed when the FFA content is in the range 3×8% (expressed as oleic acid).

It is not unusual (especially in the case of S0) for raw materials, having FAA content exceeding 8–10% to be refined; in this instance, a technology based upon a two-solvent system may be used whilst still employing sodium hydroxide as a neutralizing agent (Bernardini 1973).

The solvents, one of which is lipidic (usually hexane) and the other hydrophilic (isopropanol or acetone), are mixed with the oil and sodium hydroxide respectively, and then conveyed to a settling tank from which three streams originate. The oil-hexane stream is washed with hydrophilic solvent and pumped to recovery; the alkaline stream is washed with hexane and pumped to FFA recovery. The intermediate layer (third stream) is returned to the mixing stage.

The recovery claimed for the system described is very high (refining ratios 1.1 : 1.3) although the purchasing and maintenance costs and operating expenses (solvents losses, heat dissipation, ect.) are relatively high.

2.4.2 *Physical Refining (Fig. 3)*

Physical refining starts with the only chemical step in the entire process, degumming, which is usually accomplished using strong acids (sulfuric acid, phosphoric acid), since thorough cleansing of the oil to remove impurities is required (Bernardini 1973).

The degummed oil is submitted to bleaching similarly to the process described previously and subsequently to FFA distillation to achieve deodorization of the oil. Continuous and semi-continuous apparatuses are utilized for the operation,

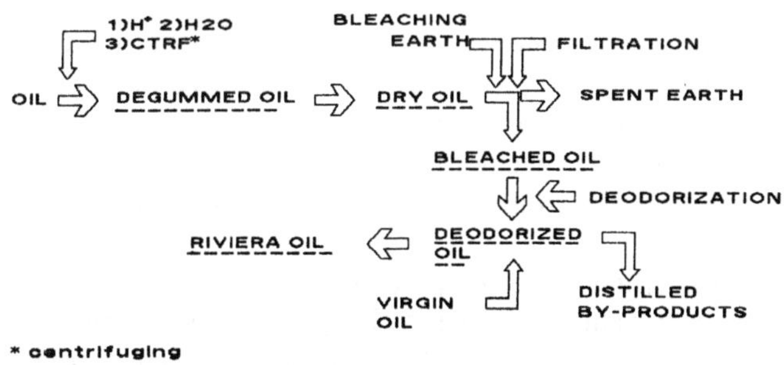

Fig. 3. Physical refining

Table 5. Italian production of olive oil (t) (SSOG)

Year	Total	Physical extrac.	Residue oil
1980	698400	686300	12100
1981	621700	606500	15200
1982	440700	430000	10700
1983	844700	634000	210700
1984	399800	343200	56600

which is effected at temperatures in the range 220–240 °C and under residual pressures (1–3 mm Hg). Physical refining is applicable more to VOO than to SO, even though some SO's may be worked when they have been submitted to suitable proparatory treatment. After refining, SO has to be submitted to winterizing to remove the greater part of the wax content.

2.4.3 Consumption

Production during the years 1980–1984 is shown in Table 5 with reference to the Italian situation.

3 In Vitro Approaches

3.1 General Account and Significance of in Vitro Techniques

A large number of woody fruit trees, rootstocks and varieties have been propagated in vitro over the last few years. At present, however, only a few among the self-rooted tissue cultures of trees have been commercially used, except for some rootstocks. Most of them are under field investigation, but in the future the situation could change (Zimmermann 1986). The common cultivars of most species usually require rootstocks to give them particular characteristics. At present, olive does not belong to this category, although there are some indications of beneficial effects of rootstocks, not yet available, therefore it could be self-rooted in vitro for mass propagation.

The aim of the in vitro technique is, however, only marginally concerned with rapid propagation. Genetic improvement, production of disease-free plants and cryopreservation of valuable germplasm, are the main objectives of using this technique. The development of in vitro techniques could greatly help the traditional methods of cross-breeding, which up to now have not given satisfactory results.

Our work is particularly concerned with short, medium and long-range objectives. The aim of the first two is to achieve rapid and cheap standard propagation of cultivars, new selections, and rootstocks difficult to propagate through conventional methods; furthermore, to obtain new cultivars through induced mutation

Table 6. Summary of in vitro studies on olive

Cultivar	Explant source (age in years)	Response	Media				Light condition	Reference
			Salts		Hormones and other additives (mg/l or as indicated)	Sugar (g/l)		
			Macro	Micro				
Manzanillo	Shoots (1)	Callus	(21-MW)		IAA(2) + kin(0.2)	S(20)	L	Lavee and Messer (1969)
Manzanillo	Shoots (1)	Callus	(21-MW)		kin(0.2) + IAA(2) + ABA(10^{-6}M)		L	Lavee and Adiri (1974)
Ogliarola di Monopoli	Internodes (1, 2) and	Root and callus	Kp	H	NAA(1)	S(50)	D	Scaramuzzi and de Gaetano (1974)
	Sections (3, 6, 12)	Callus	Kp or N	H				
Manzanillo	Young mesocarp (0.4)	Callus	(21-MW)		IAA(2) + kin(2×10^{-6}M) or Zr(2×10^{-6}M)	S(20)	L	Lavee (1977)
Picholine	Naked embryos	Embryo germination	H	H	–	Lc(20) or S(30)	D	Instanbouli and Neville (1977c)
Frantoio	Internodal (1)	Good rooting and callus	1/2Kp	H	NAA(1)	S(40)	D	Grossoni (1979)
		Poor rooting and callus	MS	MS	NAA(1 – 10)			
Leccino and Cipressino	Stem	Root primordia and tracheids	WH	WH	BA(2) + NAA(1) + CW(10%)		?	Wang et al. (1979a)
Giarraffa seedlings	Nodes	Axillary buds development	MS	MS	kin(5) + IBA(0.4) + GA_3(0.5) + GSH(50)	S(20)	L/D	Rugini et al. (1979)
		Roots	1/2Kp	H	IBA(1)			
Leccino and Cipressino	Apical twigs	Roots and shoots from callus	WH	WH	BA(2) + NAA(1) + CW(10%) or BA(2) + NAA(0.2)	S(20)	?	Wang et al. (1979b)
Seedlings	Hypocotyls	Shoots	WH	WH	NAA(0.01) + BA(0.5) + St. (0.7%)	S(20)	L/D	Bao et al. (1980)
		Roots	WH	WH	IBA(0.5) or NAA(2)			
Dolce Agogia	Shoots grown in vitro from nodal suckers (1)	Axillary bud development	1/2MS	1/2MS	Zr(10) + IBA(0.5) + S(30) GA_3(0.5) + GSH(200)	S(30)	L/D	Rugini and Fontanazza (1981)
		Roots	1/2Kp	H	NAA(1)	S(30)	L/D	

Table 6 (continued)

Cultivar	Explant source (age in years)	Response	Media				Light condition	Reference
			Salts		Hormones and other additives (mg/l or as indicated)	Sugar (g/l)		
			Macro	Micro				
Picholine marocaine	Seedlings	Roots	1/2Kp	M	NAA(1)	S(30)	L/D	Boulouha (1981)
	nodal explants (1)	Shoots from nodal explants	1/2Kp	M	?	S(30)	L/D	
Dolce Agogia	Shoots grown in vitro from	Axillary bud development	1/2MS + $Ca(NO_3)_2$	1/2MS	Zr(4) or 2.iPr(4)	S(30)	L/D	Rugini (1981)
	nodal sukers (1)	Roots	1/2Kp	H	NAA(1)	S(30)	L/D	
Manzanillo	Shoots	Callus	(21-MW)		IAA(1) + 2-iPr(1) or 2-iPr(1) + CHA(10^{-3}) or 2-iPr(1) + CIA(10^{-4}M)	S(20)	L	Lavee and Avidan (1982)
Giarraffa seeds	Naked embryos	Embryo development	1/3MS	1/3MS	2-iP(2) or Z(2)	S(10)	L/D	Rugini (1986)
Dolce Agogia, Frantoio and Morziolo	Shoots grown in vitro from nodal mature shoots, suckers (1) and	Axillary bud development and	OM	OM	Zr(4) or 2-iPr(4)	S(30)	L/D	Rugini (1984)
	Mature apices	Roots	1/2Kp	H	NAA(1)	S(20)	L/D	
Wild plants (Oleaster)	Dormant seeds	Germination	MS	MS	SAN 9789	S(30)	D	Mitrakos and Diamantoglou (1984)
Several cultivars	In vitro shoots	Rooting in high %	1/2BN	BN	NAA 1 + Putrescine (160)	S(20)	L/D	Rugini (1990)
Maurino seedlings	Nodes	Shoots (normal)	MS_1	MS	Zr(10)	S(30)	L/D	Fiorino and Leva (1986)
		Shoots (apical necrosis)	B_5	B_5	Zr(10)	S(30)	L/D	
		Roots (high %)	MS_1	MS	IBA(3)	S(30)	L/D	
		Roots (low %)	B_5	B_5	IBA(3)	S(30)	L/D	

Spanish Varieties	Naked Embryos	Embryo development	OMg	OM	BAP(0.8) + NAA(0.01)	S(30)	L/D	Cañas et al. (1987)
		Shoots induction from axillary buds	OM	OM	Zr(4) or 2-iP(4)	S(30)	L/D	
		Rooting	1/2OM	1/2OM	IBA(1)	S(30)	L/D	
Several italian cultivars	Immature zygotic embryos (75 day old)	Somatic embryogenesis	1/2MS or 1/2OM mod	1/2MS or 1/2OM	BAP(0.5)	S(20)	D	Rugini (1988)
Spanish cultivars	Mature cotyledons	Callus from protoplasts	BN	OM + glutamine (500) + ornitine (0.03mM)	2.4-D(0.5) + NAA(1) + Zr(0.5)	S(20)	L/D	Cañas et al. (1987)
Spanish cultivars	Callus of cotyledons	Shoot organogenesis	BN	OM	2-iP or Zr(0.5 – 4) + CH(1000)			Cañas and Benbadis (1988)
Several cultivars	In vitro shoots	Roots in high %	BN	BN	NAA(1) + basal blanching	S(20)	L/D	Rugini et al. (1987b)

Abbreviations: ABA = abscisic acid; BA = benzylaminopurine; B = Gamborg et al. (1968) (Ca/N = 1/31); CH = casein hydrolysate; CHA = chlorogenic acid; CIA = cinnamic acid; CW = coconut water; D = dark; Dikegulac = 2,4 : 4,6 Gi-O-isopropylidene-2-keto-L.gulonate; BN = Bourgin and Nitsch (1967), GSH = reduced glutathione; GA = gibberellic acid; H = Heller (1953); IBA = indole-3-butyric-acid; IAA = indole-3-acetic acid; Kp = Knop; Lc = lactose; L = light; L/D = 16 h light; M = Miller; MS = Murashige and Skoog (1962); MS1 = MS having 940 mg/l of $CaCl_2$ (Ca/N = 1/9.4); kin = kinetin; 21MW = 2 × macro White plus micro White; N = Nitsch; NAA = naphthaleneacetic acid; OM = olive medium (Rugini 1984); OMg, OMs, OMr, OMe = variants of OM (see Cañas et al. 1987); 2.4-D = 2,4-Dichlorophenoxyacetic acid; PP333 = paclobutrazol; 2iP = dimethylallylaminopurine (+ r = riboside); S = sucrose; St = stearic acid; SADH = succinic acid 2.2-dimethylhydrazide; SAN 9789 = 4-chloro-5-(methylamine)-2-(a,a,a,trifluoro-m-tolyl)-3-(2H)-pyridazinone; TIBA = 2,3,6,triodobenzoic acid; WH = White; Z = zeatin (r = riboside) (e.g., ionizing radiation, etc.) to cryopreserve the numerous existing cultivars, and to produce disease-free plants. The long-range objectives consist of producing highly productive hybrids with high photosynthetic rates, disease-, drought-, salt-resistant, through protoplast fusion, genetic engineering or stress selection, and to produce homozygous plants from haploid tissues.

Little work has been done in the field of in vitro culture of olive even if research has been under way for about 20 years (cf. Table 6).

3.2 Methods for Sterilization

Heavy systemic contamination occurs in olive particularly when the material, including seeds, is collected from field-grown plants; therefore, greenhouse plants are recommended for collecting explants. Sterile woody internodal olive explants were obtained by Scaramuzzi and de Gaetano (1974), Rugini and Fontanazza (1981), cleaned woody nodal explants by dipping them in ethanol (95%) followed by soaking in calcium hypochloride (9%) for 20 min, rinsing in sterile distilled water and cleaning the necrotic bark, again dipping them in ethanol, and then rinsing in sterile water. Grossoni (1979) used a solution of sodium hypochloride for 22–24 min then washed the woody explants with sterile water (eight to ten times), and subsequently dipped in ethanol for 10 s, and rinsed them in water. Lavee (1977) obtained sterile olive fruits using 70% ethanol. Diamantoglou and Mitrakos (1979) found that the antibacterial Serezan (Bayer) used for 10 min was more effective than calcium-, or sodium hypochloride when used on olive seeds. Sterile, tender shoots are much easier to obtain by prewashing in water plus Tween 80 (5 drops/100 ml), then sterilizing in a $HgCl_2$ solution of 677 mg/l (2.5 mM) plus 30 drops/l of Tween 80 for 3 min, and subsequently rinsing three times in sterile H_2O (Rugini 1984). At present in Rugini's laboratory, any explants, including contaminated shoots growing in in vitro cultures, are easily disinfected with 120 ml Domestos (Na hypochloride with 5.5% active Cl) mixed with 350 mg of $HgCl_2$ (1.3 mM) in 880 ml H_2O, shaken frequently under vacuum for 3 to 5 min, then rinsed under vacuum in abundant sterile water for 30–40 min. Then cuts of nodal explants are renewed before placing them in the medium of shoot tips are collected if they come from in vitro cultures. Shoot tips or meristems, which usually warrant sterility, collected from field or greenhouse olive plants, rarely give rise to shoots, in contrast to shoot tips collected from in vitro-growing shoots. Disinfection of contaminated apical shoots growing in vitro by this method and subsequent collection of shoot tips could thus be the sole measure by which it is possible to obtain sterile culture in woody species which require woody nodal explants or buds to start the culture. Normally they do not give rise to clean shoots, because the inner part is infected.

3.3 Micropropagation

Explant Source. Meristem and shoot tip cultures, often used in many species to obtain disease-free plants (Boxus 1976, 1978), are very difficult to use on olive. Various attempts with meristems or shoot tips from field-grown or greenhouse plants were unsuccessful, due to oxidation of tissues after collection, even with the use of active anti-oxidants (Rugini 1984). Therefore, nodal explants, possibly collected from greenhouse plants, are recommended (Rugini and Fontanazza 1981). However, the use of apical herbaceous twigs, buds collected from vigorous shoots, immediately after sprouting (10–30 mm long), located on trunks or large branches could be used (Rugini 1984, 1986).

Culture Initiation. Cultures were initiated by inoculating buds or tender apical twigs (1 to 2 cm long) in a low concentration of mineral salt agar medium plus zeatin (mixed isomers or better trans) at 0.5 mg/l (Rugini 1984). In addition, tender shoots have been obtained by culturing woody node explants, collected in winter, in 1/2 Knop macro plus Heller micro, sucrose 3% and IBA 2 mg/l (Rugini and Fontanazza 1981).

3.4 Shoot Proliferation

In order to achieve rapid shoot growth, various attempts have been made (Rugini 1984) using all the common media reported in the literature, but the new medium, OM (olive medium), was found to be more efficient (Rugini 1984). This medium has been devised by using data from the analysis of the main mineral elements found in field-growing olive and almond shoots apices (4–5 cm long) during the rapid growth phase, and in embryos during the final stage of maturation. By comparing the concentrations in the tissues, the medium has been formulated on the basis of an adequate quantity of elements according to the degree in which they differed from the MS medium, where almond grows very well (Rugini and Verma 1983). Olive medium is richer in Ca, Mg, S, P, Cu, and Zn than MS medium. For satisfactory growth, olive requires both NO_3^- and NH_4^+, inorganic nitrogen sources, and glutamine (Table 2). However, glutamine could be replaced, but with less effectiveness, in decreasing order, by allantoic acid (K salt) (2140 mg/l) allantoin (474 mg/l) or urea (1200 mg/l). Arginine and asparagine (500 mg/l), on their own or together, slightly improve growth, while alanine delays it, even using 500 mg/l (Rugini 1984). Regarding the quality of shoots, the Ca/N ratio is very important. A ratio of about 1:11, as used in OM (Rugini 1984), allows for both good multiplication rate and good shoot quality. This has been confirmed by Fiorino and Leva (1986), who carried out specific experiments using shoots of Maurino seedlings. They observed a high frequency of apex necrosis on using B_5 medium, having a Ca/N ratio = 1:31, while no apex necrosis was observed on modified MS medium with a Ca/N ratio = 1:7.6 to 9. On OM, it is possible to maintain cultures for a longer time with a relatively more rapid growth rate than on other already tested media. In olive, however, growth is not as rapid as in other crop species, and it depends on the cultivar and its age; e.g., cvs. Moraiolo and Dolce Agogia grow faster than Frantoio, and the culture grows faster only after many subcultures. The multiplication rate is about 8 to 9 (nodes for each cultured node) in 35–40 days at 23 °C, with 3000 lx for 16/h/day (Fig. 4).

Cytokinin quality, agar concentration, and aeration are critical factors. Olive shows satisfactory growth only with zeatin (mixed or trans-isomer). The other

Fig. 4. Olive shoots cv Moraiolo during proliferation phase in disposable polystirole jars, covered with polyethylene film, on OM medium

natural hormone, 2-iP, gives rise to thin shoots, while the synthetic ones, BAP, PAB, and kinetin give rise to thin and short shoots and large basal callus. Agar 0.5–0.6% allows satisfactory growth without any vitrification problem, while at a higher concentration, growth decreases due to low water availability, but above all because agar decreases invertase activity (Rugini et al. 1987a). Jar lids should not be tight, in order to permit the exchange of air, otherwise shoots do not grow, but only form basal calli.

Olive has strong apical dominance: from one uninodal explant, it is possible to obtain no more than two shoots, even with spray treatments or by incorporating SADH, Dikegulak, PP 333, thiourea, and TIBA in the medium or by inserting the explants upside down in the medium.

A significant improvement in growth rate, particularly for old shoots, has been obtained by soaking the explants in sterile distilled water or with aqueous GSH solution (50–100 mg/l) or GA_{4-7} (50–100 mg/l), before transferring them to the new medium; GA_3 (0.1–500 mg/l) incorporated in the medium was less effective (Rugini 1984).

3.5 Rooting Phase

More than 80% of in vitro rooting has been observed with bi- or trinodal of both mature and seedling explants, on media having low mineral salt content (1/3 or 1/4 MS, BN or Knop macro plus Heller micro), 2% sucrose, and 1 mg/l of NAA or IBA. Cañas et al. (1987a) obtained more than 90% rooting in seedling explants in half-strength OM (olive medium), 3% sucrose and 1 mg/l of IBA. Generally, high rooting percentage is obtained by using any explant source (juvenile or mature) if kept in vitro for many subcultures (Rugini 1984). However, olive gives inconsistent rooting even though the same technique is used. To solve this problem, two alternatives are suggested (Rugini et al. 1987b): rooting with basal etiolation or with addition of polyamines.

3.5.1 Rooting with Basal Etiolation

This method consists of maintaining the base of the shoots in darkness during the rooting phase, for 16 h light per day, by painting the outside of the jars black, and by covering the agarized rooting medium with black polycarbonate granules. This method does not increase rooting any more than by keeping the cultures in darkness for 1 week, but it prevents senescence, which happens in the dark, and may affect growth restart of plantlets in pots (Fig. 5). Medium stained with China ink has recently given the same results of above method (Rugini unpubl.).

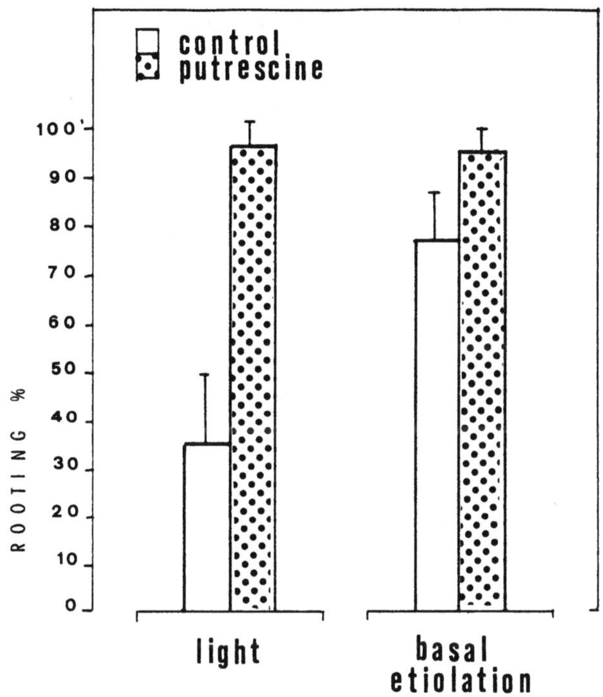

Fig. 5. Effect of putrescine treatment (1 mM) on rooting of olive unetiolated (light) and etiolated (*basal etiolation*) explants at their base by painting the outside of the jars black and by placing polycarbonate granules on top of the solidified medium. Both treatments were done under 16-h *light* days at 60 $\mu E/m^2/s$

Fig. 6. Olive explants in presence of putrescine (*right*) rooted faster than control (*left*)

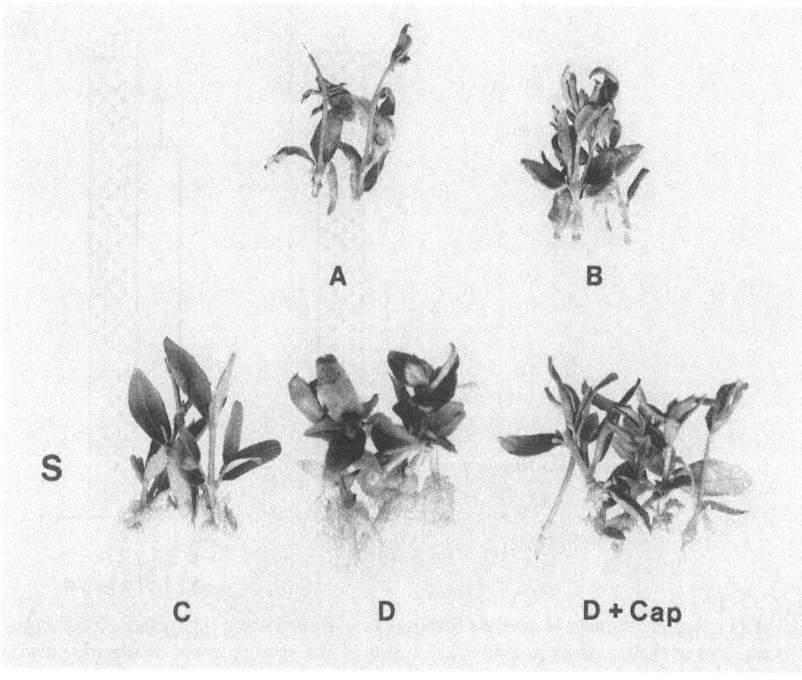

Fig. 7. *Top* Olive explants from rooting medium lacing carbohydrate (**A**), and (**B**) same medium plus putrescine. Note the more healthy explants with putrescine treatment and some of them rooted. *Bottom* Explants from medium with 2% sucrose (*S*), (**C**) control, putrescine (**D**), (**D+Cap**) putrescine plus Captan. Note in the last treatment the absence of basal callus and higher rooting percentage

3.5.2 Rooting with Addition of Polyamines

In the presence of auxin, polyamines, in particular putrescine, promote early (Fig. 6), higher and stable rooting percentage, increase the number and length of roots per explant; they also promote rooting in the absence of sucrose. They are found to be particularly effective when explants are not suitable for rooting (e.g., too old) or when rooted under high light intensity (more than 30 $\mu E/m^2/s$) (Figs. 5, 6).

Rooting percentage has also been improved by soaking explants in a suspension of Captan brand (50% p.a.) at 1% for 30 min, or by incorporating it in the rooting medium, when the agar is warm (Rugini unpubl.) (Fig. 7).

Attempts to root olive in vivo have given poor results, so the use of polyamines which delay sensescence could be helpful.

Concerning auxins, NAA (1 mg/l) incorporated in the medium is found to be more effective, in woody explants (Scaramuzzi and de Gaetano 1974), in embryonal material (Rugini et al. 1979; Bao et al. 1980) and even in shoots coming from mature branches of cultivars (Rugini and Fontanazza 1981; Rugini 1984). Cañas et al. (1987 a) reported that IBA was more effective than NAA in the Spanish cultivars. High rooting may also be obtained by dipping the basal part or the entire explant in a solution of 100–200 mg/l for 10–20 s. IAA is able to stimulate rooting efficiently only in embryonal explants (Rugini 1981).

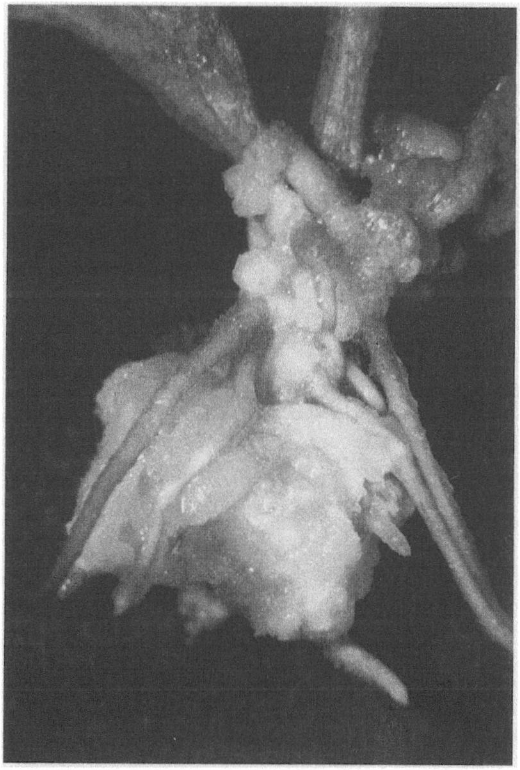

Fig. 8. Olive explants rooted with *Agrobacterium rhizogenes*

In rooting medium, 2% sucrose is generally used with tender shoots. Scaramuzzi and de Gaetano (1974) found better rooting in woody internodal explants with 4–8% sucrose.

Vitamins are not essential for the rooting phase; on the contrary, they may even reduce rooting (Rugini et al. 1979).

Root elongation medium, as used in the past, is not really necessary.

3.5.3 Rooting with Agrobacterium rhizogenes

Olive explants of several cultivars have been rooted with *A. rhizogenes* strain NCPBB 1855 by wounding the shoot base with a knife blade infected with the bacteria or by inoculating the middle of the explants with a microglass Pasteur pipette (Rugini 1986, 1990; Rugini and Wang 1986). Both methods gave the same rooting percentage sometimes reaching 80% (Fig. 8). However, roots almost always arise from the base of shoots and frequently they are agropine-negative, even though they arise from the inoculation point and in the presence of the ipomethylant 5-azacytidine (Rugini and Wang 1986). Also in this case, the polyamines promote early root emission and higher rooting percentage (Figs. 9, 10).

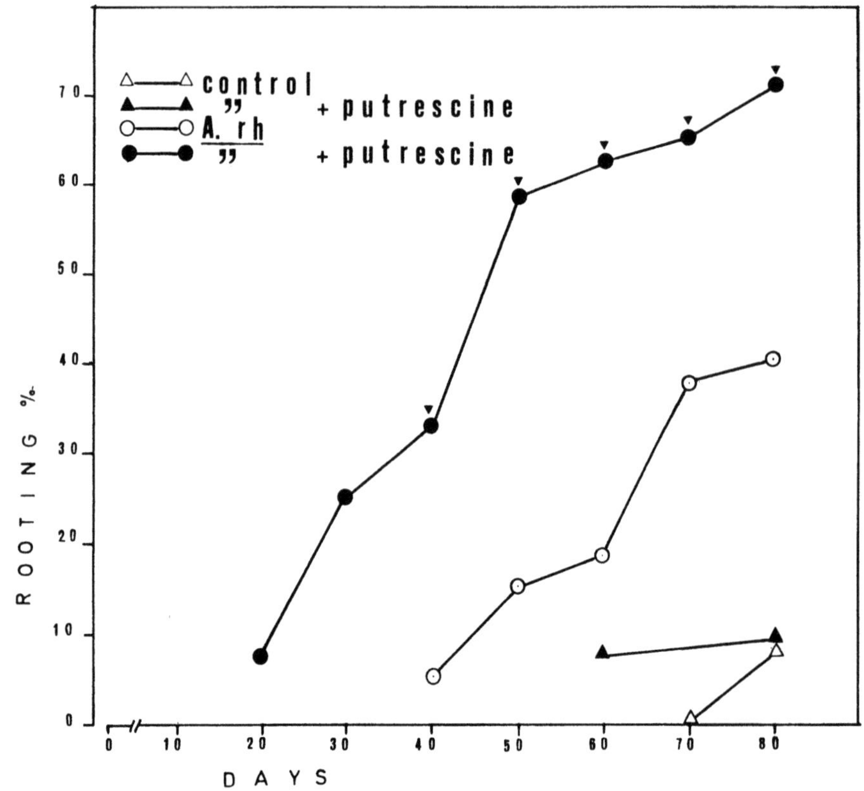

Fig. 9. Effect of putrescine at 1 mM, on rooting of Moraiolo explants with *A. rhizogenes*, and on untreated explants (control)

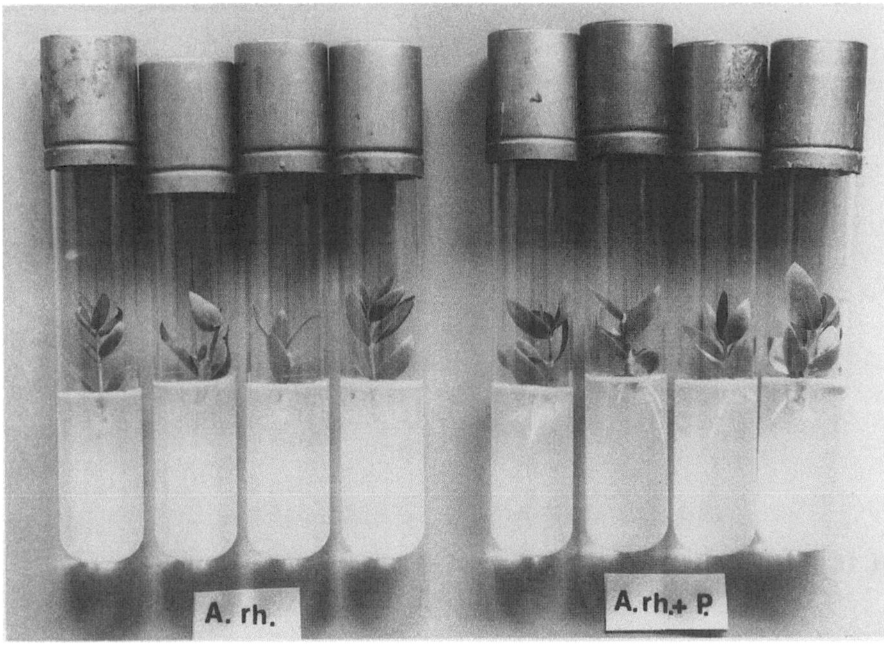

Fig. 10. Olive shoots rooted only with *Agrobacterium rhizogenes* (*left*) and with bacteria plus 1 mM putrescine incorporated in the medium (*right*). Note the early rooting with putrescine treatment

3.6 Transfer of Plantlets to Pots

Acclimatation for olive is a particularly critical phase if transferal is not followed by a rapid growth restart. Good results have been obtained by using a mixture of peat moss, perlite, and polystirole granules (1 : 1 : 0.5) and by covering the plantlets with transparent covers (Fig. 11). Survival is usually around 70%. Initial dormancy may be interrupted by spraying leaves with 300 mg/l GA_3 solution. Plants obtained with *A. rhizogenes* showed less vigor compared with those obtained with auxin treatment. To date, many plants have been obtained in our laboratory from in vitro cultivars, but unfortunately field performance has not been possible due to the frost of 1985, which destroyed all the plants in the field.

3.7 Callus Culture, Relative Physiological Studies

Callus induction is relatively easy in olive and it may be obtained using any part of the plant (see Table 6). Lavee and Messer (1969) established callus culture with 1-year-old olive stems of cv. Manzanillo by using modified White's medium (21 – MW) plus 2 mg/l IAA and 0.2 mg/l kinetin. Additionally, they pointed out that auxin is not necessary initially for early development, but after several subcultures both auxin and cytokinin are required for growth. They also found that IAA, followed by 2,4-D, are the most efficient auxins for inducing callus growth.

Fig. 11. Two-month olive plantlets in vitro rooted with *Agrobacterium rhizogenes*

Successively, Lavee (1977) reported that kinetin (2 mg/l) followed by zeatin-riboside (2 mg/l) are the best cytokinins for inducing growth in the fruit mesocarp, combined with 2 mg/l IAA. Concerning the olive mesocarp, the initiation of callus and growth depends on the age of the fruit. The author found a double sigmoid growth curve where rapid growth occurs when oil droplets start to appear in the cells (about 80 days after fruit set) and continues for about 20–30 days; after this period, oil increases up to 10–12%, and development potential is lost. Lavee and Avidan (1982), using stem calli which require both auxin (1 mg/l IAA) and cytokinin (2-iP 1 mg/l) for growing, observed that chlorogenic acid (CHA) (10^{-3} M) or cinnamic acid (CIA) (10^{-4} M) could replace IAA in stimulating growth. This growth induction by chlorogenic acid replacing IAA, seems a peculiarity of olive callus. Lavee and Avidan (1986) postulated the possibility, in olive tissue, of a reverse pathway leading back from CHA to IAA. This could be explained by the possibility of the substitution of IAA with CHA, but also by the necessity of a period of time of 96 h, compared to IAA, for inducing active growth, which also coincides with the reaction of cathodic isoperoxidases.

In our laboratory, friable callus is usually initiated on BN medium supplemented with 2,4-D (0.1–5 mg/l) or NAA (5 mg/l) or IBA (5 mg/l) plus zeatin riboside (0.1 mg/l) in the dark from any part of the in vitro-germinated seedling. However, with 2,4-D, calli are more subjected to browning than when other auxins are used. This problem may be eliminated, without influencing growth, by adding a few drops of GSH (500 mg/l) on callus inoculum during the phase of subculturing (Rugini 1986). This may indicate that the cellular antioxidant, GSH, which was found at a high concentration during the 2,4-D treatment in carrot cell suspensions (Earnshaw and Johnson 1985), is not produced sufficiently in olive to avoid oxidation. After initiation, callus can be kept for a long time in MS medium

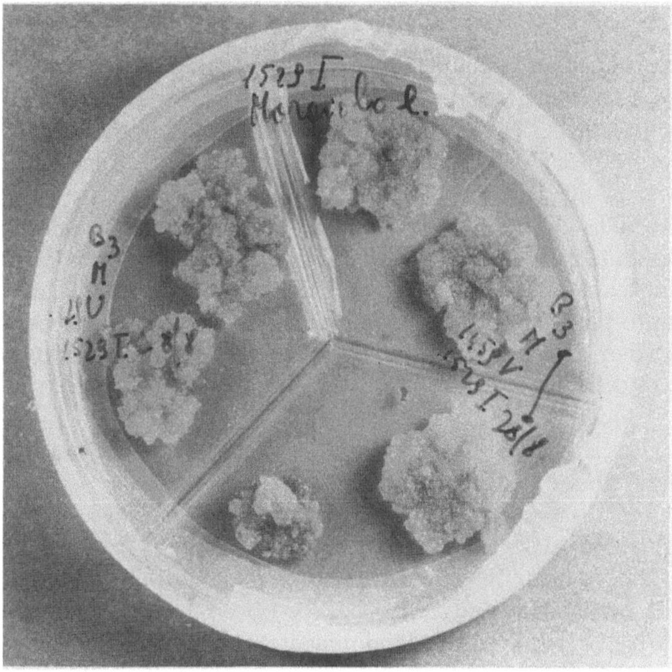

Fig. 12. Moraiolo leaf callus maintained on MS medium plus 5 mg/l NAA 0.1 mg/l BAP or Zr

plus NAA (2–5 mg/l) and zeatin riboside (0.1 mg/l) under 16 h light per day at 23 to 25 °C (Fig. 12).

Lavee and Adiri (1974) found that ABA (10^{-6}, 10^{-7} and 10^{-8} M) had a poor effect on the initiation of bark callus in the cv. Manzanillo, while it caused early senescence at concentrations above 10^{-5} M, expressed as sudden browning and cessation of growth; by using 10^{-6} to 10^{-8} M ABA, it continued to grow and grow better than the control in subsequent subcultures.

GA_3, on the contrary, had an inhibitory effect on olive bark growth. No clear interaction has been found between ABA and GA_3, although in some cases a slight synergistic effect on growth is evident. Furthermore, they demonstrated that ABA had poor or no effect on the uptake of ^{3}HGA.

Epstein (1982) compared the changes in the levels of free and conjugated indole-acetic acid (IAA), in field-grown olive leaves and Manzanillo bark callus, resulting from ethylene treatments. In both olive leaves and olive callus, the levels of bound IAA increased almost three times more than in the untreated ones, while free IAA did not change after ethylene treatment.

Lavee and Epstein (1976) showed that metabolism of IAA in in vitro olive callus does not correspond to metabolism in olive leaves. In fact, the callus converts IAA into indolacetylaspartic acid, while in leaves the major metabolite is indolecarboxyl acid.

3.8 Cell Suspension

In our laboratory we produce and maintain suspensions starting from soft calli obtained from any type of tissue (micropropagated leaves, roots, and stems of cultivars and seedlings) on agar BN medium or even better in MS in the dark or in the light). We have produced small and overfed cell suspensions in modified OM macro as follows: in mg/l: KNO_3 1772; NH_4NO_3 1058; $Ca(NO_3)_2 \cdot 4H_20$ 1300; Mg $SO_4 \cdot 7H_2O$ 1500; KH_2PO_4 340 by omitting glutamine and by adding IBA or NAA (5 mg/l) and zeatin riboside (0.1 mg/l), in autophyton or in flask or multiwells on the shaker. At the beginning growth is slow and requires the addition of a new medium to the batch rather than subculturing a sample to a fresh one. Furthermore, high density is required for growth, and the eventual precipitate in the medium after autoclaving is to not use.

3.9 Direct and Indirect Shoot Organogenesis

Bao et al. (1980) obtained efficient direct shoot organogenesis from olive hypocotyl sections in White's medium with NAA (0.01 mg/l) and BAP (0.5 mg/l) at pH 6 to 6.5, 1000 lx per 8 h/day and at 25 °C.

Wang et al. (1979a, b, 1981) studied the developmental stages of a callus from twigs of cvs. Leccino and Cipressino in correlation with nucleus, nucleolus, cell size, and RNA accumulation. They succeeded in obtaining buds and roots from calli by using NAA (0.2 mg/l) and BAP (2 mg/l) and CW (10%). In addition, they observed two types of meristematic tissues: meristematic cellular masses and meristematic nodules. Roots differentiated from meristematic masses, while adventitious buds derived from meristematic nodules.

Cañas and Benbadis (1988) induced callus in mature cotyledon fragments from the cvs. Tanche and Picual, in modified OM medium (OMc), where OM macro elements were replaced with BN macro elements, glutamine was omitted, and 1000 mg/l of hydrolyzed casein, 5 mg/l IBA and 0.2–0.5 mg/l zeatin riboside or 2-iP were added. Subsequently, highest shoot regeneration frequency was obtained by using the same basal medium plus 2-iP (from 0.5 to 4 mg/l) in the absence of or at low auxin concentrations. Shoot regeneration was higher in calli from cotyledon segments proximal to the embryo axes than in the distal ones. Shoots were then rooted and transplanted in pots.

3.10 Somatic Embryogenesis

Rugini and Tarini (1986) reported somatic embryogenesis from both root seedlings and immature embryos. Somatic embryos from callus root seedlings of cv. S. Agostino still attached to the shoot were obtained in MS medium with 0.1 mg/l NAA and 0.1 mg/l BAP. Previously, callus was induced in roots purposely damaged by a knife and placed in MS medium plus 5 mg/l NAA and 0.1 mg/l BAP. However, the number of embryos obtained was very low and strictly linked to the genotypes. On the contrary, high embryogenetic capacity is possessed by imma-

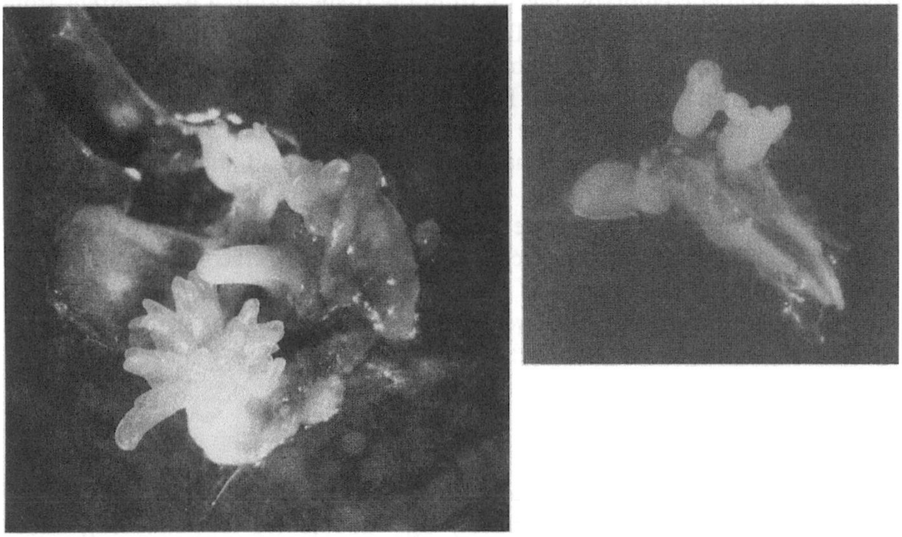

Fig. 13. Somatic embryogenesis in 75-day-old zygotic embryos on half-strength MS medium plus 0.5 mM BAP with details on the *right side*

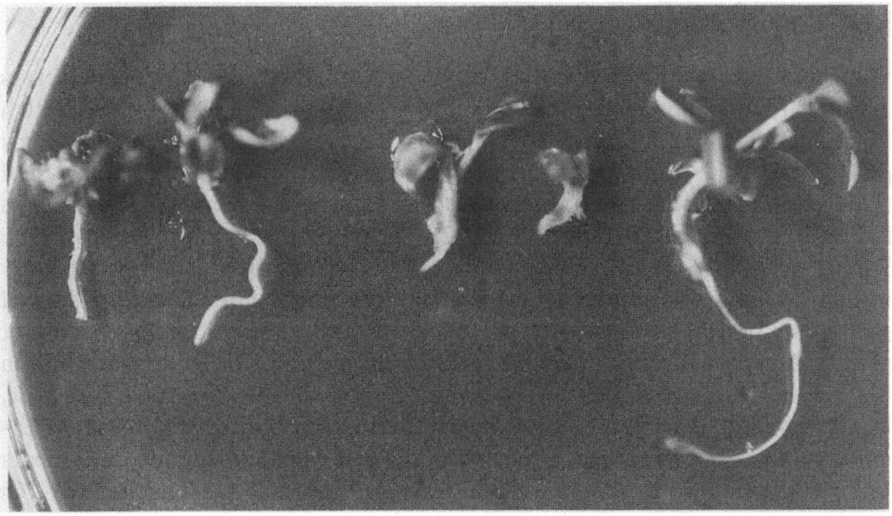

Fig. 14. Germinating somatic embryos – 2 weeks after transferring in hormone-free Olive Medium

ture embryos of cvs. Frantoio, Moraiolo, and Dolce Agogia, as shown by Rugini (1988). Date of embryo collection, growth regulator quality and light requirement have been the major critical factors. Only embryos collected 75 days after fertilization were able to give rise to embryos, whereas younger and older ones were unable to. Zygotic embryos, placed on half-strength MS agarized medium with low BAP concentration (0 to 2.5 μM) in the presence or absence of NAA (0 to 12.5 μM), gave rise to somatic embryos directly from the tissue and to a lesser extent from the callus formed in the same medium (Figs. 13, 14); 2,4-D alone or in combination with BAP hampered any embryo formation, but gave rise to yellow callus, while NAA, when combined with BAP, never hampered embryogenesis even at a high concentration (12.5 μM). The highest number of zygotic embryos capable of giving to somatic embryogenesis has been obtained in half-strength MS medium plus 0.5 μM BAP after 6 weeks in the dark. Light inhibited embryogenesis completely. Attempts to regenerate shoots or somatic embryos from leaf discs of shoots grown in vitro were unsuccessful. Orinos and Mitracos (pers. commun.) obtained somatic embryogenesis from root seedling callus of *Oleaster* on modified OM (macro elements were replaced by BN macro) with 1 mg/l IBA without any hormone.

3.11 Variability

Olive seedlings or mature plants are usually characterized by some uncommon features which appear and disappear, e.g., double leaves (unfolded), shoot dichotomy, and triple whorl (verticals with three leaves). Cañas and Benbadis (1988) observed that the frequency of these features was higher in regenerated plants from cotyledon calli than in seedlings from the same stock of seeds.

3.12 Protoplast Isolation and Cultures

Viable protoplasts from hypocotyls, cotyledons, or from micropropagated leaves of cultivars, were obtained in our laboratory by using 1.5% Driselase after 10 h of incubation in 0.6 M mannitol, 500 mg/l $CaCl_2 \cdot 2H_2O$, $2H_2O$ and 500 mg/l KCl. Many of the common enzymatic combinations of Cellulase, Pectinase, and Hemicellulase gave only poor protoplast yield. Some divisions were observed in BN medium supplemented with NAA, 2,4-D and zeatin riboside (Rugini 1986). Subsequently, Cañas et al. (1987b) obtained protoplasts from leaves of seedlings from naked embryos germinated in vitro and from cotyledons. They also isolated protoplasts by using Driselase (1 to 1.5%), 3.5 mM $CaCl_2$ and 0.5 M sucrose at pH 5.6. After purification, they diluted and cultivated 3×10^4 protoplasts/ml, in BN macro, OM micro and OM vitamins, 500 mg/l glutamine, 20 g/l sucrose, 91 g/l Mannitol, 0.5 mg/l, 2,4-D, 0.5 mg/l zeatin, 1 mg/l NAA and 0.03 mM Ornitine, at pH 5.7. After 6 weeks, the microcalli were transferred to the same media without Mannitol, plus 0.5 mg/l zeatin, 2.5 mg/l IBA or NAA, and 0.6% (W/V) Sea Plaque Agarose, for further callus development.

By comparing the two protoplast sources, it was shown that only cell colonies from cotyledon protoplasts developed visible microcalli, giving rise to two kinds of calli: white friable callus and green compact callus.

3.13 Embryo Culture

The main purpose of embryo culture and embryo rescue is to shorten or make possible the breeding cycle.

Olive embryos are characterized by light dormancy, which disappears progressively when seeds are kept in a dry place and at room temperature. The optimal germination temperature of naked dormant embryos and nondormant embryos, inhibited by the seed coats (mainly endosperm) is 13 °C, while the best for nondormant naked embryos is 25 °C (Instanbouli and Neville 1977a). Dormant naked embryos showed positive photosensitivity, like all olive seeds, being both dormant (after harvesting) or nondormant (22 months after harvesting), while nondormant embryos were indifferent to light. However, if nondormant embryos are cultivated in the presence of 3% glucose, which inhibits embryo germination, they become photosensitive again (Instanbouli and Neville 1977b).

Embryo germination is inhibited by hexose (glucose, galactose, levulose) and mannitol at concentrations of 2 to 5%. On the contrary, with higher or equal molarities of hexoses, the disaccharides (sucrose and lactose) played either a small or noninhibitory role (Instanbouli and Neville 1977c).

According to Instanbouli and Neville (1977c), hexoses play the same inhibitory role on the germination of olive seeds, but the inhibitory effect is reversible, so that seeds can germinate when shifted to hexose-free medium.

Glucose inhibits chlorophyll synthesis in the light, perhaps because it is not a natural reserve in seeds; reserves are largely fats (Sondheimer et al. 1970).

Mitrakos and Diamantoglou (1984) reported that dormant olive endocarp-free seeds have not germinated either at 15 or at 25 °C even though treated with GA_3 and cytokinins, while 70% germinated at 25 °C in the presence of 16.5 to 33 μM SAN 9789 (4-chloro-5-(methylamine)-2 (a, a, a,-trifluoro-m-tolyl)-3-2H9-puridazinone). They attributed to SAN 9789 the property of decreasing inhibitors in the endosperm. However, the inhibitory effect of a treatment with ABA cannot be overcome.

The presence of mineral elements in the medium is not essential for naked embryo germination (Istanbouli and Neville 1977c). However, rapid germination and seedling development with one or two nodes and long roots is usually reached in 4 weeks by culturing the naked embryos in the following medium: 1/3 MS+Ca $(NO_3)_2 \cdot 4H_2O$ 500 mg/l+sucrose 1%+agar 0.7%+2 mg/l zeatin or 2-iP in the light at 3 klx, 16 h light per day at 23 °C. Under these conditions, germination is strictly dependent on fruit harvesting; less mature embryos germinate more slowly (Rugini et al. 1979).

According to Lagarda and Martin (1983), embryo germination was inhibited by 10 ppm ABA. GA_3, GA_{4+7} and BA did not affect embryo germination; on the contrary 100 mg/l GA_3 reduced whole seed germination. In addition, they observed that young developing embryos contained up to 600 ng/g fresh weight ABA, which declined at seed maturity. ABA was not detected in endosperm and seed coats until seed maturity.

The embryos are more sensitive to ABA, and 8 ng/g (dry weight) in the coats and endosperm can inhibit embryo germination.

Naked embryo germination depends on seed, harvest date, and storage. Embryos from seeds collected in September germinate quite well if seeds are stored for not more than 1 month; germination increases with increasing storage time, when seeds are harvested in late October (Lagarda et al. 1983).

4 Conclusions and Perspectives

Economically, olive is one of the most important woody fruit trees for most countries bordering the Mediterranean Basin, but its price is not competitive with other vegetable fats, so research is needed in order to improve productivity and to reduce production costs. More information is required to better define the systematic classification of the genus *Olea* and to make genetic improvement easier and to improve cultivation methods (shape-training, pruning, fertilization, irrigation, harvesting, and the clonal selection of cvs. etc.). International cooperation would be very useful in order to enlarge and preserve germplasm collections, and to carry out selection programs for the best adaptibility to the environment. Clonal selection and genetic improvement by traditional crossing methods have not yet given satisfactory results. The use of tissue culture methods may offer new perspectives of improvement even though up to now, research has given only preliminary results. This is partly due to low morphogenetic capacity, which actually is linked only to zygotic embryos. This also hampers obtaining haploids, which are extremely important in breeding programs. They would speed up selection in this highly heterozygous and prevalently self-incompatible species characterized by a long juvenile phase. It is advisable, however, to try to obtain haploids by irradiated pollen and subsequent embryo rescue by using in vitro embryo cultures as already obtained in apple by James (1987). Presently, good perspectives for improving this species are more likely to be obtained by stimulating axillary buds, which are used to induce mutations with ionizing irradiation or chemicals and subsequent isolation of mutated parts, by avoiding chimeras. Another possibility, with the stimulation of axillary buds, is to obtain disease-free plants in the near future using the apical meristem from in vitro-cultured shoots in combination with thermotherapy, as suggested by Walkey (1978) for other species. Sufficient evidence is available to support the hypothesis that studies should be carried out to create a germplasm bank of shoot apices which represent the ideal material for cryopreservation (Bajaj 1979). In fact, there are large number of important cultivars to use for breeding programs which otherwise would be lost due to the current practice of using only a few cvs. in intensive cultivation.

References

AA.VV. (1982) Lavori della Commissione Tecnica presentati in inchiesta pubblica RISG 59:103

AA.VV. (1984) Olive acclimatation and breeding. Shanon H, Ying G (eds) Najing PRC (in Chinese)

Altamura Betti MM, Altamura L, Mazzolani G (1985) Elements for the revision of the genus *Olea* (Tourn.) L. IV The taxa present in *Oceania* which can be ascribed to *Olea* and allied genera. Ann Bot 43:45–52

Antognozzi E, Catalano F (1985) Risposta varietale dell'olivo ai danni da freddo. Ann Fac Agr Perugia 39:185–198

Antognozzi E, Preziosi P (1986) Effects of paclobutrazol (PP 333) on nursery trees of olive. Acta Hortic 179:583–586

Bajaj YPS (1979) Establishment of germplasm banks through freeze-storage of plant tissue culture and their implications in agriculture. In: Sharp WR, Larsen PO, Paddock EF, Raghavan V (eds), Plant cell and tissue culture: principles and applications. Ohio State Univ Press, Columbus, pp 745–774
Baldini E, Guccione E (1952) Osservazioni su una razza di olivo con antere sterili. Ann Sper Agrar 6:1205–1216
Bao Z-H, Ma Y-F, Liu J-F, Wang K-J, Zhang P-F, Ni D-X, Yang W-Q (1980) Induction of plantlets from the hypocotyl of *Olea europaea* L. in vitro. Acta Bot Sin 2:96–97 (in Chinese)
Bartolini G, Tattini M, Fabbri A (1986) The effects of regulators of ethylene synthesis on rooting of *Olea europaea* L. cuttings. Acta Hortic 179:841–843
Ben-Tal Y, Lavee S (1985) Girdling olive trees, a partial solution to biennal bearing. III. Chemical girdling: its influence on flowering and yield. Riv, Ortoflorofrutt, Ital, 69.1–11
Berenguer AG (1978) Seleccion clonal en Oliva (*Olea europaea* L.). Olea, Juno 7–15
Bernardini E (1973) Oil and fat technology. Part II Rome
Bongi G (1986) Oleuropein: an *Olea europaea* secoiridioid biologically active on growth regulation. Acta Hortic 178:245–249
Bongi G, Long S (1987) Light-dependent damage to photosynthesis in olive leaves during chilling and high temperature stress. Plant Cell Environ 10:241–249
Bongi G, Mencuccini M, Fontanazza G (1987a) Photosynthesis of olive leaves: effect of light flux density, leaf age, temperature, peltates and H_2O vapour pressure deficit on gas exchange. J Am Soc Hortic Sci 112:143–148
Bongi G, Soldatini GF, Hubick KT (1987b) The mechanism of photosynthesis in the olive tree (*Olea europaea* L.). Photosynthetica 21:572–578
Bouat A (1960) Fertilization of the olive tree. Fertilite 10:13–25
Boulouha B (1981) Recherches sur la micropropagation de la Picholine marocaine. Olea 6:45–56
Bourgin JP, Nitsch JP (1967) Obtention de *Nicotiana* haploides à partir d'ètamines cultivèes in vitro. Ann Physiol Veg 9:377–382
Boxus Ph (1976) Rapid production of virus-free strawberry by in vitro culture. Acta Hortic 66:35–38
Boxus Ph (1978) In vitro multiplication of woody species. Round Table Conf, June, 6–8 Gembloux, Belgium, p 147
Briccoli-Bati C (1986) The effect of boron on the rooting of olive cuttings. Ann Ist Sper Olivicoltura, Cosenza 7:13–20 (in Italian)
Caballero JM (1981) Multiplication de l'olivo par estaquillado semilenoso bojo nebulizacion. Comunicaciones INIA Ser Prod Veg 31
Caballero JM, Nahlawi N (1979) Influence of carbohydrates and washing on rooting in semi-hardwood cuttings of the olive (*Olea europaea* L.) cultivar Gordales. An Inst Nat Invest Agr 11:219–230
Cañas LA, Carramolino L, Vicente V (1987a) Vegetative propagation of the olive tree from in vitro-cultured embryos. Plant Sci 50:85–90
Cañas LA, Wyssmann AM, Benbadis MC (1987b) Isolation, culture and division of olive (*Olea europaea* L.) protoplasts. Plant Cell Rep 5:369–371
Cañas LA, Benbadis A (1988) Plant regeneration from cotyledon fragments of the olive tree (*Olea europaea* L.) Plant Sci 54:65–74
Cantarelli C (1962) Cultivar e pigmentazione delle olive: esame analitico dei polifenoli delle drupe di diverse cultivar dell'Italia Centrale. Atti 1 Conf Naz Olivicolo-Oleario, Spoleto, pp 283–293
Carr DJ, Oates K, Carr SGM (1980) Studies on intercellular pectin strands of leaf palisade parenchyma. Ann Bot 45:403–413
Charlet M (1965) Observation sur le comportament au froid de certaines varietes et de porte-graffes d'oliviers en France. Inf Oleic Int 3:13–35
Chevalier A (1948) L'origine de l'Olivier cultivé et ses variations: Rev Bot Appl 303–304:1–25
Ciccarone A (1964) Orientamenti sui problemi patologici di maggior interesse per il futuro dell'olivicoltura. Inf Fitopatol 14:430–436
Ciferri R (1942) Recenti progressi degli Studi botanico-agrari sull'olivo. Convegno di Studi Olivicoli. R Accad Georgofili, Firenze, pp 49–95
Ciferri R (1950) Dati ed ipotesi sull'origine e l'evoluzione dell'olivo. Olearia 1:115–122
Ciferri R, Marinucci M, Morettini A (1942) Dati preliminari per una sistemazione delle razze di olivo in coltura. L'Olivicoltura 1:3–37

Cimato A, Fiorino P (1981) Propagazione. In: Baldini E, Scaramuzzi F, (eds). L'Olivo. Frutticoltura anni '80 Roma, Reda, pp 53–71
Cirulli M, Montemurro G (1975) Comparison of pathogenic isolate of *Verticillium dahliae* and source of resistence in olive. 4th Congr Mediterranean Phytopathol Union, Yugoslavia
Cirulli M, Laviola C, Roberti D (1981) Avversità e difesa. In: Baldini E, Scaramuzzi F (eds). L'Olivo. Frutticoltura anni '80. Reda, Roma, pp 142–187
Cortesi N, Fedeli E (1983) I composti polari di oli di oliva vergine. Nota Ia. RISG 60:341
Cortesi N, Fedeli R, Tiscornia E (1978) Indagine sulla composizione di insaponificabili di oli vegetali mediante HPLC RISG, 55:168
Cortesi N, Mariani C, Fedeli E (1973) Analisi degli steroli presenti negli insaponificabili degli oli mediante GLC. RISG 50:411
Cortesi N, Fedeli R, Gasparoli A, Tiscorni E (1977) Cromatografia liquida ad alta pressione (HPLC) nella analisi degli insaponificabili degli oli vegetali RISG 54:16
Crescimanno FG, Sottile I (1971) Osservazioni su sette combinazioni di innesto delle cultivar dell'olivo Frantoio. Sci Tech Agr 9–10:197–204
Crescimanno FG, Sottile I (1976) Ricerche sulla nutrizione minerale dell'olivo. Agric Ital 104:65–80
Cronquis A (1981) An integrated system of classification of flowering plants. Columbia Univ Press, NY
De Candolle A (1883) Origine des plantes cultivèes. Germer Baillière, Paris, pp 222–227
Diamantoglou S, Mitrakos K (1979) Sur la culture in vitro de l'embrion d'olivier (*Olea europaea* L. var. Oleaster). CR Acad Sci Paris Ser D 288:1537–1540
Diana G (1986) The transmission of *A. rhizogenes* plasmids in the propagation of olive cuttings. Ann Ist Sper Olivicoltura, Cosenza, 7:1–11 (in Italian)
Donini R, Roselli G (1972) Mutazioni indotte nell'olivo per irraggiamento di talee autoradicate. Genet Agrar 26:149–160
Earnshaw BA, Johnson MA (1985) The effect of glutathione on development in wild carrot suspension culture. Biochem Biophys Res Commun 133:988–993
El-Deen SAS, Wassis MM, Fadl MS (1979) Effect of salinity on olive seedling growth. Agric Res Rev Hort 57:259–269
Epstein E (1982) Levels of free and conjugated indole-3-acetic acid in ethylene-treated leaves and callus of olive. Physiol Plant 56:371–373
Ercolani GL (1971) Presenza epifitica di *Pseudomonas savastanoi*. Smith EF, Stevens, sull'olivo in Puglia. Phytopat, Mediterr 10:130–132
FAO (1980) China: development of olive production. 23 FAO, Rome
FAO (1985) Production yearbook, FAO Statistic, Series N.76 Rome
Fadl MS, El-Deen SAS (1980) Effect of N6-benzyl adenine on photosynthetic pigments and total soluble sugars of olive seedlings grown under saline conditions. Egypt J Hort 6:169–183
Fedeli E (1968) Recherches sur les composants mineurs des huiles vegetales RFCG 15:281
Fedeli E (1983) Les composants mineurs des lipides RFCG 30:51
Fedeli E, Brillo A (1975a) Meccanismo dei fenomeni di autossidazione delle sostanze grasse. Nota I. Olio di oliva. Meccanismo di saturazione con ossigeno. RISG 52:88
Fedeli E, Brillo A (1975b) Meccanismo dei fenomeni di autossidazione delle sostanze grasse. Nota II: Olio di oliva RISG 52:109
Fedeli E, Jacini G (1967) La distribuzione degli acidi grassi nei trigliceridi. RISG 44:393
Fedeli E, Jacini G (1970) Sui componenti odorosi dell'olio di oliva. Nota preliminare. Chim Ind 52:161
Fedeli and Mariani C (1973) Indentificazione dei polialcoli terpenici nell' insaponificabile di olio di oliva. RISG 50:164
Fedeli E, Mariani C (1974) Complessita' della frazione sterolica e triterpenica negli oli vegetali. Nota III RISG 51:129
Fedeli E, Lanzani A, Capella P, Jacini G (1966) Triterpenic alcohols and sterols of vegetables oils JAOCS 43:254
Fedeli E, Baroni D, Jacini G (1973a) Sui componenti odorosi dell'olio di oliva. Nota III RISG 50:38
Fedeli E, Favini G, Baroni D, Jacini G (1973b) Sui componenti odorosi dell'olio di oliva. Nota II Chim Ind 55:681
Fedeli E, Cortesi N, Mariani C, Baroni D, Jacini G (1974) Complessita' della frazione sterolica degli oli vegetali. Nota II Sci Tec Alim 3:143
Ferreira Llamas J (1984) Results of fertilization in olive. Olea 6:11–28

Fiorino P, Natali S (1985) Proposte tecniche per una olivicoltura moderna. In: Aspetti tecnici ed economici dell'olivicoltura Viterbese. Mannino P. (ed). ENEA Roma, pp 125–160

Fiorino F, Leva AR (1986) Investigation on the micropropagation of the olive (*Olea europaea* L.). Influence of some mineral elements on the proliferation and rooting of explants. Olea 17:101–104

Flath R, Forrey R, Guadagni GD (1973) Olive oil volatiles J Agric Food Chem 21:948

Fontanazza G, Rugini E (1977) Sull attitudine rizogena di talee di olivo provenienti da piante madri trattate con GA_3 e SADH Sci Tech Agr 3:75–83

Fontanazza G, Rugini E (1978) Le nuove tecniche di propagazione per talea. Seminario sul vivaismo e controllo della rizogenesi mediante fitoregolatori, Pistoia, pp 57–70

Fontanazza G (1987) Presentiamo la cultivar '1–77' – Terra e Vita 46 (Speciale Olio e Olivo) 10–11

Gabr NF, Bakr FI, Wally YA (1982) Tissue culture of olive (*Olea europaea* L.) using stem-internode section explants. Egypt J Hort 9:87–92

Gallitelli D, Savino V (1985) Olive latent virus-1, an isometric virus with a single RNA species isolated from olive in Apulia, southern Italy. Ann Appl Biol 106:295–303

Gamborg OL, Miller RA and Ojima K (1968) Nutrient requirement of suspension cultures of soybean root cells. Exp Cell Res 50:151–158

Gonzales F, Chaves M, Mazuelos C and Troncoso A (1973) Aspectos Fisiologicos en la Nutricion del Olivar, Variedad 'Manzanillo' de Mesa. Ciclo de Nutrientes en Hojas, Organos de Crecimiento y de Produccion. Ann Edaf Agr Nos 7–8:615–634

Griggs WH (1975) Olive pollination in California. Calif Agr Exp Stu Bull 869

Grossoni P (1979) Problem concerning the vitro culture of *Olea europaea* L. G. Bot Ital 113:75–88

Guerbaa H (1987) Relazione sulla situazione e sulle tendenze dell'offerta e della domanda dei principali prodotti dell'olivo. Olivae 16:6–13

Guerriero R, Scaramuzzi F (1979) Osservazioni sul comportamento di alcune combinazioni di innesto della cv di olivo 'Frantoio' in un impianto ad elevata densità. Riv Ortoflorafrutt It 63:137–146

Guerriero R, Scaramuzzi F, Crescimanno FG, Sottile I (1974) Confronto tra diverse combinazioni di innesto dell'olivo in piante autoradicate. Tech Agr 4:979–1000

Hartmann HT (1953) Effect of winter chilling on fruitfulness and vegetative growth in the olive. Proc Am Soc Hortic Sci 62:184–190

Hartmann HT (1958) Rootstock effects in the olive. Proc Am Soc Hortic Sci 72:242–251

Hartmann HT, Loreti F (1965) Seasonal variation in the rooting of olive cuttings. Proc Am Soc Hort Sci 87:194–198

Hartmann HT, Porlingis I (1957) Effect of different amount of winter chilling on fruitfulness of several olive varieties. Bot Gaz 119:102–104

Hartmann HT, Uriu K, Lilleland O (1966) Olive nutrition. In: Childer NF (ed) Temperate to tropical fruit nutrition. Horticul Rutgers, State Univ NY, pp 252–261

Hartmann HT, Whisler JE (1975) Some rootstocks and interstock influences in the olive (*Olea europaea* L.) cv 'Seviliano'. J Am Soc Hort Sci 95:670–674

Hartmann HT, Schnathorst WC, Whisler J (1971) "Oblonga" a clonal olive rootstock resistant to *Verticillium* wilt. Calif Agric 25:12

Hegazi ES, Stino GR (1982) Chemical regulation of sex expression in certain olive cultivars. Acta Agrobot 35:185–190

Heller R (1953) Recherches sur la nutrition minérale de tissus végétaux cultivé in vitro. Ann Sci Nat Bot Biol Veg 14:1–223

Hutzinger O, Kosuge T (1968) 3-Indole-acetyl-L-lysine, a new conjugate of 3-indolacetic acid produced by *Pseudomonas savastanoi*. Biochemistry 7:183–194

Inserra R, Vovlas N, Lamberti F, Zacheo B (1976) Plant parasitic nematodes associated with declining olive trees in southern Italy. In: Poljoprivredna znanstvena smotra. Agric Consp Sci Fasc 39:419–424

Inserra RN, Vovlas N, Fontanazza G, La Casta G (1981) Comportamento di alcune cultivar di olivo alle infestazioni di quattro specie di nematodi. Riv Ortoflorofrutt Ital 65:143–148

Instanbouli A, Neville P (1977a) Etude de la 'dormance' des semences d'olivier (*Olea europaea* L.). Mise en evidence d'une dormance embrionnaire. C R Acad Sci Paris Ser D 2503–2506

Instanbouli A, Neville P (1977b) Etude de la 'dormance' des semence d'olivier (*Olea europaea* L.). Influence favorable de la lumiere en presence d'obstacles à la germination. C R Acad Sci Paris Ser D 285:41–44

Instanbouli A, Neville P (1977c) Influences de quelques glucides sur la germination de l'olivier. Rev Genet Bot 84:305–317
Itoh T, Tamura T, Matsumoto T (1973a) Sterol composition of 19 vegetables oils. JAOCS 50:122
Itoh T, Tamura T, Matsumoto T (1973b) Methyl sterols composition of 19 vegetables oils. JAOCS 50:
Jacini G, Fedeli E (1977) Acidi triterpenici dell'olio di oliva. R Ist Lomb Sci Lett 106:446
Jacini G, Fedeli E, Lanzani A (1967) Research on the nonglyceridic substances of vegetables oils (preliminary note). J Ass Off Anal Chem 50:84
Jacoboni N, Fontanazza G (1976) Un nuovo tipo di propagazione dell'olivo: l'innesto talea in cassone riscaldato alla base. Ital Agr 113:104–112
Jacoboni N, Fontanazza G (1981) Le cultivar di olivo In: Baldini E, Scaramuzzi F (eds) l'Olivo. Frutticultura anni '80. Reda, Roma pp 9–52
James DJ (1987) Cell and tissue culture technology for the genetic manipulation of temperate fruit trees. Biotechnology and genetic engineering Rev, Vol 5. Intercept, Newcastle-upon-Tyne, pp 33–79
Johnson LAS (1957) A rewiew of the family Oleaceae. Contribution from the National Herbarium 2:395–418
Klein I, Lavee S (1977) The effect of nitrogen and potassium fertilizers on olive production. In: Fertilizer use and production of carbohydrates and lipids. Proc 13th Colloq Potash Inst Coll York, England pp 295–303
Knoblauch E (1895) Oleaceae: Die natürlichen Pflanzenfamilien. In: Engler A, Prantl K (eds) 4:1–16
Lagarda A, Martin G (1983) Manzanillo olive seed dormancy and influence by exogenous hormone application and endogenous abscisic acid content. Hortscience 18:869–871
Lagarda A, Martin GC, Kester DE (1983) Influence of environmental, seed tissue and seed maturity on Manzanillo olive seed germination. Hortscience 18:868–869
Lamberti F, Boines RC (1969) Pathogenecity of four species of Meloidogyne on three varieties of olive tree. J Nematol 1:111–115
Larcher W (1981) Low temperature effects on Mediterranean sclerophylls: an unconventional viewpoint. In: Margaries NS, Mooney HA (eds) Components of productivity of Mediterranean climate regions. Junk, The Hague, pp 259–266
Larcher W, De Moraes JAPV, Bauer H (1981) Adaptive response of leaf water potential, CO_2-gas exchange and water use efficiency of *Olea europaea* during dryng and rewatering. In: Margaris NS, Mooney HA (eds) Components of productivity of Mediterranean climate regions. Junk, The Hague, pp 77–84
Lavee S (1977) The growth potential of olive fruit mesocarp in vitro (*Olea europaea*). Tissue culture for horticultural purposes. Acta Hortic 78:115–122
Lavee S (1985) *Olea europaea*. In: Halevy A (ed) Handbook of flowering. Vol III. CRC Boca Raton:423–433
Lavee S, Adiri N (1974) The effects of abscisic acid and gibelleric acid on the development of apple and olive callus. In: Plant growth substances, part VI. Tiss Cul Proc 18th Int Conf, Hirokawa, Tokyo, pp 1141–1148
Lavee S, Avidan N (1982) Growth responses of tree callus to chlorogenic acid and related phenolic substances. In: Fujiwara A (ed) Plant tissue culture 1982. Maruzen, Tokyo, pp 165–168
Lavee S, Avidan N (1986) Involvement of cathodic isoperoxidases in growth processes. In: Greppin H, Penel C and Gaspar TH (eds) Molecular and physiological aspects of plants peroxidase. Cent Bot Univ Geneve, pp 305–316
Lavee S, Epstein E (1976) Metabolic differences in degradation of ^{14}C-1AA in apple and olive tissue culture. 9th Int Conf Plant Growth Subst, Lausanne, pp 216–218
Lavee S, Haskal A, Vodner M (1986) *Barnea*: a new olive cultivar from first breeding generation. Olea 17:95–99
Lavee S, Messer G (1969) The effect of growth-regulating substances and light on olive callus growth in vitro. J Exp Bot 20:604–614
Lavee S, Tanne E (1984) Spherosis: a virus disease of olive production. Olea 6:71–75
Lingelsheim A (1920) Oleaceae – Oleoideae – Fraxineae and Oleaceae – Oleoideae – Syringe: In: Egler A (ed). Pflanzenreich 4:1–25
Loreto F, Bongi G (1987) Control of photosynthesis under salt stress in the olive. Int Conf on Agrometereology, Cesena, pp 411–414
Loukas M, Krimbas CB (1983) History of olive cultivars based on their genetic distances. J Hortic Sci, 58:121–127

Loussert R, Brousse G (eds) (1978) L'olivier: Maisonneuve and Larose, Paris
Mariani C, Fedeli E (1985) I gliceridi parziali degli oli vegetali RISG 62:129
Marte M, Gadani F, Savino V, Rugini E (1985) Strawberry latent ringspot virus associated with a new disease of olive in central Italy. Plant Dis. 70:171–172
Martelli GP (1981) Le virosi dell'olivo esistono? Inf Fitopatol 1–2:97–100
Mazzolani G, Altamura Betti MM (1976–77) Elementi per la revisione del genere *Olea* (Tourn) Linn. 1-Nota introduttiva. Ann Bot 35–36:463–469
Mazzolani G, Altamura Betti MM (1979) Elementi per la revisione del genere *Olea* (Tourn). Linn. III – *Olea Crysophylla* Lam: forme affini. Ann Bot 28:15–31
Mazzolani G, Altamura Betti MM (1980) Element per la revisione del genere *Olea* (Tourn) Linn. IV: Ciclo di *Olea capensis* L. Ann Bot 28:149–172
Milella A (1983) Alternate bearing in olives: origins; causes and possible treatment. Studi Sassaresi 30:155–160
Mitrakos K, Diamantoglou S (1984) Endosperm dormancy breakage in olive seeds. Physiol Plant. 62:8–10
Morchio G, De Andreis R, Fedeli E (1987) Indagine sul contenuto di steroli totali in oli di oliva e sulle variazioni nel ciclo di raffinazione. RISG 64:185
Morettini A (1954) Ricerche sull'autonomia delle foglie delle più note varietà degli olivi toscani in relazione alle loro resistenza al *Cicloconium oleaginum*. Not Mal Piante 28:3–11
Morettini A (1961) Influenza della defogliazione anticipata sulla fioritura e la fruttificazione dell'oliva. Ann Sper Ago 3
Morettini A (ed) (1972) Olivicoltura: Reda, Roma
Murashige T, Skoog F (1962) A revised medium for rapid growth and bioassay with tobacco tissue cultures. Physiol Plant 15:473–497
Natali S, Xiloyannis C, Angelini P (1985) Water consumption use of olive trees (*Olea europaea* L.) and effect of water stress on leaf water potential and diffusive resistance. Acta Hortic 171:341–350
Pacini E, Cresti M (1977) Viral particles in developing pollen grains of *Olea europaea* L.. Planta 137:1–4
Pannelli G, Filippucci B (1982) The effect of shading the mother plants on the rooting of olive cuttings cv. Frantoio. Inf Agr 38:21613–21615 (in Italian)
Petri L (1927) Le malattie dell'olivo. Ist Microrg Ital, Firenze
Poli M (1979) Etude bibliographique de la physiologie de l'alternance de production chez l'olivier (*Olea europaea* L.) Fruits 34:687–695
Pontikis CA, Loukas M, Kousounis G (1980) The use of biochemical markers to distinguish olive cultivars. J Hortic Sci 55:333–343
Porlingis IC, Voyiatzis DG (1986) Influence of paclobutrazol plant growth regulator on vegetative and reproductive growth of olive (*Olea europaea* L.). Acta Hortic 179:587–588
Roselli G, Donini (1982) "Briscola" Nuova cultivar di olivo a sviluppo compatto. Riv Ortoflorofrutt Ital 66:103–114
Roselli G, Scaramuzzi F (1974) Definizione dei termini di cultivar e clone. Atti Accad Agr Bologna 2:119
Rugini E (1981) Propagazione in vitro di una cultivar di olivo (*Olea europaea* L.). Valutazione di varie citochinine ed auxine. Atti Congr Fitofarmaci in Agricoltura, Firenze, pp 171–180
Rugini E (1984) In vitro propagation of some olive (*Olea europaea* L.) cultivars with different rootability, and medium development using analytical data from developing shoots and embryos. Sci Hortic 24:123–134
Rugini E (1986) Olive (*Olea europaea* L.). In: Bajaj YPS (ed) Biotechnology in agriculture and forestry. Trees I. Springer, Berlin Heidelberg New York Tokyo, pp 253–267
Rugini E (1988) Somatic embryogenesis and plant regeneration in Olive (*Olea europaea* L.). Plant Cell, Tissue and Organ Culture, 14:207–214
Rugini E (1990) The involvement of polyamines on in vitro rooting of fruit trees induced by auxins or *Agrobacterium rhizogenes*. J. Am Hort Sci (in press)
Rugini E, Fontanazza G (1981) In vitro propagation of Dolce Agogia olive. Hort Science 16:492–493
Rugini E, Tarini P (1986) Somatic embryogenesis in olive (*Olea europaea* L.). Conf Fruit tree Biotech, Oct 14–15, Paris, p 62
Rugini E, Verma DC (1983) Micropropagation of a difficult to propagate almond (*Prunus amygdalus* Batsch). Plant Sci Lett 28:273–281

Rugini E, Wang XS (1986) Effect of polyamines, 5-azacytidine and growth regulators on rooting in vitro of fruit trees, treated and untreated with *Agrobacterium rhizogenes*. In: 6th Int Congr plant tissues Cell Cult. Univ Minnesota, Minneapolis, p 374
Rugini E, Fontanazza G, Bongi G (1979) Ricerche preliminari sulla coltura in vitro nella specie Olea europaea L. Atti Incontro sulle tecniche di colture in vitro per la propagazione su vasta scala delle specie ortoflorofrutticole, Pistoia, pp 193–202
Rugini E, Bongi G, Fontanazza G (1982) Effect of Ethephon on olive ripening. J Am Soc Hortic Sci 107:835–838
Rugini E, Bongi G, Mencuccini M (1986) Effect of putrescine L-arginine and cobalt on fruit-set, ethylene content and apparent parthenocarpy in olive. Acta Hortic 179:369–375
Rugini E, Tarini P, Rossodivita ME (1987a) Control of shoot vitrification of almond and olive grown in vitro. Acta Hortic 212:177–183
Rugini E, Jacoboni A, Bazzoffia A (1987b) A simple in vitro method to avoid the initial dark period and to increase rooting in woody species. Acta Hortic 227:438–440
Rugini E, Politi V, Bignami C, De Agazio M, Grepo S (1990) Preliminary results on the effect of polyemine treatment on rooting of three Italian olive cultivar cuttings. Acta Hortic (in press)
Salleo S, Lo Gullo MA (1983) Water transparent pathways in nodes and internodes of 1-year-old twigs of *Olea europaea* L. Giorn Bot Ital 117:63–74
Savastano L (1886) Les maladies de L'olivier: hyperplasies et tumeurs. CR Acad agr France 103:1278
Savino V, Gallitelli D (1981) Cherry leafroll virus in olive. Phytopathol Mediterr 20:202–203
Savino V, Gallitelli D (1983) Isolation of cucumber mosaic virus from olive in Italy. Phytopathol Mediterr 22:76–77
Savino V, Barba M, Gallitelli D, Martelli GP (1979) Two nepovirus isolated from olive in Italy. Phytopathol Mediterr 18:135–142
Savino V, Gallitelli D, Barba M (1983) Olive latent ringspot virus, a newly recognized virus infecting olive in Italy. Ann Appl Biol 103:243–249
Savino V, Piazzolla P, Di Franco A, Martelli GP (1984) Olive latent virus, 2, a newly recognized virus with apparently shaped particles. Proc 5th Congr Mediterr Phytopathol Un, Cairo
Sax K, Abbe EC (1932) Chromosome numbers and the anatomy of the secondary xylem in the Oleaceae. J Arnold Arbor 23:37–48
Scaramuzzi F (1962) Gradiente di vegetazione e nuove forme di allevamento dell'olivo. Ital Agr 1:633–640
Scaramuzzi F, Gaetano de A (1974) Ricerche sulla rizogenesi di espianti di fusto di *Olea europaea* L. cv. Ogliarola di Monopoli, coltivati in vitro. Riv Ortoflorafrutt It 58:419–436
Scaramuzzi F, Roselli G (1986) Olive genetic improvement. Olea 17:1–17
Shamtsyan SM (1985) Dynamic of sugar content in leaves of some olive cultivars in relation to frost resistance. Hortic Abstr 1986 56:786
Shulman Y, Lavee S (1973) The effect of cytokinins and auxins on anthocyanin accumulation in green Manzanillo olive. J Exp Bot 81:655–661
Sondheimer E, Blank GE, Galson EC, Sheets FM (1970) Metabolically active glucosides in Oleaceae seeds. I. The effect of germination growth and hormone treatments. Plant Physiol 45:658–662
Stutte GW, Martin GC (1986) Effect of light intensity and carbohydrate reserves on flowering in olive. J Am Soc Hortic Sci 111:27–31
Surico G (1986) Indolacetic acid and cytokinins in the olive knot disease. An overview of their role and their genetic determinants. In: Baily JA (ed) Biology and molecular biology of plant-pathogen interactions. Springer, Berlin Heidelberg New York Tokyo, pp 315–329
Taylor H (1945) Cyto-taxonomy and phylogeny of the Oleaceae. Brittonia 5:337–367
Taufel T and Sevzisko R (1961) Ernährungsforschung 6:323
Therios I (1984) Kinetics and energetics of nitrate adsorption by olive plants (cultivar Chondrolia Chalkidikis). Riv Ortoflorofrutt Ital 68:255–263
Therios I, Sakellariadis SD (1982) Some effects of varied magnesium nutrition on the growth and composition of olive plants (cultivar, Chondrolia Chalkidikis). Sci Hort 17:33–41
Thompson RG, Tyree MT, Lo Gullo MA, Salleo S (1983) The water relations of young olive trees in a Mediterranean winter: measurements of evaporation from leaves and water conduction in wood. Ann Bot 52:399–406
Tombesi A and Standardi A (1977) Effetti della illuminazione sulla fruttificazione dell'olivo. Riv Ortoflorafrutt It 6:368–380

Troncoso A, Cerda A, Rallo L (1984) A system of tabular containers for the propagations of cuttings. Olea 6:33–37
Varvaro L, Surico G (1978) The responses of various olive cultivars to artificial inoculum with *Pseudomonas savastanoi*. Phytopathol Mediterr 17:174–177
Vioque E, Maza MP (1961) Minor components of olive oils JAOCS 38:485
Vitagliano C, Sonnoli A (1979) Indagini preliminari sulla fruttificazione di piante di olivo allevate in contenitore. Atti L'irrigazione delle colture arboree, 28 Sept Bologna
Vitagliano M, Turri E (1958) Componenti minori dell'olio di oliva. Olearia 12:145
Vitagliano M (1960) Componenti minori dell'olio di oliva RISG 37:136
Walkey DGA (1978) In vitro methods for virus elimination. In: Thorpe TA (ed) Frontiers of plants tissue culture. Univ Press, Calgary, pp 245–254
Wallace A, Mueller RT, Hartsock T (1979) Effect of irrigating small olive trees grown in a glasshouse with different percentages of seawater. Alex. J. Agric Res 27:221–224
Wang KJ, Zhang PF, Ni DX, Zhu XZ, Yang WQ, Bao ZH (1979a) Cytohistological studies on the tissue culture of *Olea europaea* L. I. Development of callus in vitro. Acta Bot Sin 21:127–134 (in Chinese)
Wang KJ, Zhang PF, Ni DX, Zhu XZ, Yang WQ, Bao ZH (1979b) Cytohistological studies on the tissue culture of *Olea europaea* L. II. Histogenesis and organogenesis. Acta Bot Sin 21(3):225–230 (in Chinese)
Wang KJ, Zhang PF, Ni DX, Zhu XZ, Yang WQ, Bao ZH (1981) Callus formation and organ regeneration in tissue culture of woody plants. Acta Bot Sin 23:97–108 (in Chinese)
White PR (1963) The cultivation of animal and plant cells. Ronald, New York
Wigodsky DE, Philipps A (1937) L'epoca della differenziazione delle gemme fiorali nell'olivo. Nuovo G Bot Ital 46:484–487
Wilhelm S, Taylor JB (1965) Control of *Verticillium* wilt of olive through natural recovery and resistance. Phytopathology 55:310–316
Ying G, Shu-zhi H, Ziu-jun S, Hui-chan B, Hou-jun W, Shan-an H (1984) Studies on variation and selection of olive seedlings for cold resistance. In: Shan-an H, Ying G (eds) Olive acclimatation and breeding. Nanjing, PRC p 247–331 (in Chinese)
Yuce B (1979) Studies on the germination of olive seed. Tarimsal Arastirma Dergisi 1:189–203 (in Turkish)
Zarrouk M, Cherif A (1981) Effect of sodium chloride on lipid content of olive trees (*Olea europaea* L.) Pflanzenphysiol 105:85–92
Zhigarevitch IA (1986) Varietal frost-resistance of the olive tree in Azerbajian. Bujul Inst Subtrop Cult 3:37
Zimmermann RH (1986) Propagation of fruit, nut, and vegetable crops – overview. In: Zimmermann RH, Griesbach RJ, Hammerschlag FA, Lawson RH (eds). Tissue culture as a plant production system for horticultural crops. Nijhoff, Dordrecht, pp 183–200
Zui-jun S, Shu-zhi H, Kan-xi L, Chui-hua L, Liv L, Ying G (1984) Breeding of olive trees for freezing resistance. In: Shan-an H, Ying G (eds) Olive acclimatation and breeding. Najing PRC pp 429–445 (in Chinese)

Section V Miscellaneous

V.1 *Indigofera* spp.

P. K. EVANS[1]

1 Introduction

Indigofera is a large genus containing some 800 species within the family Leguminosae. As a genus it is best known for the indigo plant, from which the dye indigo is obtained. Many of the species are annual or perennial herbs, but some are shrubs, whilst others grow to small trees. Members of the genus are widespread in tropical and subtropical areas, particularly the Sino-Indian region, but they are also found in the tropics of both America and Africa. Indeed, the genus *Indigofera* is one of the largest and most widely distributed in South Africa. Many of the species are adapted to semi-dry areas, and species are characteristically found in open and waste places, along roadsides and in grass and brushwood (Allen and Allen 1981).

1.1 Traditional Uses

At one time, species of *Indigofera* were extensively planted as a source of indigo. The quest to obtain supplies of this dye had a major impact on world history and on the development of international trade. The use of indigo as a dye goes back at least 4000 years, with evidence that the Egyptians used it to dye mummy cloths. The peoples of Central Africa and the Peruvians are thought to have used the dye since prehistoric times. It was also used extensively in India and the Far East. Although in the 16th century the Portugese had established trade links in indigo with India, the dye was effectively kept out of much of Europe until the end of the 16th century by the activities of the woad growers and distributors, even though woad produces a dye which is more difficult to prepare and inferior to indigo. However, by the middle of the 17th century, a considerable trade in indigo had been built up between Europe and India, and it was with the aim of obtaining a more secure supply of the dye that the Dutch East India company was formed in 1631. Cultivation of indigo was also undertaken by the American Colonists and, helped by restrictive trading practices in Europe, the West Indies came to surpass India as the main centre of cultivation. But towards the end of the 18th century, cotton and sugar became more lucrative crops for the colonists, and the centre of indigo cultivation shifted back to India. In the latter half of the 19th

[1] Department of Biology, University of Southampton, Southampton, SO9 5NH, United Kingdom

Biotechnology in Agriculture and Forestry, Vol. 10
Legumes and Oilseed Crops I (ed. by Y. P. S. Bajaj)

century, events in Europe were again to have a profound effect on the cultivation of indigo. William Henry Perkin in 1856 discovered that the addition of caustic potash to anile sulphate resulted in a deep purple staining precipitate. This anile sulphate is actually an oil derived by combining indigo with sulphur. It was already known that aniline could be derived from the distillation of coal tar. Subsequently, it was found that aniline from coal tar was identical to the substance derived from indigo. Indigo was synthesized by Baeyer in 1878 and the first commercial process for its production was developed in 1890. Eight years later, synthetic indigo was first marketed on a large scale. Cultivation of indigo then rapidly declined such that by the late 1930's export of indigo from India represented less than 0.1% of the level just prior to the development of a method for its manufacture. Currently, *Indigofera* is no longer planted widely as a source of dye, but it is still grown for local usage. In West Africa it is used in the art of tie-dyeing and Nigerian tie-dyed cloth has found a market in Europe and America (for an account of the history of indigo see Peattie 1936; Rembert 1979; Allen and Allen 1981).

Over the years some six species of *Indigofera* have been cultivated as major sources of the dye. *Indigofera suffruticosa* (= *I. anil* L.) from tropical America was widely grown in both the New and Old World, whilst *I. tinctoria* was the main source of indigo in the Far East. *Indigofera sumatrana* was introduced by the Dutch to India from Malaysia because it produced a higher yield of indigo, but this species is now considered to be synonymous with *I. tinctoria* and to have been produced in cultivation. It was *I. sumatrana* which was taken from the Old World to the West Indies to establish the crop there. In addition, *I. articulata*, and *I. arrecta* from tropical Africa and *I. longeracemosa* from Madagascar have been cultivated. Both *I. arrecta* and *I. longeracemosa* produce high yields of indigo, with the former at one time being an important source of the dye in Abyssinia.

The actual coloring agent in indigo dye is indigotin, which as such is not present in the indigo plant. Instead the plant contains indican, which is a colorless glycoside (indoxyl-β-D-glycoside). On hydrolysis, indican yields glucose and indoxyl. Subsequent oxidation of indoxyl in air results in the precipitation of blue particles of indigotin. In practise the hydrolysis is brought about by harvesting the plant just before flowering. The plant tissues are crushed and then allowed to steep in water for half a day, whereupon the activity of β-glucosidase in the tissue cleaves the glucose residue. The liquid is drawn off and aerated, often by beaters, until the color develops. When the agitation stops, the blue precipitate settles out. It is then heated to prevent any further enzyme activity. Indigo dye is actually insoluble and, therefore, before cloth can be dyed, the indigo must first be reduced to indigo white, which is soluble in alkaline solution. After being soaked in the indigo white solution, the fabric is hung up to dry, whereupon oxidation in the air produces the deep blue color which is deposited in the fabric (Martin-Leake 1975).

In addition to cultivation as a source of dye, *Indigofera* species have been used traditionally for other purposes. Some species are valued as local medicines. A decoction of the roots of *I. pascuorum* is used in parts of tropical America to treat fevers and as a stimulant, and *I. lespedezioides* is used in Guatemala to treat stomach complaints. This latter species is also considered to have insecticidal properties and to deter lice and mites. A water extract of the plant parts is applied to domestic animals for this purpose in some parts of Central and South America. In India the seeds of *I. trita* have been used as a tonic (Usher 1974).

1.2 Current Uses

A number of species of *Indigofera* are considered to have potential as grazing forage and as feed supplements for ruminants and non-ruminants. Like other members of the Leguminosae, species of *Indigofera* have the capacity, in association with *Rhizobium*, to fix nitrogen, and the plants are rich in protein. Indeed, nodulation studies have been carried out on a wide range of *Indigofera* species (see Allen and Allen 1981). In addition, the ability of some species to tolerate semi-arid conditions or salinity increases their agronomic desirability. In fact, *I. pauciflora* has been grown for some time in Asia and Africa as fodder for camels. Recently, a considerable number of *Indigofera* species have been evaluated as a forage crop. There are, however, reservations about the use of *Indigofera* species as forage because it is known that there are deleterious substances present in a number of species. As a tropical forage legume *I. spicata* shows some potential, but canavanine, a non-protein amino acid, is present in the seeds, and this substance is toxic to mice. More importantly, it has been shown that *I. spicata* in the diet causes the death of chicks and is hepatotoxic to cattle, rabbits and mice (see Aylward et al. 1987). Indospicine, a free non-protein amino acid analogue of arginine, has been detected in the leaves and seeds of several *Indigofera* species including *I. spicata* (Hegarty and Peterson 1973), and when purified indospicine was included in the diets of mice and rats, it caused similar toxicity (Christie et al. 1969).

Another use for some species of *Indigofera* is as a cover crop to protect the soil from erosion or improve the quality of the soil. *Indigofera arrecta* is grown for this purpose because it produces a deep root system with a plentiful supply of nodules. In addition, it gives a dense shade, which smothers weeds. It has been suggested as an intercrop to provide shade and protection for young cocoa trees and for oil palms. Recently, evidence has begun to accumulate that the growth of some *Indigofera* species as a cover crop, notably *I. hirsuta*, leads to a reduction in the soil population of nematodes (Rhoades 1976; Rhoades and Forbes 1986). This capacity would further increases the agronomic interest in *Indigofera*. As well as acting to protect the soil as cover crops, *Indigofera* species have been used as a green manure to enrich impoverished tropical soils.

About a dozen species of *Indigofera*, among them *I. amblyantha*, *I. potaninii* and *I. delsiana*, are valued in Britain and Northern Europe as ornamental garden shrubs (Bean 1970). These plants are not native to this region, but have been introduced from parts of China, Japan and the Himalayas.

2 In Vitro Studies (Table 1)

Indigofera has received only limited attention from the viewpoint of in vitro culture. Where work has been carried out it has usually been part of a larger exercise to assess the regenerative capacity of various legumes.

2.1 Callus Formation

Excised leaflets from in vitro-grown *I. enneaphyll* plants produced small patches of green callus when cultured on B5 medium (Gamborg et al. 1968) with IAA and BAP (Bharal and Rashid 1979). Similarly, Howell et al. (1987) were able to induce callus formation on leaf explants of *I. potaninii* on B5 medium using IAA and BAP, with an average total volume of callus of somewhat less than 0.5 cm^3 being obtained per 1 cm^2 explant after 5 weeks in culture. They achieved rather higher yields of callus when NAA or 2,4-D was used instead of IAA as the auxin in combination with BAP. However, when B5 medium was compared with MS (Murashige and Skoog 1962) and UM (Uchimiya and Murashige 1974) medium,

Fig. 1. *Indigofera potaninii* leaf explant producing friable callus and some roots after 5-weeks culture on MS medium with 2,4-D at 2 mg/l and kin at 0.25 mg/l

Table 1. In vitro culture of various species of *Indigofera*

Species	Inoculum	Medium (mg/l)	Growth response	Reference
I. enneaphylla	Leaf	B5 + BAP (1) + IAA (0.5)	Callus and buds	Bharal and Rashid (1979)
	Cotyledon	B5 + BAP (1) + IAA (0.5)	Callus and buds and shoots	Bharal and Rashid (1979)
	Cotyledon	B5 + BAP (1) + 2,4-D (0.5)	Callus	Bharal and Rashid (1979)
	Cotyledon	B5 + BAP (1) + NAA (0.5)	Callus and a few shoot buds	Bharal and Rashid (1979)
	Cotyledon	B5 + zeatin (1) + IAA (0.5)	Little response	Bharal and Rashid (1979)
	Cotyledon	B5 + kin (1) + IAA (0.5)	Some callus	Bharal and Rashid (1979)
	Immature seed	B5 + BAP (1) + IAA (0.5)	Callus and buds and shoots	Bharal and Rashid (1979)
	Callus tissue	Liquid B5 + BAP (1) + 2,4-D (0.5)	Cell suspension culture	Bharal and Rashid (1984)
	Suspension cells	plated in B5 + BAP (1)	Shoots	Bharal and Rashid (1984)
I. potaninii	Leaf	MS + BAP (2) + NAA (0.2)	Callus and buds	Howell et al. (1987)
	Leaf	B5 + BAP (2) + NAA (0.2)	Callus and buds and shoots	Howell et al. (1987)
	Leaf	MS + BAP (0.25) + 2,4-D (2)	Callus and roots	Howell et al. (1987)
	Leaf	UM + NAA (20)	Callus and roots	Howell et al. (1987), Evans et al. (1986)
	Leaf	UM + BAP (5) + NAA (10)	Callus and buds and shoots	Howell et al. (1987), Evans et al. (1986)
I. amblyantha	Leaf	MS + K (0.25) + 2,4-D (2)	Some callus and roots	Howell et al. (1987)
	Leaf	UM + BAP (5) + NAA (0.2)	Some callus	Howell et al. (1987)
I. delsiana	Leaf	UM + BAP (5) + NAA (0.2)	Small amount of callus	Howell et al. (1987)

it was found that these latter two media supported a greater development of callus than B5 when 2,4-D was used as the auxin in combination with either kin or BAP (Fig. 1). An average volume of callus of 2 cm^3 was obtained per explant after 5 weeks in culture. In contrast, when NAA was the auxin in combination with BAP, no difference in terms of callus development was found between the three media. When Howell et al. (1987) cultured leaf explants of *I. amblyantha* and *I. delsiana* on medium of the same composition as that which supported callus development with *I. potaninii*, only very small amounts of callus were formed on *I. amblyantha* explants, and virtually none on *I. delsiana*. Bharal and Rashid (1979) produced callus on cotyledon explants of *I. enneaphylla*. Green compact callus formed on cotyledons cultured on B5 medium with IAA and BAP, but on medium with zeatin instead of BAP the cotyledons enlarged but did not produce callus. Substitution of NAA or 2,4-D for IAA also led to callus formation and with 2,4-D, this callus was light brown and fragile. When immature seeds with green cotyledons were cultured on B5 medium with IAA and BAP, they initially expanded to form a small root, hypocotyl, and main shoot with the hypocotyl subsequently producing callus.

2.2 Cell Suspension Culture

When friable callus produced on leaf explants of *I. potaninii* cultured on MS medium containing 2 mg/l 2,4-D and 0.25 mg/l kin was transferred to an agitated liquid UM medium containing the same levels of growth regulators, a suspension culture was established (Howell et al. 1987). After several monthly subcultures, a highly dispersed fine cell suspension was obtained. Similarly, Bharal and Rashid (1984) reported the formation of a cell suspension culture on transfer to liquid medium of callus tissue formed on explants of *I. enneaphylla* cultured on a medium containing BAP at 1 mg/l and 2,4-D at 0.5 mg/l.

2.3 Plating of Cells from Suspension Cultures

Bharal and Rashid (1984) filtered cells from a cell suspension culture of *I. enneaphylla* through muslin to obtain a population of 95% free cells and 5% cell aggregates of two to six cells. These cells were mixed with an equal volume of medium containing 1.2% agar kept molten at 40 °C. This was overlaid onto solidified medium containing 0.6% agar in Petri dishes. After some 25 days, small colonies had developed on the plates. A plating efficiency was determined by counting the total number of colonies formed per Petri dish, dividing by the number of viable cells plated in the Petri dish, and expressing this figure as a percentage. Determinations of cell viability at the time of plating were made using fluorescein diacetate (Widholm 1972) and only about 50% of the cells were found to be viable. The plating efficiency was influenced by the plating density, media composition, and the level of CO_2 present. Only a very low plating efficiency (1.24%) was obtained when medium suitable for maintenance of callus was used. However, if this medium was supplemented with casein hydrolysate, the plating efficiency increased. Inclusion of casein hydrolysate at 500 mg/l resulted in plating efficiencies of up to 20%. This level was obtained with cells plated at 2.5×10^4 per ml. As the plating density was reduced, there was a reduction in plating efficiency, with only very few colonies being formed at a plating density of 5×10^2 cells/ml. The level of CO_2 was found to be a critical factor for colony development. Absorption of CO_2 with KOH caused the failure of colonies to develop at a plating density of 2.5×10^3 cells/ml, a level which supported considerable colony formation in the controls. Even at 5×10^3 cells/ml very few colonies formed in a CO_2-free atmosphere, and those that did were slow-growing. In contrast, light intensity did not appear to have any influence on the plating efficiency.

2.4 Protoplast Isolation

Only a very limited amount of work has been done using *Indigofera* for protoplast isolation. Stewart (unpubl.) attempted to isolate protoplasts from *I. potaninii*. She was unable to peel the leaflets, but cells could be released from the leaves by grinding in a pestle and mortar with sand and water or by using a vortex impinger (Schwenk 1980). However, conversion of these isolated cells into protoplasts proved difficult. Subsequently, thinly sliced pieces of leaf tissue incubated for 6 h in an enzyme solution containing 1% Cellulase RS, 0.5% Rhozyme and 0.1% Pectolyase with 11% mannitol led to the release of workable quantities of protoplasts. Yields of up to 1.7×10^6/g of leaf tissue were obtained. Reasonable yields could also be obtained by replacing Pectolyase with 0.4% Macerozyme. Attempts were also made to isolate protoplasts from cell suspension cultures by treatment with 2% Cellulase R10 and 0.4% Macerozyme in either 9% or 11% mannitol, but only small numbers of protoplasts were recovered. Some attempt was made to culture the mesophyll protoplasts but no division was obtained.

2.5 Shoot Organogenesis

The development of adventitious buds occurred on excised green cotyledons of *I. enneaphylla* when they were cultured on a medium containing IAA at 0.5 mg/l and BAP at 1 mg/l (Bharal and Rashid 1979). At first the explants expanded and produced some callus. But after some 6 weeks in culture 90% of the explants had produced shoot bud-like structures, and of these 50% had shoots with well-developed leaves. Between 5 and 30 shoots and 50 to 150 shoot buds occurred on each explant. Immature seeds with green cotyledons also produced shoot buds. Callus formed on the base of the hypocotyl and this differentiated to produce shoot buds, which grew into shoots with some 40 to 100 shoot buds on each hypocotyl. Cotyledons still attached to the embryo axis also produced buds and shoots. Interestingly, substitution of BAP with either zeatin or kin at 1 mg/l failed to lead to shoot bud formation on the cotyledon explants. Similarly, substitution of IAA with 2,4-D failed to lead to shoots, but instead resulted in callus formation. Transfer of this callus to medium with IAA and BAP induced shoot organogenesis. Replacement of IAA with NAA also caused callus formation but some buds did

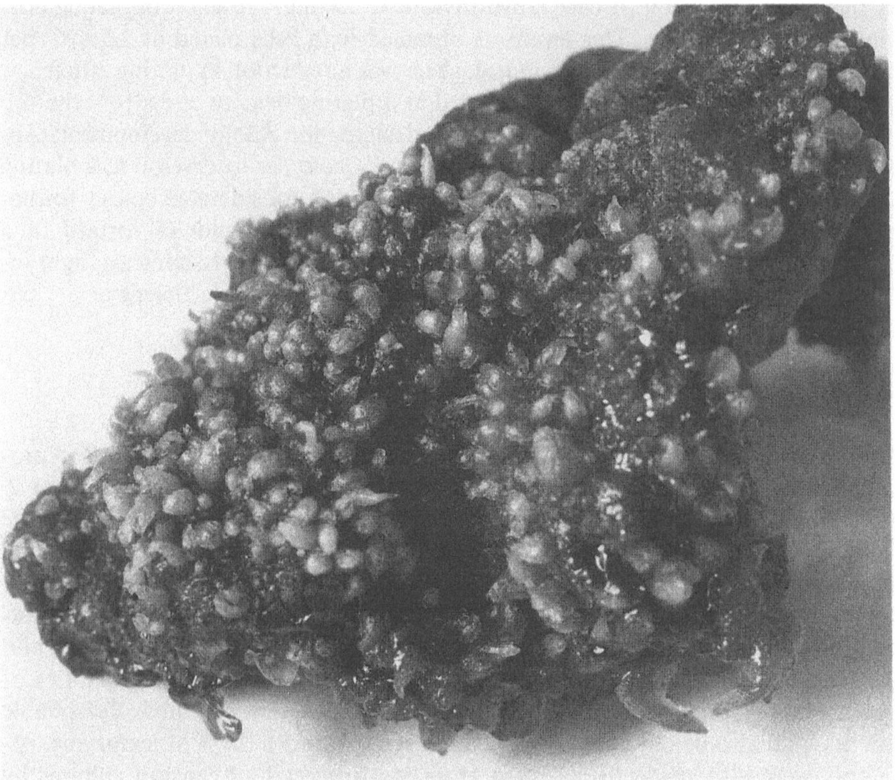

Fig. 2. Callus derived from *I. potaninii* leaf tissue producing a large number of shoot buds 4 weeks after subculture on UM medium with NAA at 0.2 mg/l and BAP at 5 mg/l

form; however, these soon callused. Shoot bud-like structures also developed on leaf explants when they were cultured on medium containing IAA and BAP. Some 50%–60% of the cultures showed this response and from 10–50 buds formed on each culture. However, the shoot buds formed on the leaf explants failed to develop any further. Howell et al. (1987) were also able to induce bud formation on leaf explants in culture. However, in these experiments the leaf explants were taken from *I. potaninii.* When B5 medium was used, buds and shoots formed in response to IAA and BAP and to NAA and BAP but not to 2,4-D and BAP. However, when B5 medium was replaced with either MS or UM medium, shoots and buds were not obtained in response to IAA, but only to NAA (Fig. 2). In agreement with the findings of Bharal and Rashid (1979), buds formed in response to IAA did not expand. These buds could be induced to expand properly only when transferred to media where NAA replaced IAA. Subculture of tissue produced on UM medium with NAA at 0.2 mg/l and BAP at 5 mg/l at monthly intervals resulted in the production of large numbers of healthy shoots. These shoots were produced 7 to 14 days after transfer of the callus to fresh medium. All shoot-producing callus was dark brown on the outside with a compact bright green central core. Tissue derived from one leaf explant continued to produce shoots for over

2 years in culture. A more detailed study of the response of leaf explants to various concentrations of NAA and BAP revealed that, in the absence of NAA, bud formation occurred on callused leaf explants of *I. potaninii* at BAP concentrations of 5, 10, and 20 mg/l, but not at lower concentrations. In the presence of NAA, buds were also formed at the lower concentrations of BAP. However, at low concentrations of BAP (0.5 mg/l), the budding response was poor except in the presence of high levels of NAA. In contrast, no buds were formed when NAA was present in the absence of BAP. The buds expanded to form shoots on media containing high levels of both BAP and NAA.

It has also proved possible to induce shoot organogenesis on callus colonies derived from cell suspension cultures. Bharal and Rashid (1984) transferred colonies which had formed from cells derived from a suspension culture plated in agar to a medium containing BAP but no auxin. After 3–4 weeks shoots appeared which, after transfer to a medium to induce rooting, subsequently developed into plants. Howell et al. (1987) also obtained shoot organogenesis from cell suspension cultures. When a suspension culture was transferred from a medium containing 2,4-D and kin to one containing NAA and BAP, small brown/green cell aggregates formed. On transfer to agar these aggregates developed into callus colonies producing buds and shoots. However, these shoots often died back and those that survived could not be induced to expand beyond 1–1.5 cm.

2.6 Root Formation

Unlike cultures of a number of legumes, root production on cultured explants and callus of *Indigofera* is not prolific. On cultured leaf explants of *I. potaninii* (Howell et al. 1987) roots formed when media lacked cytokinin but contained relatively high levels of NAA (5 mg/l and above). However, when the medium contained 2,4-D, roots were formed in the presence of BAP. Leaf explants of *I. amblyantha* also produced roots on media containing BAP and 2,4-D but, in addition, roots were produced on media containing NAA above 2 mg/l with or without the presence of BAP at concentrations of 2 mg/l and above.

Rooting of excised shoots of *I. potaninii* could be achieved fairly readily (Howell et al. 1987). When incubated in the dark at 27 °C, a small number of shoots produced roots in half-strength B5 medium solidified with 0.4% agar either with 1 mg/l IBA or without growth regulators. This number could be increased by culturing them on medium with 1 mg/l IBA for 2–3 days and then transferring them to growth regulator-free medium. A further improvement in the proportion of shoots producing roots could be obtained by incubation in the light at 3000 to 4000 lx. Although a contact time with IBA of 4 days increased the proportion of rooting shoots, callus was also produced at the base of the shoots and this was regarded as detrimental. Bharal and Rashid (1979) found that in order to obtain profuse rooting and good growth of *I. enneaphylla* shoots it was necessary to transfer them with some callus still attached to medium containing NAA and BAP and then subsequently excise the shoots and place them on basal medium.

2.7 Weaning of Regenerated Plants

No difficulty was experienced in weaning *Indigofera* plantlets from in vitro conditions (Howell et al. 1987). After short roots had developed on the excised shoots they were transferred to a sand/Levingtons compost mixture and placed in a mist propagator (Fig. 3). When new leaves had formed, the plants could be removed from the mist and grown on in the glasshouse in John Innes No. 2 compost (Fig. 4). A high proportion of the rooted shoots negotiated this process.

3 Summary and Conclusions

Although only a relatively small amount of work has been published on in vitro culture of *Indigofera*, it is established that callus can be induced to form on a range of explants, including leaf tissue. Under appropriate conditions this callus can be induced to produce adventitious shoots with a high frequency. Further-

Fig. 3. Small plants of *I. potaninii* derived from shoots formed on leaf callus at the time of removal from the mist propagator

Fig. 4. *I. potaninii* plant derived from leaf callus

more, callus cultures will continue to produce shoots even when passaged through a considerable number of subcultures. Subsequently, these shoots can be rooted and established in soil fairly easily. In addition, cell suspension cultures can be obtained and when these suspension culture cells are plated into solid medium they will proliferate to produce colonies from which shoots can be recovered. Some species of *Indigofera*, therefore, offer a legume system in which adventitious shoot formation can be induced readily. It is particularly interesting that shoots can be induced to form on leaf explants in a manner somewhat reminiscent of the situation in tobacco and other members of the Solanaceae. However, the effect of exogenously supplied growth regulators appears to be rather different to that in tobacco. Normally, shoots and buds are induced in tobacco when cytokinin is present, provided the levels of auxin are low with high auxin levels being inhibitory. Whereas for buds and shoots to form on *Indigofera* cytokinin is still required, the presence of auxin is not inhibitory. In contrast, the requirement for the formation of roots appears to be similar in tobacco and *Indigofera*, with auxin being necessary and the presence of cytokinin being detrimental (Howell et al. 1987). The fact that shoots can be regenerated from colonies derived from single cells is also encouraging, suggesting that recovery of whole plants from isolated protoplasts might also be possible in due course.

4 Protocol

For the production of large quantities of friable callus leaf explants of *I. potaninii* can be cultured on MS medium containing 2 mg/l 2,4-D and 0.25 mg/l kin. For the production of shoots it is best to culture the leaf explants on UM medium with NAA at 0.2 mg/l and BAP at 5 mg/l. On this medium the callus can be subcultured at 5-week intervals and it will continue to produce large quantities of buds, many of which will expand to produce shoots. Excised shoots can be rooted with a high frequency by placing them in contact with MS medium containing IBA at 1 mg/l for 2 days in the light and then transferring the shoots to growth regulator-free medium and continuing to incubate in the light. Once short roots have been produced, the plantlets can be weaned by transferring them to a sand/ Levingtons compost mixture and maintaining them in a mist propagator until new leaves form, whereupon they can be moved to the greenhouse bench.

References

Allen ON, Allen EK (1981) The Leguminosae. A source book of characteristics, uses and nodulation. Univ Press, Wisconsin

Aylward JH, Court RD, Haydock KP, Strickland RW, Hegarty MP (1987) *Indigofera* species with agronomic potential in the tropics; rat toxicity studies. Aust J Agric Res 38:177–186

Bean WJ (1970) Trees and shrubs hardy in the British Isles. 8th Edition, Murray, London

Bharal S, Rashid A (1979) Regeneration of plants from tissue cultures of the legume *Indigofera enneaphylla* Linn. Z Pflanzenphysiol 92:443–447

Bharal S, Rashid A (1984) Growth of free-cell suspension and plantlet regeneration in the legume *Indigofera enneaphylla* Linn. Biol Plant Praha 26:202–205

Christie GS, Madsen NP, Hagerty MP (1969) Acute biochemical changes in rat liver induced by the naturally occurring amino acid indospicine. Biochem Pharmacol 18:693–700

Evans PK, Taylor E, Stewart CE (1986) Organogenesis on cultured leaf explants of a legume, *Indigofera potaninii*. In: 6th Int Congr plant tissue and cell culture. Abstr, Univ Minnesota, Minneapolis, p 178

Gamborg OL, Miller RA, Ojima K (1968) Nutrient requirements of suspension cultures of soybean root cells. Exp Cell Res 50:151–158

Hegarty MP, Peterson PJ (1973) Free amino acids, bound amino acids, amines and ureides. In: Butler GW, Bailey RW (eds) Chemistry and biochemistry of herbage, vol 1. Academic Press, London, pp 1–62

Howell EC, Stewart CB, Evans PK (1987) Tissue culture and plant regeneration of *Indigofera potaninii* Craib. J Plant Physiol 128:259–269

Martin-Leake H (1975) An historical memoir of the indigo industry of Bihar. Econ Bot 29:361–371

Murashige T, Skoog F (1962) A revised medium for rapid growth and bioassay with tobacco tissue cultures. Physiol Plant 15:473–497

Peattie DC (1936) Cargoes and harvests. Appleton-Century, New York

Rembert DH (1979) The indigo of commerce in colonial North America. Econ Bot 33:128–134

Rhoades HL (1976) Effect of *Indigofera hirsuta* on *Belonolaimus longigaudatus*, *Meloidogyne incognita*, and *M. javanica* and subsequent crop yields. Plant Dis Rep 60:384–386

Rhoades HL, Forbes RB (1986) Effects of fallow, cover crops, organic mulches, and fenamiphos on nematode populations, soil nutrients, and subsequent crop growth. Nematropica 16:141–151
Schwenk FW (1980) Callus formation from mechanically isolated soybean cotyledonary cells. Plant Sci Lett 17:437–442
Uchimiya H, Murashige T (1974) Evaluation of parameters in the isolation of viable protoplasts from cultured tobacco cells. Plant Physiol 54:936–944
Usher G (1974) A dictionary of plants used by man. Constable, London
Widholm JM (1972) The use of fluorescein diacetate and phenosafranine for determining viability of cultured plant cells. Stain Technol 47:189–194

V.2 Carob (*Ceratonia siliqua* L.)

M. A. MARTINS-LOUÇÃO[1]

1 Introduction

Carob (*Ceratonia siliqua* L.) is a dioecious tree (Fig. 1) of the family Leguminosae, subfamily Caesalpinoideae, noted for its great morphogenetic plasticity and its ability to produce and resist in drought stress conditions (Catarino et al. 1981; Nunes and Matos 1986). Carob has a great potential as a tree crop for restoring vegetation and improving the productivity of marginal drylands in many parts of the world. It is a multipurpose tree crop whose fruit may be utilized for food, fodder, sugar and syrup, industrial gums, and alcohol (Winer 1980).

1.1 Economic Importance

Carob has historically been linked to the lives of the mediterranean peoples. Raw carob pods, sometimes called St. John's bread, have been used since ancient times, mainly as livestock food, but also for human consumption in times of food scarcity (Coit 1967). The pods were also distilled into wine and brandy and processed for tannins. Due to their consistent weight and size, carob seeds were used by jewellers as scale weights and ground into powder for glues and gums.

More recently, as a result of modern research in chemistry, carob pods are applied in various forms of human food, animal feed, and pharmaceuticals. They are being used extensively as a substitute and an extender for cocoa and chocolate (Carlson 1980). Except when pods are eaten by animals as forage, it is customary for feed manufacturers to "kibble" or chop the pods into particle size range suitable for use in mixed feeds. In the process, the kernels (seeds) are easily separated for sale to manufacturers of carob bean gum (also known as locust bean gum or tragasol). The most developed industry of raw carob pods and kernels is located in Portugal (Figs. 2, 3).

The most valuable product from carob is carob gum. By weight, 40% of the carob seed is endosperm, rich in galactomannans. The gum extracted is commercially valuable due to its properties of viscosity and low percentage of starch and other impurities. It is used in human food industry as an emulsion stabilizer and thickener in ice creams, cheese, soups, and salad dressings (Fig. 3). It improves

[1] Departamento de Biologia Vegetal, Faculdade de Ciências, Campo Grande Bloco C2, 1700 Lisboa, Portugal

Biotechnology in Agriculture and Forestry, Vol. 10
Legumes and Oilseed Crops I (ed. by Y. P. S. Bajaj)

Fig. 1. Carob tree, more than 200 years old (Livramento, Luz de Tavira, Portugal)

the water retention properties of doughs, resulting in bakery products with improved textures. Carob gum and its derivatives are excellent film-formers and are used alone or in combination with starch as textile sizing agents and dye thickeners in textile printing, as fiber bonding and beater additives in the papermaking industry. Carob gum is also used for cosmetics, pharmaceuticals, pet food, and for plugging drill holes in oil wells (Martins-Loução and Brito de Carvalho 1989).

The embryo accounts for 20%–25% of the total seed dry weight. It is very rich in proteins and is a good source of vitamin B_2. In spite of these interesting qualities for human food, most of it is misused as fodder and only small amounts are incorporated in dietetic food and beverages in Japan (Martins-Loução and Brito de Carvalho 1989).

While producing a commercial crop, carob trees are a source of decorative hardwood, and contribute to soil conservation and reclamation. Due to their dense evergreen canopies, carob trees are efficient for noise reduction, being used in landscaping, along trafficways and parks (Schaudinischky et al. 1982).

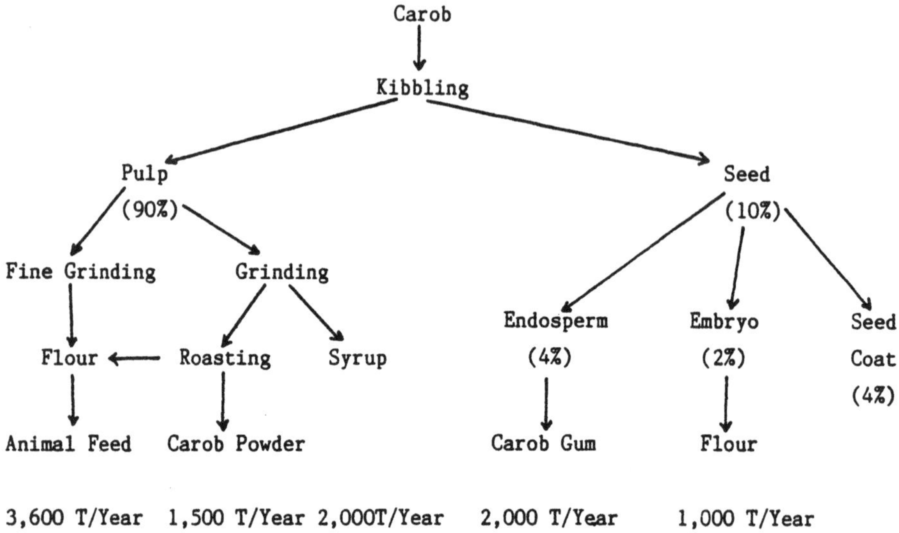

Fig. 2. Carob industrialization in Portugal (Martins-Loução and Brito de Carvalho 1989)

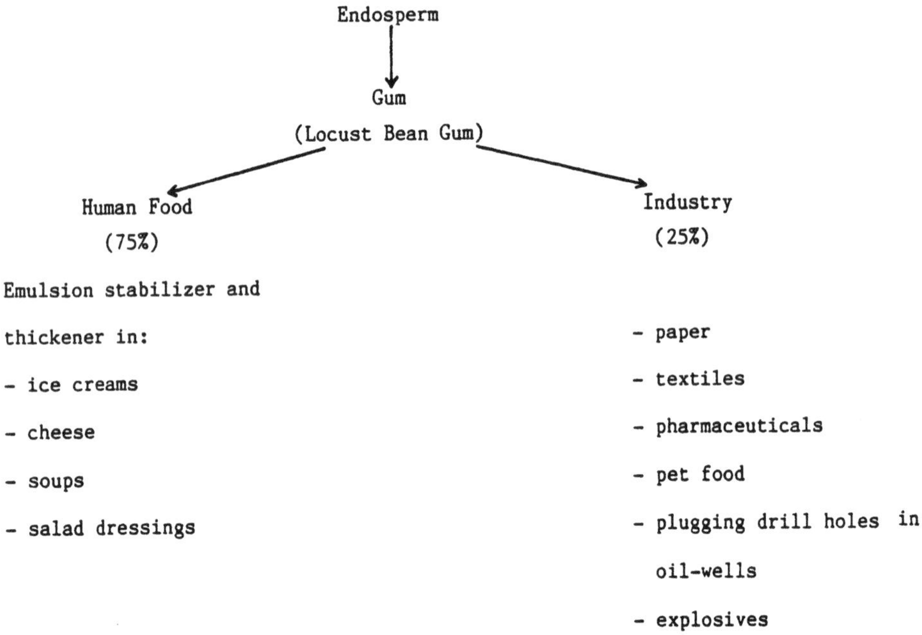

Fig. 3. Industrial uses of carob endosperm (Martins-Loução and Brito de Carvalho 1989)

1.2 Geographical Distribution and Cultivation

Ceratonia siliqua has been cultivated along the shores of the Mediterranean Sea since ancient times (Ribeiro 1980). The exact place of its origin has been obscured by extensive cultivation. The discovery in the Somali Republic of the second species of *Ceratonia* – *Ceratonia oreothauma* (Hillcoat et al. 1979) – supports the hypothesis that *Ceratonia* is part of a tropical flora, probably a xeric Indo-Malesian relic (Zohary 1973).

Carob is believed to have been spread by the Arabs along the Mediterranean coast (Ribeiro 1980). Today this tree is widely cultivated in zones with a mediterranean climate. It grows in the eastern Mediterranean, from Egypt to Greece, in southern Europe, in the Mediterranean islands, and throughout North Africa. Carob has also been introduced in the southwest of the United States, Hawaii, India, Australia, and South Africa (Vardar et al. 1980).

In Mediterranean countries the cultivation of this crop was neglected for many years for two main reasons. First, given the post-war decrease in the use of traction animals, the carob feed market was greatly reduced both in volume and monetary value; secondly, intensive cash crops – irrigated citrus as well as potatoes – expanded dramatically, replacing many carob orchards.

Recently, however, the demand for carob gum for food processing has increased enormously, opening a new hope for the crop. On a world scale, some 500000 metric tonnes (mt) of pods are processed annually, yielding 450000 mt of kibbled pods ("kibbles") and 50000 mt of kernels (Carlson 1986). The UK and the USA are the major importers of kibbles. Japan, the Netherlands, and Denmark import mostly carob gum (Martins-Louçāo and Brito de Carvalho 1989). According to INEC (Institut Européen des Industries de la Gomme de Caroube), the annual demand for carob gum on the European market is 32/35000 mt, although the world production attains only 12/14000 mt. Seventy percent of world production comes from the Mediterranean Basin.

Spain is the largest producer (Table 1), but 80% of the crop is processed and consumed domestically (Marti 1985). Portugal and Cyprus are the world's largest exporters of carob. In both countries only 10% remain for national consumption (Orphanus 1980; Martins-Louçāo and Brito de Carvalho 1989). Recently, due to

Table 1. World carob production (Martins-Louçāo and Brito de Carvalho 1989)

Location	Area under cultivation (ha)	Production (mt)
Spain	109000	120000
Italy	75000	45000
Portugal	18334	30000
Morocco	–	28000
Cyprus	[a]	20000
Greece	–	15000
Turkey	[b]	10000

[a], [b] Roughly 2 and 1 million dispersed trees, respectively

the industrial development of carob pods, Portugal is importing kernels from Morocco to process carob bean gum, which is totally exported (Graça 1986).

The largest commercial orchard in the western Hemisphere is located near Ensenada, Baja California, Mexico. Scattered small plantings still exist today on horse ranches in southern California. Small plantations intended for commercialization were tried during the 1940's in the low desert near Palm Springs (USA) (Merwin 1980).

Several other countries are also interested in carob cultivation. For instance, in Australia there is much interest in using the tree as a valuable resource for reflorestation of degraded zones where erosion and desertification are a threat (Anonymous 1979). Carob trees are also promoted for farm and village plantings in South Africa and India (Loock 1947; Greenstock 1978).

1.3 Pests and Diseases

In most countries, rats and carob moth (*Ectomyelois ceratoniae* Zell.) are the most serious pests of carob. Damage caused on developing pods by this insect facilitates further attack by other pests during storage. Other insect pests of carob are the moth *Zeuzera pyrina* L., that enters branches and trunk, and the cochineal *Aspidiotus heredae* L., that causes premature abscision of leaves and pods (Marti 1984). The red fungus, *Oidium ceratoniae* C., attacks leaves, buds and pods, especially during protracted periods of high humidity. These attacks often cause delay of pod maturation, fermentation, and facilitate insect infestation of the ripening pods, decreasing their commercial value (Davies 1970; Anonymous 1979; Feio and Almeida 1980).

1.4 Conventional Propagation

Carob trees are generally grown from seeds. Once the seedling rootsock has attained 2–4 years it can be successfully grafted with buds taken from productive female trees. Once established carob trees require little maintenance. However, regular pruning of carob trees is essential for sustained production.

Vegetative propagation by rooted cuttings is difficult; however, interesting results have recently been obtained (Medrano 1986). The procedure includes an acid pretreatment prior to the application of IBA, accompanied by a fungicide treatment. Success rate varies greatly with different varieties and with the season of the year. Some plants respond very weakly to the treatment, while rare hermaphrodite plants exhibit the best results. The plants propagated by this method have shown quite successful and promising results on transplantation to the field.

1.5 Breeding and Crop Improvement Problems

Centuries of cultivation all over the Mediterranean countries have given rise to a large number of varieties that are still waiting for evaluation regarding the agronomic and industrial value of their traits. There is a broad variation in size and quality of pods, vigor, productivity, and adaptability to various soils. Trial plantings of clones and selection of promising varieties are basic research requisites for evaluating the carob's potential in any specific area.

Although most carob varieties are dioecious, occasionally hermaphrodite trees can occur, at least in Majorca, where they seem to present a higher production than the more common female trees (Medrano pers. commun). Probably the hermaphroditism is a result of the successive graftings performed in order to obtain as many fruit-bearing plants as possible (Linskens and Scholten 1980).

Carob, like most tree crops, matures slowly, with a pronounced juvenile phase. Most cultivars do not begin to bear commercial quantities of fruits before 6–7 or more years after planting.

Mature carob trees grow in almost any soil that is well drained and aerated, including sands, clay loams, limestones, alkaline, or moderately acid soils. Carob thrives even on rocky or stony slopes, and is therefore well suited to grow on marginal lands where few if any other agricultural crops will survive without intensive care. Nevertheless, good yields are not always achieved. Generally, this tree does not need direct fertilization during maturity, but application of ammonium sulphate or ammonium nitrate during the juvenile phase has proved beneficial (Martins-Loução and Duarte 1987).

Unfortunately, there is a lack of information about the fertilization requirements to yield improvement. This is due to the relatively unimproved cultural practices and to the tolerance of the plant to nutrient deficiencies. Like most Caesalpinoideae, carob does not nodulate, and therefore it is not a nitrogen-fixing plant (Martins-Loução 1985). The nitrogenous nutrient pathways need further testing and development in order to understand better the capacity of this plant to survive and produce under apparently deficient soil conditions.

Main goals of carob improvement would include breeding and selection of plants that have high yield, a relatively short juvenile phase, and easy harvest. Pod quality, with a high content of kernels, while maintaining a sweet flavor and low tannin content, would be important, since 90% of the pod is pulp.

The long juvenile phase constitutes the main obstacle for rapid genetic improvement. Partial asynchrony in flowering of staminate and pistillate trees may represent other limitations that in part can be overcome by adequate pollen storage.

To circumvent the difficulties of classic vegetative propagation, tissue culture methods are promising ways, making it possible to obtain a large number of propagules in a short time (see Bajaj 1986). In vitro methods are usually less season-dependent and they need less mother plants than propagation of cuttings. Shoot multiplication and root induction have already been obtained by methods of micropropagation with carob axillary buds (Thomas and Mehta 1983; Sebastian and McComb 1986; Alorda and Medrano 1986). Detailed investigation about the viability of such propagules and their field testing are, unfortunately, still missing.

2 In Vitro Culture Studies

2.1 Literature Review

The first attempts to initiate callus of carob were difficult due to poor callus development and rapid necrotization. Pioneer work with carob tissues was published by Schröder (1961), who studied the regenerative capacities of different fruit tissues. Later, Scaramuzzi et al. (1971) compared seasonal effects on callus induction in different shoot trees grown in vitro. In both studies, carob cultures were used only as a comparative material and not as an objective for itself.

During the last decade a great deal of attention has focused on the commercial potential of this plant as a tree crop for the mediterranean areas. There have been two main trends. One was the development of cell and tissue culture techniques for the induction of callus cultures as an in vitro system to study symbiotic relationships. The other was to utilize the technique of shoot and meristem culture for large-scale clonal propagation. Table 2 lists the published work on callus formation and shoot propagation from several carob explants.

2.2 Methods of Sterilization

Carob seedlings have responded quite well as a source of plant tissue (Martins-Loução and Rodriguez-Barrueco 1981; Martins-Loução 1985; Sebastian and McComb 1986; Alorda and Medrano 1986). In some cases seedlings were germinated under controlled conditions inside a glasshouse (Sebastian and McComb 1986; Alorda and Medrano 1986). Nevertheless, germination of seeds in vitro has proved to provide a convenient source of sterile material (Martins-Loução and Rodriguez-Barrueco 1981). Prior to germination, the seeds were surface sterilized by immersing them for 30 s in alcohol and in a 2% mercuric chloride solution for 5 min. After a rinse in sterile water, embryos were removed and placed on a solid medium containing half-strength macronutrients of Johnson et al. (1957), micronutrients of Fortin and Piche (1979), and glucose (0.5%), as carbon source. Glucose proved to be beneficial, in contrast to sucrose, which inhibited further growth and development.

Carob embryos germinate both in dark and light, although a source of light must be provided for a good development. All embryos were maintained at $26° \pm 1\,°C$ constant temperature in a 12-h photoperiod. After 8 days, roots, cotyledons, hypocotyls, and apical meristems could be used as first explants (Martins-Loução and Rodriguez-Barrueco 1981; Martins-Loução 1985).

The advantage of seedlings for tissue culture is that they are a good source of sterile material. Juvenile tissues are also the most responsive explants. In spite of this advantage, seedling tissues have also limited usefulness for the purpose of selection procedures applicable to mature plants. These are of obvious major practical interest, particularly when they exhibit advantageous agronomic characteristics.

Unfortunately, organs from trees are very difficult to put in sterile conditions. Under field conditions, the plants have their organs routinely contaminated both

Table 2. In vitro studies on *Ceratonia siliqua* L.

Inoculum	Medium (mg l^{-1})	Growth response	Reference
Root	MS+Nvit+2,4-D(2)+C.M.(15%)	Callusing	Martins-Loução and Rodriguez-Barrueco (1981)
Cotyledon	MS+Nvit+NAA(1)+K(2)	Callusing	Martins-Loução and Rodriguez-Barrueco (1981)
Cotyledon	MS+Nvit+IAA(1)+2-iP(1)	Roots and shoots	Martins-Loução and Rodriguez-Barrueco (1981)
Cotyledon	MS+Nvit+IAA(1)+K(1)	Roots and shoots	Martins-Loução and Rodriguez-Barrueco (1981)
Cotyledon	MS+Nvit+IAA(2)	Rhizogenesis	Martins-Loução (1985)
Hypocotyl	MS+Nvit+NAA(1)+BAP(2)	Vigorous callus for nitrogen fixation studies	Martins-Loução and Rodriguez-Barrueco (1981)
Hypocotyl	MS+Nvit+NAA(1)+BAP(2)	Organogenesis	Schwab and Martins-Loução (1987)
Apical meristems	MS $NO_3/2$+Nvit+IAA(0.02)+BAP(5)	Callusing and bud burst	Martins-Loução (1985)
Shoot tips+nodal segments	MS+NAA(1)+BAP(2)	Callusing along with bud burst	Thomas and Mehta (1983)
Shoot tips+nodal segments	MS+BAP(2)+IBA(0.1)+GA_3(2)+PG(162)	Shoot growth and multiplication	Thomas and Mehta (1983)
Shoot tips+nodal segments	MS+IBA(1)+PG(630)	Root induction	Thomas and Mehta (1983)
Shoot tips+nodal segments	MS+PG(162)	Root development	Thomas and Mehta (1983)
Shoot tips+nodal segments	MS+Suc(2%)+BAP(0.1)	Callusing along with bud burst	Alorda and Medrano (1986)
Shoot tips+nodal segments	MS+Suc(2%)+IAA(2)	Root induction	Alorda and Medrano (1986)
Shoot tips	MS+Suc(2%)+Z(1)	Bud burst	Sebastian and McComb (1986)
Shoot tips	$\frac{1}{2}$MS+Suc(2%)+IBA(1)	Root induction	Sebastian and McComb (1986)
Shoot	K+H+Suc(2%)+IAA(0.1)+K(0.5)	Callusing with necrotization	Scaramuzzi et al. (1971)
Fruit	N+Nvit+IAA(10)+Suc(2%)	Callusing with necrotization	Schröder (1961)

MS = Murashige and Skoog medium; Nvit = Nitsch vitamins; C.M. = Coconut milk; PG = Phloroglucinol; K = Knop; H = Heller externally and internally by microorganisms and, hence, they are often difficult to control in vitro.

The usual method of tissue sterilization is to soak plant parts in a known concentration of sodium hypochlorite (usually a diluted commercial bleach). A dilution of 5% sodium hypochlorite for 10 min after surface sterilization in 50% ethanol for 1 min resulted in 5% and 40% contamination of seedlings and mature trees, respectively (Sebastian and McComb 1986). Alorda and Medrano (1986) also reported a high contamination of mature tree material, although they have used 20% sodium hypochlorite mixed with Tween 20 for 15 min. In both studies no damage was reported. The tissue was also washed three to five times with sterile water to ensure dilution of bleach.

Many researchers have found that a single bleach treatment is not sufficient to ensure contaminant-free cultures. In our laboratory many different procedures have been assayed. All of them were unsuccessful, with high contamination rates.

The best procedure includes a pre-washed step in running water for 4 h followed by 1–2 h wash in 1% captan, aerated continuously and two rinses in distilled water. Under aseptic conditions, nodal cross-sections of the stems, which contained two, the maximum three lateral buds, were cut. The next sterilization step included a flowing wash system during 10–15 min in 10% commercial bleach (with 5% active chloride) solution, containing 0.1% Tween 80 or 1% normal liquid detergent, followed by four rinses in distilled water. Contamination was avoided on 80%–90% of the explants. A good development and regeneration of the explants was achieved. It was observed that aeration during both steps of sterilization was very important in increasing the percentage of sterilized explants.

Thomas and Mehta (1983) developed a different procedure for explants removed from 4-year-old trees. They used 0.1% $HgCl_2$ solution for 2 min. The percentage of sterile material obtained was not given, but in our laboratory it was very low and the development and regeneration of the explants were affected.

2.3 Shoot Cultures

Micropropagation technique has been applied as an attempt to multiply and propagate carob plants, for the last 3 years (Table 2). The use of shoot tips as functional explants was first adopted by Thomas and Mehta (1983). Bud burst was obtained in 4 to 5 weeks in culture medium containing MS salts, supplement with BAP (2 mg/l) and NAA (1 mg/l). However, no growth of the sprouted buds was observed on this shoot induction medium because the conditions favored more callus growth than organized development. Replacement of NAA by IBA (0.1 mg/l) reduced callus growth in the subsequent subcultures. Moreover, incorporation of phloroglucinol (PG, 162 mg/l) and GA_3 enhanced general growth and leaf expansion.

It is known that either exogenous auxin or cytokinin may be sufficient to bring about morphogenesis (Thorpe 1980). Induction of buds from carob shoot tips could be achieved in 8 weeks by the addition of either BAP (1 mg/l) (Alorda and Medrano 1986) or zeatin (1 mg/l) alone (Sebastian and McComb 1986). It was moreover, observed that, in contrast to Thomas and Mehta (1983), the presence

Table 3. Effect of hormones in basal MS + shoot multiplication media on subsequent root formation in rooting medium for carob shoots (Sebastian and McComb 1986)

Shoot multiplication media	Rooting medium (% rooting)	No. of shoots tested
Z	92	63
Z + G	21	58
Z + G→Z	54	24
Z + G→A	90	52
Z + A	48	34

Z = 5 μM zeatin; G = 2.5 μM GA_3; A = 5 μM ancymidol; Z + G→Z = shoots cultured in a medium with zeatin and GA_3, were then grown on passage on a medium with zeatin alone before testing rooting ability. The rooting medium contained 10 μM IBA

of GA_3 in the shoot induction medium inhibited posterior root induction (Table 3).

GA_3 repression could be reversed by ancymidol, an antigibberellin substance. These contrasting results are probably due to the interaction between endogenous phytohormones and phloroglucinol. Evidence has been gathered that this compound stimulates organogenesis by altering the effective endogenous level of auxin and thus the auxin-cytokinin balance (Leopold and Kriedemann 1975). In addition, gibberellins may then be active per se causing stimulation of rooting (Thorpe 1980).

In a manner similar to other fruit trees (James et al. 1979), a synergistic effect of PG in controlling carob root induction was detected in the presence of IBA (Thomas and Mehta 1983). IBA proved to be the best auxin to promote root induction after 3 to 4 weeks (Thomas and Mehta 1983; Sebastian and McComb 1986). However, IAA seems to induce the development of more vigorous roots, an important trait for a successful transplant to the soil (Alorda and Medrano 1986). According to Sebastian and McComb (1986), rooted plants potted in a mixture of peat : moss : vermiculite (2 : 1 : 1) and kept under plastic cover at 26 °C for 2–4 weeks can be successfully transferred to normal glasshouse conditions, after a controlled period of 4 weeks.

2.4 Callus Cultures

Martins-Loução and Rodriguez-Barrueco (1981) were the first to be successful with carob callus cultures. As in other woody plants, the most responsive carob explant material is juveniles. In fact, carob tissue cultures were established from any portion of a seedling (Fig. 4). Attempts to obtain callus from mature plant organs were unsuccessful (Schröder 1961; Scaramuzzi et al. 1971).

About 3 weeks are necessary to obtain developed callus from explants of cotyledons, hypocotyls, and roots, in a MS medium (1962) supplemented with Nitsch vitamins (1965). Martins-Loução and Rodriguez-Barrueco (1981) consider the vitamin solution to be very important. They obtained 95% success with these vitamins, but only 58% with MS ones. The main advantage of these Nitsch vitamins was to provide both biotin and folic acid.

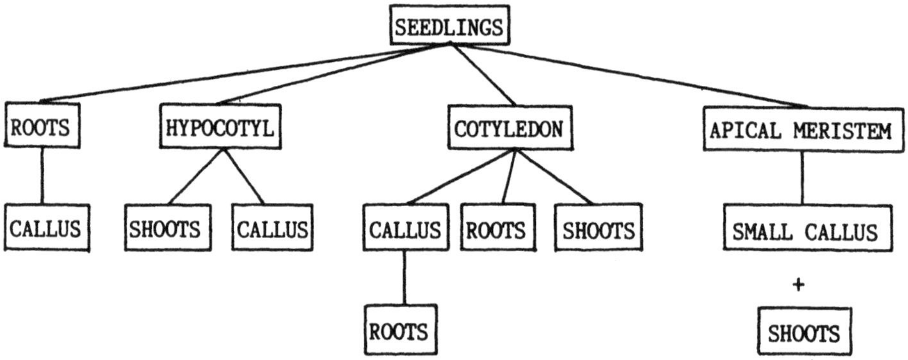

Fig. 4. Scheme for callus formation and dediferentiation in carob seedlings

The in vitro morphogenetic responses of explants have shown that the increase in callus mass was influenced more by the explant origin than the hormone concentration (Martins-Loução and Rodriguez-Barrueco 1981; Martins-Loução 1985). In cotyledons, callus initiation started on the edges and on the main rib region. White, rather hard, callus proliferated from the cut ends and edges of hypocotyl explants (Fig. 5 A). Roots showed a smaller percentage of callus formation with many necrotization problems. Nevertheless, those calli formed developed rapidly in a vigorous mass of undifferentiated cells.

The experiments have indicated that the best callus initiation and growth occurred on hypocotyl segments placed in culture medium supplemented with NAA (1 mg/l) and BAP (2 mg/l). All callus grew better on this medium under both light and dark conditions, independent of callus origin. Improvement of callus growth was obtained only by supplementing the media with coconut milk (Fig. 5 B).

Shoot formation in hypocotyl segments was obtained on the same callus induction medium passing through a less-developed callus stage formation (Fig. 5 C). It was proved that the quality of light is the most important factor for the induction of morphogenetic events. Daylight-grown seedlings develop hypocotyls which induce only callus formation. Radiant energy near red ($\lambda_{max} = 660$ nm) during germination induce short and green hypocotyls which have the potentiality to undergo organogenesis after 15–20 days of culture both on daylight and near-red light conditions (Schwab and Martins-Loução 1987). It seems that the seedlings require more quanta of red than fluorescent light to promote hypocotyls with a higher regenerative capacity, independent of light quality where they are later cultured. Such "carry-over" effects are probably the result of the phytochrome photoequilibrium established within the embryo during in vitro germination of seedlings (Morgan and Smith 1981).

Dark germination induces long, white hypocotyls which have the potentiality to undergo organogenesis under near-red light conditions. Darkness induces only callus formation. It seems that the spectral quality of light used has significant effects on carob hypocotyl induction of organogenesis. Hypocotyls become "competent" for shoot induction since they have a special light stimulus either in

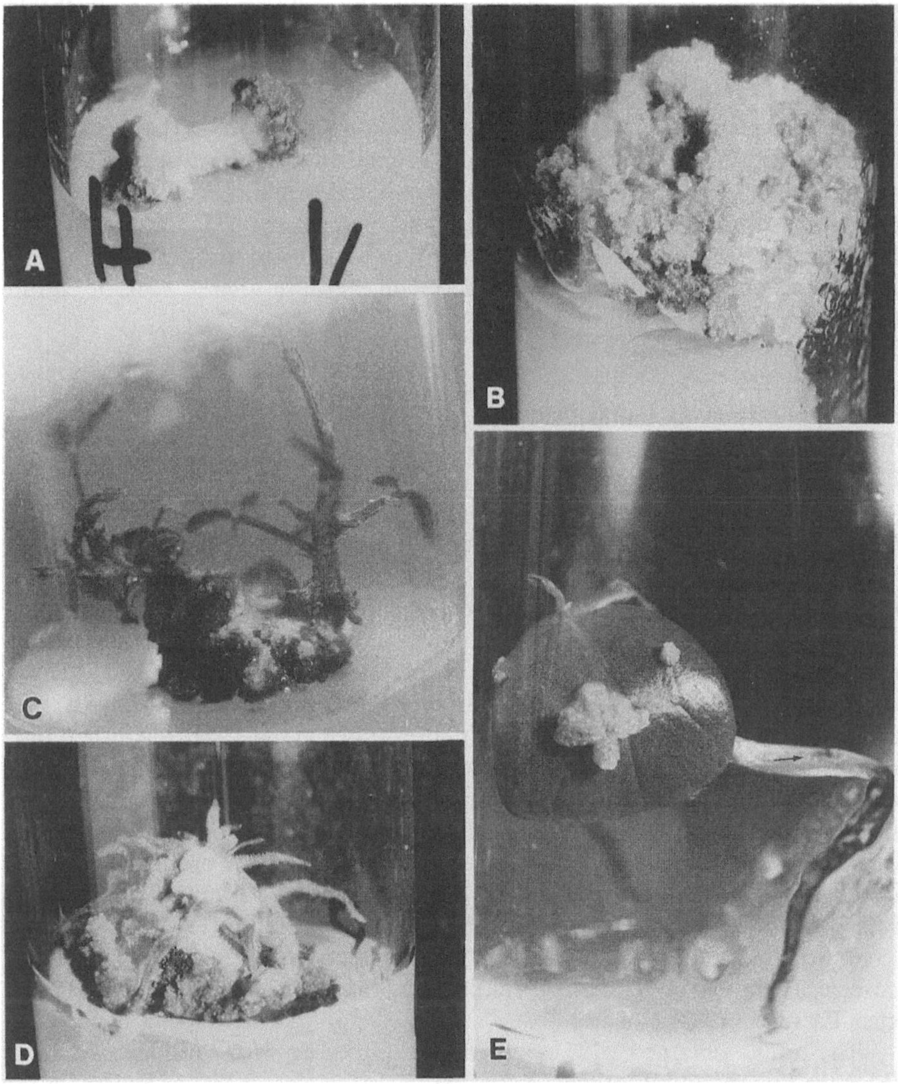

Fig. 5. A Hypocotyl explant on a MS medium supplemented with NAA (1 mg/l)+BAP (2 mg/l), after 10 days in culture. **B** Callus from hypocotyl explants. Medium: MS+NAA (1 mg/l)+BAP (2 mg/l) +CM (15%). **C** Shoot formation on hypocotyl callus explants placed on a MS medium supplemented with NAA (1 mg/l)+BAP (2 mg/l), under near-red ($\lambda_{max} = 660$ nm) light conditions. Age: 20 days. **D** Root differentiation from cotyledon explants after 25 days in culture. Medium: MS+IAA (2 mg/l). **E** Tumor-like crown-gall on carob cotyledons after 15 days in culture. Aseptic germination on germination medium developed by Martins-Loução and Rodriguez-Barrueco (1981)

the seedling during germination or in the explant during differentiation (Schwab and Martins-Loução 1987).

These findings suggest that the induction of bud formation on carob hypocotyls is stimulated by radiant energy near-red ($\lambda_{max} = 660$ nm) in a phytochrome-mediated response similar to what happens in both embryo-derived callus of pine (Kadkade et al. 1978) and in lettuce cotyledon cultures (Kadkade and Seibert 1977).

Root differentiation was obtained in a medium supplemented with IAA from cotyledon explants (Fig. 5D). Nevertheless the roots necrotize as they contact the medium (Martins-Loução 1985). In order to prevent root necrosis, cotyledon explants were cultured on filter paper soaked in liquid medium. Fast growth of root clusters was obtained on agar, while in liquid medium cotyledon growth was inhibited and no roots differentiated. Attempts to use in vitro roots for both physiological and nitrogen studies were unsuccessful, because after 1 week of culture they stopped growing and necrotized.

2.5 Applications of Callus Cultures

2.5.1 Experiments with Rhizobium

Carob cultures have been useful as an endeavor to study symbiotic relationships. With the high cost of energy and subsequent high cost of nitrogen fertilizer, the interest to improve N_2 fixation has been revitalized.

The establishment of carob callus cultures permitted our group to try a biological approach in order to throw some light on the inability of carob plants to form root nodules (Martins-Loução and Rodriguez-Barrueco 1982). Following the procedure of Scowcroft and Gibson (1975), a separated system was established on a specific medium (GLH), supplemented with glutamine (5 mM), NAA (1 mg/l) and BAP (2 mg/l). The initiation of nitrogenase activity by *Rhizobium leguminosarum* associated with carob callus cultures was stimulated by providing both a carbon source and a citric acid substance into the medium, as previously reported in other systems (Gibson et al. 1976). Although marked variation from assay to assay has already been reported for different systems (Phillips 1974; Gibson et al. 1976; Schetter and Hess 1977), linear rates of acetylene reduction up to 5 days have been achieved under defined culture conditions (Martins-Loução and Rodriguez-Barrueco 1982).

The results obtained with carob callus cultures or with carob excised roots (Martins-Loução and Rodriguez-Barrueco 1983a) suggest that the barriers for the establishment of a functional symbiotic association can be manifest either at the stage of infection or nodule formation rather than at the nitrogenase expression.

Endogenous ethylene production was detected during the above experiments. This result led to the study of the nutritional and environmental influence on the production of ethylene (Martins-Loução and Rodriguez-Barrueco 1983b). The amounts of ethylene produced by carob callus is variable, as found in other tissue cultures (LaRue and Gamborg 1971; Huxter et al. 1979). Ethylene production seems to be mainly dependent on the composition of the medium and to be associated with the callus growth cycle. Ethylene production seems to be light-dependent. Particularly, no correlation was found between growth rate and ethylene production under dark conditions, as has been observed for tobacco callus cultures (Huxter et al. 1979). The results seem to indicate that the ethylene produced by carob callus cultures is a metabolic byproduct, although ethylene may function as a growth regulator during the initial phases of callus development.

2.5.2 Experiments with Agrobacterium

Carob tissue cultures were also used as an approach to the process of transformation of a normal intact cell into a tumor cell. During the past few years it has become apparent that the induction of the plant tumor crown gall by *Agrobacterium tumefaciens* is a natural example of genetic transformation (Schilperoort et al. 1978), with great potentials in the genetic engineering of plant cells (Bryant 1986; also see Bajaj 1989). Furthermore, crown gall induction is a valuable model to study neoplastic transformation, although it has a wide range only in dicotyledonous plants (DeCleene and DeLey 1976).

Since all attempts to induce symbiotic associations between carob cultures and several species of *Rhizobium* proved unsuccessful (Martins-Loução 1985), the response to infection by *A. tumefaciens* was studied to test the receptivity of carob tissues to another member of the Rhizobiaceae. Both cotyledons and hypocotyls from in vitro seedlings have been infected with *A. tumefaciens* strain B_6 (Martins-Loução 1985). The seedlings were cultivated on the germination medium and were infected after 8 days in culture. Incubation was under light at 24 °C. Observations were made after 15 days (Fig. 5E).

The results showed that only carob cotyledons exhibited response to the infection, developing a small tumor-like crown gall. The difference between cotyledons and hypocotyls may reflect qualitative or quantitative differences in the endogenous phytohormone production (Schilperoort et al. 1978). Since *A. tumefaciens* is able to transform carob plant cells, it would be of interest to see if this tool can be utilized for genetic engineering purposes on *Ceratonia siliqua.*

3 Protocols

Callus Formation

1. Surface sterilize imbibed seeds in alcohol, 2% mercuric chloride solution for 5 min and rinse three times with sterile water.
2. Aseptically remove embryos and plant them on a simple medium containing glucose, a solution 1 : 2 of macronutrients defined by Johnson et al. (1957) and a solution of micronutrients and ferric citrate, as used by Fortin and Piche (1979).
3. At 8 – 10 days, excise hypocotyl, cotyledons, and roots. Cut sections of 6 – 8 mm long and culture them on agar. Medium for callus induction is outlined in Table 2. Culture in dark or daylight conditions, 10 $\mu E/m^2/s$, 26° ± 1 °C, 12-h photoperiod.
4. Subculture callus following 1 month of growth. The subculture medium is the same as the hypocotyl callus induction medium for all callus, independent of its origin.

Differentiation

1. Germinate embryos under dark or near-red (λ_{max} = 660 nm) light conditions.
2. At 8–10 days excise hypocotyls, place on the same callus induction medium and culture under near-red light conditions, 10 $\mu E/m^2/s$, 26° ± 1 °C, 12-h photoperiod.
3. After 2 weeks excise meristemoids with a piece of callus and place them on a MS medium without phytohormones. Culture in the same light conditions.
4. After 20–30 days remove plants with three normal leaves to root induction medium.

4 Summary and Prospects

Callus cultures can be obtained from any part of carob seedlings. In addition, shoot propagation can also be achieved, although it needs some improvement. Serious problems of mature plant material sterilization and elucidation of plant genetic integrity may seriously limit this practical utilization for the near future.

Ceratonia siliqua callus cultures can be established and maintained for years by periodic transfer to fresh media (Martins-Loução and Rodriguez-Barrueco 1981). This aspect facilitates the development of research programs directed towards the elucidation of plant-microorganism interactions. Emerging tissue culture techniques offer the promise of an indirect but broadly applicable approach to this objective. Fungus mycorrhizal and nitrogen-fixing symbiotic associations would be important to bring about crop yield increases. Partial success with carob cultures is so recent that techniques such as protoplast culture and genetic manipulation have yet to be tried. Protoplasts have been isolated from hypocotyl-root segments of germinated seedlings in our laboratory (Martins-Loução 1986). Unfortunately, at present, all attempts to culture them have failed due to the low rate of divisions.

Micropropagation of selected carob plants will continue to be of major importance. The best-developed method applies zeatin as the most effective hormone for shoot induction and multiplication (Sebastian and McComb 1986).

Carob has a marked morphogenetic plasticity (Catarino et al. 1981) which may be related to great genetic variability. In this way, the selection of varieties for carob crop improvement can be expected to be highly rewarding. In this respect the experience obtained with apical meristems will be most profitable (Martins-Loução 1985), in spite of the time involved for adequate evaluation. The improvement of meristem tip culture techniques and the methods of long-term storage of valuable germplasm will provide a means of rapid propagation of elite material.

The protocols that are urgently needed are those for readily controlled explant sterilization and improvement of culture media in order to increase the possibility of large-scale propagation. The uncertainty regarding the genetic integrity of regenerates causes limitations for extended rapid use. Since no genetic control has yet been tried, research programs are needed to elucidate the conditions allowing for clonal fidelity. However, the success already obtained indicates that this approach will be an efficient method for rapid in vitro clonal propagation.

The application of cell culture techniques to crop improvement of *Ceratonia siliqua* is a valuable undertaking. With this research tool we also expect to make substantial contribution to understanding the genetics and physiology of this tree. Carob, a very important economic crop for mediterranean countries, still represents a large and untouched field for investigations on the practical use of cell culture techniques.

Acknowledgments. I would like to express my thanks to Prof. F.M. Catarino, who originally suggested my study on this fascinating tree species and helped me in the revision of this manuscript.

References

Bajaj YPS (ed) (1986) Biotechnology in agriculture and forestry. 1. Trees I. Springer, Berlín Heidelberg New York Tokyo

Bajaj YPS (ed) (1989) Biotechnology in agriculture and forestry. 9. Protoplasts and genetic engineering II. Springer, Berlin Heidelberg New York Tokyo

Alorda M, Medrano H (1986) Micropropagacion del algarrobo (*Ceratonia siliqua* L.) a partir de yemas caulinares de planta joven. 2nd Congr Soc Esp C Hor, pp 28–31

Anonymous (1979) Tropical legumes. Resources for the future. Natl Acad Sci USA, Washington DC

Bryant J (1986) The role of host-bacterium interactions in the *Agrobacterium* mediated transformation of plant cells. Biotechnology 4(6):131–132

Carlson WA (1980) Carob as a food. Historical review. Lebensm Wiss Technol 13:51–52

Carlson WA (1986) The carob: evaluation of trees, pods and kernels. Int Tree Crops J 3:281–290

Catarino FM, Correia OA, Webb E, David M (1981) Morphological and physiological responses of the mediterranean evergreen sclerophyll, *Ceratonia siliqua*, to different light intensities. In: Margaris NS, Mooney HA (eds) Components of productivity of mediterranean climate regions – basic and applied aspects. Junk, The Hague

Coit JE (1967) Carob varieties for the semi-arid southwest. Fruit Var Hort Dig 21:5–9

Davies WNL (1970) The carob tree and its importance in the agricultural economy of Cyprus. Econ Bot 24(4):460–470

DeCleene M, DeLey J (1976) The host range of crown-gall. Bot Rev 42:389–466

Feio M, Almeida G (1980) The carob tree climate in Portugal. Port Acta Biol (A) XVI (1–4):19–32

Fortin A, Piche Y (1979) Cultivation of *Pinus strobus* root-hypocotyl explants for synthesis of ectomycorrhizaea. New Phytol 83:109–119

Gibson AH, Child JJ, Pagan JD, Scowcroft WR (1976) The induction of nitrogenase activity in *Rhizobium* by non-legume plant cells. Planta 128:233–239

Graça JM (1986) A alfarrobeira no Algarve. 1° Encontro Farroba. AIDA Comissão Coordenadora Região Algarve

Greenstock DW (1978) Food from locust plant. New Ecol 2:55–56

Hillcoat D, Lewis G, Verdcourt B (1979) A new species of *Ceratonia* (Leguminosae-Caesalpinoideae) from Arabia and Somali Republic. Kew Bull 35(2):261–271

Huxter TJ, Reid DM, Thorpe TA (1979) Ethylene production by tobacco (*Nicotiana tabacum*) callus. Physiol Plant 46:374–380

James OJ, Knight VH, Turbon IJ (1979) Micropropagation of the red raspberry and the influence of phloroglucinol. Sci Hortic 12:313–319

Johnson CM, Stout PR, Broyer TC, Carlton AB (1957) Comparative chlorine requirements of different plant species. Plant Soil VIII:337–353

Kadkade PG, Seibert M (1977) Phytochrome-regulated organogenesis in lettuce tissue culture. Nature (Lond) 270:49–50

Kadkade PG, Wetherbee P, Jopson H, Botticelli C (1978) Photoregulation of organogenesis in pine embryo and seedling. In: 4th Int Congr plant tissue and cell culture. Abstr, Calgary 108:29

LaRue TAG, Gamborg OL (1971) Ethylene production by plant cell cultures. Variations in production during growing cycle and in different plant species. Plant Physiol 48:394–398

Leopold AC, Kriedemann PE (1975) Plant growth and development. McGraw-Hill, NY

Linskens HF, Scholten W (1980) The flower of carob. Port Acta Biol (A) XVI (1–4):95–102

Loock EEM (1974) Three useful leguminous fodder trees. Farm S Africa 22 (250):7–12
Marti JT (1984) El algarrobo. Su importancia en la cuenca mediterranica. Agricultura 5:474–478
Marti JT (1985) Comercializacion y variedades de algarroba HDMin Agric Pescas Alimentacion 1/85:1–24
Martins-Loução MA (1985) Estudos fisiologicos e microbiologicos da associação da alfarrobeira (*Ceratonia siliqua* L.) com bacterias de Rhizobiaceae. Thesis, Univ Lisboa
Martins-Loução MA (1986) Isolation of protoplasts from hypocotyl-root segments of carob germinated seedlings. In: 6th Int Congr plant tissue and cell culture. Abstr, Univ Minnesota, Minneapolis
Martins-Loução MA, Duarte P (1987) Effect of ammonium and nitrate nutrition on the growth of carob (*Ceratonia siliqua* L.) plants. In: Ulrich W, Aparicio PJ, Syrett PJ, Castillo F (eds) Inorganic Nitrogen Metabolism. Springer, Berlin Heidelberg New York, pp 250–252
Martins-Loução MA, Rodriguez-Barrueco C (1981) Establishment of proliferating callus from roots, cotyledons and hypocotyls of carob (*Ceratonia siliqua* L.) seedlings. Z Pflanzenphysiol 103:297–303
Martins-Loução MA, Rodriguez-Barrueco C (1982) Studies on nitrogenase activity of carob (*Ceratonia siliqua* L.) callus cultures associated with *Rhizobium*. In: Fujiwara A (ed) Plant tissue culture 1982. Maruzen, Tokyo, pp 671–672
Martins-Loução MA, Rodriguez-Barrueco C (1983a) In vitro nitrogenase activity of *Rhizobium* associated with carob excised roots. In: Veeger C, Newton WE (eds) Advances in nitrogen fixation, pp 514
Martins-Loução MA, Rodriguez-Barrueco C (1983b) Ethylene production by carob (*Ceratonia siliqua* L.) callus cultures on varying media. Physiol Plant 58:204–208
Martins-Loução MA, Brito de Carvalho J (1989) A Alfarrobeira. Riqueza actual e potencial para os paises mediterranicos. DGPA Min Agr Pali
Medrano H (1986) Propagacion mediante estanquillado. 1° Encontro Farroba. Aida, Comissão Coordenadora Região Algarve
Merwin ML (1980) The culture of carob (*Ceratonia siliqua*) L.) for food, fodder and fuel in semi-arid environments. In: YMCA (ed) Tree Crops for Energy Co-Production on Farms. Ester Park, Colorado
Mitrakos K (1968) The carob. Tate and Lyle, Keston, UK
Morgan DC, Smith H (1981) Non-photosynthetic responses to light quality. Encyclop Plant Physiol 12A:109–135
Nitsch JP, Nitsch C (1965) Néoformation des fleurs in vitro chez une espèce de jours courts: *Plumbago indica* L., Ann Physiol Veg 7:251–256
Nunes A, Matos C (1986) Comportamento fisiológico da alfarrobeira em relação com a água disponivel. 1° Encontro Farroba. AIDA, Comissão Coordenadora Região Algarve
Orphanos PI (1980) Practical aspects of carob cultivation in Cyprus. Port Acta Biol (A) XVI (1–4):221–228
Phillips DA (1974) Factors affecting the reduction of acetylene by *Rhizobium* soybean cell associations in vitro. Plant Physiol 53:67–72
Ribeiro O (1980) Le caroubier. Ses conditions naturelles, son expansion, ses rapports avec l'agriculture. Port Acta Biol (A) XVI (1–4):3–10
Scaramuzzi F, Porcelli-Armenise V, Gaetano A (1971) Recherches sur le comportement et le rythme annuel de la calogenèse de fragments de tiges de certains arbres. CR Acad Sci Paris 272:2544–2547
Schaudinischsky LH, Schiller G, Keller Y (1982) Acoustic properties of individual trees and forest strips. Forstarchiv 53 (2):52–57
Schetter C, Hess D (1977) Nitrogenase activity in in vitro associations between callus tissues of non-leguminous horticultural plants and *Rhizobium*. Plant Sci Lett 9:1–5
Schilperoort RA, Klapwijk PM, Hooykaas PJJ, Koekman BP, Ooms G, Otten LAM, Wurzer-Figurelli EM, Wullems GJ, Rorsch A (1978) *A. tumefaciens* plasmids as vectors for genetic transformation of plant cells. In: Thorpe TA (ed) Frontiers of plant tissue culture. International Association for Plant Tissue Culture, Calgary Univ, pp 85–94
Schröder CA (1961) Some morphological aspects of fruit tissues grown in vitro. Bot Gaz 122:198–204
Schwab L, Martins-Loução MA (1987) Shoot formation in *Ceratonia siliqua* hypocotyl callus. In: Fito P, Mulet A (eds) Generalitat Valenciana Con. Agr Pescas, pp 245–254, 2nd Carob Symp
Scowcroft WR, Gibson AH (1975) Nitrogen fixation by *Rhizobium* associated with tobacco and cowpea cell cultures. Nature (Lond) 253:351–352

Sebastian KT, McComb JA (1986) A micropropagation system for carob (*Ceratonia siliqua* L.). Sci Hortic 28:127–131
Thomas V, Mehta AR (1983) Effect of phloroglucinol on shoot growth and initiation of roots in carob tree cultures grown in vitro. In: Giles KL, Sen SK (eds) Plant Cell Culture in Crop Improvement. Plenum Press, New York, pp 451–457
Thorpe TA (1980) Organogenesis in vitro: structural, physiological and biochemical aspects. Int Rev Cyt Suppl 11A:71–111
Vardar Y, Seçmen O, Özturk M (1980) Some distributional problems and biological characteristics of *Ceratonia* in Turkey. Port Acta Biol (A) XVI (1–4):75–86
Winer N (1980) The potential of carob (*Ceratonia siliqua* L.). Int Tree Crops J 1:15–26
Zohary M (1973) Geobotanical foundations of the Middle-East. Gustav Fischer Verlag, Stuttgart

Subject Index

MIX
Papier aus verantwortungsvollen Quellen
Paper from responsible sources
FSC® C105338

If you have any concerns about our products,
you can contact us on
ProductSafety@springernature.com

In case Publisher is established outside the EU,
the EU authorized representative is:
Springer Nature Customer Service Center GmbH
Europaplatz 3, 69115 Heidelberg, Germany

Printed by Libri Plureos GmbH
in Hamburg, Germany